Lecture Notes in Statistics

Volume 212

For further volumes:
http://www.springer.com/series/694

To Barbara, Lilas, Zélie and Nora,

to Tatiana, Miriam, René, Klára, Kristína, Emma, Filip, David,

and all our friends.

Preface

The final form of this volume differs a lot from our initial project that was born in 2003. Our collaboration that was initiated about 10 years before was then at a pace and we formulated the objective of reviving it through a project that could be conducted over several years (we thought three or four) and could serve as a motivation for maintaining regular exchanges between us. Our initial idea was to write a short monograph that would expose the connections between the asymptotic properties of estimators and experimental design. This corresponds basically to parts which are covered by Chaps. 2–4. The deviation from this initial project was progressive. First, we realized that we had more to say about regression models with heteroscedastic errors than what we expected initially and we found that the investigation of asymptotic normality in the case of singular designs required a rather fundamental revision (Chap. 3). We then quickly agreed that we could not avoid writing a chapter on optimality criteria and optimum experimental design based on asymptotic normality (Chap. 5). Up to that point, the presentation was rather standard, although we gave more emphasis than usual to some particular aspects, like the estimation of a nonlinear function of the model parameters and models with heteroscedastic errors. The results obtained during our collaboration in the 1990s encouraged us to write a chapter on non-asymptotic design approaches (Chap. 6). The motivation for exposing our views on the specific difficulties caused by nonlinear models in LS estimation had always been present in our mind; this project gave us the opportunity to develop and present some of these ideas (Chap. 7). Since this book focused on nonlinear models, having a chapter devoted to the problem raised by the dependency of an optimal experiment on the value of the parameters to be estimated appeared to be essential (Chap. 8). Here some kind of prior for design purposes is unavoidable, with the property that an incorrect prior causes less damage when used for design than for estimation. Finally, we hesitated about indicating or not algorithms for the optimization of the different design criteria that are presented throughout the chapters. This could have led us quite far from the initial project but at the same time was essential for the practical use of the methods suggested. We reached a compromise solution where all algorithms are gathered in a specific chapter (Chap. 9), where the principles are indicated, in connection with more classical optimization methods. The result is thus much different from what we planned in 2003 and this book covers the following aspects.

Asymptotic Normality. The first three chapters expose the necessary background on asymptotic properties of estimators in nonlinear models. The presentation is mathematically rigorous, with detailed proofs indicated in an appendix to improve readability. The stress here is on deriving asymptotic properties of estimators from properties of the experimental design, in particular the "design measure" which is a basic notion in classical experimental design since the pioneering work of Jack Kiefer in the early 1960s. For nonlinear models, this is not covered in other books on design and considered in a few research papers only; in general, the published proofs of asymptotic properties of estimators require many assumptions of different types, which are usually rather technical and not directly related to the design. Besides that, some results in Chap. 3 are new, e.g., on singular designs and on models with misspecification or with parameterized variance.

Optimality Criteria. The next chapters concern optimum design more directly. Readers only interested in the application of optimal design methodology can possibly start by reading Chap. 5, where the classical theory is presented together with several new aspects and results. The optimality criteria considered in Chap. 5 are related to the asymptotic behavior of estimators. Optimality criteria obtained under non-asymptotic considerations (small-sample situation) are considered in Chap. 6, while Chap. 7 concerns the connection between design and identifiability/estimability issues, including new extensions of some classical optimality criteria. Nonlinear models have the particularity that an optimal design depends on the value of the parameters to be estimated; this is considered in detail in Chap. 8. Once an optimality criterion is chosen, we still need to optimize it; algorithms are presented in Chap. 9 that cover all situations presented in Chaps. 5–8.

Small-sample Properties. The small-sample differential-geometric approach to the subject is considered in Chap. 6; it mainly corresponds to results obtained by the second author in a series of research papers. Chapter 7 contains results on situations when the nonlinearity of the model is such that the estimator can be totally erroneous because of too small a sample at hand. This subject is not much considered in the literature, even in research papers, although it may be of crucial importance in applications.

Several academic friends gave a substantial support through exchanges of different forms; a special mention goes to Radoslav Harman, Werner Müller, Éric Thierry, Henry Wynn, and Anatoly Zhigljavsky. We thank Jean-Pierre Gauchi, Rainer Schwabe, and particularly Henry Wynn, for their help and encouragement in reviewing our manuscript. Of course all possible mistakes remain ours. Our friends and families gave a remarkable level of support. The first author expresses his deepest thanks to the Bittner clan in St. Peter in der Au (Austria) whose kind hospitality provided a perfect environment for precious periods of intensive work during winter and summer holidays. The second author thanks the I3S Laboratory for the friendly working atmosphere during his many visits at Sophia Antipolis.

A long-term project like this one necessarily involves supports and grants of different sources. In particular, the research of Andrej Pázman has been partly supported by the VEGA grants Nb. 1/3016/06, 1/0077/09, and 2/0038/12. The work of Luc Pronzato was partially supported by the IST Programme of the European Community, under the PASCAL Network of Excellence, IST-2002-506778; he also benefited from invitations by the Isaac Newton Institute for Mathematical Sciences, Cambridge, UK, in summer 2008 and 2011. Andrej Pázman benefited from several invitations from the University of Nice Sophia Antipolis. The institutions of both authors, CNRS for Luc Pronzato and the Comenius University for Andrej Pázman, are gratefully acknowledged for the freedom they allowed through the years.

Sophia Antipolis, France — Luc Pronzato
Bratislava, Slovakia — Andrej Pázman

Contents

1

Introduction

1.1 Experiments and Their Designs

This book is about *experiments*; it concerns situations where we have to organize an experiment in order to gain some information about an object of interest. Fragments of this information can be obtained by making observations within some elementary experiments called *trials*. We shall confound the action of making an experiment with the variables that characterize this action and use the term *experimental design* for both. The set of all trials which can be incorporated in a prepared experiment will be denoted by $\mathscr{X}$, which we shall call the *design space*. The problem to be solved in experimental design is how to choose, say N trials $x_i \in \mathscr{X}$, $i = 1, \ldots, N$, called the *support points of the design*, or eventually how to choose *the size N of the design*, to gather enough information about the object of interest. *Optimum experimental design* corresponds to the maximization, in some sense, of this information. Throughout this monograph we shall generally assume that all the trials have to be chosen before the collection of information starts; that is, the experiment is designed *nonsequentially*. A brief exposition on sequential design is given in Sect. 8.5.

There are almost no restrictions on the definition of the set $\mathscr{X}$. For example, $\mathscr{X}$ can be the set of all scientific instruments that could be used in the experiment and is then a finite set. In general, we can suppose that in each trial d control variables are chosen that specify the experimental conditions in the trial. Each point $x \in \mathscr{X}$, called *design point*, is then a vector in a d-dimensional space, with components given by the chosen values of the controlled variables. If these can be changed continuously, then $\mathscr{X}$ can be considered as a subset of $\mathbb{R}^d$.

By design of (fixed) size N (sometimes called *exact design*), we understand the choice of N trials $X = (x_1, \ldots, x_N)$. Usually, replications (or repetitions) are permitted; that is, we can have $x_i = x_j$ for $i \neq j$. We suppose that in each trial x_i we can observe a random variable $y(x_i)$. The *observation* $y(x_i)$ may eventually denote a vector of random variables, or a realization of a

L. Pronzato and A. Pázman, *Design of Experiments in Nonlinear Models*, Lecture Notes in Statistics 212, DOI 10.1007/978-1-4614-6363-4_1,

random process, etc. In the present setup we suppose that observations in different trials are *independent*. This is a key assumption. It does hold in most designed experiments, which justifies that the restriction to this case is rather standard. Only this case will be considered in this volume.[1] Note that it does not mean that observations inside one trial (in case there are several) must be independent or uncorrelated; see Sect. 5.6.

When observations are independent, it is convenient to reformulate the concept of a design. The design $X = (x_1, \ldots, x_N)$ can be equivalently defined by its size N and the associated *design measure*

$$\xi(x) = \frac{N(x)}{N}, \quad x \in \mathscr{X},$$

which corresponds to the relative frequency of the trial x in the N-tuple $x_1, \ldots, x_N$: $N(x)$ denotes the number of replications of the trial x in the sequence $x_1, \ldots, x_N$. Alternatively, if the total cost C is fixed instead of the size N of the design, one may define the design measure as

$$\xi(x) = \frac{c(x)\,N(x)}{C}, \quad x \in \mathscr{X},$$

where $c(x)$ denotes the cost of one replication of the trial x. One may go further and take ξ as any probability measure supported on some finite subset of $\mathscr{X}$, or even any probability measure on $\mathscr{X}$, with $\mathscr{X}$ a subset of an Euclidian space. One then speaks of *approximate* (or *continuous*) *design*. For a given $\xi \in \Xi$, with Ξ the set of design measures on $\mathscr{X}$, and a given size N (or cost C), implementing the experiment generally requires approximating $N(x)$ by an integer (hence the name approximate design), which does not cause difficulties when N is large. This is briefly considered in Sect. 9.2.[2]

A major consequence of the independence of observations entering into the experiment is that, in some situations, the information obtained from the experiment is the sum of informations from the individual trials. It depends on how this information is measured, but the property typically holds when measuring information by the Fisher information matrix: the information matrix

[1]Situations where the assumption of independence for different trials does not hold require a special treatment, and the methods to be used differ very much according to the type of prior knowledge about the dependence structure of the observations; see, e.g., Fedorov and Hackl (1997, Sect. 5.3), Pázman and Müller (2001, 2010), Müller and Pázman (2003), and Zhu and Zhang (2006). However, replacing the assumption of independent errors by that of errors forming a martingale difference sequence does not modify the situation very much for regression models, in particular in the context of sequential design; see, e.g., Lai and Wei (1982) and Pronzato (2009a).

[2]There exist situations, however, where a continuous design can be implemented without any approximation; this is the case, for example, when designing the experiment corresponds to choosing the power spectral density of the input signal for a dynamical system; see Goodwin and Payne (1977, Chap. 6), Zarrop (1979), Ljung (1987, Chap. 14), and Walter and Pronzato (1997, Chap. 6).

for the whole experiment is the sum of the information matrices for the trials $x_1, \dots, x_N$; when a design measure ξ is used, it corresponds to the weighted mean of the matrices for trials at x. This simple structure is extremely useful in optimum experimental design and makes the concept of design measure ξ extremely effective. It is central in the developments presented in most monographs on optimum experimental design; see Fedorov (1972), Silvey (1980), Pázman (1986), Atkinson and Donev (1992), Pukelsheim (1993), and Fedorov and Hackl (1997). We shall see, however, that in many circumstances the information cannot be measured in such a simple form, even asymptotically, so that the additivity of information for independent trials is lost; see Chaps. 3 and 4 and Sect. 5.5.

1.2 Models

Designing an experiment requires a model. Indeed, since the design is anterior to observations, it requires the possibility to predict somehow the amount of useful information that will be obtained from observations in the experiment. This cannot be done without some model assumptions on the probability distributions of the observed variables $y(x_1), \dots, y(x_N)$. Naturally, the method to be used to design the experiment, and the design itself, will much depend on these assumptions.

Unless otherwise stated, the observations $y(x_i)$ will be assumed to be scalar real variables. A very common (although rather strong) assumption corresponds to *regression modeling*. In this case, one usually supposes that all useful information to be obtained from the experiment is contained in the mean $\nu(x) = \mathbb{E}[y(x)]$ of the observed variable. We shall call the function $\nu(\cdot)$ the (mean or expected) *response function*. We then have

$$y(x_i) = \nu(x_i) + \varepsilon_i\,, \quad i = 1, \dots, N\,, \tag{1.1}$$

with $\mathbb{E}(\varepsilon_i) = 0$. In this context, $\varepsilon_i = \varepsilon(x_i)$ is a nuisance random term, containing no useful information (excepted for testing the correctness of the model), and is called the *random error*. When nothing precise is known about $\nu(\cdot)$, excepted, for instance, some smoothness and regularity properties, one may use a *nonparametric* approach; see, e.g., Tsybakov (2004). This is related to the rather fashionable framework of statistical learning, for which one may refer, e.g., to the books (Vapnik 1998, 2000; Hastie et al. 2001) and the surveys (Cucker and Smale 2001; Bartlett 2003). Typically, based on so-called training data $\{[x_1, y(x_1)], \dots, [x_N, y(x_N)]\}$, the objective is then to predict the response $y(x)$ at some unsampled site x using Nadaraya–Watson regression (1964), Radial Basis Functions (RBF), Support Vector Machine (SVM) regression, or kriging (Gaussian process). All these approaches can be casted in the class of kernel methods; see Vazquez (2005), Vazquez and Walter (2003), Schaback (2003), Berlinet and Thomas-Agnan (2004) for a detailed exposition. Due to the lack of precise knowledge on $\nu(\cdot)$, the design points x_i should then be somewhat spread in $\mathscr{X}$ and have a space-filling property; see, e.g., Morris and Mitchell (1995), Müller (2007), and Pronzato and Müller (2012). When $\mathscr{X}$ is a compact subset of $\mathbb{R}^d$, the optimal design in such a nonparametric setting corresponds to a measure having a density with respect to the Lebesgue measure (see, for instance, Müller (1984), and Cheng et al. (1998)), in contrast with the more standard parametric situation where the optimal design concentrates at a few locations only. The optimal density depends on the unknown function $\nu(\cdot)$, and, when considering minimax-optimality in terms of integrated mean-squared error over some Sobolev classes of functions, the uniform distribution may turn out to be optimal; see, e.g., Biedermann and Dette (2001).

We shall restrict our attention to the parametric situation, with the exception of Sect. 3.4 where modeling error is considered. In the case of a regression model, the mean response function is then *parameterized* as

$$\nu(x) = \eta(x, \theta)\,,$$

where θ is a vector of unknown parameters, a priori restricted to a feasible parameter space $\Theta \subset \mathbb{R}^p$. In such models, the information about the response function $\eta(\cdot, \theta)$, in terms of prediction for instance, is obtained through the information about the parameters θ. In some cases these parameters may receive a strong interpretation by themselves (for instance, they may correspond to some physical quantities of interest) and not only in terms of the response function. The situation is simple when $\eta(x, \theta)$ is *linear* in θ, i.e., when

$$\eta(x, \theta) = \mathbf{f}^\top(x)\theta\,.$$

The regression is then termed *linear*. Notice that a linear regression can still be nonlinear in x, depending on the form of the vector function $x \in \mathscr{X} \longrightarrow \mathbf{f}(x) \in \mathbb{R}^p$. Design methods for linear regression models with uncorrelated observations are very well elaborated and are exposed in many textbooks and papers. Here we shall focus our attention on *nonlinear* models.

Other types of models, becoming more and more popular, can be considered as "nearly regression models." In these models, also the variance of the observation errors contains some useful information. Usually, this variance is also parameterized, but these parameters may differ from those appearing in the response function. Such models are usually termed *variance-component models* or *mixed regression models* since the information about the response, and the variance components is mixed when the same parameters appear in the variance and the response function.

Regression modeling is a very powerful tool in statistical analysis since most often, we have little knowledge about the exact distribution of the observed variables. Some important exceptions exist, however. For example, in many dichotomic "yes–no" experiments there is a strong evidence that the underlying distribution is binomial. In some other situations, it may be multinomial, Poisson, etc. All these distributions belong to the *exponential family*, and, when the so-called canonical variable is linearly parameterized, they are known as *generalized linear model*. In these models, the mean of the observed variables is still a nonlinear function of the parameters, and, concerning experimental design, we shall consider them as nonlinear models. Design theory for such models is much less elaborated than for regression models. We shall see however (Sect. 4.3) that the information carried by an experiment in such models can be measured (asymptotically) in a similar way to nonlinear regression models.

In many circumstances, the analytic form of the response $\eta(x, \theta)$ as a function of x is not available and can only be obtained through the simulation of a differential or difference equation. The computation of derivatives with respect to θ (sensitivity functions), which are required for most developments throughout the book, is considered in Appendix B.

1.3 Parameters

As indicated above, we shall consider *parametric models* and suppose that the objective of the experiment is to obtain a "precise" estimator for the parameters, or parametric functions, which are considered as useful.

Parameter estimation is of wider statistical interest than the estimation of the values of parameters per se. This may include tests for the validity of a model, model discrimination, tests for homoscedasticity of the data, rejection of outliers, etc., a variety of situations where parameter estimation can be put at work to get the solution. For example, when testing the validity of a regression model, we can incorporate it in a larger model by adding some extra parameters and accept the original model if the estimates of these new parameters are close to zero; when testing for homoscedasticity of the data, we can estimate the trend of the variance of the residuals; robust estimators can be used to neutralize the influence of outliers, which again leads to a parameter estimation problem; etc. Therefore, the restriction of experimental design to parameter estimation problems is not as severe as it might seem, and applications to other statistical problems are often natural if not straightforward.

We give special attention to least-squares (LS) estimation (Chaps. 3 and 6) but also consider maximum likelihood (ML) and to a certain extent also Bayesian estimation (Chap. 4). In all circumstances, we consider that there exists a true value $\bar{\theta}$ for the parameters, which is used to generate the observations; θ denotes a generic value for the parameters, and estimators based on N observations will be denoted by $\hat{\theta}^N$. Although in Bayesian estimation the main concepts are traditionally the prior and posterior distributions for θ, we only consider the situation where one can safely assume that a "true" $\bar{\theta}$ does exist. We can then consider, say the convergence of the maximum a posteriori estimator $\hat{\theta}^N$ to $\bar{\theta}$ as $N \to \infty$.

The primary objective of experimental design should be to ensure that the parameters are *estimable*. In linear regression models, this means that, under the given design, there exists at least one linear unbiased estimator of the parameters of interest. The existence of a design making the parameters estimable corresponds to the requirement of *identifiability*. By identifiability of a regression model, we mean that the mapping $\theta \longrightarrow \eta(\cdot, \theta)$ is one-to-one; see, e.g., Walter (1982, 1987), and Walter and Pronzato (1995, 1997, Chap. 2). With a large and rich enough set of trials $x_1, x_2, \ldots, x_N$, one should then be able to approach the true value $\bar{\theta}$ of the model parameters with arbitrary precision. There is a large amount of literature on how to construct designs of reasonable size N that ensure estimability in simple linear models, such as ANOVA models (linear models for which the components of the vectors $\mathbf{f}(x_i)$ can only take the values 0 and 1), or factorial models (for which the components of $\mathbf{f}(x_i)$ correspond to different levels of the controlled factors, which may be categorial, or qualitative, variables). Such problems are solved in the so-called classical experimental design theory and are outside the scope of this book; see, e.g., Cochran and Cox (1957), Montgomery (1976), and Casella (2008) for references.

In nonlinear regression models the estimability of θ, or of some function of θ, cannot be reduced to the issue of identifiability. For example, when θ is estimated by least squares (LS), the numerical estimability of θ means that the function

$$\theta \in \Theta \longrightarrow \sum_{i=1}^{N} [y(x_i) - \eta(x_i, \theta)]^2$$

(the sum of squares) has a unique global minimum and no other important local minima. Similarly, for maximum likelihood (ML) estimation in a nonlinear model, we require that the likelihood function does not have several important local maxima. This type of statistical identifiability, related to the sensitivity of the estimates with respect to variations in the values of the observations for a finite horizon N, raises difficult issues, too often overlooked, that are considered in Chap. 7.

1.4 Information and Design Criteria

Since our interest is in parameter estimation, the information carried by an experiment will be related to the precision of the estimation it allows. The precision of an estimator $\hat{\theta}$ can be expressed in different ways, in particular through the mean-squared error (MSE) matrix

$$\mathrm{MSE}_\theta(\hat{\theta}) = \mathbb{E}_\theta \left\{ (\hat{\theta} - \theta)(\hat{\theta} - \theta)^\top \right\},$$

which forms a standard and widely accepted characteristic. Here, $\mathbb{E}_\theta$ denotes the conditional mean, given θ. Different functionals can be used to express some aspects of this information. For example, the i-th diagonal element of $\mathrm{MSE}_\theta(\hat{\theta})$ is the MSE of the estimator $\{\hat{\theta}\}_i$. We have

$$\mathrm{MSE}_\theta(\hat{\theta}) = \mathrm{Var}_\theta(\hat{\theta}) + \mathbf{b}_\theta(\hat{\theta})\mathbf{b}_\theta^\top(\hat{\theta}),$$

where $\mathrm{Var}_\theta(\hat{\theta})$ is the variance–covariance matrix and $\mathbf{b}_\theta(\hat{\theta}) = \mathbb{E}_\theta(\hat{\theta}) - \theta$ the bias vector of $\hat{\theta}$.

In linear regression models where uniformly minimum variance unbiased estimators $\hat{\theta}$ (UMVUE) of θ exist, we have

$$\mathrm{MSE}_\theta(\hat{\theta}) = \mathrm{Var}_\theta(\hat{\theta})$$

for such an estimator. Optimality criteria are then usually based on the variance–covariance matrix of the estimator, which is often available even for a finite sample size N. Moreover, in linear situations $\mathrm{Var}_\theta(\hat{\theta})$ does not depend on the unknown true value of θ. The situation is more complex in nonlinear regression, or more generally in nonlinear models, essentially for two reasons which form the core of this book.

First, neither $\mathrm{Var}_\theta(\hat{\theta})$ nor $\mathbf{b}_\theta(\hat{\theta})$ may be known *exactly*, so that approximations are required. The most widely used relies on the asymptotic normality of the estimator $\hat{\theta}$, so that the most classical approach to optimum experimental design relies on criteria that are scalar functions of the covariance matrix of the estimator (linear models) or of the covariance matrix of the asymptotic normal distribution of the estimator (nonlinear models). In some situations this matrix coincides with that in the Cramér–Rao bound, which gives a further justification for this asymptotic approach. This leads to measuring information through the Fisher information matrix which has the attractive property of being additive for independent trials, with important consequences concerning properties of optimal designs and methods for their construction.

However, it is only under some specific assumptions on the experimental design that information matrices characterize the asymptotic precision of estimators, which creates a sort of circular argument: optimal experiments are designed on the basis of asymptotic properties, which, as we shall see, may not hold depending on how the optimal design is approached or implemented.

One of the objectives of Chaps. 2–4 is to expose the connections between general properties of estimators, such as (strong) consistency, asymptotic normality, efficiency, and Cramér–Rao inequality on one side and properties of designs on the other side. In general, the correct proofs of asymptotic properties require many assumptions of different types, which are usually very technical and not directly related to the design; see, e.g., the assumptions on finite tail products of the regression function and its derivatives used in the classical paper of Jennrich (1969) on the least-squares (LS) estimator or the Lipschitz and growth conditions in (Wu, 1981). See also Gallant (1987) and Ivanov (1997). A second objective here is to obtain simple but still rigorous proofs by formulating assumptions just in terms of designs. In particular, clear results can be obtained in the case when the design measure ξ has a finite support, which covers most situations where the design space $\mathscr{X}$ is finite, or when the design is randomly generated according to a probability measure ξ. We detail the proofs of asymptotic theorems for these cases, indicate the method, and give references in more general situations.

Experimental design based on information matrices is considered in detail in Chap. 5. Although it covers classical approaches to optimum design, new results and proofs are presented. In particular, we shall see in Chaps. 3 and 4 that for many estimators the asymptotic covariance matrix takes a more complicated expression than the inverse of a standard information matrix, this simple form being obtained under particular circumstances only: maximum likelihood estimation, weighted least squares with optimum weights, etc. The design of optimal experiments based on such covariance matrices is considered in Sect. 5.5.

More accurate small-sample characteristics of the precision of the estimation than the asymptotic normal approximation can be used to design experiments in a nonlinear situation, such as the volumes of confidence regions (Hamilton and Watts 1985), the mean-squared error (Pázman and Pronzato 1992; Gauchi and Pázman 2006) or the entropy of the distribution of the LS estimator (Pronzato and Pázman 1994b). They are considered in Chap. 6, which also contains a survey on the geometry of nonlinear regression models and on approximations of the density of the LS estimator as developed in (Pázman, 1993b) and related papers.

Chapter 7 discusses a topic which is usually overlooked in the statistical literature: how can we take the issues of identifiability and estimability into account at the design stage? Neither the Fisher information matrix nor the approximate densities of Chap. 6 can be used to solve this problem. This forms a difficult area which is still in evolution; some extended notions of measures of nonlinearity and optimality criteria are presented.

Another important difficulty raised by nonlinear models comes from the fact that the information carried by the experiment, for instance, through $\mathrm{Var}_\theta(\hat{\theta})$ or $\mathbf{b}_\theta(\hat{\theta})$, may very much depend on the unknown θ, creating a second circular argument: the optimal experiment for estimating θ depends on the value of θ to be estimated. The knowledge of a prior distribution for θ can be used to overcome this conundrum, through the use of average-optimum (Bayesian) or maximin-optimum design, or probability-level criteria. These approaches are considered in Chap. 8, which also gives a brief exposition of sequential design in two particular situations: the experiment only contains two stages or is multistage but the design space $\mathscr{X}$ is finite.

Chapter 9 gives an overview of algorithms that can be used to construct optimal experiments. Local design criteria based on the information matrix evaluated at a given nominal value of the model parameters (Chap. 5) are considered, as well as average and maximin-optimum designs of Chap. 8.

Basic notions on subdifferentials and subgradients are presented in Appendix A. Appendix B indicates how to compute derivatives with respect to parameters (sensitivity functions) in a model given by differential or difference equations. The proofs of auxiliary lemmas are given in Appendix C.

Theorems, lemmas, remarks, examples, etc., are numbered consecutively inside a chapter, so that, for instance, Lemma 5.16 follows Theorem 5.15. The symbol □ closes remarks and examples. A list of symbols and notations is given at the end of the volume together with the list of labeled assumptions that are used.

2

Asymptotic Designs and Uniform Convergence

2.1 Asymptotic Designs

In order to study the asymptotic properties of estimators, we need to indicate how the sequence of design points $x_1, x_2, \ldots$ in $\mathscr{X} \subset \mathbb{R}^d$ is generated, i.e., specify some properties of the experimental design. Throughout the monograph (excepted in Sect. 8.5), we suppose that the design is not sequential; that is, each point x_i is chosen independently of the observations already collected. It means in particular that the design can be considered as constructed in advance, prior to the collection of observations. To each truncated subsequence $x_1, \ldots, x_N$, we associate the empirical design measure ξ_N and its (cumulative) distribution function (d.f.)

$$\mathbb{F}_{\xi_N}(x) = \sum_{i=1,\ x_i \leq x}^{N} \frac{1}{N},$$

where the inequality $x_i \leq x$ must be understood componentwise.

We shall denote by $\xi_N \Rightarrow \xi$ the weak convergence of the empirical design measure ξ_N to ξ and by $\mathbb{F}_{\xi_N} \Rightarrow \mathbb{F}_\xi$ the weak convergence of the associated d.f. $\mathbb{F}_{\xi_N}$ (see, e.g., Shiryaev (1996, Sect. III.1) and Billingsley (1995, Sect. 25)), that is,

$$\lim_{N\to\infty} \mathbb{F}_{\xi_N}(x) = \mathbb{F}_\xi(x) \ \text{ for every continuity point } x \text{ of } \mathbb{F}_\xi(\cdot)\,.$$

We shall see through several examples (see, e.g., Example 2.4) that weak convergence is not always enough to ensure that an estimator using the design corresponding to ξ_N has the same asymptotic properties (same asymptotic variance–covariance matrix in particular) as an estimator that uses ξ.

We shall consider two special cases for designs on a given set $\mathscr{X}$. A first motivation is that those cases cover most practical situations in experimental design and that they allow us to easily obtain uniform strong laws of large

L. Pronzato and A. Pázman, *Design of Experiments in Nonlinear Models*, Lecture Notes in Statistics 212, DOI 10.1007/978-1-4614-6363-4_2,

numbers (SLLN) which are essential for deriving asymptotic properties of estimators. Moreover, they permit to avoid the aforementioned difficulty with weak convergence of design measures.

The first one concerns the case where ξ_N *converges strongly*, i.e., in variation (see Shiryaev 1996, p. 360) to a *discrete limiting design* ξ. Asymptotically the design points thus belong to a finite set, each point being sampled with a given frequency; this covers in particular the case of designs consisting of *repetitions* of a given set of trials.

Definition 2.1 (Asymptotically discrete design). *Let ξ be a discrete probability measure on $\mathscr{X}$, with finite support*

$$\mathcal{S}_\xi = \{x \in \mathscr{X} : \xi(\{x\}) > 0\} = \{x^{(1)}, \ldots, x^{(k)}\}. \tag{2.1}$$

We say that the design sequence $\{x_i\}$ is asymptotically discrete when ξ_N converges strongly to ξ:

$$\lim_{N\to\infty} \xi_N(\{x\}) = \xi(\{x\}) \text{ for any } x \in \mathscr{X}.$$

Since $\mathscr{X}$ usually corresponds to a subset of $\mathbb{R}^d$, $\xi(\cdot)$ is defined on the Borel algebra of subsets of $\mathscr{X}$. With a somewhat abusive notation, we shall write $\xi(x) = \xi(\{x\})$. Also, for any function $f(\cdot)$ on $\mathscr{X}$ we shall write $\int_{\mathscr{X}} f(x)\xi(\mathrm{d}x) = \sum_{x\in\mathscr{X}} f(x)\xi(x) = \sum_{i=1}^k w_i f(x^{(i)})$ with $w_i = \xi(x^{(i)})$. In the literature the discrete design ξ is often denoted by

$$\xi = \left\{ \begin{matrix} x^{(1)} & \cdots & x^{(k)} \\ w_1 & \cdots & w_k \end{matrix} \right\}.$$

In the second situation considered, the limiting design measure ξ is not necessary discrete but *the design sequence is a random sample from* ξ; that is, the design is randomized and the points are independently sampled according to the probability measure ξ.

Definition 2.2 (Randomized design). *We call randomized design with measure ξ on $\mathscr{X}$, $\int_{\mathscr{X}} \xi(\mathrm{d}x) = 1$, a sequence $\{x_i\}$ of design points independently sampled from the measure ξ on $\mathscr{X}$.*

Naturally, $\xi_N \Rightarrow \xi$ with probability one (w.p.1) as $N \to \infty$ for a randomized design, but the asymptotic properties are much stronger than that: from the SLLN, $\xi_N(\mathcal{A}) \to \xi(\mathcal{A})$ w.p.1 as $N \to \infty$ for any ξ-measurable set $\mathcal{A}$, and, when $\mathscr{X} \subset \mathbb{R}$, then $\sup_x |\mathbb{F}_{\xi_N}(x) - \mathbb{F}(x)| \to 0$ w.p.1 as $N \to \infty$ from Glivenko–Cantelli theorem; see, e.g., Billingsley (1995, p. 269)—with possible extension to other sets $\mathscr{X}$ with empirical measures $\xi_N(\mathcal{A})$ indexed by measurable sets $\mathcal{A}$ in some suitable class $\mathcal{C}$ and uniform convergence of ξ_N on this class.

When considering consistency of estimators, almost sure (a.s.) convergence, that is, strong consistency, or convergence w.p.1, will be with respect to both the sequences of errors $\{\varepsilon_i\}$ and design points $\{x_i\}$. We shall see that the random fluctuations induced by a randomized design are asymptotically negligible, in the sense that they have no effect on the asymptotic variance–covariance matrix of estimators; see Remark 3.9-(*iii*).

Notice the difference between the two definitions above: there is no requirement of randomness in Definition 2.1, but the support of the asymptotic design is finite. The support is not necessarily finite in Definition 2.2, but the sequence of design points is an i.i.d. sample from ξ. Of course, not all situations are covered by these definitions. The examples below, however, illustrate the difficulties that can be encountered when the assumptions on the design are too relaxed and thus stress the importance of the design for the asymptotic properties of the estimator, its asymptotic normality in particular. In the first example we construct a nonrandom sequence $\{x_i\}$ of design points such that the sample distribution $\mathbb{F}_{\xi_N}$ does not converge. Of course, such sequences are excluded by Definition 2.1 and cannot be generated randomly according to Definition 2.2.

Example 2.3. We construct a non-converging sequence on the finite set $\mathscr{X} = \{x^{(1)}, x^{(2)}\}$. Define $n_1(N) = N(x^{(1)})$ and $n_2(N) = N - n_1(N)$ with $N(x^{(1)})$ the number of times $x^{(1)}$ is used in the sequence $x_1, \ldots, x_N$. The construction is as follows: take $n_1(0) = 0$ and $n_1(N+1) = n_1(N) + c(N+1)$ where $c(n) = 0$ for $n \leq N_0$, $c(N_0) = 1$, and then

$$c(N+1) = \begin{cases} 1 & \text{if } n_1(N) \leq \alpha N \\ c(N) & \text{if } \alpha N < n_1(N) \leq \beta N \\ 0 & \text{otherwise}\,; \end{cases}$$

that is, $c(N+1) = c(N)\,\mathbb{I}_{[0,\beta N]}[n_1(N)] + [1 - c(N)]\,\mathbb{I}_{[0,\alpha N]}[n_1(N)]$, for some α, β satisfying $0 < \alpha < \beta < 1$. $n_1(N)/N$ then oscillates between α and β with $\liminf_{N\to\infty} n_1(N)/N = \alpha \neq \limsup_{N\to\infty} n_1(N)/N = \beta$, and the design does not converge to a discrete design; that is, it does not correspond to Definition 2.1.

This has obvious consequences on the asymptotic properties of estimators. Even in a model as simple as the linear regression with response $\eta(x,\theta) = \theta_1 + \theta_2 x$ and observations $y_i = \eta(x_i, \bar{\theta}) + \varepsilon_i$, where $\bar{\theta} = (\bar{\theta}_1, \bar{\theta}_2)^\top$ denotes the true value of the parameters θ and the errors ε_i are normal $\mathscr{N}(0, \sigma^2)$, if $\hat{\theta}^N_{LS}$ denotes the ordinary least-squares (LS) estimator of θ (see Sect. 3.1) the covariance matrix of $\sqrt{N}(\hat{\theta}^N_{LS} - \bar{\theta})$ does not converge to a limit as $N \to \infty$. □

Consider now a deterministic sequence supported on an interval in $\mathbb{R}^d$, a rather standard situation not covered by Definitions 2.1 and 2.2, and, in order to avoid the difficulty met in the example above, suppose that the sample distribution $\mathbb{F}_{\xi_N}$ converges weakly to some d.f. $\mathbb{F}$. Denote by ξ the probability measure on $\mathscr{X}$ corresponding to $\mathbb{F}$. In that case, most often the asymptotic results do not differ from the situation where the sequence $\{x_i\}$ is random and generated according to ξ. Heuristically, this can be explained as follows: for any N, a random permutation of the sequence of the independent observations $y(x_1), \ldots, y(x_N)$ does not modify the estimate $\hat{\theta}^N$. Since the design points are chosen nonsequentially, this is equivalent to a random permutation of the design points. Although this is not a formal proof, it indicates why deterministic design sequences often yield the same asymptotic properties as randomized sequences. One may refer to Lemma 2.6' of Sect. 2.3 to obtain asymptotic properties of estimators under nonrandomized designs; different assumptions are used in (Jennrich, 1969) for the LS estimator.

The following example, however, shows that in some situations $\xi_N \Rightarrow \xi$ is not enough to yield the same asymptotic properties for the estimators obtained with ξ_N and with a randomized design with measure ξ. In this example we construct a sequence $\{x_i\}$ of design points such that $\mathbb{F}_{\xi_N} \Rightarrow \mathbb{F}_{\xi}$, with ξ a discrete design allocating mass 1 to some point x_*, but $\xi_N(x_*)$ does not converge to $\xi(x_*) = 1$. Notice that, again, such a sequence is excluded by Definition 2.1 and that, w.p.1, it is also excluded by Definition 2.2.

Example 2.4. Consider a linear regression model with $p = 2$ parameters and observations

$$y(x_i) = \bar{\theta}_1 x_i + \bar{\theta}_2 x_i^2 + \varepsilon_i\,,\ i = 1, 2\ldots$$

where true value $\bar{\theta}$ of the model parameters $\theta = (\theta_1,\ \theta_2)^\top$ is assumed to satisfy $\bar{\theta}_1 \geq 0$, $\bar{\theta}_2 < 0$ and the errors ε_i are i.i.d., with zero mean and variance 1. The LS estimator of $\theta = (\theta_1,\ \theta_2)^\top$ for N observations, see Sect. 3.1, is $\hat{\theta}^N_{LS} = \mathbf{M}_N^{-1} \sum_{i=1}^N y(x_i)(x_i,\ x_i^2)^\top$ with

$$\mathbf{M}_N = \sum_{i=1}^N \begin{pmatrix} x_i^2 & x_i^3 \\ x_i^3 & x_i^4 \end{pmatrix}.$$

Define $\xi_* = \delta_{x_*}$, the delta measure that puts mass 1 at some point $x_* \neq 0$. When ξ_* is used, $\theta_1 x_* + \theta_2 x_*^2$ is estimable since $\mathbf{c}_* = (x_*,\ x_*^2)^\top$ is in the range of

$$\mathbf{M}(\xi_*) = \begin{pmatrix} x_*^2 & x_*^3 \\ x_*^3 & x_*^4 \end{pmatrix}.$$

The variance of $\mathbf{c}_*^\top \hat{\theta}^N_{LS}$ for ξ_*, which we denote by $\mathrm{var}(\mathbf{c}_*^\top \hat{\theta}^N_{LS}|\xi_*)$, satisfies

$$N \,\mathrm{var}(\mathbf{c}_*^\top \hat{\theta}^N_{LS}|\xi_*) = \mathbf{c}_*^\top \mathbf{M}^-(\xi_*)\mathbf{c}_* = 1\,,$$

with $\mathbf{M}^-$ any g-inverse of $\mathbf{M}$ (i.e., such that $\mathbf{M}\mathbf{M}^-\mathbf{M} = \mathbf{M}$).

We construct now design sequences such that $\xi_N \Rightarrow \xi_*$ by considering design points that satisfy

$$x_i = \begin{cases} x_* & \text{if } i = 2k-1 \\ x_* + (1/k)^\alpha & \text{if } i = 2k \end{cases} \tag{2.2}$$

for some $\alpha > 0$, $i = 1, 2, \ldots$ with $\sqrt{2} - 1 < x_* \leq 1$. From Corollary 1 of Wu (1980), $\mathbf{c}^\top \hat{\theta}^N_{LS} \overset{\text{a.s.}}{\to} \mathbf{c}^\top \bar{\theta}$ for any $\mathbf{c} \in \mathbb{R}^2$ when $S^\infty(\mathbf{w}) = \sum_{i=1}^\infty [\mathbf{w}^\top \mathbf{f}(x_i)]^2 = \infty$

for all $\mathbf{w} = (w_1,\ w_2)^\top \neq \mathbf{0}$, with $\stackrel{\text{a.s.}}{\rightarrow}$ denoting almost sure convergence, i.e., convergence with probability one (w.p.1), with respect to the random sequence $\{\varepsilon_i\}$; $\mathbf{c}^\top\hat{\theta}^N_{LS}$ is then a strongly consistent estimator of $\mathbf{c}^\top\theta$. Here we have

$$S^\infty(\mathbf{w}) = \sum_{k=1}^{\infty} x_*^2(w_1 + x_* w_2)^2 + \sum_{k=1}^{\infty} \left[w_1(x_* + 1/k^\alpha) + w_2(x_* + 1/k^\alpha)^2\right]^2$$

so that $S^\infty(\mathbf{w}) = \infty$ when $w_1 + x_* w_2 \neq 0$. When $w_1 + x_* w_2 = 0$,

$$S^\infty(\mathbf{w}) = \frac{w_1^2}{x_*^2}\sum_{k=1}^{\infty}\frac{(1+x_* k^\alpha)^2}{k^{4\alpha}} > w_1^2\sum_{k=1}^{\infty} k^{-2\alpha},$$

and $w_1 \neq 0$ since $\mathbf{w} \neq 0$, so that $S^\infty(\mathbf{w}) = \infty$ when $\alpha \leq 1/2$. The LS estimator $\hat{\theta}^N_{LS}$ is thus strongly consistent when ξ_N converges to ξ_* slowly enough, i.e., when $\alpha \leq 1/2$, which we suppose in the rest of the example.

The variance of $\mathbf{c}_*^\top\hat{\theta}^N_{LS}$, with $\mathbf{c}_* = (x_*,\ x_*^2)^\top$, for the design ξ_N satisfies $N\,\mathrm{var}(\mathbf{c}_*^\top\hat{\theta}^N_{LS}|\xi_N) = \mathbf{c}_*^\top\mathbf{M}^{-1}(\xi_N)\mathbf{c}_*$ with,

$$\mathbf{M}(\xi_N) = \mathbf{M}_N/N = \begin{pmatrix} \mu_2(N) & \mu_3(N) \\ \mu_3(N) & \mu_4(N) \end{pmatrix}$$

and, for $N = 2M$, $\mu_i(N) = x_*^i/2 + (1/N)\sum_{k=1}^{M}[x_* + (1/k)^\alpha]^i$, $i = 2, 3, 4$. We then obtain

$$\lim_{N\to\infty} N\,\mathrm{var}(\mathbf{c}_*^\top\hat{\theta}^N_{LS}|\xi_N) = V(\alpha) = \frac{2(1-\alpha)^2}{\alpha^2 + (1-\alpha)^2}, \tag{2.3}$$

which is monotonically decreasing in α for $0 < \alpha < 1/2$, with $V(0) = 2$ and $V(1/2) = 1$. For any $\alpha \in [0, 1/2)$ we thus have

$$\lim_{N\to\infty} N\,\mathrm{var}(\mathbf{c}_*^\top\hat{\theta}^N_{LS}|\xi_N) = V(\alpha) > N\,\mathrm{var}(\mathbf{c}_*^\top\hat{\theta}^N_{LS}|\xi_*) = 1\,.$$

That is, the limiting variance for ξ_N is always larger than the variance for the limiting design ξ_*. This is due to the discontinuity of the function $\mathbf{M}(\xi) \longrightarrow N\,\mathrm{var}(\mathbf{c}_*^\top\hat{\theta}^N_{LS}|\xi)$ at $\mathbf{M}(\xi_*)$; see Pázman (1980, 1986, p. 67) and Sects. 5.1.6 and 5.1.7.

Moreover, one can show that Lindeberg's condition is satisfied for any linear combination of θ (see, e.g., Shiryaev 1996) and for any $\mathbf{c} \neq \mathbf{0}$,

$$\sqrt{N}\frac{\mathbf{c}^\top(\hat{\theta}^N_{LS} - \bar{\theta})}{(\mathbf{c}^\top\mathbf{M}^{-1}(\xi_N)\mathbf{c})^{1/2}} \stackrel{\text{d}}{\rightarrow} \zeta \sim \mathcal{N}(0,1)\,,\ N \to \infty\,.$$

For $\mathbf{c} = \mathbf{c}_*$, $\mathbf{c}_*^\top\mathbf{M}^{-1}(\xi_N)\mathbf{c}_*$ tends to $V(\alpha)$ and

$$\sqrt{N}\mathbf{c}_*^\top(\hat{\theta}^N_{LS} - \bar{\theta}) \stackrel{\text{d}}{\rightarrow} \zeta_* \sim \mathcal{N}(0, V(\alpha))\,,\ N \to \infty\,,$$

whereas $\mathbf{c}^\top \mathbf{M}^{-1}(\xi_N)\mathbf{c}$ grows as $N^{2\alpha}$ for $\mathbf{c}$ not parallel to $\mathbf{c}_*$. Hence, the rate of convergence of $\mathbf{c}^\top \hat{\theta}^N_{LS}$ to $\mathbf{c}^\top \bar{\theta}$ depends on the direction of the vector $\mathbf{c}$. For instance, for $\mathbf{c} = \mathbf{c}_0 = (1,\ 0)^\top$, we obtain

$$N^{1/2-\alpha}\mathbf{c}_0^\top (\hat{\theta}^N_{LS} - \bar{\theta}) \stackrel{\mathrm{d}}{\to} \zeta_0 \sim \mathscr{N}(0, W(\alpha))\,,\ N \to \infty\,, \tag{2.4}$$

with

$$W(\alpha) = 2^{2(1-\alpha)}\,\frac{(1-2\alpha)(1-\alpha)^2}{\alpha^2 + (1-\alpha)^2}\,. \tag{2.5}$$

□

As previous example shows, the asymptotic properties of the estimator obtained for a design with empirical measure ξ_N converging weakly to some ξ and of the estimator for a randomized design with measure ξ may differ. This is true in particular for the estimation of scalar functions of the parameters; see Sect. 3.2. Using designs that obey Definitions 2.1 and 2.2 allow us to avoid the difficulties raised by Examples 2.3 and 2.4. In particular, we shall give a rigorous proof (Sect. 3.1) that both definitions give the same asymptotic properties for the LS estimator. For other estimators considered in the rest of the book only randomized designs will be used, but the results can easily be extended to the asymptotically discrete designs of Definition 2.1.

2.2 Uniform Convergence

Our proofs are based on uniform convergence with respect to θ of the function $J_N(\theta)$, called the estimation criterion, that defines the estimator $\hat{\theta}^N$ though $\hat{\theta}^N = \arg\min_\theta J_N(\theta)$. We shall thus need uniform SLLN.

In the case of an asymptotically discrete design we shall use the following result to establish asymptotic properties of the LS estimator.

Lemma 2.5 (Uniform SLLN 1). *Let $\{x_i\}$ be an asymptotically discrete design with measure ξ. Assume that $a(x,\theta)$ is a bounded function on $\mathscr{X} \times \Theta$ and that to every $x \in \mathscr{X}$ we can associate a random variable $\varepsilon(x)$. Let $\{\varepsilon_i\}$ be a sequence of independent random variables, with ε_i distributed like $\varepsilon(x_i)$, and assume that for all $x \in \mathscr{X}$,*

$$\mathbb{E}\{b[\varepsilon(x)]\} = m(x)\,,\ |m(x)| < \bar{m} < \infty\,,$$
$$\mathrm{var}\{b[\varepsilon(x)]\} = V(x) < \bar{V} < \infty\,,$$

with $b(\cdot)$ a Borel function on $\mathbb{R}$. Then we have

$$\frac{1}{N}\sum_{k=1}^{N} a(x_k,\theta)b(\varepsilon_k) \stackrel{\theta}{\rightsquigarrow} \sum_{x\in S_\xi} a(x,\theta)m(x)\xi(x)$$

as N tends to ∞, where $\overset{\theta}{\rightsquigarrow}$ means uniform convergence with respect to $\theta \in \Theta$, and the convergence is almost sure (a.s.), i.e., with probability one, with respect to the random sequence $\{\varepsilon_i\}$.

In the case of randomized designs we shall use the following result. An extension to more general situations is given in Lemma 2.6' of Sect. 2.3.

Lemma 2.6 (Uniform SLLN 2). *Let $\{z_i\}$ be a sequence of i.i.d. random vectors of $\mathbb{R}^r$ and $a(z,\theta)$ be a Borel measurable real function on $\mathbb{R}^r \times \Theta$, continuous in $\theta \in \Theta$ for any z, with Θ a compact subset of $\mathbb{R}^p$. Assume that*

$$\mathbb{E}\{\max_{\theta\in\Theta} |a(z_1,\theta)|\} < \infty\,, \tag{2.6}$$

then $\mathbb{E}\{a(z_1,\theta)\}$ is continuous in $\theta \in \Theta$ and

$$\frac{1}{N}\sum_{i=1}^{N} a(z_i,\theta) \overset{\theta}{\rightsquigarrow} \mathbb{E}[a(z_1,\theta)] \text{ a.s. when } N \to \infty\,.$$

In the proofs of the theorems of Chaps. 3 and 4, we shall have $z_i = (x_i, \varepsilon_i)$ for regression models, with x_i the design variable and ε_i the error for the i-th observation y_i or more generally $z_i = (x_i, y_i)$. The condition (2.6) will then be fulfilled when $\mathbb{E}_{x_1}\{\max_{\theta\in\Theta}|a(z_1,\theta)|\}$ is bounded for $x_1 \in \mathscr{X}$, i.e., in particular when the design space $\mathscr{X}$ is a compact set and $\mathbb{E}_{x_1}\{\max_{\theta\in\Theta}|a(z_1,\theta)|\}$ is a continuous function of x_1. More generally, a simple sufficient condition for (2.6) to be satisfied is when the continuity of $a(z,\theta)$ with respect to θ is uniform in z, as shown in the following lemma. In the case of randomized designs this assumption thus allows us to avoid any reference to uniform convergence properties; see, e.g., Fourgeaud and Fuchs (1967, p. 214) for maximum likelihood estimation.

Lemma 2.7. *Let $\{z_i\}$, θ, Θ and $a(z,\theta)$ be defined as in Lemma 2.6. Assume that*

$$\sup_{\theta\in\Theta} \mathbb{E}\{|a(z_1,\theta)|\} < \infty$$

and that $a(z,\theta)$ is continuous in $\theta \in \Theta$ uniformly in z. Then the conclusions of Lemma 2.6 apply.

For the case of asymptotically discrete designs and other estimators than LS, we shall use the following property.

Lemma 2.8 (Uniform SLLN 3). *Let $\{x_i\}$ be an asymptotically discrete design with measure ξ. Assume that to every $x \in \mathscr{X}$ we can associate a random variable $\varepsilon(x)$. Let $\{\varepsilon_i\}$ be a sequence of independent random variables, with ε_i distributed like $\varepsilon(x_i)$. Let $a(x,\varepsilon,\theta)$ be a Borel measurable function of ε for any $(x,\theta) \in \mathscr{X} \times \Theta$, continuous in $\theta \in \Theta$ for any x and ε, with Θ a compact subset of $\mathbb{R}^p$. Assume that*

$$\forall x \in \mathcal{S}_\xi\,, \quad \mathbb{E}\{\max_{\theta\in\Theta} |a[x,\epsilon(x),\theta]|\} < \infty\,,$$
$$\forall x \in \mathscr{X} \setminus \mathcal{S}_\xi\,, \quad \mathbb{E}\{\max_{\theta\in\Theta} |a[x,\epsilon(x),\theta]|^2\} < \infty\,,$$

with $\mathcal{S}_\xi$ defined by (2.1). Then we have

$$\frac{1}{N}\sum_{k=1}^{N} a(x_k,\epsilon_k,\theta) \overset{\theta}{\rightsquigarrow} \sum_{x\in S_\xi} \mathbb{E}\{a[x,\epsilon(x),\theta]\}\,\xi(x) \text{ a.s. when } N\to\infty\,,$$

where the function on the right-hand side is continuous in θ on Θ.

The uniform SLLN properties given in the lemmas above will be used to obtain the almost sure uniform convergence of the criterion function $J_N(\cdot)$, but first, we need the estimator to be properly defined as a random variable. This is ensured by the following lemma, taken from Jennrich (1969); see also Bierens (1994, p. 16). For the sake of completeness the proof is given in Appendix C.

Lemma 2.9. *Let Θ be a compact subset of $\mathbb{R}^p$, $\mathscr{Z}$ be a measurable subset of $\mathbb{R}^m$, and $J(z,\theta)$ be a Borel measurable real function on $\mathscr{Z}\times\Theta$, continuous in $\theta\in\Theta$ for any $z\in\mathscr{Z}$. Then there exists a mapping $\hat\theta$ from $\mathscr{Z}$ into Θ with Borel measurable components such that $J[z,\hat\theta(z)] = \min_{\theta\in\Theta} J(z,\theta)$, which therefore is also Borel measurable. If, moreover, $J(z,\theta)$ is continuous on $\mathscr{Z}\times\Theta$, then $\min_{\theta\in\Theta} J(z,\theta)$ is a continuous function on $\mathscr{Z}$.*

In the following, $\hat\theta^N$ will always denote the measurable choice from $\arg\min_\theta J_N(\theta)$ according to Lemma 2.9. The almost sure convergence of the estimator will then follow from the next lemma.

Lemma 2.10. *Assume that the sequence of functions $\{J_N(\theta)\}$ converges uniformly on Θ to the function $J_{\bar\theta}(\theta)$, with $J_N(\theta)$ continuous with respect to $\theta\in\Theta$ for any N, Θ a compact subset of $\mathbb{R}^p$ and $J_{\bar\theta}(\theta)$ such that*

$$\forall\theta\in\Theta\,,\ \theta\neq\bar\theta \Longrightarrow J_{\bar\theta}(\theta) > J_{\bar\theta}(\bar\theta)\,.$$

Then $\lim_{N\to\infty}\hat\theta^N = \bar\theta$, where $\hat\theta^N \in \arg\min_{\theta\in\Theta} J_N(\theta)$. When the functions $J_N(\cdot)$ are random and the uniform convergence to $J_{\bar\theta}(\cdot)$ is almost sure, the convergence of $\hat\theta^N$ to $\bar\theta$ is also almost sure.

This lemma admits a straightforward extension to the case where the function $J_{\bar\theta}(\theta)$ has several global minimizers. We then denote $\Theta^\#{=}\arg\min_{\theta\in\Theta} J_{\bar\theta}(\theta)$ the set of these minimizers and say that θ converges to $\Theta^\#$ when $d(\theta,\Theta^\#) = \min_{\theta'\in\Theta^\#}\|\theta-\theta'\|$ converges to zero. We have the following.

Lemma 2.11. *Assume that the sequence of functions $\{J_N(\theta)\}$ converges uniformly on Θ to the function $J_{\bar\theta}(\theta)$, with $J_N(\theta)$ continuous with respect to $\theta\in\Theta$ for any N, Θ a compact subset of $\mathbb{R}^p$. Let $\Theta^\# = \arg\min_{\theta\in\Theta} J_{\bar\theta}(\theta)$*

denote the set of minimizers of $J_{\bar{\theta}}(\theta)$. *Then* $\lim_{N\to\infty} d(\hat{\theta}^N, \Theta^{\#}) = 0$, *where* $\hat{\theta}^N \in \arg\min_{\theta\in\Theta} J_N(\theta)$. *When the functions* $J_N(\cdot)$ *are random and the uniform convergence to* $J_{\bar{\theta}}(\cdot)$ *is almost sure, the convergence of* $d(\hat{\theta}^N, \Theta^{\#})$ *to* 0 *is also almost sure.*

We shall also need to perform Taylor developments and apply a random version of the mean value theorem. This is possible through the following Lemma, taken from Jennrich (1969).

Lemma 2.12. *Let* Θ *be a convex compact subset of* $\mathbb{R}^p$, $\mathscr{Z}$ *be a measurable subset of* $\mathbb{R}^m$, *and* $J(z,\theta)$ *be a Borel measurable real function on* $\mathscr{Z} \times \Theta$, *continuously differentiable in* $\theta \in \mathrm{int}(\Theta)$ *for any* $z \in \mathscr{Z}$. *Let* $\theta^1(z)$ *and* $\theta^2(z)$ *be measurable functions from* $\mathscr{Z}$ *into* Θ. *There exists a measurable function* $\tilde{\theta}$ *from* $\mathscr{Z}$ *into* $\mathrm{int}(\Theta)$ *such that for all* $z \in \mathscr{Z}$, $\tilde{\theta}(z)$ *lies on the segment joining* $\theta^1(z)$ *and* $\theta^2(z)$ *and*

$$J[z,\theta^1(z)] - J[z,\theta^2(z)] = \left.\frac{\partial J(z,\theta)}{\partial \theta^\top}\right|_{\tilde{\theta}(z)} [\theta^1(z) - \theta^2(z)]\,.$$

2.3 Bibliographic Notes and Further Remarks

A general SLLN

Lemma 2.6 can be extended to situations where the z_k are not i.i.d., which, concerning the applications considered in this book, is of interest to establish asymptotic properties of estimators when the experiment *does not* correspond to a randomized design as defined by Definition 2.2. This extension is as follows; see Bierens (1994, Theorem 2.7.1).

Lemma 2.6'. *Let* $z_1, z_2 \ldots$ *be a sequence of independent random vectors with distribution functions* $\mathbb{F}_1, \mathbb{F}_2, \ldots$ *and* $a(z,\theta)$ *be a continuous real function of* $(z,\theta) \in \mathbb{R}^r \times \Theta$, *with* Θ *a compact subset of* $\mathbb{R}^p$. *Suppose that*

$$\frac{1}{N}\sum_{k=1}^{N} \mathbb{F}_k \Rightarrow \mathbb{F} \text{ when } N \to \infty$$

$$\limsup_{N\to\infty} \frac{1}{N}\sum_{k=1}^{N} \mathbb{E}\{\max_{\theta\in\Theta} |a(z_k,\theta)|^{2+\alpha}\} < \infty \text{ for some } \alpha > 0\,,$$

then $\mathbb{E}\{a(z,\theta)\}$ *is continuous in* $\theta \in \Theta$ *and*

$$\frac{1}{N}\sum_{i=1}^{N} a(z_i,\theta) \overset{\theta}{\rightsquigarrow} \mathbb{E}[a(z,\theta)] \text{ a.s. when } N \to \infty$$

with z *having the distribution* $\mathbb{F}$.

Uniform Convergence and Stochastic Equicontinuity

A general approach for obtaining uniform convergence results over classes of functions (possibly parameterized) is to use properties of empirical processes; see, e.g., van de Geer (2000), van der Vaart and Wellner (1996), and van der Vaart (1998). Also, although not considered here, stochastic equicontinuity is a powerful tool to derive uniform laws of large numbers; see Andrews (1987, 1992), and Newey (1991), for several extensions, in particular to strong (a.s.) convergence and totally bounded parameter spaces instead of compact sets.[1]

The standard definition of equicontinuity is as follows; see, e.g., Andrews (1992). A family of functions $\{J_N(\cdot) : N \geq 1\}$ is stochastically equicontinuous on Θ if for any $\epsilon > 0$ there exists $\delta > 0$ such that

$$\limsup_{N\to\infty} \operatorname{Prob}\left\{ \sup_{\theta\in\Theta} \sup_{\theta'\in\mathscr{B}(\theta,\delta)} |J_N(\theta') - J_N(\theta)| > \epsilon \right\} < \epsilon .$$

Andrews (1992, Theorem 1) then shows that when Θ is totally bounded and $J_N(\theta) \xrightarrow{\text{p}} J(\theta)$ as $N \to \infty$ for any $\theta \in \Theta$ (pointwise convergence in probability), then the equicontinuity of $\{J_N(\cdot) : N \geq 1\}$ implies the uniform convergence of $J_N(\cdot)$ to $J(\cdot)$ in probability; that is, $\sup_{\theta\in\Theta} |J_N(\theta) - J(\theta)| \xrightarrow{\text{p}} 0$, $N \to \infty$. Conversely, the uniform convergence in probability is shown to imply the pointwise convergence and stochastic equicontinuity of $J_N(\cdot)$; see also Newey (1991, Theorem 2.1). In the case where $J_N(\theta) = (1/N)\sum_{i=1}^N j_i(\theta)$, this yields uniform laws of large numbers under growth conditions for the random functions $j_i(\cdot)$; see, e.g., Newey (1991, Corollary 3.1).

[1] A subset of a metric space is totally bounded if it can be covered by a finite number of balls with radius ϵ for any ϵ, which is weaker than compactness. A subset of an Euclidean space Θ is totally bounded if and only if it is bounded, which compared to compactness relaxes the condition of Θ being closed. In the next chapters, the assumption of compactness of Θ will force us to work in the interior of Θ when differentiability will be required, in particular, to obtain the asymptotic normality of estimators.

3

Asymptotic Properties of the LS Estimator

3.1 Asymptotic Properties of the LS Estimator in Regression Models

We consider asymptotic properties ($N \to \infty$) of the (ordinary) LS estimator $\hat{\theta}^N_{LS}$ for a model defined by the mean (or expected) response $\eta(x, \theta)$. For observations $y(x_1), \ldots, y(x_N)$, $\hat{\theta}^N_{LS}$ is obtained by minimizing

$$J_N(\theta) = \frac{1}{N} \sum_{i=1}^{N} [y(x_i) - \eta(x_i, \theta)]^2 , \tag{3.1}$$

with respect to $\theta \in \Theta$. We assume that the true model is

$$y(x_i) = \eta(x_i, \bar{\theta}) + \varepsilon_i , \quad \text{with } \bar{\theta} \in \Theta \text{ and } \mathbb{E}\{\varepsilon_i\} = 0 \text{ for all } i , \tag{3.2}$$

where $\{\varepsilon_i\}$ is a sequence of independent random variables. The errors can be (second-order) stationary (or homoscedastic)

$$\mathbb{E}\{\varepsilon_i^2\} = \sigma^2 < \infty, \tag{3.3}$$

or nonstationary (heteroscedastic)

$$\mathbb{E}\{\varepsilon_i^2\} = \sigma^2(x_i) , \tag{3.4}$$

where $\sigma^2(x)$ is defined for every $x \in \mathscr{X}$ with the property $0 < a < \sigma^2(x) < b < \infty$. In the nonstationary case, we assume that the distribution of the errors ε_i in (3.2) only depends on the design point x_i. Notice that the random variables (vectors) $z_i = (x_i, \varepsilon_i)$ are then i.i.d. when the x_i are i.i.d. (randomized design).

We shall also consider weighted LS estimation, where $\hat{\theta}^N_{WLS}$ minimizes

$$J_N(\theta) = \frac{1}{N} \sum_{i=1}^{N} w(x_i)[y(x_i) - \eta(x_i, \theta)]^2 , \tag{3.5}$$

L. Pronzato and A. Pázman, *Design of Experiments in Nonlinear Models*, Lecture Notes in Statistics 212, DOI 10.1007/978-1-4614-6363-4_3,

with $w(x) \geq 0$ the weighting function. The following hypotheses will be used throughout this book:

H$_\Theta$: Θ is a compact subset of $\mathbb{R}^p$ such that $\Theta \subset \overline{\text{int}(\Theta)}$.
H1$_\eta$: $\eta(x, \theta)$ is bounded on $\mathscr{X} \times \Theta$ and $\eta(x, \theta)$ is continuous on Θ, $\forall x \in \mathscr{X}$.
H2$_\eta$: $\bar{\theta} \in \text{int}(\Theta)$ and, $\forall x \in \mathscr{X}$, $\eta(x, \theta)$ is twice continuously differentiable with respect to $\theta \in \text{int}(\Theta)$, and its first two derivatives are bounded on $\mathscr{X} \times \text{int}(\Theta)$.

We first prove the strong consistency of the ordinary LS estimator $\hat{\theta}^N_{LS}$ (Sect. 3.1.1). The extension to weighted LS (WLS) is straightforward. Next, in Sect. 3.1.2, we relax the estimability condition on the support of ξ and show that consistency can still be obtained when $\mathscr{X}$ is finite, by taking into account the information provided by design points that asymptotically receive zero mass, i.e. such that $\xi(x) = 0$. Again, the proof is given for $\hat{\theta}^N_{LS}$, but the extension to WLS and nonstationary errors is immediate. The asymptotic normality of the WLS estimator is considered in Sect. 3.1.3, both for randomized and nonrandomized (asymptotically discrete) designs. The estimation of a scalar function of $\hat{\theta}^N_{LS}$ is considered in Sects. 3.1.4 and 3.2 for situations where $\hat{\theta}^N_{LS}$ is not consistent.

3.1.1 Consistency

Theorem 3.1 (Consistency of the LS estimator). *Let $\{x_i\}$ be an asymptotically discrete design (Definition 2.1) or a randomized design (Definition 2.2) on $\mathscr{X} \subset \mathbb{R}^d$. Consider the estimator $\hat{\theta}^N_{LS}$ that minimizes (3.1) in the model (3.2), (3.4). Assume that H_Θ and $H1_\eta$ are satisfied and that the parameters of the model are LS estimable for the design ξ at $\bar{\theta}$, that is:*

$$\forall \theta \in \Theta\,, \quad \int_{\mathscr{X}} [\eta(x, \theta) - \eta(x, \bar{\theta})]^2 \xi(\mathrm{d}x) = 0 \Longleftrightarrow \theta = \bar{\theta}\,. \tag{3.6}$$

Then, w.p.1 the observed sequence $y(x_1), y(x_2), \ldots$ is such that

$$\lim_{N \to \infty} \hat{\theta}^N_{LS} = \bar{\theta} \quad \text{and} \quad \lim_{N \to \infty} \left[\hat{\sigma}^2\right]^N = \int_{\mathscr{X}} \sigma^2(x) \xi(\mathrm{d}x)$$

where

$$\left[\hat{\sigma}^2\right]^N = \frac{1}{N-p} \sum_{k=1}^{N} \left[y(x_k) - \eta(x_k, \hat{\theta}^N_{LS})\right]^2 \tag{3.7}$$

with $p = \dim(\theta)$.

Proof. For any $\theta \in \Theta$

$$J_N(\theta) = \frac{1}{N}\sum_{k=1}^{N}\varepsilon_k^2 + \frac{2}{N}\sum_{k=1}^{N}\left[\eta\left(x_k,\bar{\theta}\right) - \eta\left(x_k,\theta\right)\right]\varepsilon_k$$
$$+\frac{1}{N}\sum_{k=1}^{N}\left[\eta\left(x_k,\bar{\theta}\right) - \eta\left(x_k,\theta\right)\right]^2 . \tag{3.8}$$

The first term in (3.8) converges a.s. to $\int_{\mathscr{X}}\sigma^2(x)\xi(\mathrm{d}x)$ as $N\to\infty$ (SLLN).

Consider first the case where $\{x_i\}$ is an asymptotically discrete design. Define $a(x,\theta) = [\eta(x,\bar{\theta}) - \eta(x,\theta)]$, $b(\varepsilon) = \varepsilon$. Lemma 2.5 implies that the second term of (3.8) converges to zero uniformly in θ and a.s. The third term converges to $\int_{\mathscr{X}}[\eta(x,\bar{\theta}) - \eta(x,\theta)]^2\xi(\mathrm{d}x)$ uniformly in θ.

Consider now the case where $\{x_i\}$ is a randomized design. Define $a(z,\theta) = [\eta(x,\bar{\theta}) - \eta(x,\theta)]\varepsilon$ with $z=(x,\varepsilon)$. We have

$$\begin{aligned}\mathbb{E}\{\max_{\theta\in\Theta}|a(z,\theta)|\} &= \int_{\mathscr{X}}\max_{\theta\in\Theta}|\eta(x,\bar{\theta}) - \eta(x,\theta)|\,\mathbb{E}_x\{|\varepsilon|\}\,\xi(\mathrm{d}x)\\ &\le \int_{\mathscr{X}}\sigma(x)\max_{\theta\in\Theta}|\eta(x,\bar{\theta}) - \eta(x,\theta)|\xi(\mathrm{d}x) < \infty\,,\end{aligned}$$

so that Lemma 2.6 implies that the second term in (3.8) converges to $\mathbb{E}\{a(z,\theta)\} = 0$ as $N\to\infty$ and the convergence is uniform in θ and a.s. with respect to x and ε. Take now $a(z,\theta) = [\eta(x,\bar{\theta}) - \eta(x,\theta)]^2$ with $z=x$. We have $\mathbb{E}\{\max_{\theta\in\Theta}|a(z,\theta)|\} = \int_{\mathscr{X}}\max_{\theta\in\Theta}[\eta(x,\bar{\theta})-\eta(x,\theta)]^2\xi(\mathrm{d}x) < \infty$, and Lemma 2.6 implies for the third term in (3.8):

$$\frac{1}{N}\sum_{k=1}^{N}\left[\eta\left(x_k,\bar{\theta}\right) - \eta\left(x_k,\theta\right)\right]^2 \overset{\theta}{\rightsquigarrow} \int_{\mathscr{X}}[\eta(x,\bar{\theta}) - \eta(x,\theta)]^2\xi(\mathrm{d}x)\,,\ \text{a.s.}$$

Therefore, for both types of designs,

$$J_N(\theta) \overset{\theta}{\rightsquigarrow} J_{\bar{\theta}}(\theta) = \int_{\mathscr{X}}\sigma^2(x)\xi(\mathrm{d}x) + \int_{\mathscr{X}}[\eta(x,\bar{\theta}) - \eta(x,\theta)]^2\xi(\mathrm{d}x)\,,\ \text{a.s.}$$

as $N\to\infty$. The conditions of Lemma 2.10 are satisfied and $\hat{\theta}^N_{LS} \overset{\text{a.s.}}{\to} \bar{\theta}$ as $N\to\infty$.

Finally, the strong consistency of $\hat{\theta}^N_{LS}$ and $J_N(\theta) \overset{\theta}{\rightsquigarrow} J_{\bar{\theta}}(\theta)$ a.s. imply that the estimator (3.7) converges to $\int_{\mathscr{X}}\sigma^2(x)\xi(\mathrm{d}x)$ a.s. ■

Remark 3.2.

(*i*) When $\{x_i\}$ is a randomized design, the assumption of boundedness of $\eta(x,\theta)$ can be replaced by

$$\int_{\mathscr{X}}\sup_{\theta\in\Theta}[\eta(x,\theta) - \eta(x,\bar{\theta})]^2\,\xi(\mathrm{d}x) < \infty\,. \tag{3.9}$$

(*ii*) One can readily check that the WLS estimator that minimizes (3.5) is strongly consistent when $w(x)$ is bounded and the LS estimability condition (3.6) in Theorem 3.1 is replaced by

$$\forall\theta \in \Theta\,, \int_{\mathscr{X}} w(x)[\eta(x,\theta) - \eta(x,\bar{\theta})]^2\xi(\mathrm{d}x) = 0 \Longleftrightarrow \theta = \bar{\theta}\,. \qquad (3.10)$$

When $\{x_i\}$ is a randomized design, the condition of boundedness of $w(x)$ and of $\eta(x,\theta)$ in $\mathrm{H1}_\eta$ can be replaced by

$$\int_{\mathscr{X}} w(x) \sup_{\theta\in\Theta}[\eta(x,\theta) - \eta(x,\bar{\theta})]^2\, \xi(\mathrm{d}x) < \infty\,. \qquad (3.11)$$

(*iii*) When $\sigma^2(x) = \sigma^2$ for any x, $[\hat{\sigma}^2]^N$ given by (3.7) is a strongly consistent estimator of σ^2. It remains strongly consistent when $\hat{\theta}^N_{LS}$ is replaced by any strongly consistent estimator of θ.
(*iv*) When $\{x_i\}$ is asymptotically discrete, with $\mathcal{S}_\xi = (x^{(1)},\ldots,x^{(k)})$ the support of the limiting measure ξ, under H_Θ and $\mathrm{H1}_\eta$, the WLS criterion (3.5) satisfies

$$J_N(\theta) \overset{\theta}{\rightsquigarrow} \frac{1}{k}\sum_{i=1}^{k} w(x^{(i)})\xi(x^{(i)})\sigma^2(x^{(i)}) + \frac{1}{k}\sum_{i=1}^{k} w(x^{(i)})\xi(x^{(i)})[\eta(x^{(i)},\theta) - \eta(x^{(i)},\bar{\theta})]^2\,, \text{ a.s.}$$

and therefore

$$J_N(\theta) - J_N(\hat{\theta}^N_{WLS}) \overset{\theta}{\rightsquigarrow} \frac{1}{k}\sum_{i=1}^{k} w(x^{(i)})\xi(x^{(i)})[\eta(x^{(i)},\theta) - \eta(x^{(i)},\bar{\theta})]^2\,, \text{ a.s.} \;\square$$

Even when the LS estimability condition (3.6) of Theorem 3.1 is not satisfied, the estimator of a parametric function $h(\theta)$ can still be strongly consistent. This is expressed in the following theorem. Notice that the case of a vector function $\mathbf{h}(\cdot) : \Theta \longrightarrow \mathbb{R}^q$ need not be considered separately since the consistency of an estimator of $\mathbf{h}(\theta)$ is equivalent to that of its individual components.

Theorem 3.3 (Consistency of a function of the LS estimator). *Suppose that the condition (3.6) of Theorem 3.1 is replaced by*

$$\forall\theta \in \Theta\,, \int_{\mathscr{X}} [\eta(x,\theta) - \eta(x,\bar{\theta})]^2\, \xi(\mathrm{d}x) = 0 \Longrightarrow h(\theta) = h(\bar{\theta}) \qquad (3.12)$$

for some scalar function $h(\cdot)$ continuous on Θ. Assume that the other assumptions of Theorem 3.1 remain valid, then $h(\hat{\theta}^N_{LS}) \overset{\text{a.s.}}{\to} h(\bar{\theta})$ as $N \to \infty$.

Proof. Take $\Theta^{\#} = \{\theta \in \Theta : \int_{\mathscr{X}} [\eta(x,\theta) - \eta(x,\bar{\theta})]^2 \xi(dx) = 0\}$. Following the same arguments as in the proof of Theorem 3.1, according to Lemma 2.11, $\hat{\theta}^N_{LS}$ converges to $\Theta^{\#}$; that is, all limit points of $\hat{\theta}^N_{LS}$ belong to $\Theta^{\#}$ a.s. From the continuity of $h(\cdot)$, it follows that all limit points of $h(\hat{\theta}^N_{LS})$ belong to $h(\Theta^{\#})$; that is, they are equal to $h(\bar{\theta})$, a.s. ■

This property can be generalized as follows. Let $\hat{\theta}^N$ be a measurable choice from $\arg\min_{\theta \in \Theta} J_N(\theta)$ for some criterion $J_N(\cdot)$ and assume that $J_N(\theta) \overset{\theta}{\rightsquigarrow} J_{\bar{\theta}}(\theta)$ a.s. when $N \to \infty$. Suppose that $h(\cdot)$ is continuous on Θ and that $\theta^* \in \arg\min_{\theta \in \Theta} J_{\bar{\theta}}(\theta)$ implies $h(\theta^*) = h(\bar{\theta})$. Then, $h(\hat{\theta}^N) \overset{\text{a.s.}}{\to} h(\bar{\theta})$ when $N \to \infty$.

3.1.2 Consistency Under a Weaker LS Estimability Condition

Suppose that $\{x_i\}$ is an asymptotically discrete design; see Definition 2.1. The LS estimability condition (3.6) in Theorem 3.1 concerns the asymptotic support $\mathcal{S}_\xi$ of the design. For a linear model with $\eta(x,\theta) = \mathbf{f}^\top(x)\theta$, it implies that $\lim_{N\to\infty} \lambda_{\min}(\mathbf{M}_N/N) > \epsilon > 0$, with $\mathbf{M}_N = \sum_{k=1}^N \mathbf{f}(x_k)\mathbf{f}^\top(x_k)$, which is clearly over-restrictive. Indeed, using martingale convergence results, Lai et al. (1978, 1979) show that $\lambda_{\min}(\mathbf{M}_N) \to \infty$ is sufficient for the strong consistency of the LS estimator[1] (provided that the design is not sequential).

The condition (3.6) was used in order to apply Lemma 2.10, which supposes that $NJ_N(\theta)$ grows to infinity at rate N when $\theta \neq \bar{\theta}$, which is also used in the classic reference (Jennrich 1969). On the other hand, using this condition amounts to ignoring the information provided by design points $x \in \mathscr{X}$ with a relative frequency $N(x)/N$ tending to zero, which therefore do not appear in the support of ξ. In order to acknowledge the information carried by such points, we shall follow the same approach as in (Wu, 1981), based on an idea originated from Wald's proof of the strong consistency of the maximum likelihood (ML) estimator (Wald 1949). Since the points x such that $N(x)/N \to 0$ do not contribute to the (discrete) design measure ξ, the following asymptotic results cannot be expressed in terms of ξ. However, they are of importance for sequential design of experiments; see Sect. 8.5.

We denote

$$S_N(\theta) = NJ_N(\theta) = \sum_{k=1}^{N} [y(x_k) - \eta(x_k,\theta)]^2 \,. \tag{3.13}$$

We shall need the following lemma of Wu (1981); see Appendix C.

---

[1]In fact, they consider the more general situation where the errors ε_k form a martingale difference sequence with respect to an increasing sequence of σ-fields $\mathcal{F}_k$ such that $\sup_k(\varepsilon_k^2|\mathcal{F}_{k-1}) < \infty$. When the errors ε_k are i.i.d. with zero mean and variance $\sigma^2 > 0$, they also show that $\lambda_{\min}(\mathbf{M}_N) \to \infty$ is both necessary and sufficient for the strong consistency of $\hat{\theta}^N_{LS}$.

Lemma 3.4. *If for any $\delta > 0$*

$$\liminf_{N\to\infty} \inf_{\|\theta-\bar{\theta}\|\geq\delta} [S_N(\theta) - S_N(\bar{\theta})] > 0 \text{ a.s.} \tag{3.14}$$

then $\hat{\theta}^N_{LS} \overset{\text{a.s.}}{\to} \bar{\theta}$ as $N \to \infty$. If for any $\delta > 0$,

$$\text{Prob}\left\{ \inf_{\|\theta-\bar{\theta}\|\geq\delta} [S_N(\theta) - S_N(\bar{\theta})] > 0 \right\} \to 1\,,\ N \to \infty\,, \tag{3.15}$$

then $\hat{\theta}^N_{LS} \overset{\text{p}}{\to} \bar{\theta}$ as $N \to \infty$.

One can then prove the convergence of the LS estimator, in probability and a.s., when the sum $\sum_{k=1}^N [\eta(x_k,\theta) - \eta(x_k,\bar{\theta})]^2$ tends to infinity fast enough for $\|\theta - \bar{\theta}\| \geq \delta > 0$ and the design space $\mathscr{X}$ for the x_k is finite.

Theorem 3.5 (Consistency of the LS estimator with finite $\mathscr{X}$). *Let $\{x_i\}$ be a design sequence on a finite set $\mathscr{X}$. Assume that*

$$D_N(\theta,\bar{\theta}) = \sum_{k=1}^N [\eta(x_k,\theta) - \eta(x_k,\bar{\theta})]^2 \tag{3.16}$$

satisfies

$$\forall \delta > 0\,,\ \left[\inf_{\|\theta-\bar{\theta}\|\geq\delta} D_N(\theta,\bar{\theta}) \right] / (\log\log N) \overset{\text{a.s.}}{\to} \infty\,,\ N \to \infty\,. \tag{3.17}$$

Then, $\hat{\theta}^N_{LS} \overset{\text{a.s.}}{\to} \bar{\theta}$ as $N \to \infty$, where $\hat{\theta}^N_{LS}$ minimizes $S_N(\theta)$ in the model (3.2), (3.3). When $D_N(\theta,\bar{\theta})$ simply satisfies $\inf_{\|\theta-\bar{\theta}\|\geq\delta} D_N(\theta,\bar{\theta}) \to \infty$ as $N \to \infty$, then $\hat{\theta}^N_{LS} \overset{\text{p}}{\to} \bar{\theta}$.

Proof. The proof is based on Lemma 3.4. Denote $\mathcal{I}_N(x) = \{k \in \{1,\ldots,N\} : x_k = x\}$. We have

$$\begin{aligned} &S_N(\theta) - S_N(\bar{\theta}) \\ &= D_N(\theta,\bar{\theta}) \left[1 + 2\, \frac{\sum_{x\in\mathscr{X}} \left(\sum_{k\in\mathcal{I}_N(x)} \varepsilon_k \right) [\eta(x,\bar{\theta}) - \eta(x,\theta)]}{D_N(\theta,\bar{\theta})} \right] \\ &\geq D_N(\theta,\bar{\theta}) \left[1 - 2\, \frac{\sum_{x\in\mathscr{X}} \left| \sum_{k\in\mathcal{I}_N(x)} \varepsilon_k \right| |\eta(x,\bar{\theta}) - \eta(x,\theta)|}{D_N(\theta,\bar{\theta})} \right]. \end{aligned}$$

From Lemma 3.4, under the condition (3.17) it suffices to prove that

$$\sup_{\|\theta-\bar{\theta}\|\geq\delta} \frac{\sum_{x\in\mathscr{X}} \left| \sum_{k\in\mathcal{I}_N(x)} \varepsilon_k \right| |\eta(x,\bar{\theta}) - \eta(x,\theta)|}{D_N(\theta,\bar{\theta})} \overset{\text{a.s.}}{\to} 0 \tag{3.18}$$

for any $\delta > 0$ to obtain the strong consistency of $\hat{\theta}_{LS}^N$. Since $D_N(\theta, \bar{\theta}) \to \infty$, only the design points such that $N(x) \to \infty$ have to be considered, where $N(x)$ denotes the number of times x appears in the sequence $x_1, \ldots, x_N$. Define $\beta(n) = \sqrt{n \log\log n}$. We have

$$\forall x \in \mathscr{X}\,,\ \limsup_{N(x)\to\infty} \left| \frac{1}{\beta[N(x)]} \sum_{k\in\mathcal{I}_N(x)} \varepsilon_k \right| < B\,,\ \text{a.s.} \tag{3.19}$$

for some $B > 0$ from the Law of the Iterated Logarithm; see, e.g., Shiryaev (1996, p. 397). Next, Cauchy–Schwarz inequality gives

$$\sum_{x\in\mathscr{X}} \frac{\beta[N(x)]|\eta(x,\bar{\theta}) - \eta(x,\theta)|}{D_N(\theta,\bar{\theta})} \leq$$

$$\frac{1}{D_N(\theta,\bar{\theta})} \left(\sum_{x\in\mathscr{X}} \log\log N(x) \right)^{1/2} \left(\sum_{x\in\mathscr{X}} N(x)[\eta(x,\bar{\theta}) - \eta(x,\theta)]^2 \right)^{1/2}$$
$$= \left(\frac{\sum_{x\in\mathscr{X}} \log\log N(x)}{D_N(\theta,\bar{\theta})} \right)^{1/2}.$$

Using the concavity of the function $\log\log(\cdot)$, we obtain

$$\sum_{x\in\mathscr{X}} \log\log N(x) \leq \ell[\log\log(N/\ell)]$$

where ℓ is the number of elements of $\mathscr{X}$, so that

$$\sum_{x\in\mathscr{X}} \frac{\beta[N(x)]|\eta(x,\bar{\theta}) - \eta(x,\theta)|}{D_N(\theta,\bar{\theta})} \leq \left(\frac{\ell[\log\log(N/\ell)]}{D_N(\theta,\bar{\theta})} \right)^{1/2},$$

which, together with (3.17) and (3.19), gives (3.18).

When $\inf_{\|\theta-\bar{\theta}\|\geq\delta} D_N(\theta,\bar{\theta}) \to \infty$ as $N \to \infty$, we only need to prove that

$$\sup_{\|\theta-\bar{\theta}\|\geq\delta} \frac{\sum_{x\in\mathscr{X}} \left|\sum_{k\in\mathcal{I}_N(x)} \varepsilon_k\right| |\eta(x,\bar{\theta}) - \eta(x,\theta)|}{D_N(\theta,\bar{\theta})} \xrightarrow{\text{p}} 0 \tag{3.20}$$

for any $\delta > 0$ to obtain the weak consistency of $\hat{\theta}_{LS}^N$. We proceed as above and only consider the design points such that $N(x) \to \infty$, with now $\beta(n) = \sqrt{n}$. From the central limit theorem for i.i.d. random variables, for any $x \in \mathscr{X}$, $\left(\sum_{k\in\mathcal{I}_N(x)} \varepsilon_k\right) / \sqrt{N(x)} \xrightarrow{\text{d}} \zeta_x \sim \mathscr{N}(0,\sigma^2)$ as $N \to \infty$ and is thus bounded in probability. From Cauchy–Schwarz inequality

$$\sum_{x\in\mathscr{X}} \frac{\sqrt{N(x)}|\eta(x,\bar{\theta}) - \eta(x,\theta)|}{D_N(\theta,\bar{\theta})} \leq \frac{1}{D_N^{1/2}(\theta,\bar{\theta})},$$

which gives (3.20). ■

Remark 3.6.

(*i*) The condition

$$\forall \theta \neq \bar{\theta}\,,\ D_N(\theta, \bar{\theta}) = \sum_{k=1}^{N} [\eta(x_k, \theta) - \eta(x_k, \bar{\theta})]^2 \to \infty \text{ as } N \to \infty \quad (3.21)$$

is proved in (Wu, 1981) to be necessary for the existence of a weakly consistent estimator of $\bar{\theta}$ in the regression model $y(x_i) = \eta(x_i, \bar{\theta}) + \varepsilon_i$ with i.i.d. errors ε_i having a distribution with density $\bar{\varphi}(\cdot)$ with respect to the Lebesgue measure, positive almost everywhere, and with finite Fisher information for location ($\int [\bar{\varphi}'(x)]^2/\bar{\varphi}(x)dx < \infty$). Notice that the LS estimability condition (3.6) corresponds to $D_N(\theta, \bar{\theta}) = \mathcal{O}(N)$, which is much stronger than (3.17).

(*ii*) The condition (3.21) is also sufficient for the strong consistency of $\hat{\theta}^N_{LS}$ when the parameter set Θ is finite; see Wu (1981). From Theorem 3.5, when $\mathscr{X}$ is finite, this condition is sufficient for the weak consistency of $\hat{\theta}^N_{LS}$ without restriction on Θ.

(*iii*) As for Theorem 3.1, the theorem remains valid if the errors ε_i in (3.2) are no longer stationary but satisfy (3.4) and/or if the LS estimator $\hat{\theta}^N_{LS}$ is replaced by the WLS estimator $\hat{\theta}^N_{WLS}$ that minimizes (3.5) with $w(x) > 0$ and bounded on $\mathscr{X}$. The extension to $\{\varepsilon_i\}$ forming a martingale difference sequence is considered in (Pronzato, 2009a, Theorem 1). In that case, the a.s. convergence of $\hat{\theta}^N_{LS}$ is obtained under the condition $\forall \delta > 0$, $\left[\inf_{\|\theta - \bar{\theta}\| \geq \delta} D_N(\theta, \bar{\theta})\right] / (\log N)^\rho \overset{\text{a.s.}}{\to} \infty$ as $N \to \infty$ for some $\rho > 1$.

(*iv*) The continuity of $\eta(x, \theta)$ with respect to θ is not required in Theorem 3.5.

(*v*) Wu (1981) also considers the asymptotic normality of $\hat{\theta}^N_{LS}$ and shows that, under suitable conditions, when $D_N(\theta, \bar{\theta})$ tends to infinity at a rate slower than N for $\theta \neq \bar{\theta}$, then $\tau_N(\hat{\theta}^N_{LS} - \bar{\theta})$ tends to be normally distributed as $N \to \infty$, with τ_N tending to infinity more slowly than the usual $\sqrt{N}$; compare with the results in Sect. 3.1.3. See also Pronzato (2009a, Theorem 2). Theorem 3.5 will be used in Sect. 3.2.2 to obtain the asymptotic normality of a scalar function of $\hat{\theta}^N_{LS}$. □

It is important to notice that we never used the condition that the x_i were nonrandom constants, so that Theorem 3.5 also applies for sequential design provided that $\mathscr{X}$ is finite; this will be used in Sect. 8.5.2. In this context, it is interesting to compare the results of the theorem with those obtained without the assumption of a finite design space $\mathscr{X}$. For linear regression, the condition (3.17) takes the form $\log\log N = o[\lambda_{\min}(\mathbf{M}_N)]$. Noticing that $\lambda_{\max}(\mathbf{M}_N) = \mathcal{O}(N)$, we thus get a condition weaker than the sufficient condition $\{\log[\lambda_{\max}(\mathbf{M}_N)]\}^{1+\alpha} = o[\lambda_{\min}(\mathbf{M}_N)]$ derived by Lai and Wei (1982) for the strong convergence of the LS estimator in a linear regression model under a sequential design. Also, in nonlinear regression, (3.17) is much less restrictive than the condition obtained by Lai (1994) for the strong consistency of the LS estimator under a sequential design.[2]

3.1.3 Asymptotic Normality

The following lemma will be useful to compare the asymptotic covariance matrices of different estimators. Its proof is given in Appendix C.

Lemma 3.7. *Let* $\mathbf{u}, \mathbf{v}$ *be two random vectors of* $\mathbb{R}^r$ *and* $\mathbb{R}^s$ *respectively defined on a probability space with probability measure* μ*, with* $\mathbb{E}(\|\mathbf{u}\|^2) < \infty$ *and* $\mathbb{E}(\|\mathbf{v}\|^2) < \infty$*. We have*

$$\mathbb{E}(\mathbf{u}\mathbf{u}^\top) \succeq \mathbb{E}(\mathbf{u}\mathbf{v}^\top)[\mathbb{E}(\mathbf{v}\mathbf{v}^\top)]^+\mathbb{E}(\mathbf{v}\mathbf{u}^\top), \tag{3.22}$$

where $\mathbf{M}^+$ *denotes the Moore–Penrose g-inverse of* $\mathbf{M}$ *(i.e., is such that* $\mathbf{M}\mathbf{M}^+\mathbf{M} = \mathbf{M}$, $\mathbf{M}^+\mathbf{M}\mathbf{M}^+ = \mathbf{M}^+$, $(\mathbf{M}\mathbf{M}^+)^\top = \mathbf{M}\mathbf{M}^+$ *and* $(\mathbf{M}^+\mathbf{M})^\top = \mathbf{M}^+\mathbf{M}$*) and* $\mathbf{A} \succeq \mathbf{B}$ *means that* $\mathbf{A} - \mathbf{B}$ *is nonnegative definite. Moreover, the equality is obtained in (3.22) if and only if* $\mathbf{u} = \mathbf{A}\mathbf{v}$ μ*-a.s. for some nonrandom matrix* $\mathbf{A}$*.*

We consider directly the case of WLS estimation with the criterion (3.5).

Theorem 3.8 (Asymptotic normality of the WLS estimator). *Let* $\{x_i\}$ *be an asymptotically discrete design (Definition 2.1) or a randomized design (Definition 2.2) on* $\mathscr{X} \subset \mathbb{R}^d$*. Consider the estimator* $\hat{\theta}^N_{WLS}$ *that minimizes (3.5) in the model (3.2), (3.4). Assume that* H_Θ, $H1_\eta$, $H2_\eta$*, and the LS estimability condition (3.10) are satisfied and that the matrix*

$$\mathbf{M}_1(\xi, \bar{\theta}) = \int_{\mathscr{X}} w(x) \left.\frac{\partial \eta(x,\theta)}{\partial \theta}\right|_{\bar{\theta}} \left.\frac{\partial \eta(x,\theta)}{\partial \theta^\top}\right|_{\bar{\theta}} \xi(\mathrm{d}x) \tag{3.23}$$

is nonsingular, with $w(x)$ *bounded on* $\mathscr{X}$*. Then,* $\hat{\theta}^N_{WLS}$ *satisfies*

$$\sqrt{N}(\hat{\theta}^N_{WLS} - \bar{\theta}) \xrightarrow{\mathrm{d}} \mathbf{z} \sim \mathscr{N}(\mathbf{0}, \mathbf{C}(w, \xi, \bar{\theta})), \ N \to \infty,$$

where

$$\mathbf{C}(w, \xi, \theta) = \mathbf{M}_1^{-1}(\xi, \theta)\mathbf{M}_2(\xi, \theta)\mathbf{M}_1^{-1}(\xi, \theta) \tag{3.24}$$

with

$$\mathbf{M}_2(\xi, \theta) = \int_{\mathscr{X}} w^2(x)\sigma^2(x) \frac{\partial \eta(x,\theta)}{\partial \theta} \frac{\partial \eta(x,\theta)}{\partial \theta^\top} \xi(\mathrm{d}x). \tag{3.25}$$

---

[2] His proof, based on properties of Hilbert space valued martingales, requires a condition that gives for linear regression $\lambda_{\max}(\mathbf{M}_N) = \mathcal{O}\{[\lambda_{\min}(\mathbf{M}_N)]^\rho\}$ for some $\rho \in (1,2)$, to be compared to the condition $\log\log \lambda_{\max}(\mathbf{M}_N) = o[\lambda_{\min}(\mathbf{M}_N)]$.

Moreover, $\mathbf{C}(w,\xi,\bar{\theta}) - \mathbf{M}^{-1}(\xi,\bar{\theta})$ *is nonnegative definite for any choice of* $w(x)$*, where*

$$\mathbf{M}(\xi,\bar{\theta}) = \int_{\mathscr{X}} \sigma^{-2}(x) \left.\frac{\partial \eta(x,\theta)}{\partial \theta}\right|_{\bar{\theta}} \left.\frac{\partial \eta(x,\theta)}{\partial \theta^\top}\right|_{\bar{\theta}} \xi(\mathrm{d}x) \tag{3.26}$$

and $\mathbf{C}(w,\xi,\bar{\theta}) = \mathbf{M}^{-1}(\xi,\bar{\theta})$ *for* $w(x) = c\,\sigma^{-2}(x)$ *with* c *a positive constant.*

Proof. The criterion $J_N(\cdot)$ satisfies

$$\nabla_\theta J_N(\theta) = -\frac{2}{N}\sum_{k=1}^N w(x_k)[y(x_k) - \eta(x_k,\theta)]\frac{\partial \eta(x_k,\theta)}{\partial \theta},$$

$$\nabla^2_\theta J_N(\theta) = \frac{2}{N}\sum_{k=1}^N w(x_k)\frac{\partial \eta(x_k,\theta)}{\partial \theta}\frac{\partial \eta(x_k,\theta)}{\partial \theta^\top} - \frac{2}{N}\sum_{k=1}^N w(x_k)[y(x_k) - \eta(x_k,\theta)]\frac{\partial^2 \eta(x_k,\theta)}{\partial \theta \partial \theta^\top}.$$

Since $\bar{\theta} \in \mathrm{int}(\Theta)$ and $\hat{\theta}^N_{WLS} \overset{\text{a.s.}}{\to} \bar{\theta}$ as $N \to \infty$, $\hat{\theta}^N_{WLS} \in \mathrm{int}(\Theta)$ for N larger than some N_0 (which exists w.p.1), and thus, since $J_N(\theta)$ is differentiable for $\theta \in \mathrm{int}(\Theta)$, $\nabla_\theta J_N(\hat{\theta}^N_{WLS}) = \mathbf{0}$ for $N > N_0$. Using a Taylor development of the i-th component of $\nabla_\theta J_N(\cdot)$, we get

$$\{\nabla_\theta J_N(\hat{\theta}^N_{WLS})\}_i = \{\nabla_\theta J_N(\bar{\theta})\}_i + \{\nabla^2_\theta J_N(\beta^N_i)(\hat{\theta}^N_{WLS} - \bar{\theta})\}_i$$

for some $\beta^N_i = (1-\alpha_{N,i})\bar{\theta} + \alpha_{N,i}\,\hat{\theta}^N_{WLS}$, $\alpha_{N,i} \in (0,1)$ (and β^N_i is measurable; see Lemma 2.12). Since $\hat{\theta}^N_{WLS} \overset{\text{a.s.}}{\to} \bar{\theta}$, $\beta^N_i \overset{\text{a.s.}}{\to} \bar{\theta}$. For $N > N_0$, the previous equation can be written

$$\{\nabla^2_\theta J_N(\beta^N_i)(\hat{\theta}^N_{WLS} - \bar{\theta})\}_i = -\{\nabla_\theta J_N(\bar{\theta})\}_i\,. \tag{3.27}$$

Consider first the case where $\{x_i\}$ is asymptotically discrete with measure ξ. Since $\eta(x,\theta)$ is twice continuously differentiable with respect to $\theta \in \Theta$ for any $x \in \mathscr{X}$, Lemma 2.5 gives

$$\frac{1}{N}\sum_{k=1}^N w(x_k)\,\frac{\partial \eta(x_k,\theta)}{\partial \theta}\frac{\partial \eta(x_k,\theta)}{\partial \theta^\top} \overset{\theta}{\rightsquigarrow} \mathbf{M}_1(\xi,\theta) \;\text{ a.s.}$$

and

$$\frac{1}{N}\sum_{k=1}^N w(x_k)[y(x_k) - \eta(x_k,\theta)]\frac{\partial^2 \eta(x_k,\theta)}{\partial \theta \partial \theta^\top} \overset{\theta}{\rightsquigarrow} \int_{\mathscr{X}} w(x)[\eta(x,\bar{\theta}) - \eta(x,\theta)]\frac{\partial^2 \eta(x,\theta)}{\partial \theta \partial \theta^\top}\,\xi(\mathrm{d}x) \;\text{ a.s.}$$

as $N \to \infty$. Therefore, $\nabla^2_\theta J_N(\beta_i^N) \stackrel{\text{a.s.}}{\to} 2\,\mathbf{M}_1(\xi,\bar{\theta})$, and since $\mathbf{M}_1(\xi,\bar{\theta})$ is nonsingular,

$$\mathbf{M}_1^{-1}(\xi,\bar{\theta})\,\nabla^2_\theta J_N(\beta_i^N) \stackrel{\text{a.s.}}{\to} 2\,\mathbf{I}_p\,. \tag{3.28}$$

We shall now consider the distribution of

$$-\sqrt{N}\nabla_\theta J_N(\bar{\theta}) = \frac{2}{\sqrt{N}} \sum_{k=1,\,x_k\notin\mathcal{S}_\xi}^{N} w(x_k)\varepsilon_k \left.\frac{\partial\eta(x_k,\theta)}{\partial\theta}\right|_{\bar{\theta}} + \frac{2}{\sqrt{N}} \sum_{k=1,\,x_k\in\mathcal{S}_\xi}^{N} w(x_k)\varepsilon_k \left.\frac{\partial\eta(x_k,\theta)}{\partial\theta}\right|_{\bar{\theta}}. \tag{3.29}$$

Let $N(\mathscr{X}\setminus\mathcal{S}_\xi)$ denote the number of points of the sequence $x_1,\ldots,x_N$ that do not belong to $\mathcal{S}_\xi$. The first term can be written as

$$2\,\frac{\sqrt{N(\mathscr{X}\setminus\mathcal{S}_\xi)}}{\sqrt{N}} \quad \frac{1}{\sqrt{N(\mathscr{X}\setminus\mathcal{S}_\xi)}} \sum_{k=1,\,x_k\notin\mathcal{S}_\xi}^{N} w(x_k)\varepsilon_k \left.\frac{\partial\eta(x_k,\theta)}{\partial\theta}\right|_{\bar{\theta}} = 2\,\frac{\sqrt{N(\mathscr{X}\setminus\mathcal{S}_\xi)}}{\sqrt{N}}\,\mathbf{t}_N$$

where, for all $i=1,\ldots,p$, $\mathbb{E}\{[\mathbf{t}_N]_i\}=0$ and

$$\begin{aligned}\mathrm{var}\{[\mathbf{t}_N]_i\} &= \frac{1}{N(\mathscr{X}\setminus\mathcal{S}_\xi)} \sum_{k=1,\,x_k\notin\mathcal{S}_\xi}^{N} \left(\left.\frac{\partial\eta(x_k,\theta)}{\partial\theta_i}\right|_{\bar{\theta}}\right)^2 w^2(x_k)\sigma^2(x_k) \\ &\le \bar{V}_i = \max_{x\in\mathscr{X},\theta\in\Theta} \left(\left.\frac{\partial\eta(x,\theta)}{\partial\theta_i}\right|_{\bar{\theta}}\right)^2 \max_{x\in\mathscr{X}} w^2(x) \max_{x\in\mathscr{X}} \sigma^2(x)\,.\end{aligned}$$

Chebyshev inequality then gives $\forall\epsilon>0$, $\mathrm{Prob}\{|[\mathbf{t}_N]_i| > A_i\} < \epsilon$, with $A_i = \sqrt{\bar{V}_i/\epsilon}$, so that $[\mathbf{t}_N]_i$ is bounded in probability[3] and $N(\mathscr{X}\setminus\mathcal{S}_\xi)/N \to 0$ implies that the first term on the right-hand side of (3.29) tends to zero in probability. The second term can be written as

$$2\sum_{x\in\mathcal{S}_\xi} \sqrt{\frac{N(x)}{N}} \left(\frac{1}{\sqrt{N(x)}} \sum_{k=1,\,x_k=x}^{N} \varepsilon_k\right) \left.\frac{\partial\eta(x,\theta)}{\partial\theta}\right|_{\bar{\theta}} w(x)$$

with $\sqrt{N(x)/N} \to \sqrt{\xi(x)}$ and $(\sum_{k=1,\,x_k=x}^{N}\varepsilon_k)/\sqrt{N(x)} \stackrel{\text{d}}{\to} u(x) \sim \mathscr{N}(0,\sigma^2(x))$. Therefore,

$$-\sqrt{N}\nabla_\theta J_N(\bar{\theta}) \stackrel{\text{d}}{\to} 2\sum_{x\in\mathcal{S}_\xi} \sqrt{\xi(x)} \left.\frac{\partial\eta(x,\theta)}{\partial\theta}\right|_{\bar{\theta}} w(x)u(x)\,;$$

---

[3] A sequence of random variables z_n is bounded in probability if for any $\epsilon > 0$, there exist A and n_0 such that $\forall n > n_0$, $\mathrm{Prob}\{|z_n| > A\} < \epsilon$.

that is

$$-\sqrt{N}\nabla_\theta J_N(\bar{\theta}) \stackrel{\mathrm{d}}{\to} \mathbf{v} \sim \mathscr{N}(\mathbf{0}, 4\,\mathbf{M}_2(\xi,\bar{\theta}))\,.$$

Finally, (3.27) and (3.28) show that $\sqrt{N}(\hat{\theta}^N_{WLS} - \bar{\theta}) \stackrel{\mathrm{d}}{\to} \mathbf{z} \sim \mathscr{N}(\mathbf{0}, \mathbf{C}(w,\xi,\bar{\theta}))$ as $N \to \infty$.

Concerning the consequences of the choice of $w(x)$ for the matrix $\mathbf{C}(w,\xi,\bar{\theta})$, denote $a(x) = w(x)\sigma^2(x)$, $x \in \mathcal{S}_\xi$ and consider the vectors

$$\mathbf{v}(x) = \frac{\sqrt{\xi(x)}}{\sigma(x)} \left.\frac{\partial\eta(x,\theta)}{\partial\theta}\right|_{\bar{\theta}} e(x)\,,\ \ x \in \mathcal{S}_\xi\,,$$

where the $e(x)$ are independent normal random variables $\mathscr{N}(0,1)$. Denote

$$\bar{\mathbf{M}}(a) = \begin{pmatrix} \mathbf{M}_2(\xi,\bar{\theta}) & \mathbf{M}_1(\xi,\bar{\theta}) \\ \mathbf{M}_1(\xi,\bar{\theta}) & \mathbf{M}(\xi,\bar{\theta}) \end{pmatrix}.$$

We have

$$\bar{\mathbf{M}}(a) = \mathbb{E}\left\{ \begin{pmatrix} \sum_{x\in\mathcal{S}_\xi} a(x)\mathbf{v}(x) \\ \sum_{x\in\mathcal{S}_\xi} \mathbf{v}(x) \end{pmatrix} \begin{pmatrix} \sum_{x\in\mathcal{S}_\xi} a(x)\mathbf{v}(x) \\ \sum_{x\in\mathcal{S}_\xi} \mathbf{v}(x) \end{pmatrix}^\top \right\}$$

and from Lemma 3.7, $\mathbf{M}_2(\xi,\bar{\theta}) \succeq \mathbf{M}_1(\xi,\bar{\theta})\mathbf{M}^{-1}(\xi,\bar{\theta})\mathbf{M}_1(\xi,\bar{\theta})$, or equivalently

$$\mathbf{C}(w,\xi,\bar{\theta}) = \mathbf{M}_1^{-1}(\xi,\bar{\theta})\mathbf{M}_2(\xi,\bar{\theta})\mathbf{M}_1^{-1}(\xi,\bar{\theta}) \succeq \mathbf{M}^{-1}(\xi,\bar{\theta})\,.$$

One can readily check that $\mathbf{C}(w,\xi,\bar{\theta}) = \mathbf{M}^{-1}(\xi,\bar{\theta})$ for $w(x) = c\,\sigma^{-2}(x)$.

Consider now the case of a randomized design. Lemma 2.6 with $z = x$ gives

$$\frac{1}{N}\sum_{k=1}^N w(x_k)\frac{\partial\eta(x_k,\theta)}{\partial\theta}\frac{\partial\eta(x_k,\theta)}{\partial\theta^\top} \stackrel{\theta}{\rightsquigarrow} \mathbf{M}_1(\xi,\theta) \ \text{ a.s.}$$

as $N \to \infty$. Similarly, we can write

$$\frac{1}{N}\sum_{k=1}^N w(x_k)[y(x_k) - \eta(x_k,\theta)]\frac{\partial^2\eta(x_k,\theta)}{\partial\theta\partial\theta^\top} = \frac{1}{N}\sum_{k=1}^N w(x_k)\varepsilon(x_k)\frac{\partial^2\eta(x_k,\theta)}{\partial\theta\partial\theta^\top}$$
$$+\frac{1}{N}\sum_{k=1}^N w(x_k)[\eta(x_k,\bar{\theta}) - \eta(x_k,\theta)]\frac{\partial^2\eta(x_k,\theta)}{\partial\theta\partial\theta^\top}\,. \tag{3.30}$$

Define $a(z,\theta) = \partial^2\eta(x,\theta)/(\partial\theta_i\partial\theta_j)\varepsilon(x)$ with $z = (x,\varepsilon)$. We have

$$\mathbb{E}\{\max_{\theta\in\Theta}|a(z,\theta)|\} = \int_{\mathscr{X}} w(x)\,\max_{\theta\in\Theta}\left|\frac{\partial^2\eta(x_k,\theta)}{\partial\theta_i\partial\theta_j}\right|\,\mathbb{E}_x\{|\varepsilon|\}\,\xi(\mathrm{d}x)$$
$$\leq \int_{\mathscr{X}} w(x)\,\sigma(x)\,\max_{\theta\in\Theta}\left|\frac{\partial^2\eta(x_k,\theta)}{\partial\theta_i\partial\theta_j}\right|\xi(\mathrm{d}x) < \infty\,,$$

so that Lemma 2.6 implies that the first term in (3.30) converges to $\mathbb{E}\{a(z,\theta)\} = 0$ when $N \to \infty$ and the convergence is uniform in θ and almost sure with respect to x and ε. Similarly, Lemma 2.6 implies that

$$\frac{1}{N}\sum_{k=1}^{N} w(x_k)\,[y(x_k) - \eta(x_k,\theta)]\,\frac{\partial^2\eta(x_k,\theta)}{\partial\theta\partial\theta^\top} \overset{\theta}{\rightsquigarrow} \int_{\mathscr{X}} w(x)\,[\eta(x,\bar{\theta}) - \eta(x,\theta)]\frac{\partial^2\eta(x,\theta)}{\partial\theta\partial\theta^\top}\,\xi(\mathrm{d}x) \text{ a.s.}$$

as $N \to \infty$. Therefore,

$$\nabla^2_\theta J_N(\theta) \overset{\theta}{\rightsquigarrow} 2\,\mathbf{M}_1(\xi,\theta) - 2\int_{\mathscr{X}} w(x)\,[\eta(x,\bar{\theta}) - \eta(x,\theta)]\frac{\partial^2\eta(x,\theta)}{\partial\theta\partial\theta^\top}\,\xi(\mathrm{d}x) \text{ a.s.}$$

and $\nabla^2_\theta J_N(\beta_i^N) \overset{\text{a.s.}}{\to} 2\,\mathbf{M}_1(\xi,\bar{\theta})$. Since $\mathbf{M}_1(\xi,\bar{\theta})$ is nonsingular, we have again

$$\mathbf{M}_1^{-1}(\xi,\bar{\theta})\,\nabla^2_\theta J_N(\beta_i^N) \overset{\text{a.s.}}{\to} 2\,\mathbf{I}_p\,. \tag{3.31}$$

We consider now the distribution of

$$-\sqrt{N}\nabla_\theta J_N(\bar{\theta}) = \frac{2}{\sqrt{N}}\sum_{k=1}^{N} w(x_k)\,\varepsilon_k\,\frac{\partial\eta(x_k,\theta)}{\partial\theta}\bigg|_{\bar{\theta}}.$$

We have

$$-\sqrt{N}\nabla_\theta J_N(\bar{\theta}) \overset{\text{d}}{\to} \mathbf{v} \sim \mathscr{N}(\mathbf{0}, 4\,\mathbf{M}_2(\xi,\bar{\theta}))\,.$$

Finally, (3.27) and (3.31) give $\sqrt{N}(\hat{\theta}^N_{WLS} - \bar{\theta}) \overset{\text{d}}{\to} \mathbf{z} \sim \mathscr{N}(\mathbf{0}, \mathbf{C}(w,\xi,\bar{\theta}))$.

The rest of the proof is similar to the case of a nonrandomized design, with now

$$\mathbf{v}(x) = \frac{1}{\sigma(x)}\frac{\partial\eta(x,\theta)}{\partial\theta}\bigg|_{\bar{\theta}},$$

so that the vectors $\mathbf{v}(x)$ are i.i.d. ■

Remark 3.9.

(*i*) Theorem 3.8 indicates that, in general, the best choice of weighting factors in terms of asymptotic covariance of the WLS estimator is $w(x) = \sigma^{-2}(x)$; see also Sects. 4.2 and 4.4. However, the choice of $w(x)$ is indifferent when ξ is supported on p point only. Indeed, suppose that ξ has support $\{x^{(1)},\ldots,x^{(p)}\}$, that

$$\mathbf{F}_\theta^\top(\xi) = \left[\frac{\partial\eta(x^{(1)},\theta)}{\partial\theta},\ldots,\frac{\partial\eta(x^{(p)},\theta)}{\partial\theta}\right]$$

has full rank, and that $w(x^{(i)}) > 0$ for all $i = 1,\ldots,p$. Define $\mathbf{W}_1 = \mathrm{diag}\{\xi(x^{(i)})w(x^{(i)}), i = 1,\ldots,p\}$, $\mathbf{W}_2 = \mathrm{diag}\{\xi(x^{(i)})w^2(x^{(i)})\sigma^2(x^{(i)}),$

$i = 1, \ldots, p\}$, and $\mathbf{W} = \mathrm{diag}\{\xi(x^{(i)})/\sigma^2(x^{(i)}), i = 1, \ldots, p\}$ so that $\mathbf{M}_1(\xi, \theta) = \mathbf{F}_\theta^\top(\xi)\mathbf{W}_1\mathbf{F}_\theta(\xi)$, $\mathbf{M}_2(\xi, \theta) = \mathbf{F}_\theta^\top(\xi)\mathbf{W}_2\mathbf{F}_\theta(\xi)$, and $\mathbf{M}(\xi, \theta) = \mathbf{F}_\theta^\top(\xi)\mathbf{W}\mathbf{F}_\theta(\xi)$. We then obtain

$$\begin{aligned}\mathbf{C}(w, \xi, \theta) &= \mathbf{M}_1^{-1}(\xi, \theta)\mathbf{M}_2(\xi, \theta)\mathbf{M}_1^{-1}(\xi, \theta) \\ &= \mathbf{F}_\theta^{-1}(\xi)\mathbf{W}_1^{-1}\mathbf{W}_2\mathbf{W}_1^{-1}(\mathbf{F}_\theta^\top)^{-1} \\ &= \mathbf{M}^{-1}(\xi, \theta)\end{aligned}$$

and all WLS estimators have the same asymptotic covariance matrix.

(*ii*) In the case of a randomized design, the boundedness assumption of $w(x)$ and $\eta(x, \theta)$ and its derivatives in $\mathrm{H1}_\eta$ and $\mathrm{H2}_\eta$ can be replaced by (3.11) and

$$\int_{\mathscr{X}} w(x) \sup_{\theta \in \Theta} \left| \frac{\partial \eta(x, \theta)}{\partial \theta_i} \frac{\partial \eta(x, \theta)}{\partial \theta_j} \right| \xi(\mathrm{d}x) < \infty\,,$$

$$\int_{\mathscr{X}} w(x)\, \sigma(x) \sup_{\theta \in \Theta} \left| \frac{\partial^2 \eta(x, \theta)}{\partial \theta_i \partial \theta_j} \right| \xi(\mathrm{d}x) < \infty\,,$$

$$\int_{\mathscr{X}} w(x) \sup_{\theta \in \Theta} \left[|\eta(x, \bar{\theta}) - \eta(x, \theta)| \left| \frac{\partial^2 \eta(x, \theta)}{\partial \theta_i \partial \theta_j} \right| \right] \xi(\mathrm{d}x) < \infty\,.$$

(*iii*) The random fluctuations due to the use of a randomized design according to Definition 2.2 have (asymptotically) no effect on the covariance matrix of the WLS estimator. We shall see later that this is also true for other estimators. However, the situation is different in presence of modeling error; see Remark 3.38. □

When the design is such that $\mathbf{M}_1(\xi, \bar{\theta})$ is singular, asymptotic normality may still hold for $\hat{\theta}^N_{WLS}$, but with a norming constant smaller than $\sqrt{N}$; see Wu (1981) for nonsequential design and Lai and Wei (1982) and Pronzato (2009a) for respectively linear and nonlinear models under a sequential design.

Consider the case where $w(x) \equiv 1$ (ordinary LS) and $\sigma^2(x) \equiv \sigma^2$ (stationary errors). Theorem 3.8 then gives $\sqrt{N}(\hat{\theta}^N_{LS} - \bar{\theta}) \overset{\mathrm{d}}{\to} \mathbf{z} \sim \mathscr{N}(\mathbf{0}, \mathbf{M}^{-1}(\xi, \bar{\theta}))$ with

$$\mathbf{M}(\xi, \theta) = \frac{1}{\sigma^2} \int_{\mathscr{X}} \frac{\partial \eta(x, \theta)}{\partial \theta} \frac{\partial \eta(x, \theta)}{\partial \theta^\top} \xi(\mathrm{d}x)\,, \tag{3.32}$$

provided that $\mathbf{M}(\xi, \bar{\theta})$ has full rank. The situation is totally different when $\mathbf{M}(\xi, \bar{\theta})$ is singular.

We shall say that *the design ξ is singular* (for the matrix $\mathbf{M}$ and the parameter space Θ) when $\mathbf{M}(\xi, \theta)$ is singular for some $\theta \in \Theta$. It may be the case in particular because ξ is supported on less than $p = \dim(\theta)$ points, $\mathbf{M}(\xi, \theta)$ being thus singular for all $\theta \in \Theta$. Then, linear functions $\mathbf{c}^\top \hat{\theta}^N_{LS}$ of the LS estimator can be asymptotically normal, but with a normalizing constant different from $\sqrt{N}$; see Examples 2.4 and 5.39, respectively, for a linear and nonlinear model; see also Examples 3.13 and 3.17 for the estimation of a nonlinear function $h(\theta)$ in a linear model. Another situation, more atypical, is when $\mathbf{M}(\xi,\theta)$ is nonsingular for almost all θ but is singular for the particular parameter value $\theta=\bar{\theta}$. The following example illustrates the situation.

Example 3.10. Take $\eta(x,\theta) = x\,\theta^3$ in the model (3.2), (3.3). For $\bar{\theta}=0$, $\mathbf{M}(\xi,\bar{\theta})=0$ for any ξ, whereas $\int_{\mathscr{X}}[\eta(x,\theta)-\eta(x,\bar{\theta})]^2\xi(dx)=0$ implies $\theta=\bar{\theta}$ for any $\xi\neq\delta_0$, the delta measure with weight one at 0. Take, for instance, $x_i=x_*$ for $i=1,2\ldots$ with $x_*\neq 0$. Then $(\hat{\theta}^N_{LS})^3=(1/x_*)\left(\sum_{i=1}^N\varepsilon_i\right)/N$ is strongly consistent and $\sqrt{N}(\hat{\theta}^N_{LS})^3$ is asymptotically normal $\mathscr{N}(0,\sigma^2/x_*^2)$. More generally, when $\{x_i\}$ is an asymptotically discrete design (Definition 2.1) or a randomized design (Definition 2.2), we can use the same approach as in the proof of Theorem 3.8 and construct a Taylor development of the LS criterion $J_N(\theta)$. The difference is that here the first two nonzero derivatives at $\bar{\theta}$ are the third and the sixth, which gives

$$0=\nabla_\theta J_N(\hat{\theta}^N_{LS})=(1/2)(\hat{\theta}^N_{LS})^2\nabla^3_\theta J_N(\bar{\theta})+(1/5!)(\hat{\theta}^N_{LS})^5\nabla^6_\theta J_N$$

with $\nabla^3_\theta J_N(\theta)=120\,\theta^3\,(\sum_{i=1}^N x_i^2)/N-12\,(\sum_{i=1}^N x_i\varepsilon_i)/N$, so that $\nabla^3_\theta J_N(\bar{\theta})=-12\,(\sum_{i=1}^N x_i\varepsilon_i)/N$ and $\nabla^6_\theta J_N = 720\,(\sum_{i=1}^N x_i^2)/N$. Therefore, $\nabla^6_\theta J_N \stackrel{\text{a.s.}}{\to} 720\,\mathbb{E}_\xi(x^2)$ and $\sqrt{N}\nabla^3_\theta J_N(\bar{\theta}) \stackrel{\text{d}}{\to} z\sim\mathscr{N}(0,\,144\,\sigma^2\mathbb{E}_\xi(x^2))$ when $N\to\infty$, with $\mathbb{E}_\xi(x^2)=\int_{\mathscr{X}}x^2\,\xi(dx)$, which gives

$$\sqrt{N}(\hat{\theta}^N_{LS})^3 \stackrel{\text{d}}{\to} \zeta\sim\mathscr{N}(0,\sigma^2/\mathbb{E}_\xi(x^2))\,,\ N\to\infty\,.$$

Straightforward calculation then gives $N^{1/6}\,\hat{\theta}^N_{LS} \stackrel{\text{d}}{\to} [\sigma/\mathbb{E}_\xi(x^2)]^{1/3}\,t$ as $N\to\infty$, where t has the p.d.f.

$$f(t)=\frac{3}{\sqrt{2\pi}}\,t^2\,\exp(-t^6/2)\,. \tag{3.33}$$

The LS estimator thus converges as slowly as $N^{-1/6}$ with a bimodal limiting distribution; see Fig. 3.1.

In the case of normal errors, i.e., when $\varepsilon_i\sim\mathscr{N}(0,\sigma^2)$ for all i, the distribution above is exact for any N when all the x_i coincide. The derivation of the exact finite sample distribution of the LS estimator with normal errors in such a situation where the design consists of repetitions of observations at p points only is considered in Sect. 6.2.1. Finite sample approximations are also presented in the same section for the case of designs supported on more than p points. □

Previous example illustrates the importance of considering designs ξ such that $\mathbf{M}(\xi,\theta)$ is nonsingular for all $\theta\in\Theta$. The model is then said to be *regular* for ξ. A model which is not regular for ξ is called singular; it is such that $\mathbf{M}(\xi,\theta)$ is singular for some $\theta\in\Theta$. Under the assumption H2$_\eta$ (p. 22), when $\mathbf{M}(\xi,\theta)$ is nonsingular at some θ it is also nonsingular in some neighborhood of

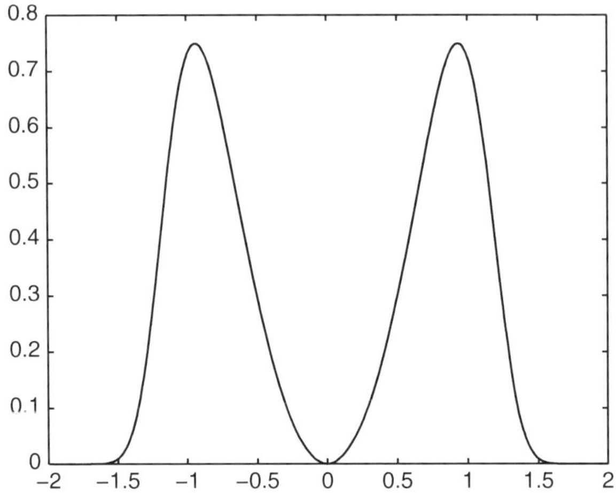

Fig. 3.1. Probability density function (3.33)

θ; in such cases one may sometimes suppose that Θ is chosen such that $\mathbf{M}(\xi, \theta)$ is nonsingular for all $\theta \in \Theta$. On the other hand, the model is often singular simply because ξ is a design supported on less than $p = \dim(\theta)$ points. In that case $\mathbf{M}(\xi, \theta)$ is singular for every $\theta \in \Theta$. This is what we consider mainly in our examples with singular designs in the rest of the book.

3.1.4 Asymptotic Normality of a Scalar Function of the LS Estimator

For the sake of simplicity, only the case $w(x) \equiv 1$ (ordinary LS) and $\sigma^2(x) \equiv \sigma^2$ (stationary errors) is considered, so that $\mathbf{C}(w, \xi, \bar{\theta}) = \mathbf{M}^{-1}(\xi, \bar{\theta})$ in Theorem 3.8 with $\mathbf{M}(\xi, \theta)$ given by (3.32). We shall use the following assumption:

H1$_h$: The function $h(\cdot) : \Theta \longrightarrow \mathbb{R}$ is continuous and has continuous second-order derivatives in $\mathrm{int}(\Theta)$.

The so-called delta method (Lehmann and Casella 1998, p. 61) then gives the following.

Theorem 3.11 (The delta method). *Let $\{\hat{\theta}^N\}$ be a sequence of random vectors of Θ satisfying $\hat{\theta}^N \xrightarrow{\text{p}} \bar{\theta} \in \mathrm{int}(\Theta)$ and $\sqrt{N}(\hat{\theta}^N - \bar{\theta}) \xrightarrow{\text{d}} \mathbf{z} \sim \mathscr{N}(\mathbf{0}, \mathbf{V}(\xi, \bar{\theta}))$, $N \to \infty$, for some matrix $\mathbf{V}(\xi, \bar{\theta})$. When H1$_h$ holds and $\partial h(\theta)/\partial\theta\big|_{\bar{\theta}} \neq \mathbf{0}$, then*

$$\sqrt{N}[h(\hat{\theta}^N) - h(\bar{\theta})] \xrightarrow{\text{d}} \zeta \sim \mathscr{N}\left(0, \left.\frac{\partial h(\theta)}{\partial \theta^\top}\right|_{\bar{\theta}} \mathbf{V}(\xi, \bar{\theta}) \left.\frac{\partial h(\theta)}{\partial \theta}\right|_{\bar{\theta}}\right), \quad N \to \infty .$$

Proof. We can write

$$h(\hat{\theta}^N) = h(\bar{\theta}) + \left.\frac{\partial h(\theta)}{\partial \theta^\top}\right|_{\bar{\theta}} (\hat{\theta}^N - \bar{\theta}) + \frac{1}{2}(\hat{\theta}^N - \bar{\theta})^\top \left.\frac{\partial^2 h(\theta)}{\partial \theta \partial \theta^\top}\right|_{\beta^N} (\hat{\theta}^N - \bar{\theta})$$

for some β^N on the segment connecting $\hat{\theta}^N$ and $\bar{\theta}$. Since $\hat{\theta}^N \overset{\text{p}}{\to} \bar{\theta}$, $\beta^N \overset{\text{p}}{\to} \bar{\theta}$ too and $\partial^2 h(\theta)/\partial\theta\partial\theta^\top\big|_{\beta^N}(\hat{\theta}^N - \bar{\theta}) \overset{\text{p}}{\to} \mathbf{0}$. Hence $\sqrt{N}[h(\hat{\theta}^N) - h(\bar{\theta})]$ converges in distribution to the same limit as $\partial h(\theta)/\partial\theta^\top\big|_{\bar{\theta}}\sqrt{N}(\hat{\theta}^N - \bar{\theta})$, i.e., to $\partial h(\theta)/\partial\theta^\top\big|_{\bar{\theta}}\mathbf{z}$. ∎

The theorem is not valid when $\partial h(\theta)/\partial\theta\big|_{\bar{\theta}} = \mathbf{0}$. We have in that case, supposing that $h(\cdot)$ is three times continuously differentiable,

$$N[h(\hat{\theta}^N) - h(\bar{\theta})] = \frac{1}{2}\sqrt{N}(\hat{\theta}^N - \bar{\theta})^\top \left.\frac{\partial^2 h(\theta)}{\partial\theta\partial\theta^\top}\right|_{\bar{\theta}} \sqrt{N}(\hat{\theta}^N - \bar{\theta})$$
$$+\frac{1}{6}\sum_{i,j,k=1}^{p} \sqrt{N}\{\hat{\theta}^N - \bar{\theta}\}_i \sqrt{N}\{\hat{\theta}^N - \bar{\theta}\}_j \left[\{\hat{\theta}^N - \bar{\theta}\}_k \left.\frac{\partial^3 h(\theta)}{\partial\theta_i \partial\theta_j \theta_k}\right|_{\beta^N}\right].$$

The term between brackets converges to zero in probability; hence, $N[h(\hat{\theta}^N) - h(\bar{\theta})] \overset{\text{d}}{\to} \zeta$ where ζ is distributed as $(1/2)\mathbf{z}^\top \partial^2 h(\theta)/\partial\theta\partial\theta^\top\big|_{\bar{\theta}}\mathbf{z}$, which is not normal. On the other hand, $\sqrt{N}[h(\hat{\theta}^N) - h(\bar{\theta})]$ may be asymptotically normal in situations where $\hat{\theta}^N$ is not consistent, or even not unique, and not asymptotically normal. We investigate this situation more deeply in the next section.

3.2 Asymptotic Properties of Functions of the LS Estimator Under Singular Designs

Again, for the sake of simplicity we shall assume that $\sigma^2(x) \equiv 1$ and take $w(x) \equiv 1$. When applied to LS estimation, under the conditions of Theorem 3.8, Theorem 3.11 gives

$$\sqrt{N}[h(\hat{\theta}^N_{LS}) - h(\bar{\theta})] \overset{\text{d}}{\to} \zeta \sim \mathscr{N}\left(0, \left.\frac{\partial h(\theta)}{\partial\theta^\top}\right|_{\bar{\theta}} \mathbf{M}^{-1}(\xi,\bar{\theta}) \left.\frac{\partial h(\theta)}{\partial\theta}\right|_{\bar{\theta}}\right), \quad N \to \infty,$$

with $\mathbf{M}(\xi,\theta)$ given by (3.32). We shall call this property *regular asymptotic normality*, with $\mathbf{M}^{-1}$ replaced by a g-inverse $\mathbf{M}^-$ when $\mathbf{M}$ is singular.

Definition 3.12 (Regular asymptotic normality). *We say that* $h(\hat{\theta}^N_{LS})$ *satisfies the property of* regular asymptotic normality *when*

$$\sqrt{N}[h(\hat{\theta}^N_{LS}) - h(\bar{\theta})] \overset{\text{d}}{\to} \zeta \sim \mathscr{N}\left(0, \left.\frac{\partial h(\theta)}{\partial\theta^\top}\right|_{\bar{\theta}} \mathbf{M}^{-}(\xi,\bar{\theta}) \left.\frac{\partial h(\theta)}{\partial\theta}\right|_{\bar{\theta}}\right), \quad N \to \infty.$$

Singular designs cause no special difficulty in linear regression (which partly explains why singularity issues have been somewhat disregarded in the design literature): a linear combination $\mathbf{c}^\top\theta$ of the parameters is either estimable or not, depending on the direction of $\mathbf{c}$.

3.2.1 Singular Designs in Linear Models

Consider an exact design of size N formed by N points $x_1,\ldots,x_N$ from $\mathscr{X}$, with associated observations $y(x_1),\ldots,y(x_N)$ modeled by

$$y(x_i) = \mathbf{f}^\top(x_i)\bar{\theta} + \varepsilon_i\,,\ i=1,\ldots,N\,, \tag{3.34}$$

where the errors ε_i are independent and $\mathbb{E}(\varepsilon_i)=0$, $\mathrm{Var}(\varepsilon_i)=\sigma^2$ for all i. If the information matrix

$$\mathbf{M}_N = \mathbf{M}(x_1,\ldots,x_N) = \sum_{i=1}^N \mathbf{f}(x_i)\mathbf{f}^\top(x_i)$$

is nonsingular, then the least-squares estimator of θ,

$$\hat{\theta}^N \in \arg\min_\theta \sum_{i=1}^N \left[y(x_i) - \mathbf{f}^\top(x_i)\theta\right]^2 \tag{3.35}$$

is unique, and its variance is

$$\mathrm{Var}(\hat{\theta}^N) = \sigma^2\mathbf{M}_N^{-1}\,.$$

On the other hand, if $\mathbf{M}_N$ is singular, then $\hat{\theta}^N$ is not defined uniquely. However, $\mathbf{c}^\top\hat{\theta}^N$ does not depend on the choice of the solution $\hat{\theta}^N$ of (3.35) if (and only if) $\mathbf{c}\in\mathcal{M}(\mathbf{M}_N)$; that is, $\mathbf{c}=\mathbf{M}_N\mathbf{u}$ for some $\mathbf{u}\in\mathbb{R}^p$. Then

$$\mathrm{Var}(\mathbf{c}^\top\hat{\theta}^N) = \sigma^2\mathbf{c}^\top\mathbf{M}_N^-\mathbf{c} \tag{3.36}$$

where the choice of the g-inverse $\mathbf{M}_N^-$ is arbitrary. This last expression can be used as a criterion (the c-optimality criterion, see Chap. 5) for an optimal choice of the N-point design $x_1,\ldots,x_N$, and the design minimizing this criterion may be singular; see Silvey (1980) and Pázman (1980) for some properties and Wu (1980, 1983) for a detailed investigation of the consistency of $\mathbf{c}^\top\hat{\theta}^N$ when $N\to\infty$. As shown below, the situation is much more complicated in the nonlinear case, and we shall see in particular how the condition $\mathbf{c}\in\mathcal{M}(\mathbf{M}_N)$ has then to be modified; see (3.39) and assumption H2$_h$.

3.2.2 Singular Designs in Nonlinear Models

In a nonlinear situation with $\mathbf{M}(\xi,\bar{\theta})$ singular, regular asymptotic normality relies on precise conditions concerning the model $\eta(x,\theta)$, the true value $\bar{\theta}$ of its parameters, the function $h(\cdot)$, and the convergence of the empirical design measure ξ_N associated with the design sequence $x_1, x_2 \ldots$ to the design measure ξ. The following example illustrates the difficulties that may occur when ξ_N converges weakly to a singular ξ and $h(\cdot)$ is a nonlinear function, even if the regression model is linear. In this example, $\hat{\theta}^N_{LS}$ is consistent, but in general $h(\hat{\theta}^N_{LS})$ converges more slowly than $1/\sqrt{N}$; it may be not asymptotically normal, and, in the very specific situation where it is asymptotically normal and converges as $1/\sqrt{N}$, its limiting variance is larger than that obtained when using the limiting design ξ.

Example 3.13. We consider the same situation as in Example 2.4 and are now interested in the estimation of the point x where $\eta(x,\theta) = \theta_1 x + \theta_2 x^2$ is maximum, i.e., in the estimation of $h(\theta) = -\theta_1/(2\theta_2)$, with $h \geq 0$ and

$$\frac{\partial h(\theta)}{\partial \theta} = -\frac{1}{2\theta_2}\begin{pmatrix} 1 \\ 2h \end{pmatrix}.$$

Let θ^* be a prior guess for θ with $\theta_1^* \geq 0$, $\theta_2^* < 0$; let $h_* = -\theta_1^*/(2\theta_2^*)$ denote the corresponding prior guess for h, and define $x_* = 2h_*$. The c-optimum design ξ_* supported in $\mathscr{X} = [0,1]$ that minimizes $[\partial h(\theta)/\partial\theta^\top \mathbf{M}^-(\xi)\, \partial h(\theta)/\partial\theta]_{\theta^*}$ is easily computed from Elfving's theorem (1952), see Sect. 5.3.1, and is given by

$$\xi_* = \begin{cases} \gamma_*\delta_{\sqrt{2}-1} + (1-\gamma_*)\delta_1 & \text{if } 0 \leq x_* \leq \sqrt{2}-1 \text{ or } 1 \leq x_*\,, \\ \delta_{x_*} \text{ otherwise}\,, \end{cases} \tag{3.37}$$

with δ_x the delta measure that puts weight 1 at x and

$$\gamma_* = \frac{\sqrt{2}}{2}\,\frac{1-x_*}{2(\sqrt{2}-1)-x_*}.$$

Suppose that the prior guess θ^* is such that $\sqrt{2}-1 < x_* \leq 1$ so that the c-optimum design puts mass 1 at x_*; that is, it coincides with ξ_* considered in Example 2.4. We consider the design given by (2.2) for which the empirical design measure ξ_N converges weakly to ξ_*, with $\alpha \leq 1/2$ to ensure the consistency of $\hat{\theta}^N_{LS}$.

We show the following in the rest of the example: (i) $h(\hat{\theta}^N_{LS})$ is generally asymptotically normal but converges as slowly as $N^{\alpha-1/2}$; (ii) in the particular case when $x_* = \arg\max_x \eta(x,\bar{\theta})$, we obtain that $h(\hat{\theta}^N_{LS})$ is not asymptotically normal supposing that $1/4 \leq \alpha \leq 1/2$, although such a choice of x_* may be considered as optimal for the estimation of the maximum of the regression function. On the other hand, for the same choice of x_* but with $\alpha < 1/4$ the estimator $h(\hat{\theta}^N_{LS})$ is asymptotically normal and converges as $1/\sqrt{N}$, but its limiting variance is larger than that obtained with the limiting design ξ_*.

(i) When $\bar{\theta}_1 + x_*\bar{\theta}_2 \neq 0$, i.e., when $h(\bar{\theta}) \neq h_*$, which corresponds to the typical situation, we have

$$h(\hat{\theta}^N_{LS}) = h(\bar{\theta}) + (\hat{\theta}^N_{LS} - \bar{\theta})^\top \left[\frac{\partial h(\theta)}{\partial \theta}\bigg|_{\bar{\theta}} + o_p(1) \right],$$

with $\partial h(\theta)/\partial\theta\big|_{\bar{\theta}} = -1/(2\bar{\theta}_2)[1,\ 2h(\bar{\theta})]^\top$ not parallel to $\mathbf{c}_* = (x_*,\, x_*^2)^\top$, and

$$N^{1/2-\alpha}\,[h(\hat{\theta}^N_{LS}) - h(\bar{\theta})] \overset{\mathrm{d}}{\to} \zeta \sim \mathscr{N}(0, v_{\bar{\theta}}),\ N \to \infty,$$

with $v_{\bar{\theta}} = W(\alpha)[x_* - 2h(\bar{\theta})]^2/(4\bar{\theta}_2^2 x_*^2)$ where $W(\alpha)$ is given by (2.5); $h(\hat{\theta}^N_{LS})$ is thus asymptotically normal but converges as $N^{\alpha-1/2}$.

(*ii*) In the particular situation where the prior guess h_* coincides with the true value $h(\bar{\theta})$, $\bar{\theta}_1 + x_*\bar{\theta}_2 = 0$ and we write

$$\begin{aligned} h(\hat{\theta}^N_{LS}) &= h(\bar{\theta}) + (\hat{\theta}^N_{LS} - \bar{\theta})^\top \frac{\partial h(\theta)}{\partial \theta}\bigg|_{\bar{\theta}} \\ &\quad + \frac{1}{2}(\hat{\theta}^N_{LS} - \bar{\theta})^\top \left[\frac{\partial^2 h(\theta)}{\partial\theta\partial\theta^\top}\bigg|_{\bar{\theta}} + o_p(1)\right](\hat{\theta}^N_{LS} - \bar{\theta}), \end{aligned} \tag{3.38}$$

with

$$\frac{\partial h(\theta)}{\partial \theta}\bigg|_{\bar{\theta}} = -\frac{1}{2\bar{\theta}_2 x_*}\mathbf{c}_* \ \text{ and } \ \frac{\partial^2 h(\theta)}{\partial\theta\partial\theta^\top}\bigg|_{\bar{\theta}} = \frac{1}{2\bar{\theta}_2^2}\begin{pmatrix} 0 & 1 \\ 1 & 2x_* \end{pmatrix}.$$

Define $\Delta_N = \hat{\theta}^N_{LS} - \bar{\theta}$ and $E_N = 2\bar{\theta}_2^2\, \Delta_N^\top\, [\partial^2 h(\theta)/\partial\theta\partial\theta^\top]\big|_{\bar{\theta}}\, \Delta_N$. The eigenvector decomposition of $[\partial^2 h(\theta)/\partial\theta\partial\theta^\top]\big|_{\bar{\theta}}$ gives

$$E_N = \beta\left[(\mathbf{v}_1^\top \Delta_N)^2 - (\mathbf{v}_2^\top \Delta_N)^2\right]$$

with $\mathbf{v}_{1,2} = (1,\ x_* \pm \sqrt{1+x_*^2})$ and $\beta = (x_* + \sqrt{1+x_*^2})/[2(1 + x_*^2 + x_*\sqrt{1+x_*^2})]$. Similarly to (2.4) we then obtain

$$N^{1/2-\alpha}\mathbf{v}_{1,2}^\top \Delta_N \overset{\mathrm{d}}{\to} \zeta_{1,2} \sim \mathscr{N}(0, [1 + 1/x_*^2]W(\alpha)),\ N \to \infty,$$

with $W(\alpha)$ given by (2.5). From (3.38), the limiting distribution of $h(\hat{\theta}^N_{LS})$ is not normal when $\alpha \geq 1/4$; when $\alpha > 1/4$, $N^{1-2\alpha}[h(\hat{\theta}^N_{LS}) - h(\bar{\theta})]$ tends to be distributed as $[1/(4\bar{\theta}_2^2)]\beta W(\alpha)(1 + 1/x_*^2)$ times the difference of two independent chi-square random variables. When $\alpha < 1/4$ we have $\sqrt{N}\, E_N = o_p(1)$, $N \to \infty$, and (3.38) implies

$$\sqrt{N}[h(\hat{\theta}^N_{LS}) - h(\bar{\theta})] \overset{\mathrm{d}}{\to} \zeta \sim \mathscr{N}(0, V(\alpha)/(4\bar{\theta}_2^2 x_*^2)),\ N \to \infty,$$

with $V(\alpha)$ given by (2.3). The limiting variance is thus larger than

$$[\partial h(\theta)/\partial\theta^\top\, \mathbf{M}^-(\xi_*)\, \partial h(\theta)/\partial\theta]_{\bar{\theta}} = 1/(4\bar{\theta}_2^2 x_*^2). \qquad \square$$

As previous example shows, regular asymptotic normality may fail to hold when the empirical design measure ξ_N converges weakly to a singular design. Stronger types of convergence are thus required, and we shall focus our attention on the sequences defined in Sect. 2.1. Besides conditions on the design sequence, we shall see that regular asymptotic normality also relies on conditions concerning:

(i) The true value $\bar{\theta}$ of the model parameters in relation to the geometry of the model
(ii) The function $h(\cdot)$

We first consider the situation when the design sequence is such that the empirical measure ξ_N converges strongly to a singular discrete design, but the convergence is slow enough to ensure the consistency of $\hat{\theta}^N_{LS}$. Such a condition on the design sequence can be obtained by invoking Theorem 3.5. The more general situation of design sequences obeying to Definition 2.1 or 2.2 is considered next.

Regular Asymptotic Normality when $\mathscr{X}$ is Finite

When the design space $\mathscr{X}$ is finite, we can invoke Theorem 3.5 to ensure the consistency of $\hat{\theta}^N_{LS}$, and regular asymptotic normality follows for suitable functions $h(\cdot)$.

Theorem 3.14. *Let $\{x_i\}$ be an asymptotically discrete design (Definition 2.1) on the finite design space $\mathscr{X} \subset \mathbb{R}^d$, the limiting design ξ being possibly singular. Suppose that the assumptions H_Θ, $H1_\eta$, $H2_\eta$, and $H1_h$ given in Sects. 3.1 and 3.1.4 and the condition (3.17) of Theorem 3.5 are satisfied, so that $\hat{\theta}^N_{LS}$ is strongly consistent. If, moreover, $\partial h(\theta)/\partial\theta\big|_{\bar{\theta}} \neq \mathbf{0}$ and*

$$\left.\frac{\partial h(\theta)}{\partial \theta}\right|_{\bar{\theta}} \in \mathcal{M}[\mathbf{M}(\xi,\bar{\theta})]\,, \tag{3.39}$$

then $h(\hat{\theta}^N_{LS})$ satisfies

$$\sqrt{N}\,[h(\hat{\theta}^N_{LS}) - h(\bar{\theta})] \overset{\mathrm{d}}{\to} \zeta \sim \mathscr{N}\left(0, \left.\frac{\partial h(\theta)}{\partial \theta^\top}\right|_{\bar{\theta}} \mathbf{M}^-(\xi,\bar{\theta}) \left.\frac{\partial h(\theta)}{\partial \theta}\right|_{\bar{\theta}}\right),\ N \to \infty\,, \tag{3.40}$$

where $\mathbf{M}(\xi,\theta)$ is given by (3.32) and the choice of the g-inverse is arbitrary.

Proof. Similarly to the proof of Theorem 3.8, we can write for N larger than some N_0,

$$\{\nabla_\theta J_N(\hat{\theta}^N_{LS})\}_i = 0 = \{\nabla_\theta J_N(\bar{\theta})\}_i + \{\nabla^2_\theta J_N(\beta^N_i)(\hat{\theta}^N_{LS} - \bar{\theta})\}_i\,,\ i = 1,\ldots,p\,,$$

with $J_N(\cdot)$ the LS criterion (3.1) and β^N_i between $\hat{\theta}^N_{LS}$ and $\bar{\theta}$. Moreover, $-\sqrt{N}\nabla_\theta J_N(\bar{\theta}) \overset{\mathrm{d}}{\to} \mathbf{v} \sim \mathscr{N}(\mathbf{0}, 4\,\mathbf{M}(\xi,\bar{\theta}))$ and $\nabla^2_\theta J_N(\beta^N_i) \overset{\mathrm{a.s.}}{\to} 2\,\mathbf{M}(\xi,\bar{\theta})$ as $N \to \infty$. From that, we obtain

$$\sqrt{N}\mathbf{c}^\top \mathbf{M}(\xi,\bar{\theta})(\hat{\theta}^N_{LS} - \bar{\theta}) \overset{\mathrm{d}}{\to} z \sim \mathscr{N}(0, \mathbf{c}^\top \mathbf{M}(\xi,\bar{\theta})\mathbf{c})\,,\ N \to \infty\,,$$

for any $\mathbf{c} \in \mathbb{R}^p$. Applying the Taylor formula again we can write

$$\sqrt{N}[h(\hat{\theta}^N_{LS}) - h(\bar{\theta})] = \sqrt{N}\, \frac{\partial h(\theta)}{\partial \theta^\top}\bigg|_{\alpha^N} (\hat{\theta}^N_{LS} - \bar{\theta})$$

for some α^N between $\hat{\theta}^N_{LS}$ and $\bar{\theta}$ and $\partial h(\theta)/\partial\theta\big|_{\alpha^N} \stackrel{\text{a.s.}}{\to} \partial h(\theta)/\partial\theta\big|_{\bar{\theta}}$ as $N \to \infty$. When (3.39) is satisfied, we can write $\partial h(\theta)/\partial\theta\big|_{\bar{\theta}} = \mathbf{M}(\xi,\bar{\theta})\mathbf{u}$ for some $\mathbf{u} \in \mathbb{R}^p$, which gives (3.40). ■

Remark 3.15.

(*i*) When the limiting design ξ is singular, the property of regular asymptotic normality in Theorem 3.14 relies on all the design sequence and is not a property of ξ. Even more importantly, the sequence itself, not ξ, is responsible for the consistency of $h(\hat{\theta}^N_{LS})$. Using a c-optimal experiment ξ^*_c that minimizes $\left[\partial h(\theta)/\partial\theta^\top\, \mathbf{M}^-(\xi,\theta)\, \partial h(\theta)/\partial\theta\right]_{\theta^*}$ for some θ^* (see Chap. 5) may thus raise difficulties when ξ^*_c is singular.

(*ii*) It is instructive to consider the case of a linear function $h(\cdot)$, i.e., $h(\theta) = \mathbf{c}^\top\theta$, in Theorem 3.14. The choice of $\mathbf{c}$ is then not arbitrary when the limiting design ξ is singular: from (3.39), the vectors $\mathbf{c}$ for which regular asymptotic normality holds depend on the (unknown) value of $\bar{\theta}$; see Example 5.39 in Sect. 5.4. This means in particular that regular asymptotic normality does not hold in general for the estimation of a component $\{\theta\}_i$ of θ in a nonlinear regression model when the limiting design ξ is singular.

(*iii*) The conclusion of the theorem remains valid when $D_N(\theta,\bar{\theta})$ in Theorem 3.5 only satisfies $\inf_{\|\theta-\bar{\theta}\|\geq\delta} D_N(\theta,\bar{\theta}) \to \infty$ as $N \to \infty$ (which ensures $\hat{\theta}^N_{LS} \stackrel{\text{p}}{\to} \bar{\theta}$) with Θ a convex set; see Bierens (1994, Theorem 4.2.2); see also Theorem 4.15. □

Regular Asymptotic Normality when $\hat{\theta}^N_{LS}$ is not Consistent

The situation is much more complicated than above when $\hat{\theta}^N_{LS}$ does not converge to $\bar{\theta}$. Indeed, all possible limit points of the sequence $\{\hat{\theta}^N_{LS}\}$ in $\Theta^\# = \{\theta \in \Theta : \int_{\mathscr{X}} [\eta(x,\theta) - \eta(x,\bar{\theta})]^2\, \xi(\mathrm{d}x) = 0\}$ should be considered; see Theorem 3.3. This can be investigated through the geometry of the model under the design measure ξ.

For ξ any design measure, we define $\mathscr{L}_2(\xi)$ as the Hilbert space of real-valued functions $f(\cdot)$ on $\mathscr{X}$ which are square integrable with the norm

$$\|f\|_\xi = \left[\int_{\mathscr{X}} f^2(x)\, \xi(\mathrm{d}x)\right]^{1/2} < \infty\,. \tag{3.41}$$

Under H2$_\eta$, the functions $\eta(\cdot,\theta)$ and

$$\{\mathbf{f}_\theta\}_i(\cdot) = \partial\eta(\cdot,\theta)/\partial\theta_i\,,\ i = 1,\ldots,p\,, \tag{3.42}$$

belong to $\mathscr{L}_2(\xi)$. We shall write

$$\theta \stackrel{\xi}{\equiv} \theta^*$$

when the values θ and θ^* satisfy $\|\eta(\cdot,\theta)-\eta(\cdot,\theta^*)\|_\xi = 0$. Note that $\theta \stackrel{\xi}{\equiv} \bar{\theta}$ is equivalent to $\theta \in \Theta^{\#}$ as defined in the proof of Theorem 3.3.

We shall need the following technical assumptions on the geometry of the model:

H3$_\eta$: Let $\mathscr{S}_\epsilon$ denote the set $\left\{\theta \in \mathrm{int}(\Theta) : \left\|\eta(\cdot,\theta)-\eta(\cdot,\bar{\theta})\right\|_\xi^2 < \epsilon\right\}$, then there exists $\epsilon > 0$ such that for every $\theta^{\#}$ and θ^* in $\mathscr{S}_\epsilon$, we have

$$\left[\frac{\partial}{\partial\theta}\left\|\eta(\cdot,\theta)-\eta(\cdot,\theta^{\#})\right\|_\xi^2\right]_{\theta=\theta^*} = \mathbf{0} \implies \theta^{\#} \stackrel{\xi}{\equiv} \theta^* \,.$$

H4$_\eta$: For any point $\theta^* \stackrel{\xi}{\equiv} \bar{\theta}$, there exists a neighborhood $\mathcal{V}(\theta^*)$ such that

$$\forall \theta \in \mathcal{V}(\theta^*)\,,\ \mathrm{rank}[\mathbf{M}(\xi,\theta)] = \mathrm{rank}[\mathbf{M}(\xi,\theta^*)]\,.$$

The assumptions H3$_\eta$ and H4$_\eta$ admit a straightforward geometrical and statistical interpretation in the case where the measure ξ is discrete. Let $X = \mathcal{S}_\xi = \left\{x^{(1)},\ldots,x^{(k)}\right\}$ denote the support of ξ and define $\eta(\theta) = (\eta(x^{(1)},\theta),\ldots, \eta(x^{(k)},\theta))^\top$. The set $\mathbb{S}_\eta = \{\eta(\theta) : \theta \in \Theta\}$ is then the *expectation surface* of the model under the design ξ, and $\theta^* \stackrel{\xi}{\equiv} \bar{\theta}$ is equivalent to $\eta(\theta^*) = \eta(\bar{\theta})$. When H3$_\eta$ is not satisfied, it means that the surface $\mathbb{S}_\eta$ intersects itself at the point $\eta(\bar{\theta})$, and therefore, there are points $\eta(\theta)$ arbitrary close to $\eta(\bar{\theta})$ with θ far from $\bar{\theta}$. The asymptotic distribution of $\sqrt{N}(\hat{\theta}_{LS}^N - \bar{\theta})$ is then not normal, even if $\mathbf{M}(\xi,\bar{\theta})$ has full rank. When H4$_\eta$ is not satisfied, it means that the surface $\mathbb{S}_\eta$ possesses edges, and the point $\eta(\bar{\theta})$ belongs to such an edge, although $\bar{\theta} \in \mathrm{int}(\Theta)$. Again, the asymptotic distribution of $\sqrt{N}(\hat{\theta}_{LS}^N - \bar{\theta})$ is not normal.

Besides the geometrical assumptions above on the model, we need an assumption that generalizes (3.39) to the case where the set $\Theta^{\#}$ is not reduced to $\{\bar{\theta}\}$.

H2$_h$: The function $h(\cdot)$ is defined and has a continuous nonzero vector of derivatives $\partial h(\theta)/\partial\theta$ on $\mathrm{int}(\Theta)$. Moreover, for any $\theta \stackrel{\xi}{\equiv} \bar{\theta}$, there exists a linear mapping A_θ from $\mathscr{L}_2(\xi)$ to $\mathbb{R}$ (a continuous linear functional on $\mathscr{L}_2(\xi)$), such that $A_\theta = A_{\bar{\theta}}$ and that

$$\frac{\partial h(\theta)}{\partial \theta_i} = A_\theta\left[\{\mathbf{f}_\theta\}_i\right]\,,\ i = 1,\ldots,p\,,$$

where $\{\mathbf{f}_\theta\}_i$ is defined by (3.42).

When $\Theta^{\#} = \{\bar{\theta}\}$, H2$_h$ corresponds to (3.39) used in Theorem 3.14. In a linear model $\eta(x,\theta) = \mathbf{f}^\top(x)\theta$ with $h(\theta) = \mathbf{c}^\top\theta$, H2$_h$ is equivalent to the classical condition $\mathbf{c} \in \mathcal{M}[\mathbf{M}(\xi)]$; see Pázman and Pronzato (2009). More generally, it receives a simple interpretation when ξ is a discrete design measure with support $X = \mathcal{S}_\xi = \{x^{(1)}, \ldots, x^{(k)}\}$. Suppose that $h(\cdot)$ is continuously differentiable, then H2$_h$ and (3.12) are satisfied under the following assumption:

H2$'_h$: There exists a function $\Psi(\cdot)$, with continuous gradient, such that $h(\theta) = \Psi[\eta(\theta)]$, with $\eta(\theta) = (\eta(x^{(1)},\theta), \ldots, \eta(x^{(k)},\theta))^\top$.

We then obtain

$$\frac{\partial h(\theta)}{\partial \theta^\top} = \left.\frac{\partial \Psi(\mathbf{t})}{\partial \mathbf{t}^\top}\right|_{\mathbf{t}=\eta(\theta)} \frac{\partial \eta(\theta)}{\partial \theta^\top} .$$

H2$_h$ thus holds for every $\bar{\theta} \in \mathrm{int}(\Theta)$ with $A_\theta = \partial\Psi(\mathbf{t})/\partial\mathbf{t}^\top\big|_{\mathbf{t}=\eta(\theta)}$.

When ξ is a continuous design measure, an example where H2$_h$ holds and (3.12) is satisfied is when the following assumption is satisfied.

H2$''_h$: $h(\theta) = \Psi[h_1(\theta), \ldots, h_k(\theta)]$ with $\Psi(\cdot)$ a continuously differentiable function of k variables and with

$$h_i(\theta) = \int_{\mathscr{X}} g_i[\eta(x,\theta), x]\, \xi(\mathrm{d}x)\,,\ i = 1, \ldots, k\,,$$

for some functions $g_i(t,x)$ differentiable with respect to t for any x in the support of ξ.

Indeed, supposing that we can interchange the order of derivatives and integrals, we obtain

$$\frac{\partial h(\theta)}{\partial \theta_i} = \sum_{j=1}^{k} \left[\frac{\partial \Psi(v)}{\partial v_j}\right]_{v_j=h_j(\theta)} \int_{\mathscr{X}} \left[\frac{\partial g_j(t,x)}{\partial t}\right]_{t=\eta(x,\theta)} \{\mathbf{f}_\theta\}_i(x)\, \xi(\mathrm{d}x)\,,$$

and, for any $f \in \mathscr{L}_2(\xi)$,

$$A_\theta(f) = \sum_{j=1}^{k} \left[\frac{\partial \Psi(v)}{\partial v_j}\right]_{v_j=h_j(\theta)} \int_{\mathscr{X}} \left[\frac{\partial g_j(t,x)}{\partial t}\right]_{t=\eta(x,\theta)} f(x)\, \xi(\mathrm{d}x)\,,$$

so that H2$_h$ holds.

The following result on regular asymptotic normality without consistency of $\hat{\theta}^N_{LS}$ is proved in (Pázman and Pronzato, 2009).

Theorem 3.16. *Let $\{x_i\}$ be an asymptotically discrete design (Definition 2.1) or a randomized design (Definition 2.2) on $\mathscr{X} \subset \mathbb{R}^d$. Suppose that H$_\Theta$, H1$_\eta$, and H2$_\eta$ are satisfied and that the function of interest $h(\cdot)$ is continuous and satisfies (3.12) and H2$_h$. Let $\{\hat{\theta}^N_{LS}\}$ be a sequence of LS estimators. Then, H3$_\eta$ and H4$_\eta$ imply regular asymptotic normality: the sequence*

$\sqrt{N}\left\{h(\hat{\theta}_{LS}^N) - h(\bar{\theta})\right\}$ *converges in distribution as* $N \to \infty$ *to a random variable distributed*

$$\mathscr{N}\left(0, \left[\frac{\partial h(\theta)}{\partial \theta^\top}\mathbf{M}^-(\xi, \theta)\frac{\partial h(\theta)}{\partial \theta}\right]_{\theta=\bar{\theta}}\right),$$

where the choice of the g-inverse is arbitrary.

The importance of the assumption $H2_h$ in a nonlinear situation is illustrated in Example 3.17 below. In this example, the empirical design measure ξ_N converges strongly to a singular ξ_*, but regular asymptotic normality only holds in the special situation when $H2_h$ is satisfied.

Example 3.17. We consider the same linear regression model as in Example 3.13, but now the design is such that $N - m$ observations are taken at $x = x_* = 2h_* \in (0,1]$, with $h_* = -\theta_1^*/(2\theta_2^*)$ a prior guess for the location of the maximum of the function $\theta_1 x + \theta_2 x^2$, and m observations are taken at $x = z \in (0,1]$, $z \neq x_*$. We shall suppose that either m is fixed[4] or $m \to \infty$ with $m/N \to 0$ as N tends to infinity. In both cases the sequence $\{x_i\}$ is such that the empirical measure ξ_N converges strongly to δ_{x_*} as $N \to \infty$, in the sense of Definition 2.1. Note that $\delta_{x_*} = \xi_*$, the c-optimum design measure for $h(\theta_*)$, when $\sqrt{2} - 1 < x_* \leq 1$; see (3.37).

The LS estimator $\hat{\theta}_{LS}^N$ is given by

$$\hat{\theta}_{LS}^N = \bar{\theta} + \frac{1}{x_* z(x_* - z)}\left[\frac{\beta_m}{\sqrt{m}}\begin{pmatrix} x_*^2 \\ -x_* \end{pmatrix} + \frac{\gamma_{N-m}}{\sqrt{N-m}}\begin{pmatrix} -z^2 \\ z \end{pmatrix}\right], \quad (3.43)$$

where $\beta_m = (1/\sqrt{m})\sum_{x_i=z}\varepsilon_i$ and $\gamma_{N-m} = (1/\sqrt{N-m})\sum_{x_i=x_*}\varepsilon_i$ are independent random variables that tend to be distributed $\mathscr{N}(0,1)$ as $m \to \infty$ and $N - m \to \infty$. $\hat{\theta}_{LS}^N$ is consistent if and only if $m \to \infty$. However, $h(\hat{\theta}_{LS}^N)$ is also consistent when m is finite provided that $\bar{\theta}_1 + x_*\bar{\theta}_2 = 0$. Indeed, for m finite we have

$$\hat{\theta}_{LS}^N \overset{\text{a.s.}}{\to} \hat{\theta}^\# = \bar{\theta} + \frac{1}{z(x_* - z)}\frac{\beta_m}{\sqrt{m}}\begin{pmatrix} x_* \\ -1 \end{pmatrix}, \quad N \to \infty,$$

and $h(\hat{\theta}^\#) = -\hat{\theta}_1^\#/(2\hat{\theta}_2^\#) = x_*/2 = h(\bar{\theta})$. Also,

$$\begin{aligned}\sqrt{N}[h(\hat{\theta}_{LS}^N) - h(\bar{\theta})] &= \sqrt{N}[h(\hat{\theta}_{LS}^N) - h(\hat{\theta}^\#)] \\ &= \sqrt{N}(\hat{\theta}_{LS}^N - \hat{\theta}^\#)^\top\left[\left.\frac{\partial h(\theta)}{\partial \theta}\right|_{\hat{\theta}^\#} + o_p(1)\right],\end{aligned}$$

[4]Taking only a finite number of observations at another place than x_* might seem an odd strategy; note, however, that Wynn's algorithm (Wynn, 1972) for the minimization of $[\partial h(\theta)/\partial\theta^\top\, \mathbf{M}^-(\xi)\, \partial h(\theta)/\partial\theta]_{\theta^*}$ generates such a sequence of design points when the design space is $\mathscr{X} = [-1, 1]$, see Pázman and Pronzato (2006b), or when $\mathscr{X}$ is a finite set containing x_*.

with $\partial h(\theta)/\partial\theta\big|_{\hat{\theta}^{\#}} = -1/(2\hat{\theta}_2^{\#})[1,\ x_*]^\top$, and

$$\sqrt{N}(\hat{\theta}_{LS}^N - \hat{\theta}^{\#}) = \frac{\sqrt{N}}{x_*(x_* - z)} \frac{\gamma_{N-m}}{\sqrt{N-m}} \begin{pmatrix} -z \\ 1 \end{pmatrix}.$$

Therefore, $\sqrt{N}[h(\hat{\theta}_{LS}^N) - h(\bar{\theta})] \overset{\text{d}}{\to} \nu/(2\zeta)$ with $\nu \sim \mathscr{N}(0, 1/x_*^2)$ and $\zeta \sim \mathscr{N}(\bar{\theta}_2, 1/[mz^2(x_* - z)^2])$, and $h(\hat{\theta}_{LS}^N)$ is not asymptotically normal.

Suppose now that $m = m(N) \to \infty$ with $m/N \to 0$ as $N \to \infty$. If $\bar{\theta}_1 + x_*\bar{\theta}_2 \neq 0$, we can write

$$\sqrt{m}[h(\hat{\theta}_{LS}^N) - h(\bar{\theta})] = \sqrt{m}(\hat{\theta}_{LS}^N - \bar{\theta})^\top \left[\frac{\partial h(\theta)}{\partial \theta}\bigg|_{\bar{\theta}} + o_{\text{p}}(1) \right]$$

and, using (3.43), we get

$$\sqrt{m}[h(\hat{\theta}_{LS}^N) - h(\bar{\theta})] \overset{\text{d}}{\to} \zeta \sim \mathscr{N}\left(0, \frac{(\bar{\theta}_1 + x_*\bar{\theta}_2)^2}{4\bar{\theta}_2^4 z^2 (x_* - z)^2}\right), \ N \to \infty.$$

$h(\hat{\theta}_{LS}^N)$ thus converges as $1/\sqrt{m}$ and is asymptotically normal with a limiting variance depending on z. If $\bar{\theta}_1 + x_*\bar{\theta}_2 = 0$,

$$\sqrt{N}[h(\hat{\theta}_{LS}^N) - h(\bar{\theta})] = \sqrt{N}\left(-\frac{\hat{\theta}_1^N}{2\hat{\theta}_2^N} - \frac{x_*}{2}\right) = -\frac{\sqrt{N}}{2} \frac{\gamma_{N-m}}{\sqrt{N-m}} \frac{1}{x_*\hat{\theta}_2^N}$$

and

$$\sqrt{N}[h(\hat{\theta}_{LS}^N) - h(\bar{\theta})] \overset{\text{d}}{\to} \zeta \sim \mathscr{N}(0, 1/(4\bar{\theta}_2^2 x_*^2)).$$

This is the only situation within Examples 3.13 and 3.17 where regular asymptotic normality holds: $h(\hat{\theta}_{LS}^N)$ converges as $1/\sqrt{N}$, is asymptotically normal and has a limiting variance that can be computed from the limiting design ξ_*, i.e., which coincides with $[\partial h(\theta)/\partial\theta^\top \mathbf{M}^-(\xi_*)\, \partial h(\theta)/\partial\theta]_{\bar{\theta}}$. Note that assuming that $\bar{\theta}_1 + x_*\bar{\theta}_2 = 0$ amounts to assuming that the prior guess $h_* = x_*/2$ coincides with the true location of the maximum of the model response, which is rather unrealistic.

It is instructive to discuss (3.12) and H2$_h$ in the context this example. The limiting design is $\xi_* = \delta_{x_*}$, the measure that puts mass one at x_*. Therefore, $\theta \overset{\xi_*}{\equiv} \bar{\theta} \iff \theta_1 + x_*\theta_2 = \bar{\theta}_1 + x_*\bar{\theta}_2$. It follows that $\theta \overset{\xi_*}{\equiv} \bar{\theta} \Longrightarrow h(\theta) = h(\bar{\theta})$ only if $\bar{\theta}_1 + x_*\bar{\theta}_2 = 0$, and this is the only case where the condition (3.12) is satisfied. Since we have seen above that regular asymptotic normality does not hold when $\bar{\theta}_1 + x_*\bar{\theta}_2 \neq 0$, this shows the importance of (3.12). Consider now the derivative of $h(\theta)$. We have $\partial h(\theta)/\partial\theta = -1/(2\theta_2)[1,\ -\theta_1/\theta_2]^\top$ and $\partial\eta(x_*,\theta)/\partial\theta = (x_*,\ x_*^2)^\top$. Therefore, even if $\bar{\theta}_1 + x_*\bar{\theta}_2 = 0$, we obtain $\theta \overset{\xi_*}{\equiv} \bar{\theta} \Longrightarrow \partial h(\theta)/\partial\theta = [-1/(2\theta_2)][1,\ x_*]^\top = [-1/(2x_*\theta_2)]\partial\eta(x_*,\theta)/\partial\theta$, and H2$_h$ does not hold if $\theta \neq \bar{\theta}$. When m is fixed, $\Theta^{\#} = \arg\min J_{\bar{\theta}}(\theta)$, with $J_{\bar{\theta}}(\theta) = \lim_{N\to\infty} J_N(\theta)$, contains points other than $\bar{\theta}$; H2$_h$ does not hold and there is no regular asymptotic normality for $h(\hat{\theta}^N)$. On the opposite, when $m \to \infty$, $\Theta^{\#} = \{\bar{\theta}\}$ and H2$_h$ holds: this is the only situation in the example where regular asymptotic normality holds. □

Regular Asymptotic Normality of a Multidimensional Function $\mathbf{h}(\boldsymbol{\theta})$

Let $\mathbf{h}(\theta) = [h_1(\theta), \ldots, h_q(\theta)]^\top$ be a q-dimensional function defined on Θ. We then have the following straightforward extension of Theorem 3.3.

Theorem 3.18. *Suppose that the q functions $h_i(\cdot)$ are continuous on Θ. Then, under the assumptions of Theorem 3.3, but with*

$$\theta \overset{\xi}{\equiv} \bar{\theta} \Longrightarrow \mathbf{h}(\theta) = \mathbf{h}(\bar{\theta}) \tag{3.44}$$

replacing (3.12), we have $\lim_{N\to\infty} \mathbf{h}(\hat{\theta}^N_{LS}) = \mathbf{h}(\bar{\theta})$ a.s. for $\{\hat{\theta}^N_{LS}\}$ any sequence of LS estimators.

Also, Theorem 3.14 can be extended into the following.

Theorem 3.19. *Under the assumptions of Theorem 3.14 but with*

$$\left.\frac{\partial \mathbf{h}^\top(\theta)}{\partial \theta}\right|_{\bar{\theta}} \in \mathcal{M}[\mathbf{M}(\xi, \bar{\theta})],$$

replacing (3.39), $\mathbf{h}(\hat{\theta}^N_{LS})$ satisfies

$$\sqrt{N}\left\{\mathbf{h}(\hat{\theta}^N_{LS}) - \mathbf{h}(\bar{\theta})\right\} \overset{\mathrm{d}}{\to} \mathbf{z} \sim \mathscr{N}\left(\mathbf{0}, \left[\frac{\partial \mathbf{h}(\theta)}{\partial \theta^\top}\mathbf{M}^-(\xi, \theta)\frac{\partial \mathbf{h}^\top(\theta)}{\partial \theta}\right]_{\bar{\theta}}\right)$$

as $N \to \infty$, where the choice of the g-inverse is arbitrary.

Proof. Take any $\mathbf{c} \in \mathbb{R}^q$, and define $h_c(\theta) = \mathbf{c}^\top \mathbf{h}(\theta)$. Evidently $h_c(\theta)$ satisfies the assumptions of Theorem 3.14 and $\sqrt{N}\left\{h_c(\hat{\theta}^N_{LS}) - h_c(\bar{\theta})\right\}$ converges in distribution as $N \to \infty$ to a random variable distributed

$$\mathscr{N}\left(0, \mathbf{c}^\top \left[\frac{\partial \mathbf{h}(\theta)}{\partial \theta^\top}\mathbf{M}^-(\xi, \theta)\frac{\partial \mathbf{h}^\top(\theta)}{\partial \theta}\right]_{\bar{\theta}} \mathbf{c}\right). \qquad \blacksquare$$

Consider now the following assumption; its substitution for H2$_h$ in Theorem 3.16 gives Theorem 3.20 below.

H3$_h$: The vector function $\mathbf{h}(\theta)$ has a continuous Jacobian $\partial \mathbf{h}(\theta)/\partial \theta^\top$ on $\mathrm{int}(\Theta)$. Moreover, for each $\theta \overset{\xi}{\equiv} \bar{\theta}$, there exists a continuous linear mapping B_θ from $\mathscr{L}_2(\xi)$ to $\mathbb{R}^q$ such that $B_\theta = B_{\bar{\theta}}$ and that

$$\frac{\partial \mathbf{h}(\theta)}{\partial \theta_i} = B_\theta\left[\{\mathbf{f}_\theta\}_i\right], \ i = 1, \ldots, p,$$

where $\{\mathbf{f}_\theta\}_i$ is given by (3.42).

Theorem 3.20. *Under the assumptions of Theorem 3.16, but with (3.44) and $H3_h$ replacing (3.12) and $H2_h$, respectively, for any sequence $\{\hat{\theta}^N_{LS}\}$ of LS estimators*

$$\sqrt{N}\left\{\mathbf{h}(\hat{\theta}^N_{LS}) - \mathbf{h}(\bar{\theta})\right\} \overset{\mathrm{d}}{\to} \mathbf{z} \sim \mathscr{N}\left(\mathbf{0}, \left[\frac{\partial \mathbf{h}(\theta)}{\partial \theta^\top}\mathbf{M}^-(\xi,\theta)\frac{\partial \mathbf{h}^\top(\theta)}{\partial \theta}\right]_{\bar{\theta}}\right)$$

as $N \to \infty$, where the choice of the g-inverse is arbitrary.

3.3 LS Estimation with Parameterized Variance

In some situations, more information than the mean response $\eta(x,\theta)$ can be included in the characteristics of the model, even if the full parameterized probability distribution of the observations is not available. In particular, this includes the situation where the variance function $\sigma^2(x_i)$ of the error ε_i in the model (3.2) has a known parametric form.

The (ordinary) LS estimator, which ignores this information, is still strongly consistent and asymptotically normally distributed under the conditions given in Sect. 3.1. However, when the parameters θ that enter into the mean response also enter the variance function, it is natural to expect that using the variance information will provide a more precise estimation of θ. Only the case of randomized designs will be considered, but similar developments can be obtained for asymptotically discrete designs using Lemma 2.5 instead of Lemma 2.6.

We consider the situation where the errors in (3.2) satisfy

$$\mathbb{E}\{\varepsilon_i^2\} = \sigma^2(x_i) = \bar{\beta}\,\lambda(x_i,\bar{\theta}) \geq 0 \ \text{ for all } i\,, \tag{3.45}$$

with $\bar{\beta}$ a positive scaling factor, either known or estimated (see Sect. 3.3.6). We shall use the following assumptions:

H1$_\lambda$: $\lambda(x,\bar{\theta})$ is bounded and strictly positive on $\mathscr{X}$, $\lambda^{-1}(x,\theta)$ is bounded on $\mathscr{X} \times \Theta$, and $\lambda(x,\theta)$ is continuous on Θ for all $x \in \mathscr{X}$.

H2$_\lambda$: For all $x \in \mathscr{X}$, $\lambda(x,\theta)$ is twice continuously differentiable with respect to $\theta \in \mathrm{int}(\Theta)$, and its first two derivatives are bounded on $\mathscr{X} \times \mathrm{int}(\Theta)$.

The influence of a misspecification of the variance function $\lambda(x,\theta)$, in particular of $\bar{\beta}$ in (3.45), is considered in Sect. 3.3.5, and the case of variance functions depending on parameters other than θ is treated in Sect. 3.3.6.

3.3.1 Inconsistency of WLS with Parameter-Dependent Weights

Since the optimum weights given in Theorem 3.8, $w(x) = \lambda^{-1}(x,\bar{\theta})$, cannot be used ($\bar{\theta}$ is unknown), it is tempting to use the weights $\lambda^{-1}(x,\theta)$, i.e., to choose $\hat{\theta}^N$ that minimizes the criterion

$$J_N(\theta) = \frac{1}{N} \sum_{k=1}^{N} \frac{[y(x_k) - \eta(x_k, \theta)]^2}{\lambda(x_k, \theta)} . \qquad (3.46)$$

However, this approach is not recommended since $\hat{\theta}^N$ is generally not consistent, as shown in the following theorem.

Theorem 3.21. *Let $\{x_i\}$ be a randomized design with measure ξ on $\mathscr{X} \subset \mathbb{R}^d$; see Definition 2.2. Consider the estimator $\hat{\theta}^N$ that minimizes (3.46) in the model (3.2), (3.45). Assume that H_Θ, $H1_\eta$, and $H1_\lambda$ are satisfied. Then, as $N \to \infty$, $\hat{\theta}^N$ converges a.s. to the set $\Theta^{\#}$ of values of θ that minimize*

$$J_{\bar{\theta}}(\theta) = \bar{\beta} \int_{\mathscr{X}} \lambda(x, \bar{\theta}) \lambda^{-1}(x, \theta)\, \xi(\mathrm{d}x) + \int_{\mathscr{X}} \lambda^{-1}(x, \theta) [\eta(x, \theta) - \eta(x, \bar{\theta})]^2\, \xi(\mathrm{d}x) .$$

Notice that, in general, $\bar{\theta} \notin \Theta^{\#}$.

Proof. As in Theorem 3.1, we have for every θ,

$$\begin{aligned} J_N(\theta) &= \frac{1}{N} \sum_{k=1}^{N} \lambda^{-1}(x_k, \theta) [y(x_k) - \eta(x_k, \theta)]^2 \\ &= \frac{1}{N} \sum_{k=1}^{N} \lambda^{-1}(x_k, \theta) \varepsilon_k^2 + \frac{2}{N} \sum_{k=1}^{N} \lambda^{-1}(x_k, \theta) [\eta(x_k, \bar{\theta}) - \eta(x_k, \theta)] \varepsilon_k \\ &\quad + \frac{1}{N} \sum_{k=1}^{N} \lambda^{-1}(x_k, \theta) [\eta(x_k, \bar{\theta}) - \eta(x_k, \theta)]^2 . \end{aligned}$$

Lemma 2.6 implies that, when $N \to \infty$, the first term on the right-hand side converges a.s. to $\bar{\beta} \int_{\mathscr{X}} \lambda^{-1}(x, \theta) \lambda(x, \bar{\theta}) \xi(\mathrm{d}x)$, the second to zero, and the third to $\int_{\mathscr{X}} \lambda^{-1}(x, \theta)\ [\eta(x, \theta) - \eta(x, \bar{\theta})]^2 \xi(\mathrm{d}x)$, and for each term, the convergence is uniform with respect to θ. Lemma 2.11 then gives the result. ■

3.3.2 Consistency and Asymptotic Normality of Penalized WLS

Consider now the following modification of the criterion (3.46),

$$J_N(\theta) = \frac{1}{N} \sum_{k=1}^{N} \frac{[y(x_k) - \eta(x_k, \theta)]^2}{\lambda(x_k, \theta)} + \frac{\bar{\beta}}{N} \sum_{k=1}^{N} \log \lambda(x_k, \theta) . \qquad (3.47)$$

It can be considered as a *penalized WLS estimator*, where the term $(\bar{\beta}/N) \sum_{k=1}^{N} \log \lambda(x_k, \theta)$ penalizes large variances. This construction will receive a formal justification in Sects. 4.2 and 4.3; see in particular Example 4.12: it corresponds to the maximum likelihood estimator when the errors ε_i are normally distributed. It can be obtained numerically by direct minimization of (3.47) using a nonlinear optimization method or by solving an infinite sequence[5] of weighted LS problems as suggested in (Downing et al., 2001); see also Sect. 4.3.2. Next theorems show that this estimator is strongly consistent and asymptotically normally distributed without the assumption of normal errors. Notice that the method requires that $\bar{\beta}$ is known. The consequences of a misspecification of $\bar{\beta}$ will be considered in Sect. 3.3.5 and the joint estimation of $\bar{\beta}$ and θ in Sect. 3.3.6. On the other hand, the two-stage method of Sect. 3.3.3 does not require $\bar{\beta}$ to be known.

Theorem 3.22 (Consistency of the penalized WLS estimator). *Let $\{x_i\}$ be a randomized design with measure ξ on $\mathscr{X} \subset \mathbb{R}^d$; see Definition 2.2. Consider the estimator $\hat{\theta}^N_{PWLS}$ that minimizes (3.47) in the model (3.2), (3.45). Assume that H_Θ, $H1_\eta$, and $H1_\lambda$ are satisfied and that*

$$\forall \theta \in \Theta\,, \left\{ \begin{array}{r} \int_{\mathscr{X}} \lambda^{-1}(x,\theta)[\eta(x,\theta) - \eta(x,\bar{\theta})]^2\, \xi(\mathrm{d}x) = 0 \\ \int_{\mathscr{X}} |\lambda^{-1}(x,\theta)\lambda(x,\bar{\theta}) - 1|\, \xi(\mathrm{d}x) = 0 \end{array} \right\} \Longleftrightarrow \theta = \bar{\theta}\,. \tag{3.48}$$

Then $\hat{\theta}^N_{PWLS}$ converges a.s. to $\bar{\theta}$ as $N \to \infty$.

Proof. The proof follows the same lines as for Theorem 3.1. Using Lemma 2.6, we show that $J_N(\theta) \overset{\theta}{\rightsquigarrow} J_{\bar{\theta}}(\theta)$ a.s. when $N \to \infty$, with

$$\begin{aligned} J_{\bar{\theta}}(\theta) &= \bar{\beta} \int_{\mathscr{X}} \lambda(x,\bar{\theta})\lambda^{-1}(x,\theta)\, \xi(\mathrm{d}x) + \int_{\mathscr{X}} \lambda^{-1}(x,\theta)[\eta(x,\theta) - \eta(x,\bar{\theta})]^2\, \xi(\mathrm{d}x) \\ &\quad + \bar{\beta} \int_{\mathscr{X}} \log \lambda(x,\theta)\, \xi(\mathrm{d}x) \\ &= \int_{\mathscr{X}} \lambda^{-1}(x,\theta)[\eta(x,\theta) - \eta(x,\bar{\theta})]^2\, \xi(\mathrm{d}x) \\ &\quad + \bar{\beta} \int_{\mathscr{X}} \left\{ \lambda(x,\bar{\theta})\lambda^{-1}(x,\theta) - \log\left[\lambda(x,\bar{\theta})\lambda^{-1}(x,\theta)\right] \right\} \xi(\mathrm{d}x) \\ &\quad + \bar{\beta} \int_{\mathscr{X}} \log \lambda(x,\bar{\theta})\, \xi(\mathrm{d}x)\,. \end{aligned}$$

The function $x - \log x$ is positive and minimum at $x = 1$, so that $J_{\bar{\theta}}(\theta) \geq \bar{\beta}[1 + \int_{\mathscr{X}} \log \lambda(x,\bar{\theta})\, \xi(\mathrm{d}x)]$. From (3.48) the equality is obtained only at $\theta = \bar{\theta}$. Lemma 2.10 then implies that $\hat{\theta}^N \overset{\text{a.s.}}{\to} \bar{\theta}$ as $N \to \infty$. ■

Remark 3.23. Suppose that $\bar{\beta}$ in (3.45) is an unknown positive constant that forms a nuisance parameter for the estimation of θ. Assuming that $y(x_k)$ is normally distributed $\mathscr{N}(\eta(x_k,\bar{\theta}), \bar{\beta}\lambda(x_k,\bar{\theta}))$ for all k, we obtain that the maximum likelihood estimator of θ and β minimizes

$$J_N(\theta,\beta) = \frac{1}{N} \sum_{k=1}^{N} \frac{[y(x_k) - \eta(x_k,\theta)]^2}{\beta\, \lambda(x_k,\theta)} + \frac{1}{N} \sum_{k=1}^{N} \log \lambda(x_k,\theta) + \log(\beta)$$

[5]However, we shall in Remark 3.28-(*iv*) that two steps are enough to obtain the same asymptotic behavior as the maximum likelihood estimator for normal errors.

with respect to θ and β; see Example 4.12. For any θ, this function reaches its minimum with respect to β at

$$\hat{\beta}^N = \frac{1}{N} \sum_{k=1}^{N} \frac{[y(x_k) - \eta(x_k, \theta)]^2}{\lambda(x_k, \theta)}$$

and the substitution of $\hat{\beta}^N$ for β in $J_N(\theta, \beta)$ yields the criterion

$$J_N(\theta) = \log \left\{ \frac{1}{N} \sum_{k=1}^{N} \frac{[y(x_k) - \eta(x_k, \theta)]^2}{\lambda(x_k, \theta)} \right\} + \frac{1}{N} \sum_{k=1}^{N} \log \lambda(x_k, \theta) \,, \tag{3.49}$$

to be minimized with respect to θ. Similarly to the proof of Theorem 3.22, we obtain that $J_N(\theta) \overset{\theta}{\rightsquigarrow} J_{\bar{\theta}}(\theta)$ a.s. for $\{x_k\}$ a randomized design with measure ξ, where now

$$J_{\bar{\theta}}(\theta) = \log \left\{ \bar{\beta} \int_{\mathscr{X}} \frac{\lambda(x, \bar{\theta})}{\lambda(x, \theta)} \, \xi(\mathrm{d}x) + \int_{\mathscr{X}} \frac{[\eta(x, \theta) - \eta(x, \bar{\theta})]^2}{\lambda(x, \theta)} \, \xi(\mathrm{d}x) \right\}$$
$$+ \int_{\mathscr{X}} \log \lambda(x, \theta) \, \xi(\mathrm{d}x) \,.$$

Therefore,

$$\begin{aligned} J_{\bar{\theta}}(\theta) - J_{\bar{\theta}}(\bar{\theta}) &= \log \left\{ \int_{\mathscr{X}} \frac{\lambda(x, \bar{\theta})}{\lambda(x, \theta)} \, \xi(\mathrm{d}x) + \int_{\mathscr{X}} \frac{[\eta(x, \theta) - \eta(x, \bar{\theta})]^2}{\bar{\beta} \lambda(x, \theta)} \, \xi(\mathrm{d}x) \right\} \\ &\quad + \int_{\mathscr{X}} \log \frac{\lambda(x, \theta)}{\lambda(x, \bar{\theta})} \, \xi(\mathrm{d}x) \\ &\geq \log \left\{ \int_{\mathscr{X}} \frac{\lambda(x, \bar{\theta})}{\lambda(x, \theta)} \, \xi(\mathrm{d}x) \right\} + \int_{\mathscr{X}} \log \frac{\lambda(x, \theta)}{\lambda(x, \bar{\theta})} \, \xi(\mathrm{d}x) \\ &\geq 0 \end{aligned}$$

(from Jensen's inequality), with equality if and only if

$$\int_{\mathscr{X}} \frac{[\eta(x, \theta) - \eta(x, \bar{\theta})]^2}{\lambda(x, \theta)} \, \xi(\mathrm{d}x) = 0 \tag{3.50}$$

and

$$\int_{\mathscr{X}} \left| \frac{\lambda(x, \bar{\theta})}{\lambda(x, \theta)} - C \right| \xi(\mathrm{d}x) = 0 \;\; \text{for some constant} \;\; C > 0 \,. \tag{3.51}$$

Under the estimability condition [(3.50) and (3.51) $\Longrightarrow \theta = \bar{\theta}$], the estimator that minimizes (3.49) thus converges a.s. to $\bar{\theta}$ as $N \to \infty$. □

Theorem 3.24 (Asymptotic normality of the penalized WLS estimator). *Let $\{x_i\}$ be a randomized design with measure ξ on $\mathscr{X} \subset \mathbb{R}^d$; see*

Definition 2.2. Consider the penalized WLS estimator $\hat{\theta}^N_{PWLS}$ *that minimizes the criterion (3.47) in the model (3.2), (3.45) where the errors* ε_i *have finite fourth moments* $\mathbb{E}(\varepsilon_i^4)$. *Assume that* H_Θ, $H1_\eta$, $H2_\eta$, $H1_\lambda$, *and* $H2_\lambda$ *are satisfied, that the condition (3.48) is satisfied, and that the matrix*

$$\mathbf{M}_1(\xi,\bar{\theta}) = \int_{\mathscr{X}} \lambda^{-1}(x,\bar{\theta}) \frac{\partial \eta(x,\theta)}{\partial \theta}\bigg|_{\bar{\theta}} \frac{\partial \eta(x,\theta)}{\partial \theta^\top}\bigg|_{\bar{\theta}} \xi(\mathrm{d}x) + \frac{\bar{\beta}}{2} \int_{\mathscr{X}} \lambda^{-2}(x,\bar{\theta}) \frac{\partial \lambda(x,\theta)}{\partial \theta}\bigg|_{\bar{\theta}} \frac{\partial \lambda(x,\theta)}{\partial \theta^\top}\bigg|_{\bar{\theta}} \xi(\mathrm{d}x) \tag{3.52}$$

is nonsingular. Then, $\hat{\theta}^N_{PWLS}$ *satisfies*

$$\sqrt{N}(\hat{\theta}^N_{PWLS} - \bar{\theta}) \overset{\mathrm{d}}{\to} \mathbf{z} \sim \mathscr{N}(\mathbf{0}, \mathbf{M}_1^{-1}(\xi,\bar{\theta})\mathbf{M}_2(\xi,\bar{\theta})\mathbf{M}_1^{-1}(\xi,\bar{\theta}))\,,\ N \to \infty\,,$$

with

$$\mathbf{M}_2(\xi,\bar{\theta}) = \bar{\beta}\mathbf{M}_1(\xi,\bar{\theta}) + \frac{\bar{\beta}^{3/2}}{2} \int_{\mathscr{X}} \lambda^{-3/2}(x,\bar{\theta}) \left[\frac{\partial \eta(x,\theta)}{\partial \theta}\bigg|_{\bar{\theta}} \frac{\partial \lambda(x,\theta)}{\partial \theta^\top}\bigg|_{\bar{\theta}} + \frac{\partial \lambda(x,\theta)}{\partial \theta}\bigg|_{\bar{\theta}} \frac{\partial \eta(x,\theta)}{\partial \theta^\top}\bigg|_{\bar{\theta}} \right] s(x)\xi(\mathrm{d}x) + \frac{\bar{\beta}^2}{4} \int_{\mathscr{X}} \lambda^{-2}(x,\bar{\theta}) \frac{\partial \lambda(x,\theta)}{\partial \theta}\bigg|_{\bar{\theta}} \frac{\partial \lambda(x,\theta)}{\partial \theta^\top}\bigg|_{\bar{\theta}} \kappa(x)\,\xi(\mathrm{d}x)\,, \tag{3.53}$$

where $s(x) = \mathbb{E}_x\{\varepsilon^3(x)\}\sigma^{-3}(x)$ *is the skewness and* $\kappa(x) = \mathbb{E}_x\{\varepsilon^4(x)\}\sigma^{-4}(x) - 3$ *the kurtosis of the distribution of* $\varepsilon(x)$.

Proof. The proof is similar to that of Theorem 3.8. Using Lemma 2.6 with the conditions stated in the theorem, the calculation of the derivatives of $J_N(\theta)$ gives $\nabla^2_\theta J_N(\hat{\theta}^N_{PWLS}) \overset{\text{a.s.}}{\to} 2\,\mathbf{M}_1(\xi,\bar{\theta})$ as $N \to \infty$. Also,

$$-\sqrt{N}\nabla_\theta J_N(\bar{\theta}) = \frac{2}{\sqrt{N}} \sum_{k=1}^N \left\{ \lambda^{-1}(x_k,\bar{\theta})\varepsilon_k \frac{\partial \eta(x_k,\theta)}{\partial \theta}\bigg|_{\bar{\theta}} + \frac{1}{2}\lambda^{-2}(x_k,\bar{\theta})\varepsilon_k^2 \frac{\partial \lambda(x_k,\theta)}{\partial \theta}\bigg|_{\bar{\theta}} - \frac{\bar{\beta}}{2}\lambda^{-1}(x_k,\bar{\theta}) \frac{\partial \lambda(x_k,\theta)}{\partial \theta}\bigg|_{\bar{\theta}} \right\} = \frac{2}{\sqrt{N}} \sum_{k=1}^N \mathbf{w}_k\,,$$

where the $\mathbf{w}_k$ are independent. Direct calculation gives $\mathbb{E}\{\mathbf{w}_k\} = \mathbf{0}$ and $\mathbb{E}\{\mathbf{w}_k\mathbf{w}_k^\top\} = \mathbf{M}_2(\xi,\bar{\theta})$. Therefore, $\sqrt{N}(\hat{\theta}^N_{PWLS} - \bar{\theta}) \overset{\mathrm{d}}{\to} \mathbf{z}$ as $N \to \infty$, with $\mathbf{z}$ distributed $\mathscr{N}(\mathbf{0}, \mathbf{M}_1^{-1}(\xi,\bar{\theta})\mathbf{M}_2(\xi,\bar{\theta})\mathbf{M}_1^{-1}(\xi,\bar{\theta}))$. ■

Remark 3.25. When the errors ε_k are normally distributed, $s(x) = \kappa(x) = 0$, $\mathbf{M}_2(\xi,\bar{\theta}) = \bar{\beta}\mathbf{M}_1(\xi,\bar{\theta})$, and $\sqrt{N}(\hat{\theta}^N_{PWLS} - \bar{\theta}) \overset{\mathrm{d}}{\to} \mathbf{z} \sim \mathscr{N}(\mathbf{0}, \bar{\beta}\mathbf{M}_1^{-1}(\xi,\bar{\theta}))$ as $N \to \infty$. □

3.3.3 Consistency and Asymptotic Normality of Two-stage LS

By two-stage LS, we mean using first some estimator $\hat{\theta}_1^N$ and then plugging this estimate into the weight function $\lambda(x, \theta)$. The second-stage estimator $\hat{\theta}_{TSLS}^N$ is then obtained by minimizing

$$J_N(\theta, \hat{\theta}_1^N) = \frac{1}{N} \sum_{k=1}^{N} \frac{[y(x_k) - \eta(x_k, \theta)]^2}{\lambda(x_k, \hat{\theta}_1^N)} \tag{3.54}$$

with respect to $\theta \in \Theta$. We shall write

$$\nabla_\theta J_N(\theta, \theta') = \frac{\partial J_N(\theta, \theta')}{\partial \theta}, \; \nabla^2_{\theta,\theta} J_N(\theta, \theta') = \frac{\partial^2 J_N(\theta, \theta')}{\partial \theta \partial \theta^\top},$$

and

$$\nabla^2_{\theta,\theta'} J_N(\theta, \theta') = \frac{\partial^2 J_N(\theta, \theta')}{\partial \theta \partial \theta'^\top}.$$

We first show that $\hat{\theta}_{TSLS}^N$ is consistent when $\hat{\theta}_1^N$ converges to some $\bar{\theta}_1 \in \Theta$; $\hat{\theta}_1^N$ does not need to be consistent, i.e., its convergence to $\bar{\theta}$ is not required. Then, we show that when $\hat{\theta}_1^N$ is $\sqrt{N}$-consistent, i.e., when $\sqrt{N}(\hat{\theta}_1^N - \bar{\theta})$ is bounded in probability,[6] $\hat{\theta}_{TSLS}^N$ is asymptotically normally distributed with the best possible covariance matrix among WLS estimators. Note that we implicitly assume that, excepted for the scaling factor $\bar{\beta}$, the variance function λ does not depend on any unknown parameter other than θ. The more general situation where such other parameters enter λ will be considered in Sect. 3.3.6. Also note that, in opposition to previous section, $\bar{\beta}$ does not need to be known.

Theorem 3.26 (Consistency of the two-stage LS estimator). *Let $\{x_i\}$ be a randomized design with measure ξ on $\mathscr{X} \subset \mathbb{R}^d$; see Definition 2.2. Consider the estimator $\hat{\theta}_{TSLS}^N$ that minimizes (3.54) in the model (3.2), (3.45). Assume that H_Θ, $H1_\eta$, and $H1_\lambda$ are satisfied, that $\hat{\theta}_1^N$ converges a.s. to some $\bar{\theta}_1 \in \Theta$, and that*

$$\forall \theta \in \Theta, \; \int_{\mathscr{X}} \lambda^{-1}(x, \bar{\theta}_1)[\eta(x, \theta) - \eta(x, \bar{\theta})]^2 \, \xi(\mathrm{d}x) = 0 \Longleftrightarrow \theta = \theta. \tag{3.55}$$

Then, $\hat{\theta}_{TSLS}^N$ converges a.s. to $\bar{\theta}$ as $N \to \infty$.

Proof. Using Lemma 2.6, one can easily show that $J_N(\theta, \theta')$ converges a.s. and uniformly in θ, θ' to

$$J_{\bar{\theta}}(\theta, \theta') = \bar{\beta} \int_{\mathscr{X}} \lambda(x, \bar{\theta}) \lambda^{-1}(x, \theta') \, \xi(\mathrm{d}x) + \int_{\mathscr{X}} \lambda^{-1}(x, \theta')[\eta(x, \theta) - \eta(x, \bar{\theta})]^2 \, \xi(\mathrm{d}x),$$

so that $J_{\bar{\theta}}(\theta, \hat{\theta}_1^N)$ converges a.s. and uniformly in θ to $J_{\bar{\theta}}(\theta, \bar{\theta}_1)$ as $N \to \infty$. Lemma 2.10 and (3.55) then give the result. ∎

[6] See page 31 for the definition.

Theorem 3.27 (Asymptotic normality of the two-stage LS estimator). *Let $\{x_i\}$ be a randomized design with measure ξ on $\mathscr{X} \subset \mathbb{R}^d$; see Definition 2.2. Consider the two-stage LS estimator $\hat{\theta}^N_{TSLS}$ that minimizes the criterion (3.54) in the model (3.2), (3.45). Assume that H_Θ, $H1_\eta$, $H2_\eta$, $H1_\lambda$, and $H2_\lambda$ are satisfied, that the condition (3.55) is satisfied, and that the matrix*

$$\mathbf{M}(\xi,\bar{\theta}) = \int_{\mathscr{X}} \lambda^{-1}(x,\bar{\theta}) \left.\frac{\partial \eta(x,\theta)}{\partial \theta}\right|_{\bar{\theta}} \left.\frac{\partial \eta(x,\theta)}{\partial \theta^\top}\right|_{\bar{\theta}} \xi(\mathrm{d}x) \qquad (3.56)$$

is nonsingular. Assume that the first-stage estimator $\hat{\theta}^N_1$ plugged in (3.54) is $\sqrt{N}$-consistent. Then, $\hat{\theta}^N_{TSLS}$ satisfies

$$\sqrt{N}(\hat{\theta}^N_{TSLS} - \bar{\theta}) \overset{\mathrm{d}}{\to} \mathbf{z} \sim \mathscr{N}(\mathbf{0}, \bar{\beta}\mathbf{M}^{-1}(\xi,\bar{\theta}))\,,\ N \to \infty\,.$$

Proof. We use the same approach as in Theorem 3.8. Since $\hat{\theta}^N_1$ is strongly consistent, $\hat{\theta}^N_{TSLS}$ is strongly consistent too from Theorem 3.26 and, since $\bar{\theta} \in \mathrm{int}(\Theta)$, $\hat{\theta}^N_{TSLS} \in \mathrm{int}(\Theta)$ for N large enough. $J_N(\theta, \hat{\theta}^N_1)$ is differentiable with respect to θ and $\nabla_\theta J_N(\hat{\theta}^N_{TSLS}, \hat{\theta}^N_1) = \mathbf{0}$ for large N. A Taylor series development of $\{\nabla_\theta J_N(\theta,\theta')\}_i$ for $i = 1, \ldots, p$ then gives

$$\begin{aligned} 0 = \{\nabla_\theta J_N(\hat{\theta}^N_{TSLS}, \hat{\theta}^N_1)\}_i &= \{\nabla_\theta J_N(\bar{\theta}, \hat{\theta}^N_1)\}_i + \{\nabla^2_{\theta,\theta} J_N(\beta^N_i, \hat{\theta}^N_1)(\hat{\theta}^N_{TSLS} - \bar{\theta})\}_i \\ &= \{\nabla_\theta J_N(\bar{\theta},\bar{\theta})\}_i + \{\nabla^2_{\theta,\theta'} J_N(\bar{\theta},\gamma^N_i)(\hat{\theta}^N_1 - \bar{\theta})\}_i \\ &\quad + \{\nabla^2_{\theta,\theta} J_N(\beta^N_i, \hat{\theta}^N_1)(\hat{\theta}^N_{TSLS} - \bar{\theta})\}_i \end{aligned}$$

for some $\beta^N_i = (1-\alpha_{1,i,N})\bar{\theta} + \alpha_{1,i,N}\hat{\theta}^N_{TSLS}$ and $\gamma^N_i = (1-\alpha_{2,i,N})\bar{\theta} + \alpha_{2,i,N}\hat{\theta}^N_1$, $\alpha_{1,i,N}, \alpha_{2,i,N} \in (0,1)$ (and β^N_i, γ^N_i are measurable; see Lemma 2.12). Previous equation can be written

$$\begin{aligned} \{\nabla^2_{\theta,\theta} J_N(\beta^N_i, \hat{\theta}^N_1)[\sqrt{N}(\hat{\theta}^N_{TSLS} - \bar{\theta})]\}_i &= -\sqrt{N}\{\nabla_\theta J_N(\bar{\theta},\bar{\theta})\}_i \\ &\quad -\{\nabla^2_{\theta,\theta'} J_N(\bar{\theta},\gamma^N_i)[\sqrt{N}(\hat{\theta}^N_1 - \bar{\theta})]\}_i\,. \end{aligned}$$

Using Lemma 2.6, we can easily show that $\nabla^2_{\theta,\theta} J_N(\theta,\theta')$ tends a.s. and uniformly in θ,θ' to the matrix $2\,\mathbf{M}(\xi,\theta,\theta') - 2\int_{\mathscr{X}} \lambda^{-1}(x,\theta')[\eta(x,\bar{\theta}) - \eta(x,\theta)]$ $\partial^2\eta(x,\theta)/\partial\theta\partial\theta^\top\, \xi(\mathrm{d}x)$ as $N \to \infty$, with

$$\mathbf{M}(\xi,\theta,\theta') = \int_{\mathscr{X}} \lambda^{-1}(x,\theta') \frac{\partial \eta(x,\theta)}{\partial \theta} \frac{\partial \eta(x,\theta)}{\partial \theta^\top} \xi(\mathrm{d}x)\,.$$

Therefore, $\nabla^2_{\theta,\theta} J_N(\beta^N_i, \hat{\theta}^N_1) \overset{\text{a.s.}}{\to} 2\,\mathbf{M}(\xi,\bar{\theta})$ when $N \to \infty$. Similarly, Lemma 2.6 implies that $\nabla^2_{\theta,\theta'} J_N(\bar{\theta},\gamma^N_i) \overset{\text{a.s.}}{\to} \mathbf{O}$ when $N \to \infty$. Since $\hat{\theta}^N_1$ is $\sqrt{N}$-consistent, $\nabla^2_{\theta,\theta'} J_N(\bar{\theta},\gamma^N_i)[\sqrt{N}(\hat{\theta}^N_1 - \bar{\theta})] \overset{\mathrm{p}}{\to} \mathbf{0}$. Next, we obtain

$$-\sqrt{N}\nabla_\theta J_N(\bar{\theta},\bar{\theta}) \overset{\mathrm{d}}{\to} \mathbf{v} \sim \mathscr{N}(\mathbf{0}, 4\,\bar{\beta}\mathbf{M}(\xi,\bar{\theta}))$$

which finally gives $\sqrt{N}(\hat{\theta}^N_{TSLS} - \bar{\theta}) \overset{\mathrm{d}}{\to} \mathbf{z} \sim \mathscr{N}(\mathbf{0}, \bar{\beta}\mathbf{M}^{-1}(\xi,\bar{\theta}))\,,\ N \to \infty\,.$ ■

Remark 3.28.

(*i*) A natural candidate for the first-stage estimator $\hat{\theta}_1^N$ is the WLS estimator $\hat{\theta}_{WLS}^N$, which is $\sqrt{N}$-consistent under the assumptions of Theorem 3.8. In particular, we can use the ordinary LS estimator $\hat{\theta}_{LS}^N$ for which $w(x) \equiv 1$.

(*ii*) $\bar{\beta}\mathbf{M}^{-1}(\xi, \bar{\theta})$ gives the asymptotic covariance matrix of the WLS estimator of θ when the optimum weights $\lambda^{-1}(x, \bar{\theta})$ are used; see Sect. 3.1.3.

(*iii*) The sequence of weights in (3.54) can also be computed recursively by replacing $\lambda(x_k, \hat{\theta}_1^N)$ by $\lambda(x_k, \hat{\theta}_1^k)$, where $\hat{\theta}_1^k$ is constructed from the k first observations and design points only. Under suitable assumptions, see Pronzato and Pázman (2004), the corresponding recursively reweighted estimator $\hat{\theta}_{RWLS}^N$ has the same asymptotic properties as $\hat{\theta}_{TSLS}^N$; that is, the recursive estimation of the weights does not increase the asymptotic variance of the estimates. When the regression model is linear[7] in θ, i.e., $\eta(x, \theta) = \mathbf{f}^\top(x)\theta$, the ordinary LS estimator $\hat{\theta}_{LS}^N$ can be computed recursively through

$$\mathbf{C}_{k+1} = \mathbf{C}_k - \frac{\mathbf{C}_k \mathbf{f}(x_{k+1})\mathbf{f}^\top(x_{k+1})\mathbf{C}_k}{1 + \mathbf{f}^\top(x_{k+1})\mathbf{C}_k \mathbf{f}(x_{k+1})},$$
$$\hat{\theta}_{LS}^{k+1} = \hat{\theta}_{LS}^k + \frac{\mathbf{C}_k \mathbf{f}(x_{k+1})}{1 + \mathbf{f}^\top(x_{k+1})\mathbf{C}_k \mathbf{f}(x_{k+1})}[y(x_{k+1}) - \mathbf{f}^\top(x_{k+1})\hat{\theta}_{LS}^k],$$

with $\mathbf{C}_k$ initialized by

$$\mathbf{C}_{k_0} = \left[\sum_{i=1}^{k_0} \mathbf{f}(x_i)\mathbf{f}^\top(x_i)\right]^{-1}$$

for k_0 the first integer such that $\mathbf{f}(x_1), \ldots, \mathbf{f}(x_{k_0})$ span $\mathbb{R}^p$. When the auxiliary estimator $\hat{\theta}_1^N$ is taken equal to $\hat{\theta}_{LS}^N$, $\hat{\theta}_{RWLS}^N$ can be computed simultaneously by a similar recursion

$$\mathbf{C}'_{k+1} = \mathbf{C}'_k - \frac{\mathbf{C}'_k \mathbf{f}(x_{k+1})\mathbf{f}^\top(x_{k+1})\mathbf{C}'_k}{\lambda(x_{k+1}, \hat{\theta}_{LS}^{k+1}) + \mathbf{f}^\top(x_{k+1})\mathbf{C}'_k \mathbf{f}(x_{k+1})},$$
$$\hat{\theta}_{RWLS}^{k+1} = \hat{\theta}_{RWLS}^k + \frac{\mathbf{C}'_k \mathbf{f}(x_{k+1})}{\lambda(x_{k+1}, \hat{\theta}_{LS}^{k+1}) + \mathbf{f}^\top(x_{k+1})\mathbf{C}'_k \mathbf{f}(x_{k+1})} \times [y(x_{k+1}) - \mathbf{f}^\top(x_{k+1})\hat{\theta}_{RWLS}^k],$$

with the initialization $\mathbf{C}'_{k_0} = \mathbf{C}_{k_0}$ and $\hat{\theta}_{RWLS}^{k_0} = \hat{\theta}_{LS}^{k_0}$. Note that $\hat{\theta}_{LS}^N$ is linear in the observations $y_1, \ldots, y_N$ but $\hat{\theta}_{RWLS}^N$ is not.

(*iv*) $\mathbf{M}_1(\xi, \bar{\theta}) \succeq \mathbf{M}(\xi, \bar{\theta})$, see (3.52) and (3.56), so that, from Remark 3.25, $\hat{\theta}_{PWLS}^N$ has a smaller asymptotic covariance matrix than $\hat{\theta}_{TSLS}^N$ for normal errors. However, when the two-stage procedure is modified such that at the second stage $\hat{\theta}_{TSLS'}^N$ minimizes

[7]The variance function $\lambda(x, \theta)$ may be nonlinear.

$$J_N(\theta, \hat{\theta}_1^N) = \frac{1}{N} \sum_{k=1}^{N} \frac{[y(x_k) - \eta(x_k, \theta)]^2}{\bar{\beta}\lambda(x_k, \hat{\theta}_1^N)}$$
$$+ \frac{1}{2N} \sum_{k=1}^{N} \frac{\left\{[y(x_k) - \eta(x_k, \hat{\theta}_1^N)]^2 - \bar{\beta}\lambda(x_k, \theta)\right\}^2}{\bar{\beta}^2\lambda^2(x_k, \hat{\theta}_1^N)} \tag{3.57}$$

with respect to $\theta \in \Theta$, then, under the assumptions of Theorem 3.22, $\hat{\theta}^N_{TSLS'} \overset{\text{a.s.}}{\to} \bar{\theta}$ when $N \to \infty$, and, under the assumptions of Theorem 3.24, $\sqrt{N}(\hat{\theta}^N_{TSLS'} - \bar{\theta})$ has the same asymptotic normal distribution as $\sqrt{N}(\hat{\theta}^N_{PWLS} - \bar{\theta})$. See also Example 4.12. □

3.3.4 Consistency and Asymptotic Normality of Iteratively Reweighted LS

Iteratively reweighted LS estimation relies on sequence of estimators constructed as follows:

$$\hat{\theta}_k^N = \arg\min_{\theta \in \Theta} J_N(\theta, \hat{\theta}_{k-1}^N) \,, \; k = 2, 3 \ldots \tag{3.58}$$

where $J_N(\theta, \theta')$ is defined by (3.54) and where $\hat{\theta}_1^N$ can be taken equal to the LS estimator. One may refer to Green (1984) and del Pino (1989) for iteratively reweighted LS procedures in more general situations than the nonlinear regression considered here.

Using Theorems. 3.26 and 3.27, a simple induction shows that, for any fixed k, $\hat{\theta}_k^N$ is strongly consistent and has the same asymptotic normal distribution as the TSLS estimator of previous section.

The following property states that the recursion (3.58) converges a.s. for any fixed N large enough.

Theorem 3.29. *Under the conditions of Theorem 3.27, the iteratively reweighted LS estimator defined by (3.58) in the model (3.2), (3.45) converges a.s. when $k \to \infty$ for N fixed but large enough:*

$$\lim_{N_0 \to \infty} \text{Prob}\{\forall N > N_0 \,, \; \lim_{k \to \infty} \hat{\theta}_k^N \; \text{ exists}\} = 1 \,.$$

Proof. Define $\tau_N(\theta') = \arg\min_{\theta \in \Theta} J_N(\theta, \theta')$, so that the recursion (3.58) corresponds to $\hat{\theta}_k^N = \tau_N(\hat{\theta}_{k-1}^N)$, $k = 1, 2, \ldots$ with $\hat{\theta}_1^N$ $\sqrt{N}$-consistent. $\tau_N(\theta')$ is solution of the stationary equation $\nabla J_N[\tau_N(\theta'), \theta'] = \mathbf{0}$; that is

$$-\frac{2}{N} \sum_{k=1}^{N} \lambda^{-1}(x_k, \theta') \left\{y(x_k) - \eta[x_k, \tau_N(\theta')]\right\} \left.\frac{\partial \eta(x_k, \theta)}{\partial \theta}\right|_{\tau_N(\theta')} = \mathbf{0} \,,$$

which implicitly relates θ' to $\tau_N(\theta')$. The implicit function theorem then gives

$$\nabla^2_{\theta,\theta} J_N[\tau_N(\theta'),\theta'] \frac{\partial \tau_N(\theta')}{\partial \theta'^\top} + \frac{2}{N}\sum_{k=1}^N \lambda^{-2}(x_k,\theta')\left\{[y(x_k)-\eta[x_k,\tau_N(\theta')]]\right\} \left.\frac{\partial \eta(x_k,\theta)}{\partial \theta}\right|_{\tau_N(\theta')} \frac{\partial \lambda(x_k,\theta')}{\partial \theta'^\top} = \mathbf{O}\,.$$

Lemma 2.6 implies that second term tends to zero a.s. and uniformly in θ' when $N \to \infty$. Also, $\nabla^2_{\theta,\theta} J_N[\tau_N(\theta'),\theta']$ tends a.s. and uniformly in θ' to $2\mathbf{M}(\xi,\bar\theta,\theta')$ when $N\to\infty$; see the proof of Theorem 3.27. This implies that

$$\limsup_{N\to\infty} \max_i \sup_{\theta\in\Theta} \left|\frac{\partial \tau_N(\theta)}{\partial \theta_i}\right| = 0 \text{ a.s.}$$

and, for any $\epsilon > 0$, $\text{Prob}\{\forall N > N_0\,,\ \forall \theta,\theta' \in \Theta\,,\ \|\tau_N(\theta)-\tau_N(\theta')\| \le \epsilon\|\theta-\theta'\|\}$ tends to 1 when $N_0 \to \infty$. Fix ϵ to some value smaller than 1. By a classical theorem in fixed-point theory, see, e.g., Stoer and Bulirsch (1993, p. 267), this implies that the probability P_{N_0} that the recursion (3.58) converges to a fixed point for all $N > N_0$ tends to 1 as $N_0 \to \infty$. ∎

Remark 3.30.

(*i*) The value of ϵ in the preceding proof can be chosen arbitrarily small, which indicates that the convergence of the recursion to a fixed point will accelerate as N increases.

(*ii*) The convergence of the recursion (3.58) implies that $\hat\theta^N_\infty$ tends to $\bar\theta$ a.s. (Theorem 3.26) and $\hat\theta^N_\infty$ is asymptotically normal according to Theorem 3.27. Apparently, there is thus no gain in pushing this recursion to its limit, rather than using simply the two-stage LS estimator, with, for instance, $\hat\theta^N_1 = \hat\theta^N_{LS}$, the ordinary LS estimator. However, these are only asymptotic results, and the finite sample behaviors of both methods may differ.

(*iii*) Replacing in (3.58) $J_N(\theta,\theta')$ by the expression given by (3.57), we obtain the iterative procedure suggested in (Downing et al., 2001), which in case of normal errors has the same asymptotic behavior as penalized WLS estimation.

(*iv*) By letting the weights $\lambda(x_k,\theta)$ depend on the observations $y(x_k)$, we can use iteratively reweighted LS iterations to determine estimators of other types than LS. For instance, the $\mathscr{L}_1$ estimator $\hat\theta^N_{\mathscr{L}_1}$ that minimizes $(1/N)\sum_{i=1}^N |y(x_i)-\eta(x_i,\theta)|$ can be obtained by using $\lambda(x_i,\theta) = |y(x_i)-\eta(x_i,\theta)|$; see Schlossmacher (1973). The method then possesses strong links with the EM algorithm; see Phillips (2002). □

3.3.5 Misspecification of the Variance Function

The penalized WLS estimator of Sect. 3.3.2 requires $\bar\beta$ to be known in (3.45), whereas the asymptotic properties of $\hat\theta^N_{TSLS}$ (Sect. 3.3.3) do not depend on $\bar\beta$.

More generally, the functional relation between σ^2 and x, θ is typically at best an approximation, even in situations where the model for the mean response can be considered as fairly accurate. The robustness of the estimator with respect to misspecification of the variance function $\lambda(x,\theta)$ is therefore an important issue.

Suppose that the true variance of the measurement errors ε_i in (3.2) satisfies

$$\mathbb{E}\{\varepsilon_i^2\} = \sigma^2(x_i) = \bar{\lambda}(x_i, \bar{\theta})$$

whereas the variance function $\lambda(x_i,\theta)$ is used for estimation. We suppose that both λ and $\bar{\lambda}$ satisfy $\mathrm{H1}_\lambda$ and $\mathrm{H2}_\lambda$.

The only consequence on TSLS estimation is the use of wrong weights at the second stage, $\lambda^{-1}(x, \hat{\theta}_1^N)$ instead of $\bar{\lambda}^{-1}(x, \hat{\theta}_1^N)$. Under conditions similar to those of Theorem 3.26, we still have $\hat{\theta}_{TSLS}^N \overset{\text{a.s.}}{\rightarrow} \bar{\theta}$, $N \to \infty$, and, using Theorem 3.8 and 3.27, we obtain

$$\sqrt{N}(\hat{\theta}_{TSLS}^N - \bar{\theta}) \overset{\text{d}}{\rightarrow} \mathbf{z} \sim \mathscr{N}(\mathbf{0}, \mathbf{C}(w,\xi,\bar{\theta}))\,, \ N \to \infty\,,$$

where $\mathbf{C}(w,\xi,\bar{\theta})$ is given by (3.24) with the weight function $w(x) = \lambda^{-1}(x,\bar{\theta})$ and the variance $\sigma^2(x) = \bar{\lambda}(x,\bar{\theta})$.

The situation is much different for the penalized WLS estimator. Assume, for instance, that

$$\mathbb{E}\{\varepsilon_i^2\} = \bar{\beta}\lambda(x_i, \bar{\theta})$$

whereas $\beta\lambda(x,\theta)$ is used for estimation, with $\beta \neq \bar{\beta}$; that is, the error is only in the scaling factor β in (3.47). Under the same assumptions as in Theorem 3.22, we obtain $J_N(\theta) \overset{\theta}{\rightsquigarrow} J_{\bar{\theta}}(\theta)$ a.s., $N \to \infty$, with now

$$\begin{aligned} J_{\bar{\theta}}(\theta) = & \int_{\mathscr{X}} \lambda^{-1}(x,\theta)[\eta(x,\theta) - \eta(x,\bar{\theta})]^2\, \xi(\mathrm{d}x) \\ & + \beta \int_{\mathscr{X}} \left\{ \left[\frac{\bar{\beta}}{\beta} \lambda(x,\bar{\theta})\lambda^{-1}(x,\theta) \right] - \log \left[\frac{\bar{\beta}}{\beta} \lambda(x,\bar{\theta})\lambda^{-1}(x,\theta) \right] \right\} \xi(\mathrm{d}x) \\ & + \beta \int_{\mathscr{X}} \log \lambda(x,\bar{\theta})\, \xi(\mathrm{d}x) + \beta \log(\bar{\beta}/\beta)\,, \end{aligned}$$

which, in general, is minimum for some $\hat{\theta}$ different from $\bar{\theta}$ when $\beta \neq \bar{\beta}$. Then, $\hat{\theta}_{PWLS}^N \overset{\text{a.s.}}{\rightarrow} \hat{\theta} \neq \bar{\theta}$ when $N \to \infty$, and $\hat{\theta}_{PWLS}^N$ is not consistent.[8]

Following the same approach as in (Carroll and Ruppert, 1982), consider small deviations from $\lambda(x,\theta)$, in the form

$$\mathbb{E}\{\varepsilon_i^2\} = \lambda(x_i,\bar{\theta}) \left[1 + \frac{2B}{\sqrt{N}}\, h(x_i,\bar{\theta}) \right]\,, \tag{3.59}$$

---

[8]It seems therefore more reasonable to consider β an unknown nuisance parameter for the estimation of θ; this approach will be considered in the next section. See also Remark 3.23.

with B some positive constant and $h(x,\bar{\theta})$ bounded on $\mathscr{X}$. This may account for $\sqrt{N}$ deviation of the assumed (e.g., estimated) variance from its true expression. In this case, the asymptotic properties of $\hat{\theta}^N_{TSLS}$ remain identical to those given in Sect. 3.3.3. Concerning $\hat{\theta}^N_{PWLS}$, we still obtain $\hat{\theta}^N_{PWLS} \overset{\text{a.s.}}{\to} \bar{\theta}$ under the conditions of Theorem 3.22, but the asymptotic distribution of $\sqrt{N}(\hat{\theta}^N_{PWLS} - \bar{\theta})$ is now different. Following the same lines as in the proof of Theorem 3.24, for $J_N(\theta)$ given by (3.47) (with $\bar{\beta}=1$), we obtain

$$-\sqrt{N}\nabla_\theta J_N(\bar{\theta}) = \frac{2}{\sqrt{N}} \sum_{k=1}^{N} \left\{ \lambda^{-1}(x_k,\bar{\theta})\varepsilon_k \frac{\partial \eta(x_k,\theta)}{\partial \theta}\bigg|_{\bar{\theta}} + \frac{1}{2}\lambda^{-2}(x_k,\bar{\theta})\varepsilon_k^2 \frac{\partial \lambda(x_k,\theta)}{\partial \theta}\bigg|_{\bar{\theta}} - \frac{1}{2}\lambda^{-1}(x_k,\bar{\theta}) \frac{\partial \lambda(x_k,\theta)}{\partial \theta}\bigg|_{\bar{\theta}} \right\};$$

that is,

$$-\sqrt{N}\nabla_\theta J_N(\bar{\theta}) = \frac{2}{N} \sum_{k=1}^{N} B\lambda^{-1}(x_k,\bar{\theta})h(x_k,\bar{\theta}) \frac{\partial \lambda(x_k,\theta)}{\partial \theta}\bigg|_{\bar{\theta}} + \frac{2}{\sqrt{N}} \sum_{k=1}^{N} \mathbf{w}_k \tag{3.60}$$

with

$$\mathbf{w}_k = \lambda^{-1}(x_k,\bar{\theta})\varepsilon_k \frac{\partial \eta(x_k,\theta)}{\partial \theta}\bigg|_{\bar{\theta}} + \frac{1}{2}\left[\varepsilon_k^2\lambda^{-2}(x_k,\bar{\theta}) - \lambda^{-1}(x_k,\bar{\theta}) - \frac{2B}{\sqrt{N}}\lambda^{-1}(x_k,\bar{\theta})h(x_k,\bar{\theta})\right] \frac{\partial \lambda(x_k,\theta)}{\partial \theta}\bigg|_{\bar{\theta}}.$$

As $N \to \infty$, the first sum in (3.60) tends to $2B\mathbf{b}(\xi,\bar{\theta})$ a.s., with

$$\mathbf{b}(\xi,\bar{\theta}) = \int_{\mathscr{X}} \lambda^{-1}(x,\bar{\theta})h(x,\bar{\theta}) \frac{\partial \lambda(x,\theta)}{\partial \theta}\bigg|_{\bar{\theta}} \xi(\mathrm{d}x). \tag{3.61}$$

The $\mathbf{w}_k$ are i.i.d., and again

$$\frac{2}{\sqrt{N}} \sum_{k=1}^{N} \mathbf{w}_k \overset{\text{d}}{\to} \mathbf{z} \sim \mathscr{N}(\mathbf{0}, 4\,\mathbf{M}_2(\xi,\bar{\theta}))\,,\ N \to \infty\,,$$

with $\mathbf{M}_2(\xi,\bar{\theta})$ given by (3.53) (with $\bar{\beta}=1$). Also, $\nabla^2_\theta J_N(\beta^N) \overset{\text{a.s.}}{\to} 2\,\mathbf{M}_1(\xi,\bar{\theta})$ when $\beta^N \overset{\text{a.s.}}{\to} \bar{\theta}$, with $\mathbf{M}_1(\xi,\bar{\theta})$ given by (3.52) (with $\bar{\beta}=1$). The Taylor series developments

$$\{\sqrt{N}\nabla_\theta J_N(\hat{\theta}^N_{PWLS})\}_i = 0 = \sqrt{N}\{\nabla_\theta J_N(\bar{\theta})\}_i + \{\nabla^2_\theta J_N(\beta^N_i)\sqrt{N}(\hat{\theta}^N_{PWLS} - \bar{\theta})\}_i$$

for some $\beta_i^N = (1 - \alpha_{i,N})\bar{\theta} + \alpha_{i,N}\hat{\theta}^N_{PWLS} \overset{\text{a.s.}}{\rightarrow} \bar{\theta}$, $N \to \infty$, $i = 1, \ldots, p$, give

$$\sqrt{N}(\hat{\theta}^N_{PWLS} - \bar{\theta}) \overset{\text{d}}{\rightarrow} \mathbf{z} \sim \mathscr{N}(B\mathbf{M}_1^{-1}(\xi, \bar{\theta})\mathbf{b}(\xi, \bar{\theta}), \mathbf{M}_1^{-1}(\xi, \bar{\theta})\mathbf{M}_2(\xi, \bar{\theta})\mathbf{M}_1^{-1}(\xi, \bar{\theta})) \tag{3.62}$$

as $N \to \infty$. Notice the presence of the asymptotically vanishing bias term $B\mathbf{M}_1^{-1}(\xi, \bar{\theta})\mathbf{b}(\xi, \bar{\theta})$, where $\mathbf{b}(\xi, \bar{\theta})$ is given by (3.61). In the particular case where the misspecification only concerns the scaling factor $\bar{\beta}$ in (3.45), $h(x, \bar{\theta}) = 1$ in (3.59) and

$$\mathbf{b}(\xi, \bar{\theta}) = \int_{\mathscr{X}} \left. \frac{\partial \log \lambda(x, \theta)}{\partial \theta} \right|_{\bar{\theta}} \xi(\mathrm{d}x)\,. \tag{3.63}$$

Instead of incurring the risk of a bias in penalized WLS estimation, one may wish to include additional parameters, besides θ, in the variance function. For instance, the constant $\bar{\beta}$ in (3.45) might be considered as a nuisance parameter for the estimation of θ, as in the next section.

3.3.6 Different Parameterizations for the Mean and Variance

Suppose that the true variance of the measurement errors ε_i in (3.2) satisfies

$$\mathbb{E}\{\varepsilon_i^2\} = \sigma^2(x_i) = \lambda(x_i, \bar{\alpha}, \bar{\beta}) \tag{3.64}$$

with α a subset of the parameters θ of the mean function $\eta(x, \theta)$ and β a vector of q additional parameters entering only the variance function. We shall denote by γ the full vector of unknown parameters, $\gamma = (\theta^\top, \beta^\top)^\top$, with $\gamma \in \Gamma \subset \mathbb{R}^{p+q}$, and $\bar{\gamma} = (\bar{\theta}^\top, \bar{\beta}^\top)^\top$ its true value.

Penalized WLS Estimation

Denote $\tilde{\eta}(x, \gamma) = \eta(x, \theta)$ and $\tilde{\lambda}(x, \gamma) = \lambda(x, \alpha, \beta)$. Under conditions similar to those of Theorem 3.22, with now

$$\forall \gamma \in \Gamma, \left\{ \begin{array}{r} \int_{\mathscr{X}} \tilde{\lambda}^{-1}(x, \gamma)[\tilde{\eta}(x, \gamma) - \tilde{\eta}(x, \bar{\gamma})]^2\, \xi(\mathrm{d}x) = 0 \\ \int_{\mathscr{X}} |\tilde{\lambda}^{-1}(x, \gamma)\tilde{\lambda}(x, \bar{\gamma}) - 1|\, \xi(\mathrm{d}x) = 0 \end{array} \right\} \Longleftrightarrow \gamma = \bar{\gamma}\,, \tag{3.65}$$

the penalized WLS estimator $\hat{\gamma}^N_{PWLS}$ that minimizes

$$J_N(\gamma) = \frac{1}{N} \sum_{k=1}^{N} \frac{[y(x_k) - \tilde{\eta}(x_k, \gamma)]^2}{\tilde{\lambda}(x_k, \gamma)} + \frac{1}{N} \sum_{k=1}^{N} \log \tilde{\lambda}(x_k, \gamma) \tag{3.66}$$

converges strongly to $\bar{\gamma}$.

Consider in particular the situation where $\alpha = \theta$ and β is a positive scalar that corresponds to the scaling factor in (3.45), i.e., $\lambda(x, \alpha, \beta) = \tilde{\lambda}(x, \gamma) = \beta\lambda(x, \theta)$. The condition (3.65) then becomes

$$\forall\beta>0, \theta\in\Theta\,, \left\{\begin{array}{r}\int_{\mathscr{X}} \lambda^{-1}(x,\theta)[\eta(x,\theta)-\eta(x,\bar{\theta})]^2\,\xi(\mathrm{d}x)=0\\ \int_{\mathscr{X}} |[\beta\lambda(x,\theta)]^{-1}\bar{\beta}\lambda(x,\bar{\theta})-1|\,\xi(\mathrm{d}x)=0\end{array}\right\} \Longleftrightarrow \theta=\bar{\theta}\,,\ \beta=\bar{\beta}\,.$$

Note that it is equivalent to the estimability condition in Remark 3.23. When β is considered as a nuisance parameter for the estimation of θ, the asymptotic distribution of the estimator of θ can be obtained in two different ways: either we use developments similar to those in the proof of Theorem 3.24 but for the criterion (3.49) of Remark 3.23 or we use the asymptotic distribution given in Theorem 3.24 and marginalize out β, that is, we integrate the joint p.d.f. of θ and β over β. Suppose for simplicity that the errors are normally distributed, the second approach then gives

$$\sqrt{N}(\hat{\gamma}^N_{PWLS}-\bar{\gamma}) \stackrel{\mathrm{d}}{\rightarrow} \mathbf{z}\sim\mathscr{N}(\mathbf{0},\mathbf{M}_1^{-1}(\xi,\bar{\gamma}))\,,\quad N\rightarrow\infty\,, \tag{3.67}$$

with

$$\begin{aligned}\mathbf{M}_1(\xi,\bar{\gamma}) &= \bar{\beta}^{-1}\int_{\mathscr{X}} \lambda^{-1}(x,\bar{\theta})\frac{\partial\eta(x,\theta)}{\partial\gamma}\Big|_{\bar{\gamma}}\frac{\partial\eta(x,\theta)}{\partial\gamma^\top}\Big|_{\bar{\gamma}}\,\xi(\mathrm{d}x)\\ &\quad+\frac{1}{2\bar{\beta}^2}\int_{\mathscr{X}} \lambda^{-2}(x,\bar{\theta})\frac{\partial[\beta\lambda(x,\theta)]}{\partial\gamma}\Big|_{\bar{\gamma}}\frac{\partial[\beta\lambda(x,\theta)]}{\partial\gamma^\top}\Big|_{\bar{\gamma}}\,\xi(\mathrm{d}x)\,.\end{aligned}$$

Notice that compared to (3.47) the scaling factor $\bar{\beta}$ equals one in (3.66). The matrix $\mathbf{M}_1(\xi,\bar{\gamma})$ can be partitioned into

$$\mathbf{M}_1(\xi,\bar{\gamma}) = \begin{bmatrix}\mathbf{M}_{1,\theta}(\xi,\bar{\gamma}) & \mathbf{v}_1(\xi,\bar{\gamma})\\ \mathbf{v}_1^\top(\xi,\bar{\gamma}) & m_1(\bar{\beta})\end{bmatrix}$$

where

$$\begin{aligned}\mathbf{M}_{1,\theta}(\xi,\bar{\gamma}) &= \bar{\beta}^{-1}\int_{\mathscr{X}} \lambda^{-1}(x,\bar{\theta})\frac{\partial\eta(x,\theta)}{\partial\theta}\Big|_{\bar{\theta}}\frac{\partial\eta(x,\theta)}{\partial\theta^\top}\Big|_{\bar{\theta}}\,\xi(\mathrm{d}x)\\ &\quad+\frac{1}{2}\int_{\mathscr{X}} \lambda^{-2}(x,\bar{\theta})\frac{\partial\lambda(x,\theta)}{\partial\theta}\Big|_{\bar{\theta}}\frac{\partial\lambda(x,\theta)}{\partial\theta^\top}\Big|_{\bar{\theta}}\,\xi(\mathrm{d}x)\,,\\ \mathbf{v}_1(\xi,\bar{\gamma}) &= \frac{1}{2\bar{\beta}}\int_{\mathscr{X}} \lambda^{-1}(x,\bar{\theta})\frac{\partial\lambda(x,\theta)}{\partial\theta}\Big|_{\bar{\theta}}\,\xi(\mathrm{d}x) = \frac{1}{2\bar{\beta}}\mathbf{b}(\xi,\bar{\theta})\,,\end{aligned}$$

see (3.63), and

$$m_1(\bar{\beta}) = \frac{1}{2\bar{\beta}^2}\,.$$

When marginalizing out β, we obtain

$$\sqrt{N}(\hat{\theta}^N_{PWLS}-\bar{\theta}) \stackrel{\mathrm{d}}{\rightarrow} \mathbf{z}\sim\mathscr{N}(\mathbf{0},\mathbf{C}_\theta(\xi,\bar{\gamma}))\,,\quad N\rightarrow\infty\,, \tag{3.68}$$

with

$$\mathbf{C}_\theta(\xi,\bar{\gamma}) = \left[\mathbf{M}_{1,\theta}(\xi,\bar{\gamma}) - \frac{\mathbf{v}_1(\xi,\bar{\gamma})\mathbf{v}_1^\top(\xi,\bar{\gamma})}{m_1(\bar{\beta})}\right]^{-1}\,. \tag{3.69}$$

Similarly, marginalizing out θ, we get

$$\sqrt{N}(\hat{\beta}^N_{PWLS} - \bar{\beta}) \stackrel{\mathrm{d}}{\to} \mathbf{z} \sim \mathscr{N}(0, C_\beta(\xi, \bar{\gamma})) \,, \quad N \to \infty \,,$$

where

$$C_\beta(\xi, \bar{\gamma}) = \frac{1}{m_1(\bar{\beta}) - \mathbf{v}_1^\top(\xi, \bar{\gamma})\mathbf{M}_{1,\theta}^{-1}(\xi, \bar{\gamma})\mathbf{v}_1(\xi, \bar{\gamma})} \,.$$

Remark 3.31. The asymptotic result (3.68) can be compared with that obtained when β is assumed to be known but the relative misspecification is of order $2B/\sqrt{N}$; that is, $\mathbb{E}\{\varepsilon_i^2\} = \beta\lambda(x_i, \bar{\theta})[1 + 2Bh(x_i, \bar{\theta})/\sqrt{N}]$; see (3.59). In that case (still for normal errors), similarly to (3.62), we obtain

$$\sqrt{N}(\hat{\theta}^N_{PWLS} - \bar{\theta}) \stackrel{\mathrm{d}}{\to} \mathbf{z} \sim \mathscr{N}(B\mathbf{M}_{1,\theta}^{-1}(\xi, \bar{\theta})\mathbf{b}(\xi, \bar{\theta}), \mathbf{M}_{1,\theta}^{-1}(\xi, \bar{\theta})) \,, \quad N \to \infty \,, \tag{3.70}$$

where $\mathbf{b}(\xi, \bar{\theta})$ is given by (3.61). Hence, the mean-squared error satisfies

$$N\mathbb{E}\{(\hat{\theta}^N_{PWLS} - \bar{\theta})^\top(\hat{\theta}^N_{PWLS} - \bar{\theta})\} \to B^2\mathbf{b}^\top(\xi, \bar{\theta})\mathbf{M}_{1,\theta}^{-2}(\xi, \bar{\theta})\mathbf{b}(\xi, \bar{\theta}) + \mathrm{trace}[\mathbf{M}_{1,\theta}^{-1}(\xi, \bar{\theta})] \,. \tag{3.71}$$

When β is considered as a nuisance parameter that we marginalize out, the asymptotic mean-squared error is

$$\begin{aligned} \mathrm{trace}[\mathbf{C}_\theta(\xi, \bar{\gamma})] &= C_\beta(\xi, \bar{\gamma})\mathbf{v}_1^\top(\xi, \bar{\gamma})\mathbf{M}_{1,\theta}^{-2}(\xi, \bar{\theta})\mathbf{v}_1(\xi, \bar{\gamma}) + \mathrm{trace}[\mathbf{M}_{1,\theta}^{-1}(\xi, \bar{\theta})] \\ &= \frac{C_\beta(\xi, \bar{\gamma})}{4\bar{\beta}^2}\mathbf{b}^\top(\xi, \bar{\theta})\mathbf{M}_{1,\theta}^{-2}(\xi, \bar{\theta})\mathbf{b}(\xi, \bar{\theta}) + \mathrm{trace}[\mathbf{M}_{1,\theta}^{-1}(\xi, \bar{\theta})] \,, \end{aligned}$$

which is smaller than (3.71) as soon as $2B > \sqrt{C_\beta(\xi, \bar{\gamma})}/\bar{\beta}$. □

Two-stage LS Estimation

The developments of Sect. 3.3.3 must be modified, since the estimation of θ by ordinary LS at the first stage does not give an estimate of $\bar{\beta}$ in (3.64), which is required to compute the weights used at the second stage. We thus introduce an intermediate stage 1' for the estimation of $\bar{\beta}$ by $\hat{\beta}^N_1$. We shall denote by $\delta = (\alpha^\top, \beta^\top)^\top$ the parameters of the variance function and write $\lambda(x, \delta) = \bar{\lambda}(x, \alpha, \beta)$; $\bar{\delta} = (\bar{\alpha}^\top, \bar{\beta}^\top)^\top$ will denote the true value of δ. Both δ and θ are assumed to belong to compact sets. The procedure is as follows.

First, we estimate $\hat{\theta}^N_1 = \hat{\theta}^N_{LS}$ which minimizes

$$J_{N,1}(\theta) = \frac{1}{N}\sum_{k=1}^{N}[y(x_k) - \eta(x_k, \theta)]^2$$

with respect to $\theta \in \Theta$; this is similar to stage 1 of Sect. 3.3.3. Next, we estimate $\hat{\delta}^N_1$ that minimizes

$$J_{N,1'}(\delta) = J_N(\delta, \hat{\theta}_1^N) = \frac{1}{N}\sum_{k=1}^{N}\left\{[y(x_k) - \eta(x_k, \hat{\theta}_1^N)]^2 - \lambda(x_k, \delta)\right\}^2 \quad (3.72)$$

with respect to δ; this is the new intermediate step, similar to LS estimation for the model $\lambda(x, \delta)$ with parameters δ and "observations" $[y(x_k) - \eta(x_k, \hat{\theta}_1^N)]^2$. Finally, the second stage is similar to that of Sect. 3.3.3, and $\hat{\theta}_{TSLS}^N$ minimizes

$$J_{N,2}(\theta) = \frac{1}{N}\sum_{k=1}^{N}\frac{[y(x_k) - \eta(x_k, \theta)]^2}{\lambda(x_k, \hat{\delta}_1^N)}$$

with respect to $\theta \in \Theta$. Following the same lines as in Sect. 3.3.3, one can show that $\hat{\theta}_{TSLS}^N$ is strongly consistent when $\hat{\delta}_1^N$ converges to some $\bar{\delta}_1$ a.s. Also, when $\hat{\delta}_1^N$ is $\sqrt{N}$-consistent,

$$\sqrt{N}(\hat{\theta}_{TSLS}^N - \bar{\theta}) \xrightarrow{\text{d}} \mathbf{z} \sim \mathscr{N}(\mathbf{0}, \mathbf{M}^{-1}(\xi, \bar{\gamma}))\,, \quad N \to \infty\,, \quad (3.73)$$

with

$$\mathbf{M}(\xi, \bar{\gamma}) = \int_{\mathscr{X}} \lambda^{-1}(x, \bar{\delta}) \left.\frac{\partial \eta(x, \theta)}{\partial \theta}\right|_{\bar{\theta}} \left.\frac{\partial \eta(x, \theta)}{\partial \theta^\top}\right|_{\bar{\theta}} \xi(\mathrm{d}x)\,.$$

Note that $\mathbf{M}^{-1}(\xi, \bar{\gamma})$ is the asymptotic covariance matrix obtained for WLS estimation with the optimum weights $\bar{\lambda}^{-1}(x, \bar{\alpha}, \bar{\beta})$. The rest of the section is devoted to the asymptotic properties of $\hat{\delta}_1^N$ obtained by the minimization of $J_{N,1'}(\delta)$.

Consider $J_N(\delta, \theta)$ defined by (3.72). Under conditions similar to those of Theorem 3.26 (plus the additional assumption that the errors ε_i have finite fourth moments $\mathbb{E}(\varepsilon_i^4)$), $J_N(\delta, \theta)$ tends a.s. and uniformly in δ and θ to

$$\begin{aligned} J_{\bar{\theta}}(\delta, \theta) = & \int_{\mathscr{X}} [\kappa(x) + 2]\lambda^2(x, \bar{\delta})\, \xi(\mathrm{d}x) \\ & + 4 \int_{\mathscr{X}} \lambda^{3/2}(x, \delta)[\eta(x, \bar{\theta}) - \eta(x, \theta)]s(x)\, \xi(\mathrm{d}x) \\ & + 4 \int_{\mathscr{X}} \lambda(x, \delta)[\eta(x, \bar{\theta}) - \eta(x, \theta)]^2\, \xi(\mathrm{d}x) \\ & + \int_{\mathscr{X}} \left\{[\eta(x, \bar{\theta}) - \eta(x, \theta)]^2 + \lambda(x, \bar{\delta}) - \lambda(x, \delta)\right\}^2 \xi(\mathrm{d}x)\,, \end{aligned}$$

with $s(x)$ and $\kappa(x)$, respectively, the skewness and kurtosis of the distribution of $\varepsilon(x)$; see Theorem 3.24. Therefore, when $\hat{\theta}_1^N \xrightarrow{\text{a.s.}} \bar{\theta}$, $J_{N,1'}(\delta) = J_N(\delta, \hat{\theta}_1^N)$ tends a.s. and uniformly in δ to

$$\begin{aligned} J_{\bar{\theta}}(\delta, \bar{\theta}) = & \int_{\mathscr{X}} [\kappa(x) + 2]\lambda^2(x, \bar{\delta})\, \xi(\mathrm{d}x) \\ & + \int_{\mathscr{X}} [\lambda(x, \bar{\delta}) - \lambda(x, \delta)]^2\, \xi(\mathrm{d}x)\,, \end{aligned}$$

and, under the estimability condition

$$\forall \delta\,, \int_{\mathscr{X}} [\lambda(x,\delta) - \lambda(x,\bar{\delta})]^2 \,\xi(\mathrm{d}x) = 0 \Longleftrightarrow \delta = \bar{\delta}\,, \tag{3.74}$$

$\hat{\delta}_1^N \stackrel{\text{a.s.}}{\rightarrow} \bar{\delta}$ when $N \rightarrow \infty$. The asymptotic normality of $\hat{\delta}_1^N$ can be proved following the same lines as in Theorem 3.27. Using notations for derivatives similar to those of Sect. 3.3.3, we can write

$$\begin{aligned}\{\nabla_\delta J_N(\hat{\delta}_1^N, \hat{\theta}_1^N)\}_j = 0 &= \{\nabla_\delta J_N(\bar{\delta}, \hat{\theta}_1^N)\}_j + \{\nabla^2_{\delta,\delta} J_N(\mathbf{u}_j^N, \hat{\theta}_1^N)(\hat{\delta}_1^N - \bar{\delta})\}_j \\ &= \{\nabla_\delta J_N(\bar{\delta}, \bar{\theta})\} + \{\nabla^2_{\delta,\theta} J_N(\bar{\delta}, \mathbf{v}_j^N)(\hat{\theta}_1^N - \bar{\theta})\}_j \\ &\quad + \{\nabla^2_{\delta,\delta} J_N(\mathbf{u}_j^N, \hat{\theta}_1^N)(\hat{\delta}_1^N - \bar{\delta})\}_j \end{aligned} \tag{3.75}$$

for some $\mathbf{u}_j^N \stackrel{\text{a.s.}}{\rightarrow} \bar{\delta}$ and $\mathbf{v}_j^N \stackrel{\text{a.s.}}{\rightarrow} \bar{\theta}$, $N \rightarrow \infty$, for $j = 1, \ldots, \dim(\delta)$. Direct calculation gives

$$\nabla^2_{\delta,\delta} J_N(\mathbf{u}_j^N, \hat{\theta}_1^N) \stackrel{\text{a.s.}}{\rightarrow} 2\mathbf{N}_1(\xi, \bar{\delta}) = 2 \int_{\mathscr{X}} \left.\frac{\partial \lambda(x,\delta)}{\partial \delta}\right|_{\bar{\delta}} \left.\frac{\partial \lambda(x,\delta)}{\partial \delta^\top}\right|_{\bar{\delta}} \xi(\mathrm{d}x) \tag{3.76}$$

$$\nabla^2_{\delta,\theta} J_N(\bar{\delta}, \mathbf{v}_j^N) \stackrel{\text{a.s.}}{\rightarrow} \mathbf{O}$$

as $N \rightarrow \infty$ and

$$\begin{aligned} -\sqrt{N} \nabla_\delta J_N(\bar{\delta}, \bar{\theta}) &= \frac{2}{\sqrt{N}} \sum_{k=1}^N \left[\varepsilon_k^2 - \lambda(x_k, \bar{\delta})\right] \left.\frac{\partial \lambda(x_k, \delta)}{\partial \delta}\right|_{\bar{\delta}} \\ &\stackrel{\text{d}}{\rightarrow} \mathbf{z} \sim \mathscr{N}(\mathbf{0}, 4\mathbf{N}_2(\xi, \bar{\gamma})) \end{aligned}$$

with

$$\mathbf{N}_2(\xi, \bar{\delta}) = \int_{\mathscr{X}} [\kappa(x) + 2]\lambda^2(x, \bar{\delta}) \left.\frac{\partial \lambda(x,\delta)}{\partial \delta}\right|_{\bar{\delta}} \left.\frac{\partial \lambda(x,\delta)}{\partial \delta^\top}\right|_{\bar{\delta}} \xi(\mathrm{d}x)\,. \tag{3.77}$$

Together with (3.75), this gives

$$\sqrt{N}(\hat{\delta}_1^N - \bar{\delta}) \stackrel{\text{d}}{\rightarrow} \mathbf{z} \sim \mathscr{N}(\mathbf{0}, \mathbf{N}_1^{-1}(\xi, \bar{\delta}) \mathbf{N}_2(\xi, \bar{\delta}) \mathbf{N}_1^{-1}(\xi, \bar{\delta}))\,. \tag{3.78}$$

The estimator $\hat{\delta}_1^N$ minimizing $J_{N,1'}(\delta)$ is thus $\sqrt{N}$-consistent[9] and the TSLS estimator $\hat{\theta}_{TSLS}^N$ satisfies (3.73).

Remark 3.32.

(*i*) A modification of the second stage similar to (3.57) yields an estimator $\hat{\theta}_{TSLS'}^N$ with the same asymptotic properties as $\hat{\theta}_{PWLS}^N$; see Remark 3.28-(*iv*).

[9] The asymptotic normality mentioned above for $\hat{\delta}_1^N$ extends Theorem 1 of Jobson and Fuller (1980) which concerns the case where $\eta(x, \theta)$ is linear in θ and the errors ε_k are normally distributed.

(*ii*) The estimability condition (3.74) involves some parameters α that are present in θ and are thus already estimated at the first stage by $\hat{\alpha}_1^N$, a part of $\hat{\theta}_1^N$. A less restrictive condition would be

$$\forall\beta\,, \int_{\mathscr{X}} [\bar{\lambda}(x,\bar{\alpha},\beta) - \bar{\lambda}(x,\bar{\alpha},\bar{\beta})]^2\, \xi(\mathrm{d}x) = 0 \Longleftrightarrow \beta = \bar{\beta}\,,$$

to be applied when only β is estimated at stage 1' by

$$\hat{\beta}_1^N = \arg\min_{\beta} \frac{1}{N}\sum_{k=1}^{N} \left\{[y(x_k) - \eta(x_k,\hat{\theta}_1^N)]^2 - \bar{\lambda}(x_k,\hat{\alpha}_1^N,\beta)\right\}^2 .$$

Again, one can prove that β_1^N is strongly consistent and asymptotically normal as $N \to \infty$, but with a covariance matrix different from that obtained by marginalizing out α in the normal distribution (3.78). □

3.3.7 Penalized WLS or Two-Stage LS?

Normal Errors

Suppose that the variance of the errors satisfies (3.45). When the kurtosis and skewness are zero (e.g., for normal errors), $\mathbf{M}_2(\xi,\bar{\theta})$, given by (3.53), equals $\bar{\beta}\mathbf{M}_1(\xi,\bar{\theta})$, see (3.52), with $\mathbf{M}_1(\xi,\bar{\theta}) \succeq \mathbf{M}(\xi,\bar{\theta})$ (see (3.56)), and penalized WLS should be preferred to TSLS estimation.

This remains true when the scaling factor $\bar{\beta}$ in (3.45) is unknown and estimated by penalized WLS. In that case, the estimator $\hat{\theta}_{PWLS}^N$ satisfies

$$\sqrt{N}(\hat{\theta}_{PWLS}^N - \bar{\theta}) \overset{\mathrm{d}}{\to} \mathbf{z} \sim \mathscr{N}(\mathbf{0}, \mathbf{C}_\theta(\xi,\bar{\gamma}))\,, \quad N \to \infty\,,$$

see (3.68), with $\mathbf{C}_\theta(\xi,\bar{\gamma})$ given by (3.69), whereas $\hat{\theta}_{TSLS}^N$ satisfies

$$\sqrt{N}(\hat{\theta}_{TSLS}^N - \bar{\theta}) \overset{\mathrm{d}}{\to} \mathbf{z} \sim \mathscr{N}(\mathbf{0}, \mathbf{M}^{-1}(\xi,\bar{\gamma}))\,, \quad N \to \infty\,,$$

with

$$\mathbf{M}(\xi,\bar{\gamma}) = \bar{\beta}^{-1} \int_{\mathscr{X}} \lambda^{-1}(x,\bar{\theta}) \frac{\partial\eta(x,\theta)}{\partial\theta}\bigg|_{\bar{\theta}} \frac{\partial\eta(x,\theta)}{\partial\theta^\top}\bigg|_{\bar{\theta}} \xi(\mathrm{d}x)\,,$$

see (3.73). Direct calculation gives

$$\mathbf{C}_\theta^{-1}(\xi,\bar{\gamma}) - \mathbf{M}(\xi,\bar{\gamma}) = \frac{1}{2}\int_{\mathscr{X}} \mathbf{z}(x,\bar{\theta})\mathbf{z}^\top(x,\bar{\theta})\,\xi(\mathrm{d}x)$$

where

$$\mathbf{z}(x,\theta) = \frac{\partial \log\lambda(x,\theta)}{\partial\theta} - \int_{\mathscr{X}} \frac{\partial \log\lambda(x,\theta)}{\partial\theta}\,\xi(\mathrm{d}x)$$

and thus $\mathbf{C}_\theta^{-1}(\xi,\bar{\gamma}) \succeq \mathbf{M}(\xi,\bar{\gamma})$.

No Common Parameters in the Mean and Variance Functions

Suppose again that the distribution of errors has no kurtosis and no skewness, and that no parameter of the mean function enter the variance function, i.e., $\mathbb{E}\{\varepsilon_i^2\} = \lambda(x_i, \bar{\beta})$; see (3.64).

The asymptotic covariance matrix for penalized WLS estimation is then block diagonal, with $\mathbf{M}_\theta^{-1}(\xi, \bar{\theta})$ the block corresponding to the parameters θ and

$$\mathbf{M}_\theta(\xi, \bar{\gamma}) = \int_{\mathscr{X}} \lambda^{-1}(x, \bar{\beta}) \frac{\partial \eta(x, \theta)}{\partial \theta}\bigg|_{\bar{\theta}} \frac{\partial \eta(x, \theta)}{\partial \theta^\top}\bigg|_{\bar{\theta}} \xi(\mathrm{d}x),$$

see (3.52). It coincides with the information matrix obtained for TSLS estimation; see (3.73). The two estimation methods thus have the same asymptotic behavior.

Remark 3.33. Consider the estimation of the parameters β of the variance function. The asymptotic covariance matrix is $\mathbf{M}_\beta^{-1}(\xi, \bar{\beta})$ for penalized WLS estimation, with

$$\mathbf{M}_\beta(\xi, \bar{\beta}) = \frac{1}{2} \int_{\mathscr{X}} \lambda^{-2}(x, \bar{\beta}) \frac{\partial \lambda(x, \beta)}{\partial \beta}\bigg|_{\bar{\beta}} \frac{\partial \lambda(x, \beta)}{\partial \beta^\top}\bigg|_{\bar{\beta}} \xi(\mathrm{d}x).$$

The asymptotic covariance matrix for $\hat{\beta}_1^N$ estimated at stage 1' of the TSLS method is $\mathbf{C}(\xi, \bar{\beta}) = \mathbf{N}_1^{-1}(\xi, \bar{\beta})\mathbf{N}_2(\xi, \bar{\beta})\mathbf{N}_1^{-1}(\xi, \bar{\beta})$; see (3.78) and (3.76), (3.77). Lemma 3.7 with

$$\mathbf{u} = \lambda(x, \bar{\beta}) \frac{\partial \lambda(x, \beta)}{\partial \beta}\bigg|_{\bar{\beta}} \quad \text{and } \mathbf{v} = \lambda^{-1}(x, \bar{\beta}) \frac{\partial \lambda(x, \beta)}{\partial \beta}\bigg|_{\bar{\beta}}$$

gives $\mathbf{C}(\xi, \bar{\beta}) \succeq \mathbf{M}_\beta^{-1}(\xi, \bar{\beta})$. This is due to the use of ordinary LS in the minimization of (3.72). Indeed, consider a second stage, similar to WLS, where β is estimated by

$$\hat{\beta}_{TSLS}^N = \arg\min_\beta \frac{1}{N} \sum_{k=1}^N \frac{\left\{[y(x_k) - \eta(x_k, \hat{\theta}_1^N)]^2 - \lambda(x_k, \beta)\right\}^2}{\lambda^2(x_k, \hat{\beta}_1^N)}.$$

The asymptotic covariance matrix is then the same as for penalized WLS estimation. □

Non-normal Errors

The advantage of $\hat{\theta}_{PWLS}^N$ over $\hat{\theta}_{TSLS}^N$ when the mean and variance functions have common parameters may disappear when the distribution of the errors is not normal, e.g., when it has a large positive kurtosis. In general, the conclusion depends on the design ξ, which raises the issue of choosing simultaneously the method of estimation and the design. In some cases, however, as illustrated by the following example, the conclusion does not depend on ξ, one estimation method being uniformly better.

Example 3.34. Suppose that in the model (3.2) the variance of the errors satisfies

$$\mathbb{E}\{\varepsilon_k^2\} = \bar{\beta}_1 \lambda(x_k, \bar{\theta}) = \bar{\beta}_1 [\eta(x_k, \bar{\theta}) + \bar{\beta}_2]^2 \text{ for all } k\,,$$

which in particular, when $\bar{\beta}_2 = 0$, corresponds to the situation where the relative precision of the observations is constant. Suppose, moreover, that the distribution of the errors is symmetric (so that the skewness $s(x)$ equals zero) and has a constant kurtosis κ (which is the case when the distributions of errors at different x are similar and only differ by a scaling factor).

Direct calculation then gives for the asymptotic covariance matrix of the penalized WLS estimator with $\bar{\beta}_1$ known:

$$\mathbf{M}_1^{-1}(\xi, \bar{\theta})\mathbf{M}_2(\xi, \bar{\theta})\mathbf{M}_1^{-1}(\xi, \bar{\theta}) = \frac{1 + 2\bar{\beta}_1 + \kappa\bar{\beta}_1}{(1 + 2\bar{\beta}_1)^2}\, \bar{\beta}_1\, \mathbf{M}^{-1}(\xi, \bar{\theta})\,,$$

see (3.52), (3.53), with $\bar{\beta}_1 \mathbf{M}^{-1}(\xi, \theta)$ the asymptotic covariance matrix of the two-stage LS estimator; see (3.56).

Two-stage LS should thus be preferred to penalized WLS when $\kappa > 2$ and $\bar{\beta}_1 < \beta_1^* = (\kappa - 2)/4$ and vice versa otherwise. Generally speaking, it means that the two-stage procedure is always preferable when $\kappa > 2$, provided that the errors are small enough. For instance, when the errors have the exponential distribution $\bar{\varphi}(\varepsilon) = [\sqrt{2}/(2\sigma)] \exp(-|\varepsilon|\sqrt{2}/\sigma)$, $\kappa = 3$ and the limiting value for $\bar{\beta}_1$ is $\beta_1^* = 1/4$. Figure 3.2 gives the evolution of the ratio

$$\rho(\kappa, \beta_1) = \frac{1 + 2\beta_1 + \kappa\beta_1}{(1 + 2\beta_1)^2}$$

as a function of β_1 for $\kappa = 3$. □

In more general situations the estimator and the design must be chosen simultaneously. The following approach is suggested in (Pázman and Pronzato, 2004) for the case of errors with constant kurtosis κ (not depending on x) and no skewness: (a) determine the optimum design ξ_{PWLS}^* for the penalized WLS estimator under the assumption of zero kurtosis and the optimum design ξ_{TSLS}^* for the two-stage LS estimator; (b) compare the values of the design criteria for both estimators at different values of the kurtosis κ. Note that the asymptotic covariance matrix of $\hat{\theta}_{PWLS}$ is linear in κ. Therefore, for any isotonic design criterion $\Phi(\cdot)$, see Definition 5.3, a value κ^* exists such that $(\xi_{TSLS}^*, \hat{\theta}_{TSLS})$ should be preferred to $(\xi_{PWLS}^*, \hat{\theta}_{PWLS})$ for $\kappa > \kappa^*$.

Bernoulli Experiments

Consider the case of qualitative binary observations, indicating for instance a success or failure, and take $y(x) = 1$ for a successful trial at the design point x and $y(x) = 0$ otherwise. When the probability of success is parameterized, $\text{Prob}\{y(x) = 1\} = \pi(x, \bar{\theta})$, with $\pi(x, \theta)$ a known function of x and θ, we get

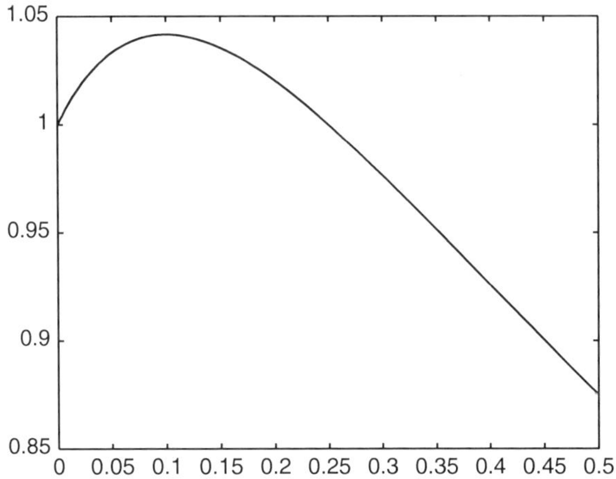

Fig. 3.2. Evolution of $\rho(\kappa, \beta_1)$ as a function of β_1 in Example 3.34: $\kappa = 3$ and $\hat{\theta}^N_{TSLS}$ should be preferred to $\hat{\theta}^N_{PWLS}$ for $\beta_1 < 1/4$

$\mathbb{E}\{y(x)\} = \pi(x, \bar{\theta})$ and $\mathrm{var}\{y(x)\} = \pi(x, \bar{\theta})[1 - \pi(x, \bar{\theta})]$. Forgetting the binary character of $y(x)$, we may consider this as a regression model with parameterized variance and take $\eta(x, \theta) = \pi(x, \theta)$ and $\lambda(x, \theta) = \pi(x, \theta)[1 - \pi(x, \theta)]$; see also Sect. 4.3.1. In that case, the asymptotic covariance matrix for TSLS estimation is $\mathbf{M}^{-1}(\xi, \bar{\theta})$ with

$$\mathbf{M}(\xi, \theta) = \int_{\mathscr{X}} \frac{1}{\pi(x, \theta)[1 - \pi(x, \theta)]} \frac{\partial \pi(x, \theta)}{\partial \theta} \frac{\partial \pi(x, \theta)}{\partial \theta^\top} \, \xi(\mathrm{d}x) \,,$$

i.e., the same as for maximum likelihood estimation; see (4.38). See also Green (1984). From the results in Sect. 4.4, TSLS estimation is thus preferable (asymptotically) to penalized LS estimation for Bernoulli experiments. This can also be checked directly. The asymptotic covariance matrix $\mathbf{C}(\xi, \theta)$ for penalized WLS estimation is as indicated in Theorem 3.24, with the skewness

$$s(x) = \frac{1 - 2\pi(x, \theta)}{\sqrt{\pi(x, \theta)[1 - \pi(x, \theta)]}}$$

and kurtosis

$$\kappa(x) = \frac{1 - 6\pi(x, \theta) + 6\pi^2(x, \theta)}{\pi(x, \theta)[1 - \pi(x, \theta)]} \,.$$

Substitution in (3.52), (3.53) gives

$$\mathbf{M}_1(\xi, \theta) = \int_{\mathscr{X}} \frac{2\pi^2(x, \theta) - 2\pi(x, \theta) + 1}{2\pi^2(x, \theta)[1 - \pi(x, \theta)]^2} \frac{\partial \pi(x, \theta)}{\partial \theta} \frac{\partial \pi(x, \theta)}{\partial \theta^\top} \, \xi(\mathrm{d}x) \,,$$

$$\mathbf{M}_2(\xi, \theta) = \int_{\mathscr{X}} \frac{[2\pi^2(x, \theta) - 2\pi(x, \theta) + 1]^2}{4\pi^3(x, \theta)[1 - \pi(x, \theta)]^3} \frac{\partial \pi(x, \theta)}{\partial \theta} \frac{\partial \pi(x, \theta)}{\partial \theta^\top} \, \xi(\mathrm{d}x) \,.$$

Denote

$$\mathbf{v} = \begin{pmatrix} \frac{2\pi^2(x,\theta)-2\pi(x,\theta)+1}{2\pi^{3/2}(x,\theta)[1-\pi(x,\theta)]^{3/2}} \frac{\partial \pi(x,\theta)}{\partial \theta} \\ \frac{1}{\sqrt{\pi(x,\theta)[1-\pi(x,\theta)]}} \frac{\partial \pi(x,\theta)}{\partial \theta} \end{pmatrix},$$

so that

$$\mathbb{E}\{\mathbf{v}\mathbf{v}^\top\} = \begin{pmatrix} \mathbf{M}_2(\xi,\theta) & \mathbf{M}_1(\xi,\theta) \\ \mathbf{M}_1(\xi,\theta) & \mathbf{M}(\xi,\theta) \end{pmatrix},$$

where the expectation is with respect to $\xi(\mathrm{d}x)$. Using Lemma 3.7, we get $\mathbf{M}_2(\xi,\theta) \succeq \mathbf{M}_1(\xi,\theta)\mathbf{M}^{-1}(\xi,\theta)\mathbf{M}_1(\xi,\theta)$, and thus, $\mathbf{C}(\xi,\theta) \succeq \mathbf{M}^{-1}(\xi,\theta)$; see Theorem 3.24.

3.3.8 Variance Stabilization

A parametric transformation $T_\omega(\cdot)$ can be applied to the regression model and the observations when the errors satisfy (3.45), in order to stabilize the variance and use ordinary LS. In the seminal paper (Box and Cox, 1964), it is assumed that the transformed observations $z(x_i) = T_\omega[y(x_i)]$ are independently normally distributed with constant variance σ^2 for some unknown parameters ω. The transformation $T_\omega(\cdot)$ is obtained as follows. Assume that $\lambda(x,\theta)$ in (3.45) takes the form

$$\lambda(x,\theta) = [\eta(x,\theta) + c]^{2(1-\alpha)}.$$

Then, neglecting terms of power in $\bar{\beta}$ larger than 1, we have

$$\mathrm{var}[z(x_i)] \approx \bar{\beta}\,[\eta(x_i,\bar{\theta}) + c]^{2(1-\alpha)} \left[\left.\frac{\mathrm{d}T_\omega(t)}{\mathrm{d}t}\right|_{\eta(x_i,\bar{\theta})} \right]^2$$

which becomes constant when $\mathrm{d}T_\omega(t)/\mathrm{d}t = (t+c)^{\alpha-1}$, i.e., when

$$T_\omega(t) = \begin{cases} \frac{(t+c)^\alpha - 1}{\alpha} & \text{if } \alpha \neq 0 \\ \log(t+c) & \text{if } \alpha = 0 \end{cases} \tag{3.79}$$

where $\omega = (\alpha, c)$. The choice $T_\omega(t) = (t+c)^\alpha$ when $\alpha \neq 0$ is equally valid, but (3.79) presents the advantage of being continuous at $\alpha = 0$. In Box and Cox (1964), the unknown ω is estimated by maximum likelihood, see Sect. 4.2, or by maximum a posteriori using a suitable prior. A regression model is formed for the transformed observations $z(x_i)$ directly, in the form $\mathbb{E}[z(x_i)] = \nu(x_i,\theta)$ and $\mathrm{var}[z(x_i)] = \sigma^2$ for all i, independently of any model for the original variables $y(x_i)$.

On the other hand, one may wish to start from the model (3.2), (3.45) of the original observations $y(x_i)$. This entails several difficulties and approximations.

First, the transformation (3.79) makes sense only if $y(x_i) - c > 0$, i.e., $\varepsilon_i > -\eta(x_i,\bar{\theta}) + c$, for all i. For a given set of observations, this is easily overcome by choosing a suitable value for c. However, for asymptotic considerations concerning the estimation of θ, it requires that the support of the distribution of the ε_i is bounded from below or simply bounded if the distribution is symmetric.

Second, when the original observations $y(x_i)$ are independently distributed with mean $\eta(x_i, \bar{\theta})$ and variance $\bar{\beta}\lambda(x_i, \bar{\theta}) = \bar{\beta}[\eta(x_i, \bar{\theta}) + c]^{2(1-\alpha)}$, in general the transformed observations $z(x_i)$ do not have mean $T_\omega[\eta(x_i, \bar{\theta})]$ and constant variance. This can be checked by calculating the Taylor expansion of $z(x_i) = T[\eta(x_i, \bar{\theta}) + \varepsilon_i]$ at $\varepsilon_i = 0$. For instance, for $\alpha = 0$, we obtain

$$\mathbb{E}[z(x_i)] = \log[\eta(x_i, \bar{\theta}) + c] + \sum_{k=1}^{\infty} (-1)^{k+1} \frac{\mu_k}{k}$$

with μ_k the k-th moment of the random variable $u_i = \varepsilon_i / [\eta(x_i, \bar{\theta}) + c]$. In general, $\mathbb{E}[z(x)]$ is thus difficult to express as a function of θ and x, so that a transformed nonlinear regression model (with constant variance) for the transformed observations $z(x_i)$ is not directly available from $\eta(x, \theta)$. Whereas the effect of an error in the variance of the $z(x_i)$ will asymptotically disappear when replicating observations at x_i, the bias will remain. The regression model

$$z(x_i) = T_\omega[\eta(x_i, \bar{\theta})] + \varepsilon_i' \text{ with } \mathbb{E}(\varepsilon_i') = 0 \text{ and } \mathbb{E}[(\varepsilon_i')^2] = \sigma^2 \text{ for all } i \quad (3.80)$$

may thus be rather inaccurate and the asymptotic properties given in Sect. 3.1 for the LS estimator in such models may therefore not be valid here. The asymptotic variance of the LS estimator for the model (3.80) may nevertheless be used to design experiments; see, e.g., Atkinson (2003, 2004). Alternatively, assuming that the observations transformed by $T_\omega(\cdot)$ are independently normally distributed with constant variance σ^2, the parameters ω, σ^2 and θ can be estimated by maximum likelihood, see Sect. 4.2. The asymptotic covariance matrix of the estimator can then be used for experimental design, considering σ^2 and ω as nuisance parameters for the estimation of θ; see Atkinson and Cook (1996).

3.4 LS Estimation with Model Error

So far, we assumed that there is no modeling error, see (3.2), a condition rarely satisfied in practice. It is thus important to investigate the asymptotic behavior of the estimator in presence of modeling errors, i.e. when the true response function $\nu(\cdot)$ is not of the form $\eta(\cdot, \theta)$ for some θ. Only the case of randomized designs with second-order stationary measurement errors will be considered. The results can easily be extended to nonstationary errors.

Assume that

$$y(x_k) = \nu(x_k) + \varepsilon_k\,, \quad \text{with } \mathbb{E}\{\varepsilon_k\} = 0 \text{ for all } k\,, \qquad (3.81)$$

where $\nu(\cdot)$ is some unknown function of x, bounded on $\mathscr{X}$. Theorem 3.1 can then be modified as follows.

Theorem 3.35. *Let $\{x_i\}$ be a randomized design on $\mathscr{S}_\xi$; see Definition 2.2. Consider the estimator $\hat{\theta}^N_{LS}$ that minimizes (3.1) in the model (3.81) with second-order stationary errors, that is,* $\mathrm{var}(\varepsilon_k) = \sigma^2$ *for all k. Assume that H_Θ and $H1_\eta$ are satisfied. Then, as $N \to \infty$, $\hat{\theta}^N_{LS}$ converges a.s. to the set $\Theta^\# \subset \Theta$ of values of θ that minimize*

$$J_\nu(\theta) = \int_{\mathscr{X}} [\nu(x) - \eta(x,\theta)]^2 \,\xi(\mathrm{d}x)\,. \tag{3.82}$$

Moreover, the estimator

$$\left[\hat{\sigma}^2\right]^N = \frac{1}{N-p} \sum_{k=1}^{N} \left[y(x_k) - \eta(x_k, \hat{\theta}^N_{LS})\right]^2$$

converges a.s. to $\sigma^2 + J_\nu(\bar{\theta})$, $\bar{\theta} \in \Theta^\#$.

The proof is similar to that of Theorem 3.1. In particular, when $\Theta^\#$ is reduced to a singleton $\{\bar{\theta}\}$, $\hat{\theta}^N_{LS}$ converges a.s. to that value $\bar{\theta}$ that gives the closest approximation to $\nu(x)$ in terms of $\mathscr{L}_2$ norm. The presence of modeling error can be detected by comparing the value of $\left[\hat{\sigma}^2\right]^N$, which converges a.s. to $\sigma^2 + J_\nu(\bar{\theta})$, to the estimated value of σ^2 obtained by replicating observations at a given design point x.

For nonlinear models, modeling errors affect the asymptotic distribution of the LS estimator as shown by the next theorem. We first introduce some notations. Define the matrices

$$\begin{aligned}
\mathbf{M}(\xi,\theta) &= \int_{\mathscr{X}} \frac{\partial \eta(x,\theta)}{\partial \theta} \frac{\partial \eta(x,\theta)}{\partial \theta^\top} \,\xi(\mathrm{d}x)\,,\\
\mathbf{D}_\nu(\xi,\theta) &= \int_{\mathscr{X}} [\eta(x,\theta) - \nu(x)] \,\frac{\partial^2 \eta(x,\theta)}{\partial\theta \partial \theta^\top} \,\xi(\mathrm{d}x)\,,\\
\mathbf{M}_\nu(\xi,\theta) &= \int_{\mathscr{X}} [\eta(x,\theta) - \nu(x)]^2 \,\frac{\partial \eta(x,\theta)}{\partial \theta} \frac{\partial \eta(x,\theta)}{\partial \theta^\top} \,\xi(\mathrm{d}x)\,.
\end{aligned}$$

Let P_θ denote the orthogonal projector onto $\mathcal{L}_\theta = \{\alpha^\top \partial\eta(\cdot,\theta)/\partial\theta : \alpha \in \mathbb{R}^p\}$,

$$(P_\theta f)(x') = \frac{\partial \eta(x',\theta)}{\partial \theta^\top} \mathbf{M}^{-1}(\xi,\theta) \int_{\mathscr{X}} \frac{\partial \eta(x,\theta)}{\partial \theta} \,f(x)\,\xi(\mathrm{d}x)\,. \tag{3.83}$$

We define the *intrinsic curvature* of the model at θ for the design measure by

$$C_{int}(\xi,\theta) = \sup_{\mathbf{u} \in \mathbb{R}^p - \{\mathbf{0}\}} \frac{\|[I - P_\theta]\sum_{i,j=1}^{p} u_i [\partial^2\eta(\cdot,\theta)/\partial\theta_i\partial\theta_j] u_j\|_\xi}{\mathbf{u}^\top \mathbf{M}(\xi,\theta)\mathbf{u}}\,, \tag{3.84}$$

where I denotes the identity operator and $\|f\|_\xi = [\int_{\mathscr{X}} f^2(x)\,\xi(\mathrm{d}x)]^{1/2}$ for any $f(\cdot)$ in $\mathscr{L}_2(\xi)$. Notice that replacing ξ by the empirical measure ξ_N associated with the N design points $X_N = (x_1, \ldots, x_N)$, we obtain

$$C_{int}(\xi_N, \theta) = \sqrt{N}\, C_{int}(X_N, \theta)$$

with $C_{int}(X_N, \theta)$ the intrinsic curvature of the model at θ for the exact design X_N; see, e.g., Bates and Watts (1988), Pázman (1993b) and Remark 6.1.

Theorem 3.36. *Let $\{x_i\}$ be a randomized design on $\mathscr{X}$; see Definition 2.2. Consider the LS estimator $\hat{\theta}^N_{LS}$ that minimizes (3.1) in the model (3.81) with second-order stationary errors ($\mathrm{var}(\varepsilon_k) = \sigma^2$ for all k). Assume that H_Θ, $H1_\eta$, and $H2_\eta$ are satisfied, that $\Theta^\# = \{\bar{\theta}\}$, that the matrix $\mathbf{M}(\xi, \bar{\theta})$ is nonsingular, and that*

$$C_{int}(\xi, \bar{\theta}) \|\nu(\cdot) - \eta(\cdot, \bar{\theta})\|_\xi < 1\,.$$

Then, the matrix $\mathbf{M}(\xi, \bar{\theta}) + \mathbf{D}_\nu(\xi, \bar{\theta})$ is nonsingular, and $\hat{\theta}^N_{LS}$ satisfies

$$\sqrt{N}(\hat{\theta}^N_{LS} - \bar{\theta}) \stackrel{\mathrm{d}}{\to} \mathbf{z} \sim \mathscr{N}(\mathbf{0}, \sigma^2 \mathbf{C}_\nu(\xi, \bar{\theta}))\,,\ N \to \infty\,, \tag{3.85}$$

with

$$\begin{aligned} \mathbf{C}_\nu(\xi, \theta) = {} & [\mathbf{M}(\xi, \theta) + \mathbf{D}_\nu(\xi, \theta)]^{-1} [\mathbf{M}(\xi, \theta) + (1/\sigma^2) \mathbf{M}_\nu(\xi, \theta)] \\ & \times [\mathbf{M}(\xi, \theta) + \mathbf{D}_\nu(\xi, \theta)]^{-1}\,. \end{aligned} \tag{3.86}$$

Proof. We first show that under the conditions of the theorem, the matrix $\mathbf{M}(\xi, \bar{\theta}) + \mathbf{D}_\nu(\xi, \bar{\theta})$ is nonsingular. Since $\bar{\theta}$ minimizes $J_\nu(\theta)$ given by (3.82), we have

$$\int_{\mathscr{X}} [\eta(x, \bar{\theta}) - \nu(x)] \left. \frac{\partial \eta(x, \theta)}{\partial \theta} \right|_{\bar{\theta}} \xi(\mathrm{d}x) = \mathbf{0}$$

so that $P_{\bar{\theta}}[\eta(\cdot, \bar{\theta}) - \nu(\cdot)] = 0$ and

$$\mathbf{D}_\nu(\xi, \bar{\theta}) = \int_{\mathscr{X}} ([I - P_{\bar{\theta}}][\eta(\cdot, \bar{\theta}) - \nu(\cdot)])(x) \left. \frac{\partial^2 \eta(x, \theta)}{\partial \theta \partial \theta^\top} \right|_{\bar{\theta}} \xi(\mathrm{d}x)\,.$$

Cauchy–Schwarz inequality gives for any $\mathbf{u} \in \mathbb{R}^p - \{\mathbf{0}\}$

$$\begin{aligned} \mathbf{u}^\top [\mathbf{M}(\xi, \bar{\theta}) + \mathbf{D}_\nu(\xi, \bar{\theta})] \mathbf{u} &= \mathbf{u}^\top \mathbf{M}(\xi, \bar{\theta}) \mathbf{u} \\ & \times \left(1 + \frac{\int_{\mathscr{X}} ([I - P_{\bar{\theta}}][\eta(\cdot, \bar{\theta}) - \nu(\cdot)])(x) \left[\mathbf{u}^\top \left. \frac{\partial^2 \eta(x, \theta)}{\partial \theta \partial \theta^\top} \right|_{\bar{\theta}} \mathbf{u}\right] \xi(\mathrm{d}x)}{\mathbf{u}^\top \mathbf{M}(\xi, \bar{\theta}) \mathbf{u}} \right) \\ & \geq \mathbf{u}^\top \mathbf{M}(\xi, \bar{\theta}) \mathbf{u} \left[1 - \|\eta(\cdot, \bar{\theta}) - \nu(\cdot)\|_\xi\, C_{int}(\xi, \bar{\theta}) \right] > 0\,. \end{aligned}$$

The rest of proof is similar to that of Theorem 3.8. The second-order derivative $\nabla^2_\theta J_N(\theta)$ of the LS criterion (3.1) tends a.s. and uniformly in θ to $2[\mathbf{M}(\xi, \theta) + \mathbf{D}_\nu(\xi, \theta)]$ when $N \to \infty$. The first-order derivative evaluated at $\bar{\theta}$ satisfies

$$-\sqrt{N}\nabla_\theta J_N(\bar{\theta}) = \frac{2}{\sqrt{N}}\left\{\sum_{k=1}^{N}\varepsilon_k \frac{\partial \eta(x_k,\theta)}{\partial \theta}\bigg|_{\bar{\theta}} + \sum_{k=1}^{N}[\nu(x_k)-\eta(x_k,\theta)]\frac{\partial \eta(x_k,\bar{\theta})}{\partial \theta}\bigg|_{\bar{\theta}}\right\} \quad (3.87)$$

and converges in distribution to a vector distributed $\mathscr{N}(\mathbf{0}, 4[\sigma^2\mathbf{M}(\xi,\bar{\theta}) + \mathbf{M}_\nu(\xi,\bar{\theta})])$. A Taylor development of $\nabla_\theta J_N(\theta)$ at $\bar{\theta}$ gives the result; see the proof of Theorem 3.8. ■

The asymptotic distribution of the estimator is thus generally affected by the presence of modeling errors. As shown below, one can easily construct examples where the precision can be either deteriorated or improved.

Example 3.37. Take $\nu(x) = \exp(-x)$ and $\eta(x,\theta) = 1/(x+\theta)$, with $\Theta = [0,\infty)$, $\mathscr{X} = [0,\infty)$, and $\sigma^2 = 1$. The design is obtained by equal replications of observations at the three points $\{0,\ 1,\ 2\}$.

Direct calculation gives $\Theta^\# = \{\bar{\theta}\}$ with $\bar{\theta} \simeq 1.0573$, and $\mathbf{M}(\xi,\bar{\theta}) \simeq 0.2892$, $\mathbf{D}_\nu(\xi,\bar{\theta}) \simeq -0.0170$ and $\mathbf{M}_\nu(\xi,\bar{\theta}) \simeq 0.0010$. We get $\mathbf{M}^{-1}(\xi,\bar{\theta}) \simeq 3.4577$, while $\mathbf{C}_\nu(\xi,\bar{\theta}) \simeq 3.9181$, and the modeling error increases the asymptotic variance of the LS estimator.

Take now $\eta(x,\theta) = 1/(x+\theta) - 1/2$. We get $\Theta^\# = \{\bar{\theta}\}$ with $\bar{\theta} \simeq 0.6431$, and $\mathbf{M}(\xi,\bar{\theta}) \simeq 2.001$, $\mathbf{D}_\nu(\xi,\bar{\theta}) \simeq 0.0894$ and $\mathbf{M}_\nu(\xi,\bar{\theta}) \simeq 0.0088$ so that $\mathbf{M}^{-1}(\xi,\bar{\theta}) \simeq 0.4997$, while $\mathbf{C}_\nu(\xi,\bar{\theta}) \simeq 0.4599$, and the modeling error decreases the asymptotic variance of the LS estimator. □

When the model response $\eta(x,\theta)$ is linear in θ, i.e., when $\eta(x,\theta) = \mathbf{f}^\top(x)\theta$, the presence of modeling errors always increases the asymptotic variance of the LS estimator. Indeed, when the matrix $\mathbf{M}(\xi) = \int_{\mathscr{X}} \mathbf{f}(x)\mathbf{f}^\top(x)\,\xi(\mathrm{d}x)$ is nonsingular, the set $\Theta^\#$ of Theorem 3.35 only contains the vector

$$\bar{\theta} = \mathbf{M}^{-1}(\xi)\int_{\mathscr{X}} \mathbf{f}(x)\,\nu(x)\,\xi(\mathrm{d}x)\,,$$

$\mathbf{D}_\nu(\xi,\bar{\theta})$ is the null matrix, $\mathbf{M}_\nu(\xi,\theta) = \int_{\mathscr{X}}[\mathbf{f}^\top(x)\theta - \nu(x)]^2\,\mathbf{f}(x)\mathbf{f}^\top(x)\,\xi(\mathrm{d}x)$ and $\mathbf{C}_\nu(\xi,\bar{\theta}) = \mathbf{M}^{-1}(\xi) + (1/\sigma^2)\mathbf{M}^{-1}(\xi)\mathbf{M}_\nu(\xi,\bar{\theta})\mathbf{M}^{-1}(\xi) \succeq \mathbf{M}^{-1}(\xi)$ in Theorem 3.36.

The expression of the asymptotic covariance matrix $\mathbf{C}_\nu(\xi,\theta)$ given by (3.86) cannot be used directly for optimum design since the true mean function $\nu(x)$ is unknown. For $\Phi(\cdot)$, a positively homogeneous design criterion, see Definition 5.3, bounds on $\Phi[\mathbf{C}_\nu^{-1}(\xi,\theta)]$ are given in Sect. 5.5.3, to be used for experimental design; see also Pázman and Pronzato (2006a).

Remark 3.38. The covariance $\sigma^2\mathbf{C}_\nu(\xi,\bar{\theta})$ in (3.85) contains a term that does not disappear when the variance of the observations tends to zero:

$$\lim_{\sigma^2\to 0}\sigma^2\mathbf{C}_\nu(\xi,\bar{\theta}) = [\mathbf{M}(\xi,\bar{\theta}) + \mathbf{D}_\nu(\xi,\bar{\theta})]^{-1}\mathbf{M}_\nu(\xi,\bar{\theta})[\mathbf{M}(\xi,\bar{\theta}) + \mathbf{D}_\nu(\xi,\bar{\theta})]^{-1}\,.$$

This term comes from the second sum in (3.87); its presence is due to the fact that the design is considered as randomly generated with the probability measure ξ.

When the design sequence is deterministic and such that the empirical measure ξ_N converges to ξ according to Definition 2.1, the other conditions of Theorem 3.36 being unchanged, then $\hat{\theta}^N_{LS}$ fluctuates around $\bar{\theta}$ and the fluctuations decrease as $1/\sqrt{N}$, with a random component due to the observation errors and a deterministic component due to the variability of the sequence of design points $x_1, x_2 \ldots$ When there are no observation errors ($\sigma^2 = 0$), the deterministic fluctuations remain. □

3.5 LS Estimation with Equality Constraints

Suppose that Θ is defined by the q equations $\mathbf{c}(\theta) = (c_1(\theta), \ldots, c_q(\theta))^\top = \mathbf{0}$, $q < p$. The constrained LS estimator $\hat{\theta}^N_{LS}$ in the model (3.2), (3.3) is obtained by minimizing the criterion $J_N(\theta)$ given by (3.1) under the constraint $\mathbf{c}(\theta) = \mathbf{0}$. Assume that $\mathbf{c}(\bar{\theta}) = \mathbf{0}$ —the situation gets more complicated when $\bar{\theta}$ does not satisfy the constraints.[10] Theorem 3.1 then remains valid and it is enough to consider a neighborhood of $\bar{\theta}$ when investigating the asymptotic distribution of $\hat{\theta}^N_{LS}$.

The following construction is used in (Pázman, 2002a), assuming suitable regularity conditions on $\mathbf{c}(\cdot)$ and $\eta(x,\cdot)$. Denote $\mathbf{L}(\theta) = \partial\mathbf{c}(\theta)/\partial\theta^\top \in \mathbb{R}^{q\times p}$ and assume that $\mathbf{L}(\bar{\theta})$ has rank q; without any loss of generality, we may assume that the first q columns of $\mathbf{L}(\bar{\theta})$ are linearly independent. We then denote $\alpha = (\theta_1, \ldots, \theta_q)^\top$ and $\beta = (\theta_{q+1}, \ldots, \theta_p)^\top$ and use β to reparameterize the regression model. Since $\mathbf{c}(\theta) = \mathbf{c}(\alpha, \beta) = \mathbf{0}$, by the implicit function theorem, there exist a neighborhood $\mathcal{V}$ of $\bar{\theta}$, an open set $\mathscr{B}$ or $\mathbb{R}^{p-q}$ containing $\bar{\beta} = (\bar{\theta}_{q+1}, \ldots, \bar{\theta}_p)^\top$, and a mapping $\mathbf{g}(\cdot)$ continuously differentiable on $\mathscr{B}$ such that $\mathbf{L}(\theta)$ has rank q for all $\theta \in \mathcal{V}$, $\mathbf{g}(\bar{\beta}) = \bar{\alpha} = (\bar{\theta}_1, \ldots, \bar{\theta}_q)^\top$, $\mathbf{c}(\mathbf{g}(\beta), \beta) = \mathbf{0}$ on $\mathscr{B}$, and

$$\frac{\partial\mathbf{g}(\beta)}{\partial\beta^\top} = -\left[\left.\frac{\partial\mathbf{c}(\alpha,\beta)}{\partial\alpha^\top}\right|_{\alpha=\mathbf{g}(\beta)}\right]^{-1} \frac{\partial\mathbf{c}(\alpha,\beta)}{\partial\beta^\top}, \ \forall\beta \in \mathscr{B}.$$

Now, $\phi(\beta) = (\mathbf{g}^\top(\beta), \beta^\top)^\top$ defines a reparameterization for the regression model such that $\phi(\bar{\beta}) = \bar{\theta}$ and $\mathbf{c}[\phi(\beta)] = \mathbf{0}$ for all $\beta \in \mathscr{B}$. Moreover,

$$\mathbf{D}(\beta) = \frac{\partial\phi(\beta)}{\partial\beta^\top} = \begin{pmatrix} \frac{\partial\mathbf{g}(\beta)}{\partial\beta^\top} \\ \mathbf{I}_{p-q} \end{pmatrix}, \ \beta \in \mathscr{B}, \tag{3.88}$$

[10]By enforcing constraints $\mathbf{c}(\theta) = \mathbf{0}$ in the estimation in a situation where $\mathbf{c}(\bar{\theta}) \neq \mathbf{0}$, we introduce a modeling error, the effect of which on the asymptotic properties of the LS estimator $\hat{\theta}^N_{LS}$ could be taken into account by combining the developments below with those in Sect. 3.4.

which has rank $p-q$. The regression model (3.2), (3.3) with the constraints $\mathbf{c}(\theta)=0$ is thus asymptotically equivalent to

$$y(x_i) = \eta[x_i, \phi(\bar{\beta})] + \varepsilon_i\,,\ \bar{\beta} \in \mathscr{B} \subset \mathbb{R}^{p-q}$$

with i.i.d. errors ε_i satisfying (3.3). This means in particular that the asymptotic normality property of Theorem 3.8 remains valid and, under the regularity assumptions mentioned in this theorem,

$$\sqrt{N}(\hat{\beta}_{LS}^N - \bar{\beta}) \xrightarrow{\mathrm{d}} \mathbf{z} \sim \mathscr{N}(\mathbf{0},\, \sigma^2 \mathbf{M}_\beta^{-1}(\xi, \bar{\beta}))\,,\ N \to \infty\,,$$

where

$$\begin{aligned}\mathbf{M}_\beta(\xi,\beta) &= \int_{\mathscr{X}} \frac{\partial \eta[x,\phi(\beta)]}{\partial \beta} \frac{\partial \eta[x,\phi(\beta)]}{\partial \beta^\top}\, \xi(\mathrm{d}x) \\ &= \mathbf{D}^\top(\beta)\mathbf{M}[\xi,\phi(\beta)]\mathbf{D}(\beta)\,, \end{aligned} \tag{3.89}$$

with $\mathbf{D}(\beta)$ given by (3.88) and

$$\mathbf{M}(\xi,\theta) = \int_{\mathscr{X}} \frac{\partial \eta(x,\theta)}{\partial \theta} \frac{\partial \eta(x,\theta)}{\partial \theta^\top}\, \xi(\mathrm{d}x)\,,$$

the matrix given in (3.32) for $\sigma^2 = 1$.

The conditions of Theorem 3.8 require in particular that $\mathbf{M}_\beta(\xi,\bar{\beta})$ be nonsingular. It is worthwhile to notice that we do not need to have $\mathbf{M}(\xi,\bar{\theta})$ nonsingular. Indeed, it is enough if the $p \times p$ matrix

$$\mathbf{H}(\xi,\bar{\theta}) = \mathbf{M}(\xi,\bar{\theta}) + \mathbf{L}^\top(\bar{\theta})\mathbf{L}(\bar{\theta}) \tag{3.90}$$

has full rank: $\mathbf{c}[\phi(\beta)] = \mathbf{0}$ on $\mathscr{B}$ implies that

$$\left.\frac{\partial \mathbf{c}[\phi(\beta)]}{\partial \beta^\top}\right|_{\bar{\beta}} = \mathbf{L}(\bar{\beta})\mathbf{D}(\bar{\beta}) = \mathbf{O}_{q,p-q}\,, \tag{3.91}$$

the $q \times (p-q)$-dimensional null matrix, so that the matrix $\mathbf{M}_\beta(\xi,\bar{\beta}) = \mathbf{D}^\top(\bar{\beta})\mathbf{M}(\xi,\bar{\theta})\mathbf{D}(\bar{\beta}) = \mathbf{D}^\top(\bar{\beta})\mathbf{H}(\xi,\bar{\theta})\mathbf{D}(\bar{\beta})$ has the same rank as $\mathbf{D}(\bar{\beta})$, i.e., $p-q$, see (3.88), and is nonsingular as required by Theorem 3.8.

Consider now the asymptotic distribution of $\hat{\theta}_{LS}^N = \phi(\beta_{LS}^N)$. The delta method (see Theorem 3.11) gives

$$\sqrt{N}(\hat{\theta}_{LS}^N - \bar{\theta}) \xrightarrow{\mathrm{d}} \mathbf{z}' \sim \mathscr{N}(\mathbf{0},\, \sigma^2 \mathbf{V}_{\theta,\xi}(\bar{\beta}))\,,\ N \to \infty\,,$$

where

$$\begin{aligned}\mathbf{V}_{\theta,\xi}(\bar{\beta}) &= \mathbf{D}(\bar{\beta})\mathbf{M}_\beta^{-1}(\xi,\bar{\beta})\mathbf{D}^\top(\bar{\beta}) \\ &= \mathbf{D}(\bar{\beta})[\mathbf{D}^\top(\bar{\beta})\mathbf{H}(\xi,\bar{\theta})\mathbf{D}(\bar{\beta})]^{-1}\mathbf{D}^\top(\bar{\beta})\,. \end{aligned} \tag{3.92}$$

However, this expression depends on $\mathbf{D}(\bar{\beta})$, which is not known explicitly. We can write $\mathbf{V}_{\theta,\xi}(\bar{\beta}) = \mathbf{H}^{-1/2}(\xi,\bar{\theta})\mathbf{Z}(\xi,\bar{\theta})\mathbf{H}^{-1/2}(\xi,\bar{\theta})$, where $\mathbf{H}(\xi,\bar{\theta})$ is given by (3.90) and

$$\mathbf{Z}(\xi,\bar{\theta}) = \mathbf{H}^{1/2}(\xi,\bar{\theta})\mathbf{D}(\bar{\beta})[\mathbf{D}^\top(\bar{\beta})\mathbf{H}(\xi,\bar{\theta})\mathbf{D}(\bar{\beta})]^{-1}\mathbf{D}^\top(\bar{\beta})\mathbf{H}^{1/2}(\xi,\bar{\theta})$$

is the orthogonal projector onto the column space of $\mathbf{H}^{1/2}(\xi,\bar{\theta})\mathbf{D}(\bar{\beta})$. Since $\mathbf{L}(\bar{\beta})\mathbf{D}(\bar{\beta}) = \mathbf{O}_{q,p-q}$, see (3.91), $\mathbf{H}^{1/2}(\xi,\bar{\theta})\mathbf{D}(\bar{\beta})$ and $\mathbf{H}^{-1/2}(\xi,\bar{\theta})\mathbf{L}^\top(\bar{\theta})$ are mutually orthogonal, with $\text{rank}[\mathbf{H}^{1/2}(\xi,\bar{\theta})\mathbf{D}(\bar{\beta})] = \text{rank}[\mathbf{D}(\bar{\beta})] = p - q$, and $\text{rank}[\mathbf{H}^{-1/2}(\xi,\bar{\theta})\mathbf{L}^\top(\bar{\theta})] = \text{rank}[\mathbf{L}(\bar{\theta})] = q$. Therefore, $\mathbf{Z}(\xi,\bar{\theta}) = \mathbf{I} - \mathbf{U}(\xi,\bar{\theta})$, with $\mathbf{U}(\xi,\bar{\theta})$ the complementary orthogonal projector of $\mathbf{Z}(\xi,\bar{\theta})$:

$$\mathbf{U}(\xi,\bar{\theta}) = \mathbf{H}^{-1/2}(\xi,\bar{\theta})\mathbf{L}^\top(\bar{\theta})[\mathbf{L}(\bar{\theta})\mathbf{H}^{-1}(\xi,\theta)\mathbf{L}^\top(\bar{\theta})]^{-1}\mathbf{L}(\bar{\theta})\mathbf{H}^{-1/2}(\xi,\bar{\theta})\,.$$

We thus obtain

$$\mathbf{V}_{\theta,\xi}(\bar{\beta}) = \mathbf{H}^{-1}(\xi,\bar{\theta}) - \mathbf{H}^{-1}(\xi,\bar{\theta})\mathbf{L}^\top(\bar{\theta})[\mathbf{L}(\bar{\theta})\mathbf{H}^{-1}(\xi,\bar{\theta})\mathbf{L}^\top(\bar{\theta})]^{-1}\mathbf{L}(\bar{\theta})\mathbf{H}^{-1}(\xi,\bar{\theta}) \tag{3.93}$$

with $\mathbf{H}(\xi,\bar{\theta})$ given by (3.90). Optimum design for the estimation of θ in this context will be briefly considered in Sect. 5.6 where a simpler form of $\mathbf{V}_{\theta,\xi}(\bar{\beta})$ will be used.

3.6 Bibliographic Notes and Further Remarks

Slow rates of convergence, cube-root asymptotics

The asymptotic properties considered in this book concern $\sqrt{N}$-consistency: we consider estimators $\hat{\theta}^N$ which, under suitable conditions, satisfy $\sqrt{N}(\hat{\theta}^N - \bar{\theta}) \xrightarrow{\text{d}} \mathbf{z} \sim \mathscr{N}(\mathbf{0},\mathbf{C})$ as $N \to \infty$, with $\mathbf{C}$ some positive-definite matrix; $\sqrt{N}(\hat{\theta}^N - \bar{\theta})$ is thus bounded in probability. When other asymptotic convergence rates are mentioned, it is in relation with the design sequence $\{x_i\}$; see Remark 3.6-(v) and Examples 2.4, 3.13, 3.17, and 5.39.[11]

There are, however, situations where $\sqrt{N}$-consistency cannot be obtained, independently of the design. For instance, this is the standard situation when the estimation of the parameters involves the estimation of a nonparametric component;[12] see, e.g., Parzen (1962) for the estimation of the mode of a density. There are also situations where the slow rate of convergence is only

[11] We only pay attention to rates slower than $\sqrt{N}$ because $\mathscr{X}$ is compact, but notice that by allowing the design points to expand to infinity, we might easily generate convergence rates faster than $\sqrt{N}$.

[12] However it is not always so: adaptive estimation precisely concerns efficient parameter estimation for models involving a nonparametric component; see the references in Sect. 4.4.2.

due to the parametric estimation procedure itself, as it is the case for some robust estimation methods, e.g., the least median of squares estimator; see Rousseeuw (1984) and Rousseeuw and Leroy (1987). A convergence rate of $N^{1/3}$ is obtained in this case, as well as in many other situations as shown in (Kim and Pollard, 1990). See also van der Vaart (1998, p. 77).

4

Asymptotic Properties of M, ML, and Maximum A Posteriori Estimators

4.1 M Estimators in Regression Models

We consider the regression model (3.2) where the errors ε_i are independently distributed, ε_i having the p.d.f. $\bar{\varphi}_{x_i}(\cdot)$. Notice that the vectors (ε_i, x_i) are i.i.d. in the case of a randomized design. An M estimator $\hat{\theta}^N_M = \hat{\theta}^N_M(\mathbf{y})$ is obtained by minimizing a function

$$J_N(\theta) = \frac{1}{N} \sum_{k=1}^{N} \rho[y(x_k) - \eta(x_k, \theta)] \tag{4.1}$$

with respect to θ, with $\rho(z)$ minimum at $z = 0$. This includes LS estimation, for which $\rho(z) = z^2$. Choosing a function $\rho(\cdot)$ with slower increase than quadratic conveys some robustness to the estimation; see in particular Huber (1981). Here we shall briefly consider the asymptotic properties of M estimators under randomized designs when $\rho(\cdot)$ is a smooth function. This is by no way exhaustive since it does not cover situations as simple as $\rho(z) = |z|$. However, the techniques required in such cases are more advanced and beyond the scope of this monograph; one may refer, for instance, to the books (van de Geer 2000; van der Vaart 1998, Chap. 5) for methods based on empirical processes.

The presentation used below follows the same lines as in Sects. 3.1.1 and 3.1.3 and corresponds to randomized designs; remember that almost sure properties are then over the product measure for the errors ε_i and design points x_i. Similar developments can be obtained for asymptotically discrete designs, using Lemma 2.8 instead of Lemma 2.6. One may notice that when $\hat{\theta}^N_M \in \text{int}(\Theta)$ and $J_N(\cdot)$ is differentiable in $\text{int}(\Theta)$, then $\hat{\theta}^N_M$ satisfies the estimating equation $\nabla_\theta J_N(\hat{\theta}^N_M) = \mathbf{0}$; see Sect. 4.6.

Theorem 4.1 (Consistency of M estimators). *Let $\{x_i\}$ be a randomized design with measure ξ on $\mathscr{X} \subset \mathbb{R}^d$; see Definition 2.2. Assume that H_Θ and $H1_\eta$ are satisfied, that*

L. Pronzato and A. Pázman, *Design of Experiments in Nonlinear Models*, Lecture Notes in Statistics 212, DOI 10.1007/978-1-4614-6363-4_4,

$$\int_{\mathscr{X}} \left\{ \int_{-\infty}^{\infty} \max_{\theta \in \Theta} \left| \rho[\eta(x, \bar{\theta}) - \eta(x, \theta) + \varepsilon] \right| \bar{\varphi}_x(\varepsilon) \, d\varepsilon \right\} \xi(dx) < \infty \tag{4.2}$$

with $\rho(\cdot)$ a continuous function, and that the estimability condition (3.6) is satisfied, together with

$$\forall x \in \mathscr{X}, \quad \forall z \neq 0,$$
$$J_x(z) = \int_{-\infty}^{\infty} \rho(\varepsilon - z) \, \bar{\varphi}_x(\varepsilon) \, d\varepsilon > \int_{-\infty}^{\infty} \rho(\varepsilon) \, \bar{\varphi}_x(\varepsilon) \, d\varepsilon = J_x(0) \,. \tag{4.3}$$

Then $\hat{\theta}_M^N \stackrel{\text{a.s.}}{\to} \bar{\theta}$, $N \to \infty$, where $\hat{\theta}_M^N$ minimizes (4.1) in the model (3.2) with errors $\varepsilon_k = \varepsilon(x_k) \sim \bar{\varphi}_{x_k}(\cdot)$.

Proof. The proof is similar to that of Theorem 3.1. We get from Lemma 2.6

$$J_N(\theta) \stackrel{\theta}{\rightsquigarrow} J_{\bar{\theta}}(\theta) = \int_{\mathscr{X}} \left\{ \int_{-\infty}^{\infty} \rho[\eta(x, \bar{\theta}) - \eta(x, \theta) + \varepsilon] \, \bar{\varphi}_x(\varepsilon) \, d\varepsilon \right\} \xi(dx)$$

a.s. as $N \to \infty$. In order to use Lemma 2.10, we only need to prove that for all $\theta \in \Theta$, $\theta \neq \bar{\theta}$, $J_{\bar{\theta}}(\theta) > J_{\bar{\theta}}(\bar{\theta})$. First note that $J_{\bar{\theta}}(\bar{\theta}) = \int_{\mathscr{X}} J_x(0) \xi(dx)$. From the estimability condition (3.6), $\theta \neq \bar{\theta}$ implies that $\eta(x, \bar{\theta}) \neq \eta(x, \theta)$ on some set $\mathcal{A}$ with $\xi(\mathcal{A}) > 0$. From (4.3), this implies

$$\int_{\mathcal{A}} \left\{ \int_{-\infty}^{\infty} \rho[\eta(x, \bar{\theta}) - \eta(x, \theta) + \varepsilon] \, \bar{\varphi}_x(\varepsilon) \, d\varepsilon \right\} \xi(dx) > \int_{\mathcal{A}} J_x(0) \, \xi(dx)$$

and thus $J_{\bar{\theta}}(\theta) > J_{\bar{\theta}}(\bar{\theta})$. Lemma 2.10 implies $\hat{\theta}_M^N \stackrel{\text{a.s.}}{\to} \bar{\theta}$ as $N \to \infty$. ■

Remark 4.2.

(*i*) Suppose that $\rho(\cdot)$ is twice continuously differentiable and that conditions allowing differentiation under the integral in $J_x(\cdot)$ are fulfilled; see (4.4). Then, (4.3) implies that the function $J_x(\cdot)$ is convex at $z = 0$ so that $\int_{-\infty}^{\infty} \rho''(\varepsilon) \, \bar{\varphi}_x(\varepsilon) \, d\varepsilon \geq 0$.
The condition (4.3) is satisfied, for instance, when $\bar{\varphi}_x(\cdot)$ and $\rho(\cdot)$ are symmetric, i.e., $\forall x \in \mathscr{X}$, $\forall z \in \mathbb{R}$, $\bar{\varphi}_x(-z) = \bar{\varphi}_x(z)$, $\rho(-z) = \rho(z)$, and, respectively, decreasing and increasing for $z > 0$, i.e., $\forall x \in \mathscr{X}$, $\forall z_2 > z_1 \geq 0$, $\bar{\varphi}_x(z_2) < \bar{\varphi}_x(z_1)$, $\rho(z_2) > \rho(z_1)$. Indeed, we have in this case, for any $x \in \mathscr{X}$ and any $z \neq 0$,

$$J_x(z) - J_x(0) = \int_{-\infty}^{0} [\rho(\varepsilon - z) - \rho(\varepsilon)][\bar{\varphi}_x(\varepsilon) - \bar{\varphi}_x(\varepsilon - z)] \, d\varepsilon > 0 \,.$$

(*ii*) When $\mathscr{X}$ is compact and $\eta(x, \theta)$ is continuous in (x, θ) on $\mathscr{X} \times \Theta$, $|\eta(x, \theta)|$ is bounded on $\mathscr{X} \times \Theta$, say by A. The condition (4.2) can then be replaced by $\int_{-\infty}^{\infty} \sup_{|z|<2A} |\rho(\varepsilon + z)| \bar{\varphi}_x(\varepsilon) \, d\varepsilon$ is bounded for $x \in \mathscr{X}$. □

Theorem 4.3 (Asymptotic normality of M estimators). *Let $\{x_i\}$ be a randomized design with measure ξ on $\mathscr{X} \subset \mathbb{R}^d$; see Definition 2.2. Assume that the conditions of Theorem 4.1 and H2$_\eta$ are satisfied, that $\rho(\cdot)$ is twice continuously differentiable, with derivatives $\rho'(\cdot)$ and $\rho''(\cdot)$ satisfying for all $x \in \mathscr{X}$*

$$\int_{-\infty}^{\infty} \rho'(\varepsilon)\bar{\varphi}_x(\varepsilon)d\varepsilon = 0\,, \quad \int_{-\infty}^{\infty} [\rho'(\varepsilon)]^2\bar{\varphi}_x(\varepsilon)d\varepsilon < \infty\,, \quad \int_{-\infty}^{\infty} |\rho''(\varepsilon)|\bar{\varphi}_x(\varepsilon)d\varepsilon < \infty\,, \tag{4.4}$$

that the matrix

$$\mathbf{M}_1(\xi,\bar{\theta}) = \int_{\mathscr{X}} \left\{ \int_{-\infty}^{\infty} \rho''(\varepsilon)\,\bar{\varphi}_x(\varepsilon)\,\mathrm{d}\varepsilon \right\} \left.\frac{\partial\eta(x,\theta)}{\partial\theta}\right|_{\bar{\theta}} \left.\frac{\partial\eta(x,\theta)}{\partial\theta^\top}\right|_{\bar{\theta}} \xi(\mathrm{d}x) \tag{4.5}$$

is nonsingular and that for all $i,j = 1,\ldots,p$,

$$\int_{\mathscr{X}} \left[\int_{-\infty}^{\infty} \max_{\theta\in\Theta} \left\{ \left|\rho''[\eta(x,\bar{\theta}) - \eta(x,\theta) + \varepsilon]\right| \left| \frac{\partial\eta(x,\theta)}{\partial\theta_i}\frac{\partial\eta(x,\theta)}{\partial\theta_j} \right| \right\} \bar{\varphi}_x(\varepsilon)\,\mathrm{d}\varepsilon \right] \times \xi(\mathrm{d}x) < \infty\,, \tag{4.6}$$

$$\int_{\mathscr{X}} \left[\int_{-\infty}^{\infty} \max_{\theta\in\Theta} \left\{ \left|\rho'[\eta(x,\bar{\theta}) - \eta(x,\theta) + \varepsilon]\right| \left| \frac{\partial^2\eta(x,\theta)}{\partial\theta_i\partial\theta_j} \right| \right\} \bar{\varphi}_x(\varepsilon)\,\mathrm{d}\varepsilon \right] \times \xi(\mathrm{d}x) < \infty\,. \tag{4.7}$$

Then, the M estimator $\hat{\theta}^N_M$ that minimizes (4.1) in the model (3.2) with errors $\varepsilon_k = \varepsilon(x_k) \sim \bar{\varphi}_{x_k}(\cdot)$ satisfies

$$\sqrt{N}(\hat{\theta}^N_M - \bar{\theta}) \overset{\mathrm{d}}{\to} \mathbf{z} \sim \mathscr{N}(\mathbf{0}, \mathbf{C}(\xi,\bar{\theta}))\,,\ N \to \infty\,,$$

where

$$\mathbf{C}(\xi,\theta) = \mathbf{M}_1^{-1}(\xi,\theta)\mathbf{M}_2(\xi,\theta)\mathbf{M}_1^{-1}(\xi,\theta)$$

with

$$\mathbf{M}_2(\xi,\theta) = \int_{\mathscr{X}} \left\{ \int_{-\infty}^{\infty} [\rho'(\varepsilon)]^2\,\bar{\varphi}_x(\varepsilon)\,\mathrm{d}\varepsilon \right\} \frac{\partial\eta(x,\theta)}{\partial\theta}\frac{\partial\eta(x,\theta)}{\partial\theta^\top}\,\xi(\mathrm{d}x)\,.$$

Proof. The proof is similar to that of Theorem 3.8. Lemma 2.6, and Theorem 4.1 imply that $\nabla^2_\theta J_N(\hat{\theta}^N_M) \overset{\mathrm{a.s.}}{\to} \mathbf{M}_1(\xi,\bar{\theta})$, and since $\mathbf{M}_1(\xi,\bar{\theta})$ is nonsingular,

$$[\nabla^2_\theta J_N(\hat{\theta}^N_M)]^{-1} \overset{\mathrm{a.s.}}{\to} \mathbf{M}_1^{-1}(\xi,\bar{\theta})$$

as $N \to \infty$. Also,

$$-\sqrt{N}\nabla_\theta J_N(\bar{\theta}) = \frac{1}{\sqrt{N}}\sum_{k=1}^{N} \rho'(\varepsilon_k) \left.\frac{\partial\eta(x_k,\theta)}{\partial\theta}\right|_{\bar{\theta}} \overset{\mathrm{d}}{\to} \mathbf{v} \sim \mathscr{N}(\mathbf{0}, \mathbf{M}_2(\xi,\bar{\theta}))$$

for $N \to \infty$, which gives $\sqrt{N}(\hat{\theta}^N_M - \bar{\theta}) \overset{\mathrm{d}}{\to} \mathbf{z} \sim \mathscr{N}(\mathbf{0}, \mathbf{C}(\xi,\bar{\theta}))$. ■

Remark 4.4. Suppose that $\mathscr{X}$ is compact and that $\eta(x,\theta)$ and its first two derivatives in θ are continuous in (x,θ) on $\mathscr{X} \times \Theta$. Then, $|\eta(x,\theta)| < A$ for some $A > 0$, and the conditions (4.6), (4.7) can be replaced by

$$\int_{-\infty}^{\infty} \sup_{|z|<2A} |\rho''(\varepsilon+z)| \bar{\varphi}_x(\varepsilon)\, \mathrm{d}\varepsilon \quad \text{and} \quad \int_{-\infty}^{\infty} \sup_{|z|<2A} |\rho'(\varepsilon+z)| \bar{\varphi}_x(\varepsilon)\, \mathrm{d}\varepsilon$$

are bounded for any $x \in \mathscr{X}$. □

When the errors ε_k are i.i.d. with a p.d.f. $\bar{\varphi}(\cdot)$, previous theorem gives

$$\sqrt{N}(\hat{\theta}_M^N - \bar{\theta}) \overset{\mathrm{d}}{\to} \mathbf{z} \sim \mathscr{N}(\mathbf{0}, \mathbf{M}^{-1}(\xi,\bar{\theta}))\,, \; N \to \infty\,,$$

with

$$\mathbf{M}(\xi,\theta) = \mathcal{I}(\rho,\bar{\varphi}) \int_{\mathscr{X}} \frac{\partial \eta(x,\theta)}{\partial \theta} \frac{\partial \eta(x,\theta)}{\partial \theta^\top}\, \xi(\mathrm{d}x)\,, \tag{4.8}$$

where

$$\mathcal{I}(\rho,\bar{\varphi}) = \frac{\left[\int_{-\infty}^{\infty} \rho''(\varepsilon)\, \bar{\varphi}(\varepsilon)\, \mathrm{d}\varepsilon\right]^2}{\int_{-\infty}^{\infty} [\rho'(\varepsilon)]^2\, \bar{\varphi}(\varepsilon)\, \mathrm{d}\varepsilon}\,. \tag{4.9}$$

Choosing $\rho(z) = z^2$, which corresponds to LS estimation, gives $\mathcal{I}(\rho,\bar{\varphi}) = 1/\sigma^2$ in (4.8), a result already obtained in Sect. 3.1.3.

Assume that the *Fisher information for location* for $\bar{\varphi}$ exists; that is,

$$\mathcal{I}_{\bar{\varphi}} = \int_{-\infty}^{\infty} \left[\frac{\bar{\varphi}'(\varepsilon)}{\bar{\varphi}(\varepsilon)}\right]^2 \bar{\varphi}(\varepsilon)\, \mathrm{d}\varepsilon < \infty$$

with $\bar{\varphi}'(\cdot)$ the derivative of $\bar{\varphi}(\cdot)$. Then, integration by parts and Cauchy–Schwarz inequality give

$$\begin{aligned}
\left[\int_{-\infty}^{\infty} \rho''(\varepsilon)\, \bar{\varphi}(\varepsilon)\, \mathrm{d}\varepsilon\right]^2 &= \left[\int_{-\infty}^{\infty} \rho'(\varepsilon)\, \frac{\bar{\varphi}'(\varepsilon)}{\bar{\varphi}(\varepsilon)}\, \bar{\varphi}(\varepsilon)\, \mathrm{d}\varepsilon\right]^2 \\
&\le \int_{-\infty}^{\infty} [\rho'(\varepsilon)]^2\, \bar{\varphi}(\varepsilon)\, \mathrm{d}\varepsilon \times \int_{-\infty}^{\infty} \left[\frac{\bar{\varphi}'(\varepsilon)}{\bar{\varphi}(\varepsilon)}\right]^2 \bar{\varphi}(\varepsilon)\, \mathrm{d}\varepsilon\,;
\end{aligned}$$

that is,

$$\mathcal{I}(\rho,\bar{\varphi}) \le \mathcal{I}_{\bar{\varphi}} \tag{4.10}$$

for any function $\rho(\cdot)$ twice continuously differentiable and such that

$$\int_{-\infty}^{\infty} [\rho'(\varepsilon)]^2\, \bar{\varphi}(\varepsilon)\, \mathrm{d}\varepsilon < \infty\,.$$

The equality is achieved in (4.10) for $\rho'(z) = K_1\, \bar{\varphi}'(z)/\bar{\varphi}(z)$, i.e., $\rho(z) = K_1 \log \bar{\varphi}(z) + K_2$, with $K_1 < 0$ since $\rho(\cdot)$ is minimum at zero. This choice is legitimate if $-\log \bar{\varphi}(z)$ satisfies the conditions in Theorems. 4.1 and 4.3;

it corresponds to maximum likelihood estimation, considered in the following section. We can already notice that the condition (4.2) then requires that the support of the p.d.f. $\bar{\varphi}(\cdot)$ is infinite.[1] The inequality (4.10) shows that the minimum asymptotic variance, in the class of estimators defined by (4.1) and for regression models with i.i.d. errors, is obtained for maximum likelihood estimation. A similar result will be presented below in a more general context, see Theorem 4.7, which corresponds to extending the class of estimators considered by allowing $\rho(\cdot)$ to depend on x; that is, we shall consider estimators obtained by the minimization of

$$J_N(\theta) = \frac{1}{N} \sum_{k=1}^{N} \rho_{x_k}[y(x_k) - \eta(x_k, \theta)] \,. \tag{4.11}$$

The properties stated in Theorems. 4.1 and 4.3 remain valid for such estimators, provided that the assumptions mentioned are satisfied when $\rho(\cdot)$ is replaced by $\rho_x(\cdot)$.

4.2 The Maximum Likelihood Estimator

Let $\varphi_{x,\theta}(y)$ denote the probability density function of the observation $y(x)$ at the design point x for some postulated true value θ of the model parameters. The observations $y(x_1), y(x_2), \ldots$ are assumed to be independent.[2] The maximum likelihood (ML) estimator $\hat{\theta}^N_{ML}$ maximizes the *likelihood function*, which is defined as being proportional to the density of the observations given θ, that is,

$$\mathrm{L}_{X,\mathbf{y}}(\cdot) : \theta \longrightarrow \mathrm{L}_{X,\mathbf{y}}(\theta) = C(\mathbf{y}) \prod_{k=1}^{N} \varphi_{x_k,\theta}(y_k) \,,$$

for some positive measurable function $C(\cdot)$ of $\mathbf{y} = [y(x_1), \ldots, y(x_N)]^\top$, not depending on θ. Notice that the value of $C(\mathbf{y})$ is arbitrary in the sense that it has no influence on $\hat{\theta}^N_{ML}$. The choice $C(\mathbf{y}) = 1$ is usual; see, e.g., Cox and Hinkley (1974, p. 11), Lehmann and Casella (1998, p. 238). Equivalently, $\hat{\theta}^N_{ML}$ minimizes

$$J_N(\theta) = -\frac{1}{N} \sum_{k=1}^{N} \log \varphi_{x_k,\theta}(y_k) \,, \tag{4.12}$$

with $y_k = y(x_k)$. For instance, when $\varphi_{x,\theta}(\cdot)$ is the normal density

---

[1] A more standard condition, used in more general situations than regression models, is that the support of the density of the observations should not depend on the value θ of the parameters in the model generating these observations.

[2] See Sect. 4.6 for a brief discussion on the application of maximum likelihood estimation to dynamical systems, for which the independence assumption does not hold.

$$\varphi_{x,\theta}(y) = \frac{1}{\sigma(x)\sqrt{2\pi}} \exp\left\{-\frac{[y-\eta(x,\theta)]^2}{2\sigma^2(x)}\right\},$$

then the ML estimator $\hat{\theta}^N_{ML}$ coincides with the WLS estimator minimizing (3.5) for the optimal weights $w(x_i) = 1/\sigma^2(x_i)$.

4.2.1 Regression Models

Denote $\eta(x,\theta) = \mathbb{E}_{x,\theta}(y) = \int_{-\infty}^{\infty} y\,\varphi_{x,\theta}(y)\,\mathrm{d}y$ and $\bar{\varphi}_x(\cdot)$ the p.d.f. of $\varepsilon(x) = y(x) - \eta(x,\bar{\theta})$, with $\bar{\theta} \in \Theta$ the true value of the model parameters, used to generate the observations y_k. The model thus takes the usual form (3.2); $\hat{\theta}^N_{ML}$ is an M estimator with $\rho_x(\cdot) = -\log\bar{\varphi}_x(\cdot)$, see (4.11), and Theorems. 4.1 and 4.3 apply. We repeat these properties below for the special case of ML estimation. Again, similar developments can be obtained for asymptotically discrete designs, using Lemma 2.8 instead of Lemma 2.6.

Theorem 4.5 (Consistency of ML estimators in regression models). *Let $\{x_i\}$ be a randomized design with measure ξ on $\mathscr{X} \subset \mathbb{R}^d$; see Definition 2.2. Consider the ML estimator $\hat{\theta}^N_{ML}$ that minimizes (4.12) in the model (3.2) with errors $\varepsilon_k = \varepsilon(x_k) \sim \bar{\varphi}_{x_k}(\cdot)$. Assume that H_Θ and $H1_\eta$ are satisfied, that $\bar{\varphi}_x(\cdot)$ is continuous for any x with*

$$\int_{\mathscr{X}} \left\{\int_{-\infty}^{\infty} \max_{\theta\in\Theta} |\log\bar{\varphi}_x[\eta(x,\bar{\theta}) - \eta(x,\theta) + \varepsilon]|\,\bar{\varphi}_x(\varepsilon)\,\mathrm{d}\varepsilon\right\}\xi(\mathrm{d}x) < \infty\,, \quad (4.13)$$

and that the estimability condition (3.6) is satisfied, with the function

$$J_x(z) = -\int_{-\infty}^{\infty} \log[\bar{\varphi}_x(\varepsilon - z)]\,\bar{\varphi}_x(\varepsilon)\,\mathrm{d}\varepsilon \quad (4.14)$$

having a unique minimum at $z = 0$ for any $x \in \mathscr{X}$. Then $\hat{\theta}^N_{ML}$ satisfies $\hat{\theta}^N_{ML} \overset{\text{a.s.}}{\to} \bar{\theta}$ as $N \to \infty$.

Note that $J_x(z) - J_x(0) = D(P_{x,0}\|P_{x,z})$, the Kullback–Leibler divergence (or information divergence, or relative entropy) between the probability distributions having densities $\varphi_{x,0}(\varepsilon) = \bar{\varphi}_x(\varepsilon)$ and $\varphi_{x,z}(\varepsilon) = \bar{\varphi}_x(\varepsilon - z)$. Also note that (4.13) implies that the support of $\bar{\varphi}_x(\cdot)$ must be infinite.

Theorem 4.6 (Asymptotic normality of ML estimators in regression models). *Let $\{x_i\}$ be a randomized design with measure ξ on $\mathscr{X} \subset \mathbb{R}^d$; see Definition 2.2. Assume that the conditions of Theorem 4.5 and $H2_\eta$ are satisfied, that $\bar{\varphi}_x(\cdot)$ is twice continuously differentiable for any $x \in \mathscr{X}$, and such that the Fisher information for location exists at any x,*

$$\forall x \in \mathscr{X}\,,\ \mathcal{I}_{\bar{\varphi}}(x) = \int_{-\infty}^{\infty} \left[\frac{\bar{\varphi}'_x(\varepsilon)}{\bar{\varphi}_x(\varepsilon)}\right]^2 \bar{\varphi}_x(\varepsilon)\,\mathrm{d}\varepsilon < \infty\,, \quad (4.15)$$

with $\bar{\varphi}'_x(\varepsilon) = \mathrm{d}\bar{\varphi}_x(\varepsilon)/\mathrm{d}\varepsilon$, *that for all* $i, j = 1, \ldots, p$,

$$\int_{\mathscr{X}} \left[\int_{-\infty}^{\infty} \max_{\theta \in \Theta} \left\{ \left| \frac{\mathrm{d}^2 \log \bar{\varphi}_x(z)}{\mathrm{d}z^2} \right|_{\eta(x,\bar{\theta}) - \eta(x,\theta) + \varepsilon} \right| \left| \frac{\partial \eta(x,\theta)}{\partial \theta_i} \frac{\partial \eta(x,\theta)}{\partial \theta_j} \right| \right\} \bar{\varphi}_x(\varepsilon)\, \mathrm{d}\varepsilon \right] \times \xi(\mathrm{d}x) < \infty\,,$$

$$\int_{\mathscr{X}} \left[\int_{-\infty}^{\infty} \max_{\theta \in \Theta} \left\{ \left| \frac{\mathrm{d} \log \bar{\varphi}_x(z)}{\mathrm{d}z} \right|_{\eta(x,\bar{\theta}) - \eta(x,\theta) + \varepsilon} \right| \left| \frac{\partial^2 \eta(x,\theta)}{\partial \theta_i \partial \theta_j} \right| \right\} \bar{\varphi}_x(\varepsilon)\, \mathrm{d}\varepsilon \right] \times \xi(\mathrm{d}x) < \infty\,,$$

and that the average Fisher information matrix *per sample (for the parameters θ)*

$$\mathbf{M}(\xi, \bar{\theta}) = \int_{\mathscr{X}} \mathcal{I}_{\bar{\varphi}}(x) \left. \frac{\partial \eta(x,\theta)}{\partial \theta} \right|_{\bar{\theta}} \left. \frac{\partial \eta(x,\theta)}{\partial \theta^\top} \right|_{\bar{\theta}} \xi(\mathrm{d}x) \tag{4.16}$$

is nonsingular. Then, the ML estimator $\hat{\theta}^N_{ML}$ *that minimizes (4.12) in the model (3.2) with the errors* $\varepsilon_k = \varepsilon(x_k) \sim \bar{\varphi}_{x_k}(\cdot)$ *satisfies*

$$\sqrt{N}(\hat{\theta}^N_{ML} - \bar{\theta}) \xrightarrow{\mathrm{d}} \mathbf{z} \sim \mathscr{N}(\mathbf{0}, \mathbf{M}^{-1}(\xi, \bar{\theta}))\,,\ N \to \infty\,.$$

Moreover, we have the following property which states that in a regression model, any estimator defined by the minimization of a function of the form (4.11), with suitable regularity properties for $\rho_x(\cdot)$, has an asymptotic covariance matrix at least as large as that obtained for ML estimation. This corresponds to a Cramér–Rao-type inequality; see also Theorem 4.10 and Sect. 4.4.

Theorem 4.7. *Assume that the conditions of Theorem 4.6 and those of Theorem 4.3 with* $\rho(\cdot)$ *replaced by* $\rho_x(\cdot)$ *are satisfied. Then, the estimator* $\hat{\theta}^N$ *that minimizes (4.11) satisfies*

$$\sqrt{N}(\hat{\theta}^N - \bar{\theta}) \xrightarrow{\mathrm{d}} \mathbf{z} \sim \mathscr{N}(\mathbf{0}, \mathbf{C}(\xi, \bar{\theta}))\,,\ N \to \infty\,,$$

where $\mathbf{C}(\xi, \theta)$ *is defined as in Theorem 4.3, with* $\rho(\cdot)$ *replaced by* $\rho_x(\cdot)$, *and, for any* $\theta \in \Theta$,

$$\mathbf{C}(\xi, \theta) \succeq \mathbf{M}^{-1}(\xi, \theta) \tag{4.17}$$

with $\mathbf{M}(\xi, \theta)$ *defined by (4.16). The equality is obtained when*

$$\rho_x(z) = K_1 \log \bar{\varphi}_x(z) + K_2\,,\ K_1 < 0\,.$$

Proof. We have $\mathbf{C}(\xi, \theta) = \mathbf{M}_1^{-1}(\xi, \theta)\mathbf{M}_2(\xi, \theta)\mathbf{M}_1^{-1}(\xi, \theta)$ with

$$\mathbf{M}_1(\xi, \bar{\theta}) = \int_{\mathscr{X}} \left\{ \int_{-\infty}^{\infty} \rho''_x(\varepsilon)\, \bar{\varphi}_x(\varepsilon)\, \mathrm{d}\varepsilon \right\} \left. \frac{\partial \eta(x,\theta)}{\partial \theta} \right|_{\bar{\theta}} \left. \frac{\partial \eta(x,\theta)}{\partial \theta^\top} \right|_{\bar{\theta}} \xi(\mathrm{d}x) \tag{4.18}$$

$$\mathbf{M}_2(\xi, \theta) = \int_{\mathscr{X}} \left\{ \int_{-\infty}^{\infty} [\rho'_x(\varepsilon)]^2\, \bar{\varphi}_x(\varepsilon)\, \mathrm{d}\varepsilon \right\} \frac{\partial \eta(x,\theta)}{\partial \theta} \frac{\partial \eta(x,\theta)}{\partial \theta^\top}\, \xi(\mathrm{d}x)\,,$$

where $\rho'_x(\varepsilon) = \mathrm{d}\rho_x(\varepsilon)/\mathrm{d}\varepsilon$ and $\rho''_x(\varepsilon) = \mathrm{d}^2\rho_x(\varepsilon)/\mathrm{d}\varepsilon^2$. We use the same approach as in the proof of Theorem 3.8, and define

$$\mathbf{v} = \begin{pmatrix} \left\{\int_{-\infty}^{\infty} [\rho'_x(\varepsilon)]^2\, \bar{\varphi}_x(\varepsilon)\,\mathrm{d}\varepsilon\right\}^{1/2} \frac{\partial\eta(x,\theta)}{\partial\theta} \\ \left\{\int_{-\infty}^{\infty} \rho''_x(\varepsilon)\,\bar{\varphi}_x(\varepsilon)\,\mathrm{d}\varepsilon\right\} \left\{\int_{-\infty}^{\infty} [\rho'_x(\varepsilon)]^2\, \bar{\varphi}_x(\varepsilon)\,\mathrm{d}\varepsilon\right\}^{-1/2} \frac{\partial\eta(x,\theta)}{\partial\theta} \end{pmatrix}.$$

It gives

$$\mathbb{E}\{\mathbf{v}\mathbf{v}^\top\} = \begin{pmatrix} \mathbf{M}_2(\xi,\theta) & \mathbf{M}_1(\xi,\theta) \\ \mathbf{M}_1(\xi,\theta) & \tilde{\mathbf{M}}(\xi,\theta) \end{pmatrix},$$

where the expectation is with respect to $\xi(\mathrm{d}x)$ and

$$\tilde{\mathbf{M}}(\xi,\theta) = \int_{\mathscr{X}} \mathcal{I}(\rho_x,\bar{\varphi}_x)\, \frac{\partial\eta(x,\theta)}{\partial\theta}\frac{\partial\eta(x,\theta)}{\partial\theta^\top}\,\xi(\mathrm{d}x)$$

with $\mathcal{I}(\rho,\bar{\varphi})$ defined by (4.9). We have $\mathcal{I}(\rho_x,\bar{\varphi}_x) \leq \mathcal{I}_{\bar{\varphi}}(x)$ for any x, with $\mathcal{I}_{\bar{\varphi}}(x)$ given by (4.15); see the discussion following the proof of Theorem 4.3. Therefore, $\tilde{\mathbf{M}}(\xi,\theta) \preceq \mathbf{M}(\xi,\theta)$ and $\mathbf{M}_2(\xi,\theta) \succeq \mathbf{M}_1(\xi,\theta)\tilde{\mathbf{M}}^{-1}(\xi,\theta)\mathbf{M}_1(\xi,\theta)$ from Lemma 3.7, which together give (4.17). ■

4.2.2 General Situation

We are interested in the properties of the estimator $\hat{\theta}^N_{ML}$ that minimizes (4.12) when the observations $y_k = y(x_k)$ are independent random variables, $y(x)$ having the density (or Radon–Nikodým derivative) $\varphi_{x,\bar{\theta}}(y)$ with respect to a σ-finite measure μ_x on the set $\mathcal{Y}_x \subset \mathbb{R}$ to which $y(x)$ belongs, with $\bar{\theta} \in \Theta$.

In parallel, we shall consider estimators defined by the minimization of a criterion given by

$$J_N(\theta) = \frac{1}{N}\sum_{k=1}^{N} \varrho_{x_k}(y_k,\theta) \tag{4.19}$$

for some functions $\varrho_{x_k}(y,\theta)$. We proceed as previously: we first present the results for an estimator defined by the minimization of (4.19), then we particularize these results to the case of ML estimation by taking $\varrho_x(y,\theta) = -\log\varphi_{x,\theta}(y)$ and compare the asymptotic variances.

Theorem 4.8 (Consistency). *Let $\{x_i\}$ be a randomized design with measure ξ on $\mathscr{X} \subset \mathbb{R}^d$; see Definition 2.2. Assume that H_Θ is satisfied, that $\varrho_x(y,\theta)$ is continuous in $\theta \in \Theta$ for any (x,y) in $\mathscr{X} \times \mathbb{R}$, that*

$$\int_{\mathscr{X}} \left\{\int_{-\infty}^{\infty} \max_{\theta\in\Theta} |\varrho_x(y,\theta)|\, \varphi_{x,\bar{\theta}}(y)\,\mu_x(\mathrm{d}y)\right\} \xi(\mathrm{d}x) < \infty\,, \tag{4.20}$$

and that the function

$$J_{\bar{\theta}}(\theta) = \int_{\mathscr{X}} \left\{ \int_{-\infty}^{\infty} \varrho_x(y,\theta)\, \varphi_{x,\bar{\theta}}(y)\, \mu_x(\mathrm{d}y) \right\} \xi(\mathrm{d}x) \tag{4.21}$$

has a unique minimum at $\theta = \bar{\theta} \in \Theta$. *Then the estimator* $\hat{\theta}^N$ *that minimizes (4.19), with independent observations* $y_k \sim \varphi_{x_k,\bar{\theta}}(\cdot)$, *satisfies* $\hat{\theta}^N \overset{\text{a.s.}}{\to} \bar{\theta}$ *as* $N \to \infty$.

Proof. The proof directly follows from the application of Lemmas 2.6 and 2.10. ∎

Theorem 4.9 (Asymptotic normality). *Let* $\{x_i\}$ *be a randomized design with measure* ξ *on* $\mathscr{X} \subset \mathbb{R}^d$*; see Definition 2.2. Assume that* $\varrho_x(y,\theta)$ *is twice continuously differentiable in* $\theta \in \mathrm{int}(\Theta)$ *for any* (x,y) *in* $\mathscr{X} \times \mathbb{R}$ *and that* $\bar{\theta} \in \mathrm{int}(\Theta)$. *Assume moreover that the conditions of Theorem 4.8 are satisfied, that for all* $i,j = 1,\ldots,p$,

$$\int_{\mathscr{X}} \left[\int_{-\infty}^{\infty} \max_{\theta\in\Theta} \left| \frac{\partial \varrho_x(y,\theta)}{\partial \theta_i} \right| \varphi_{x,\bar{\theta}}(y)\, \mu_x(\mathrm{d}y) \right] \xi(\mathrm{d}x) < \infty\,, \tag{4.22}$$
$$\int_{\mathscr{X}} \left[\int_{-\infty}^{\infty} \max_{\theta\in\Theta} \left| \frac{\partial^2 \varrho_x(y,\theta)}{\partial \theta_i \partial \theta_j} \right| \varphi_{x,\bar{\theta}}(y)\, \mu_x(\mathrm{d}y) \right] \xi(\mathrm{d}x) < \infty\,,$$

and that the matrix

$$\mathbf{M}_1(\xi,\bar{\theta}) = \int_{\mathscr{X}} \left[\int_{-\infty}^{\infty} \left. \frac{\partial^2 \varrho_x(y,\theta)}{\partial \theta \partial \theta^\top} \right|_{\bar{\theta}} \varphi_{x,\bar{\theta}}(y)\, \mu_x(\mathrm{d}y) \right] \xi(\mathrm{d}x) \tag{4.23}$$

is nonsingular. Then, the estimator $\hat{\theta}^N$ *that minimizes (4.19), with independent observations* $y_k \sim \varphi_{x_k,\bar{\theta}}(\cdot)$, *satisfies*

$$\sqrt{N}(\hat{\theta}^N - \bar{\theta}) \overset{\mathrm{d}}{\to} \mathbf{z} \sim \mathscr{N}(\mathbf{0}, \mathbf{C}(\xi,\bar{\theta}))\,,\ N \to \infty\,,$$

where $\mathbf{C}(\xi,\theta) = \mathbf{M}_1^{-1}(\xi,\theta)\mathbf{M}_2(\xi,\theta)\mathbf{M}_1^{-1}(\xi,\theta)$, *with*

$$\mathbf{M}_2(\xi,\theta) = \int_{\mathscr{X}} \left[\int_{-\infty}^{\infty} \frac{\partial \varrho_x(y,\theta)}{\partial \theta} \frac{\partial \varrho_x(y,\theta)}{\partial \theta^\top} \varphi_{x,\bar{\theta}}(y)\, \mu_x(\mathrm{d}y) \right] \xi(\mathrm{d}x)\,.$$

Proof. The proof follows the same lines as that of Theorem 4.3. An additional requirement is that $\mathbb{E}\{\nabla_\theta J_N(\bar{\theta})\}$ should be proved to be equal to zero, where $\mathbb{E}\{\cdot\}$ denotes expectation with respect to y and x. Under the conditions (4.20), (4.22), we can write $\mathbb{E}\{\nabla_\theta J_N(\theta)\} = \nabla_\theta \mathbb{E}\{J_N(\theta)\} = \nabla_\theta J_{\bar{\theta}}(\theta)$, with $J_{\bar{\theta}}(\theta)$ given by (4.21), which is indeed differentiable with respect to θ and satisfies $\nabla_\theta J_{\bar{\theta}}(\bar{\theta}) = \mathbf{0}$ since $J_{\bar{\theta}}(\theta)$ is minimum at $\bar{\theta} \in \mathrm{int}(\Theta)$. ∎

The choice $\varrho_x(y,\theta) = -\log \varphi_{x,\theta}(y)$ corresponds to ML estimation, with the following asymptotic properties. Notice that the condition (4.20) implies that the support of $\varphi_{x,\theta}(\cdot)$ does not depend on θ.

Theorem 4.10 (Asymptotic normality of ML estimators–II). *Assume that the conditions of Theorems. 4.8 and 4.9 are satisfied with $\varrho_x(y,\theta) = -\log \varphi_{x,\theta}(y)$ and that for all $i,j = 1,\ldots,p$,*

$$\int_{\mathscr{X}} \left[\int_{-\infty}^{\infty} \max_{\theta\in\Theta} \varphi_{x,\theta}(y)\, \mu_x(\mathrm{d}y) \right] \xi(\mathrm{d}x) < \infty\,, \tag{4.24}$$

$$\int_{\mathscr{X}} \left[\int_{-\infty}^{\infty} \max_{\theta\in\Theta} \left| \frac{\partial \varphi_{x,\theta}(y)}{\partial\theta_i} \right| \mu_x(\mathrm{d}y) \right] \xi(\mathrm{d}x) < \infty\,, \tag{4.25}$$

$$\int_{\mathscr{X}} \left[\int_{-\infty}^{\infty} \max_{\theta\in\Theta} \left| \frac{\partial^2 \varphi_{x,\theta}(y)}{\partial\theta_i \partial\theta_j} \right| \mu_x(\mathrm{d}y) \right] \xi(\mathrm{d}x) < \infty\,. \tag{4.26}$$

Then the ML estimator $\hat\theta^N_{ML}$ that minimizes (4.12), with independent observations y_k having the p.d.f. $\varphi_{x_k,\bar\theta}(\cdot)$, satisfies

$$\sqrt{N}(\hat\theta^N_{ML} - \bar\theta) \overset{\mathrm{d}}{\to} \mathbf{z} \sim \mathscr{N}(\mathbf{0}, \mathbf{M}^{-1}(\xi,\bar\theta))\,,\ N \to \infty\,, \tag{4.27}$$

with

$$\mathbf{M}(\xi,\theta) = \int_{\mathscr{X}} \left[\int_{-\infty}^{\infty} \frac{\partial \log \varphi_{x,\theta}(y)}{\partial\theta} \frac{\partial \log \varphi_{x,\theta}(y)}{\partial\theta^\top} \varphi_{x,\theta}(y)\, \mu_x(\mathrm{d}y) \right] \xi(\mathrm{d}x) \tag{4.28}$$

the average Fisher information matrix per sample for the parameters θ. Moreover, for any estimator $\hat\theta^N$ that minimizes (4.19), with $\varrho_x(\cdot,\cdot)$ satisfying the conditions of Theorem 4.8 and 4.9 and such that for all $i,j = 1,\ldots,p$,

$$\int_{\mathscr{X}} \left[\int_{-\infty}^{\infty} \max_{\theta\in\Theta} \left| \frac{\partial \varrho_x(y,\theta)}{\partial\theta_i} \right| \varphi_{x,\theta}(y)\, \mu_x(\mathrm{d}y) \right] \xi(\mathrm{d}x) < \infty\,, \tag{4.29}$$

$$\int_{\mathscr{X}} \left[\int_{-\infty}^{\infty} \max_{\theta\in\Theta} \left| \frac{\partial^2 \varrho_x(y,\theta)}{\partial\theta_i \partial\theta_j} \right| \varphi_{x,\theta}(y)\, \mu_x(\mathrm{d}y) \right] \xi(\mathrm{d}x) < \infty\,, \tag{4.30}$$

$$\int_{\mathscr{X}} \left[\int_{-\infty}^{\infty} \max_{\theta\in\Theta} \left| \frac{\partial \varrho_x(y,\theta)}{\partial\theta_i} \frac{\partial \varphi_{x,\theta}(y)}{\partial\theta_j} \right| \mu_x(\mathrm{d}y) \right] \xi(\mathrm{d}x) < \infty\,, \tag{4.31}$$

the asymptotic covariance matrix $\mathbf{C}(\xi,\bar\theta)$ of $\sqrt{N}(\hat\theta^N - \bar\theta)$ satisfies

$$\mathbf{C}(\xi,\bar\theta) \succeq \mathbf{M}^{-1}(\xi,\bar\theta)\,. \tag{4.32}$$

Proof. To prove (4.27) we only need to compute $\mathbf{C}(\xi,\bar\theta)$ as defined in Theorem 4.9. Direct calculation for $\varrho_x(y,\theta) = -\log \varphi_{x,\theta}(y)$ gives $\mathbf{M}_2(\xi,\theta) = \mathbf{M}(\xi,\theta)$ and

$$\mathbf{M}_1(\xi,\bar\theta) = -\int_{\mathscr{X}} \left[\int_{-\infty}^{\infty} \left. \frac{\partial^2 \varphi_{x,\theta}(y)}{\partial\theta\partial\theta^\top} \right|_{\bar\theta} \mu_x(\mathrm{d}y) \right] \xi(\mathrm{d}x) + \mathbf{M}(\xi,\bar\theta)\,.$$

We prove now $\mathbf{C}(\xi,\bar\theta) \succeq \mathbf{M}^{-1}(\xi,\bar\theta)$ for any arbitrary $\varrho_x(\cdot,\cdot)$. The conditions (4.24)–(4.26) imply that the first term on the right-hand side of the equation above is zero, so that $\mathbf{M}_1(\xi,\bar{\theta}) = \mathbf{M}(\xi,\bar{\theta})$ and $\mathbf{C}(\xi,\bar{\theta}) = \mathbf{M}^{-1}(\xi,\bar{\theta})$ for $\varrho_x(y,\theta) = -\log \varphi_{x,\theta}(y)$.

The conditions (4.20), (4.22) give

$$\int_{\mathscr{X}} \left[\int_{-\infty}^{\infty} \left. \frac{\partial \varrho_x(y,\theta)}{\partial \theta} \right|_{\bar{\theta}} \varphi_{x,\bar{\theta}}(y)\, \mu_x(\mathrm{d}y) \right] \xi(\mathrm{d}x) = \nabla_\theta J_{\bar{\theta}}(\bar{\theta}) = \mathbf{0}\,,$$

where $J_{\bar{\theta}}(\theta)$ is given by (4.21). Using (4.29)–(4.31), differentiation with respect to $\bar{\theta}$ gives

$$\int_{\mathscr{X}} \left[\int_{-\infty}^{\infty} \left. \frac{\partial^2 \varrho_x(y,\theta)}{\partial \theta \partial \theta^\top} \right|_{\bar{\theta}} \varphi_{x,\bar{\theta}}(y)\, \mu_x(\mathrm{d}y) \right] \xi(\mathrm{d}x) = \\ - \int_{\mathscr{X}} \left[\int_{-\infty}^{\infty} \left. \frac{\partial \varrho_x(y,\theta)}{\partial \theta} \right|_{\bar{\theta}} \left. \frac{\partial \varphi_{x,\theta}(y)}{\partial \theta^\top} \right|_{\bar{\theta}} \mu_x(\mathrm{d}y) \right] \xi(\mathrm{d}x)$$

and thus the matrix $\mathbf{M}_1(\xi,\bar{\theta})$ defined by (4.23) is also given by

$$\mathbf{M}_1(\xi,\bar{\theta}) = - \int_{\mathscr{X}} \left[\int_{-\infty}^{\infty} \left. \frac{\partial \varrho_x(y,\theta)}{\partial \theta} \right|_{\bar{\theta}} \left. \frac{\partial \log \varphi_{x,\theta}(y)}{\partial \theta^\top} \right|_{\bar{\theta}} \varphi_{x,\bar{\theta}}(y)\, \mu_x(\mathrm{d}y) \right] \xi(\mathrm{d}x)\,.$$

Define

$$\mathbf{v} = \begin{pmatrix} \left. \frac{\partial \varrho_x(y,\theta)}{\partial \theta} \right|_{\bar{\theta}} \\ \left. \frac{\partial \log \varphi_{x,\theta}(y)}{\partial \theta^\top} \right|_{\bar{\theta}} \end{pmatrix},$$

we have

$$\mathbb{E}\{\mathbf{v}\mathbf{v}^\top\} = \begin{pmatrix} \mathbf{M}_2(\xi,\bar{\theta}) & \mathbf{M}_1(\xi,\bar{\theta}) \\ \mathbf{M}_1(\xi,\bar{\theta}) & \mathbf{M}(\xi,\bar{\theta}) \end{pmatrix},$$

and $\mathbf{C}(\xi,\bar{\theta}) \succeq \mathbf{M}^{-1}(\xi,\bar{\theta})$ from Lemma 3.7. ■

The properties (4.17) and (4.32) of Theorems. 4.7 and 4.10 correspond to Cramér–Rao-type inequalities, which, together with the related notion of efficiency, will form the subject of Sect. 4.4.

4.3 Generalized Linear Models and Exponential Families

Models from exponential families have received much attention due in particular to their interest for practical applications; see, e.g., Barndorff-Nielsen (1978), Jørgensen (1997), and McCullagh and Nelder (1989). They correspond to the situation where the observations $y_k = y(x_k)$ are independent and the variable $y(x)$, the observation at $x \in \mathscr{X}$, is distributed according to an exponential family. This means that for any $x \in \mathscr{X}$, there is a measurable set $\mathcal{Y}_x \subset \mathbb{R}$ with a σ-finite measure μ_x on $\mathcal{Y}_x$, and the p.d.f. $\varphi_{x,\bar{\theta}}(\cdot)$ of $y(x)$ (with respect to μ_x) is of the exponential form. We consider first the case where the sufficient statistic is one dimensional and then the multidimensional case in Sect. 4.3.2.

4.3.1 Models with a One-Dimensional Sufficient Statistic

In this case, the p.d.f. $\varphi_{x,\theta}(\cdot)$ has the form

$$\varphi_{x,\theta}(y) = \exp\{-\psi_x(y) + y\gamma(x,\theta) - \zeta_x[\gamma(x,\theta)]\}\,, \tag{4.33}$$

where $\psi_x(y)$ is a measurable function on $\mathcal{Y}_x$ and $\gamma(x,\theta)$ is continuous on Θ and twice continuously differentiable with respect to $\theta \in \mathrm{int}(\Theta)$. When $\gamma(x,\theta)$ is linear in θ, $\gamma(x,\theta) = \mathbf{f}^\top(x)\theta$, the model is called a *generalized linear model.*

Example 4.11. In the so-called Bernoulli experiments, $y(x) \in \mathcal{Y}_x = \{0,1\}$, and the success probability $\mathrm{Prob}\{y(x) = 1\}$ of a trial at x is given by $\mathrm{Prob}\{y(x) = 1\} = \pi(x,\bar{\theta})$, with $\pi(x,\theta)$ a known function of the design point x and parameter vector θ. With $\mu_x\{0\} = \mu_x\{1\} = 1$ (this choice is somewhat arbitrary[3]), we obtain $\varphi_{x,\theta}(y) = [\pi(x,\theta)]^y[1-\pi(x,\theta)]^{1-y}$.

If n_x independent trials are made at x, the number of successes is a sufficient statistic. Let now $y(x)$ denote the observed value of this number, the set $\mathcal{Y}_x$ becoming $\mathcal{Y}_x = \{0,1,\ldots,n_x\}$. We take again $\mu_x\{y\} = 1$ for every $y \in \mathcal{Y}_x$, and obtain $y(x) \sim Bi(\pi(x,\bar{\theta}), n_x)$; that is,

$$\begin{aligned}\varphi_{x,\theta}(y) &= \binom{n_x}{y}[\pi(x,\theta)]^y[1-\pi(x,\theta)]^{n_x-y}\\ &= \exp\left\{\log\binom{n_x}{y} + y\log\left[\frac{\pi(x,\theta)}{1-\pi(x,\theta)}\right] + n_x\log[1-\pi(x,\theta)]\right\},\end{aligned}$$

which is of the form (4.33), with $\gamma(x,\theta) = \log\{\pi(x,\theta)/[1-\pi(x,\theta)]\}$, or

$$\pi(x,\theta) = \frac{e^{\gamma(x,\theta)}}{1+e^{\gamma(x,\theta)}}\,,$$

and $\zeta_x(\gamma) = n_x\log(1+e^\gamma)$.

The mean $\mathbb{E}_{x,\theta}\{y(x)\} = n_x\pi(x,\theta)$ is bounded by n_x, which is not appropriate for a linear regression. On the other hand, $\gamma(x,\theta) \in (-\infty,\infty)$, so that one often takes $\gamma(x,\theta) = \mathbf{f}^\top(x)\theta$ which gives

$$\eta(x,\theta) = \mathbb{E}_{x,\theta}\{y(x)\} = n_x\frac{e^{\mathbf{f}^\top(x)\theta}}{1+e^{\mathbf{f}^\top(x)\theta}}\,,$$

the well-known logistic regression model. □

Coming back to the general density (4.33), from

$$\int_{\mathcal{Y}_x}\varphi_{x,\theta}(y)\mu_x(\mathrm{d}y) = 1 \tag{4.34}$$

we obtain, for $\gamma = \gamma(x,\theta)$ with x,θ fixed,

[3] μ_x and $\varphi_{x,\theta}(\cdot)$ only need to satisfy the condition (4.34).

$$\zeta_x(\gamma) = \log \int_{\mathscr{Y}_x} \exp\{y\gamma - \psi_x(y)\}\mu_x(\mathrm{d}y)\,.$$

Define $\Gamma = \{t \in \mathbb{R} : \int_{\mathscr{Y}_x} \exp\{yt - \psi_x(y)\}\mu_x(\mathrm{d}y) < \infty\}$. Supposing that $\gamma \in \mathrm{int}(\Gamma)$, we can differentiate $\zeta_x(\gamma)$ and

$$\frac{\mathrm{d}\zeta_x(\gamma)}{\mathrm{d}\gamma} = \frac{\int_{\mathscr{Y}_x} y \exp\{y\gamma - \psi_x(y)\}\mu_x(\mathrm{d}y)}{\int_{\mathscr{Y}_x} \exp\{y\gamma - \psi_x(y)\}\mu_x(\mathrm{d}y)} = \int_{\mathscr{Y}_x} y\varphi_{x,\theta}(y)\mu_x(\mathrm{d}y) = \mathbb{E}_{x,\theta}\{y\}\,.$$

Similarly, we obtain $d^2\zeta_x(\gamma)/d\gamma^2 = \mathrm{var}_{x,\theta}\{y\}$. Denote

$$\eta(x,\theta) = \mathbb{E}_{x,\theta}\{y\} = \left.\frac{\mathrm{d}\zeta_x(\gamma)}{\mathrm{d}\gamma}\right|_{\gamma(x,\theta)}, \tag{4.35}$$

$$\lambda(x,\theta) = \mathrm{var}_{x,\theta}\{y\} = \left.\frac{\mathrm{d}^2\zeta_x(\gamma)}{\mathrm{d}\gamma^2}\right|_{\gamma(x,\theta)},$$

which gives

$$\frac{\partial\eta(x,\theta)}{\partial\theta} = \left.\frac{\mathrm{d}^2\zeta_x(\gamma)}{\mathrm{d}\gamma^2}\right|_{\gamma(x,\theta)} \frac{\partial\gamma(x,\theta)}{\partial\theta}\,;$$

that is,

$$\frac{\partial\gamma(x,\theta)}{\partial\theta} = \lambda^{-1}(x,\theta)\,\frac{\partial\eta(x,\theta)}{\partial\theta}\,.$$

From (4.33), (4.35) we obtain

$$\frac{\partial \log\varphi_{x,\theta}(y)}{\partial\theta} = [y - \eta(x,\theta)]\frac{\partial\gamma(x,\theta)}{\partial\theta}\,, \tag{4.36}$$

$$\frac{\partial^2 \log\varphi_{x,\theta}(y)}{\partial\theta\partial\theta^\top} = -\frac{\partial\gamma(x,\theta)}{\partial\theta}\frac{\partial\eta(x,\theta)}{\partial\theta^\top} + [y - \eta(x,\theta)]\frac{\partial^2\gamma(x,\theta)}{\partial\theta\partial\theta^\top}\,. \tag{4.37}$$

Hence the Fisher information matrix at x is

$$\mathbb{E}_{x,\theta}\left\{\frac{\partial \log\varphi_{x,\theta}(y)}{\partial\theta}\frac{\partial \log\varphi_{x,\theta}(y)}{\partial\theta^\top}\right\} = \lambda(x,\theta)\,\frac{\partial\gamma(x,\theta)}{\partial\theta}\frac{\partial\gamma(x,\theta)}{\partial\theta^\top}\,,$$
$$= \lambda^{-1}(x,\theta)\,\frac{\partial\eta(x,\theta)}{\partial\theta}\frac{\partial\eta(x,\theta)}{\partial\theta^\top}$$

and the average Fisher information matrix per sample (4.28) for a design ξ is equal to

$$\mathbf{M}(\xi,\theta) = \int_{\mathscr{X}} \lambda^{-1}(x,\theta)\,\frac{\partial\eta(x,\theta)}{\partial\theta}\frac{\partial\eta(x,\theta)}{\partial\theta^\top}\,\xi(\mathrm{d}x)\,. \tag{4.38}$$

In particular, the validity of the assumptions of Theorem 4.10 can be checked by using (4.36), (4.37). The matrix $\mathbf{M}(\xi,\theta)$ coincides with the matrix obtained in Sect. 3.3.3 for the TSLS estimator in a regression model with parameterized variance; see (3.56). See also Sect. 3.3.7. Note, however, that the situation concerned here does not correspond to nonlinear regression, which will be considered in Sect. 4.3.2.

This connection with two-stage LS (or iteratively reweighted LS; see Sect. 3.3.4) is even stronger, as revealed by considering the normal equation of the ML estimator $\hat{\theta}^N_{ML}$ that minimizes (4.19). We have

$$\hat{\theta}^N_{ML} = \arg\min_{\theta\in\Theta} \sum_{k=1}^{N} \{\zeta_{x_k}[\gamma(x_k,\theta)] - y(k)\gamma(x_k,\theta)\}$$

and the normal equation for $\hat{\theta}^N_{ML}$ is

$$\begin{aligned}\mathbf{0} &= \sum_{k=1}^{N} \left[\left.\frac{\mathrm{d}\zeta_{x_k}(\gamma)}{\mathrm{d}\gamma}\right|_{\gamma(x_k,\hat{\theta}^N_{ML})} - y(x_k) \right] \left.\frac{\partial\gamma(x_k,\theta)}{\partial\theta}\right|_{\hat{\theta}^N_{ML}} \\ &= \sum_{k=1}^{N} \lambda^{-1}(x_k,\hat{\theta}^N_{ML})[\eta(x_k,\hat{\theta}^N_{ML}) - y(x_k)] \left.\frac{\partial\eta(x_k,\theta)}{\partial\theta}\right|_{\hat{\theta}^N_{ML}} .\end{aligned}$$

Exactly the same expression is obtained if in the normal equation for the WLS estimator $\hat{\theta}^N_{WLS}$ that minimizes (3.5) the weights $w(x)$ are replaced by $\lambda^{-1}(x,\hat{\theta}^N_{WLS})$. This forms the basis for the iteratively reweighted LS method of Sect. 3.3.4; see (3.58).

4.3.2 Models with a Multidimensional Sufficient Statistic

For a model from the general exponential family, the p.d.f. $\varphi_{x,\bar{\theta}}(\cdot)$ has the form

$$\varphi_{x,\theta}(y) = \exp\{-\psi_x(y) + \mathbf{t}^\top(y)\gamma(x,\theta) - \zeta_x[\gamma(x,\theta)]\} . \qquad (4.39)$$

We obtain as in Sect. 4.3.1

$$\begin{aligned}\left.\frac{\mathrm{d}\zeta_x(\gamma)}{\mathrm{d}\gamma}\right|_{\gamma(x,\theta)} &= \mathbb{E}_{x,\theta}\{\mathbf{t}(y)\} = \nu(x,\theta) , \\ \left.\frac{\mathrm{d}^2\zeta_x(\gamma)}{\mathrm{d}\gamma^2}\right|_{\gamma(x,\theta)} &= \mathrm{Var}_{x,\theta}\{\mathbf{t}(y)\} = \mathbf{V}(x,\theta) ,\end{aligned}$$

and

$$\mathbf{M}(\xi,\theta) = \int_{\mathscr{X}} \frac{\partial\nu^\top(x,\theta)}{\partial\theta} \mathbf{V}^{-1}(x,\theta) \frac{\partial\nu(x,\theta)}{\partial\theta^\top}\, \xi(\mathrm{d}x) ,$$

for the average Fisher information matrix per sample for the parameters θ. The normal equation for the ML estimator $\hat{\theta}^N_{ML}$ is then

$$\mathbf{0} = \sum_{k=1}^{N} \{\nu(x_k,\hat{\theta}^N_{ML}) - \mathbf{t}[y(x_k)]\}^\top \mathbf{V}^{-1}(x_k,\hat{\theta}^N_{ML}) \left.\frac{\partial\nu(x_k,\theta)}{\partial\theta}\right|_{\hat{\theta}^N_{ML}} .$$

The regression model with normal errors is a typical example.

Example 4.12. Consider a regression model with parameterized variance and normal errors, $y(x_k) \sim \mathscr{N}(\eta(x_k, \bar{\theta}), \lambda(x_k, \bar{\theta}))$; that is,

$$\varphi_{x,\theta}(y) = \exp\left\{ -\frac{1}{2}\log(2\pi) + y\frac{\eta(x,\theta)}{\lambda(x,\theta)} - y^2 \frac{1}{2\lambda(x,\theta)} - \left[\frac{1}{2}\log\lambda(x,\theta) + \frac{\eta^2(x,\theta)}{2\lambda(x,\theta)}\right]\right\}. \tag{4.40}$$

Notice that the ML criterion (4.12) then coincides with the penalized WLS criterion (3.47). If we compare (4.40) to a model from the general exponential family (4.39), we obtain

$$\mathbf{t}(y) = \begin{pmatrix} y \\ y^2 \end{pmatrix}, \quad \gamma(x,\theta) = \lambda^{-1}(x,\theta)\begin{pmatrix} \eta(x,\theta) \\ -1/2 \end{pmatrix}$$

and

$$\zeta(\gamma) = \frac{1}{2}\log(-2\{\gamma\}_2) - \frac{\{\gamma\}_1^2}{4\{\gamma\}_2}.$$

We also obtain directly

$$\mathbb{E}_{x,\theta}\begin{pmatrix} y \\ y^2 \end{pmatrix} = \begin{pmatrix} \eta(x,\theta) \\ \lambda(x,\theta) + \eta^2(x,\theta) \end{pmatrix},$$

$$\mathrm{Var}_{x,\theta}\begin{pmatrix} y \\ y^2 \end{pmatrix} = \begin{pmatrix} \lambda(x,\theta) & 2\eta(x,\theta)\lambda(x,\theta) \\ 2\eta(x,\theta)\lambda(x,\theta) & 2\lambda^2(x,\theta) + 4\eta^2(x,\theta)\lambda(x,\theta) \end{pmatrix}.$$

This gives the normal equation

$$\sum_{k=1}^{N}\left(\frac{y(x_k) - \eta(x_k, \hat{\theta}_{ML}^N)}{\lambda(x_k, \hat{\theta}_{ML}^N)} \left.\frac{\partial \eta(x_k,\theta)}{\partial\theta}\right|_{\hat{\theta}_{ML}^N} + \frac{[y(x_k) - \eta(x_k, \hat{\theta}_{ML}^N)]^2 - \lambda(x_k, \hat{\theta}_{ML}^N)}{2\lambda^2(x_k, \hat{\theta}_{ML}^N)} \left.\frac{\partial \lambda(x_k,\theta)}{\partial\theta}\right|_{\hat{\theta}_{ML}^N} \right) = \mathbf{0}.$$

It exactly coincides with the normal equation of the WLS problem defined by the criterion

$$\frac{1}{N}\sum_{k=1}^{N} \frac{[y(x_k) - \eta(x_k,\theta)]^2}{\lambda(x_k, \hat{\theta}_{ML}^N)} + \frac{1}{2N}\sum_{k=1}^{N} \frac{\left\{[y(x_k) - \eta(x_k, \hat{\theta}_{ML}^N)]^2 - \lambda(x_k,\theta)\right\}^2}{\lambda^2(x_k, \hat{\theta}_{ML}^N)},$$

hence, the iteratively reweighted LS procedure

$$\hat{\theta}_k^N = \arg\min_{\theta\in\Theta} \frac{1}{N}\sum_{k=1}^{N} \frac{[y(x_k) - \eta(x_k,\theta)]^2}{\lambda(x_k, \hat{\theta}_{k-1}^N)} + \frac{1}{2N}\sum_{k=1}^{N} \frac{\left\{[y(x_k) - \eta(x_k, \hat{\theta}_{k-1}^N)]^2 - \lambda(x_k,\theta)\right\}^2}{\lambda^2(x_k, \hat{\theta}_{k-1}^N)}, \quad k = 2, 3, \ldots$$

suggested in (Downing et al., 2001). See also Green (1984) and del Pino (1989) for more general procedures. □

4.4 The Cramér–Rao Inequality: Efficiency of Estimators

4.4.1 Efficiency

Consider N observations $\mathbf{y} = [y(x_1), \ldots, y(x_N)]^\top$ obtained for the design $X = (x_1, \ldots, x_N)$. The likelihood of θ at $\mathbf{y}$ can be written as $\mathrm{L}_{X,\mathbf{y}}(\theta) = C(\mathbf{y})\, \varphi_{X,\theta}(\mathbf{y})$, with $C(\cdot)$ an arbitrary positive measurable function of $\mathbf{y}$ and $\varphi_{X,\theta}(\cdot)$ the density (Radon–Nikodým derivative) of $\mathbf{y}$ supposed to be generated by a model with parameters θ, with respect to a σ-finite measure μ_X on $\mathcal{Y}_X = \mathcal{Y}_{x_1} \times \cdots \times \mathcal{Y}_{x_N}$. Note that

$$\frac{\partial \log \mathrm{L}_{X,\mathbf{y}}(\theta)}{\partial \theta} = \frac{\partial \log \varphi_{X,\theta}(\mathbf{y})}{\partial \theta} \tag{4.41}$$

when the derivatives exist. Under suitable regularity conditions, any unbiased estimator $\hat{\theta}^N = \hat{\theta}^N(\mathbf{y})$ based on the N observations $\mathbf{y}$ satisfies the Cramér–Rao inequality, which can be formulated as follows.

Theorem 4.13. *Let $\hat{\theta}^N$ be an unbiased estimator of θ (i.e., such that $\mathbb{E}_{X,\theta}\{\hat{\theta}^N\} = \theta$), based on N observations $\mathbf{y}$, satisfying $\mathbb{E}_{X,\theta}[\|\hat{\theta}^N - \theta)\|^2] < \infty$. Assume that the support of $\varphi_{X,\theta}(\cdot)$, $\{\mathbf{y} \in \mathbb{R}^N : \varphi_{X,\theta}(\mathbf{y}) > 0\}$, does not depend on θ and that $\varphi_{X,\theta}(\mathbf{y})$, considered as a function of θ, is continuously differentiable in θ and satisfies $\mathbb{E}_{X,\theta}\left[\|\partial\varphi_{X,\theta}(\mathbf{y})/\partial\theta\|^2\right] < \infty$ and, for all $i, j = 1, \ldots, p$, for all $\theta \in \Theta$,*

$$\int_{\mathcal{Y}_X} \left|\{\hat{\theta}^N(\mathbf{y})\}_i\right| \varphi_{X,\theta}(\mathbf{y})\, \mu_X(\mathrm{d}\mathbf{y}) < \infty\,, \tag{4.42}$$

$$\int_{\mathcal{Y}_X} \left|\frac{\partial \varphi_{X,\theta}(\mathbf{y})}{\partial \theta_i}\right| \mu_X(\mathrm{d}\mathbf{y}) < \infty\,, \tag{4.43}$$

$$\int_{\mathcal{Y}_X} \left|\{\hat{\theta}^N(\mathbf{y})\}_i \frac{\partial \varphi_{X,\theta}(\mathbf{y})}{\partial \theta_j}\right| \mu_X(\mathrm{d}\mathbf{y}) < \infty\,. \tag{4.44}$$

The variance–covariance matrix $\mathrm{Var}_{X,\theta}\{\hat{\theta}^N\}$ of $\hat{\theta}^N$ then satisfies

$$\mathrm{Var}_{X,\theta}\{\hat{\theta}^N\} = \mathbb{E}_{X,\theta}\{(\hat{\theta}^N - \theta)(\hat{\theta}^N - \theta)^\top\} \succeq \frac{\mathbf{M}^+(X,\theta)}{N} \tag{4.45}$$

where

$$\mathbf{M}(X,\theta) = \frac{1}{N}\, \mathbb{E}_{X,\theta}\left\{\frac{\partial \log \mathrm{L}_{X,\mathbf{y}}(\theta)}{\partial \theta}\, \frac{\partial \log \mathrm{L}_{X,\mathbf{y}}(\theta)}{\partial \theta^\top}\right\} \tag{4.46}$$

is the average Fisher information matrix per sample for the parameters θ and design X and $\mathbf{M}^+$ denotes the Moore–Penrose g-inverse of $\mathbf{M}$.

Proof. We apply Lemma 3.7 with $\mathbf{u} = \hat{\theta}^N - \theta$ and $\mathbf{v} = \partial \log \varphi_{X,\theta}(\mathbf{y})/\partial\theta$. Under (4.42)–(4.44), we can differentiate under the integral sign, which gives

$$\mathbb{E}_{X,\theta}\{\mathbf{u}\mathbf{v}^\top\} = \int_{\mathscr{Y}_X} \mathbf{u}(\mathbf{y})\mathbf{v}^\top(\mathbf{y})\,\varphi_{X,\theta}(\mathbf{y})\,\mu_X(\mathrm{d}\mathbf{y})$$
$$= \int_{\mathscr{Y}_X} [\hat{\theta}^N(\mathbf{y}) - \theta]\,\frac{\partial \varphi_{X,\theta}(\mathbf{y})}{\partial \theta^\top}\mu_X(\mathrm{d}\mathbf{y})$$
$$= \frac{\partial \mathbb{E}_{X,\theta}\{\hat{\theta}^N\}}{\partial \theta^\top} - \theta\,\frac{\partial \mathbb{E}_{X,\theta}\{1\}}{\partial \theta^\top} = \mathbf{I}_p\,,$$

the p-dimensional identity matrix. Lemma 3.7 and (4.41) give (4.45), (4.46). ■

Remark 4.14.

(*i*) Within the setup of Sect. 4.2.2 the conditions (4.24), (4.25) used for the asymptotic normality of the ML estimator imply (4.42)–(4.44).

(*ii*) An estimator that achieves equality in (4.45) is said to be *efficient*. From Lemma 3.7, an efficient unbiased estimator of θ, if it exists (which is seldom the case in nonlinear situations), satisfies

$$\frac{\partial \log \mathrm{L}_{X,\mathbf{y}}(\theta)}{\partial \theta} = \mathbf{C}(X,\theta)(\hat{\theta}^N - \theta)$$

for some nonrandom matrix $\mathbf{C}(X,\theta)$; it therefore coincides with the ML estimator.

(*iii*) Relaxing the constrained that $\hat{\theta}^N$ is unbiased in Theorem 4.13, one can use the same approach as in the proof above, with now $\mathbf{u} = \hat{\theta}^N - \mathbb{E}_{X,\theta}\{\hat{\theta}^N\}$. We then obtain

$$\mathrm{Var}_{X,\theta}\{\hat{\theta}^N\} \succeq \frac{\partial \mathbb{E}_{X,\theta}\{\hat{\theta}^N\}}{\partial \theta^\top}\,\frac{\mathbf{M}^+(X,\theta)}{N}\,\frac{\partial [\mathbb{E}_{X,\theta}\{\hat{\theta}^N\}]^\top}{\partial \theta}\,.$$

From Lemma 3.7, the equality is obtained if and only if

$$\frac{\partial \log \mathrm{L}_{X,\mathbf{y}}(\theta)}{\partial \theta} = \mathbf{C}(X,\theta)(\hat{\theta}^N - \mathbb{E}_{X,\theta}\{\hat{\theta}^N\})$$

for some matrix $\mathbf{C}(X,\theta)$, which means that the model is from the exponential family; see Lehmann and Casella (1998, p. 128).

(*iv*) Under suitable regularity conditions (ensuring the existence of second-order derivatives $\partial^2 \mathrm{L}_{X,\mathbf{y}}(\theta)/\partial\theta_i\partial\theta_j$, $i,j = 1,\ldots,p$, and allowing differentiation under the integral sign), $\mathbf{M}(X,\theta)$ given by (4.46) can also be written as

$$\mathbf{M}(X,\theta) = -\frac{1}{N}\,\mathbb{E}_{X,\theta}\left\{\frac{\partial^2 \log \mathrm{L}_{X,\mathbf{y}}(\theta)}{\partial\theta\partial\theta^\top}\right\}. \qquad \square$$

4.4.2 Asymptotic Efficiency

Consider the situation of Sect. 4.2.2, where the observations $y_k = y(x_k)$ are independent random variables, $y(x)$ having the density (or Radon–Nikodým derivative) $\varphi_{x,\bar{\theta}}(\cdot)$ with respect to a σ-finite measure μ_x on the set $\mathscr{Y}_x \subset \mathbb{R}$. Suppose that the conditions of Theorem 4.10 are satisfied. Then, the matrix $\mathbf{M}(X,\theta)$ given by (4.46) can also be written as

$$\begin{aligned}
\mathbf{M}(X,\theta) = \frac{1}{N} \int_{\mathscr{Y}_X} & \left(\sum_{j=1}^N \frac{\partial \log \varphi_{x_j,\theta}(y_j)}{\partial \theta} \sum_{k=1}^N \frac{\partial \log \varphi_{x_k,\theta}(y_k)}{\partial \theta^\top} \right) \\
& \times \prod_{l=1}^N [\varphi_{x_l,\theta}(y_l) \mu_{x_l}(\mathrm{d}y_l)] \\
= \frac{1}{N} \sum_{k=1}^N & \int_{\mathscr{Y}_{x_k}} \frac{\partial \log \varphi_{x_k,\theta}(t)}{\partial \theta} \frac{\partial \log \varphi_{x_k,\theta}(t)}{\partial \theta^\top} \varphi_{x_k,\theta}(t)\, \mu_{x_k}(\mathrm{d}t) \\
& + \frac{1}{N} \sum_{j,k=1,\, j\neq k}^N \left(\int_{\mathscr{Y}_{x_j}} \frac{\partial \log \varphi_{x_j,\theta}(t)}{\partial \theta} \varphi_{x_j,\theta}(t)\, \mu_{x_j}(\mathrm{d}t) \right) \\
& \quad \times \left(\int_{\mathscr{Y}_{x_k}} \frac{\partial \log \varphi_{x_k,\theta}(t)}{\partial \theta^\top} \varphi_{x_k,\theta}(y_l) \mu_{x_k}(\mathrm{d}t) \right).
\end{aligned}$$

Now, for any $k = 1, \ldots, N$,

$$\int_{\mathscr{Y}_{x_k}} \frac{\partial \log \varphi_{x_k,\theta}(t)}{\partial \theta} \varphi_{x_k,\theta}(t) \mu_{x_k}(\mathrm{d}t) = \frac{\partial \left[\int_{\mathscr{Y}_{x_k}} \varphi_{x_k,\theta}(t) \mu_{x_k}(\mathrm{d}t) \right]}{\partial \theta} = \frac{\partial 1}{\partial \theta} = \mathbf{0}$$

and therefore $\mathbf{M}(X,\theta) = \mathbf{M}(\xi_N,\theta)$ given by (4.28) for the discrete design measure ξ_N that allocates the mass $1/N$ to each x_k, $k = 1, \ldots, N$. From Theorem 4.10 the ML estimator $\hat{\theta}^N_{ML}$ is asymptotically unbiased and asymptotically normally distributed,

$$\sqrt{N}(\hat{\theta}^N_{ML} - \bar{\theta}) \xrightarrow{\mathrm{d}} \mathbf{z} \sim \mathscr{N}(\mathbf{0}, \mathbf{M}^{-1}(\xi,\bar{\theta}))\,, \tag{4.47}$$

see (4.27). When ξ_N converges to ξ, $\mathbf{M}^{-1}(X,\theta)$ converges to $\mathbf{M}^{-1}(\xi,\theta)$, and $\hat{\theta}^N_{ML}$ is *asymptotically efficient*, in the sense that it asymptotically attains the bound of Theorem 4.13; see Lehmann and Casella (1998, p. 439).

There are alternatives to ML estimation that achieve asymptotic efficiency under milder conditions than the ML estimator. For instance, there exist adaptive estimators that are asymptotically efficient in the regression model when the distribution of errors is only known to be symmetric and thus forms an infinite-dimensional nuisance component in a semi-parametric model; see

Bickel (1982) and Manski (1984).[4] Also, some estimators may be easier to compute than $\hat{\theta}^N_{ML}$ and still be asymptotically efficient. This is the case for the following one-step estimator, see, e.g., Bierens (1994, p. 81) and van der Vaart (1998, p. 71). We only consider the general situation of Sect. 4.2.2, which can easily be particularized to the case of regression models.

Theorem 4.15 (Asymptotic efficiency of one-step estimators). *Suppose that Θ is convex, that the conditions of Theorem 4.10 are satisfied, and that $\hat{\theta}^N_1$ is a $\sqrt{N}$-consistent[5] estimator of θ for independent observations y_k having the p.d.f. $\varphi_{x_k,\bar{\theta}}(\cdot)$. Consider the one-step estimator $\hat{\theta}^N_2$ obtained by performing one single Newton step for the minimization of the ML criterion $J_N(\theta)$ given by (4.12), starting from $\hat{\theta}^N_1$,*

$$\hat{\theta}^N_2 = \hat{\theta}^N_1 - \left[\nabla^2_\theta J_N(\hat{\theta}^N_1)\right]^{-1} \nabla_\theta J_N(\hat{\theta}^N_1) \,. \tag{4.48}$$

It is asymptotically efficient,

$$\sqrt{N}(\hat{\theta}^N_2 - \bar{\theta}) \xrightarrow{\text{d}} \mathbf{z} \sim \mathscr{N}(\mathbf{0}, \mathbf{M}^{-1}(\xi, \bar{\theta}))\,,\ N \to \infty\,, \tag{4.49}$$

with $\mathbf{M}(\xi, \bar{\theta})$ the average Fisher information matrix per observation; see (4.28).

Proof. Using Lemma 2.6, we can prove that $\nabla^2_\theta J_N(\theta) \overset{\theta}{\rightsquigarrow} \mathbf{M}(\xi, \theta)$ a.s., so that $\nabla^2_\theta J_N(\hat{\theta}^N_1) \xrightarrow{\text{p}} \mathbf{M}(\xi, \bar{\theta})$, which is nonsingular. Since Θ is convex, we can consider the following Taylor expansion of $\{\nabla_\theta J_N(\hat{\theta}^N_1)\}_i$ at $\bar{\theta}$ for $i = 1, \ldots, p$:

$$\{\nabla_\theta J_N(\hat{\theta}^N_1)\}_i = \{\nabla_\theta J_N(\bar{\theta})\}_i + \{\nabla^2_\theta J_N(\beta^N_i)(\hat{\theta}^N_1 - \bar{\theta})\}_i$$

for some $\beta^N_i = (1 - \alpha_{i,N})\bar{\theta} + \alpha_{i,N}\hat{\theta}^N_1$, $\alpha_{i,N} \in (0, 1)$, with β^N_i measurable; see Lemma 2.12. Using (4.48), we obtain

$$\sqrt{N}\{\nabla^2_\theta J_N(\hat{\theta}^N_1)(\hat{\theta}^N_2 - \bar{\theta})\}_i = \sqrt{N}\left\{[\nabla^2_\theta J_N(\hat{\theta}^N_1) - \nabla^2_\theta J_N(\beta^N_i)](\hat{\theta}^N_1 - \bar{\theta})\right\}_i \\ -\sqrt{N}\{\nabla_\theta J_N(\bar{\theta})\}_i \,,\ i = 1, \ldots, p \,. \tag{4.50}$$

Since $\hat{\theta}^N_1 \xrightarrow{\text{p}} \bar{\theta}$, $\beta^N_i \xrightarrow{\text{p}} \bar{\theta}$, and we have $\nabla^2_\theta J_N(\hat{\theta}^N_1) \xrightarrow{\text{p}} \mathbf{M}(\xi, \bar{\theta})$, $\nabla^2_\theta J_N(\beta^N) \xrightarrow{\text{p}} \mathbf{M}(\xi, \bar{\theta})$. Since $\sqrt{N}(\hat{\theta}^N_1 - \bar{\theta})$ is bounded in probability, the first term on the right-hand side of (4.50) tends to zero in probability. As in Theorem 4.10, the second one is asymptotically distributed $\mathscr{N}(\mathbf{0}, \mathbf{M}(\xi, \bar{\theta}))$, which gives (4.49). ■

[4]The notion of adaptive estimation originated in (Stein, 1956); one can refer to Beran (1974), Stone (1975), and Bickel et al. (1993) for the main steps in the developments.

[5]That is, $\sqrt{N}(\hat{\theta}^N_1 - \bar{\theta})$ is bounded in probability; see page 31.

4.5 The Maximum A Posteriori Estimator

Suppose that N observations $\mathbf{y} = [y(x_1), \ldots, y(x_N)]^\top$ have been obtained with the design $X = (x_1, \ldots, x_N)$ and let $\pi(\cdot)$ denote a prior p.d.f. for θ on $\mathbb{R}^p$ or a subset of it. Bayesian estimation relies on the construction of the posterior p.d.f. for θ,

$$\pi_{X,\mathbf{y}}(\theta) = \frac{\varphi_{X,\theta}(\mathbf{y})\pi(\theta)}{\varphi_X^*(\mathbf{y})}, \tag{4.51}$$

with $\varphi_X^*(\cdot)$ the p.d.f. of the marginal distribution of the observations $\mathbf{y}$ and $\varphi_{X,\theta}(\cdot)$ the p.d.f. of their conditional distribution given θ.

When the y_k are independent, $y_k = y(x_k)$ having the density $\varphi_{x_k,\bar{\theta}}(y)$ with respect to the measure μ_{x_k} on $\mathscr{Y}_{x_k} \subset \mathbb{R}$, we have $\varphi_{X,\theta}(\mathbf{y}) = \prod_{k=1}^N \varphi_{x_k,\theta}(y_k)$ and thus

$$\log \pi_{X,\mathbf{y}}(\theta) = \sum_{k=1}^N \log \varphi_{x_k,\theta}(y_k) + \log \pi(\theta) - \log \varphi_X^*(\mathbf{y}) .$$

The maximum a posteriori estimator $\hat{\theta}^N$ maximizes $\log \pi_{X,\mathbf{y}}(\theta)$ with respect to $\theta \in \Theta$. We shall investigate the behaviors of $\hat{\theta}^N$ and $\pi_{X,\mathbf{y}}(\cdot)$ as $N \to \infty$ in a classical sense, that is, assuming that the observations $\mathbf{y}$ are generated with a fixed but unknown parameter value $\bar{\theta}$.

The maximum a posteriori estimator then satisfies the following.

Theorem 4.16. *Assume that the conditions of Theorem 4.8 are satisfied for $\varrho_x(y,\theta) = -\log \varphi_{x,\theta}(y)$ and that $\pi(\cdot)$ is continuous on Θ with $\pi(\theta) > \epsilon > 0$ for any $\theta \in \Theta$. Then $\hat{\theta}^N = \arg\max_{\theta \in \Theta} \log \pi_{X,\mathbf{y}}(\theta) \overset{\text{a.s.}}{\to} \bar{\theta}$ as $N \to \infty$. If, moreover, $\pi(\cdot)$ is twice continuously differentiable on $\text{int}(\Theta)$ and the conditions of Theorem 4.10 are satisfied, then $\hat{\theta}^N$ satisfies*

$$\sqrt{N}(\hat{\theta}^N - \bar{\theta}) \overset{\text{d}}{\to} \mathbf{z} \sim \mathscr{N}(\mathbf{0}, \mathbf{M}^{-1}(\xi, \bar{\theta})) , \ N \to \infty ,$$

with $\mathbf{M}(\xi,\theta)$ given by (4.28).

Proof. Under the conditions of Theorem 4.8, with $\varrho_x(y,\theta) = -\log \varphi_{x,\theta}(y)$, we obtain

$$\frac{1}{N} [\log \pi_{X,\mathbf{y}}(\theta) + \log \varphi_X^*(\mathbf{y})] \overset{\theta}{\rightsquigarrow} J_{\bar{\theta}}(\theta) \ \text{ a.s.}, \ N \to \infty ,$$

with

$$J_{\bar{\theta}}(\theta) = -\int_{\mathscr{X}} \left\{ \int_{-\infty}^{\infty} \log[\varphi_{x,\theta}(y)] \, \varphi_{x,\bar{\theta}}(y) \mu_x(\mathrm{d}y) \right\} \xi(\mathrm{d}x) .$$

This means that the influence of the prior term $\log \pi(\theta)$ asymptotically vanishes in $(1/N) \log \pi_{X,\mathbf{y}}(\theta)$. This influence also vanishes in the first- and second-order derivatives with respect to θ, and the asymptotic normality follows from Theorem 4.10. ∎

More generally, under some regularity conditions, for large N, the posterior density of θ is approximately normal with mean tending to $\bar{\theta}$ and variance tending to $(1/N)\mathbf{M}^{-1}(\xi,\bar{\theta})$, and the asymptotic properties of a Bayes' estimator based on $\pi_{X,\mathbf{y}}(\theta)$, for instance, the posterior expectation $\int_\Theta \theta\, \pi_{X,\mathbf{y}}(\theta)\, \mathrm{d}\theta$, are the same as those of the ML estimator based on the likelihood $\mathrm{L}_{X,\mathbf{y}}(\theta)$; see, e.g., Lehmann and Casella (1998, pp. 487–496) for details. In terms of asymptotic characteristics of the precision of the estimation, there is therefore no modification with respect to the results in previous sections.

The situation is different for finite N, where a Bayesian form of the Cramér–Rao inequality (4.45) can be derived.

Theorem 4.17. *Let $\hat{\theta}^N$ be an estimator of θ based on N observations $\mathbf{y}$. Assume that $\varphi_{X,\theta}(\mathbf{y})$ and $\pi(\theta)$ are continuously differentiable in θ and that the conditions (4.42)–(4.44) are satisfied. Assume, moreover, that Θ is unbounded, that the support of $\varphi_{X,\theta}(\cdot)$, $\{\mathbf{y} \in \mathbb{R}^N : \varphi_{X,\theta}(\mathbf{y}) > 0\}$ does not depend on θ, that $\lim_{\|\theta\|\to\infty}[\mathbb{E}_{X,\theta}\{\hat{\theta}^N\} - \theta]\pi(\theta) = \mathbf{0}$, and that $\mathbb{E}[\|\hat{\theta}^N - \theta\|^2] < \infty$ and $\mathbb{E}[\|\partial \log \pi_{X,\mathbf{y}}(\theta)/\partial\theta\|^2] < \infty$ with $\mathbb{E}(\cdot) = \mathbb{E}_\pi[\mathbb{E}_{X,\theta}(\cdot)]$. The mean-squared error matrix $\mathbb{E}\{(\hat{\theta}^N - \theta)(\hat{\theta}^N - \theta)^\top\}$ then satisfies*

$$\mathbb{E}\{(\hat{\theta}^N - \theta)(\hat{\theta}^N - \theta)^\top\} \succeq \frac{[\tilde{\mathbf{M}}(X)]^+}{N} \tag{4.52}$$

where

$$\tilde{\mathbf{M}}(X) = \frac{1}{N}\, \mathbb{E}\left\{\frac{\partial \log \pi_{X,\mathbf{y}}(\theta)}{\partial\theta}\, \frac{\partial \log \pi_{X,\mathbf{y}}(\theta)}{\partial\theta^\top}\right\} \tag{4.53}$$

and $\mathbf{M}^+$ denotes the Moore–Penrose g-inverse of $\mathbf{M}$.

Proof. The proof is similar to that of Theorem 4.13. Again we apply Lemma 3.7 with now $\mathbf{u} = \hat{\theta}^N - \theta$ and $\mathbf{v} = \partial \log[\varphi_{X,\theta}(\mathbf{y})\pi(\theta)]/\partial\theta$. It gives

$$\begin{aligned}
\mathbb{E}\left\{(\hat{\theta}^N - \theta)\frac{\partial \log[\varphi_{X,\theta}(\mathbf{y})\pi(\theta)]}{\partial\theta^\top}\right\} &= \int_\Theta \int_{\mathscr{Y}_X} (\hat{\theta}^N - \theta)\frac{\partial[\varphi_{X,\theta}(\mathbf{y})\pi(\theta)]}{\partial\theta^\top}\mu_X(\mathrm{d}\mathbf{y})\mathrm{d}\theta \\
&= \int_\Theta \left[\int_{\mathscr{Y}_X} (\hat{\theta}^N - \theta)\, \frac{\partial\varphi_{X,\theta}(\mathbf{y})}{\partial\theta^\top}\, \mu_X(\mathrm{d}\mathbf{y})\right] \pi(\theta)\, \mathrm{d}\theta \\
&\quad + \int_\Theta \left[\int_{\mathscr{Y}_X} (\hat{\theta}^N - \theta)\, \varphi_{X,\theta}(\mathbf{y})\, \mu_X(\mathrm{d}\mathbf{y})\right] \frac{\partial\pi(\theta)}{\partial\theta^\top}\, \mathrm{d}\theta\,.
\end{aligned}$$

By differentiation under the integral sign, which is allowed from (4.42)–(4.44), the first integral on the right-hand side is equal to $\int_\Theta \{\partial\mathbb{E}_{X,\theta}(\hat{\theta}^N)/\partial\theta^\top\}\, \pi(\theta)\, \mathrm{d}\theta$; the second is equal to $\int_\Theta [\mathbb{E}_{X,\theta}(\hat{\theta}^N) - \theta]\, \{\partial\pi(\theta)/\partial\theta^\top\}\, \mathrm{d}\theta$, which gives

$$\mathbb{E}(\mathbf{u}\mathbf{v}^\top) = \int_\Theta \frac{\partial\{[\mathbb{E}_{X,\theta}(\hat{\theta}^N) - \theta]\pi(\theta)\}}{\partial\theta^\top}\, \mathrm{d}\theta + \mathbf{I}_p = \mathbf{I}_p\,,$$

the p-dimensional identity matrix. Using (4.51) and Lemma 3.7 we obtain (4.52), (4.53). ∎

Remark 4.18.

(*i*) Under the conditions (4.42)–(4.44),

$$\mathbb{E}_{X,\theta}\left\{\frac{\partial \log \varphi_{X,\theta}(\mathbf{y})}{\partial \theta}\right\} = \int_{\mathscr{Y}_X} \frac{\partial \varphi_{X,\theta}(\mathbf{y})}{\partial \theta}\, \mu_X(d\mathbf{y})$$
$$= \frac{\partial\left[\int_{\mathscr{Y}_X} \varphi_{X,\theta}(\mathbf{y})\, \mu_X(d\mathbf{y})\right]}{\partial \theta} = \mathbf{0}$$

and the matrix $\tilde{\mathbf{M}}(X)$ given by (4.53) can also be written as

$$\tilde{\mathbf{M}}(X) = \mathbb{E}_\pi\{\mathbf{M}(X,\theta)\} + \frac{1}{N}\, \mathbb{E}_\pi\left\{\frac{\partial \log \pi(\theta)}{\partial \theta} \frac{\partial \log \pi(\theta)}{\partial \theta^\top}\right\}$$

with $\mathbf{M}(X,\theta)$ the average Fisher information matrix per sample for the parameters θ and design X, given by (4.46). In a linear model, $\mathbf{M}(X,\theta) = \mathbf{M}(X)$ does not depend on θ and

$$\tilde{\mathbf{M}}(X) = \mathbf{M}(X) + \frac{1}{N}\, \mathbb{E}_\pi\left\{\frac{\partial \log \pi(\theta)}{\partial \theta} \frac{\partial \log \pi(\theta)}{\partial \theta^\top}\right\}.$$

In particular, if the prior is normal $\mathscr{N}(\hat{\theta}^0, \mathbf{\Omega})$ then $\tilde{\mathbf{M}}(X) = \mathbf{M}(X) + \mathbf{\Omega}^{-1}/N$. Note that the influence of the prior vanishes as N increases.

(*ii*) From Lemma 3.7, equality in (4.52) is equivalent to $\partial \log \pi_{X,\mathbf{y}}(\theta)/\partial\theta = \mathbf{A}(\hat{\theta}^N - \theta)$ for some nonrandom matrix $\mathbf{A}$. Moreover, $\mathbf{A} = N\tilde{\mathbf{M}}(X)$, see the proof of Lemma 3.7, and when $\tilde{\mathbf{M}}(X)$ is nonsingular one can check that the posterior is the normal $\mathscr{N}(\hat{\theta}^N, [N\tilde{\mathbf{M}}(X)]^{-1})$; that is,

$$\pi_{X,\mathbf{y}}(\theta) = \sqrt{\det[\tilde{\mathbf{M}}(X)]}\, \frac{N^{p/2}}{(2\pi)^{p/2}} \exp\left[-\frac{N}{2}(\theta - \hat{\theta}^N)\tilde{\mathbf{M}}(X)(\theta - \hat{\theta}^N)\right].$$

It implies that if the equality in (4.52) is attained for some estimator $\hat{\theta}^N$, $\hat{\theta}^N$ corresponds to the maximum a posteriori estimator. From the normal form of $\pi_{X,\mathbf{y}}(\theta)$, it is also the posterior mean for θ. This is the case for linear regression with normal errors and a normal prior. □

4.6 Bibliographic Notes and Further Remarks

M Estimators Without Smoothness

Only the case of a smooth function $\rho(\cdot)$ has been considered in Sect. 4.1 and also only the case of a smooth density $\bar{\varphi}(\cdot)$ in Sect. 4.2. The extension of Theorem 4.3 to the case where $\rho(\cdot)$ is not a smooth function requires specific developments; see, e.g., van der Vaart (1998, Chap. 5). A heuristic justification of the results can be given in situations where $\rho(\cdot)$ is twice continuously differentiable almost everywhere, i.e., when the probability that $\rho(\cdot)$ is twice continuously differentiable at ϵ_k equals one when ϵ_k has the p.d.f. $\bar{\varphi}_{x_k}(\cdot)$ and x_k is distributed with the probability measure ξ.

Estimating Functions and Quasi-Likelihood

The approach we used in Chaps. 3 and 4 to show that an estimator $\hat{\theta}^N$ is asymptotically normal relies on the fact that $\hat{\theta}^N \in \mathrm{int}(\Theta)$ for N large enough so that the vector of derivatives $\nabla J_N(\theta)$ equals $\mathbf{0}$ at $\hat{\theta}^N$ when the criterion $J_N(\cdot)$ that we minimize is differentiable in $\mathrm{int}(\Theta)$. More generally, an equation $\mathbf{d}_N(\theta) = \mathbf{d}(\theta; \mathbf{y}, \xi_N) = \mathbf{0}$, with $\mathbf{d}_N(\theta) \in \mathbb{R}^p$, $\mathbf{y} = (y_1, \ldots, y_N)^\top$ the vector of observations and ξ_N the empirical design measure, is called an estimating equation for $\theta \in \mathbb{R}^p$, and $\mathbf{d}_N(\cdot)$ is an unbiased estimating function when $\mathbb{E}_\theta\{\mathbf{d}_N(\theta)\} = \mathbf{0}$ for all θ. A typical example is when $\mathbf{d}_N(\cdot)$ is the score function,

$$\mathbf{u}_N(\theta) = \frac{\partial \log \mathrm{L}_{X,\mathbf{y}}(\theta)}{\partial \theta} = \frac{\partial \log \varphi_{X,\theta}(\mathbf{y})}{\partial \theta}, \tag{4.54}$$

which satisfies $\mathbb{E}_{X,\theta}\{\mathbf{u}_N(\theta)\} = \mathbf{0}$ under (4.42)–(4.44); see Remark 4.18-(*i*). Under conditions similar to those used in Chaps. 3 and 4, one can show that estimators $\hat{\theta}^N$ defined by suitable estimating equations $\mathbf{d}_N(\hat{\theta}^N) = \mathbf{0}$ are strongly consistent and asymptotically normal.

The following standardized form of an estimating function is often used,

$$\mathbf{d}_N^{(\mathrm{s})}(\theta) = -\mathbb{E}_\theta\left\{\frac{\partial \mathbf{d}_N(\theta)}{\partial \theta^\top}\right\}^\top \left[\mathbb{E}_\theta\{\mathbf{d}_N(\theta)\mathbf{d}_N^\top(\theta)\}\right]^{-1} \mathbf{d}_N(\theta),$$

so that

$$\mathbb{E}_\theta\left\{\frac{\partial \mathbf{d}_N(\theta)}{\partial \theta^\top}\right\}^\top \left[\mathbb{E}_\theta\{\mathbf{d}_N(\theta)\mathbf{d}_N^\top(\theta)\}\right]^{-1} \mathbb{E}_\theta\left\{\frac{\partial \mathbf{d}_N(\theta)}{\partial \theta^\top}\right\} = \mathbb{E}_\theta\left\{\mathbf{d}_N^{(\mathrm{s})}(\theta)\, \mathbf{d}_N^{(\mathrm{s})\top}(\theta)\right\}.$$

Note that the score function (4.54) satisfies $\mathbf{u}_N^{(\mathrm{s})}(\theta) = \mathbf{u}_N(\theta)$; see Remark 4.14-(*iv*). The matrix $\mathbb{E}_\theta\{\mathbf{d}_N^{(\mathrm{s})}(\theta)\, \mathbf{d}_N^{(\mathrm{s})\top}(\theta)\}$ forms a natural generalization of the Fisher information matrix to other estimating functions than the score function.

When $\mathbf{d}_N(\cdot)$ belongs to some class $\mathscr{D}$ of functions which have zero mean and are square integrable, one may choose a function $\mathbf{d}_N^*(\cdot)$ such that

$$\mathbb{E}_\theta\left\{\mathbf{d}_N^{*(\mathrm{s})}(\theta)\, \mathbf{d}_N^{*(\mathrm{s})\top}(\theta)\right\} \succeq \mathbb{E}_\theta\left\{\mathbf{d}_N^{(\mathrm{s})}(\theta)\, \mathbf{d}_N^{(\mathrm{s})\top}(\theta)\right\}$$

for all $\mathbf{d}_N(\cdot)$ in $\mathscr{D}$ and all θ, or, equivalently, such that

$$\mathbb{E}_\theta\left\{[\mathbf{u}_N(\theta) - \mathbf{d}_N^{(\mathrm{s})}(\theta)]\,[\mathbf{u}_N(\theta) - \mathbf{d}_N^{(\mathrm{s})}(\theta)]^\top\right\} \succeq \mathbb{E}_\theta\left\{[\mathbf{u}_N(\theta) - \mathbf{d}_N^{*(\mathrm{s})}(\theta)]\,[\mathbf{u}_N(\theta) - \mathbf{d}_N^{*(\mathrm{s})}(\theta)]^\top\right\}$$

for all $\mathbf{d}_N(\cdot)$ in $\mathscr{D}$ and all θ; see Heyde (1997, Chap. 2). Such a $\mathbf{d}_N^*(\cdot)$ is called a quasi-score estimating function; estimators associated with quasi-score estimating functions are called quasi-likelihood estimators.

The construction of estimating functions (or estimating equations) yields a rich set of tools for parameter estimation in stochastic models. They can provide simple estimators for dynamical systems, at the expense, in general, of a loss of precision compared to LS or ML estimation. One may refer to Heyde (1997) for a general exposition of the methodology; see also the discussion paper (Liang and Zeger, 1995). Instrumental variable methods, see, e.g., Söderström and Stoica (1981, 1983) and Söderström and Stoica (1989, Chap. 8), used in dynamical systems as an alternative to LS estimation when the regressors and errors are correlated and the LS estimator is biased, can be considered as methods for constructing unbiased estimating functions.

Superefficiency

An estimator $\hat{\theta}^N$ satisfying $\sqrt{N}(\hat{\theta}^N - \bar{\theta}) \sim \mathscr{N}(\mathbf{0}, \mathbf{C}(\xi, \bar{\theta}))$ is said to be asymptotically efficient when $\mathbf{C}(\xi, \bar{\theta}) = \mathbf{M}^{-1}(\xi, \bar{\theta})$, the inverse of the average Fisher information matrix per observation; see Sect. 4.4.2. Under suitable conditions, this is the case for the ML estimator; see (4.47). One might think that the Cramér–Rao inequality (4.45) would imply that in general $\mathbf{C}(\xi, \bar{\theta}) \succeq \mathbf{M}^{-1}(\xi, \bar{\theta})$, at least under some conditions on the probability model generating the observations. However, there typically exist estimators $\hat{\theta}^N$ such that $\sqrt{N}(\hat{\theta}^N - \bar{\theta}) \sim \mathscr{N}(\mathbf{0}, \mathbf{C}(\xi, \bar{\theta}))$ but with $\mathbf{C}(\xi, \bar{\theta}) \prec \mathbf{M}^{-1}(\xi, \bar{\theta})$ at some particular points $\bar{\theta}$, which are called *points of superefficiency*. A well-known example is due to J.L. Hodges; see Lehmann and Casella (1998, p. 440), Ibragimov and Has'minskii (1981, p. 91). However, Le Cam (1953) has shown that the set of points of superefficiency has zero Lebesgue measure, so that superefficiency is not really significant from a statistical point of view, and asymptotic efficiency remains an important notion.

The LAN Property

A most important step in the construction of asymptotically efficient estimators is the concept of local asymptotic normality (LAN), due to Le Cam; see Le Cam (1960) and the review by van der Vaart (2002) of his contributions to statistics. Under minimal assumptions (e.g., replication of independent experiments), the LAN condition holds, and an asymptotically efficient estimator can be constructed from an initial estimator that is simply $\sqrt{N}$ consistent. This extends the approach of Sect. 4.4.2 to situations where the ML estimator cannot be computed. For instance, it is a key feature for the construction of adaptive estimators in (Bickel, 1982) and (Manski, 1984).

Dynamical Systems and Dependent observations

It is assumed throughout the monograph that the observations y_k form an independent sequence of random variables. However, it may happen that this assumption of independence does not hold. This is the case in particular for dynamical systems where y_k can be written as

$$y_k = F(k, Y^{k-1}, U^{k-1}, \bar{\theta}) + \varepsilon_k$$

with Y^{k-1} the observations $y_1, \ldots, y_{k-1}$ available at time k, U^{k-1} the values of inputs (controls) $u_1, \ldots, u_k$ used up to time k and $\{\varepsilon_k\}$ a sequence of independent random variables, and ε_k being distributed with a p.d.f. $\bar{\varphi}_k(\cdot)$ which possibly depends on u_k. Here $F(\cdot)$ simply expresses some dependence of y_k on the passed values of y and u, as in time-series models, for instance. The observations y_k are not independent, but the prediction errors

$$\hat{\varepsilon}(\theta, k) = y_k - F(k, Y^{k-1}, U^{k-1}, \theta) \tag{4.55}$$

form a sequence of independent random variables for $\theta = \bar{\theta}$. The results presented in Chaps. 3 and 4 for regression models can then be extended to that situation; see, e.g., Goodwin and Payne (1977) and Ljung (1987) for a detailed presentation including more complex models. For instance, the LS criterion now takes the form

$$J_N(\theta) = \frac{1}{N} \sum_{k=1}^{N} \hat{\varepsilon}^2(\theta, k)\,,$$

and for a deterministic sequence of inputs u_k, the likelihood can be written as

$$\mathrm{L}_{U^N, Y^N}(\theta) = C(Y^N) \prod_{k=1}^{N} \bar{\varphi}_k[\hat{\varepsilon}(\theta, k)]\,, \tag{4.56}$$

with $C(\cdot)$ some positive measurable function of Y^N not depending on θ. The corresponding estimation procedure is then called a *prediction-error method.* Notice that (4.56) contains no approximation when the predictor $F(k, Y^{k-1}, U^{k-1}, \theta)$ can be calculated exactly[6] in (4.55). Under suitable assumptions, the ML estimator is still asymptotically efficient in this situation. For some particular models, e.g., those given by linear differential or recurrence equations, the Fisher information matrix can be computed explicitly in closed form; see, for instance, Zarrop (1979), Goodwin and Payne (1977), Ljung (1987) and Walter and Pronzato (1997, Chap. 6). Notice that the sequence of inputs u_k plays here the role of the experimental design.

Data-Recursive Estimation Methods

It may be necessary, for dynamical systems in particular, to have an estimate of the model parameters available on line; that is, $\hat{\theta}^{N-1}$ should be available immediately after y_{N-1} has been observed. Data-recursive methods concern

[6]Typically, this implies that unknown initial values have been replaced by zero in the dynamical systems; the ML method is then called conditional ML, where conditional refers to this choice of initial values.

estimation procedures where $\hat{\theta}^N$ is constructed from $\hat{\theta}^{N-1}$ and y_N (and the design point x_N associated with y_N, or the N-th control for a dynamical system); see, for instance, Remark 3.28-(*iii*) for recursive LS and recursively reweighted LS in a linear regression model. We shall not develop this point but only mention that when $\hat{\theta}^N$ is obtained by a Newton-type step similar to (4.48) with $\hat{\theta}_1^N$ replaced by $\hat{\theta}^{N-1}$, the asymptotic properties of the off-line (nonrecursive) estimator are preserved.[7] In particular, data-recursive ML estimation is still efficient under suitable conditions. The proofs are rather technical; see, e.g., Ljung (1987), Caines (1988), and Söderström and Stoica (1989).

Cramér–Rao Inequality for Estimation with Constraints

Assume that $\bar{\theta} \in \Theta'$, with $\Theta' \subset \Theta$ a continuous m-dimensional manifold of $\mathbb{R}^p$, $m < p$, and consider constrained parameter estimation where $\hat{\theta}^N$ is forced to belong to Θ'. Then, under the conditions of Theorem 4.13, with the addition that $\mathbf{M}(X, \bar{\theta})$ is nonsingular, any constrained estimator $\hat{\theta}^N \in \Theta'$ with finite variance satisfies

$$\mathrm{Var}_{X,\bar{\theta}}\{\hat{\theta}^N\} \succeq \frac{\partial \mathbb{E}_{X,\bar{\theta}}\{\hat{\theta}^N\}}{\partial \bar{\theta}^\top} \mathbf{P}_{\bar{\theta}} \frac{\mathbf{M}^{-1}(X, \bar{\theta})}{N} \frac{\partial [\mathbb{E}_{X,\bar{\theta}}\{\hat{\theta}^N\}]^\top}{\partial \bar{\theta}} \tag{4.57}$$

where $\mathbf{P}_{\bar{\theta}}$ is the projector

$$\mathbf{P}_{\bar{\theta}} = \mathbf{M}^{-1}(X, \bar{\theta})\mathbf{Q}[\mathbf{Q}^\top \mathbf{M}^{-1}(X, \bar{\theta})\mathbf{Q}]^{-1}\mathbf{Q}^\top$$

and $\mathbf{Q} = [\mathbf{q}_1, \ldots, \mathbf{q}_m]$ is formed by m arbitrary linearly independent tangent directions on Θ' at $\bar{\theta}$; see Gorman and Hero (1990). The case where $\mathbf{M}(X, \bar{\theta})$ is singular is considered in (Stoica, 1998, 2001). A geometrical investigation of the small sample properties in nonlinear models with parameter constraints has been done in (Pázman, 2002b). Since equality constraints as considered in Sect. 3.5 also define a restriction of the parameter space Θ to a subset $\Theta' \subset \Theta$, the generalized Cramér–Rao inequality (4.57) can also be applied to that case.

[7]Many methods exist, they receive different names (recursive ML, recursive pseudo-linear regression, recursive generalized LS, extended LS...) depending on the type of model to which they are applied and on the type of approximations used in the implementation of the Newton step.

5

Local Optimality Criteria Based on Asymptotic Normality

The approach considered in this chapter is probably the most common for designing experiments in nonlinear situations. It consists in optimizing a scalar function of the asymptotic covariance matrix of the estimator and thus relies on asymptotic normality, as considered in Chaps. 3 and 4. Design based on more accurate characterizations of the precision of the estimation will be considered in Chap. 6. Additionally to asymptotic normality, the approach also supposes that the asymptotic covariance matrix takes the form of the inverse of an information matrix,[1] as it is the case, for instance, for weighted LS with optimum weights, see Sect. 3.1.3, or maximum likelihood estimation, see Sect. 4.2. The case of asymptotic covariance matrices that are products of information matrices and their inverses, as those encountered, for instance, in Sects. 3.1.3, 3.3.2, 3.4, 4.1, is considered in Sect. 5.5.

In this chapter we thus consider design criteria that can be written as

$$\phi(\xi) = \Phi[\mathbf{M}(\xi, \theta)]$$

with $\mathbf{M}(\xi, \theta)$ an information matrix of the form

$$\mathbf{M}(\zeta, \theta) = \int_{\mathscr{X}} \mathbf{M}_\theta(x)\, \xi(\mathrm{d}x)\,. \tag{5.1}$$

Here, $\mathbf{M}_\theta(x)$ denotes the symmetric nonnegative-definite $p \times p$ matrix $\mathbf{M}(\delta_x, \theta)$ with δ_x the delta measure putting mass one at x.

In regression models with scalar observations $\mathbf{M}_\theta(x)$ has often rank one and

$$\mathbf{M}_\theta(x) = \mathbf{g}_\theta(x)\mathbf{g}_\theta^\top(x)\,,\ \mathbf{g}_\theta(x) \in \mathbb{R}^p\,.$$

For instance, in the model (3.2), (3.4) the asymptotic variance–covariance matrix of the WLS estimator with weights proportional to $\sigma^{-2}(x)$ is the inverse

[1]Another interpretation is that locally optimum design is based on the Cramér–Rao bound and the Fisher information matrix; see Sect. 4.4.2. Note, however, that the Cramér–Rao inequality gives a lower bound, whereas an upper bound would be more suitable, but unfortunately is not available.

L. Pronzato and A. Pázman, *Design of Experiments in Nonlinear Models*, Lecture Notes in Statistics 212, DOI 10.1007/978-1-4614-6363-4_5,

of the matrix given by (3.26); see Theorem 3.8. In the model (3.2), (3.45) with parameterized variance, the asymptotic variance–covariance matrix of the two-stage LS estimator is proportional to the inverse of (3.56); see Theorem 3.27. See also Theorem 4.6 for the case of ML estimation. Note, however, that the matrix $\mathbf{M}_\theta(x)$ may have rank larger than one although the observations are scalar; see, e.g., Remark 3.25 for the penalized WLS estimator with normal errors and Sect. 4.3.2. Naturally, $\mathbf{M}_\theta(x)$ has generally rank larger than one in the case of multivariate observations; see Sect. 5.6.

We shall denote by $\mathcal{M}_\theta(\mathscr{X})$ and $\mathcal{M}_\theta(\Xi)$ the sets

$$\mathcal{M}_\theta(\mathscr{X}) = \{\mathbf{M}_\theta(x) : x \in \mathscr{X}\}\,, \tag{5.2}$$

$$\mathcal{M}_\theta(\Xi) = \{\mathbf{M}(\xi,\theta) : \xi \in \Xi\}\,, \tag{5.3}$$

with Ξ the set of probability measures on $\mathscr{X}$. One may already notice that $\mathcal{M}_\theta(\Xi)$ is the convex hull of $\mathcal{M}_\theta(\mathscr{X})$ and that $\mathcal{M}_\theta(\Xi) \subset \mathbb{M}^{\geq}$, the set of symmetric nonnegative-definite $p \times p$ matrices, which forms a closed cone in the set $\mathbb{M}$ of symmetric $p \times p$ matrices. The set of symmetric positive-definite $p \times p$ matrices, an open cone included in $\mathbb{M}^{\geq}$, will be denoted by $\mathbb{M}^{>}$. We shall always assume that $\mathcal{M}_\theta(\mathscr{X})$ is compact, which is not too restrictive since it holds, for instance, when $\mathscr{X}$ is finite or when $\mathscr{X}$ is a compact set with nonempty interior and $\mathbf{M}_\theta(x)$ is continuous on $\mathscr{X}$. We shall also assume that $\mathcal{M}_\theta(\Xi)$ contains at least a nonsingular information matrix.

The term *locally* in locally optimum design is due to the dependence of the criterion, and thus of the optimal design, on the value of the parameters that we precisely intend to estimate. This phenomenon is typical in nonlinear situations. The idea is then to assume a nominal value θ^0 for θ and to design for θ^0, with the hope that the optimal design for that θ^0 will not differ too much from the optimal one for the unknown true value $\bar{\theta}$. Since in this chapter θ^0 will be kept fixed, we shall most often omit the dependence in θ and simply write $\mathbf{M}(\xi) = \mathbf{M}(\xi, \theta^0)$ for the information matrix (5.1) computed at θ^0. Also, the sets (5.2), (5.3) should be considered at the value $\theta = \theta^0$. Approaches that aim at achieving some robustness with respect to the choice of θ^0 in a nonlinear situation will be considered in Chap. 8.

When $\Phi(\cdot)$ is such that the information matrix $\mathbf{M}^* = \mathbf{M}(\xi^*)$ associated with an optimal design ξ^* has full rank, the local design problem is similar to that encountered in a linear situation where $\mathbf{M}$ does not depend on θ. However, there are situations where no such similarity exists. Indeed, some of the functions $\Phi(\cdot)$ that are classical in experimental design may lead to optimal designs that are singular, i.e., such that $\mathbf{M}(\xi)$ is singular. This does not raise any special difficulty in a linear situation: θ is not estimable for a singular design ξ, and the linear combination $\mathbf{c}^\top\theta$ is estimable if and only if $\mathbf{c} = \mathbf{M}\mathbf{u}$ for some $\mathbf{u} \in \mathbb{R}^p$. We have seen in Sect. 3.2 that singular designs cause difficulties in nonlinear situations. Typically, the conditions (3.39) and the assumption H2$_h$ (p. 43) for the asymptotic normality of the estimator of a function of θ indicate that the use of a singular design requires some knowledge on the true value of the model parameters, see Example 3.17 and Remark 3.15-(*ii*). Singular designs should thus be handled with great care in nonlinear situations. These issues will be considered in Sect. 5.4 for the case where we are interested into the estimation of a scalar function of θ.

We shall consider criteria $\phi(\xi) = \Phi[\mathbf{M}(\xi)]$ that measure the amount of information provided by the experiment characterized by ξ; they will therefore be *maximized*; see in particular Pukelsheim (1993) for a justification of this choice. Concavity[2] will thus play an important role in terms of the properties of a ϕ-optimal design ξ^* and in terms of the construction of algorithms for the determination of ξ^*. This is a rather universal observation, and we shall keep the presentation general enough to be able to refer to some of the results presented in this chapter when we shall consider average, maximin, and probabilistic optimum design in Chap. 8.

The developments in this chapter involve a lot of linear algebra (due to the similarity between locally optimum design and optimum design for linear models), convex analysis, and geometry. We keep the presentation as simple as possible, giving the full proof of results when they are strongly design oriented, but only indicating references for more general properties. As in some other work on optimum design, the presentation goes through a list of optimality criteria (Sect. 5.1), each with its own merits. Their main properties are indicated, which should motivate the choice of a particular criterion in most situations.

The success of optimum design theory is largely due to the equivalence theorem, first established by Kiefer and Wolfowitz (1960) for D-optimality and then extended to any concave criterion, and to its consequences for the development of globally convergent algorithms for the optimization of design measures. The equivalence theorem forms the core of Sect. 5.2. The presentation is based on directional derivatives. An alternative introduction would be through subdifferentials and subgradients. This is briefly considered in Appendix A where the connection between the two approaches is indicated.

The case of c-optimum design, motivated by the estimation of a scalar function of the parameters, receives special attention in Sects. 5.3 and 5.4 in the light of the developments of Sect. 3.2. Several methods are proposed to overcome the difficulties raised by singular c-optimal designs in nonlinear situations.

[2]When necessary, we shall extend the definition of a concave criterion $\Phi(\cdot)$ over the whole set $\mathbb{M}$ of symmetric $p \times p$ matrices by allowing $\Phi(\cdot)$ to take the value $-\infty$. All the criteria considered are proper functions; that is, they satisfy $\Phi(\mathbf{M}) < \infty$ for all $\mathbf{M} \in \mathbb{M}$, and their effective domain $\mathrm{dom}(\Phi) = \{\mathbf{M} \in \mathbb{M} : \Phi(\mathbf{M}) > -\infty\}$ is non empty.

5.1 Design Criteria and Their Properties

5.1.1 Ellipsoid of Concentration

Consider the situation where an estimator $\hat{\theta}^N$ satisfies $\sqrt{N}(\hat{\theta}^N - \bar{\theta}) \overset{\mathrm{d}}{\to} \mathbf{z} \sim \mathscr{N}(\mathbf{0}, \mathbf{M}^{-1}(\xi, \bar{\theta}))$ as $N \to \infty$, with $\bar{\theta}$ the unknown true value of the parameters and $\mathbf{M}(\xi, \theta)$ an information matrix of the form (5.1). It means that asymptotically the density of $\hat{\theta}^N - \bar{\theta}$ can be approximated by the normal density

$$n_{\bar{\theta}}(\mathbf{t}) = \frac{N^{p/2}}{(2\pi)^{p/2} \det^{-1/2} \mathbf{M}(\xi, \bar{\theta})} \exp\left[-\frac{N}{2}\mathbf{t}^\top \mathbf{M}(\xi, \bar{\theta})\mathbf{t}\right] . \qquad (5.4)$$

The concentration of this density can be expressed by the extend of the set

$$\mathcal{E}_{\bar{\theta}} = \{\mathbf{t} \in \mathbb{R}^p : \mathbf{t}^\top \mathbf{M}(\xi, \bar{\theta})\mathbf{t} \leq 1\} = \left\{\mathbf{t} \in \mathbb{R}^p : \log\left[\frac{\max_{\mathbf{t} \in \mathbb{R}^p} n_{\bar{\theta}}(\mathbf{t})}{n_{\bar{\theta}}(\mathbf{t})}\right] \leq \frac{N}{2}\right\}$$

which is called the (asymptotic) ellipsoid of concentration; see, e.g., Cramér (1946, Chap. 22) and Fedorov and Pázman (1968). Some geometrical properties of this ellipsoid are summarized in the following lemma. Its proof is given in Appendix C.

Lemma 5.1. *Let* $\mathbf{A}$ *be a* $p \times p$ *positive-definite matrix and let* $\mathcal{E}_A = \{\mathbf{t} \in \mathbb{R}^p : \mathbf{t}^\top \mathbf{A}\mathbf{t} \leq 1\}$. *Then*

(i) $\mathrm{vol}(\mathcal{E}_A) = V_p \det^{-1/2} \mathbf{A}$, *with* $V_p = \pi^{p/2}/\Gamma(p/2+1) = \mathrm{vol}[\mathscr{B}(\mathbf{0},1)]$, *the volume of the unit ball* $\mathscr{B}(\mathbf{0},1)$ *in* $\mathbb{R}^p$.

(ii) For any vector $\mathbf{c} \in \mathbb{R}^p$ *we have*

$$\max_{\mathbf{t} \in \mathcal{E}_A} (\mathbf{c}^\top \mathbf{t})^2 = \mathbf{c}^\top \mathbf{A}^{-1} \mathbf{c} ;$$

in particular, when $\|\mathbf{c}\| = 1$, *then* $\mathbf{c}^\top \mathbf{A}^{-1}\mathbf{c}$ *is the squared half-length of the orthogonal projection of* $\mathcal{E}_A$ *onto the straight line defined by* $\mathbf{c}$.

(iii) $\max_{\|\mathbf{c}\|=1} \mathbf{c}^\top \mathbf{A}^{-1}\mathbf{c} = 1/\lambda_{\min}(\mathbf{A}) = R^2(\mathcal{E}_A)$, *with* $\lambda_{\min}(\mathbf{A})$ *the minimum eigenvalue of* $\mathbf{A}$ *and* $R(\mathcal{E}_A)$ *the radius of the smallest ball containing* $\mathcal{E}_A$*; the length of a principal axis of* $\mathcal{E}_A$ *equals* $2/\sqrt{\lambda_i(\mathbf{A})}$ *with* $\lambda_i(\mathbf{A})$ *an eigenvalue of* $\mathbf{A}$.

(iv) The squared length of the half-diagonal of the parallelepiped containing $\mathcal{E}_A$ *and parallel to the coordinate axes of the Euclidean space* $\mathbb{R}^p$ *equals the sum of the squared half-lengths of the principal axes of* $\mathcal{E}_A$ *and is given by* $\mathrm{trace}(\mathbf{A}^{-1})$.

(v) Let $\mathcal{E}_B$ *be defined similarly to* $\mathcal{E}_A$ *but for the* $p \times p$ *positive-definite matrix* $\mathbf{B}$, *then the following statements are equivalent:*

(a) $\mathcal{E}_A \subseteq \mathcal{E}_B$.

(b) $\mathbf{A} \succeq \mathbf{B}$, *i.e., the matrix* $\mathbf{A} - \mathbf{B}$ *is nonnegative definite.*

(c) For any $\mathbf{c} \in \mathbb{R}^p$, $\mathbf{c}^\top \mathbf{A}^{-1}\mathbf{c} \leq \mathbf{c}^\top \mathbf{B}^{-1}\mathbf{c}$, *i.e.,* $\mathbf{B}^{-1} \succeq \mathbf{A}^{-1}$.

Since $\hat{\theta}^N$ is asymptotically normal and $\mathbf{M}(\xi, \hat{\theta}^N)$ converges to $\mathbf{M}(\xi, \bar{\theta})$, the ellipsoid $\{\theta \in \mathbb{R}^p : N(\hat{\theta}^N - \theta)^\top \mathbf{M}(\xi, \hat{\theta}^N)(\hat{\theta}^N - \theta) \leq q\}$ can be considered as an approximate (asymptotically correct) confidence region for $\bar{\theta}$; see Sect. 6.5 for exact regions. Here, q is proportional to a quantile of some distribution. For instance, in the case of a regression model as considered in Sect. 3.1, an approximate 95% confidence region is obtained for q equal to the 95% quantile of the χ_p^2 distribution (chi-square with p degrees of freedom) when the variance σ^2 of the errors is known. Hence, ellipsoids of concentration $\mathcal{E}_\theta$ are homothetic to (asymptotic) confidence ellipsoids. We also have the following property concerning asymptotic confidence regions for the response function in a regression model; see Appendix C.

Lemma 5.2. *Suppose that the estimator $\hat{\theta}^N$ in the regression model (3.2) satisfies $\sqrt{N}(\hat{\theta}^N - \bar{\theta}) \xrightarrow{\mathrm{d}} \mathbf{z} \sim \mathscr{N}(\mathbf{0}, \mathbf{M}^{-1}(\xi, \bar{\theta}))$ as $N \to \infty$. Then, for N large we have approximately*

$$\mathrm{Prob}\left\{y(x_1), \ldots, y(x_N) : \forall x \in \mathscr{X}, |\eta(x, \hat{\theta}^N) - \eta(x, \bar{\theta})| \leq \right.$$
$$\left. \frac{1}{\sqrt{N}} \left[\chi_p^2(1-\alpha) \frac{\partial \eta(x,\theta)}{\partial \theta^\top}\bigg|_{\bar{\theta}} \mathbf{M}^{-1}(\xi, \bar{\theta}) \frac{\partial \eta(x,\theta)}{\partial \theta}\bigg|_{\bar{\theta}}\right]^{1/2} \right\} \geq 1 - \alpha \quad (5.5)$$

where $\chi_p^2(1-\alpha)$ is the $(1-\alpha)$ quantile of the χ_p^2 distribution.

Most of the classical design criteria presented below can be related to characteristics of the ellipsoids $\mathcal{E}_\theta$. Indeed, a rather natural motivation for designing an experiment is to have a "small" ellipsoid $\mathcal{E}_\theta$, where several meanings can be attached to the adjective small. Asymptotic confidence regions for the regression function $\eta(\cdot, \bar{\theta})$ are based on (5.5) and can also be related to criteria based on $\mathcal{E}_\theta$. The situation in thus similar to that in linear models.

5.1.2 Classical Design Criteria

D-Optimality

The criterion of D-optimality is defined by $\Phi_D(\mathbf{M}) = \det^{1/p}(\mathbf{M})$. According to Lemma 5.1-(i), maximizing $\Phi_D(\mathbf{M})$ amounts to minimizing the volume of the ellipsoid of concentration. The negative (Shannon) entropy of a distribution is a measure of its concentration. For the asymptotic normal distribution (5.4) it is given by

$$H_1(n_{\bar{\theta}}) = -\int_{\mathbb{R}^p} n_{\bar{\theta}}(\mathbf{t}) \log[n_{\bar{\theta}}(\mathbf{t})] \, \mathrm{d}\mathbf{t} \quad (5.6)$$
$$= \frac{p}{2}[1 + \log(2\pi) - \log(N)] - \frac{1}{2} \log \det \mathbf{M}(\xi, \bar{\theta}), \quad (5.7)$$

so that a D-optimal experiment minimizes the Shannon entropy of the asymptotic distribution of the estimator. Since the Rényi (1961) entropy of order q of the normal distribution (5.4) is given by

$$H_q(n_{\bar{\theta}}) = \frac{1}{1-q} \log \int_{\mathbb{R}^p} [n_{\bar{\theta}}(\mathbf{t})]^q \, d\mathbf{t} = H_1(n_{\bar{\theta}}) - \frac{p}{2}\left(1 + \frac{\log q}{1-q}\right), \; q > 0, \; q \neq 1$$

(with $H_q(n_{\bar{\theta}}) \to H_1(n_{\bar{\theta}})$ as $q \to 1$), a D-optimal experiment more generally minimizes the Rényi entropy of any order $q > 0$ of the distribution (5.4).

An important property of D-optimum design is its invariance with respect to reparameterization: let $\beta = \mathbf{A}\theta$ define another parameterization of the model,[3] with $\mathbf{A}$ a full-rank $p \times p$ matrix. Then, $\det[\mathbf{M}(\xi, \beta)] = \det[\mathbf{M}(\xi, \theta)] \det^{-2}(\mathbf{A})$, so that the ordering of designs and the optimal designs will be the same for both parameterizations.

D_s-Optimality

It corresponds to D-optimum design for the estimation of a subset of θ. By reordering the parameters, we may always suppose that the first s components of θ are of interest. Denote by $\theta_{[1]}$ this part of θ and $\theta = (\theta_{[1]}^\top, \theta_{[2]}^\top)^\top$. Also denote by

$$\mathbf{M} = \begin{pmatrix} \mathbf{M}_{11} & \mathbf{M}_{12} \\ \mathbf{M}_{21} & \mathbf{M}_{22} \end{pmatrix} \quad \text{and} \quad \mathbf{M}^- = \begin{pmatrix} \mathbf{M}^{11} & \mathbf{M}^{12} \\ \mathbf{M}^{21} & \mathbf{M}^{22} \end{pmatrix}$$

the corresponding partitions for the information matrix and a g-inverse of it. A D_s-optimal design then maximizes $\det^{1/s}(\mathbf{M}^*)$, or equivalently

$$\phi_{D_s}(\mathbf{M}) = \log\det(\mathbf{M}^*) \,, \tag{5.8}$$

where $\mathbf{M}^* = \mathbf{M}_{11} - \mathbf{M}_{12}\mathbf{M}_{22}^-\mathbf{M}_{21}$. The value of $\mathbf{M}^*$ does not depend on the choice of the g-inverse $\mathbf{M}_{22}^-$; see Karlin and Studden (1966, Lemma 6.2). If $\mathbf{M}$ is nonsingular, so is $\mathbf{M}_{22}$, $\mathbf{M}^- = \mathbf{M}^{-1}$, and θ is estimable; however, $\theta_{[1]}$ may remain estimable in situations where $\mathbf{M}$ is singular. In fact, $\theta_{[1]}$ is estimable if and only if $\mathbf{M}^*$ is nonsingular; in that case $\mathbf{M}^{11} = (\mathbf{M}^*)^{-1}$, and under the conditions for asymptotic normality considered in Chaps. 3 and 4, $(\mathbf{M}^*)^{-1}$ is proportional to the inverse of the (asymptotic) covariance matrix for the estimation of $\theta_{[1]}$. This justifies the name information matrix for $\theta_{[1]}$ given to $\mathbf{M}^*$; see, e.g., Atwood (1980).

A-Optimality

The criterion of A-optimality is defined by $\Phi_A(\mathbf{M}) = -\text{trace}(\mathbf{M}^{-1})$ when $\mathbf{M}$ is invertible and $\Phi_A(\mathbf{M}) = -\infty$ otherwise. A-optimum design corresponds to minimizing the sum of the asymptotic variances of the estimators of the components of θ and, from Lemma 5.1-(iv), to minimizing the squared length of the diagonal of the parallelepiped parallel to the coordinate axes of the Euclidean space and containing the ellipsoid of concentration.

---

[3] The reparameterization may also be nonlinear, with $\mathbf{A} = \partial\beta/\partial\theta^\top|_{\theta^0}$, where θ^0 denotes the nominal value for θ used in locally optimum design.

c-Optimality

The criterion of c-optimality is defined by

$$\Phi_c(\mathbf{M}) = \begin{cases} -\text{trace}(\mathbf{c}\mathbf{c}^\top\mathbf{M}^-) = -\mathbf{c}^\top\mathbf{M}^-\mathbf{c} & \text{if } \mathbf{c} \in \mathcal{M}(\mathbf{M}) \\ -\infty & \text{otherwise,} \end{cases} \tag{5.9}$$

with $\mathbf{M}^-$ a g-inverse of $\mathbf{M}$ (i.e., such that $\mathbf{M}\mathbf{M}^-\mathbf{M} = \mathbf{M}$) and $\mathbf{c}$ a vector of $\mathbb{R}^p$. According to Theorem 3.11, a c-optimal design ξ^* such that $\mathbf{M}(\xi^*,\bar{\theta})$ is nonsingular minimizes the asymptotic variance of a parametric function $h(\theta)$ such that $\partial h(\theta)/\partial\theta\big|_{\bar{\theta}} = \mathbf{c}$. Notice, however, that $\Phi_c(\mathbf{M})$ remains finite for $\mathbf{M}$ singular provided that $\mathbf{c} \in \mathcal{M}(\mathbf{M})$, i.e., $\mathbf{c} = \mathbf{M}\mathbf{u}$ for some $\mathbf{u} \in \mathbb{R}^p$. In that case, $\Phi_c(\mathbf{M}) = -\mathbf{u}^\top\mathbf{M}\mathbf{M}^-\mathbf{M}\mathbf{u} = -\mathbf{u}^\top\mathbf{M}\mathbf{u}$, and the value of $\Phi_c(\mathbf{M})$ is thus independent of the choice of the g-inverse. Also note that the set $\mathbb{M}^{\geq}_{\mathbf{c}}$ of symmetric nonnegative-definite matrices such that $\mathbf{c} \in \mathcal{M}(\mathbf{M})$ forms a convex cone such that $\mathbb{M}^{>} \subset \mathbb{M}^{\geq}_{\mathbf{c}} \subset \mathbb{M}^{\geq}$, see Pukelsheim (1993, Chap. 2). An extension of the notion of c-optimality to the case where $h(\theta)$ is a homogeneous polynomial of θ is presented in (Pázman, 1986, Sect. 7.4).

For a design such that $\mathbf{M}(\xi,\bar{\theta})$ is singular, the asymptotic distribution of a function $h(\theta)$ is normal with a variance related to $\Phi_c[\mathbf{M}(\xi,\bar{\theta})]$ under very particular circumstances only; see Sect. 3.2. These difficulties are considered in Sect. 5.4.

L-Optimality

L-optimum design forms a generalization of A and c-optimum design, see, e.g., Fedorov (1972, p. 122), and concerns criteria that are linear in $\mathbf{M}^{-1}$. A particular case that covers most applications is $\Phi_L(\mathbf{M}) = -\text{trace}[\mathbf{Q}\mathbf{Q}^\top\mathbf{M}^{-1}]$ with $\mathbf{Q}$ some $p \times m$ matrix, which generalizes to semi-definite matrices $\mathbf{M}$ as follows:

$$\Phi_L(\mathbf{M}) = \begin{cases} -\text{trace}(\mathbf{Q}\mathbf{Q}^\top\mathbf{M}^-) & \text{if } \mathcal{M}(\mathbf{Q}) \subseteq \mathcal{M}(\mathbf{M}) \\ -\infty & \text{otherwise.} \end{cases} \tag{5.10}$$

Notice that if $\mathbf{c}_i$ denotes the i-th column of $\mathbf{Q}$, $\text{trace}(\mathbf{Q}\mathbf{Q}^\top\mathbf{M}^-)$ can be written as $\sum_{i=1}^m \mathbf{c}_i^\top\mathbf{M}^-\mathbf{c}_i$. As for c-optimum design, the feasibility set $\mathbb{M}^{\geq}_{\mathbf{Q}}$ of symmetric nonnegative-definite matrices $\mathbf{M}$ such that $\mathcal{M}(\mathbf{Q}) \subseteq \mathcal{M}(\mathbf{M})$ forms a convex cone such that $\mathbb{M}^{>} \subset \mathbb{M}^{\geq}_{\mathbf{Q}} \subset \mathbb{M}^{\geq}$, see Pukelsheim (1993, Chap. 3). The choice of $\mathbf{Q}$ can be motivated by the relative importance given to different components of θ. For instance, $\mathbf{Q} = \text{diag}\{1/\theta_i^0,\ i = 1,\ldots,p\}$ standardizes the asymptotic variances of the estimates.

E and MV-Optimality

E-optimum design aims at maximizing the minimum eigenvalue of $\mathbf{M}$, i.e., $\Phi_E(\mathbf{M}) = \lambda_{\min}[\mathbf{M}]$. According to Lemma 5.1-(*iii*), it minimizes the radius of the smallest ball containing the ellipsoid of concentration. Note that maximizing $\Phi_E(\mathbf{M})$ is equivalent to minimizing $\lambda_{\max}(\mathbf{M}^{-1}) = \max_{\{\mathbf{c}:\|\mathbf{c}\|_2=1\}} \mathbf{c}^\top\mathbf{M}^{-1}\mathbf{c}$;

that is, E-optimum design minimizes the maximum (asymptotic) variance of linear combinations of parameters $\mathbf{c}^\top\theta$ for vectors $\mathbf{c}$ of $\mathscr{L}_2$ norm one. If a nonsingular design exists, then a E-optimal design is nonsingular.

Using the $\mathscr{L}_1$ norm $\|\mathbf{c}\|_1 = \sum_{i=1}^p |c_i|$ instead of the $\mathscr{L}_2$ norm gives another criterion

$$\max_{\{\mathbf{c}:\|\mathbf{c}\|_1=1\}} \mathbf{c}^\top\mathbf{M}^{-1}\mathbf{c} = \max_{i=1,\ldots,p} \mathbf{e}_i^\top\mathbf{M}^{-1}\mathbf{e}_i\,,$$

with $\mathbf{e}_i$ the i-th basis vector. This is called MV-optimality and is considered, for instance, in (López-Fidalgo et al., 1998). MV-optimum design minimizes the maximum of the (asymptotic) variances of individual parameters, which is given by the maximum diagonal element of $\mathbf{M}^{-1}$. Notice that when a nonsingular design exists, a MV-optimal design is necessarily nonsingular. Indeed, $\max_{i=1,\ldots,p} \mathbf{e}_i^\top\mathbf{M}^{-1}\mathbf{e}_i$ finite implies that $\mathbf{e}_i \in \mathcal{M}(\mathbf{M})$ for all $i = 1,\ldots,p$ so that $\mathbf{M}$ has full rank.

c-Maximin and G-Optimality

c-, E-, and MV-optimality form particular cases of what we shall call c-maximin-optimum design, which aims at maximizing $\min_{\mathbf{c}\in\mathcal{C}} \Phi_c(\mathbf{M})$ for some compact set $\mathcal{C} \subset \mathbb{R}^p$. For instance, the set $\mathcal{C}$ is given by $\{\mathbf{c}\}$ (respectively, $\{\mathbf{c} \in \mathbb{R}^p : \|\mathbf{c}\| = 1\}$ and $\{\mathbf{c} \in \mathbb{R}^p : \|\mathbf{c}\|_1 = 1\}$) for c-optimum design (respectively, E- and MV-optimum design).

G-optimum design is another important particular case of c-maximin-optimum design. Consider a regression model $\eta(x,\theta)$ with homoscedastic errors, see (3.2), (3.3), and define $\mathcal{C} = \{\mathbf{f}_\theta(x) : x \in \mathscr{X}\}$, with $\mathbf{f}_\theta(x) = \partial\eta(x,\theta)/\partial\theta$. Then, maximizing $\min_{\mathbf{c}\in\mathcal{C}} -\mathbf{c}^\top\mathbf{M}^-(\xi)\mathbf{c}$ with $\mathbf{M}(\xi) = \sigma^{-2}\int_{\mathscr{X}} \mathbf{f}_\theta(x)\mathbf{f}_\theta^\top(x)\,\xi(\mathrm{d}x)$ amounts to minimizing the maximum value of the (asymptotic) variance of the model prediction $\eta(x,\hat\theta^N_{LS})$ over the design space $\mathscr{X}$ and is called G-optimum design. This is equivalent to minimizing the maximum width of the asymptotic confidence region (5.5) in Lemma 5.2. One of the consequences of the equivalence theorem of Sect. 5.1.4 is that G-optimum design is equivalent to D-optimum design; see Remark 5.22-(ii). One may also notice that minimizing the average (asymptotic) variance of the model prediction $\eta(x,\hat\theta^N_{LS})$ over $\mathscr{X}$, i.e., $\int_{\mathscr{X}} \mathbf{f}_\theta^\top(x)\mathbf{M}^-(\xi)\mathbf{f}_\theta(\xi)\,\mu(\mathrm{d}x)$, for some probability measure μ on $\mathscr{X}$, corresponds to L-optimum design (5.10) with $\mathbf{Q}\mathbf{Q}^\top = \mathbf{M}(\mu)$.

When the set $\mathcal{C}$ is large enough to span $\mathbb{R}^p$, $\min_{\mathbf{c}\in\mathcal{C}} -\mathbf{c}^\top\mathbf{M}^-\mathbf{c}$ is infinite for any singular $\mathbf{M}$, so that $\mathbf{M}$ is nonsingular at the optimum—provided that at least one nonsingular matrix exists in $\mathcal{M}_\theta(\Xi)$. In that case, the attention can be restricted to matrices in $\mathbb{M}^>$, as for D-, A-, E-, and MV-optimum design.

Φ_q-Class of Criteria

An important class of criteria defined by Kiefer (1974) includes[4]

[4]Up to a change of sign, since Kiefer considered criteria that should be minimized.

$$\Phi_{q,\mathbf{Q}}(\mathbf{M}) = \begin{cases} -\left\{\frac{1}{m}\text{trace}[(\mathbf{Q}^\top\mathbf{M}^-\mathbf{Q})^q]\right\}^{1/q} & \text{if } \mathcal{M}(\mathbf{Q}) \subseteq \mathcal{M}(\mathbf{M}) \\ -\infty & \text{otherwise} \end{cases} \tag{5.11}$$

with $q \geq -1$ and $\mathbf{Q}$ some $(p \times m)$-dimensional matrix. The classical criteria considered above correspond to particular cases in this class: $q = 1$ gives L-optimum design; if moreover $\mathbf{Q} = \mathbf{I}_p$, the p-dimensional identity matrix, it gives A-optimum design; taking $\mathbf{Q}$ as a p-dimensional vector $\mathbf{c}$ and $q = 1$, we get c-optimum design; taking the limit when q tends to zero with $\mathbf{Q} = \mathbf{I}_p$, we obtain $\Phi_{0,\mathbf{I}}(\mathbf{M}) = -\det^{-1/p}(\mathbf{M})$, the maximization of which is equivalent to that of $\det^{1/p}(\mathbf{M})$, i.e., to D-optimum design; and taking the limit when q tends to $+\infty$ with $\mathbf{Q} = \mathbf{I}_p$, we obtain $\Phi_{\infty,\mathbf{I}}(\mathbf{M}) = -1/\lambda_{\min}(\mathbf{M})$, which corresponds to E-optimum design. Apart from the special cases of c- and D_s-optimality (and at a lesser extend of L-optimality), we shall only consider the case $\mathbf{Q} = \mathbf{I}_p$ in what follows. The influence of the function $\mathbf{C}_\mathbf{Q}(\cdot) : \mathbf{M} \in \mathbb{M}^{\geq}_{\mathbf{Q}} \longrightarrow (\mathbf{Q}^\top\mathbf{M}^-\mathbf{Q})^{-1}$ (called information mapping) on the properties of $\Phi_{q,\mathbf{Q}}(\cdot)$ is detailed in (Pukelsheim, 1993, Chap. 3). One may notice that the use of information mappings covers the situation where we are interested into the estimation of a subset of θ only: take $\mathbf{Q}$ diagonal with $\{\mathbf{Q}\}_{ii} = 0$ or 1. In particular, for $q = 0$ it corresponds to D_s-optimum design.

Orthogonally Invariant Criteria

When $\mathbf{Q} = \mathbf{I}_p$, the criteria in the $\Phi_{q,\mathbf{I}}$-class have the property of being orthogonally invariant; that is,

$$\Phi(\mathbf{U}^\top\mathbf{M}\mathbf{U}) = \Phi(\mathbf{M}) \text{ for } \mathbf{U} \text{ a } p \times p \text{ orthogonal matrix}$$

(i.e., such that $\mathbf{U}^\top\mathbf{U} = \mathbf{I}_p$). This property of orthogonal invariance is equivalent to the assumption that $\Phi(\mathbf{M})$ only depends on the eigenvalues of $\mathbf{M}$. In terms of ellipsoid of concentration, it means that the criterion depends on the shape of the ellipsoid, but not on its orientation. Less common examples of orthogonally invariant criteria are given by the coefficients of the characteristic polynomial of $\mathbf{M}^{-1}$, as considered in (López-Fidalgo and Rodríguez-Díaz, 1998), which gives A- and D-optimality as particular cases and the criteria of E_k-optimality of Harman (2004a,b),

$$\Phi_{E_k}(\mathbf{M}) = \sum_{i=1}^{k} \lambda_i(\mathbf{M}), \quad 1 \leq k \leq p, \tag{5.12}$$

where $\lambda_1(\mathbf{M}) \leq \lambda_2(\mathbf{M}) \leq \cdots \leq \lambda_p(\mathbf{M})$ denote the eigenvalues of $\mathbf{M}$. Another example of orthogonally invariant criterion is the condition number of $\mathbf{M}$ for the Frobenius norm $\|\mathbf{M}\|_F = \text{trace}^{1/2}(\mathbf{M}^2)$, which is given by $\rho_F(\mathbf{M}) = \|\mathbf{M}\|_F \, \|\mathbf{M}^{-1}\|_F$ and measures how well conditioned the estimation problem is.[5]

[5]From Cauchy–Schwarz inequality, $\rho_F(\mathbf{M}) \geq p$ with equality if and only if $\mathbf{M}$ is proportional to the identity matrix.

5.1.3 Positive Homogeneity, Concavity, and Isotonicity

Among the desirable properties of a design criterion, positive homogeneity, concavity, and monotonicity play particularly important and imbricated roles. Here we shall simply state the definitions, mention the connections with other properties, and give some examples. One may refer in particular to Pukelsheim (1993, Chap. 5) for a detailed exposition. Also, only Loewner's partial ordering is considered[6]; one may refer, e.g., to Pázman (1986, Chap. 3) for developments concerning Schur's ordering.[7]

Definition 5.3. *An optimality criterion $\Phi(\cdot)$ defined on the closed set $\mathbb{M}^{\geq}$ of symmetric nonnegative-definite $p \times p$ matrices is:*

(i) Positively homogeneous when

$$\Phi(\alpha\mathbf{M}) = \alpha\Phi(\mathbf{M}), \ \forall\alpha > 0 \ \text{ and } \ \forall\mathbf{M} \in \mathbb{M}^{\geq}.$$

(ii) Concave when

$$\Phi[(1-\alpha)\mathbf{M}_1 + \alpha\mathbf{M}_2] \geq (1-\alpha)\Phi(\mathbf{M}_1) + \alpha\Phi(\mathbf{M}_2), \\ \forall\alpha \in (0,1) \ \text{ and } \ \forall\mathbf{M}_1, \mathbf{M}_2 \in \mathbb{M}^{\geq}$$

and strictly concave on $\mathbb{M}^{>}$ when the inequality is strict for any $\mathbf{M}_1 \in \mathbb{M}^{>}$, $\mathbf{M}_2 \in \mathbb{M}^{\geq}$, $\mathbf{M}_2 \neq \mathbf{O}$ and $\mathbf{M}_2$ not proportional to $\mathbf{M}_1$.

(iii) Monotonic for Loewner's ordering (or isotonic) when

$$\Phi(\mathbf{M}_1) \geq \Phi(\mathbf{M}_2), \ \forall\mathbf{M}_1 \succeq \mathbf{M}_2 \in \mathbb{M}^{\geq}.$$

If, moreover, $\Phi(\mathbf{M}_1) > \Phi(\mathbf{M}_2)$ when $\mathbf{M}_1 \neq \mathbf{M}_2$, $\Phi(\cdot)$ is strictly isotonic on $\mathbb{M}^{\geq}$ (on $\mathbb{M}^{>}$ if the condition $\mathbf{M}_1, \mathbf{M}_2 \in \mathbb{M}^{>}$ is required).

The following properties are easily obtained; see, for instance, Pukelsheim (1993, Sects. 5.2, 5.4) and Appendix C. In particular, the last one shows that positively homogeneous and isotonic criteria can be compared using a scaling standardization such that $\Phi(\mathbf{I}_p) = 1$; see Sect. 5.1.4.

Lemma 5.4. *Let $\Phi(\cdot)$ be a function from $\mathbb{M}^{\geq}$ to $\mathbb{R}$. Then,*

(i) When $\Phi(\cdot)$ is positively homogeneous, it is concave if and only if it is superadditive, i.e., $\Phi(\mathbf{M}_1+\mathbf{M}_2) \geq \Phi(\mathbf{M}_1)+\Phi(\mathbf{M}_2)$ for all $\mathbf{M}_1, \mathbf{M}_2 \in \mathbb{M}^{\geq}$.
(ii) When $\Phi(\cdot)$ is superadditive, nonnegativity implies isotonicity.
(iii) When $\Phi(\cdot)$ is positively homogeneous, isotonicity implies nonnegativity (i.e., $\Phi(\mathbf{M}) \geq 0$ for all $\mathbf{M}$ in $\mathbb{M}^{\geq}$); moreover, either Φ is identically zero or $\Phi(.)$ is strictly positive on the open set $\mathbb{M}^{>}$.

---

[6]For any $\mathbf{M}_1$, $\mathbf{M}_2 \in \mathbb{M}$, $\mathbf{M}_1 \succeq \mathbf{M}_2$ ($\mathbf{M}_1 \succ \mathbf{M}_2$) if and only if $\mathbf{M}_1 - \mathbf{M}_2 \in \mathbb{M}^{\geq}$ ($\mathbb{M}^{>}$).

[7]For any $\mathbf{M}_1$, $\mathbf{M}_2 \in \mathbb{M}$, $\mathbf{M}_1$ is better that $\mathbf{M}_2$ according to Schur's ordering if and only if $\Phi_{E_k}(\mathbf{M}_1) \geq \Phi_{E_k}(\mathbf{M}_2)$ for $k = 1, \ldots, p$, see (5.12), with strict inequality for one k at least.

Monotonicity simply means that $\Phi[\mathbf{M}(\xi)]$ increases when $\mathbf{M}(\xi)$ increases. The practical interpretation of superadditivity and positive homogeneity is more easily understood when considering non-normalized information matrices $N\mathbf{M}(\xi)$, with N the number of observations performed. Superadditivity means that when N_1 observations are taken with ξ_1 in experiment E_1 and N_2 with ξ_2 in experiment E_2, the merge of the two experiments yields more information than the sum of the informations in E_1 and E_2 in terms of $\Phi(\cdot)$; that is, $\Phi[N_1\mathbf{M}(\xi_1)+N_2\mathbf{M}(\xi_2)] \geq \Phi[N_1\mathbf{M}(\xi_1)]+\Phi[N_2\mathbf{M}(\xi_2)]$. Finally, positive homogeneity corresponds to a normalization that ensures that doubling the number of observations N doubles the information measured by $\Phi(\cdot)$.

5.1.4 Equivalence Between Criteria

For $\Phi(\cdot)$, a design criterion (to be maximized) and $\psi(\cdot) : \mathbb{R} \longrightarrow \mathbb{R}$ a strictly increasing function, $\tilde{\Phi}(\cdot) = \psi[\Phi(\cdot)]$ defines an optimality criterion equivalent to $\Phi(\cdot)$, in the sense that it preserves the ordering of designs. Such a transformation preserves (strict) isotonicity and can be used, for instance, to obtain a positive homogeneous criterion and then to form an efficiency criterion, see Sect. 5.1.8, or a compound criterion, see Sect. 5.1.9. Also, equivalence with a strictly concave criterion ensures the uniqueness of the optimal information matrix. At the same time, from a more practical point of view, such transformations allow us to use simple equivalent forms of criteria for optimization, where simplifications, for instance, in terms of computations of derivatives, may be of interest.[8]

Consider, for instance, D-optimum design that maximizes $\det^{1/p}(\mathbf{M})$, which is a positively homogenous, concave, and isotonic function of $\mathbf{M}$; see Sect. 5.1.5. One might consider equivalently the maximization of $\det(\mathbf{M})$; however, this function is neither positively homogeneous nor concave. Concerning the numerical determination of a D-optimal design, it may be of interest to work with $\log\det(\mathbf{M})$, which is clearly not positively homogeneous but has simpler derivatives than $\det^{1/p}(\mathbf{M})$, see Sect. 5.2.1; it is also a concave function of $\mathbf{M}$, see Sect. 5.1.5.

We shall denote by $\Phi^+(\cdot)$ the positively homogenous form of a criterion $\Phi(\cdot)$. A typical convention is to use a normalization such that $\Phi^+(\mathbf{I}_p) = 1$. The positively homogeneous form of a strictly isotonic criterion $\Phi(\cdot)$ is easily obtained as follows. Set $\Phi^+(\mathbf{M}) = \psi[\Phi(\mathbf{M})]$; for $\mathbf{M} = \mathbf{I}$, we should have $\psi[\Phi(a\mathbf{I}_p)] = a\psi[\Phi(\mathbf{I}_p)] = a\Phi^+(\mathbf{I}_p) = a$, and $\psi(\cdot)$ is simply the inverse function of $a \longrightarrow \Phi(a\mathbf{I}_p)$. We obtain, respectively, for D-, A-, c-, L-, and E-optimality

[8]It should be stressed here that neither positive homogeneity nor concavity do necessarily matter for the optimization of a differentiable criterion. Positive homogeneity is important when we wish to compare criteria through their efficiency. Equivalence with a concave criterion is crucial for preventing the existence of local maxima, but concavity itself is required in some particular situations only (typically, when the criterion is not differentiable). One may refer, e.g., to Avriel (2003, Chap. 6), for extensions of the notion of convexity.

$$\Phi_D^+(\mathbf{M}) = \det{}^{1/p}(\mathbf{M}), \tag{5.13}$$

$$\Phi_A^+(\mathbf{M}) = \begin{cases} \left[\frac{1}{p}\text{trace}(\mathbf{M}^{-1})\right]^{-1} & \text{if } \det(\mathbf{M}) \neq 0 \\ 0 & \text{otherwise,} \end{cases}$$

$$\Phi_c^+(\mathbf{M}) = \begin{cases} (\mathbf{c}^\top\mathbf{c})\,(\mathbf{c}^\top\mathbf{M}^-\mathbf{c})^{-1} & \text{if } \mathbf{c} \in \mathcal{M}(\mathbf{M}) \\ 0 & \text{otherwise,} \end{cases} \tag{5.14}$$

$$\Phi_L^+(\mathbf{M}) = \begin{cases} \frac{\text{trace}(\mathbf{Q}\mathbf{Q}^\top)}{\text{trace}(\mathbf{Q}\mathbf{Q}^\top\mathbf{M}^-)} & \text{if } \mathcal{M}(\mathbf{Q}) \subseteq \mathcal{M}(\mathbf{M}) \\ 0 & \text{otherwise,} \end{cases}$$

$$\Phi_E^+(\mathbf{M}) = \lambda_{\min}(\mathbf{M}).$$

Similarly, for the criteria $\Phi_{q,\mathbf{I}}(\cdot)$, see (5.11), we define

$$\Phi_{q,\mathbf{I}}^+(\mathbf{M}) = \begin{cases} \left[\frac{1}{p}\text{trace}(\mathbf{M}^{-q})\right]^{-1/q} & \text{if } \det(\mathbf{M}) \neq 0 \\ 0 & \text{otherwise.} \end{cases} \tag{5.15}$$

The composition of $\Phi_{q,\mathbf{I}}^+(\cdot)$ with an information mapping $\mathbf{M} \longrightarrow \mathbf{C}_\mathbf{Q}(\mathbf{M}) = (\mathbf{Q}^\top\mathbf{M}^-\mathbf{Q})^{-1}$ yields a positively homogeneous form for the criterion (5.11); see Pukelsheim (1993, Chap. 5).

5.1.5 Concavity and Isotonicity of Classical Criteria

Concavity

The following property shows that the c-optimality criterion defined by (5.9) is concave. It will also be used in Sect. 5.2.1 to obtain its directional derivative. The proof is given in Appendix C.

Lemma 5.5. *For any $p \times p$ matrix $\mathbf{M}$ in $\mathbb{M}^\geq$ and any $\mathbf{c} \in \mathcal{M}(\mathbf{M})$ (i.e., such that $\mathbf{c} = \mathbf{M}\mathbf{u}$ for some $\mathbf{u} \in \mathbb{R}^p$), we have*

$$\Phi_c(\mathbf{M}) = -\mathbf{c}^\top\mathbf{M}^-\mathbf{c} = \min_{\mathbf{z}\in\mathbb{R}^p}[\mathbf{z}^\top\mathbf{M}\mathbf{z} - 2\mathbf{z}^\top\mathbf{c}].$$

When $\mathbf{c} \notin \mathcal{M}(\mathbf{M})$, the right-hand side equals $-\infty$.

For any $\mathbf{M} \in \mathbb{M}^\geq$, we can thus define $\Phi_c(\mathbf{M})$ as in Lemma 5.5, compare with (5.9). Since the function $\mathbf{M} \longrightarrow \mathbf{z}^\top\mathbf{M}\mathbf{z}$ is concave for any $\mathbf{z} \in \mathbb{R}^p$, it implies that $\Phi_c(\cdot)$ is concave on $\mathbb{M}^\geq_\mathbf{c}$ (as the minimum of a family of concave functions). It also implies that $\Phi_L(\cdot)$ given by (5.10) is concave on $\mathbb{M}^\geq_\mathbf{Q}$. Also, any c-maximin-optimality criterion

$$\Phi_\mathcal{C}(\mathbf{M}) = \min_{\mathbf{c}\in\mathcal{C}} -\mathbf{c}^\top\mathbf{M}^-\mathbf{c} \tag{5.16}$$

is concave. It implies in particular that

$$\Phi_E(\mathbf{M}) = -\lambda_{\max}(\mathbf{M}^{-1}) = \min_{\|\mathbf{c}\|=1} -\mathbf{c}^\top \mathbf{M}^{-1}\mathbf{c}\,,$$

$$\Phi_{MV}(\mathbf{M}) = \min_{i=1,\ldots,p} -\mathbf{e}_i^\top \mathbf{M}^{-1}\mathbf{e}_i \tag{5.17}$$

and

$$\Phi_G(\mathbf{M}) = \min_{x\in\mathscr{X}} -\mathbf{f}_\theta^\top(x)\mathbf{M}^{-1}\mathbf{f}_\theta(x) \tag{5.18}$$

are concave on $\mathbb{M}^>$.

The following lemma is the analogue of Lemma 5.5 for the criterion $\Phi_c^+(\cdot)$. The proof is given in Appendix C.

Lemma 5.6. *For any $p \times p$ matrix $\mathbf{M}$ in $\mathbb{M}^{\geq}$ and any $\mathbf{c} \in \mathcal{M}(\mathbf{M})$, we have*

$$\Phi_c^+(\mathbf{M}) = (\mathbf{c}^\top\mathbf{c})(\mathbf{c}^\top\mathbf{M}^-\mathbf{c})^{-1} = (\mathbf{c}^\top\mathbf{c}) \min_{\mathbf{z}^\top\mathbf{c}=1} \mathbf{z}^\top\mathbf{M}\mathbf{z}\,.$$

When $\mathbf{c} \notin \mathcal{M}(\mathbf{M})$, the minimum on the right-hand side equals 0.

$\Phi_c^+(\cdot)$ is thus concave on $\mathbf{M} \in \mathbb{M}^{\geq}$ as the minimum of a family of concave functions.[9] Similarly, since $\Phi_E^+(\mathbf{M}) = \lambda_{\min}(\mathbf{M}) = \min_{\|\mathbf{c}\|=1} \mathbf{c}^\top\mathbf{M}\mathbf{c}$, this criterion is concave on the set $\mathbb{M}$ of symmetric $p\times p$ matrices for the same reasons. The criteria $\Phi_{E_k}(\cdot)$ defined by (5.12) are also concave on $\mathbb{M}$ since we can write

$$\Phi_{E_k}(\mathbf{M}) = \min_{\mathbf{Q}^\top\mathbf{Q}=\mathbf{I}_k} \operatorname{trace}(\mathbf{Q}^\top\mathbf{M}\mathbf{Q})\,,$$

see, e.g., Magnus and Neudecker (1999, p. 211). The strict concavity of the criteria $\Phi_{q,\mathbf{I}}^+(\cdot)$ on $\mathbb{M}^>$ for $q \in (-1,\infty)$ is proved in (Pukelsheim, 1993, Chap. 6).[10] Notice that it directly implies the concavity of $\Phi_D^+(\cdot)$ and $\Phi_A^+(\cdot)$. Also, since the function logarithm is concave, $\log\det(\cdot) = p\log[\Phi_D^+(\cdot)]$ is strictly concave on $\mathbb{M}^>$; see also Pázman (1986, p. 81) for a direct proof.

The following lemma shows that the D_s-optimality criterion is concave; the proof is given in Appendix C.

Lemma 5.7. *For any $p \times p$ matrix $\mathbf{M}$ in $\mathbb{M}^{\geq}$ partitioned as*

$$\mathbf{M} = \begin{pmatrix} \mathbf{M}_{11} & \mathbf{M}_{12} \\ \mathbf{M}_{21} & \mathbf{M}_{22} \end{pmatrix}$$

with $\mathbf{M}_{11}$ of dimension $s \times s$ we have

$$\log\det(\mathbf{M}_{11} - \mathbf{M}_{12}\mathbf{M}_{22}^-\mathbf{M}_{21}) \leq \log\det(\mathbf{M}_{11} + \mathbf{D}^\top\mathbf{M}_{22}\mathbf{D} - \mathbf{M}_{12}\mathbf{D} - \mathbf{D}^\top\mathbf{M}_{21})$$

for any $\mathbf{D} \in \mathbb{R}^{(p-s)\times s}$, with equality if and only if $\mathbf{M}_{22}\mathbf{D} = \mathbf{M}_{12}$.

---

[9] Notice that it implies that $\log[-\Phi_c(\cdot)]$ is convex on $\mathbb{M}_{\mathbf{c}}^{\geq}$, which is a stronger results than $\Phi_c(\cdot)$ being concave on $\mathbb{M}_{\mathbf{c}}^{\geq}$.

[10] Note that $\log[\Phi_{q,\mathbf{I}}^+(\mathbf{M})]$ is therefore concave on $\mathbb{M}^>$ for $q \in (-1,\infty)$, so that $\log[\operatorname{trace}(\mathbf{M}^{-q})]^{1/q}$ is convex, which is stronger than the convexity results mentioned, e.g., in (Kiefer, 1974) for $[\operatorname{trace}(\mathbf{M}^{-q})]^{1/q}$.

Isotonicity

Lemma 5.1 directly implies that the D-, A-, c-, E-, and MV-optimality criteria are isotonic on $\mathbb{M}^>$. Since in (5.10) $\text{trace}(\mathbf{Q}\mathbf{Q}^\top\mathbf{M}^-) = \sum_{i=1}^m \mathbf{c}_i^\top\mathbf{M}^-\mathbf{c}_i$ for $\mathbf{Q} = (\mathbf{c}_1,\ldots,\mathbf{c}_m)$, it also implies that the L-optimality criterion (5.10) is isotonic. From Lemma 5.1-(iv), the squared half-length of the i-th principal axis of the ellipsoid $\mathcal{E}_M$ is $\lambda_i^{-1}(\mathbf{M})$, where $\lambda_{\min}(\mathbf{M}) = \lambda_1(\mathbf{M}) \leq \lambda_2(\mathbf{M}) \leq \cdots \leq \lambda_p(\mathbf{M}) = \lambda_{\max}(\mathbf{M})$. Therefore, $\mathbf{M}_1 \succeq \mathbf{M}_2$ implies $\mathcal{E}_{\mathbf{M}_1} \subseteq \mathcal{E}_{\mathbf{M}_2}$ and $\lambda_i(\mathbf{M}_1) \geq \lambda_i(\mathbf{M}_2)$ for any $i = 1,\ldots,p$ and any $\mathbf{M}_1, \mathbf{M}_2 \in \mathbb{M}^>$, and the criteria $\Phi_{E_k}(\cdot)$ given by (5.12) are isotonic on $\mathbb{M}^>$. Also, we can write $\Phi_{q,\mathbf{I}}(\mathbf{M}) = -\left[(1/p)\sum_{i=1}^p \lambda_i^{-q}(\mathbf{M})\right]^{1/q}$, which is an increasing function of each λ_i for any $q \in \mathbb{R}$. $\Phi_{q,\mathbf{I}}(\cdot)$ is thus isotonic on $\mathbb{M}^>$ for any q; see also Pukelsheim (1993, Chap. 6) who proves the strict isotonicity of $\Phi_{q,\mathbf{I}}(\cdot)$ on $\mathbb{M}^\geq$ for $q \in [-1, 0)$ and on $\mathbb{M}^>$ for $q \in [0, \infty)$.

5.1.6 Classification into Global and Partial Optimality Criteria

As a rule, the matrix functions $\Phi(\cdot)$ from $\mathbb{M}^\geq$ to $\mathbb{R}$ that we consider are well defined, finite, and continuous on the set $\mathbb{M}^>$ of positive-definite $p\times p$ matrices. A positively homogeneous, isotonic, and nonconstant criterion $\Phi^+(\cdot)$ is strictly positive on $\mathbb{M}^>$; see Lemma 5.4-(iii). However, $\Phi^+(\cdot)$ may also take strictly positive values for singular matrices, so that the optimum can sometimes be obtained at a singular $\mathbf{M}$. In that case the attention cannot be restricted to the set $\mathbb{M}^>$. Some of the difficulties that may result, in particular in terms of differentiability, will be mentioned later. An additional difficulty concerns the absence of continuity on $\mathbb{M}^\geq$, which we detail below for the case of c-optimum design.

The following classification, see Pázman (1986), singularizes the situations where continuity on $\mathbb{M}^\geq$ may not hold.

Definition 5.8. *An isotonic design criterion $\Phi(\cdot)$ from $\mathbb{M}^\geq$ to $\mathbb{R}$ is said to be* global *when its nonnegative positively homogeneous version $\Phi^+(\cdot)$ satisfies $\Phi^+(\mathbf{M}) > 0$ if and only if $\mathbf{M}$ is nonsingular; it is said to be* partial *(or singular) when $\Phi^+(\mathbf{M}) > 0$ also for some singular $\mathbf{M}$.*

The criteria (5.8)–(5.11) are typical examples of partial optimality criteria. As a rule, a partial optimality criterion is only upper semicontinuous on $\mathbb{M}^\geq$, where upper semicontinuity is defined as follows; one may refer to Rockafellar (1970, Chap. 7) for alternative definitions.

Definition 5.9. *A design criterion $\Phi(\cdot)$ is said to be upper semicontinuous at $\mathbf{M}_*$ when $\limsup_{n\to\infty} \Phi(\mathbf{M}_n) \leq \Phi(\mathbf{M}_*)$ for any sequence of matrices $\mathbf{M}_n$ converging to $\mathbf{M}_*$. Similarly, a design criterion $\phi(\cdot)$ is said to be upper semicontinuous at ξ_* when $\limsup_{n\to\infty} \phi(\xi_n) \leq \Phi(\xi_*)$ for any sequence of measures ξ_n such that $\xi_n \Rightarrow \xi_*$.*

A proof of upper semicontinuity for the traditional criteria of Sect. 5.1.2 can be found in (Pázman, 1986). Design criteria $\Phi^+(\cdot)$ that are positively homogeneous, concave, nonnegative, nonconstant, and upper semicontinuous are called information functions in (Pukelsheim, 1993). In the next section we consider more particularly the case of c-optimality; notice that it forms a particular case of D_s-optimality with $s = 1$ when $\mathbf{c}$ is a basis vector $\mathbf{e}_i$.

5.1.7 The Upper Semicontinuity of the c-Optimality Criterion

Next example from Pázman (1986, p. 67) shows that $\phi_c(\xi) = \Phi_c[\mathbf{M}(\xi)]$, with $\Phi_c(\mathbf{M})$ given by (5.9), may be discontinuous at a ξ_* such that $\mathbf{M}(\xi_*) \in \mathbb{M}^{\geq}_{\mathbf{c}}$ is singular. See also Pukelsheim (1993, Sect. 3.16).

Example 5.10. Consider again the two parameter regression model of Example 2.4 with $\eta(x,\theta) = \theta_1 x + \theta_2 x^2$ and let $\xi_* = \delta_{x_*}$ ($x_* \neq 0$), so that the associated information matrix $\mathbf{M}(\xi_*)$ is singular (the model being linear in θ, the information matrix does not depend on θ). Take $\mathbf{c}_* = (x_*,\ x_*^2)$, which is in the range of $\mathbf{M}(\xi_*)$. We shall see in Example 5.34 that ξ_* is c-optimal for the estimation of $\mathbf{c}_*^\top\theta$ when $\sqrt{2}-1 < x_* \leq 1$. In Example 2.4 we obtained $\lim_{N\to\infty} \mathbf{c}_*^\top \mathbf{M}^{-1}(\xi_N)\mathbf{c}_* > \mathbf{c}_*^\top \mathbf{M}^{-1}(\xi_*)\mathbf{c}_* = 1$ for some design sequences such that $\xi_N \Rightarrow \xi_*$, illustrating that the criterion $\phi(\xi) = -\mathbf{c}_*^\top \mathbf{M}^{-1}(\xi)\mathbf{c}_*$ is not continuous at ξ_* and that the criterion $\Phi(\mathbf{M}) = -\mathbf{c}_*^\top \mathbf{M}^{-1}\mathbf{c}_*$ is not continuous at $\mathbf{M}(\xi_*)$. This phenomenon is not due to the particular type of sequence that was considered, and a similar discontinuity occurs for other ξ_N converging to ξ_*. Take, for instance, $\xi_N = (1/2)\delta_{x_{1,N}} + (1/2)\delta_{x_{2,N}}$ with $x_{1,N} = x_* + 1/N$, $x_{2,N} = x_* + \beta/N$, $\beta \neq 1$. We have $\xi_N \Rightarrow \xi_{x_*}$, and the matrix $\mathbf{M}(\xi_N)$ is invertible for any N. Direct calculations give

$$\begin{aligned}\mathbf{c}_*^\top \mathbf{M}^{-1}(\xi_N)\mathbf{c}_* &= \frac{2x_*^2}{(1-\beta)^2}\left[\frac{1}{(x_*+\beta/N)^2} + \frac{\beta^2}{(x_*+1/N)^2}\right] \\ &\to \frac{2(1+\beta^2)}{(1-\beta)^2}\,,\quad N\to\infty\,,\end{aligned}$$

with

$$\frac{2(1+\beta^2)}{(1-\beta)^2} = 1 + \frac{(1+\beta)^2}{(1-\beta)^2} > 1 = \mathbf{c}_*^\top \mathbf{M}^{-1}(\xi_*)\mathbf{c}_*\,.$$

Note that $\mathbf{c}_*^\top \mathbf{M}^{-1}(\xi_N)\mathbf{c}_*$ can thus become arbitrarily large as β approaches 1 and $N\to\infty$. □

Although not continuous at every ξ where it is defined (finite), $\phi_c(\cdot)$ satisfies the following (see Appendix C).

Lemma 5.11. *The criterion $\phi_c(\cdot) = \Phi_c[\mathbf{M}(\cdot)]$, with $\Phi_c(\mathbf{M})$ given by (5.9), is upper semicontinuous at any $\xi_* \in \Xi_c = \{\xi \in \Xi : \mathbf{c} \in \mathcal{M}[\mathbf{M}(\xi)]\}$.*

This property will be used in Sect. 5.4 to justify the use of a regularized version of the criterion.

Consider a sequence of matrices $\mathbf{M}(t)$ converging to $\mathbf{M}_0 = \mathbf{M}(0)$ as $t \to 0$ and take $\mathbf{c} \in \mathcal{M}(\mathbf{M}_0)$. As we have seen in Example 5.10, $\Phi_c[\mathbf{M}(t)]$ does not necessarily converge to $\Phi_c(\mathbf{M}_0)$ as $t \to 0$ when $\mathbf{M}_0$ is singular; it is therefore of interest to investigate under which conditions on $\mathbf{M}(t)$ the function $t \longrightarrow \Phi_c[\mathbf{M}(t)]$ is continuous at $t = 0$. For sequences of matrices defined by

$$\mathbf{M}(t) = \mathbf{M}_0 + t^\alpha \mathbf{M}_\alpha + \mathbf{R}_t\,, \quad t, \alpha > 0\,,\ \mathbf{M}_0 \in \mathbb{M}^{\geq}\,,\ \mathbf{M}_\alpha, \mathbf{R}_t \in \mathbb{M}\,, \tag{5.19}$$

a partial answer is given in the following lemma and its corollary, which use the following two conditions:

C1: $\|\mathbf{R}_t\| = [\text{trace}(\mathbf{R}_t^\top \mathbf{R}_t)]^{1/2} = o(t^\alpha)$ as $t \to 0^+$.
C2: $\mathbf{M}_0 + t^\alpha \mathbf{M}_\alpha \in \mathbb{M}^{>}$ for arbitrary small $t > 0$.

The proofs are given in Appendix C.

Lemma 5.12. *Consider a sequence of matrices satisfying (5.19) and suppose that* $\mathbf{c} \in \mathcal{M}(\mathbf{M}_0)$. *Then, under the conditions C1 and C2, we have*

$$\lim_{t \to 0^+} \Phi_c[\mathbf{M}(t)] = \Phi_c(\mathbf{M}_0)\,. \tag{5.20}$$

Corollary 5.13. *For a sequence of matrices* $\mathbf{M}(t) \in \mathbb{M}^{\geq}$ *satisfying (5.19) with* $\mathbf{c} \in \mathcal{M}(\mathbf{M}_0)$ *and the condition C1, either the continuity property (5.20) is satisfied or the convergence of* $\mathbf{M}(t)$ *to* $\mathbf{M}_0$ *is along a hyperplane tangent to the cone* $\mathbb{M}^{\geq}$ *at* $\mathbf{M}_0$, *i.e.,* $\mathbf{M}_\alpha$ *belongs to a supporting hyperplane to* $\mathbb{M}^{\geq}$ *at* $\mathbf{M}_0$.

It is instructive to consider again Example 5.10 in the light of the properties above.

Example 5.14. Consider the sequence of matrices $\mathbf{M}(\xi_N)$ in Example 5.10. Setting $t = 1/N$, we obtain

$$\mathbf{M}(t) = (1/2) \begin{pmatrix} (x+t)^2 & (x+t)^3 \\ (x+t)^3 & (x+t)^4 \end{pmatrix} + (1/2) \begin{pmatrix} (x+\beta\,t)^2 & (x+\beta\,t)^3 \\ (x+\beta\,t)^3 & (x+\beta\,t)^4 \end{pmatrix}.$$

Suppose first that $\beta \neq -1$. We obtain $\mathbf{M}(t) = \mathbf{M}_0 + t\mathbf{M}_1 + \mathbf{R}_t$ with

$$\mathbf{M}_0 = \begin{pmatrix} x^2 & x^3 \\ x^3 & x^4 \end{pmatrix}, \quad \mathbf{M}_1 = (1+\beta) \begin{pmatrix} x & (3/2)\,x^2 \\ (3/2)\,x^2 & 2\,x^3 \end{pmatrix}$$

and $\|\mathbf{R}(t)\| = \mathcal{O}(t^2)$ when $t \to 0$. The sequence $\mathbf{M}(t)$ is thus of the form (5.19) and satisfies C1.

The normal to the cone $\mathbb{M}^{\geq}$ at $\mathbf{M}_0$ is one-dimensional and defined by the rank-one matrices of the form $\mathbf{A} = \mathbf{a}\mathbf{a}^\top$ with $\mathbf{a}^\top \mathbf{M}_0 \mathbf{a} = 0$, i.e., $[x\ x^2]\mathbf{a} = 0$. Writing $\mathbf{a} = (a_1,\ a_2)^\top$, we get $a_1 + x a_2 = 0$, and direct calculation gives $\mathbf{a}^\top \mathbf{M}_1 \mathbf{a} = 0$; that is, $\mathbf{M}_1$ belongs to the supporting hyperplane to $\mathbb{M}^{\geq}$ at $\mathbf{M}_0$, which explains the discontinuity observed in Example 5.10.

Suppose now that $\beta = -1$. Then, $\mathbf{M}(t) = \mathbf{M}_0 + t^2 \mathbf{M}_2 + \mathbf{R}_t$ with

$$\mathbf{M}_2 = \begin{pmatrix} 1 & 3\,x \\ 3\,x & 6\,x^2 \end{pmatrix}, \quad \mathbf{R}_t = t^4 \begin{pmatrix} 0 & 0 \\ 0 & 1 \end{pmatrix}.$$

We then obtain $\mathbf{a}^\top \mathbf{M}_2 \mathbf{a} = (x a_2)^2 \neq 0$, and (5.20) is satisfied. □

5.1.8 Efficiency

Consider a positively homogeneous and isotonic criterion $\Phi^+(\cdot)$. Let ξ^* be an optimal design for the criterion $\phi^+(\xi) = \Phi^+[\mathbf{M}(\xi)]$; i.e., $\xi^* = \arg\max_{\Xi} \phi^+(\xi)$, with Ξ the set of design measures on $\mathscr{X} \subset \mathbb{R}^d$ (we assume for the moment that such a ξ^* exists). With the criteria $\Phi^+(\cdot)$ and $\phi^+(\cdot)$, we can then, respectively, associate the efficiency criteria

$$\mathscr{E}_\Phi(\mathbf{M}) = \frac{\Phi^+(\mathbf{M})}{\Phi^+[\mathbf{M}(\xi^*)]} \quad \text{and} \quad \mathscr{E}_\phi(\xi) = \frac{\phi^+(\xi)}{\phi^+(\xi^*)},$$

with $\mathscr{E}_\phi(\xi) \in [0, 1]$ for any $\xi \in \Xi$. For instance, the D- and c-efficiency criteria are, respectively, defined by

$$\mathscr{E}_D(\xi) = \frac{\det^{1/p}[\mathbf{M}(\xi)]}{\det^{1/p}[\mathbf{M}(\xi_D^*)]}, \quad \mathscr{E}_c(\xi) = \frac{\mathbf{c}^\top \mathbf{M}^-(\xi_c^*)\mathbf{c}}{\mathbf{c}^\top \mathbf{M}^-(\xi)\mathbf{c}}, \tag{5.21}$$

with ξ_D^* and ξ_c^*, respectively, a D- and a c-optimal design measure.

The c-maximin efficiency criterion is defined by

$$\mathscr{E}_{mmc}(\xi) = \min_{\mathbf{c} \in \mathcal{C}} \mathscr{E}_c(\xi) \tag{5.22}$$

with $\mathscr{E}_c(\cdot)$ given in (5.21) and $\mathcal{C}$ a set of directions of interest. This criterion is concave (as the minimum over a family of concave criteria), and we shall see in Sect. 5.3.2 that, under suitable conditions on $\mathcal{C}$, an optimal design for $\mathscr{E}_{mmc}(\cdot)$ is D-optimal.

Maximizing Efficiency over a Class of Criteria

Designing experiments that are reasonably efficient for a variety of criteria is a quite natural objective, hence the importance of establishing lower bounds on efficiency over a large class of criteria. The next important theorem (Harman, 2004a,b) shows that minimum efficiency over the class $\mathscr{C}_\perp$ of orthogonally invariant criteria (see Sect. 5.1.2) is reached in the finite set of the E_k-optimality criteria (5.12) and that a lower bound can be constructed from a $\Phi_{q,\mathbf{I}}$-optimal design; see (5.15). We first state the theorem (without proof) and then mention some of its consequences.

Theorem 5.15. *For any design measure ξ on $\mathscr{X}$ and any orthogonally invariant design criterion $\phi(\cdot) \in \mathscr{C}_\perp$,*

$$\mathscr{E}_\phi(\xi) \geq \min_{k=1,\ldots,p} \mathscr{E}_{\phi_{E_k}}(\xi)$$

where $\mathscr{E}_{\phi_{E_k}}(\xi) = \Phi_{E_k}[\mathbf{M}(\xi)]/\Phi_{E_k}[\mathbf{M}(\xi^*_{E_k})]$, with $\Phi_{E_k}[\mathbf{M}(\xi)]$ given by (5.12) and $\xi^*_{E_k}$ an optimal design for $\Phi_{E_k}(\cdot)$. Consequently,

$$\min_{\phi \in \mathscr{C}_\perp} \mathscr{E}_\phi(\xi) = \min_{k=1,\ldots,p} \mathscr{E}_{\phi_{E_k}}(\xi)$$

for any measure ξ on $\mathscr{X}$. Moreover, for any $q \in [-1, \infty)$ the maximum value of Φ_{E_k} satisfies

$$\Phi_{E_k}[\mathbf{M}(\xi^*_{E_k})] \leq \max_{r=1,\ldots,k} \frac{r \sum_{i=1}^p 1/(\lambda^*_{i,q})^q}{\sum_{i=k+1-r}^p 1/(\lambda^*_{i,q})^{q+1}} \tag{5.23}$$

where $\lambda^*_{1,q} \leq \lambda^*_{2,q} \leq \cdots \leq \lambda^*_{p,q}$ denote the eigenvalues of $\mathbf{M}(\xi^*_q)$ with ξ^*_q an optimal design measure for $\Phi^+_{q,\mathbf{I}}[\mathbf{M}(\xi)]$; see (5.15).

One may notice that (5.23) directly gives a lower bound on the E_k-efficiency of a $\Phi_{q,\mathbf{I}}$-optimal design. This simplifies for $k = 1$, where E_k-efficiency is E-efficiency. Indeed, with the same notations as in the theorem, we obtain

$$\mathscr{E}_E(\xi^*_q) = \frac{\lambda^*_{1,q}}{\Phi_{E_1}[\mathbf{M}(\xi^*_{E_1})]} \geq \frac{\sum_{i=1}^p (\lambda^*_{1,q}/\lambda^*_{i,q})^{q+1}}{\sum_{i=1}^p (\lambda^*_{1,q}/\lambda^*_{i,q})^q}$$

for any $\Phi_{q,\mathbf{I}}$-optimal design ξ^*_q, $q \in [-1, \infty)$.

In the special case of D-optimality, for which $q = 0$, this gives

$$\mathscr{E}_E(\xi^*_D) \geq \frac{1}{p} \sum_{i=1}^p \lambda^*_{1,0}/\lambda^*_{i,0} \geq m/p$$

with m the multiplicity of the smallest eigenvalue $\lambda^*_{1,0}$ of $\mathbf{M}(\xi^*_D)$. Moreover, for a D-optimal design measure ξ^*_D, Theorem 5.15 gives

$$\begin{aligned}\mathscr{E}_\phi(\xi^*_D) &\geq \min_{k=1,\ldots,p} \frac{\sum_{i=1}^k \lambda^*_{i,0}}{\Phi_{E_k}[\mathbf{M}(\xi^*_{E_k})]} \\ &\geq \min_{k=1,\ldots,p} \min_{r=1,\ldots,k} \frac{1}{rp} \left(\sum_{i=1}^k \lambda^*_{i,0}\right) \left(\sum_{i=k+1-r}^p 1/\lambda^*_{i,0}\right) \geq \frac{1}{p}\end{aligned}$$

for any orthogonally invariant criterion $\Phi(\cdot)$, where the last inequality follows from $\sum_{i=1}^k \lambda^*_{i,0} = \sum_{j=1}^k \lambda^*_{k+1-j,0} \geq \sum_{j=1}^r \lambda^*_{k+1-j,0} \geq r\lambda^*_{k+1-r,0}$ for all $r \in \{1, \ldots, k\}$. This property of D-optimal designs is rather exceptional: for ξ^*_q a $\Phi_{q,\mathbf{I}}$-optimal measure with $q > 0$, examples can be constructed for which the D-efficiency $\mathscr{E}_D(\xi_q^*)$ is arbitrarily small; see Galil and Kiefer (1977, Theorem 5.2).

Consider finally the case where there exists an optimal measure ξ_q^* for some $q \in [-1, \infty)$ such that the associated information matrix is proportional to the identity, i.e., $\mathbf{M}(\xi_q^*) = \alpha \mathbf{I}_p$ for some $\alpha > 0$. Theorem 5.15 then gives $\mathscr{E}_\phi(\xi_q^*) = 1$ for any orthogonally invariant criterion $\phi(\cdot)$; that is, ξ_q^* is universally optimal in the class of orthogonally invariant criteria. Harman (2004a) shows that the same universal optimality property holds when $\mathbf{M}(\xi_q^*)$ has at most two distinct eigenvalues.

5.1.9 Combining Criteria

Compound Criteria

The positively homogeneous forms $\Phi^+(\cdot)$ of different design criteria can be easily combined. Possible objectives are the achievement of a reasonable efficiency for several criteria simultaneously and the regularization of the optimization of a partial design criterion $\Phi_0^+(\cdot)$ through the introduction of a second global criterion $\Phi_1^+(\cdot)$; see Definition 5.8. This may be facilitated by a prior normalization of the criteria as done in Sect. 5.1.4.

Consider, for instance, the case where two positively homogeneous criteria $\Phi_0^+(\cdot)$ and $\Phi_1^+(\cdot)$, possibly normalized such that $\Phi_i^+(\mathbf{I}_p) = 1$, $i = 0, 1$, are combined into

$$\Phi_\alpha(\cdot) = (1-\alpha)\Phi_0^+(\cdot) + \alpha \Phi_1^+(\cdot)\,,\ \ \alpha \in [0,1]\,. \tag{5.24}$$

$\Phi_\alpha(\cdot)$ is positively homogeneous, isotonic, and concave when $\Phi_0^+(\cdot)$ and $\Phi_1^+(\cdot)$ are. Moreover, the directional derivative of $\Phi_\alpha(\cdot)$ is simply the weighted sum of the directional derivatives for $\Phi_0^+(\cdot)$ and $\Phi_1^+(\cdot)$; see Sect. 5.2.1. Choosing α close to 0 will make an optimal design for $\Phi_\alpha(\cdot)$ almost optimal for $\Phi_0^+(\cdot)$. Indeed, let ξ_α^* and ξ_0^* denote two optimal design measures for $\Phi_\alpha(\cdot)$ and $\Phi_0^+(\cdot)$, respectively. Since $\Phi_\alpha[\mathbf{M}(\xi_\alpha^*)] \geq \Phi_\alpha[\mathbf{M}(\xi_0^*)]$ and $\Phi_0^+[\mathbf{M}(\xi_0^*)] \geq \Phi_0^+[\mathbf{M}(\xi_\alpha^*)]$, we have $\Phi_1^+[\mathbf{M}(\xi_\alpha^*)] \geq \Phi_1^+[\mathbf{M}(\xi_0^*)]$, with a strict inequality if one of the previous two inequalities is strict, and direct calculations give

$$\begin{aligned}\mathscr{E}_{\Phi_0}(\xi_\alpha^*) = \frac{\Phi_0^+[\mathbf{M}(\xi_\alpha^*)]}{\Phi_0^+[\mathbf{M}(\xi_0^*)]} &\geq 1 - \alpha\, \frac{\Phi_1^+[\mathbf{M}(\xi_\alpha^*)] - \Phi_1^+[\mathbf{M}(\xi_0^*)]}{\Phi_\alpha[\mathbf{M}(\xi_\alpha^*)] - \alpha\Phi_1^+[\mathbf{M}(\xi_0^*)]} \\ &\geq 1 - \frac{\alpha}{1-\alpha}\, \frac{\Phi_1^+[\mathbf{M}(\xi_\alpha^*)] - \Phi_1^+[\mathbf{M}(\xi_0^*)]}{\Phi_0^+[\mathbf{M}(\xi_0^*)]}\,,\end{aligned}$$

which tends to one as α tends to zero.

Using Design Criteria as Constraints

Another possibility for combining two criteria $\Phi_0(\cdot)$ and $\Phi_1(\cdot)$ is to maximize $\Phi_0(\mathbf{M})$ under the constraint that $\Phi_1(\mathbf{M}) \geq \Delta$ for some $\Delta \in \mathbb{R}$, see Mikulecká

(1983)—we suppose that Δ is such that there exists $\mathbf{M} \in \mathcal{M}_\theta(\xi)$ such that $\Phi_0(\mathbf{M}) > -\infty$ and $\Phi_1(\mathbf{M}) > \Delta$. See also Cook and Wong (1994) for the equivalence between constrained and compound optimum design. When $\Phi_0(\cdot)$ and $\Phi_1(\cdot)$ are concave, so is the Lagrangian

$$L(\mathbf{M}, \lambda) = \Phi_0(\mathbf{M}) + \lambda[\Phi_1(\mathbf{M}) - \Delta]$$

used in the Kuhn–Tucker theorem, see, e.g., Alexéev et al. (1987, p. 75), from which a necessary-and-sufficient condition for optimality can be obtained; see Sect. 5.2.2. The extension to the case where several constraints are present is straightforward.

Maximin Criteria

We can also combine two (or more) criteria $\Phi_0(\cdot)$ and $\Phi_1(\cdot)$ using a maximin approach through the definition of

$$\Phi(\mathbf{M}) = \min[\Phi_0(\mathbf{M}), \Phi_1(\mathbf{M})] , \tag{5.25}$$

which is concave, positively homogeneous, and isotonic when $\Phi_0(\cdot)$ and $\Phi_1(\cdot)$ are. The criteria of c-maximin optimality (5.16) and c-maximin efficiency (5.22) form examples of maximin criteria over possibly infinite classes. Section 5.1.8 provides lower bounds on c-maximin efficiency over the whole class $\mathscr{C}_\perp$ of orthogonally invariant criteria. Few general results of this type exist, and most often one should be contented with the numerical determination of a maximin-optimal design. A necessary-and-sufficient condition for maximin optimality is given in Sect. 5.2.2.

5.1.10 Design with a Cost Constraint

A related but somewhat different situation is when cost constraints are present, of the form

$$\psi_i(\xi) = \int_{\mathscr{X}} C_i(x)\, \xi(\mathrm{d}x) \leq c_i \,, \; i = 1, \ldots, n_c \,, \tag{5.26}$$

where the $C_i(\cdot)$ are continuous in x; see Cook and Fedorov (1995) and Fedorov and Hackl (1997, p. 57). The treatment is then similar to the case above where one criterion is used as constraint. A noticeable particular situation is when only one constraint is present, in the form of a total-cost constraint

$$N\, \psi_C(\xi) = N \int_{\mathscr{X}} C(x)\, \xi(\mathrm{d}x) \leq C \,, \; C(x) \geq 0 \,, \; C > 0 \,, \tag{5.27}$$

with N the number of observations, considered as a free variable. For any isotonic criterion $\Phi_0(\cdot)$, maximizing the total information $\Phi_0[N\, \mathbf{M}(\xi)]$ with respect to ξ and N gives $N = N(\xi) = C/\psi_C(\xi)$, so that ξ should maximize

$$\phi(\xi) = \Phi_0 \left[\frac{C\, \mathbf{M}(\xi)}{\psi_C(\xi)} \right] . \tag{5.28}$$

An alternative presentation is as follows. Define $\tilde{\xi} = N\,\xi$, so that $\int_{\mathscr{X}} \tilde{\xi}(\mathrm{d}x) = N$, $N\,\mathbf{M}(\xi) = \mathbf{M}(\tilde{\xi})$; $\tilde{\xi}$ should then maximize $\Phi_0[\mathbf{M}(\tilde{\xi})]$ under the constraint $\psi_C(\tilde{\xi}) \leq C$. From the isotonicity of $\Phi_0(\cdot)$, the constraint is saturated at the optimum, and defining $\xi'(\mathrm{d}x) = \tilde{\xi}(\mathrm{d}x)\, C(x)/C$, we obtain from (5.1) $\mathbf{M}(\tilde{\xi}) = \int_{\mathscr{X}} \mathbf{M}_\theta(x)\, \tilde{\xi}(\mathrm{d}x) = \int_{\mathscr{X}} \mathbf{M}_\theta(x)\, [C/C(x)]\, \xi'(\mathrm{d}x)$ with $\int_{\mathscr{X}} \xi'(\mathrm{d}x) = 1$. The design problem thus recovers a standard unconstrained form; see, e.g., Fedorov and Hackl (1997, p. 57). Notice that C only appears as a multiplicative factor in the expression of $\mathbf{M}(\tilde{\xi})$ and has thus no influence on the optimal design. The case where $\Phi_0(\cdot)$ corresponds to D-optimality is considered in (Dragalin and Fedorov, 2006) and (Pronzato, 2010b).

5.2 Derivatives and Conditions for Optimality of Designs

5.2.1 Derivatives

Definitions and Notations

For a function $f(\cdot) : \mathbb{R}^d \longrightarrow \mathbb{R}$, the one-sided directional derivative of $f(\cdot)$ at $\mathbf{x} \in \mathbb{R}^d$ in the direction $\mathbf{y} \in \mathbb{R}^d$ (when it exists) is usually defined as

$$f'(\mathbf{x}; \mathbf{y}) = \lim_{\alpha \to 0^+} \frac{f(\mathbf{x} + \alpha \mathbf{y}) - f(\mathbf{x})}{\alpha} . \tag{5.29}$$

For a function $\Phi(\cdot) : \mathbb{M}^{\geq} \longrightarrow \mathbb{R}$, this gives the definition

$$\Phi'(\mathbf{M}_1; \mathbf{M}_2) = \lim_{\alpha \to 0^+} \frac{\Phi(\mathbf{M}_1 + \alpha \mathbf{M}_2) - \Phi(\mathbf{M}_1)}{\alpha} , \tag{5.30}$$

$\mathbf{M}_1, \mathbf{M}_2 \in \mathbb{M}^{\geq}$. However, the design problem concerns information matrices of the form (5.1) with $\int_{\mathscr{X}} \xi(\mathrm{d}x) = 1$, i.e., matrices that belong to $\mathcal{M}_\theta(\Xi)$, see (5.3), whereas $\mathbf{M}_1 + \alpha \mathbf{M}_2 \notin \mathcal{M}_\theta(\Xi)$ for $\alpha > 0$ and $\mathbf{M}_1, \mathbf{M}_2 \in \mathcal{M}_\theta(\Xi)$. For that reason, it is common in design theory to define the directional derivative of $\Phi(\cdot)$ at $\mathbf{M}_1$ in the direction $\mathbf{M}_2$ as

$$F_\Phi(\mathbf{M}_1; \mathbf{M}_2) = \lim_{\alpha \to 0^+} \frac{\Phi[(1-\alpha)\mathbf{M}_1 + \alpha \mathbf{M}_2] - \Phi(\mathbf{M}_1)}{\alpha} , \tag{5.31}$$

where now $(1-\alpha)\mathbf{M}_1 + \alpha \mathbf{M}_2 \in \mathcal{M}_\theta(\Xi)$ for any $\mathbf{M}_1, \mathbf{M}_2 \in \mathcal{M}_\theta(\Xi)$ and $\alpha \in [0, 1]$. Note that $F_\Phi(\mathbf{M}_1; \mathbf{M}_2) = \Phi'(\mathbf{M}_1, \mathbf{M}_2 - \mathbf{M}_1)$, so that the existence of one type of directional derivative implies the existence of the other.

Consider in particular the case where $\Phi(\cdot)$ is differentiable on the open set of nonsingular $p \times p$ matrices, i.e., when

$$\Phi(\mathbf{M}_2) = \Phi(\mathbf{M}_1) + \mathrm{trace}[(\mathbf{M}_2 - \mathbf{M}_1)\nabla_{\mathbf{M}}\Phi(\mathbf{M}_1)] + o(\|\mathbf{M}_2 - \mathbf{M}_1\|)$$

for any $\mathbf{M}_1, \mathbf{M}_2$ nonsingular, where $\nabla_{\mathbf{M}}\Phi(\mathbf{M}_1)$, the gradient of Φ at $\mathbf{M}_1$, is the $p \times p$ matrix with elements

$$\{\nabla_{\mathbf{M}}\Phi(\mathbf{M}_1)\}_{ij} = \left.\frac{\partial \Phi(\mathbf{M})}{\partial \{\mathbf{M}\}_{ij}}\right|_{\mathbf{M}=\mathbf{M}_1} .$$

We then obtain $\Phi'(\mathbf{M}_1; \mathbf{M}_2) = \mathrm{trace}[\mathbf{M}_2 \nabla_{\mathbf{M}}\Phi(\mathbf{M}_1)]$ and

$$F_\Phi(\mathbf{M}_1; \mathbf{M}_2) = \mathrm{trace}[(\mathbf{M}_2 - \mathbf{M}_1)\nabla_{\mathbf{M}}\Phi(\mathbf{M}_1)] . \tag{5.32}$$

In this finite-dimensional situation, when $\Phi(\cdot)$ is concave, the linearity of $\Phi'(\mathbf{M}_1; \mathbf{M}_2)$ in $\mathbf{M}_2$, or of $F_\Phi(\mathbf{M}_1; \mathbf{M}_2)$ in $\mathbf{M}_2 - \mathbf{M}_1$, is a necessary-and-sufficient condition for the differentiability of $\Phi(\cdot)$ at $\mathbf{M}_1$; see Rockafellar (1970, p. 244). Note that $\nabla_{\mathbf{M}}\Phi(\mathbf{M})$ is nonnegative definite when $\Phi(\cdot)$ is isotonic, see Definition 5.3.

Slightly more generally, consider now the case of a design criterion $\phi(\cdot)$ defined on the set Ξ of design measures on $\mathscr{X} \subset \mathbb{R}^d$. We shall use the following definition for the directional derivative $F_\phi(\xi; \nu)$ of $\phi(\cdot)$ at ξ in the direction ν:

$$F_\phi(\xi; \nu) = \lim_{\alpha \to 0^+} \frac{\phi[(1-\alpha)\xi + \alpha\nu] - \phi(\xi)}{\alpha} , \tag{5.33}$$

which is also standard in design theory; see, e.g., Silvey (1980). Suppose that $\phi(\cdot)$ is concave, i.e., $\phi[(1-\alpha)\xi + \alpha\nu] \geq (1-\alpha)\phi(\xi) + \alpha\xi(\nu)$ for any $\alpha \in (0,1)$ and $\xi, \nu \in \Xi$. Take any α_1, α_2 such that $0 < \alpha_1 < \alpha_2 < 1$. We can write

$$\begin{aligned}
\phi(\xi) - \phi[(1-\alpha_1)\xi + \alpha_1\nu] &= \phi(\xi) - \phi\left\{\frac{\alpha_1}{\alpha_2}[(1-\alpha_2)\xi + \alpha_2\nu] + \left(1 - \frac{\alpha_1}{\alpha_2}\right)\xi\right\} \\
&\leq \phi(\xi) - \left\{\frac{\alpha_1}{\alpha_2}\phi[(1-\alpha_2)\xi + \alpha_2\nu] + \left(1 - \frac{\alpha_1}{\alpha_2}\right)\phi(\xi)\right\} \\
&= \frac{\alpha_1}{\alpha_2}\left\{\phi(\xi) - \phi[(1-\alpha_2)\xi + \alpha_2\nu]\right\} .
\end{aligned}$$

Therefore, the function $\alpha \in (0,1) \longrightarrow \{\phi[(1-\alpha)\xi + \alpha\nu] - \phi(\xi)\}/\alpha$ is nonincreasing, so that the limit in (5.33) exists in $\mathbb{R} \cup \{+\infty\}$. A similar formulation can be given in terms of $F_\Phi(\mathbf{M}_1; \mathbf{M}_2)$ for a concave $\Phi(\cdot)$. We thus have the following property.

Lemma 5.16. *If $\phi(\cdot)$ is concave on Ξ, then the directional derivative $F_\phi(\xi; \nu)$ exists in $\mathbb{R} \cup \{+\infty\}$ for any $\nu \in \Xi$ and any $\xi \in \Xi$ such that $\phi(\xi) > -\infty$. Similarly, if $\Phi(\cdot)$ is concave on $\mathcal{M}_\theta(\Xi)$, then $F_\Phi(\mathbf{M}_1; \mathbf{M}_2)$ exists in $\mathbb{R} \cup \{+\infty\}$ for any $\mathbf{M}_2 \in \mathcal{M}_\theta(\Xi)$ and any $\mathbf{M}_1 \in \mathcal{M}_\theta(\Xi)$ such that $\Phi(\mathbf{M}_1) > -\infty$.*

In particular, when $\phi(\xi) = \Phi[\mathbf{M}(\xi)]$, with $\mathbf{M}(\xi)$ an information matrix of the form (5.1) and $\Phi(\cdot)$ differentiable, we have

$$F_\phi(\xi; \nu) = \mathrm{trace}\left\{[\mathbf{M}(\nu) - \mathbf{M}(\xi)]\, \nabla_{\mathbf{M}}\Phi[\mathbf{M}(\xi)]\right\} , \tag{5.34}$$

which corresponds to the first-order derivative of the function $\alpha \longrightarrow \phi[(1-\alpha)\xi + \alpha\nu]$ at $\alpha = 0$. Notice that $F_\phi(\xi; \nu)$ is linear in ν.

Composition with a Differentiable Function

Let $\psi(\cdot) : \mathbb{R} \longrightarrow \mathbb{R}$ be a differentiable function with derivative $\psi'(\cdot)$. Consider a design criterion $\Phi(\cdot)$ that admits a directional derivative $F_\Phi(\mathbf{M}_1; \mathbf{M}_2)$. The criterion $\tilde{\Phi}(\cdot) = \psi[\Phi(\cdot)]$ has the directional derivative

$$F_{\tilde{\Phi}}(\mathbf{M}_1; \mathbf{M}_2) = \psi'[\Phi(\mathbf{M}_1)] F_\Phi(\mathbf{M}_1; \mathbf{M}_2) .$$

Derivatives of Compound Criteria

A compound criterion of the form (5.24) is especially useful in situations where the optimal design for $\Phi_0(\cdot)$ is difficult to determine, e.g., because it may be obtained at a singular information matrix, so that attention cannot be restricted to $\mathbb{M}^>$, and $\Phi_0(\cdot)$ is not differentiable on $\mathbb{M}^\geq$. The presence of $\Phi_1(\cdot)$ then introduces some regularization. Suppose that $\Phi_0(\cdot)$ and $\Phi_1(\cdot)$ are concave, so that their directional derivatives exist from Lemma 5.16, and denote $\phi_\alpha(\cdot) = \Phi_\alpha[\mathbf{M}(\cdot)]$. We then obtain $F_{\phi_\alpha}(\xi; \nu) = (1-\alpha) F_{\phi_0}(\xi; \nu) + \alpha F_{\phi_1}(\xi; \nu)$. More than two criteria can be combined similarly by defining $\phi_\mu(\cdot) = \sum_{i=1}^m \mu_i \, \phi_i(\cdot)$ with $\mu_i \geq 0$ for all i and $\sum_{i=1}^m \mu_i = 1$. We then have $F_{\phi_\mu}(\xi; \nu) = \sum_{i=1}^m \mu_i \, F_{\phi_i}(\xi; \nu)$. For a continuous version, one may consider a class of criteria $\phi_{\mathbf{t}}(\cdot)$ indexed by a continuous variable $\mathbf{t} \in \mathbb{R}^m$ and define $\phi(\cdot) = \int_{\mathbb{R}^m} \phi_{\mathbf{t}}(\cdot) \, \mu(d\mathbf{t})$ for some measure μ on (a subset of) $\mathbb{R}^m$. The directional derivative $F_\phi(\xi; \nu)$ is then $\int_{\mathbb{R}^m} F_{\phi_{\mathbf{t}}}(\xi; \nu) \, \mu(d\mathbf{t})$.

Derivatives of Maximin Criteria

The following properties which we state without proof, see Dem'yanov and Malozemov (1974, Chap. 3, Theorem 2.1 and Chap. 6, Theorem 2.1), are extremely useful for obtaining the expression of directional derivatives of maximin criteria.

Lemma 5.17. *Assume that the functions $f_i(\cdot) : \mathbb{R}^d \longrightarrow \mathbb{R}$ are continuously differentiable in a neighborhood $\mathscr{B}(\mathbf{x}_0, \delta)$ of $\mathbf{x}_0$, $i = 1, \ldots, n$. Then the function $\mathbf{x} \longrightarrow f_*(\mathbf{x}) = \min_{i=1,\ldots,n} f_i(\mathbf{x})$ is differentiable at $\mathbf{x}_0$ in any direction $\mathbf{y}$ and*

$$f'_*(\mathbf{x}_0; \mathbf{y}) = \min_{i \in \mathcal{I}^*(\mathbf{x}_0)} f'_i(\mathbf{x}_0; \mathbf{y})$$

where $\mathcal{I}^*(\mathbf{x}) = \{i \in \{1, \ldots, n\} : f_i(\mathbf{x}) = f_*(\mathbf{x})\}$ and $f'_i(\mathbf{x}_0; \mathbf{y}) = \mathbf{y}^\top \nabla_{\mathbf{x}} f_i(\mathbf{x}_0)$.

Lemma 5.18. *Assume that the function $f(\cdot, \cdot) : \mathcal{A} \times \mathscr{B} \longrightarrow \mathbb{R}$ is continuous in $(\mathbf{x}, \mathbf{y})$ and continuously differentiable in $\mathbf{x}$ on $\mathcal{A} \times \mathscr{B}$, with $\mathcal{A}$ and $\mathscr{B}$, respectively, an open subset of $\mathbb{R}^{d_1}$ and a compact subset of $\mathbb{R}^{d_2}$. Then the function $\mathbf{x} \longrightarrow f_*(\mathbf{x}) = \min_{\mathbf{y} \in \mathscr{B}} f(\mathbf{x}, \mathbf{y})$ is differentiable at any $\mathbf{x}_0 \in \mathcal{A}$ in any direction $\mathbf{z} \in \mathbb{R}^{d_1}$ and*

$$f'_*(\mathbf{x}_0; \mathbf{z}) = \min_{\mathbf{y} \in \mathscr{B}^*(\mathbf{x}_0)} f'_x(\mathbf{x}_0, \mathbf{y}; \mathbf{z})$$

where $\mathscr{B}^*(\mathbf{x}) = \{\mathbf{y} \in \mathscr{B} : f(\mathbf{x}, \mathbf{y}) = f_*(\mathbf{x})\}$ and $f'_x(\mathbf{x}_0, \mathbf{y}; \mathbf{z}) = \mathbf{z}^\top \nabla_{\mathbf{x}} f(\mathbf{x}_0, \mathbf{y})$.

Consider the following extension of the criterion (5.25) of Sect. 5.1.9,

$$\Phi(\mathbf{M}) = \min_{i=1,\ldots,m} \Phi_i(\mathbf{M}),$$

and denote $\phi(\xi) = \Phi[\mathbf{M}(\xi)]$ and $\phi_i(\xi) = \Phi_i[\mathbf{M}(\xi)]$, $i = 1, \ldots, m$. Suppose that all $\Phi_i(\cdot)$ are differentiable so that Lemma 5.17 applies. Then,

$$\phi'(\xi;\nu) = \Phi'[\mathbf{M}(\xi);\mathbf{M}(\nu)] = \min_{i\in\mathcal{I}^*(\xi)} \Phi_i'(\xi;\nu)$$

where $\mathcal{I}^*(\xi) = \{i \in \{0,\ldots,m\} : \phi_i(\xi) = \phi(\xi)\}$. Similarly,

$$F_\phi(\xi;\nu) = \Phi'[\mathbf{M}(\xi);\mathbf{M}(\nu) - \mathbf{M}(\xi)] = \min_{i\in\mathcal{I}^*(\xi)} F_{\phi_i}(\xi;\nu). \tag{5.35}$$

The property (5.35) will be used below to obtain the directional derivatives of the MV-, E-, c-, and D_s-optimality criteria. By substituting Lemma 5.18 for Lemma 5.17, it can be directly extended to criteria of the form

$$\phi(\xi) = \min_{\omega\in\Omega} \Phi_\omega[\mathbf{M}(\xi)]$$

where $\{\Phi_\omega(\cdot),\, \omega \in \Omega\}$ defines a parameterized family of criteria such that $\Phi_\omega(\cdot)$ is differentiable for all $\omega \in \Omega$. We have in that case

$$F_\phi(\xi;\nu) = \min_{\omega\in\Omega(\xi)} F_{\Phi_\omega}[\mathbf{M}(\xi);\mathbf{M}(\nu)] \tag{5.36}$$

with $\Omega(\xi) = \{\omega \in \Omega : \Phi_\omega[\mathbf{M}(\xi)] = \phi(\xi)\}$.

The situation will be the same for the maximin-optimality criteria of Sect. 8.2 where $\phi(\xi) = \min_{\theta\in\Theta} \Phi[\mathbf{M}(\xi,\theta)]$ when $\Phi(\cdot)$ is differentiable. We shall then have

$$F_\phi(\xi;\nu) = \min_{\theta\in\Theta(\xi)} F_\Phi[\mathbf{M}(\xi,\theta);\mathbf{M}(\nu,\theta)]$$

with $\Theta(\xi) = \{\theta \in \Theta : \Phi[\mathbf{M}(\xi,\theta)] = \phi(\xi)\}$.

Application to Classical Criteria

For D-optimality we get $\nabla_{\mathbf{M}}\Phi(\mathbf{M}) = \mathbf{M}^{-1}$ when $\Phi(\mathbf{M}) = \log\det\mathbf{M}$ and $\nabla_{\mathbf{M}}\Phi(\mathbf{M}) = [(1/p)\det^{1/p}\mathbf{M}]\mathbf{M}^{-1}$ when $\Phi(\mathbf{M}) = \det^{1/p}\mathbf{M}$; when $\mathbf{M}$ has full rank, the gradient of $\Phi(\mathbf{M}) = -\text{trace}(\mathbf{A}\mathbf{M}^{-q})$ for $q \geq 1$ is $\nabla_{\mathbf{M}}\Phi(\mathbf{M}) = \sum_{i=1}^q \mathbf{M}^{-i}\mathbf{A}^\top\mathbf{M}^{-(q+1-i)}$; see, e.g., Harville (1997, Chap. 15).

The cases of MV-optimality given by (5.17) of E-optimality with $\Phi_E^+(\mathbf{M}) = \lambda_{\min}(\mathbf{M})$ or of c- and D_s-optimality are more subtle since the criterion is not differentiable everywhere. However, they correspond to particular versions of a maximin criterion and (5.35), (5.36) can be used.

For the criterion of MV-optimality (5.17), (5.35) directly gives at a nonsingular $\mathbf{M}_1$

$$F_{\Phi_{MV}}(\mathbf{M}_1;\mathbf{M}_2) = \min_{i\in\mathcal{I}^*(\mathbf{M}_1)} \mathbf{e}_i^\top \mathbf{M}_1^{-1}(\mathbf{M}_2-\mathbf{M}_1)\mathbf{M}_1^{-1}\mathbf{e}_i$$
$$= \min_{i\in\mathcal{I}^*(\mathbf{M}_1)} \{\mathbf{M}_1^{-1}\mathbf{M}_2\mathbf{M}_1^{-1} - \mathbf{M}_1^{-1}\}_{ii}\,,$$

where $\mathcal{I}^*(\mathbf{M}_1) = \{i : -\mathbf{e}_i^\top \mathbf{M}_1^{-1}\mathbf{e}_i = \Phi_{MV}(\mathbf{M}_1)\}$. The criterion is differentiable at $\mathbf{M}$ when $\mathcal{I}^*(\mathbf{M})$ is a singleton, $\mathcal{I}^*(\mathbf{M}) = \{i^*\}$, with $\nabla_{\mathbf{M}}\Phi_{MV}(\mathbf{M}) = \mathbf{M}^{-1}\mathbf{e}_{i^*}\mathbf{e}_{i^*}^\top\mathbf{M}^{-1}$.

For the case of E-optimality, we can write $\Phi_E^+(\mathbf{M}) = \min_{\{\mathbf{c}:\|\mathbf{c}\|=1\}} \mathbf{c}^\top\mathbf{M}\mathbf{c}$, and (5.36) gives

$$F_{\Phi_E^+}(\mathbf{M}_1;\mathbf{M}_2) = \min_{\mathbf{c}\in\mathscr{B}^*(\mathbf{M}_1)} \mathbf{c}^\top(\mathbf{M}_2-\mathbf{M}_1)\mathbf{c}\,,$$

where $\mathscr{B}^*(\mathbf{M}_1) = \{\mathbf{c}\in\mathbb{R}^p : \|\mathbf{c}\| = 1 \text{ and } \mathbf{M}_1\mathbf{c} = \lambda_{\min}(\mathbf{M}_1)\mathbf{c}\}$. Here the criterion is differentiable at $\mathbf{M}$ when its minimum eigenvalue $\lambda_{\min}(\mathbf{M})$ has multiplicity of one, and in that case $\nabla_{\mathbf{M}}\Phi_E^+(\mathbf{M}) = \mathbf{c}\mathbf{c}^\top$ with $\mathbf{c}$ the associated eigenvector of unit length—unique up to multiplication by -1.

More generally, at a nonsingular $\mathbf{M}_1$, the c-maximin-optimality criterion (5.16) has the directional derivative

$$F_{\Phi_{\mathcal{C}}}(\mathbf{M}_1;\mathbf{M}_2) = \min_{\mathbf{c}\in\mathscr{B}^*(\mathbf{M}_1)} \mathbf{c}^\top\mathbf{M}_1^{-1}(\mathbf{M}_2-\mathbf{M}_1)\mathbf{M}_1^{-1}\mathbf{c}\,,$$

where $\mathscr{B}^*(\mathbf{M}) = \{\mathbf{c}\in\mathcal{C} : -\mathbf{c}^\top\mathbf{M}^{-1}\mathbf{c} = \Phi_{\mathcal{C}}(\mathbf{M})\}$. In the case of G-optimality, see (5.18), this gives

$$F_G(\mathbf{M}_1;\mathbf{M}_2) = \min_{x\in\mathscr{B}^*(\mathbf{M})} \mathbf{f}_\theta^\top(x)\mathbf{M}_1^{-1}(\mathbf{M}_2-\mathbf{M}_1)\mathbf{M}_1^{-1}\mathbf{f}_\theta(x)$$

with $\mathscr{B}^*(\mathbf{M}) = \{x\in\mathscr{X} : -\mathbf{f}_\theta^\top(x)\mathbf{M}^{-1}\mathbf{f}_\theta(x) = \Phi_G(\mathbf{M})\}$.

Consider now the case of c-optimality into more details. When $\mathbf{M}_1$ has full rank, the criterion $\Phi_c(\mathbf{M}_1)$ defined by (5.9) is differentiable at $\mathbf{M}_1$ with gradient $\nabla_{\mathbf{M}}(\Phi_c)(\mathbf{M}_1) = \mathbf{M}_1^{-1}\mathbf{c}\mathbf{c}^\top\mathbf{M}_1^{-1}$. The directional derivative $F_{\Phi_c}(\mathbf{M}_1;\mathbf{M}_2)$ can thus be written in the form

$$F_{\Phi_c}(\mathbf{M}_1;\mathbf{M}_2) = \mathbf{c}^\top\mathbf{M}_1^{-1}\mathbf{M}_2\mathbf{M}_1^{-1}\mathbf{c} + \Phi_c(\mathbf{M}_1)\,,$$

see (5.32). However, a similar formula

$$\overline{F}_{\Phi_c}(\mathbf{M}_1;\mathbf{M}_2) = \mathbf{c}^\top\mathbf{M}_1^{-\top}\mathbf{M}_2\mathbf{M}_1^{-}\mathbf{c} + \Phi_c(\mathbf{M}_1)\,, \tag{5.37}$$

cannot be used if $\mathbf{M}_1$ is singular, even when $\mathbf{c}\in\mathcal{M}(\mathbf{M}_1)$, i.e., when $\mathbf{c} = \mathbf{M}_1\mathbf{u}$ for some $\mathbf{u}\in\mathbb{R}^p$. Indeed, when $\mathbf{M}_1$ is singular, $\Phi_c(\mathbf{M}_1)$ keeps a finite value that does not depend on the choice of the g-inverse that is used provided that $\mathbf{c}\in\mathcal{M}(\mathbf{M}_1)$, but its directional derivative $F_{\Phi_c}(\mathbf{M}_1;\mathbf{M}_2)$ generally differs from $\overline{F}_{\Phi_c}(\mathbf{M}_1;\mathbf{M}_2)$ which depends on the choice of the g-inverse. This is illustrated by the following example:

Example 5.19. Take

$$\mathbf{M}_1 = \begin{pmatrix} 2 & 1 & 0 \\ 1 & 2 & 0 \\ 0 & 0 & 0 \end{pmatrix}, \quad \mathbf{M}_2 = \begin{pmatrix} 0 & 0 & 0 \\ 0 & 3 & 1 \\ 0 & 1 & 3 \end{pmatrix} \quad \text{and} \quad \mathbf{c} = (1,\ 1,\ 0)^\top \in \mathcal{M}(\mathbf{M}_1),$$

together with the g-inverse

$$\mathbf{M}_1^- = \begin{pmatrix} 2/3 & -1/3 & 0 \\ -1/3 & 2/3 & 0 \\ 0 & 0 & 1 \end{pmatrix}.$$

Then, the directional derivative computed according to the definition (5.31) is $F_{\Phi_c}(\mathbf{M}_1; \mathbf{M}_2) = -10/27 < \overline{F}_{\Phi_c}(\mathbf{M}_1; \mathbf{M}_2) = -1/3$. □

Using (5.36), one can show that $\overline{F}_{\Phi_c}(\mathbf{M}_1; \mathbf{M}_2)$ given by (5.37) forms an upper bound on $F_{\Phi_c}(\mathbf{M}_1; \mathbf{M}_2)$ for any choice of the g-inverse $\mathbf{M}_1^-$ and, more precisely, that $F_{\Phi_c}(\mathbf{M}_1; \mathbf{M}_2)$ is the minimum of $\overline{F}_{\Phi_c}(\mathbf{M}_1; \mathbf{M}_2)$ over the set of all g-inverses $\mathbf{M}_1^-$. Indeed, from Lemma 5.5, $\Phi_c(\mathbf{M}_1) = \min_{\mathbf{z}\in\mathbb{R}^p}[\mathbf{z}^\top \mathbf{M}_1 \mathbf{z} - 2\mathbf{z}^\top \mathbf{c}]$, the minimum being attained at $\mathbf{z} = \mathbf{M}_1^- \mathbf{c}$ (see the proof of Lemma 5.5 in appendix) so that from (5.36),

$$F_{\Phi_c}(\mathbf{M}_1; \mathbf{M}_2) = \min_{\{\mathbf{z}:\mathbf{M}_1\mathbf{z}=\mathbf{c}\}} \mathbf{z}^\top \mathbf{M}_2 \mathbf{z} - \mathbf{z}^\top \mathbf{c}\,.$$

Using Harville (1997, Theorem 11.5.1), $\mathbf{M}_1\mathbf{z} = \mathbf{c}$ is equivalent to $\mathbf{z} = \mathbf{M}_1^- \mathbf{c}$ for some g-inverse $\mathbf{M}_1^-$, so that

$$F_{\Phi_c}(\mathbf{M}_1; \mathbf{M}_2) = \min_{\{\mathbf{A}:\mathbf{M}_1\mathbf{A}\mathbf{M}_1=\mathbf{M}_1\}} \mathbf{c}^\top \mathbf{A}^\top \mathbf{M}_2 \mathbf{A}\mathbf{c} - \mathbf{c}^\top \mathbf{A}\mathbf{c}\,. \tag{5.38}$$

Since c-optimum design with $\mathbf{c}$ equal to a basis vector $\mathbf{e}_i$ forms a particular case of D_s-optimum design (with $s = 1$), it is not surprising that difficulties similar to those encountered for c-optimality exist for D_s-optimality.

There is no special difficulty when $\mathbf{M}$ has full rank. Denote $\mathbf{A}^\top = [\mathbf{I}_s\ \mathbf{O}]$ with $\mathbf{I}_s$ the s-dimensional identity matrix and $\mathbf{O}$ the $s \times (p-s)$ null matrix. We have $\Phi_{D_s}(\mathbf{M}) = -\log\det[\mathbf{A}^\top \mathbf{M}^{-1} \mathbf{A}]$, with directional derivative

$$\begin{aligned} F_{\Phi_{D_s}}(\mathbf{M}; \mathbf{M}') &= \mathrm{trace}[\mathbf{M}^{-1}\mathbf{A}(\mathbf{A}^\top\mathbf{M}^{-1}\mathbf{A})^{-1}\mathbf{A}^\top\mathbf{M}^{-1}(\mathbf{M}' - \mathbf{M})] \\ &= \mathrm{trace}[\mathbf{M}^{-1}\mathbf{A}(\mathbf{A}^\top\mathbf{M}^{-1}\mathbf{A})^{-1}\mathbf{A}^\top\mathbf{M}^{-1}\mathbf{M}'] - s\,. \end{aligned}$$

This expression is also valid for other choices for $\mathbf{A}$, which corresponds to the criterion of D_A-optimality; see Sibson (1974). In the general situation, where $\mathbf{M}$ may be singular, we define

$$\Phi^{\mathbf{D}}(\mathbf{M}) = \log\det\left[\mathbf{M}_{11} + \mathbf{D}^\top \mathbf{M}_{22}\mathbf{D} - \mathbf{M}_{12}\mathbf{D} - \mathbf{D}^\top \mathbf{M}_{21}\right].$$

According to Lemma 5.7, $\Phi_{D_s}(\mathbf{M}) = \min_{\mathbf{D}\in\mathbb{R}^{(p-s)\times s}} \Phi^{\mathbf{D}}(\mathbf{M})$. The directional derivative of $\Phi^{\mathbf{D}}(\cdot)$ is easily calculated as

$$F_{\Phi_{\mathbf{D}}}(\mathbf{M};\mathbf{M}') = \operatorname{trace}\left[(\mathbf{M}_{11} + \mathbf{D}^\top \mathbf{M}_{22}\mathbf{D} - \mathbf{M}_{12}\mathbf{D} - \mathbf{D}^\top \mathbf{M}_{21})^{-1} \right.$$
$$\left. \times\, (\mathbf{M}'_{11} + \mathbf{D}^\top \mathbf{M}'_{22}\mathbf{D} - \mathbf{M}'_{12}\mathbf{D} - \mathbf{D}^\top \mathbf{M}'_{21})\right] - s\,.$$

From Lemma 5.7 and (5.36), we then obtain

$$\begin{aligned} F_{\Phi_{D_s}}(\mathbf{M};\mathbf{M}') &= \min_{\{\mathbf{D}\in\mathbb{R}^{(p-s)\times s}:\mathbf{M}_{22}\mathbf{D}=\mathbf{M}_{21}\}} F_{\Phi_{\mathbf{D}}}(\mathbf{M};\mathbf{M}') \\ &= \min_{\{\mathbf{D}\in\mathbb{R}^{(p-s)\times s}:\mathbf{M}_{22}\mathbf{D}=\mathbf{M}_{21}\}} \operatorname{trace}\left[(\mathbf{M}^*)^{-1} \right. \\ &\qquad \left. \times\, (\mathbf{M}'_{11} + \mathbf{D}^\top \mathbf{M}'_{22}\mathbf{D} - \mathbf{M}'_{12}\mathbf{D} - \mathbf{D}^\top \mathbf{M}'_{21})\right] - s\,, \end{aligned}$$

with $\mathbf{M}^* = \mathbf{M}_{11} - \mathbf{M}_{12}\mathbf{M}_{22}^{-}\mathbf{M}_{21}$.

5.2.2 The Equivalence Theorem

When the set $\mathcal{M}_\theta(\mathscr{X})$ given by (5.2) is compact and $\phi(\cdot)$ is concave, a ϕ-optimum design ξ^* on $\mathscr{X}$ always exists. It may not be unique, but the set of such designs is then convex; when $\phi(\xi) = \Phi[\mathbf{M}(\xi)]$ with $\Phi(\cdot)$ equivalent to a strictly concave criterion, then the ϕ-optimal matrix $\mathbf{M}(\xi^*)$ is unique.

We first formulate precisely the rather intuitive property that for a concave criterion $\phi(\cdot)$ the maximum value of the directional derivative gives an indication of the distance to the optimum value $\phi^* = \max_{\xi\in\Xi}\phi(\xi)$.

Lemma 5.20. *Let $\phi(\cdot)$ be a concave functional on the set Ξ of design measures on $\mathscr{X}$ and let ξ be a design measure with $\phi(\xi) > -\infty$. Then,*

$$0 \le \phi^* - \phi(\xi) \le d(\xi) = \sup_{\nu\in\Xi} F_\phi(\xi;\nu)\,.$$

In particular, when $\phi(\xi) = \Phi[\mathbf{M}(\xi)]$ with $\Phi(\cdot)$ differentiable and when $\mathcal{M}_\theta(\mathscr{X})$ given by (5.2) is a compact subset of $\mathbb{M}$, then

$$d(\xi) = \max_{x\subset\mathscr{X}} F_\phi(\xi, x) \tag{5.39}$$

with

$$F_\phi(\xi,x) = \operatorname{trace}\{\mathbf{M}_\theta(x)\nabla_{\mathbf{M}}\Phi[\mathbf{M}(\xi)]\} - \operatorname{trace}\{\mathbf{M}(\xi)\nabla_{\mathbf{M}}\Phi[\mathbf{M}(\xi)]\}\,. \tag{5.40}$$

Proof. From the concavity of $\phi(\cdot)$ and the proof of Lemma 5.16 we obtain that for any $\nu\in\Xi$,

$$\phi(\nu) - \phi(\xi) \le \frac{\phi[(1-\alpha)\xi + \alpha\nu] - \phi(\xi)}{\alpha} \le F_\phi(\xi;\nu)\,. \tag{5.41}$$

In particular, when $\nu = \xi^*$, an optimum measure such that $\phi(\xi^*) = \phi^*$, $0 \le \phi^* - \phi(\xi) \le F_\phi(\xi;\xi^*) \le \sup_{\nu\in\Xi} F_\phi(\xi;\nu)$.

For the rest of the proof we use (5.34) and

$$\begin{aligned}\operatorname{trace}\{\mathbf{M}(\nu)\nabla_{\mathbf{M}}\Phi[\mathbf{M}(\xi)]\} &= \int_{\mathscr{X}} \operatorname{trace}\{\mathbf{M}_\theta(x)\nabla_{\mathbf{M}}\Phi[\mathbf{M}(\xi)]\}\, \nu(\mathrm{d}x) \\ &\leq \max_{x\in\mathscr{X}} \operatorname{trace}\{\mathbf{M}_\theta(x)\nabla_{\mathbf{M}}\Phi[\mathbf{M}(\xi)]\}.\end{aligned}$$

Let x^* be such that

$$\operatorname{trace}\{\mathbf{M}_\theta(x^*)\nabla_{\mathbf{M}}\Phi[\mathbf{M}(\xi)]\} = \max_{x\in\mathscr{X}} \operatorname{trace}\{\mathbf{M}_\theta(x)\nabla_{\mathbf{M}}\Phi[\mathbf{M}(\xi)]\}.$$

Then,

$$\begin{aligned}\sup_{\nu\in\Xi} \operatorname{trace}\{\mathbf{M}(\nu)\nabla_{\mathbf{M}}\Phi[\mathbf{M}(\xi)]\} &\leq \operatorname{trace}\{\mathbf{M}_\theta(x^*)\nabla_{\mathbf{M}}\Phi[\mathbf{M}(\xi)]\} \\ &= \operatorname{trace}\{\mathbf{M}(\delta_{x^*})\nabla_{\mathbf{M}}\Phi[\mathbf{M}(\xi)]\} \\ &\leq \sup_{\nu\in\Xi} \operatorname{trace}\{\mathbf{M}(\nu)\nabla_{\mathbf{M}}\Phi[\mathbf{M}(\xi)]\}\end{aligned}$$

which completes the proof. ■

A direct consequence of Lemma 5.20 is that $d(\xi^*) = 0$ is a sufficient condition for ξ^* to be ϕ-optimal. The next theorem shows that this condition is also necessary.

Theorem 5.21 (Equivalence theorem[11] for ϕ-optimality). *Let $\phi(\cdot)$ be a concave functional on the set Ξ of design measures on $\mathscr{X}$. A design ξ^* is ϕ-optimal if and only if $d(\xi^*) = 0$ with $d(\xi) = \sup_{\nu\in\Xi} F_\phi(\xi;\nu)$.*

Proof. Following Lemma 5.20, we only need to prove that the condition is necessary. Suppose that $d(\xi^*) > 0$. It means that there exists $\nu \in \Xi$ such that $F_\phi(\xi^*;\nu) > 0$, and therefore, there exists $\alpha > 0$ such that $\phi[(1-\alpha)\xi^* + \alpha\nu] > \phi(\xi^*)$, which contradicts the optimality of ξ^*. ■

Remark 5.22.

(*i*) The equivalence theorem only expresses that for a concave criterion defined on a convex set a necessary-and-sufficient condition for being at the optimum is that the slope is not positive in any direction. It is thus clear that concavity is not indispensable; only the absence of local maxima is really required and equivalence with a concave criterion is enough. The equivalence theorem thus remains valid for a criterion $\phi(\cdot)$ that can be written as $\phi(\cdot) = \psi[\tilde{\phi}(\cdot)]$ with $\psi(\cdot)$ strictly increasing and $\tilde{\phi}(\cdot)$ concave.

(*ii*) The equivalence theorem takes a very special form for D-optimality, which, additionally to the properties mentioned in Sect. 5.1.2, gives a

[11] The now classical denomination "equivalence theorem" can be considered as a tribute to the work of Kiefer and Wolfowitz (1960); we use this term throughout the book although "Necessary and Sufficient condition for optimality" would be appropriate too.

motivation for considering D-optimum design. For $\Phi(\mathbf{M}) = \log\det\mathbf{M}$, $\nabla_{\mathbf{M}}\Phi(\mathbf{M}) = \mathbf{M}^{-1}$. When $\mathcal{M}_\theta(\mathscr{X})$ is compact, we have

$$d(\xi) = \max_{x\in\mathscr{X}} \mathrm{trace}[\mathbf{M}_\theta(x)\mathbf{M}^{-1}(\xi)] - p$$

for any $\xi \in \Xi$ such that $\mathbf{M}(\xi)$ is nonsingular; see (5.39), (5.40). We thus obtain the following: ξ_D^* is D-optimal if and only if $\max_{x\in\mathscr{X}} \mathrm{trace}[\mathbf{M}_\theta(x)\mathbf{M}^{-1}(\xi_D^*)] \leq p = \dim(\theta)$, i.e., if and only if ξ_D^* minimizes $\max_{x\in\mathscr{X}} \mathrm{trace}[\mathbf{M}_\theta(x)\mathbf{M}^{-1}(\xi)]$. When $\mathbf{M}_\theta(x) = c\,\mathbf{f}_\theta(x)\mathbf{f}_\theta^\top(x)$ for some $c > 0$, this corresponds to G-optimality, which is related to the maximum variance of the prediction or to the maximum width of the confidence region in Lemma 5.2. □

Since $d(\xi^*) > 0$ when ξ^* is not ϕ-optimal, we also have that ξ^* is ϕ-optimal if and only if it minimizes $d(\xi)$. When $d(\xi)$ is given by (5.39), (5.40), we obtain that ξ^* is ϕ-optimum if and only if it minimizes $\max_{x\in\mathscr{X}} F_\phi(\xi, x)$, i.e., if and only if $\max_{x\in\mathscr{X}} F_\phi(\xi^*, x) = 0$. Notice that this condition is often easy to check, e.g., by computing all the values $F_\phi(\xi^*, x)$ for $x \in \mathscr{X}$ when $\mathscr{X}$ is a finite set or by plotting $F_\phi(\xi^*, x)$ as a function of x when $\dim(x) \leq 2$; see Example 5.23 below. Moreover, since $F_\phi(\xi^*; \xi^*) = 0$, see (5.40), we have $F_\phi(\xi^*, x) = 0$ ξ^*-almost everywhere, i.e., $F_\phi(\xi^*, x) = 0$ on the support of ξ^*. Also, for any $\xi \in \Xi$, $\phi(\xi^*) \leq \phi(\xi) + \max_{x\in\mathscr{X}} F_\phi(\xi, x)$; see Lemma 5.20.

Example 5.23. Consider the same regression model as in (Box and Lucas, 1959) with

$$\eta(x,\theta) = \frac{\theta_1}{\theta_1 - \theta_2}\left[\exp(-\theta_2 x) - \exp(-\theta_1 x)\right]. \tag{5.42}$$

We suppose that the observation errors are stationary and $x \in \mathscr{X} = [0, 10]$. Without any loss of generality we take $\sigma^2 = 1$ so that $\mathbf{M}(\xi) = \int_{\mathscr{X}} \mathbf{f}_{\theta^0}(x)\mathbf{f}_{\theta^0}^\top(x)\,\xi(\mathrm{d}x)$ with $\mathbf{f}_{\theta^0}(x) = \partial\eta(x,\theta)\partial\theta\big|_{\theta^0}$. When $\theta^0 = (0.7,\ 0.2)^\top$, the associated D-optimal design measure ξ_D^* on $\mathscr{X}$, which can be computed by one of the algorithms presented in Sect. 9.1, puts mass 1/2 at each of the two support points given by $x^{(1)} \simeq 1.23$, $x^{(2)} \simeq 6.86$.

Figure 5.1 presents the evolution of $F_{\phi_D}(\xi, x) = \mathbf{f}_{\theta^0}^\top(x)\mathbf{M}^{-1}(\xi)\mathbf{f}_{\theta^0}(x) - p$ (dashed line) and of $F_{\phi_D}(\xi_D^*, x)$ (solid line) as functions of x, when $\xi = (1/3)\delta_{x^{(1)}} + (2/3)\delta_{x^{(2)}}$. For any nonsingular design ξ, $F_{\phi_D}(\xi, x) + 2$ is proportional to the asymptotic variance of the LS predictor $\eta(x, \hat{\theta}_{LS}^N)$, under the assumption $\bar{\theta} = \theta^0$, when $\hat{\theta}_{LS}^N$ is estimated with a design sequence satisfying Definition 2.1 or 2.2. The D-optimality of ξ_D^* (and the non-optimality of ξ) is clear from the figure. Notice that $F_{\Phi_D}(\xi_D^*, x^{(1)}) = F_{\Phi_D}(\xi_D^*, x^{(2)}) = 0$. □

Remark 5.24. The equivalence theorem is a most useful tool for constructing optimal designs analytically in particular situations. A well-known example concerns D-optimal designs for polynomial models, see, e.g., Fedorov (1972, Sect. 2.3) and Pukelsheim (1993, pp. 418–421) for developments and

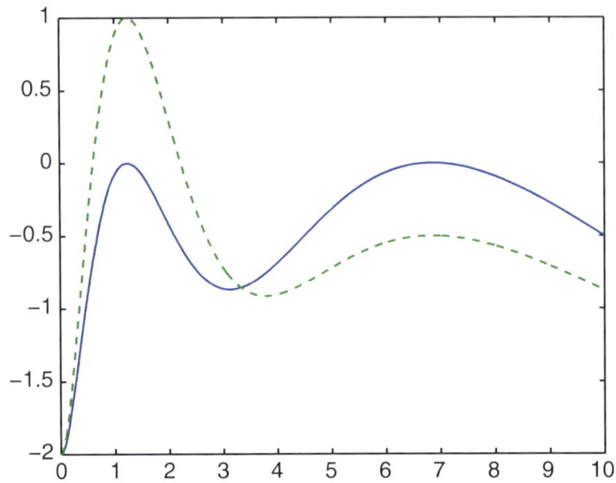

Fig. 5.1. Directional derivatives $F_{\phi_D}(\xi, x)$ (*dashed line*) and $F_{\phi_D}(\xi_D^*, x)$ (*solid line*)

references. Consider, for instance, the case of the second-order polynomial $\eta(x,\theta) = \theta_1 + \theta_2 x + \theta_3 x^2$; the D-optimal design measure on $\mathscr{X} = [a,b]$ gives weight 1/3 at each of the points a, b, and $(a+b)/2$. This is true whatever the basis chosen for the polynomial since a D-optimal design is invariant by reparameterization, see Sect. 5.1.2; for instance, the design measure above is also D-optimal for estimating $\alpha = (\alpha_1, \alpha_2, \alpha_3)^\top$ in the model $\eta(x,\alpha) = \alpha_1(x-\alpha_2)(x-\alpha_3)$ when $\alpha_1 \neq 0$ and $\alpha_2 \neq \alpha_3$, the Jacobian of the transformation $\alpha \longrightarrow \theta$ being equal to $\alpha_1^2(\alpha_2 - \alpha_3)$.

The equivalence theorem is also most useful for establishing general properties about design optimality. For instance, it is used in (Torsney, 1986) and (Pronzato et al., 2005) to obtain generalizations of various moment inequalities. Important results, much useful for simplifying the construction of optimal designs in complicated multifactor models, are proven in (Schwabe, 1996). In particular, when ξ_1 is D-optimal for the estimation of the parameters $(\alpha_0, \alpha^\top)^\top$ in the linear model with intercept $\eta_1(x, \alpha_0, \alpha) = \alpha_0 + \mathbf{f}_1^\top(x)\alpha$, $x \in \mathscr{X}_1$ and ξ_2 is D-optimal for the estimation of the parameters $(\beta_0, \beta^\top)^\top$ in the linear model with intercept $\eta_2(y, \beta_0, \beta) = \beta_0 + \mathbf{f}_2^\top(y)\beta$, $y \in \mathscr{X}_2$, then the product design[12] $\xi = \xi_1 \otimes \xi_2$ on $\mathscr{X} = \mathscr{X}_1 \times \mathscr{X}_2$ is D-optimal for the estimation of $\theta = (\theta_0\, ,\ \alpha^\top\, ,\ \beta^\top)^\top$ in the model $\eta([x,y],\theta) = \theta_0 + \mathbf{f}_1^\top(x)\alpha + \mathbf{f}_2^\top(y)\beta$. An additional condition is required when the intercept term is not present in both models; see (Schwabe, 1996, Sect. 5.2). This product design is also D-optimal in the complete product-type interaction model $\eta([x,y],\theta) = \eta_1(x,\alpha) \otimes \eta_2(y,\beta)$, see Schwabe (1996, Chap. 4), see also Schwabe and Wong (1999). For instance, it implies that the design measure allocating equal

[12] $\xi = \xi_1 \otimes \xi_2$ is the joint design measure on $\mathscr{X} = \mathscr{X}_1 \times \mathscr{X}_2$ with $\xi(\mathrm{d}x, \mathrm{d}y)$ given by the product $\xi_1(\mathrm{d}x) \times \xi_2(\mathrm{d}y)$ of the marginal measures, respectively, on $\mathscr{X}_1$ and $\mathscr{X}_2$.

weight 1/9 to each of the design points (a, c), (a, d), $(a, [c+d]/2)$, (b, c), (b, d), $(b, [c+d]/2)$, $([a+b]/2, c)$, $([a+b]/2, d)$, and $([a+b]/2, [c+d]/2)$ is D-optimal for estimating $\theta = (\theta_0, \alpha_1, \alpha_2, \beta_1, \beta_2)^\top$ in the model $\eta([x, y], \theta) = \theta_0 + \alpha_1 x + \alpha_2 x^2 + \beta_1 y + \beta_2 y^2$ and $\theta = (\theta_0, \alpha_1, \alpha_2, \beta_1, \beta_2, \theta_1, \theta_2, \theta_3, \theta_4)^\top$ in the model $\eta([x, y], \theta) = \theta_0 + \alpha_1 x + \alpha_2 x^2 + \beta_1 y + \beta_2 y^2 + \theta_1 xy + \theta_2 x^2 y + \theta_3 xy^2 + \theta_4 x^2 y^2$, with $x \in [a, b]$ and $y \in [c, d]$.

The optimality of products of D-optimal measures extends to the sum of nonlinear models with intercepts, where $\eta_1(x, \alpha_0, \alpha) = \alpha_0 + \eta_1'(x, \alpha)$, $\eta_2(y, \beta_0, \beta) = \beta_0 + \eta_1'(y, \beta)$, $\eta_1'(x, \alpha)$ and $\eta_2'(y, \beta)$ being nonlinear in α and β, respectively; see Schwabe (1995). Even more interestingly, it also applies to maximin D-optimal designs, see Sect. 8.2: the maximin D-optimal design measure over $\mathscr{X}_1 \times \mathscr{X}_2$ that maximizes $\min_{(\alpha\, \beta) \in \mathcal{A} \times \mathscr{B}} \det[\mathbf{M}(\xi, \alpha, \beta)]$ is the product $\xi_1 \otimes \xi_2$ of the maximin D-optimal measures that respectively maximize $\min_{\alpha \in \mathcal{A}} \det[\mathbf{M}_1(\xi, \alpha)]$ over $\mathscr{X}_1$ and $\min_{\beta \in \mathscr{B}} \det[\mathbf{M}_2(\xi, \beta)]$ over $\mathscr{X}_2$, with $\mathbf{M}_1(\xi, \alpha)$ the information matrix for the estimation of $(\alpha_0\,,\ \alpha^\top)^\top$ in $\eta_1(x, \alpha_0, \alpha)$, $\mathbf{M}_2(\xi, \beta)$ the information matrix for the estimation of $(\beta_0\,,\ \beta^\top)^\top$ in $\eta_2(y, \beta_0, \beta)$, and $\mathbf{M}(\xi, \alpha, \beta)$ the information matrix for the estimation of $\theta = (\theta_0\,,\ \alpha^\top\,,\ \beta^\top)^\top$ in $\eta([x, y], \theta) = \theta_0 + \eta_1'(x, \alpha) + \eta_2'(y, \beta)$. □

Equivalence Theorem for Constrained Design

Consider again the problem of maximizing $\Phi_0(\mathbf{M})$ under the constraint that $\Phi_1(\mathbf{M}) \geq \Delta$ for some $\Delta \in \mathbb{R}$, with $\Phi_0(\cdot)$ and $\Phi_1(\cdot)$ concave on $\mathcal{M}_\theta(\Xi)$. The presentation is for one constraint only but can easily be extended to several constraints of the form $\Phi_j(\mathbf{M}) \geq \Delta_j$ with $\Phi_j(\cdot)$ concave, $j = 1, \ldots, m$. The Lagrangian $L(\mathbf{M}, \lambda) = \Phi_0(\mathbf{M}) + \lambda[\Phi_1(\mathbf{M}) - \Delta]$ is also a concave function of $\mathbf{M}$ for any λ. Suppose that there exists $\mathbf{M} \in \mathcal{M}_\theta(\xi)$ with $\Phi_0(\mathbf{M}) > -\infty$ and $\Phi_1(\mathbf{M}) > \Delta$, so that Slater's condition is satisfied in the Kuhn–Tucker theorem; see Alexéev et al. (1987, p. 75). A necessary-and-sufficient condition for the optimality of $\mathbf{M}^*$ satisfying $\Phi_1(\mathbf{M}^*) \geq \Delta$ is the existence of λ such that

$$\lambda \geq 0\,,\ \lambda[\Phi_1(\mathbf{M}^*) - \Delta] = 0\,, \tag{5.43}$$

with $\mathbf{M}^*$ maximizing $L(\mathbf{M}, \lambda)$. Since $L(\mathbf{M}, \lambda)$ is concave for any λ, the equivalence theorem applies, and a necessary-and-sufficient condition for the optimality of $\mathbf{M}^*$ satisfying $\Phi_1(\mathbf{M}^*) \geq \Delta$ is the existence of λ satisfying (5.43) with

$$F_{\Phi_0}(\mathbf{M}^*; \mathbf{A}) + \lambda F_{\Phi_1}(\mathbf{M}^*; \mathbf{A}) \leq 0\,,\ \forall \mathbf{A} \in \mathcal{M}_\theta(\Xi)\,.$$

Notice that $\mathbf{M}^*$ is then optimal for the compound criterion (5.24) with $\alpha = \lambda/(1+\lambda)$, see, e.g., Cook and Fedorov (1995) and the enclosed discussion for the connection between compound and constrained designs. An optimal constrained design can thus be obtained by solving a series of compound design problems associated with increasing values of α, starting at $\alpha = 0$ and stopping at the first α such that the associated optimal matrix $\mathbf{M}^*(\alpha)$ satisfies $\Phi_1[\mathbf{M}^*(\alpha)] \geq \Delta$; see Mikulecká (1983). See also the algorithms in Sect. 9.5.

Take, for instance, $\Phi_0(\mathbf{M}) = \log\det \mathbf{M}$ and $\Phi_1(\mathbf{M}) = -\mathrm{trace}(\mathbf{M}^{-1})$. Then, ξ^* satisfying $\Phi_1[\mathbf{M}(\xi^*)] \geq \Delta$ is optimal in Ξ if and only if there exists $\lambda \geq 0$, with $\lambda\{\mathrm{trace}[\mathbf{M}^{-1}(\xi^*)] + \Delta\} = 0$ and

$$\max_{x\in\mathscr{X}} \left\{\mathrm{trace}[\mathbf{M}_\theta(x)\mathbf{M}^{-1}(\xi^*)] + \lambda\,\mathrm{trace}[\mathbf{M}_\theta(x)\mathbf{M}^{-2}(\xi^*)]\right\} \\ \leq p + \lambda\,\mathrm{trace}[\mathbf{M}^{-1}(\xi^*)]\,,$$

see Fedorov and Hackl (1997, p. 63).

Equivalence Theorem for Maximin Design

Consider the maximin criterion

$$\phi(\xi) = \min_{\omega\in\Omega} \Phi_\omega[\mathbf{M}(\xi)]$$

where $\{\Phi_\omega(\cdot),\, \omega \in \Omega\}$ defines a parameterized family of criteria such that $\Phi_\omega(\cdot)$ is concave and differentiable for all $\omega \in \Omega$. Its directional derivative is given by $F_\phi(\xi;\nu) = \min_{\omega\in\Omega(\xi)} F_{\Phi_\omega}[\mathbf{M}(\xi);\mathbf{M}(\nu)]$ with

$$\Omega(\xi) = \{\omega \in \Omega : \Phi_\omega[\mathbf{M}(\xi)] = \phi(\xi)\}\,,$$

see (5.36). The equivalence theorem 5.21 then says that ξ^* is optimal for $\phi(\cdot)$ if and only if

$$\sup_{\nu\in\Xi}\ \min_{\omega\in\Omega(\xi)} F_{\phi_\omega}(\xi^*;\nu) = 0\,, \tag{5.44}$$

where $F_{\phi_\omega}(\xi;\nu) = F_{\Phi_\omega}[\mathbf{M}(\xi);\mathbf{M}(\nu)]$. An alternative form is as follows.

Theorem 5.25. *A design ξ^* is optimal for $\phi(\cdot)$ if and only if*

$$\max_{x\in\mathscr{X}} \int_{\Omega(\xi^*)} F_{\phi_\omega}(\xi^*,x)\,\mu^*(\mathrm{d}\theta) = 0 \quad \textit{for some measure } \mu^* \in \mathscr{M}_{\xi^*}\,, \tag{5.45}$$

with $\mathscr{M}_\xi$ the set of probability measures on $\Omega(\xi)$, $F_{\phi_\omega}(\xi^*,x) = F_{\phi_\omega}(\xi^*;\delta_x)$, and δ_x the delta measure at x.

Proof. Since $\Phi_\omega(\cdot)$ is differentiable for all $\omega \in \Omega$, $F_{\phi_\omega}(\xi;\nu)$ is linear in ν, and we can write

$$F_{\phi_\omega}(\xi;\nu) = \int_{\mathscr{X}} F_{\phi_\omega}(\xi,x)\,\nu(\mathrm{d}x)\,.$$

Since

$$
\begin{aligned}
0 \leq \sup_{\nu \in \Xi} \min_{\omega \in \Omega(\xi)} \int_{\mathscr{X}} F_{\phi_\omega}(\xi, x)\, \nu(\mathrm{d}x) \\
&= \sup_{\nu \in \Xi} \min_{\mu \in \mathscr{M}_\xi} \int_{\mathscr{X}} \int_{\Omega(\xi)} F_{\phi_\omega}(\xi, x)\, \mu(\mathrm{d}\theta)\, \nu(\mathrm{d}x) \\
&= \min_{\mu \in \mathscr{M}_\xi} \sup_{\nu \in \Xi} \int_{\mathscr{X}} \int_{\Omega(\xi)} F_{\phi_\omega}(\xi, x)\, \mu(\mathrm{d}\theta)\, \nu(\mathrm{d}x) \\
&= \min_{\mu \in \mathscr{M}_\xi} \max_{x \in \mathscr{X}} \int_{\Omega(\xi)} F_{\phi_\omega}(\xi, x)\, \mu(\mathrm{d}\theta)\,, \qquad (5.46)
\end{aligned}
$$

the necessary-and-sufficient condition (5.44) can be written as (5.45). ∎

Remark 5.26.

(*i*) A necessary-and-sufficient condition for maximin c-optimality similar to (5.45) allowed Müller and Pázman (1998) to obtain explicit expressions for the optimal ξ^* for some simple polynomial models. However, the application of (5.45) is difficult in general situations.

(*ii*) The construction of $\sup_{\nu \in \Xi} F_\phi(\xi; \nu)$ in (5.44) is useful for checking the ϕ-optimality of a given design measure ξ. Also, we shall see in Chap. 9 that this construction is central in several optimization algorithms. One should then notice that $\sup_{\nu \in \Xi} F_\phi(\xi; \nu)$ is generally not obtained for ν^* equal to a one-point (delta) measure. Indeed, the minimax problem (5.46) has generally several solutions $x^{(i)}$ for x, $i = 1, \ldots, s$, and the optimal ν^* is then a linear combination $\sum_{i=1}^s w_i \delta_{x^{(i)}}$, with $w_i \geq 0$ and $\sum_{i=1}^s w_i = 1$; see Pronzato et al. (1991) for developments on a similar difficulty in T-optimum design for model discrimination. This property, due to the fact that $\phi(\cdot)$ is not differentiable, has the important consequence that the determination of a maximin-optimal design cannot be obtained via a standard vertex-direction optimization algorithm such as considered in Sect. 9.1.1. We should resort either to methods for non-differentiable optimization, see Sect. 9.3.1, or to a regularization of the design criterion, see Sect. 8.3.

(*iii*) When $\phi(\xi) = \Phi[\mathbf{M}(\xi)]$ with $\Phi(\cdot)$ differentiable, maximizing the directional derivative $F_\phi(\xi; \nu)$ defined by (5.33) is equivalent to maximizing the directional derivative defined in the usual way (5.29) since $F_\Phi(\mathbf{M}_1; \mathbf{M}_2) = \Phi'(\mathbf{M}_1; \mathbf{M}_2 - \mathbf{M}_1)$ which is linear in $\mathbf{M}_2 - \mathbf{M}_1$. This is not so obvious when $\phi(\xi) = \min_{\omega \in \Omega} \Phi_\omega[\mathbf{M}(\xi)]$, and one may wonder whether the maxima of $F_\phi(\xi; \nu)$ and $\phi'(\xi; \nu)$ are reached for the same $\nu \in \Xi$. The answer is positive at least in the following situation. Suppose that $\Phi_\omega(\cdot)$ is isotonic and differentiable for all $\omega \in \Omega$. Then, according to (5.36),

$$F_\phi(\xi; \nu) = \min_{\omega \in \Omega(\xi)} F_{\Phi_\omega}[\mathbf{M}(\xi); \mathbf{M}(\nu)]$$

where $F_{\Phi_\omega}(\mathbf{M}_1; \mathbf{M}_2) = \Phi'_\omega(\mathbf{M}_1; \mathbf{M}_2 - \mathbf{M}_1)$ is linear in $\mathbf{M}_2 - \mathbf{M}_1$. Therefore, we only need to check that $\Phi'_\omega(\mathbf{M}_1; \mathbf{M}_1)$ is a constant for all $\mathbf{M}_1 = \mathbf{M}_1(\xi)$ such that $\Phi_\omega(\mathbf{M}_1) = \phi(\xi)$. But this is the case if $\Phi_\omega(\cdot) = \psi[\Phi_\omega^+(\cdot)]$ for all

ω, with $\Phi_\omega^+(\cdot)$ the positively homogeneous form of $\Phi_\omega(\cdot)$ and $\psi(\cdot)$ continuous and strictly increasing, see Sect. 5.1.4: indeed, we have in this case $\Phi'_\omega(\mathbf{M}_1;\mathbf{M}_1) = \psi'[\Phi_\omega^+(\mathbf{M}_1)]\Phi_\omega^+(\mathbf{M}_1)$; see (5.30).

The situation will be the same for the maximin-optimality criteria of Sect. 8.2 when $\phi(\xi) = \min_{\theta\in\Theta} \Phi[\mathbf{M}(\xi,\theta)]$ and $\Phi(\cdot)$ is isotonic and differentiable. □

D_s-Optimality

As mentioned above, the equivalence theorem takes a more complicated form for criteria that are not differentiable everywhere. We detail below the case of D_s-optimum design; the case of c-optimum design will be considered in Sect. 5.3.4. We use the notations of Sect. 5.2.1. Take $\mathbf{M} = \mathbf{M}(\xi)$ for some $\xi \in \Xi$ such that $\mathbf{M}^* = \mathbf{M}_{11} - \mathbf{M}_{12}\mathbf{M}_{22}^-\mathbf{M}_{21}$ has full rank.

A sufficient condition for the D_s-optimality of ξ is as follows. If there exists $\mathbf{D} \in \mathbb{R}^{(p-s)\times s}$ such that $\mathbf{M}_{22}\mathbf{D} = \mathbf{M}_{21}$ and $F_{\phi\mathbf{D}}(\mathbf{M};\mathbf{M}') \le 0$ for all $\mathbf{M}' \in \mathcal{M}_\theta(\Xi)$, then $F_{\phi_{D_s}}(\mathbf{M};\mathbf{M}') \le 0$ for all $\mathbf{M}' \in \mathcal{M}_\theta(\Xi)$ and ξ is D_s-optimal. This condition can be slightly improved as follows. Define

$$\overline{F}_{\Phi\mathbf{D}}(\mathbf{M};\mathbf{M}') = \text{trace}\left[(\mathbf{M}^*)^{-1}(\mathbf{M}'_{11} + \mathbf{D}^\top\mathbf{M}'_{22}\mathbf{D} - \mathbf{M}'_{12}\mathbf{D} - \mathbf{D}^\top\mathbf{M}'_{21})\right] - s\,.$$

Then, $\overline{F}_{\Phi\mathbf{D}}(\mathbf{M};\mathbf{M}) = \text{trace}[(\mathbf{M}^*)^{-1}(\mathbf{C}-\mathbf{D})^\top\mathbf{M}_{22}(\mathbf{C}-\mathbf{D})]$ with $\mathbf{C}$ any matrix such that $\mathbf{M}_{22}\mathbf{C} = \mathbf{M}_{21}$; see the proof of Lemma 5.7. We thus have the following: if there exists $\mathbf{D} \in \mathbb{R}^{(p-s)\times s}$ such that and $\overline{F}_{\phi\mathbf{D}}(\mathbf{M};\mathbf{M}') \le 0$ for all $\mathbf{M}' \in \mathcal{M}_\theta(\Xi)$, then $\mathbf{D}$ necessarily satisfies $\mathbf{M}_{22}\mathbf{D} = \mathbf{M}_{21}$ (otherwise $\overline{F}_{\Phi\mathbf{D}}(\mathbf{M};\mathbf{M})$ would be strictly positive), and we have the same result as above, $F_{\phi_{D_s}}(\mathbf{M};\mathbf{M}') \le 0$ for all $\mathbf{M}' \in \mathcal{M}_\theta(\Xi)$ and ξ is D_s-optimal. When $\mathbf{M}(\xi) = \int_{\mathscr{X}} \mathbf{g}_\theta(x)\mathbf{g}_\theta^\top(x)\,\xi(\mathrm{d}x)$, denote $F_{\phi_{D_s}}(\xi;\nu) = F_{\phi_{D_s}}[\mathbf{M}(\xi);\mathbf{M}'(\nu)]$ and

$$\begin{aligned}\overline{F}_{\Phi\mathbf{D}}(\xi;\nu) &= \overline{F}_{\Phi\mathbf{D}}[\mathbf{M}(\xi);\mathbf{M}'(\nu)]\\ &= \int_{\mathscr{X}} [\mathbf{g}_{[1]}(x) - \mathbf{D}^\top\mathbf{g}_{[2]}(x)]^\top(\mathbf{M}^*)^{-1}[\mathbf{g}_{[1]}(x) - \mathbf{D}^\top\mathbf{g}_{[2]}(x)]\,\nu(\mathrm{d}x) - s,\end{aligned}$$

where $\mathbf{g}_\theta(x)$ has been partitioned as $\mathbf{g}_\theta(x) = [\mathbf{g}_{[1]}^\top(x)\ \mathbf{g}_{[2]}^\top(x)]^\top$. The sufficient condition above for the D_s-optimality of ξ becomes: there exists $\mathbf{D} \in \mathbb{R}^{(p-s)\times s}$ such that

$$\overline{F}_{\Phi\mathbf{D}}(\xi,x) = [\mathbf{g}_{[1]}(x) - \mathbf{D}^\top\mathbf{g}_{[2]}(x)]^\top(\mathbf{M}^*)^{-1}[\mathbf{g}_{[1]}(x) - \mathbf{D}^\top\mathbf{g}_{[2]}(x)] - s \le 0$$

for any $x \in \mathscr{X}$. It corresponds to the sufficient condition in (Atwood, 1969, Sect. 3)—a corrected version of a condition in (Karlin and Studden, 1966).

On the other hand, the minimax theorem applies, see, e.g., Dem'yanov and Malozemov (1974, Theorem 5.2, p. 218) and Polak (1987, Corollary 5.5.6, p. 707), and

$$\max_{\mathbf{M}'\in\mathcal{M}_\theta(\Xi)} \min_{\{\mathbf{D}\in\mathbb{R}^{(p-s)\times s}:\mathbf{M}_{22}\mathbf{D}=\mathbf{M}_{21}\}} F_{\phi\mathbf{D}}(\mathbf{M};\mathbf{M}')$$
$$= \min_{\{\mathbf{D}\in\mathbb{R}^{(p-s)\times s}:\mathbf{M}_{22}\mathbf{D}=\mathbf{M}_{21}\}} \max_{\mathbf{M}'\in\mathcal{M}_\theta(\Xi)} F_{\phi\mathbf{D}}(\mathbf{M};\mathbf{M}') \,.$$

Therefore, if ξ is D_s-optimal, it implies that there exists $\mathbf{D} \in \mathbb{R}^{(p-s)\times s}$ such that $\mathbf{M}_{22}\mathbf{D} = \mathbf{M}_{21}$ and $F_{\phi\mathbf{D}}(\mathbf{M};\mathbf{M}') \leq 0$ for all $\mathbf{M}' \in \mathcal{M}_\theta(\Xi)$. Since $F_{\phi\mathbf{D}}(\mathbf{M};\mathbf{M}') = \overline{F}_{\phi\mathbf{D}}(\mathbf{M};\mathbf{M}')$ when $\mathbf{M}_{22}\mathbf{D} = \mathbf{M}_{21}$, the D_s-optimality of ξ implies that there exists $\mathbf{D} \in \mathbb{R}^{(p-s)\times s}$ such that $\overline{F}_{\phi\mathbf{D}}(\mathbf{M};\mathbf{M}') \leq 0$ for all $\mathbf{M}' \in \mathcal{M}_\theta(\Xi)$. When $\mathbf{M}(\xi) = \int_{\mathscr{X}} \mathbf{g}_\theta(x)\mathbf{g}_\theta^\top(x)\,\xi(\mathrm{d}x)$, it implies that $\overline{F}_{\Phi\mathbf{D}}(\xi,x) \leq 0$ for all $x \in \mathscr{X}$. This corresponds to the necessary condition in (Atwood, 1969, Theorem 3.2).

5.2.3 Number of Support Points

We already noticed that $\mathcal{M}_\theta(\Xi)$ is the convex hull of the set $\mathcal{M}_\theta(\mathscr{X})$. Also, $\mathcal{M}_\theta(\Xi)$ being a set of symmetric $p \times p$ matrices, it is a subset of $\mathbb{R}^{p(p+1)/2}$. Therefore, from Caratheodory's theorem, see, e.g., Silvey (1980, p. 72), any matrix in $\mathcal{M}_\theta(\Xi)$ can be written as the linear combination of $m = p(p+1)/2+1$ elements of $\mathcal{M}_\theta(\mathscr{X})$ at most.[13] In other words, with any design measure $\xi \in \Xi$, one may associate a discrete measure ξ_d, supported on m points of $\mathscr{X}$ at most, such that $\mathbf{M}(\xi) = \mathbf{M}(\xi_d)$. This bound can be quite pessimistic. In particular, we shall see in Sect. 5.3.1 that one can always find a c-optimal design measure supported on $m \leq p$ points. Also, even if there exist situations where the bound $m = p(p+1)/2$ is reached for D-optimal measures, the number of D-optimal support points is usually much smaller, and p points are often enough. One may refer to Yang and Stufken (2009) for general results concerning the optimality of two-point designs for isotonic criteria in nonlinear models with two parameters. The generalization to models with more than two parameters is presented in (Yang, 2010) and (Dette and Melas, 2011), extending de la Garza (1954) result on polynomial regression to general nonlinear models. The duality property presented below provides a geometrical interpretation for this phenomenon.

Remark 5.27. The bound m on the number of support points of an optimal design measure is not modified by the presence of constraints defined by criteria that are functions of $\mathbf{M}(\xi)$. However, in the case of linear constraints of the form (5.26), see Sect. 5.1.10, $\{\mathbf{M}(\xi), \psi_1(\xi), \ldots, \psi_{n_c}(\xi)\}$ belongs to the convex hull of the set $\{\mathbf{M}_\theta(x), C_1(x), \ldots, C_{n_c}(x) : x \in \mathscr{X}\}$, and the bound

[13] We have $m = p(p+1)/2$ when the design criterion $\Phi(\cdot)$ is strictly isotonic since an optimum design matrix $\mathbf{M}^*$ (unique if $\Phi(\cdot)$ is strictly concave or is equivalent to a strictly concave criterion) necessarily lies on the boundary of $\mathcal{M}_\theta(\Xi)$. Indeed, when $\mathbf{M}$ is in the interior of $\mathcal{M}_\theta(\Xi)$, there exists $\alpha > 1$ such that $\alpha\mathbf{M} \in \mathcal{M}_\theta(\Xi)$ and $\Phi(\alpha\mathbf{M}) > \Phi(\mathbf{M})$. From Caratheodory's theorem, $\mathbf{M}^*$ can then be written as the linear combination of $p(p+1)/2$ elements of $\mathcal{M}_\theta(\mathscr{X})$.

on the number of support points becomes $p(p+1)/2 + n_c + 1$; see Cook and Fedorov (1995). □

5.2.4 Elfving's Set and Some Duality Properties

It is a general fact that the maximization of a concave function $\Phi(\cdot)$ over a convex set (here the set $\mathcal{M}_\theta(\Xi)$) admits a dual representation. It happens that for some design criteria this dual representation has a nice geometrical interpretation.

Consider the situation where $\mathbf{M}_\theta(x)$ in (5.1) has rank one; that is, it takes the form

$$\mathbf{M}_\theta(x) = \mathbf{g}_\theta(x)\mathbf{g}_\theta^\top(x)\,. \tag{5.47}$$

For instance, in a regression model, $\mathbf{g}_\theta(x) = \sigma^{-1}\mathbf{f}_\theta(x)$ in (3.26) and $\mathbf{g}_\theta(x) = \lambda^{-1/2}(x,\theta)\mathbf{f}_\theta(x)$ in (3.56). The Elfving's set $\mathscr{F}_\theta$ is then defined as the convex closure of the set $\{\mathbf{g}_\theta(x) : x \in \mathscr{X}\} \cup \{-\mathbf{g}_\theta(x) : x \in \mathscr{X}\}$. We first state a property concerning the support of an optimal design when $\Phi(\cdot)$ is isotonic; see Fellman (1974) and Pázman (1986, p. 56). The proof is given in Appendix C.

Lemma 5.28. *When the design criterion $\Phi(\cdot)$ is isotonic, an optimal design is supported at values of x such that $\mathbf{g}_\theta(x)$ is on the boundary of the Elfving's set $\mathscr{F}_\theta$.*

Consider now ellipsoids $\mathcal{E}_A$ defined by

$$\mathcal{E}_A = \{\mathbf{t} \in \mathbb{R}^p : \mathbf{t}^\top \mathbf{A}\mathbf{t} \le 1\}\,,$$

that contain $\mathscr{F}_\theta$, i.e., such that $\mathbf{g}_\theta^\top(x)\mathbf{A}\mathbf{g}_\theta(x) \le 1$ for all $x \in \mathscr{X}$. Note that from Lemma 5.1-(i), the volume of $\mathcal{E}_A$ is proportional to $\det^{-1/2}\mathbf{A}$.

Using Lagrangian theory, one can show that the determination of a D-optimal design measure ξ_D^* on $\mathscr{X}$ that maximizes $\det \mathbf{M}(\xi)$ is equivalent to the determination of the minimum-volume ellipsoid $\mathcal{E}_{A^*}$ that contains Elfving's set $\mathscr{F}_\theta$, with the optimum value for $\mathbf{A}$ (such that $\det \mathbf{A}$ is maximal) equal to $\mathbf{A}^* = \mathbf{M}^{-1}(\xi_D^*)/p$. Moreover, the support points of ξ_D^* satisfy $\mathbf{g}_\theta^\top(x)\mathbf{A}^*\mathbf{g}_\theta(x) = 1$; that is, they correspond to the points of contact between $\mathcal{E}_{A^*}$ and $\mathscr{F}_\theta$; see, e.g., Silvey (1980, p. 78), Pronzato and Walter (1994). When $\mathscr{X}$ is finite, $\mathscr{F}_\theta$ is a bounded convex polyhedron, and there exist fast algorithms for solving this minimal ellipsoid problem, see, e.g., Welzl (1991) and Sun and Freund (2004); see also Sect. 9.1.4. Once $\mathbf{A}^*$ is found and the support points of ξ_D^* are located, only their weights have to be determined. The connection with other ellipsoid problems is briefly considered in Sect. 5.6.

Example 5.29. Consider again the regression model of Example 5.23, with $\theta^0 = (0.7,\ 0.2)^\top$. The associated D-optimal measure is $\xi_D^* = (1/2)\delta_{x^{(1)}} + (1/2)\delta_{x^{(2)}}$, $x^{(1)} \simeq 1.23$, $x^{(2)} \simeq 6.86$.

Figure 5.2 shows the set $\{\mathbf{f}_{\theta^0}(x) : x \in \mathscr{X}\}$ (solid line), its symmetric $\{-\mathbf{f}_{\theta^0}(x) : x \in \mathscr{X}\}$ (dashed line), and Elfving's set $\mathscr{F}_{\theta^0}$ (colored region)

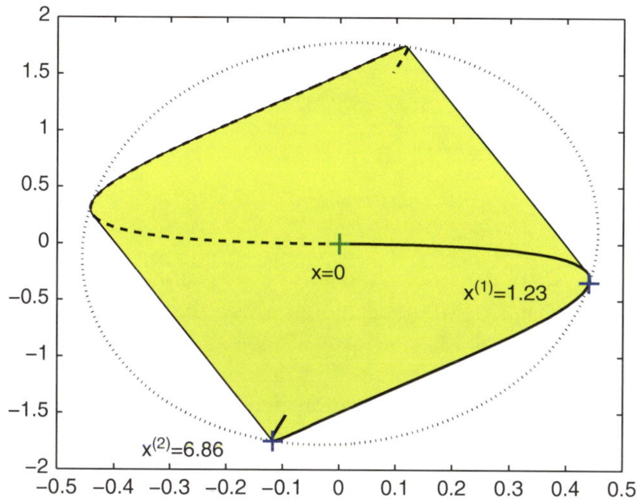

Fig. 5.2. Minimum-volume ellipsoid containing Elfving's set in Example 5.29

together with the minimum-volume ellipsoid containing $\mathscr{F}_{\theta^0}$. Note that the points of contact correspond to the support points of ξ_D^*. □

Since the support points of ξ_D^* correspond to the points of contact of the minimum-volume ellipsoid containing $\mathscr{F}_\theta$ with the set $\{\mathbf{f}_\theta(x) : x \in \mathscr{X}\}$, it is clear that a D-optimal measure has often much less than $p(p+1)/2$ support points. In fact, it often has p support points only, as in the example above, and when this happens, those p points $x^{(i)}$, $i = 1, \ldots, p$, all receive the same mass $1/p$. Indeed, we then have $\log\det \mathbf{M}(\xi_D^*) = 2\log\det[\mathbf{f}_\theta(x^{(1)}), \ldots, \mathbf{f}_\theta(x^{(p)})] + \log\prod_{i=1}^p \xi_D^*(x^{(i)})$, which, under the constraint $\sum_{i=1}^p \xi_D^*(x^{(i)}) = 1$ is maximal for $\xi_D^*(x^{(i)}) = 1/p$, $i = 1, \ldots, p$.

The duality property above, originally noticed by Sibson (1972), can be extended to other design criteria. For instance, one can show with the same Lagrangian technique that the determination of an A-optimal design measure ξ_A^* that minimizes trace$[\mathbf{M}^{-1}(\xi)]$ is equivalent to the determination of an ellipsoid $\mathcal{E}_A$ that contains $\mathscr{F}_\theta$ and such that trace$(\mathbf{A}^{1/2})$ is maximal. The corresponding value for $\mathbf{A}$ is then $\mathbf{A}^* = \mathbf{M}^{-2}(\xi_A^*)/\mathrm{trace}[\mathbf{M}^{-1}(\xi_A^*)]$.

More generally, duality can be connected with the notion of polarity as considered in (Pukelsheim, 1993, Chap. 7). The two cases above, namely, maximizing $\Phi_D^+[\mathbf{M}(\xi)]$ is equivalent to maximizing $\Phi_D^+(\mathbf{A})$, see (5.13), and maximizing $\Phi_{1,\mathbf{I}}^+[\mathbf{M}(\xi)]$ is equivalent to maximizing $\Phi_{-1/2,\mathbf{I}}^+(\mathbf{A})$, see (5.15), both under the constraint that $\mathscr{F}_\theta \subset \mathcal{E}_A$, are then special instances of polar pairs of criteria; see Pukelsheim (1993, p. 153).

The optimization of partial optimality criteria also admits a dual representation. For instance, the dual problem to D_s-optimum design (see Sect. 5.1.2) is the determination of the thinnest cylinder containing the Elfving's set $\mathscr{F}_\theta$; see Silvey and Titterington (1973). The dual problem to c-optimum design will be considered in Sect. 5.3.3. One may refer to Dette and Studden (1993) for the geometrical properties of E-optimum design.

5.3 c-Optimum Design in Linearized Nonlinear Models

In this section we consider some properties that are useful for the construction of c-optimal designs, i.e., designs that maximize the criterion $\phi_c(\xi) = \Phi_c[\mathbf{M}(\xi)]$, see (5.9), which is related to the precision of the estimation of a scalar function of θ. For a nonlinear model we write $\mathbf{M}(\xi) = \mathbf{M}(\xi, \theta^0)$, with θ^0 a nominal value for θ^0. Similarly, for a nonlinear function of interest $h(\cdot)$ we use $\mathbf{c} = \partial h(\theta)/\partial\theta_{\theta^0}$ in (5.9). In terms of design, the situation is thus similar to that encountered for a linear model with a linear function of interest.

We shall see that a c-optimal design can be singular, i.e., it can be supported on less than $p = \dim(\theta)$ points. The lack of continuity of the design criterion, see Sect. 5.1.7, then raises some specific difficulties that have been illustrated by Examples 2.4 and 5.10. The additional troubles in terms of estimation of θ and $h(\theta)$ caused by singular designs in presence of nonlinearity, of the model or of the function of interest $h(\cdot)$, will be considered in Sect. 5.4.

5.3.1 Elfving's Theorem and Related Properties

The equivalence theorem 5.21 is valid for c-optimality, but the criterion being not differentiable everywhere $d(\xi)$ does not take the simple form (5.39), (5.40). One may refer to Silvey (1980, p. 49) for a formulation of the equivalence theorem in this particular case. See also Sect. 5.3.4.

The following theorem is a much useful alternative formulation of a necessary and sufficient condition for c-optimality when $\mathbf{M}_\theta(x)$ has the form (5.47).

Theorem 5.30 (Elfving 1952). *When $\mathbf{M}_\theta(\mathbf{x})$ in (5.1) is the rank-one matrix (5.47), the design ξ_c^* is c-optimal if and only if there exists a subset $\mathcal{S}_c^*$ of the support $\mathcal{S}_{\xi_c^*}$ of ξ_c^* and a strictly positive number γ such that $\gamma\mathbf{c}$ lies on the boundary of Elfving's set $\mathscr{F}_\theta$ and*

$$\gamma\mathbf{c} = \int_{\mathcal{S}_c^*} \mathbf{g}_\theta(x)\,\xi_c^*(\mathrm{d}x) - \int_{\mathcal{S}_{\xi_c^*}\setminus\mathcal{S}_c^*} \mathbf{g}_\theta(x)\,\xi_c^*(\mathrm{d}x)\,.$$

Our proof of Elfving's theorem will rely on the following lemma which is also interesting per se.

Lemma 5.31. *When $\mathbf{M}_\theta(x)$ in (5.1) is the rank-one matrix (5.47), the design ξ_c^* is c-optimal if and only if $\mathbf{c}^\top\mathbf{M}^-(\xi_c^*)\mathbf{c} = 1/(\gamma^*)^2$ where*

$$\gamma^* = \max\{\gamma : \gamma\mathbf{c} \in \mathscr{F}_\theta\}\,. \tag{5.48}$$

Proof. We consider design measures ξ such that $\mathbf{c} = \mathbf{M}(\xi)\mathbf{u}$ for some $\mathbf{u} \in \mathbb{R}^p$. Define the measure μ_ξ by $\mu_\xi(\mathrm{d}x) = |\mathbf{u}^\top \mathbf{g}_\theta(x)|\gamma(\xi)\xi(\mathrm{d}x)$ where

$$\gamma(\xi) = \left[\int_{\mathscr{X}} |\mathbf{u}^\top \mathbf{g}_\theta(x)|\, \xi(\mathrm{d}x)\right]^{-1}$$

(so that $\int_{\mathscr{X}} \mu_\xi(\mathrm{d}x) = 1$). The constraint $\mathbf{c} = \mathbf{M}(\xi)\mathbf{u}$ can then be written as

$$\gamma(\xi)\,\mathbf{c} = \int_{\mathscr{X}} \mathrm{sign}[\mathbf{u}^\top \mathbf{g}_\theta(x)]\, \mathbf{g}_\theta(x)\, \mu_\xi(\mathrm{d}x)\,. \tag{5.49}$$

Hence, $\gamma(\xi)\mathbf{c} \in \mathscr{F}_\theta$. Moreover, from Cauchy–Schwarz inequality, we have

$$\mathbf{c}^\top \mathbf{M}^-(\xi)\mathbf{c} = \mathbf{u}^\top \mathbf{M}(\xi)\mathbf{u} = \int_{\mathscr{X}} |\mathbf{u}^\top \mathbf{g}_\theta(x)|^2\, \xi(\mathrm{d}x) \geq \frac{1}{\gamma^2(\xi)}\,, \tag{5.50}$$

with equality if and only if $|\mathbf{u}^\top \mathbf{g}_\theta(x)|$ is constant for ξ-almost every x.

We first show that the condition $\mathbf{c}^\top \mathbf{M}^-(\xi_c^*)\mathbf{c} = 1/(\gamma^*)^2$ is sufficient for the c-optimality of ξ_c^*. The right-hand side of (5.50) is minimum when ξ is such that $\gamma(\xi) = \gamma^*$ as defined in (5.48). Therefore, $\mathbf{c}^\top \mathbf{M}^-(\xi)\mathbf{c} \geq 1/(\gamma^*)^2$ for any design measure ξ on $\mathscr{X}$, and $\mathbf{c}^\top \mathbf{M}^-(\xi_c^*)\mathbf{c} = 1/(\gamma^*)^2$ implies that ξ_c^* is c-optimal.

We show now that the condition is necessary through the construction of a measure μ satisfying $\mathbf{c}^\top \mathbf{M}^-(\mu)\mathbf{c} = 1/(\gamma^*)^2$. From Caratheodory's theorem, the vector $\gamma^*\mathbf{c}$ of $\mathscr{F}_\theta$ can be represented as a finite linear combination

$$\gamma^*\mathbf{c} = \sum_{i\in\mathcal{I}^+} \mu_i \mathbf{g}_\theta(x^{(i)}) - \sum_{i\in\mathcal{I}^-} \mu_i \mathbf{g}_\theta(x^{(i)})\,,$$

where $\mu_i > 0$ and $x^{(i)} \in \mathscr{X}$ for all $i \in \mathcal{I}^+ \cup \mathcal{I}^-$, $\sum_{i\in\mathcal{I}^+\cup\mathcal{I}^-} \mu_i = 1$, and the number m of elements in $\mathcal{I}^+ \cup \mathcal{I}^-$ is bounded by p—indeed, only p points are required since $\gamma^*\mathbf{c}$ is on the boundary or $\mathscr{F}_\theta$. From that we construct a vector $\mathbf{u}_*$ such that

$$\gamma^*\mathbf{u}_*^\top \mathbf{g}_\theta(x^{(i)}) = 1 \text{ for } i \in \mathcal{I}^+ \text{ and } \gamma^*\mathbf{u}_*^\top \mathbf{g}_\theta(x^{(i)}) = -1 \text{ for } i \in \mathcal{I}^-\,.$$

Such a vector exists since there are p linear constraints at most. This gives

$$\gamma^*\mathbf{c} = \sum_{i=1}^m \mu_i \gamma^* [\mathbf{u}_*^\top \mathbf{g}_\theta(x^{(i)})]\, \mathbf{g}_\theta(x^{(i)})\,.$$

Therefore, $\mathbf{c} = \mathbf{M}(\mu)\mathbf{u}_*$ where μ is the measure $\mu = \sum_{i=1}^m \mu_i\, \delta_{x^{(i)}}$, and

$$\mathbf{c}^\top \mathbf{M}^-(\mu)\mathbf{c} = \mathbf{u}_*^\top \mathbf{M}(\mu)\mathbf{u}_* = \sum_{i=1}^m \mu_i\, [\mathbf{u}_*^\top \mathbf{g}_\theta(x^{(i)})]^2 = 1/(\gamma^*)^2\,. \qquad \blacksquare$$

Proof of Theorem 5.30.

We first show that a design measure ξ_c^* satisfying the conditions of the theorem is such that $\mathbf{c}^\top \mathbf{M}^-(\xi_c^*)\mathbf{c} = 1/(\gamma^*)^2$, with γ^* defined in (5.48), and is thus c-optimal according to Lemma 5.31. Since $\gamma\mathbf{c}$ lies on the boundary of $\mathscr{F}_\theta$, $\gamma = \gamma^*$, and $\mathbf{g}_\theta(x)$ lies on the boundary of $\mathscr{F}_\theta$ for any x in the support $\mathcal{S}_{\xi_c^*}$ of ξ_c^*. Moreover, the vectors $\mathbf{g}_\theta(x)$, $x \in \mathcal{S}_c^*$, and $-\mathbf{g}_\theta(x)$, $x \in \mathcal{S}_{\xi_c^*} \setminus \mathcal{S}_c^*$ belong to a face $\mathscr{F}$ of $\mathscr{F}_\theta$, i.e. , a m-dimensional hyperplane with $m < p$, or form a vertex of $\mathscr{F}_\theta$ if $m = 0$. Define O' as the orthogonal projection of the origin O onto this face $\mathscr{F}$ and take $\mathbf{u}_*$ as the vector $\overrightarrow{OO'}/(\gamma^*\|\overrightarrow{OO'}\|^2)$. It satisfies $\gamma^*\mathbf{u}_*^\top \mathbf{g}_\theta(x) = 1$ for $x \in \mathcal{S}_c^*$ and $\gamma^*\mathbf{u}_*^\top \mathbf{g}_\theta(x) = -1$ for $x \in \mathcal{S}_{\xi_c^*} \setminus \mathcal{S}_c^*$. The rest of the proof is as in Lemma 5.31. We have

$$\begin{aligned}\gamma^*\mathbf{c} &= \int_{\mathcal{S}_c^*} \gamma^*\, \mathbf{g}_\theta(x)\, [\mathbf{g}_\theta^\top(x)\mathbf{u}_*]\, \xi_c^*(\mathrm{d}x) + \int_{\mathcal{S}_{\xi_c^*}\setminus\mathcal{S}_c^*} \gamma^*\, \mathbf{g}_\theta(x)\, [\mathbf{g}_\theta^\top(x)\mathbf{u}_*]\, \xi_c^*(\mathrm{d}x) \\ &= \gamma^* \left[\int_{\mathcal{S}_{\xi_c^*}} \mathbf{g}_\theta(x)\mathbf{g}_\theta^\top(x)\, \xi_c^*(\mathrm{d}x)\right] \mathbf{u}_* = \gamma^*\, \mathbf{M}(\xi_c^*)\, \mathbf{u}_*\end{aligned}$$

and

$$\mathbf{c}^\top \mathbf{M}^-(\xi_c^*)\mathbf{c} = \mathbf{u}_*\mathbf{M}(\xi_c^*)\mathbf{u}_* = \int_{\mathcal{S}_{\xi_c^*}} [\mathbf{u}_*^\top \mathbf{g}_\theta(x)]^2\, \xi_c^*(\mathrm{d}x) = 1/(\gamma^*)^2$$

so that ξ_c^* is c-optimal.

Conversely, we show now that any c-optimal design measure ξ_c^* can be put in the form given in Theorem 5.30. From Lemma 5.31, the optimality of ξ_c^* implies $\mathbf{c}^\top \mathbf{M}^-(\xi_c^*)\mathbf{c} = 1/(\gamma^*)^2$ with $\mathbf{c} = \mathbf{M}(\xi_c^*)\mathbf{u}$ for some $\mathbf{u} \in \mathbb{R}^p$. From Cauchy–Schwarz inequality, see (5.50) in the proof of Lemma 5.31, we then have that $|\mathbf{u}^\top \mathbf{g}_\theta(x)|$ equals some constant β for ξ_c^*-almost any x. Defining μ_ξ and $\gamma(\xi)$ as in the proof of Lemma 5.31, we get $\mu_{\xi_c^*} = \xi_c^*$ and

$$\mathbf{c}^\top \mathbf{M}^-(\xi_c^*)\mathbf{c} = \mathbf{u}^\top \mathbf{M}(\xi_c^*)\mathbf{u} = \int_{\mathscr{X}} [\mathbf{u}^\top \mathbf{g}_\theta(x)]^2\, \xi_c^*(\mathrm{d}x) = \beta^2\,,$$

so that $\beta = 1/\gamma^*$. We also obtain

$$\gamma(\xi_c^*)\, \mathbf{c} = \int_{\mathscr{X}} \mathrm{sign}[\mathbf{u}^\top \mathbf{g}_\theta(x)]\, \mathbf{g}_\theta(x)\, \xi_c^*(\mathrm{d}x)\,, \tag{5.51}$$

see (5.49), and

$$\mathbf{c}^\top \mathbf{M}^-(\xi_c^*)\mathbf{c} = \mathbf{u}^\top \mathbf{c} = \int_{\mathscr{X}} \mathrm{sign}[\mathbf{u}^\top \mathbf{g}_\theta(x)]\, \mathbf{u}^\top \mathbf{g}_\theta(x)\, \xi_c^*(\mathrm{d}x)/\gamma(\xi_c^*) = \beta/\gamma(\xi_c^*)$$

so that $\gamma(\xi_c^*) = \gamma^*$; (5.51) then gives $\int_{\mathscr{X}} \mathrm{sign}[\mathbf{u}^\top \mathbf{g}_\theta(x)]\, \mathbf{g}_\theta(x)\, \xi_c^*(\mathrm{d}x) = \gamma^*\mathbf{c}$ and ξ_c^* can thus be put in the form indicated in the theorem. ■

Remark 5.32.

(*i*) The proof of Lemma 5.31 involves the construction of a c-optimal design measure μ supported on p points at most. Hence, a c-optimal design measure can always be put in the form $\sum_{i=1}^{m} \mu_i \, \delta_{x^{(i)}}$ where the $x^{(i)}$ belong to $\mathscr{X}$ and $m \leq p = \dim(\theta)$.

(*ii*) When the design space $\mathscr{X}$ is finite, $\mathscr{X} = \{x^{(1)}, \ldots, x^{(\ell)}\}$, the Elfving's set $\mathscr{F}_\theta$ corresponds to the convex hull of the finite set $\{\mathbf{g}_\theta(x^{(i)}),\ i = 1, \ldots, \ell\} \cup \{-\mathbf{g}_\theta(x^{(i)}),\ i = 1, \ldots, \ell\}$, and Elfving's theorem indicates that a c-optimal design can be obtained by solving the following LP problem: find some weights $w_1, \ldots, w_{2\ell}$, with $w_i \geq 0$ for all i, that maximize γ under the $p+1$ constraints $\sum_{i=1}^{\ell} w_i \, \mathbf{g}_\theta(x^{(i)}) - \sum_{i=\ell+1}^{2\ell} w_i \, \mathbf{g}_\theta(x^{(i)}) = \gamma \, \mathbf{c}$ and $\sum_{i=1}^{2\ell} w_i = 1$; see Harman and Jurík (2008). □

Using Theorem 5.30 and the representations of $\mathbf{c}^\top \mathbf{M}^- \mathbf{c}$ obtained in Lemmas 5.5 and 5.6, one can easily prove the following. We follow the approach of Pázman (2001) where graphical representations complementary to Elfving's set are also presented.

Lemma 5.33. *When $\mathbf{M}_\theta(x)$ in (5.1) is the rank-one matrix (5.47), the optimum value of the c-optimality criterion satisfies*

$$\min_{\xi \in \Xi} \mathbf{c}^\top \mathbf{M}^-(\xi)\mathbf{c} = \sup_{\mathbf{z} \in \mathbb{R}^p} \min_{x \in \mathscr{X}} \left\{ 2\mathbf{z}^\top \mathbf{c} - [\mathbf{z}^\top \mathbf{g}_\theta(x)]^2 \right\}$$
$$= \sup_{\mathbf{z} \in \mathbb{R}^p} \min_{x \in \mathscr{X}} \frac{(\mathbf{z}^\top \mathbf{c})^2}{[\mathbf{z}^\top \mathbf{g}_\theta(x)]^2} \,,$$

where $\Xi_c = \{\xi \in \Xi : \mathbf{c} \in \mathcal{M}[\mathbf{M}(\xi)]\}$.

Proof. From Lemma 5.5, for any $\xi \in \Xi_c$,

$$\mathbf{c}^\top \mathbf{M}^-(\xi)\mathbf{c} = \sup_{\mathbf{z} \in \mathbb{R}^p} \left\{ 2\mathbf{z}^\top \mathbf{c} - \mathbf{z}^\top \mathbf{M}(\xi)\mathbf{z} \right\} \geq \sup_{\mathbf{z} \in \mathbb{R}^p} \left\{ 2\mathbf{z}^\top \mathbf{c} - \max_{x \in \mathscr{X}} [\mathbf{z}^\top \mathbf{g}_\theta(x)]^2 \right\}$$
$$\geq 2\mathbf{u}_*^\top \mathbf{c} - \max_{x \in \mathscr{X}} [\mathbf{u}_*^\top \mathbf{g}_\theta(x)]^2 = \frac{1}{(\gamma^*)^2} = \mathbf{c}^\top \mathbf{M}^-(\xi_c^*)\mathbf{c}$$

with $\mathbf{u}_*$ and γ^* defined in the proof of Lemma 5.31. We have similarly,

$$\mathbf{c}^\top \mathbf{M}^-(\xi)\mathbf{c} = \sup_{\mathbf{z}^\top \mathbf{M}(\xi)\mathbf{z} \neq 0} \frac{(\mathbf{z}^\top \mathbf{c})^2}{\mathbf{z}^\top \mathbf{M}(\xi)\mathbf{z}} \,, \tag{5.52}$$

see the proof of Lemma 5.6. Take now any $\xi \in \Xi_c$, it satisfies

$$\frac{(\mathbf{z}^\top \mathbf{c})^2}{\mathbf{z}^\top \mathbf{M}(\xi)\mathbf{z}} \geq \frac{(\mathbf{z}^\top \mathbf{c})^2}{\max_{x \in \mathscr{X}} [\mathbf{z}^\top \mathbf{g}_\theta(x)]^2} = \frac{[\mathbf{z}^\top \mathbf{M}(\xi)\mathbf{u}]^2}{\max_{x \in \mathscr{X}} [\mathbf{z}^\top \mathbf{g}_\theta(x)]^2}$$

for some $\mathbf{u} \in \mathbb{R}^p$, and (5.52) gives

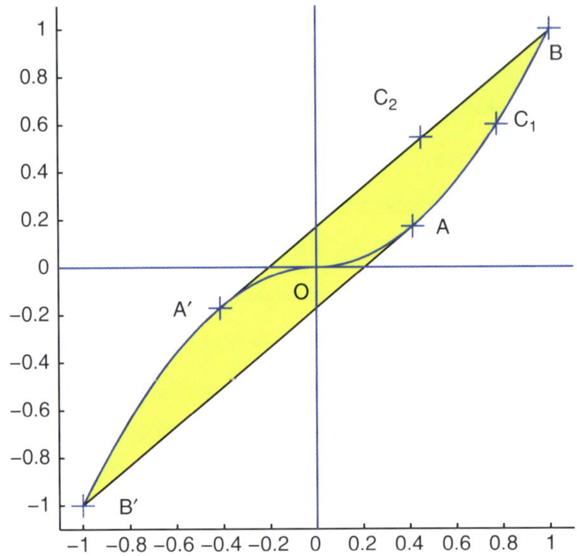

Fig. 5.3. Elfving's set and illustration of Elfving's theorem

$$\mathbf{c}^\top \mathbf{M}^-(\xi)\mathbf{c} \geq \sup_{\mathbf{z}^\top \mathbf{M}(\xi)\mathbf{z}\neq 0} \frac{[\mathbf{z}^\top \mathbf{M}(\xi)\mathbf{u}]^2}{\max_{x\in\mathscr{X}}[\mathbf{z}^\top \mathbf{g}_\theta(x)]^2}$$
$$= \sup_{\mathbf{z}\in\mathbb{R}^p} \frac{[\mathbf{z}^\top \mathbf{M}(\xi)\mathbf{u}]^2}{\max_{x\in\mathscr{X}}[\mathbf{z}^\top \mathbf{g}_\theta(x)]^2} = \sup_{\mathbf{z}\in\mathbb{R}^p} \frac{(\mathbf{z}^\top \mathbf{c})^2}{\max_{x\in\mathscr{X}}[\mathbf{z}^\top \mathbf{g}_\theta(x)]^2} .$$

As above, taking $\mathbf{z} = \mathbf{u}_*$ as defined in the proof of Lemma 5.31, we get

$$\min_{\xi\in\Xi} \mathbf{c}^\top \mathbf{M}^-(\xi)\mathbf{c} \geq \sup_{\mathbf{z}\in\mathbb{R}^p} \frac{(\mathbf{z}^\top \mathbf{c})^2}{\max_{x\in\mathscr{X}}[\mathbf{z}^\top \mathbf{g}_\theta(x)]^2}$$
$$\geq \frac{(\mathbf{u}_*^\top \mathbf{c})^2}{\max_{x\in\mathscr{X}}[\mathbf{u}_*^\top \mathbf{g}_\theta(x)]^2} = \frac{1}{(\gamma^*)^2} = \mathbf{c}^\top \mathbf{M}^-(\xi_c^*)\mathbf{c} . \qquad \blacksquare$$

A generalization of Elfving's theorem to A- and D-optimality can be found, respectively, in (Studden, 1971) and (Dette, 1993).

Example 5.34. We consider c-optimum design for the linear regression model used in Examples 2.4 and 3.13, i.e., $\eta(x,\theta) = \theta_1 x + \theta_2 x^2$, $x \in [0,1]$. Figure 5.3 presents the corresponding Elfving's set $\mathscr{F}_\theta$ (colored area). The points A and A', respectively, have coordinates (a, a^2) and $(-a, -a^2)$ with $a = \sqrt{2}-1$, B and B' are the points $(1,1)$ and $(-1,-1)$, and O is the origin $(0,0)$. The c-optimal design ξ_c^* takes a different form depending on the direction of the vector $\mathbf{c}$.

When $\mathbf{c}$ is such that the intersection C_1 of the line $\{\gamma\mathbf{c}\,,\ \gamma > 0\}$ with the boundary of $\mathscr{F}_\theta$ is between A and B, or between A' and B', ξ_c^* is supported at a single point. For instance, when $\gamma^*\mathbf{c} = \overrightarrow{OC_1} = (x_1, x_1^2)^\top$ with $a \leq x_1 \leq 1$, as shown on Fig. 5.3 for $x_1 = \sqrt{0.6}$, then $\xi_c^* = \delta_{x_1}$.

When the intersection C_2 of the line $\{\gamma\mathbf{c}\,,\ \gamma > 0\}$ with the boundary of $\mathscr{F}_\theta$ is between A' and B, then ξ_c^* is supported on the two points a and 1. Let $\gamma^*\mathbf{c} = \overrightarrow{OC_2'} = (\alpha, \beta)^\top$ (with $\alpha = 0.45$ on Fig. 5.3). Then direct calculations give $\overrightarrow{OC_2'} = w_1\overrightarrow{OA'} + (1-w_1)\overrightarrow{OB}$, i.e., $\xi_c^* = w_1\delta_a + (1-w_1)\delta_1$, with $w_1 = (1-\alpha)/(1+a)$. The situation is similar when C_2 is between A and B' since reversing $\mathbf{c}$ leaves the design problem unchanged. In particular, when $\mathbf{c} = \alpha(z, z^2)$ with $\alpha > 0$ and $0 < z \leq a$ or $z \geq 1$, we get $\xi_c^* = w_1\delta_a + (1-w_1)\delta_1$, with $w_1 = \sqrt{2}(1-z)/[2(2a-z)]$. □

Example 5.35. Consider again Example 5.23. When the function of interest is the value $x_{\max}$ where the regression function (5.42) is maximum, i.e.,

$$h(\theta) = \frac{1}{t_1 - t_2} \log\left(\frac{t_1}{t_2}\right),$$

we obtain $c_{\theta^0} = \partial h(\theta)/\partial\theta\big|_{\theta^0} \simeq (-2.1539,\ -4.9889)^\top$ at $\theta^0 = (0.7,\ 0.2)^\top$. A direct application of Elfving's theorem (e.g., through a geometrical construction on Fig. 5.2) gives the c-optimal design measure $\xi_c^* = \alpha\delta_{x_1} + (1-\alpha)\delta_{x_2}$ with $x_1 \simeq 0.9940$, $x_2 \simeq 7.1223$, and $\alpha \simeq 0.5395$.

When the function of interest is now the maximum of the response, $h(\theta) = \max_x \eta(x,\theta) = \eta(x_{\max},\theta)$, we obtain $c_{\theta^0} = \partial h(\theta)/\partial\theta\big|_{\theta^0} \simeq (0.26,\ -0.91)^\top$ at θ^0 and the c-optimal design measure is the one-point delta measure δ_{x_*} with $x_* \simeq 2.5055$. □

5.3.2 c-Maximin Efficiency and D-Optimality

c-maximin efficiency defined by the criterion (5.22) is closely related to D-optimality, as the next theorem shows; see Kiefer (1962) from whom we reproduce the proof.

Theorem 5.36. *When* $\mathbf{M}(\xi) = \int_{\mathscr{X}} \mathbf{g}_\theta(x)\mathbf{g}_\theta^\top(x)\,\xi(\mathrm{d}x)$ *and* $\mathcal{C} = \mathcal{C}_{\mathscr{X}} = \{\mathbf{g}_\theta(x) : x \in \mathscr{X}\}$*, a design measure* ξ^* *on* $\mathscr{X}$ *is c-maximin efficient (i.e.,* ξ^* *maximizes* $\mathscr{E}_{mmc}(\xi) = \min_{\mathbf{c}\in\mathcal{C}_{\mathscr{X}}} [\mathbf{c}^\top\mathbf{M}^-(\xi_c^*)\mathbf{c}]/[\mathbf{c}^\top\mathbf{M}^-(\xi)\mathbf{c}]$ *with* ξ_c^* *a c-optimal design measure minimizing* $\mathbf{c}^\top\mathbf{M}^-(\xi)\mathbf{c}$*) if and only if it is D-optimal on* $\mathscr{X}$*, i.e.,* ξ^* *maximizes* $\log\det\mathbf{M}(\xi)$.

Proof. For any $x \in \mathscr{X}$, $\mathbf{g}_\theta^\top(x)\theta$ is estimable from the delta measure δ_x and $\mathbf{g}_\theta^\top(x)\mathbf{M}^-(\delta_x)\mathbf{g}_\theta(x) = 1$. Therefore, $\min_{\xi\in\Xi} \mathbf{g}_\theta^\top(x)\mathbf{M}^-(\xi)\mathbf{g}_\theta(x) \leq 1$. When applied to D-optimality, the equivalence theorem 5.21 implies that $\max_{x\in\mathscr{X}} \mathbf{g}_\theta^\top(x)\mathbf{M}^-(\xi)\mathbf{g}_\theta(x) \geq p$ for any $\xi \in \Xi$, so that

$$\mathscr{E}_{mmc}(\xi) \leq \min_{x\in\mathscr{X}} \frac{\mathbf{g}_\theta^\top(x)\mathbf{M}^-(\delta_x)\mathbf{g}_\theta(x)}{\mathbf{g}_\theta^\top(x)\mathbf{M}^-(\xi)\mathbf{g}_\theta(x)} \leq 1/p\,.$$

Suppose first that ξ^* is D-optimal. We show that $\mathscr{E}_{mmc}(\xi^*) = 1/p$, which implies that ξ^* is c-maximin efficient. If $\mathscr{E}_{mmc}(\xi^*) < 1/p$, there exist $x \in \mathscr{X}$ and $\xi \in \Xi$ such that

$$\frac{\mathbf{g}_\theta^\top(x)\mathbf{M}^-(\xi)\mathbf{g}_\theta(x)}{\mathbf{g}_\theta^\top(x)\mathbf{M}^-(\xi^*)\mathbf{g}_\theta(x)} < 1/p\,. \tag{5.53}$$

We can always assume that $\mathbf{M}(\xi)$ is nonsingular, since otherwise we can replace ξ by $(1-\alpha)\xi + \alpha\xi^*$ with α small. Also, we can perform a linear transformation that makes $\mathbf{M}(\xi^*)$ the identity and $\mathbf{M}(\xi)$ diagonal with elements D_i, $i = 1,\ldots,p$; see, e.g., Harville (1997, p. 562). Then (5.53) gives $\sum_{i=1}^p \{\mathbf{g}_\theta(x)\}_i^2 \left(\sum_{j=1}^p \{\mathbf{g}_\theta(x)\}_j^2\right)^{-1} D_i^{-1} < 1/p$, so that at least one D_i is strictly larger than p. But then $F_{\phi_D}(\xi^*;\xi) = \mathrm{trace}\{[\mathbf{M}(\xi) - \mathbf{M}(\xi^*)]\mathbf{M}^{-1}(\xi^*)\} = \sum_{i=1}^p (D_i - 1) > 0$, which contradicts the D-optimality of ξ^*.

Conversely, suppose that ξ^* is c-maximin efficient. Since $\mathscr{E}_{mmc}(\xi_D^*) = 1/p$ for a D-optimal measure ξ_D^* and $\mathscr{E}_{mmc}(\xi) \leq 1/p$ for any ξ, we must have $\mathscr{E}_{mmc}(\xi^*) = 1/p$. But then $\min_{\xi\in\Xi} \mathbf{g}_\theta^\top(x)\mathbf{M}^-(\xi)\mathbf{g}_\theta(x) \leq 1$ implies that $\max_{x\in\mathscr{X}} \mathbf{g}_\theta^\top(x)\mathbf{M}^-(\xi^*)\mathbf{g}_\theta(x) \leq p$, so that ξ^* is D-optimal. ■

Remark 5.37.

(*i*) Kiefer (1962) also shows that, under the conditions of Theorem 5.36, a design measure ξ^* is D-optimal if and only if it maximizes the criterion $\phi_{AR}(\xi) = \min_{\mathbf{c}\in\mathcal{C}_{\mathscr{X}}} \left\{[\mathbf{c}^\top\mathbf{M}^-(\xi_c^*)\mathbf{c}] - [\mathbf{c}^\top\mathbf{M}^-(\xi)\mathbf{c}]\right\}$.

(*ii*) Schwabe (1997) extends the theorem above to situations where $\mathbf{M}(\xi) = \int_{\mathscr{X}} \mathbf{g}_\theta(x)\mathbf{g}_\theta^\top(x)\,\xi(\mathrm{d}x)$, but $\mathcal{C}$ has a more general form than $\mathcal{C}_{\mathscr{X}} = \{\mathbf{g}_\theta(x) : x \in \mathscr{X}\}$. His extension is as follows, see also Müller and Pázman (1998) for part (*a*): for $\mathcal{C}$ a subset of $\mathbb{R}^p$, let $\mathcal{C}(\lambda)$ denote the set $\{\lambda\mathbf{c} : \mathbf{c}\in\mathcal{C}\,,\ \lambda\in\mathbb{R}\}$ and let $\overline{\mathcal{C}}(\lambda)$ be its closure; then:

   (*a*) A D-optimal design measure ξ_D^* with support $\mathcal{S}_{\xi_D^*}$ satisfying $\{\mathbf{g}_\theta(x) : x \in \mathcal{S}_{\xi_D^*}\} \subseteq \overline{\mathcal{C}}(\lambda)$ is c-maximin efficient on $\mathcal{C}$.

   (*b*) A design measure ξ^* c-maximin efficient on $\mathcal{C}$ is D-optimal if $\mathcal{C}_{\mathscr{X}} \subseteq \overline{\mathcal{C}}(\lambda)$. □

5.3.3 A Duality Property for *c*-Optimality

A duality property can be formulated for c-optimum design, similarly to what was presented in Sect. 5.2.4. We only consider the case where $\mathbf{M}_\theta(x)$ has rank one; see (5.47).

Consider an ellipsoid $\mathcal{E}_A = \{\mathbf{t}\in\mathbb{R}^d : \mathbf{t}^\top\mathbf{A}\mathbf{t} \leq 1\}$, possibly degenerate, i.e., with $\mathbf{A}\in\mathbb{M}^{\geq}$. Suppose that $\mathcal{E}_A$ contains the Elfving's set $\mathscr{F}_\theta$, i.e., $\mathbf{g}_\theta^\top(x)\mathbf{A}\mathbf{g}_\theta(x) \leq 1$ for all $x\in\mathscr{X}$, which implies

$$\mathrm{trace}[\mathbf{M}(\xi)\mathbf{A}] \leq 1\,,\ \forall\xi\in\Xi\,,$$

with equality if and only if $\mathbf{g}_\theta^\top(x)\mathbf{A}\mathbf{g}_\theta(x) = 1$ for ξ-almost any x. Since $\gamma^*\mathbf{c}$ as defined in Lemma 5.31 belongs to $\mathscr{F}_\theta$, $(\gamma^*)^2\mathbf{c}^\top\mathbf{A}\mathbf{c} \leq 1$, i.e., $\mathbf{c}^\top\mathbf{A}\mathbf{c} \leq \mathbf{c}^\top\mathbf{M}^-(\xi_c^*)\mathbf{c}$. Therefore,

$$\sup_{\{\mathbf{A}\in\mathbb{M}^{\geq}:\mathscr{F}_\theta\subset\mathcal{E}_A\}} \mathbf{c}^\top\mathbf{A}\mathbf{c} \leq \mathbf{c}^\top\mathbf{M}^-(\xi_c^*)\mathbf{c} = 1/(\gamma^*)^2\,.$$

One can show, moreover, that equality is attained. Indeed, take $\mathbf{A} = (\gamma^*)^2 \mathbf{u}_* \mathbf{u}_*^\top$ with $\mathbf{u}_*$ and γ^* as in the proof of Theorem 5.30, so that $\mathbf{g}_\theta^\top(x)\mathbf{A}\mathbf{g}_\theta(x) = (\gamma^*)^2[\mathbf{g}_\theta^\top(x)\mathbf{u}_*]^2 \leq 1$ for all $x \in \mathscr{X}$; that is, $\mathscr{F}_\theta \subset \mathcal{E}_A$. We also have $\mathbf{c}^\top\mathbf{A}\mathbf{c} = (\gamma^*)^2\|\mathbf{c}^\top\mathbf{u}_*\|^2 = 1/(\gamma^*)^2$. The problem of maximizing $\mathbf{c}^\top\mathbf{A}\mathbf{c}$ under the constraint that $\mathscr{F}_\theta \subset \mathcal{E}_A$ is thus dual to c-optimum design that aims at minimizing $\mathbf{c}^\top\mathbf{M}^-(\xi)\mathbf{c}$.

5.3.4 Equivalence Theorem for c-Optimality

Elfving's theorem (Theorem 5.30) is almost constructive, in the sense that it characterizes optimality in terms of a property for ξ_c^*. One may thus wonder in which aspects it differs from the equivalence theorem 5.21 and which particular form this theorem may take in the case of c-optimality.

A direct application of Theorem 5.21 with the directional derivative given by (5.38) shows that ξ_c^* is c-optimal if and only if

$$\max_{\mathbf{M}_2\in\mathcal{M}_\theta(\Xi)} \min_{\{\mathbf{A}:\mathbf{M}^*\mathbf{A}\mathbf{M}^*=\mathbf{M}^*\}} \mathbf{c}^\top\mathbf{A}^\top\mathbf{M}_2\mathbf{A}\mathbf{c} - \mathbf{c}^\top\mathbf{A}\mathbf{c} = 0$$

where $\mathbf{M}^* = \mathbf{M}(\xi_c^*) \in \mathbb{M}_c^{\geq}$. The set of generalized inverses of $\mathbf{M}^*$ can be linearly parameterized, see Harville (1997, Theorem 9.2.7), and is thus convex. The function $(\mathbf{M}_2, \mathbf{A}) \longrightarrow \mathbf{c}^\top\mathbf{A}^\top\mathbf{M}_2\mathbf{A}\mathbf{c} - \mathbf{c}^\top\mathbf{A}\mathbf{c}$ is convex in $\mathbf{A}$ and concave in $\mathbf{M}_2$ so that the minimax theorem applies, see, e.g., Dem'yanov and Malozemov (1974, Theorem 5.2, p. 218) and Polak (1987, Corollary 5.5.6, p. 707), and

$$\begin{aligned}
&\max_{\mathbf{M}_2\in\mathcal{M}_\theta(\Xi)} \min_{\{\mathbf{A}:\mathbf{M}^*\mathbf{A}\mathbf{M}^*=\mathbf{M}^*\}} \mathbf{c}^\top\mathbf{A}^\top\mathbf{M}_2\mathbf{A}\mathbf{c} - \mathbf{c}^\top\mathbf{A}\mathbf{c} \\
&= \min_{\{\mathbf{A}:\mathbf{M}^*\mathbf{A}\mathbf{M}^*=\mathbf{M}^*\}} \max_{\mathbf{M}_2\in\mathcal{M}_\theta(\Xi)} \mathbf{c}^\top\mathbf{A}^\top\mathbf{M}_2\mathbf{A}\mathbf{c} - \mathbf{c}^\top\mathbf{A}\mathbf{c}\,.
\end{aligned}$$

The equivalence theorem for c-optimality can thus be formulated as follows; see Pukelsheim (1993, p. 52).

Theorem 5.38 (Equivalence theorem for c-optimality). *A design ξ_c^* is c-optimal if and only if $\mathbf{M}(\xi_c^*) \in \mathbb{M}_c^{\geq}$ and there exists a g-inverse $\mathbf{A}$ of $\mathbf{M}(\xi_c^*)$ such that*

$$\mathbf{c}^\top\mathbf{A}^\top\mathbf{M}\mathbf{A}\mathbf{c} \leq \mathbf{c}^\top\mathbf{A}\mathbf{c}\,,\ \forall \mathbf{M} \in \mathcal{M}_\theta(\Xi)\,.$$

Notice that this theorem is much less constructive than Elfving's theorem. On the other hand, it does not require $\mathbf{M}_\theta(x)$ to have rank one. When $\mathbf{M}_\theta(x) = \mathbf{g}_\theta(x)\mathbf{g}_\theta^\top(x)$ and $\mathbf{M}(\xi) = \int_{\mathscr{X}} \mathbf{g}_\theta(x)\mathbf{g}_\theta^\top(x)\,\xi(\mathrm{d}x)$, the necessary-and-sufficient condition above becomes $[\mathbf{c}^\top\mathbf{A}\mathbf{g}_\theta(x)]^2 \leq \mathbf{c}^\top\mathbf{A}\mathbf{c}$ for all $x \in \mathscr{X}$.

5.4 Specific Difficulties with c-Optimum Design in Presence of Nonlinearity

We assume throughout the section that $\mathbf{M}_\theta(x)$ in (5.1) is the rank-one matrix $\mathbf{g}_\theta(x)\mathbf{g}_\theta^\top(x)$; see (5.47). In Example 5.34, where we considered a linear model and a linear function of interest, the possible singularity of the optimal design does not raise any special difficulty. The situation is quite different when (i) the function of interest $h(\cdot)$ is nonlinear so that $\mathbf{c}_\theta = \partial h(\theta)/\partial\theta$ depends on θ, see Examples 3.13 and 3.17, or (ii) the model is nonlinear, so that $\mathbf{M}(\xi,\theta)$ depends on θ, as illustrated by the example below; see also Example 7.12. Indeed, a singular optimal design ξ^* cannot be used directly in such situations since it allows the estimation of the quantity of interest only in the special case where the true value $\bar{\theta}$ of the parameters exactly equals the nominal value θ^0 used to construct ξ^* as an optimal design. This is of course unrealistic in applications since we do not know $\bar{\theta}$ beforehand. When using a sequence of design points such that the empirical design measure ξ_N converges to ξ^*, then different rates of convergence are obtained for the estimator of the quantity of interest, depending on how ξ_N converges to ξ^*.

Example 5.39. Consider again Example 5.23 and take any $x_* \in [x^{(1)}, x^{(2)}]$, with $x^{(1)} \simeq 1.23$, $x^{(2)} \simeq 6.86$ (see Example 5.29). Take $\mathbf{c} = \beta\mathbf{f}_{\theta^0}(x_*)$, $\beta \neq 0$. The corresponding c-optimal design is then the delta measure δ_{x_*}; see Fig. 5.2 (this is true also for x_* slightly outside $[x^{(1)}, x^{(2)}]$). Obviously, the singular design δ_{x_*} only allows us to estimate $\eta(x_*,\theta)$.

Consider now a second design point $x^0 \neq x_*$ and suppose that when N observations are performed, m are taken at x^0 and $N-m$ at x_*, where $m/(\log\log N) \to \infty$ with $m/N \to 0$. Then, for $x^0 \neq 0$ the conditions of Theorem 3.5 are satisfied. Indeed, the design space equals $\{x^0, x_*\}$ and is thus finite, and

$$\begin{aligned} D_N(\theta,\bar{\theta}) &= \sum_{k=1}^{N} [\eta(x_k,\theta) - \eta(x_k,\bar{\theta})]^2 \\ &= (N-2m)[\eta(x_*,\theta) - \eta(x_*,\bar{\theta})]^2 \\ &\quad +m \left\{ [\eta(x_*,\theta) - \eta(x_*,\bar{\theta})]^2 + [\eta(x^0,\theta) - \eta(x^0,\bar{\theta})]^2 \right\} \end{aligned}$$

so that $\inf_{\|\theta-\bar{\theta}\|>\Delta} D_N(\theta,\bar{\theta}) \geq mC(x^0, x_*, \Delta)$, with $C(x^0, x_*, \Delta)$ a positive constant, and $\inf_{\|\theta-\bar{\theta}\|>\Delta} D_N(\theta,\bar{\theta})/(\log\log N) \to \infty$ as $N \to \infty$. Therefore, although the empirical measure ξ_N of the design points converges strongly to the singular design δ_{x_*} which does not allow the estimation of θ, this convergence is sufficiently slow to make the LS estimator $\hat{\theta}^N_{LS}$ (strongly) consistent. Moreover, for $h(\theta)$ a function satisfying the conditions of Theorem 3.14, $h(\hat{\theta}^N_{LS})$ satisfies the regular asymptotic property of Definition 3.12. In the present example, this means that when $\partial h(\theta)/\partial\theta\big|_{\bar{\theta}} = \beta\mathbf{f}_{\bar{\theta}}(x_*)$ for some $\beta \in \mathbb{R}$, with $\mathbf{f}_{\bar{\theta}}(x) = \partial\eta(x,\theta)/\partial\theta\big|_{\bar{\theta}}$, then $\sqrt{N}[h(\hat{\theta}^N_{LS}) - h(\bar{\theta})]$ converges in distribution to a variable distributed $\mathscr{N}(0, \left[\partial h(\theta)/\partial\theta^\top \mathbf{M}^-(\delta_{x_*},\theta)\,\partial h(\theta)/\partial\theta\right]_{\bar{\theta}})$. This holds, for instance, when $h(\cdot) = \eta(x_*,\cdot)$ or is a function of $\eta(x_*,\cdot)$. Notice that when $\bar{\theta}$ and x_* are such that $\mathbf{f}_{\bar{\theta}}(x_*)$ is a boundary point of the Elfving's set $\mathscr{F}_{\bar{\theta}}$ for the value $\bar{\theta}$, Theorem 5.30 implies $\left[\partial h(\theta)/\partial\theta^\top \mathbf{M}^-(\delta_{x_*},\theta)\,\partial h(\theta)/\partial\theta\right]_{\bar{\theta}} = \beta^2$.

Although this seems to be an argument for considering δ_{x_*} as an optimum design for estimating $h(\theta)$, there is a serious flaw in it, and quite severe limitations exist that restrain the application of this result in practical situations, even when the function of interest is linear, i.e., when $h(\theta) = \mathbf{c}^\top\theta$:

1. The consistency of $\hat{\theta}^N_{LS}$ and regular asymptotic normality of $h(\hat{\theta}^N_{LS})$ are due to the use of two different design points, so that ξ_N is nonsingular for any N; these asymptotic properties of the estimators are therefore not attached to the limiting design itself, here δ_{x_*}.
2. The direction $\mathbf{f}_{\bar{\theta}}(x_*)$ for which regular asymptotic normality holds is unknown since $\bar{\theta}$ is unknown. Let $\mathbf{c}$ be a direction of interest chosen in advance and θ^0 be the nominal value of the parameters used for local design. The associated c-optimal design ξ_c^* is then determined for this nominal value. For instance, when $\mathbf{c} = (0,\ 1)^\top$, which means that we are only interested into the estimation of the component θ_2, $\xi_c^* = \delta_{x_*}$ with x_* solution of $\{\mathbf{f}_{\theta^0}(x)\}_1 = 0$ (see Fig. 5.2), that is, x_* satisfies
$$\theta_2^0 = [\theta_2^0 + \theta_1^0(\theta_1^0 - \theta_2^0)x_*]\exp[-(\theta_1^0 - \theta_2^0)x_*]\,.$$
For $\theta^0 = (0.7,\ 0.2)^\top$, this gives $x_* = x_*(\theta^0) \simeq 4.28$. In general, $\mathbf{f}_{\bar{\theta}}(x_*) \neq \mathbf{f}_{\theta^0}(x_*)$ to which $\mathbf{c}$ is proportional. Therefore, $\mathbf{c} \notin \mathcal{M}[\mathbf{M}(\xi_c^*, \bar{\theta})]$, and regular asymptotic normality does not hold for $\mathbf{c}^\top\hat{\theta}^N_{LS}$.

The present example is simple enough to be able to investigate the limiting behavior of $\mathbf{c}^\top\hat{\theta}^N_{LS}$ by direct calculation. In particular, we show that different choices of $\mathbf{c}$ give different speeds of convergence for $\mathbf{c}^\top\hat{\theta}^N_{LS}$. Using a Taylor development of the LS criterion $J_N(\theta)$, see (3.1), similar to that used in the proof of Theorem 3.8, we obtain

$$0 = \{\nabla_\theta J_N(\hat{\theta}^N_{LS})\}_i = \{\nabla_\theta J_N(\bar{\theta})\}_i + \{\nabla^2_\theta J_N(\beta_i^N)(\hat{\theta}^N_{LS} - \bar{\theta})\}_i\,,\ i = 1, 2\,,$$

where $\beta_i^N \overset{\text{a.s.}}{\to} \bar{\theta}$ as $N \to \infty$. Direct calculations give

$$\begin{aligned}\nabla_\theta J_N(\bar{\theta}) &= -\frac{2}{N}\left[\sqrt{m}\,\beta_m\,\mathbf{f}_{\bar{\theta}}(x^0) + \sqrt{N-m}\,\gamma_{N-m}\,\mathbf{f}_{\bar{\theta}}(x_*)\right],\\ \nabla^2_\theta J_N(\bar{\theta}) &= \frac{2}{N}\left[m\,\mathbf{f}_{\bar{\theta}}(x^0)\mathbf{f}_{\bar{\theta}}^\top(x^0) + (N-m)\,\mathbf{f}_{\bar{\theta}}(x_*)\mathbf{f}_{\bar{\theta}}^\top(x_*)\right] + \mathcal{O}_{\rm p}(\sqrt{m}/N)\,,\end{aligned}$$

where $\beta_m = (1/\sqrt{m})\sum_{x_i=x^0}\varepsilon_i$ and $\gamma_{N-m} = (1/\sqrt{N-m})\sum_{x_i=x_*}\varepsilon_i$ are independent random variables that tend to be distributed $\mathcal{N}(0,1)$ as $m \to \infty$ and $N - m \to \infty$. We then obtain that $\sqrt{N}\mathbf{f}_{\bar{\theta}}^\top(x_*)(\hat{\theta}^N_{LS} - \bar{\theta})$ is asymptotically normal $\mathcal{N}(0,1)$, whereas for any direction $\mathbf{c}$ not parallel to $\mathbf{f}_{\bar{\theta}}(x_*)$ and not orthogonal to $\mathbf{f}_{\bar{\theta}}(x^0)$, $\sqrt{m}\mathbf{c}^\top(\hat{\theta}^N_{LS} - \bar{\theta})$ is asymptotically normal and $\mathbf{c}^\top(\hat{\theta}^N_{LS} - \bar{\theta})$ converges not faster than $1/\sqrt{m}$. In particular, $\sqrt{m}\mathbf{f}_{\bar{\theta}}^\top(x^0)(\hat{\theta}^N_{LS} - \bar{\theta})$ is asymptotically normal $\mathcal{N}(0,1)$, and $\sqrt{m}\{\hat{\theta}^N_{LS} - \bar{\theta}\}_2$ is asymptotically normal with zero mean and variance $\{\mathbf{f}_{\bar{\theta}}(x_*)\}_1^2\,\det^{-2}[\mathbf{f}_{\bar{\theta}}(x_*), \mathbf{f}_{\bar{\theta}}(x^0)]$. □

A first intuitive remedy to the difficulties encountered in Example 5.39 consists in replacing the singular design ξ_c^* by a nonsingular one, with support points close to those of ξ_c^*. As shown now in a continuation of the example, letting these support points approach those of ξ_c^* may create difficulties in the limit.

Example 5.40. We continue Example 5.39 by placing now the proportion $m = N/2$ of the observations at x^0. We thus consider the design measure $\xi_{\gamma,x^0} = (1-\gamma)\delta_{x_*} + \gamma\delta_{x^0}$ for $\gamma = 1/2$. Since the c-optimal design is δ_{x_*}, we consider the limiting behavior of $\mathbf{c}^\top(\hat{\theta}^N_{LS} - \bar{\theta})$ when N tends to infinity for x^0 approaching x_*. Note that ξ_N then converges weakly to δ_{x_*}. Since $\xi_{1/2,x^0}$ is nonsingular for $x^0 \neq x_*$ (and $x^0 \neq 0$), $\sqrt{N}\mathbf{c}^\top(\hat{\theta}^N_{LS} - \bar{\theta})$ is asymptotically normal $\mathscr{N}(0, \mathbf{c}^\top\mathbf{M}^{-1}(\xi_{1/2,x^0}, \bar{\theta})\mathbf{c})$.

A first difficulty is that the asymptotic variance $\mathbf{c}^\top\mathbf{M}^{-1}(\xi_{1/2,x^0}, \bar{\theta})\mathbf{c}$ tends to infinity as x^0 tends to x_* when $\mathbf{c}$ is not proportional to $\mathbf{f}_{\bar{\theta}}(x_*)$. This is hardly avoidable: in the limit, estimation is possible in one direction $\mathbf{c}$ only, dictated by the value of $\bar{\theta}$ due to the nonlinearity of the model.

Moreover, when $\mathbf{c} = \mathbf{f}_{\bar{\theta}}(x_*)$, $\mathbf{f}_{\bar{\theta}}^\top(x_*)\mathbf{M}^{-1}(\xi_{1/2,x^0}, \bar{\theta})\mathbf{f}_{\bar{\theta}}(x_*)$ equals 2 for any $x^0 \neq x_*$, twice more than what could be achieved with the singular design δ_{x_*} since $\mathbf{f}_{\bar{\theta}}^\top(x_*)\mathbf{M}^-(\delta_{x_*}, \bar{\theta})\mathbf{f}_{\bar{\theta}}(x_*) = 1$ (this result coincides with that of Example 5.10 for $\beta = 0$). □

Examples 3.13, 3.17, 5.39, and 5.40 show that, in a nonlinear situation, when using a design sequence such that ξ_N converges to a singular c-optimal design, the estimator of the quantity of interest may converge at a slow rate or have an excessively large variance in the limit. A possible way to circumvent these difficulties is to regularize the c-optimality criterion. Indeed, a regularization of the criterion evaluated at some θ^0 allows us to construct nonsingular designs with performance close to that of the singular optimal design for $\bar{\theta}$ when θ^0 is not too far from $\bar{\theta}$.

Such a regularization can be based, for instance, on ridge estimation; see Pázman (1986, Sect. 4.5). Another possibility is to replace $\mathbf{M}$ by $\mathbf{M} + \gamma\mathbf{I}_p$ in $\Phi_c(\cdot)$, which can be interpreted as designing for maximum a posteriori estimation with a vague prior; see Remark 4.18-(i). The regularized criterion is then

$$\Phi_c^{(\gamma)}(\mathbf{M}) = -\mathbf{c}^\top(\mathbf{M} + \gamma\mathbf{I}_p)^{-1}\mathbf{c}\,,$$

with γ a small positive number. Note that $\mathbf{M} + \gamma\mathbf{I}_p \in \mathbb{M}^>$ and that $\Phi_c^{(\gamma)}(\mathbf{M})$ is differentiable at any $\mathbf{M} \in \mathbb{M}^\geq$. The motivation for this approach is that any value $\Phi_c(\mathbf{M}_1)$ for $\mathbf{M}_1 \in \mathbb{M}^\geq$ can be obtained as a limit of values $\Phi_c(\mathbf{M}_1 + \gamma_k\mathbf{I}_p)$ with $\gamma_k > 0$ and tending to zero. Indeed, using the isotonicity property, $\Phi_c^{(\gamma_k)}(\mathbf{M}_1) \geq \Phi_c(\mathbf{M}_1)$, and from the upper semicontinuity, $\mathbf{M}_1 + \gamma_k\mathbf{I}_p \to \mathbf{M}_1$ implies $\limsup_{k\to\infty}\Phi_c^{(\gamma_k)}(\mathbf{M}_1) \leq \Phi_c(\mathbf{M}_1)$. Therefore, $\lim_{\gamma\to 0}\Phi_c^{(\gamma)}(\mathbf{M}_1) = \Phi_c(\mathbf{M}_1)$.

We may also consider the regularized criterion defined by

$$\Phi_c^{(\gamma)}(\mathbf{M}) = \Phi_c[(1-\gamma)\mathbf{M} + \gamma\mathbf{M}(\xi_0)]\,, \tag{5.54}$$

with $\gamma \in (0,1]$ and ξ_0 a design measure in Ξ such that $\mathbf{M}(\xi_0)$ has full rank; see Fedorov and Hackl (1997, p. 51). For $\mathbf{M} = \mathbf{M}(\xi)$, this equivalently defines the criterion

$$\phi_c^{(\gamma)}(\xi) = \phi_c[(1-\gamma)\xi + \gamma\xi_0]\,, \tag{5.55}$$

with $(1-\gamma)\xi + \gamma\xi_0$ a regularized version of ξ. Notice that such regularized designs are easy to implement: using the design $(1-\gamma)\xi + \gamma\xi_0$ means that when N observations are performed, approximately $(1-\gamma)\,N$ are to be taken with ξ and $\gamma\,N$ with ξ_0. Note that when the design measure ξ is discrete, then the design is asymptotically discrete in the sense of Definition 2.1 when $N \to \infty$ with $\gamma = \gamma(N)$ tending to zero. Also note that the criterion $\Phi_c^{(\gamma)}(\mathbf{M})$ is continuous in γ at $\gamma = 0$, see Lemma 5.12, and differentiable with respect to $\mathbf{M}$ at any $\mathbf{M} \in \mathbb{M}^{\geq}$ when $\gamma > 0$. Let ξ_γ^* be an optimal design for $\phi_c^{(\gamma)}(\xi) = \Phi_c^{(\gamma)}[\mathbf{M}(\xi)]$ and ξ^* be a c-optimal design. Then $(1-\gamma)\xi_\gamma^* + \gamma\xi_0$ tends to become c-optimal as $\gamma \to 0$. Indeed, from the concavity of $\phi(\cdot)$ and the proof of Lemma 5.16, we have

$$\begin{aligned}
\phi_c(\xi^*) \geq \phi_c^{(\gamma)}(\xi_\gamma^*) &= \Phi_c[(1-\gamma)\mathbf{M}(\xi_\gamma^*) + \gamma\mathbf{M}(\xi_0)] \\
&\geq \Phi_c[(1-\gamma)\mathbf{M}(\xi^*) + \gamma\mathbf{M}(\xi_0)] \\
&\geq \gamma\{\Phi_c[\mathbf{M}(\xi_0)] - \phi_c(\xi^*)\} + \phi_c(\xi^*)\,,
\end{aligned}$$

so that

$$0 \leq \phi_c(\xi^*) - \phi_c^{(\gamma)}(\xi_\gamma^*) \leq \gamma\{\phi_c(\xi^*) - \Phi_c[\mathbf{M}(\xi_0)]\}$$

where the right-hand side tends to zero as $\gamma \to 0$.

On the other hand, Examples 3.17 and 5.39 have shown the possible pitfalls caused by γ tending to zero in a nonlinear situation when the optimal design is constructed for $\theta^0 \neq \bar{\theta}$. When $\mathbf{c} = \mathbf{c}(\theta) = \partial h(\theta)/\partial\theta$, with $h(\cdot)$ the function of interest, the value of γ in the regularized criterion (5.54) can then be chosen by maximizing

$$J_{reg}(\gamma) = \min_{\theta\in\Theta^0} \Phi_c[(1-\gamma)\mathbf{M}(\xi_c^*,\theta) + \gamma\mathbf{M}(\xi_0,\theta)]\,,\ \gamma \in [0,1]\,, \tag{5.56}$$

where Θ^0 denotes a feasible set for θ and ξ_c^* maximizes $\Phi_c[\mathbf{M}(\xi,\theta^0)]$ with $\theta^0 \in \Theta^0$ a nominal value for θ. Note that $J_{reg}(\gamma)$ is a concave function of γ (as the minimum of a family of concave functions). Also note that if neither $\mathbf{c}_\theta$ nor $\mathbf{M}(\xi,\theta)$ depend on θ, the maximum of $J_{reg}(\cdot)$ is attained at $\gamma = 0$.

Other approaches for avoiding singular c-optimal designs, based on c-maximin-optimum design or on regularization through D-optimum design (using the results of Sect. 5.3.2), are suggested in (Pronzato, 2009b). See also Sect. 7.7.2.

5.5 Optimality Criteria for Asymptotic Variance–Covariance Matrices in Product Form

In Chaps. 3 and 4 we have encountered several cases when the asymptotic variance–covariance matrix of the estimator of θ is in product form $N^{-1}\mathbf{C}(\xi) = N^{-1}\mathbf{M}_1^{-1}(\xi)\mathbf{M}_2(\xi)\mathbf{M}_1^{-1}(\xi)$, where $\mathbf{M}_1(\xi) \neq \mathbf{M}_2(\xi)$ are "information-like" matrices; see, for instance, Sects. 3.1.3, 3.3.2, 3.4, and 4.1. Although from a statistical point of view it seems that design criteria should then be built on $\mathbf{C}(\xi)$, they cannot in general since $\mathbf{C}(\xi)$ depends on quantities that are unknown a priori , and moreover, $\mathbf{C}^{-1}(\xi)$ does not possess the properties required for information matrices.

In general, this raises nonstandard design problems which we do not address here.[14] In this section we consider four situations that we have met in Chaps. 3 and 4. In each case, the design is performed under idealistic assumptions (optimum weights in WLS estimation, normal errors for penalized WLS, no modeling error, etc..), and we give bounds on the loss of efficiency resulting from a violation of those assumptions. One may also refer to Sect. 7.8 for another justification, related to estimability properties, for designing under such idealistic assumptions.

5.5.1 The WLS Estimator

In Theorem 3.8 we proved that the asymptotic variance–covariance matrix of the WLS estimator is equal to $N^{-1}\mathbf{C}(w,\xi,\bar{\theta})$, with $\bar{\theta}$ the unknown true value of θ and

$$\mathbf{C}(w,\xi,\theta) = \mathbf{M}_1^{-1}(\xi,\theta)\mathbf{M}_2(\xi,\theta)\mathbf{M}_1^{-1}(\xi,\theta)\,,$$

where

$$\mathbf{M}_1(\xi,\theta) = \int_{\mathscr{X}} w(x)\,\frac{\partial\eta(x,\theta)}{\partial\theta}\frac{\partial\eta(x,\theta)}{\partial\theta^\top}\,\xi(\mathrm{d}x)\,,$$
$$\mathbf{M}_2(\xi,\theta) = \int_{\mathscr{X}} w^2(x)\sigma^2(x)\,\frac{\partial\eta(x,\theta)}{\partial\theta}\frac{\partial\eta(x,\theta)}{\partial\theta^\top}\,\xi(\mathrm{d}x)\,,$$

$w(x)$ is the (known) weight function and $\sigma^2(x)$ is the (unknown) variance of the observation at x. In the particular case where $w(x) = c\,\sigma^{-2}(x)$, $c > 0$, the matrix $\mathbf{C}(w,\xi,\theta)$ is proportional to $\mathbf{M}^{-1}(\xi,\bar{\theta})$ with

$$\mathbf{M}(\xi,\theta) = \int_{\mathscr{X}} \sigma^{-2}(x)\,\frac{\partial\eta(x,\theta)}{\partial\theta}\frac{\partial\eta(x,\theta)}{\partial\theta^\top}\,\xi(\mathrm{d}x)\,.$$

For other choices of the weight function $w(\cdot)$ we only know that $\mathbf{C}(w,\xi,\bar{\theta}) - \mathbf{M}^{-1}(\xi,\bar{\theta})$ is positive definite, with both $\mathbf{C}(w,\xi,\bar{\theta})$ and $\mathbf{M}(\xi,\bar{\theta})$ depending

[14] Possible developments may involve the use of bounds, obtained, e.g., from Kantorovich-type inequalities; see Bloomfield and Watson (1975), Pečarić et al. (1996), and Liu and Neudecker (1997).

on the variance function $\sigma^2(\cdot)$ which, in general, is unknown. The choice of the weight function $w(\cdot)$ in WLS estimation is thus always based on a prior guess $\hat{\sigma}^2(\cdot)$ for $\sigma^2(\cdot)$ so that the corresponding guessed asymptotic variance–covariance matrix of the estimator is $\mathbf{M}_1^{-1}(\xi,\bar{\theta})$. It is thus natural to build optimality criteria on the matrix $\mathbf{M}_1$ and proceed as in previous sections of this chapter. Since $\bar{\theta}$ is unknown, we use again a local design approach and substitute a nominal value θ^0 for $\bar{\theta}$ in all places where $\bar{\theta}$ appears. We then omit the dependence in θ and write $\mathbf{C}(w,\xi)$ for $\mathbf{C}(w,\xi,\theta^0)$, $\mathbf{M}(\xi)=\mathbf{M}(\xi,\theta^0)$, etc.

The loss of efficiency induced by using $\mathbf{M}_1(\xi)$ instead of $\mathbf{C}^{-1}(\xi)$ for constructing an optimal design is evaluated in the following theorem.

Theorem 5.41. *Let $\Phi^+(\cdot)$ be an isotonic and positively homogeneous global criterion, see Definitions 5.3 and 5.8. Then, for every design ξ such that $\mathbf{M}_1(\xi)$ and $\mathbf{M}_2(\xi)$ are nonsingular the efficiencies*

$$\mathscr{E}_{\Phi^+,\mathbf{M}_1}(\xi) = \frac{\Phi^+[\mathbf{M}_1(\xi)]}{\max_\nu \Phi^+[\mathbf{M}_1(\nu)]}\,, \tag{5.57}$$

$$\mathscr{E}_{\Phi^+,\mathbf{C}^{-1}}(\xi) = \frac{\Phi^+[\mathbf{M}_1(\xi)\mathbf{M}_2^{-1}(\xi)\mathbf{M}_1(\xi)]}{\max_\nu \Phi^+[\mathbf{M}_1(\nu)\mathbf{M}_2^{-1}(\nu)\mathbf{M}_1(\nu)]} \tag{5.58}$$

satisfy the inequalities

$$\frac{k}{K}\,\mathscr{E}_{\Phi^+,\mathbf{M}_1}(\xi) \le \mathscr{E}_{\Phi^+,\mathbf{C}^{-1}}(\xi) \le \frac{K}{k}\,\mathscr{E}_{\Phi^+,\mathbf{M}_1}(\xi) \tag{5.59}$$

with

$$K = \max_{x\in\mathscr{X}} w(x)\,\sigma^2(x) \quad \textit{and} \quad k = \min_{x\in\mathscr{X}} w(x)\,\sigma^2(x)\,,$$

so that $k=K$ and $\mathscr{E}_{\Phi^+,\mathbf{M}_1}(\xi)=\mathscr{E}_{\Phi^+,\mathbf{C}^{-1}}(\xi)$ when $w(x)=c\,\sigma^{-2}(x)$ for some $c>0$.

Proof. For any vector $\mathbf{u}\in\mathbb{R}^p$ we have

$$\begin{aligned} k\,\mathbf{u}^\top\mathbf{M}_1(\xi)\mathbf{u} \le \mathbf{u}^\top\mathbf{M}_2(\xi)\mathbf{u} &= \int_{\mathscr{X}} [w(x)\sigma^2(x)]\,w(x)\,\left[\mathbf{u}^\top \left.\frac{\partial\eta(x,\theta)}{\partial\theta}\right|_{\theta^0}\right]^2 \xi(\mathrm{d}x) \\ &\le K\,\mathbf{u}^\top\mathbf{M}_1(\xi)\mathbf{u}\,. \end{aligned}$$

Hence, from Lemma 5.1-(v),

$$K^{-1}\,\mathbf{M}_1(\xi) \preceq \mathbf{M}_1(\xi)\mathbf{M}_2^{-1}(\xi)\mathbf{M}_1(\xi) \preceq k^{-1}\,\mathbf{M}_1(\xi)\,.$$

From the isotonicity and positive homogeneity of $\Phi^+(\cdot)$ we obtain

$$\begin{aligned} K^{-1}\,\Phi^+[\mathbf{M}_1(\xi)] &\le \Phi^+[\mathbf{M}_1(\xi)\mathbf{M}_2^{-1}(\xi)\mathbf{M}_1(\xi)] \le k^{-1}\,\Phi^+[\mathbf{M}_1(\xi)] \\ k^{-1}\,\max_\nu \Phi^+[\mathbf{M}_1(\nu)] &\ge \max_\nu \Phi^+[\mathbf{M}_1(\nu)\mathbf{M}_2^{-1}(\nu)\mathbf{M}_1(\nu)] \ge K^{-1}\,\max_\nu \Phi^+[\mathbf{M}_1(\nu)] \end{aligned}$$

which yields the required result. ■

5.5.2 The Penalized WLS Estimator

In Theorem 3.24 we proved that the asymptotic variance–covariance matrix of the penalized WLS estimator is equal to $N^{-1}\,\mathbf{C}(\xi,\bar{\theta})$, with $\mathbf{C}(\xi,\theta)$ still given by $\mathbf{M}_1^{-1}(\xi,\theta)\mathbf{M}_2(\xi,\theta)\mathbf{M}_1^{-1}(\xi,\theta)$ where now

$$\mathbf{M}_1(\xi,\bar{\theta}) = \int_{\mathscr{X}} \lambda^{-1}(x,\bar{\theta}) \frac{\partial\eta(x,\theta)}{\partial\theta}\bigg|_{\bar{\theta}} \frac{\partial\eta(x,\theta)}{\partial\theta^\top}\bigg|_{\bar{\theta}} \xi(\mathrm{d}x)$$
$$+\frac{\bar{\beta}}{2}\int_{\mathscr{X}} \lambda^{-2}(x,\bar{\theta}) \frac{\partial\lambda(x,\theta)}{\partial\theta}\bigg|_{\bar{\theta}} \frac{\partial\lambda(x,\theta)}{\partial\theta^\top}\bigg|_{\bar{\theta}} \xi(\mathrm{d}x)\,,$$
$$\mathbf{M}_2(\xi,\bar{\theta}) = \bar{\beta}\mathbf{M}_1(\xi,\bar{\theta})$$
$$+\frac{\bar{\beta}^{3/2}}{2}\int_{\mathscr{X}} \lambda^{-3/2}(x,\bar{\theta}) \left[\frac{\partial\eta(x,\theta)}{\partial\theta}\bigg|_{\bar{\theta}} \frac{\partial\lambda(x,\theta)}{\partial\theta^\top}\bigg|_{\bar{\theta}} + \frac{\partial\lambda(x,\theta)}{\partial\theta}\bigg|_{\bar{\theta}} \frac{\partial\eta(x,\theta)}{\partial\theta^\top}\bigg|_{\bar{\theta}}\right] s(x)\,\xi(\mathrm{d}x)$$
$$+\frac{\bar{\beta}^2}{4}\int_{\mathscr{X}} \lambda^{-2}(x,\bar{\theta})\, \frac{\partial\lambda(x,\theta)}{\partial\theta}\bigg|_{\bar{\theta}} \frac{\partial\lambda(x,\theta)}{\partial\theta^\top}\bigg|_{\bar{\theta}} \kappa(x)\,\xi(\mathrm{d}x)\,,$$

with $s(x) = \mathbb{E}_x\{\varepsilon^3(x)\}\sigma^{-3}(x)$ the skewness and $\kappa(x) = \mathbb{E}_x\{\varepsilon^4(x)\}\sigma^{-4}(x) - 3$ the kurtosis of the distribution of the error $\varepsilon(x)$ and $\bar{\beta}$ the (known) scaling factor in the error variance (3.45) used in the penalized criterion (3.47). We suppose that either $\mathscr{X}$ is finite, or $s(x)$ and $\kappa(x)$ are continuous on $\mathscr{X}$ compact.

Again, $\mathbf{C}^{-1}(\xi,\bar{\theta})$ is not in the form of an information matrix, and moreover, it contains unknown functions $s(\cdot)$ and $\kappa(\cdot)$. Only in the particular case that $s(x) = \kappa(x) = 0$ for all x (normal errors, for instance), we have $\mathbf{C}^{-1}(\xi,\bar{\theta}) = \bar{\beta}^{-1}\,\mathbf{M}_1(\xi,\bar{\theta})$, see Remark 3.25, which corresponds to an information matrix. A reasonable approach consists in assuming that $s(x) = \kappa(x) = 0$ at the design stage, which amounts to using a design criterion based on $\mathbf{M}_1(\xi) = \mathbf{M}_1(\xi,\theta^0)$, with θ^0 a nominal value for θ (locally optimum design). The resulting loss of efficiency, due to the substitution of $\mathbf{M}_1(\xi)$ for $\mathbf{C}^{-1}(\xi) = \mathbf{C}^{-1}(\xi,\theta^0)$, is evaluated in the following theorem.

Theorem 5.42. *Let $\Phi^+(\cdot)$ be an isotonic and positively homogeneous global criterion, see Definitions 5.3 and 5.8, and let*

$$K = \max_{x\in\mathscr{X}} \gamma_{\max}(x)\,, \quad k = \min_{x\in\mathscr{X}} \gamma_{\min}(x)\,,$$

where $\gamma_{\max}(x) \geq \gamma_{\min}(x) > 0$ are the two solutions of the quadratic equation

$$\det[\mathbf{V}(x) - \gamma\mathbf{I}_2] = 0$$

with $\mathbf{V}(x)$ the 2×2 matrix

$$\mathbf{V}(x) = \bar{\beta}\begin{pmatrix} 1 & s(x)/\sqrt{2} \\ s(x)/\sqrt{2} & 1+\kappa(x)/2 \end{pmatrix}.$$

Then, for every design ξ such that $\mathbf{M}_1(\xi)$ and $\mathbf{M}_2(\xi)$ are nonsingular the efficiencies $\mathscr{E}_{\Phi^+,\mathbf{M}_1}(\xi)$ and $\mathscr{E}_{\Phi^+,\mathbf{C}^{-1}}(\xi)$ defined by (5.57) and (5.58) satisfy the inequalities (5.59), with equality when $s(x) = \kappa(x) = 0$ for all $x \in \mathscr{X}$.

Proof. Define

$$\mathbf{a}(x) = \lambda^{-1/2}(x,\theta^0)\,\frac{\partial\eta(x,\theta)}{\partial\theta}\bigg|_{\theta^0}\,,\quad \mathbf{b}(x) = (\bar{\beta}/2)^{1/2}\,\lambda^{-1}(x,\theta^0)\,\frac{\partial\lambda(x,\theta)}{\partial\theta}\bigg|_{\theta^0}\,,$$

so that we can write

$$\mathbf{M}_1(\xi) = \int_{\mathscr{X}} \begin{pmatrix}\mathbf{a}(x) & \mathbf{b}(x)\end{pmatrix} \mathbf{I}_2 \begin{pmatrix}\mathbf{a}^\top(x)\\ \mathbf{b}^\top(x)\end{pmatrix} \xi(\mathrm{d}x)\,,$$
$$\mathbf{M}_2(\xi) = \int_{\mathscr{X}} \begin{pmatrix}\mathbf{a}(x) & \mathbf{b}(x)\end{pmatrix} \mathbf{V}(x) \begin{pmatrix}\mathbf{a}^\top(x)\\ \mathbf{b}^\top(x)\end{pmatrix} \xi(\mathrm{d}x)\,.$$

Consider now the matrix $\mathbf{V}(x)$. From Cauchy–Schwarz inequality, $1+\kappa(x)/2 = [\mathbb{E}\{\varepsilon^4(x)\} - \sigma^4]/(2\sigma^4) \geq 0$, which implies that $\mathrm{trace}[\mathbf{V}(x)] \geq \bar{\beta}$. Also,

$$\begin{aligned}\det[\mathbf{V}(x)] &= \bar{\beta}^2\left[1 + \frac{\kappa(x)}{2} - \frac{s^2(x)}{2}\right]\\ &= \frac{\bar{\beta}^2}{2\sigma^6}\left[\mathbb{E}\{\varepsilon^2(x)\}\mathbb{E}\{\varepsilon^4(x)\} - (\mathbb{E}\{\varepsilon^3(x)\})^2 - (\mathbb{E}\{\varepsilon^2(x)\})^3\right]\\ &= \frac{\bar{\beta}^2}{2\sigma^6}\det[\mathbb{E}\{\mathbf{v}(x)\mathbf{v}^\top(x)\}]\,,\end{aligned}$$

with $\mathbf{v}^\top(x) = [1\ \varepsilon(x)\ \varepsilon^2(x)]$ (remember that $\mathbb{E}\{\varepsilon(x)\} = 0$). Hence, $\mathbf{V}(x)$ is positive definite with positive eigenvalues $\gamma_{\min}(x)$ and $\gamma_{\max}(x)$. Since $k\mathbf{I}_2 \preceq \mathbf{V}(x) \preceq K\mathbf{I}_2$, we have $k\mathbf{u}^\top\mathbf{M}_1(\xi)\mathbf{u} \leq \mathbf{u}^\top\mathbf{M}_2(\xi)\mathbf{u} \leq K\mathbf{u}^\top\mathbf{M}_1(\xi)\mathbf{u}$ for any $\mathbf{u} \in \mathbb{R}^p$. The rest of the proof is like in Theorem 5.41. ∎

5.5.3 The LS Estimator with Model Error

In Sect. 3.4 we considered the situation where the supposed regression model $y(x_k) = \eta(x_k,\theta) + \varepsilon_k$ is incorrect; see (3.81). As in Sect. 3.4, we assume that $\sigma^2 = \mathrm{var}[\varepsilon(x)]$ does not depend on $x \in \mathscr{X}$. In Theorem 3.36 we proved that in that case the asymptotic variance–covariance matrix of the LS estimator is proportional to $\mathbf{C}_\nu(\xi,\bar{\theta})$, with $\bar{\theta} = \arg\min_{\theta\in\Theta}\int_{\mathscr{X}}[\eta(x,\theta) - \nu(x)]^2\,\xi(\mathrm{d}x)$ and

$$\begin{aligned}\mathbf{C}_\nu(\xi,\theta) = {} & [\mathbf{M}(\xi,\theta) + \mathbf{D}_\nu(\xi,\theta)]^{-1}[\mathbf{M}(\xi,\theta) + (1/\sigma^2)\mathbf{M}_\nu(\xi,\theta)]\\ & \times[\mathbf{M}(\xi,\theta) + \mathbf{D}_\nu(\xi,\theta)]^{-1}\,.\end{aligned}$$

As before, we omit the dependence in θ when the evaluation is at a nominal value θ^0 and write $\mathbf{C}_\nu(\xi) = \mathbf{C}_\nu(\xi,\theta^0)$. The matrices $\mathbf{M}(\xi), \mathbf{D}_\nu(\xi)$, and $\mathbf{M}_\nu(\xi)$ are then given by

$$\mathbf{M}(\xi) = \int_{\mathscr{X}} \frac{\partial\eta(x,\theta)}{\partial\theta}\bigg|_{\theta^0} \frac{\partial\eta(x,\theta)}{\partial\theta^\top}\bigg|_{\theta^0} \xi(\mathrm{d}x)\,,$$
$$\mathbf{D}_\nu(\xi) = \int_{\mathscr{X}} [\eta(x,\theta^0) - \nu(x)]\,\frac{\partial^2\eta(x,\theta)}{\partial\theta\partial\theta^\top}\bigg|_{\theta^0} \xi(\mathrm{d}x)\,,$$
$$\mathbf{M}_\nu(\xi) = \int_{\mathscr{X}} [\eta(x,\theta^0) - \nu(x)]^2\,\frac{\partial\eta(x,\theta)}{\partial\theta}\bigg|_{\theta^0} \frac{\partial\eta(x,\theta)}{\partial\theta^\top}\bigg|_{\theta^0} \xi(\mathrm{d}x)\,,$$

see Sect. 3.4. We also define $\mathbf{N}_\nu(\xi) = \mathbf{C}_\nu^{-1}(\xi)$. In the particular case where the supposed model is correct, i.e., when $\eta(x, \theta^0) = \nu(x)$ for every x, we have $\mathbf{N}_\nu(\xi) = \mathbf{M}(\xi)$ for every ξ. In general one does not know $\bar{\theta}$, $\nu(x)$, and σ a priori , and we base optimality criteria again on the matrix $\mathbf{M}(\xi)$. The loss of efficiency due to the substitution of $\mathbf{M}(\xi)$ for $\mathbf{N}_\nu(\xi)$ in a design optimality criterion is evaluated in the following theorem.

Theorem 5.43. *Let $\Phi^+(\cdot)$ be an isotonic and positively homogeneous global criterion, see Definitions 5.3 and 5.8. Suppose that the assumptions of Theorem 3.36 hold at θ^0, i.e., $\mathbf{M}(\xi, \theta^0)$ is nonsingular and $C_{int}(\xi, \theta^0)\|\nu(\cdot) - \eta(\cdot, \bar{\theta}^0)\|_\xi < 1$, where $C_{int}(\xi, \theta)$ is defined in (3.84) and $\|\nu(\cdot) - \eta(\cdot, \theta^0)\|_\xi^2 = \int_{\mathscr{X}} [\eta(x, \theta^0) - \nu(x)]^2 \, \xi(\mathrm{d}x)$. Then,*

$$\frac{\delta^2}{b} \Phi^+[\mathbf{M}(\xi)] \leq \Phi^+[\mathbf{N}_\nu(\xi)] \leq \Delta^2 \, \Phi^+[\mathbf{M}(\xi)] \,, \tag{5.60}$$

where

$$\begin{aligned} b &= 1 + \frac{\max_{x \in \mathscr{X}} [\nu(x) - \eta(x, \theta^0)]^2}{\sigma^2} \,, \\ \delta &= 1 - C_{int}(\xi, \theta^0)\|\nu(\cdot) - \eta(\cdot, \theta^0)\|_\xi \,, \\ \Delta &= 1 + C_{int}(\xi, \theta^0)\|\nu(\cdot) - \eta(\cdot, \theta^0)\|_\xi \,. \end{aligned}$$

Moreover, the efficiencies

$$\mathscr{E}_{\Phi^+, \mathbf{M}}(\xi) = \frac{\Phi^+[\mathbf{M}(\xi)]}{\max_\mu \Phi^+[\mathbf{M}(\mu)]} \quad \text{and} \quad \mathscr{E}_{\Phi^+, \mathbf{N}_\nu}(\xi) = \frac{\Phi^+[\mathbf{N}_\nu(\xi)]}{\max_\mu \Phi^+[\mathbf{N}_\nu(\mu)]}$$

satisfy the inequalities

$$\frac{\delta^2}{b\Delta^2} \mathscr{E}_{\Phi^+, \mathbf{M}}(\xi) \leq \mathscr{E}_{\Phi^+, \mathbf{N}_\nu}(\xi) \leq \frac{b\Delta^2}{\delta^2} \mathscr{E}_{\Phi^+, \mathbf{M}}(\xi) \,, \tag{5.61}$$

which are changed to equalities when the model is correct.

Proof. Define $\tilde{\mathbf{N}}_\nu(\xi) = [\mathbf{M}(\xi) + \mathbf{D}_\nu(\xi)]\mathbf{M}^{-1}(\xi)[\mathbf{M}(\xi) + \mathbf{D}_\nu(\xi)]$. Since $\mathbf{M}_\nu(\xi)$ is positive semi-definite, we have $[\mathbf{M}(\xi) + \sigma^{-2}\mathbf{M}_\nu(\xi)] \succeq \mathbf{M}(\xi)$, and from the definition of b, $\mathbf{N}_\nu^{-1}(\xi) \succeq \tilde{\mathbf{N}}_\nu^{-1}(\xi)$ and $\mathbf{N}_\nu^{-1}(\xi) \preceq b \, \tilde{\mathbf{N}}_\nu^{-1}(\xi)$. Therefore, from Lemma 5.1-(v), $\mathbf{N}_\nu(\xi) \preceq \tilde{\mathbf{N}}_\nu(\xi) \preceq b \, \mathbf{N}_\nu(\xi)$. From the proof of Theorem 3.36 we know that for any $\mathbf{u} \in \mathbb{R}^p$

$$\begin{aligned} \mathbf{u}^\top[\mathbf{M}(\xi) + \mathbf{D}_\nu(\xi)]\mathbf{u} &\geq \mathbf{u}^\top\mathbf{M}(\xi)\mathbf{u} \, [1 - C_{int}(\xi, \theta^0)\|\nu(\cdot) - \eta(\cdot, \theta^0)\|_\xi] \\ &= \delta \, \mathbf{u}^\top\mathbf{M}(\xi)\mathbf{u} > 0 \,; \end{aligned}$$

that is, $\mathbf{M}(\xi) + \mathbf{D}_\nu(\xi) \succeq \delta \, \mathbf{M}(\xi)$. Similarly, $\mathbf{M}(\xi) + \mathbf{D}_\nu(\xi) \preceq \Delta \, \mathbf{M}(\xi)$. Hence (see Lemma 5.1-($v$)),

$$\frac{1}{\Delta} \mathbf{M}^{-1}(\xi) \preceq [\mathbf{M}(\xi) + \mathbf{D}_\nu(\xi)]^{-1} \preceq \frac{1}{\delta} \mathbf{M}^{-1}(\xi) \,.$$

Multiplying by $[\mathbf{M}(\xi) + \mathbf{D}_\nu(\xi)]$ from both sides, we obtain

$$\frac{1}{\Delta}\,\tilde{\mathbf{N}}_\nu(\xi) \preceq \mathbf{M}(\xi) + \mathbf{D}_\nu(\xi) \preceq \frac{1}{\delta}\,\tilde{\mathbf{N}}_\nu(\xi)\,.$$

Collecting and combining the results, we get

$$\mathbf{N}_\nu(\xi) \preceq \tilde{\mathbf{N}}_\nu(\xi) \preceq \Delta^2\,\mathbf{M}(\xi)\,,$$
$$b\,\mathbf{N}_\nu(\xi) \succeq \tilde{\mathbf{N}}_\nu(\xi) \succeq \delta^2\,\mathbf{M}(\xi)\,,$$

and $(\delta^2/b)\,\mathbf{M}(\xi) \preceq \mathbf{N}_\nu(\xi) \preceq \Delta^2\,\mathbf{M}(\xi)$. Applying the criterion $\Phi^+(\cdot)$ to these inequalities, we finally obtain (5.60). The inequalities (5.61) for efficiencies follow straightforwardly by maximizing with respect to ξ. ■

One may note that $\delta = \Delta = 1$ when the model is intrinsically linear with, however, $b \neq 1$ when modeling errors are present.

5.5.4 The M Estimator

Under the assumptions of Theorem 4.7 the asymptotic variance of the M estimator

$$\hat{\theta}^N_M = \arg\min_{\theta\in\Theta} \sum_{k=1}^{N} \rho_{x_k}[y(x_k) - \eta(x_k,\theta)]$$

is equal to $N^{-1}\mathbf{C}(\xi,\bar{\theta})$, with $\mathbf{C}(\xi,\theta) = \mathbf{M}_1^{-1}(\xi,\theta)\mathbf{M}_2(\xi,\theta)\mathbf{M}_1^{-1}(\xi,\theta)$ where

$$\mathbf{M}_1(\xi,\theta) = \int_{\mathscr{X}} \left\{\int_{-\infty}^{\infty} \rho''_x(\varepsilon)\,\bar{\varphi}_x(\varepsilon)\,\mathrm{d}\varepsilon\right\} \frac{\partial\eta(x,\theta)}{\partial\theta}\frac{\partial\eta(x,\theta)}{\partial\theta^\top}\,\xi(\mathrm{d}x)\,,$$
$$\mathbf{M}_2(\xi,\theta) = \int_{\mathscr{X}} \left\{\int_{-\infty}^{\infty} [\rho'_x(\varepsilon)]^2\,\bar{\varphi}_x(\varepsilon)\,\mathrm{d}\varepsilon\right\} \frac{\partial\eta(x,\theta)}{\partial\theta}\frac{\partial\eta(x,\theta)}{\partial\theta^\top}\,\xi(\mathrm{d}x)\,,$$

with $\rho'_x(\varepsilon) = \mathrm{d}\rho_x(\varepsilon)/\mathrm{d}\varepsilon$ and $\rho''_x(\varepsilon) = \mathrm{d}^2\rho_x(\varepsilon)/\mathrm{d}\varepsilon^2$. Again, we substitute a nominal value θ^0 for the unknown $\bar{\theta}$ and denote $\mathbf{M}_1(\xi) = \mathbf{M}_1(\xi,\theta^0)$, $\mathbf{M}_2(\xi) = \mathbf{M}_2(\zeta,\theta^0)$ and $\mathbf{C}(\xi) = \mathbf{C}(\xi,\theta^0)$. Since the minimum variance is obtained for $\rho_x(\cdot) = K_1 \log\bar{\varphi}_x(\cdot) + K_2$, $K_1 < 0$, see Theorem 4.7, we choose $\rho_x(\cdot)$ according to an a priori guessed p.d.f. $\hat{\varphi}_x(\cdot)$, so that

$$\rho_x(\cdot) = -\log\hat{\varphi}_x(\cdot)\,.$$

It is natural to assume that $\hat{\varphi}'_x(\varepsilon)\bar{\varphi}'_x(\varepsilon) \geq 0$ for any x and ε, with $\hat{\varphi}'_x(\varepsilon) = \mathrm{d}\hat{\varphi}_x(\varepsilon)/\mathrm{d}\varepsilon$ and $\bar{\varphi}'_x(\varepsilon) = \mathrm{d}\bar{\varphi}_x(\varepsilon)/\mathrm{d}\varepsilon$, so that

$$\int_{-\infty}^{\infty} \rho''_x(\varepsilon)\,\bar{\varphi}_x(\varepsilon)\,\mathrm{d}\varepsilon = \int_{-\infty}^{\infty} \frac{\hat{\varphi}'_x(\varepsilon)\bar{\varphi}'_x(\varepsilon)}{\hat{\varphi}_x(\varepsilon)}\,\mathrm{d}\varepsilon \geq 0$$

for all x; see also Remark 4.2-(i). The associated M estimator corresponds to the maximum likelihood estimator under the supposed error density $\hat{\varphi}_x(\cdot)$. Following this supposition, we choose

$$\mathbf{M}(\xi) = \int_{\mathscr{X}} \mathcal{I}_{\hat{\varphi}}(x) \left.\frac{\partial \eta(x,\theta)}{\partial \theta}\right|_{\theta^0} \left.\frac{\partial \eta(x,\theta)}{\partial \theta^\top}\right|_{\theta^0} \xi(\mathrm{d}x)$$

instead of $\mathbf{C}^{-1}(\xi)$ to construct optimality criteria, with

$$\mathcal{I}_{\hat{\varphi}}(x) = \int_{-\infty}^{\infty} \left[\frac{\hat{\varphi}'_x(\varepsilon)}{\hat{\varphi}_x(\varepsilon)}\right]^2 \hat{\varphi}_x(\varepsilon)\, \mathrm{d}\varepsilon\,.$$

The influence of this choice on the efficiency of designs is evaluated in the following theorem.

Theorem 5.44. *Let $\Phi^+(\cdot)$ be an isotonic and positively homogeneous global criterion; see Definitions 5.3 and 5.8. Suppose that the assumptions of Theorem 4.7 hold and that $\int_{-\infty}^{\infty} \rho''_x(\varepsilon)\, \bar{\varphi}_x(\varepsilon)\, \mathrm{d}\varepsilon \geq 0$ for all x. Then, for any design ξ such that $\mathbf{M}_1(\xi)$ and $\mathbf{M}_2(\xi)$ are nonsingular, the efficiencies $\mathscr{E}_{\phi^+,\mathbf{M}}(\xi)$ and $\mathscr{E}_{\phi^+,\mathbf{C}^{-1}}(\xi)$ given by (5.57), (5.58) satisfy the inequalities*

$$\frac{k_1^2\, k_2}{K_1^2\, K_2}\, \mathscr{E}_{\phi^+,\mathbf{M}}(\xi) \leq \mathscr{E}_{\phi^+,\mathbf{C}^{-1}}(\xi) \leq \frac{K_1^2\, K_2}{k_1^2\, k_2}\, \mathscr{E}_{\phi^+,\mathbf{M}}(\xi)$$

with

$$K_1 = \max_{x\in\mathscr{X}} \frac{\int_{-\infty}^{\infty} \rho''_x(\varepsilon)\, \bar{\varphi}_x(\varepsilon)\, \mathrm{d}\varepsilon}{\mathcal{I}_{\hat{\varphi}}(x)}\,, \quad k_1 = \min_{x\in\mathscr{X}} \frac{\int_{-\infty}^{\infty} \rho''_x(\varepsilon)\, \bar{\varphi}_x(\varepsilon)\, \mathrm{d}\varepsilon}{\mathcal{I}_{\hat{\varphi}}(x)}\,,$$
$$K_2 = \max_{x\in\mathscr{X}} \frac{\int_{-\infty}^{\infty} [\rho'_x(\varepsilon)]^2\, \bar{\varphi}_x(\varepsilon)\, \mathrm{d}\varepsilon}{\mathcal{I}_{\hat{\varphi}}(x)}\,, \quad k_2 = \min_{x\in\mathscr{X}} \frac{\int_{-\infty}^{\infty} [\rho'_x(\varepsilon)]^2\, \bar{\varphi}_x(\varepsilon)\, \mathrm{d}\varepsilon}{\mathcal{I}_{\hat{\varphi}}(x)}\,,$$

so that $k_1 = K_1 = k_2 = K_2 = 1$ and $\mathscr{E}_{\phi^+,\mathbf{M}}(\xi) = \mathscr{E}_{\phi^+,\mathbf{C}^{-1}}(\xi)$ when $\hat{\varphi}(\cdot) = \bar{\varphi}(\cdot)$.

Proof. We have $k_1\, \mathbf{M}(\xi) \preceq \mathbf{M}_1(\xi) \preceq K_1\, \mathbf{M}(\xi)$ and $k_2\, \mathbf{M}(\xi) \preceq \mathbf{M}_2(\xi) \preceq K_2\, \mathbf{M}(\xi)$. The rest of the proof is like in Theorem 5.41. ■

5.6 Bibliographic Notes and Further Remarks

Ellipsoid Problems

As mentioned in Sect. 5.2.4, the dual of a D-optimal design problem corresponds to a minimum ellipsoid problem where the center of the ellipsoid is fixed—it coincides with the origin. The determination of the ellipsoid $\mathcal{E}^*$ of free center containing a given set of points $\mathscr{X}_\ell = \{\mathbf{x}^{(i)}\,,\ i = 1, \ldots, \ell\} \subset \mathbb{R}^p$ and having minimal volume can be shown to be equivalent to a D-optimal design problem in $\mathbb{R}^{p+1}$; see Titterington (1975). The ellipsoid $\mathcal{E}^*$ is the intersection of the minimal-volume ellipsoid containing the Elfving's set associated with the linear regression model $\eta(\mathbf{x},\theta) = [\mathbf{x}^\top\ 1]\theta$, $\theta \in \mathbb{R}^{p+1}$, $\mathbf{x} \in \mathscr{X}_\ell$, and the hyperplane $\{\mathbf{z} \in \mathbb{R}^{p+1} : \mathbf{z}_{p+1} = 1\}$. A geometrical proof can be found in

(Shor and Berezovski, 1992) and (Khachiyan and Todd, 1993). One may refer, e.g., to Welzl (1991), Khachiyan (1996), Todd and Yildirim (2007), and Sun and Freund (2004) for algorithms. The construction of this minimal ellipsoid finds applications, for instance, in the robust estimation of correlation coefficients through ellipsoidal trimming, see, e.g., Cook et al. (1993), Davies (1992), and Titterington (1978); in robust control, see Raynaud et al. (2000), and quality control, see Hadjihassan et al. (1997).

The determination of the ellipsoid of maximal volume contained in a convex polyhedron can be used to form a very efficient algorithm for convex programming; see Tarasov et al. (1988). The determination of the maximum-volume ellipsoid with fixed center contained in a polyhedron is equivalent to the determination of the ellipsoid with same center containing the polar of the polyhedron and of minimal volume; see, e.g., Khachiyan and Todd (1993). This equivalence does not hold when the center is free, and one may refer, for instance, to Khachiyan and Todd (1993), Pronzato and Walter (1996), and Zhang and Gao (2003) for algorithms for the construction of maximum-volume inner ellipsoids in this more general situation. See also Pronzato et al. (2000, Chap. 6) for algorithms derived from Khachiyan (1979) algorithm for LP. Finally, one may refer to Vandenberghe et al. (1998) for a general-purpose interior-point algorithm for problems involving determinant optimization, with an overview of applications.

A short exposition on covering ellipsoids is given in Sect. 9.1.4 on algorithms for D-optimum design. See also Sect. 9.5.2.

Multidimensional Regression Models

Most of the developments in the chapter can easily be extended to the situation where the matrix $\mathbf{M}_\theta(x)$ in (5.1) has rank larger than one, see Fedorov (1971), Fedorov (1972, Chap. 5). A typical case corresponds to regression models with multidimensional observations

$$\mathbf{y}(x_i) = \eta(x_i, \bar{\theta}) + \varepsilon_i \ \in \mathbb{R}^n ,$$

where $\{\varepsilon_i\}$ is a sequence of n-dimensional independent random vectors with $\mathbb{E}\{\varepsilon_i\} = \mathbf{0}$ for all i. The errors can be (second-order) stationary

$$\mathbb{E}\{\varepsilon_i^2\} = \mathbf{\Sigma} \ \text{ for all } i ,$$

or nonstationary

$$\mathbb{E}\{\varepsilon_i^2\} = \mathbf{\Sigma}(x_i) \ \text{ for all } i . \tag{5.62}$$

The results of Chap. 3 can easily be adapted to this multidimensional case. For instance, under conditions similar to those of Theorem 3.8, when the errors satisfy (5.62), the WLS estimator $\hat{\theta}^N_{WLS}$ that minimizes

$$J_N(\theta) = \frac{1}{N} \sum_{i=1}^{N} [\mathbf{y}(x_i) - \eta(x_i, \theta)]^\top \mathbf{\Sigma}^{-1}(x_i) [\mathbf{y}(x_i) - \eta(x_i, \theta)]$$

satisfies $\sqrt{N}(\hat{\theta}^N_{WLS} - \bar{\theta}) \stackrel{\mathrm{d}}{\to} \mathbf{z} \sim \mathscr{N}(\mathbf{0}, \mathbf{M}^{-1}(\xi, \bar{\theta}))$, $N \to \infty$, where

$$\mathbf{M}(\xi, \theta) = \int_{\mathscr{X}} \frac{\partial \eta^\top(x, \theta)}{\partial \theta} \mathbf{\Sigma}^{-1}(x) \frac{\partial \eta(x, \theta)}{\partial \theta^\top} \xi(\mathrm{d}x).$$

One may refer in particular to Fedorov (1971) and Harman and Trnovská (2009) for a presentation of results concerning D-optimum design when the information matrix takes this form.

Optimum Design for LS Estimation with Equality Constraints

The presence of constraints $\mathbf{c}(\theta) = \mathbf{0}$ in nonlinear regression models has been considered in Sect. 3.5, where we have shown that the asymptotic variance–covariance matrix $\sigma^2 \mathbf{V}_{\theta,\xi}$ of the LS estimator $\hat{\theta}^N_{LS}$ is given by (3.92) or equivalently (3.93). These expressions can be used to define criteria for optimum design; see Pázman (2002a).

One should notice that $\mathbf{V}_{\theta,\xi}$ is singular since the presence of constraints makes the model over-parameterized. We can nevertheless use A-optimum design and maximize

$$\begin{aligned}\phi_A(\xi) &= -\mathrm{trace}[\mathbf{V}_{\theta,\xi}] \\ &= -\mathrm{trace}[\mathbf{H}^{-1}(\xi, \theta)] \\ &\quad + \mathrm{trace}[\mathbf{H}^{-1}(\xi, \theta)\mathbf{L}^\top(\theta)[\mathbf{L}(\theta)\mathbf{H}^{-1}(\xi, \theta)\mathbf{L}^\top(\theta)]^{-1}\mathbf{L}(\theta)\mathbf{H}^{-1}(\xi, \theta)],\end{aligned}$$

see (3.93), where we used the notations of Sect. 3.5 with $\theta = \theta^0$ some nominal value for θ (locally optimum design). The concavity of $\phi_A(\cdot)$ follows from the expression (3.92) of $\mathbf{V}_{\theta,\xi}$. Similarly, one may consider E-optimum design and maximize $\phi_E(\xi) = \min_{\mathbf{u} \in \mathbb{R}^p,\, \|\mathbf{u}\|=1} -\mathbf{u}^\top \mathbf{V}_{\theta^0,\xi} \mathbf{u}$.

D-optimum design cannot be used directly since $\mathbf{V}_{\theta,\xi}$ is singular for any ξ. To overcome the problem induced by the over-parameterization of the model, we should consider the reparameterization of the regression model with the auxiliary parameters β so that a D-optimal design can be obtained by maximizing the expression $\log\det[\mathbf{D}_0^\top \mathbf{M}(\xi, \theta^0)\mathbf{D}_0]$ with respect to $\xi \in \Xi$, with $\theta^0 = \phi(\theta^0)$; see (3.89) for the notation. Here the matrix $\mathbf{D}_0 = \mathbf{D}(\beta^0)$ given by (3.88) is not known explicitly, but any other parameterization can be substituted. We can in particular follow the recommendation in (Pázman, 2002a): since $\mathbf{L}(\theta^0)\mathbf{D}_0 = \mathbf{O}_{q,p-q}$, see (3.91), construct a QR decomposition of $\mathbf{L}^\top(\theta^0)$, see, e.g., Harville (1997, p. 66),

$$\mathbf{L}^\top(\theta^0) = (\mathbf{Q} \;\; \mathbf{T}) \begin{pmatrix} \mathbf{R} \\ \mathbf{O}_{p-q,q} \end{pmatrix}$$

where $\mathbf{R}$ is $q \times q$ upper triangular, $\mathbf{Q}^\top \mathbf{Q} = \mathbf{I}_q$, $\mathbf{T}^\top \mathbf{T} = \mathbf{I}_{p-q}$, and $\mathbf{Q}^\top \mathbf{T} = \mathbf{O}_{q,p-q}$, so that the columns of $\mathbf{Q}$ form an orthogonal basis for the range of $\mathbf{L}^\top(\theta^0)$ and those of $\mathbf{T}$ form an orthogonal basis for the column space of $\mathbf{D}_0$, and consider the maximization of $\phi_D(\xi) = \log\det[\mathbf{T}^\top \mathbf{M}(\xi, \theta^0)\mathbf{T}]$. Note that from (3.89), (3.92), the asymptotic variance–covariance matrix of $\hat{\theta}^N_{LS}$ at θ^0 is given by $\sigma^2 \mathbf{V}_{\theta^0,\xi} = \sigma^2 \mathbf{T}[\mathbf{T}^\top \mathbf{M}(\xi,\theta^0)\mathbf{T}]^{-1}\mathbf{T}^\top$. One may also notice that direct calculations give the following expression for the directional derivative of $\log\det[\mathbf{D}_0^\top \mathbf{M}(\cdot,\theta^0)\mathbf{D}_0]$ at ξ in the direction ν, see (5.33),

$$\begin{aligned} F_{\phi_D}(\xi;\nu) &= \mathrm{trace}\left\{[\mathbf{D}_0^\top \mathbf{M}(\xi,\theta^0)\mathbf{D}_0]^{-1}\mathbf{D}_0^\top[\mathbf{M}(\nu,\theta^0) - \mathbf{M}(\xi,\theta^0)]\mathbf{D}_0\right\} \\ &= \int_{\mathscr{X}} \left.\frac{\partial \eta(x,\theta)}{\partial \theta^\top}\right|_{\theta^0} \mathbf{D}_0[\mathbf{D}_0^\top \mathbf{M}(\xi,\theta^0)\mathbf{D}_0]^{-1}\mathbf{D}_0^\top \left.\frac{\partial \eta(x,\theta)}{\partial \theta}\right|_{\theta^0} \nu(\mathrm{d}x) - (p-q) \\ &= \int_{\mathscr{X}} \left.\frac{\partial \eta(x,\theta)}{\partial \theta^\top}\right|_{\theta^0} \mathbf{V}_{\theta^0,\xi} \left.\frac{\partial \eta(x,\theta)}{\partial \theta}\right|_{\theta^0} \nu(\mathrm{d}x) - (p-q) \end{aligned}$$

where the last equality follows from (3.92). Note that $F_{\phi_D}(\xi;\nu)$ does not depend on $\mathbf{D}_0$. The equivalence theorem 5.21 then indicates that a design measure ξ^* is D-optimal in the framework of Sect. 3.5 if and only if

$$\max_{x\in\mathscr{X}} \left.\frac{\partial \eta(x,\theta)}{\partial \theta^\top}\right|_{\theta^0} \mathbf{V}_{\theta^0,\xi^*} \left.\frac{\partial \eta(x,\theta)}{\partial \theta}\right|_{\theta^0} \leq p-q\,,$$

which resembles a G-optimality condition; see Remark 5.22-(*ii*).

Design for Bayesian Estimation

Consider a linear regression model $\mathbf{y} = \mathbf{F}(X)\theta + \varepsilon$, with $\varepsilon \sim \mathscr{N}(\mathbf{0},\sigma^2\mathbf{I}_N)$, and suppose that θ has the prior normal distribution $\mathscr{N}(\theta^0,\mathbf{\Omega})$. Its posterior p.d.f. is then the density of the normal distribution $\mathscr{N}(\hat{\theta}^N, [N\mathbf{M}(X)+\mathbf{\Omega}^{-1}]^{-1})$, with $\mathbf{M}(X) = \mathbf{F}^\top(X)\mathbf{F}(X)/(N\sigma^2)$ and $\hat{\theta}^N$ the maximum a posteriori estimator, $\hat{\theta}^N = [N\mathbf{M}(X) + \mathbf{\Omega}^{-1}]^{-1}\,[\mathbf{F}^\top(X)\mathbf{y}/\sigma^2 + \mathbf{\Omega}^{-1}\theta^0]$. When designing an optimal experiment for Bayesian estimation, it is then natural to apply the criteria of Sect. 5.1.2 to the (Bayesian information) matrix $\mathbf{M}(X) + \mathbf{\Omega}^{-1}/N$.

More generally, for Bayesian estimation in a nonlinear model we can replace the information matrix $\mathbf{M}(\xi,\theta)$ by $\mathbf{M}_B(\xi,\theta) = \mathbf{M}(\xi,\theta) + \mathbf{\Omega}^{-1}/N$ with $\mathbf{\Omega}$ the prior covariance matrix for θ; see Remark 4.18-(*i*). Properties such as concavity, the existence of directional derivatives, and equivalence theorems can be obtained in this context too; one may refer to Pilz (1983) for a thorough presentation and to the survey (Chaloner and Verdinelli, 1995) for a bibliography; see also Pukelsheim (1993, Chap. 11).

Design Measures Bounded from Above

Let $\mu(\cdot)$ denote a probability measure on $\mathscr{X}$ and, for a given $\alpha \in (0,1)$, denote by $\mathcal{D}(\mu,\alpha)$ the set of measures on $\mathscr{X}$ bounded by μ/α,

$$\mathcal{D}(\mu,\alpha) = \{\xi \in \Xi :\ \xi(dx) \leq \mu(dx)/\alpha\}\,.$$

Optimum design problems with such constraints occur, for instance, in sampling, see Wynn (1977, 1982), and in experiments with spatially distributed observations or with time as independent variable, see Fedorov (1989).

Suppose that $\phi(\xi) = \Phi[\mathbf{M}(\xi)]$, with $\Phi(\cdot)$ isotonic, concave, and linearly differentiable. Denote by ξ_α^* a ϕ-optimal design measure in $\mathcal{D}(\mu,\alpha)$. The theorem below (Wynn, 1982, Sahm and Schwabe, 2001) states that the design space $\mathscr{X}$ can be partitioned into three subsets $\mathscr{X}_{1,\alpha}^*$, $\mathscr{X}_{2,\alpha}^*$, and $\mathscr{X}_{3,\alpha}^* = \mathscr{X} \backslash (\mathscr{X}_{1,\alpha}^* \cup \mathscr{X}_{2,\alpha}^*)$, with $\xi_\alpha^* = 0$ on $\mathscr{X}_{1,\alpha}^*$, $\xi_\alpha^* = \mu/\alpha$ on $\mathscr{X}_{2,\alpha}^*$ and the directional derivative $F_\phi(\xi_\alpha^*, x)$ constant on $\mathscr{X}_{3,\alpha}^*$.

Theorem 5.45. *The following statements are equivalent:*

(i) ξ_α^* *is a* ϕ*-optimum constrained design measure.*
(ii) There exists a number c *such that* $F_\phi(\xi_\alpha^*, x) \geq c$ *for* ξ_α^**-almost all* x *and* $F_\phi(\xi_\alpha^*, x) \leq c$ *for* $(\mu - \alpha\xi_\alpha^*)$*-almost all* x*.*
(iii) There exist two subsets $\mathscr{X}_{1,\alpha}^*$ *and* $\mathscr{X}_{2,\alpha}^*$ *of* $\mathscr{X}$ *such that*
- $\xi_\alpha^* = 0$ *on* $\mathscr{X}_{1,\alpha}^*$ *and* $\xi_\alpha^* = \mu/\alpha$ *on* $\mathscr{X}_{2,\alpha}^*$.
- $\inf_{x\in\mathscr{X}_{2,\alpha}^*} F_\phi(\xi_\alpha^*, x) \geq c \geq \sup_{x\in\mathscr{X}_{1,\alpha}^*} F_\phi(\xi_\alpha^*, x)$.
- $F_\phi(\xi_\alpha^*, x) = c$ *on* $\mathscr{X}_{3,\alpha}^* = \mathscr{X} \backslash (\mathscr{X}_{1,\alpha}^* \cup \mathscr{X}_{2,\alpha}^*)$.

Wynn (1982), Fedorov (1989), and Fedorov and Hackl (1997) consider the case where μ has no atoms; that is, for any $\mathcal{A} \subset \mathscr{X}$ there exists $\mathcal{A}' \subset \mathscr{X}$ such that $\int_{\mathcal{A}'} \mu(dx) < \int_{\mathcal{A}} \mu(dx)$. As a consequence, $\mu(\mathscr{X}_{3,\alpha}^*) = 0$, and ξ_α^* belongs to the following subclass of $\mathcal{D}(\mu,\alpha)$:

$$\mathcal{D}^*(\mu,\alpha) = \{\xi \in \mathcal{D}(\mu,\alpha) : \exists \mathcal{A} \subset \mathscr{X}\,,\ \xi(\mathcal{A}) = \mu(\mathcal{A})/\alpha\,,\ \xi(\mathscr{X}\backslash\mathcal{A}) = 0\}\,.$$

The condition *(ii)* of Theorem 5.45 is then formulated as $F_\phi(\xi_\alpha^*, x)$ separating the two sets $\mathscr{X}_\alpha^*$ and $\mathscr{X}\backslash\mathscr{X}_\alpha^*$, with

$$\mathscr{X}_\alpha^* = \mathrm{supp}\xi_\alpha^* = \{x \in \mathscr{X} : \ \xi_\alpha^*(x) > 0\}\,.$$

Moreover, $\int_{\mathscr{X}_\alpha^*} F_\phi(\xi_\alpha^*, x)\, \mu(dx) = \int_{\mathscr{X}} F_\phi(\xi_\alpha^*, x)\xi_\alpha^*(dx) = 0$, see Fedorov (1989) and Fedorov and Hackl (1997) who also present iterative algorithms of the exchange type for the construction of ξ_α^*.

For a given ξ, consider the random variable $F_\phi(\xi, X_1)$ where X_1 is distributed with the probability measure μ, and let $\mathbb{F}_\xi$ denote the corresponding distribution function, $\mathbb{F}_\xi(s) = \mu\{x : \ F_\phi(\xi,x) \leq s\}$. Define $c_\alpha(\xi)$ as $c_\alpha(\xi) = \min\{s : \ \mathbb{F}_\xi(s) \geq 1-\alpha\}$ and

$$\begin{aligned}
\mathscr{X}_{1,\alpha}(\xi) &= \{x : \ F_\phi(\xi,x) < c_\alpha(\xi)\}\,,\\
\mathscr{X}_{2,\alpha}(\xi) &= \{x : \ F_\phi(\xi,x) > c_\alpha(\xi)\}\,,\\
\mathscr{X}_{3,\alpha}(\xi) &= \{x : \ F_\phi(\xi,x) = c_\alpha(\xi)\}\,.
\end{aligned}$$

We then obtain $\mathscr{X}_{j,\alpha}^* = \mathscr{X}_{j,\alpha}(\xi_\alpha^*)$, $j = 1,2,3$, and $c_\alpha(\xi_\alpha^*)$ is the constant c of Theorem 5.45. Consider now the transformation $T_{\phi,\alpha} : \xi \in \Xi \longrightarrow T_{\phi,\alpha}(\xi) \in \mathcal{D}(\mu,\alpha)$ defined by

$$T_{\phi,\alpha}(\xi) = \begin{cases} \mu/\alpha \text{ on } \mathscr{X}_{2,\alpha}(\xi)\,, \\ \frac{\alpha - \mu[\mathscr{X}_{2,\alpha}(\xi)]}{\mu[\mathscr{X}_{3,\alpha}(\xi)]}\, \mu/\alpha \text{ on } \mathscr{X}_{3,\alpha}(\xi)\,, \\ 0 \text{ on } \mathscr{X}_{1,\alpha}(\xi)\,. \end{cases}$$

Notice that $F_\phi[\xi; T_{\phi,\alpha}(\xi)] = \max_{\nu \in \mathcal{D}(\mu,\alpha)} F_\phi(\xi;\nu)$; indeed, $T_{\phi,\alpha}(\xi)$ distributes its mass on $\mathscr{X}$ where $F_\phi(\xi, x)$ takes its highest values. Next theorem (see Pronzato, 2004, 2006), complements Theorem 5.45 by a minimax formulation similar to the equivalence theorem 5.21.

Theorem 5.46. *The following statements are equivalent:*

(i) ξ_α^* *is a* ϕ*-optimum constrained design measure.*
(ii) $F_\phi[\xi_\alpha^*; T_{\phi,\alpha}(\xi_\alpha^*)] = 0$.
(iii) ξ_α^* *minimizes* $F_\phi[\xi; T_{\phi,\alpha}(\xi)]$ *with respect to* $\xi \in \mathcal{D}(\mu,\alpha)$.
(iv) ξ_α^* *minimizes* $\max_{\nu \in \mathcal{D}(\mu,\alpha)} F_\phi(\xi;\nu)$ *with respect to* $\xi \in \mathcal{D}(\mu,\alpha)$.

The continuity and differentiability of $\phi(\xi_\alpha^*)$ with respect to α are considered in (Pronzato, 2004). A method is presented in (Pronzato, 2006) that allows us to sample (asymptotically) from ξ_α^* by applying a simple acceptation/rejection rule to samples from μ, without requiring the construction of ξ_α^* or the knowledge of the sets $\mathscr{X}_{j,\alpha}^*$, or even the knowledge of μ itself.

6

Criteria Based on the Small-Sample Precision of the LS Estimator

This chapter deals with designs with a fixed finite size N (exact designs), of the form $X = (x_1, \ldots, x_N)$, possibly with replications; that is, we may have $x_i = x_j$ for some $i \neq j$. The number N of observations will be supposed to be small, so that the limit theorems of Chap. 3 concerning the asymptotic normal distribution of estimators cannot be used and the introduction of a design measure ξ to approximate X is not helpful. We consider regression models (3.2), (3.3); that is

$$y_i = y(x_i) = \eta(x_i, \bar{\theta}) + \varepsilon_i \,, \; i = 1, \ldots, N \,,$$

with

$$\mathbb{E}(\varepsilon_i) = 0 \,, \; \mathrm{var}(\varepsilon_i) = \sigma^2 \,, \; \mathrm{cov}(\varepsilon_i, \varepsilon_j) = 0 \text{ if } i \neq j \,, \; i, j = 1, \ldots, N \,.$$

We suppose that the assumptions H_Θ, $\mathrm{H1}_\eta$, and $\mathrm{H2}_\eta$ of Sect. 3.1 are satisfied: $\bar{\theta} \in \mathrm{int}(\Theta)$ with Θ a compact subset of $\mathbb{R}^p$ such that $\Theta \subset \overline{\mathrm{int}(\Theta)}$, and $\eta(x, \theta)$ is twice continuously differentiable with respect to $\theta \in \mathrm{int}(\Theta)$ for any $x \in \mathscr{X}$. In a vector notation, we shall write

$$\mathbf{y} = \eta(\bar{\theta}) + \varepsilon \,, \quad \text{with } \mathbb{E}(\varepsilon) = \mathbf{0} \,, \; \mathrm{Var}(\varepsilon) = \sigma^2 \mathbf{I}_N \,, \tag{6.1}$$

where $\eta(\theta) = \eta_X(\theta) = (\eta(x_1, \theta), \ldots, \eta(x_N, \theta))^\top$, $\mathbf{y} = (y_1, \ldots, y_N)^\top$ and $\varepsilon = (\varepsilon_1, \ldots, \varepsilon_N)^\top$. The more general nonstationary (heteroscedastic) case where $\mathrm{var}(\varepsilon_i) = \sigma^2(x_i)$ can easily be transformed into the model (6.1) with $\sigma^2 = 1$ via the division of y_i and $\eta(x_i, \theta)$ by $\sigma(x_i)$.

The geometrical properties of the model (6.1) are considered in Sect. 6.1 and used to obtain a classification of nonlinear regression models. The probability density $q(\cdot|\bar{\theta})$ of the LS estimator $\hat{\theta} = \hat{\theta}^N_{LS}$ is derived in Sect. 6.2; depending on the model and the design X, the expression obtained is exact or approximate. Criteria for designing optimal experiments based on $q(\cdot|\bar{\theta})$ are considered in Sect. 6.3. Since N is too small to use the asymptotic results of Chap. 3, assumptions on the distribution of ε are required to investigate

L. Pronzato and A. Pázman, *Design of Experiments in Nonlinear Models*, Lecture Notes in Statistics 212, DOI 10.1007/978-1-4614-6363-4_6,

the statistical properties of the LS estimator. In Sects. 6.2 and 6.3, we shall suppose that the errors are normal, $\varepsilon \sim \mathscr{N}(\mathbf{0}, \sigma^2 \mathbf{I}_N)$. Higher-order approximations of optimality criteria are considered in Sect. 6.4.

6.1 The Geometry of the Regression Model

6.1.1 Basic Notions

A geometrical interpretation of the model (6.1) is extremely useful to investigate the statistical properties of the LS estimator of θ. One can refer to Pázman (1993b) for details and precise proofs.

In the sample space $\mathbb{R}^N$ of the model we consider the expectation surface

$$\mathbb{S}_\eta = \{\eta(\theta) : \theta \in \Theta\}\,, \tag{6.2}$$

it corresponds to the set of all hypothetical means of the observed vectors $\mathbf{y}$ in (6.1). Since $\eta(\theta)$ is supposed to have continuous first- and second-order derivatives in $\mathrm{int}(\Theta)$, $\mathbb{S}_\eta$ is a smooth surface in $\mathbb{R}^N$ with a (local) dimension given by $r = \mathrm{rank}[\partial\eta(\theta)/\partial\theta^\top]$. If $r = p$, which means full rank, we call the model (6.1) regular. In regular models with no overlapping of $\mathbb{S}_\eta$, i.e., when $\eta(\theta) = \eta(\theta')$ implies $\theta = \theta'$, the LS estimator

$$\hat{\theta} = \hat{\theta}^N_{LS} = \arg\min_{\theta\in\Theta} \|\mathbf{y} - \eta(\theta)\|^2 \tag{6.3}$$

is defined uniquely (with probability one; see Theorem 7.2); hence, only designs corresponding to regular models will be considered. Then, $\mathbb{S}_\eta$ is a p-dimensional surface, and $\eta(\hat{\theta})$ is the orthogonal projection of $\mathbf{y}$ onto $\mathbb{S}_\eta$, unless $\hat{\theta} \notin \mathrm{int}(\Theta)$—the boundary of Θ requires special considerations; a smooth approximation will be considered in Sect. 6.2.5.

The vectors $\partial\eta(\theta)/\partial\theta_1, \ldots, \partial\eta(\theta)/\partial\theta_p$ of $\mathbb{R}^N$ are tangent to $\mathbb{S}_\eta$ at the point $\eta(\theta)$, and their linear span is the tangent space $\mathscr{T}_\theta$ to $\mathbb{S}_\eta$ at $\eta(\theta)$. Since they are linearly independent, $\mathscr{T}_\theta$ is a p-dimensional subspace of $\mathbb{R}^N$. We denote by $\mathbf{P}_\theta$ the orthogonal projector onto $\mathscr{T}_\theta$, i.e., the $N \times N$ matrix

$$\mathbf{P}_\theta = \mathbf{J}(\theta)\mathbf{M}_X^{-1}(\theta)\mathbf{J}^\top(\theta)\,, \tag{6.4}$$

where $\mathbf{J}(\theta)$ is the Jacobian matrix $\partial\eta(\theta)/\partial\theta^\top$ and

$$\mathbf{M}_X(\theta) = \mathbf{J}^\top(\theta)\mathbf{J}(\theta) = \sum_{i=1}^N \frac{\partial\eta(x_i,\theta)}{\partial\theta}\frac{\partial\eta(x_i,\theta)}{\partial\theta^\top}$$

is the information matrix for $\sigma = 1$. Note that $\mathbf{M}_X(\theta)$ is un-normalized, in the sense that $\mathbf{M}_X(\theta) = N\mathbf{M}(\xi_N,\theta)$ with ξ_N the empirical design measure associated with X that gives weight $1/N$ to each design point x_i, $i = 1, \ldots, N$. Also note that $\eta(\theta)$, $\mathbf{J}(\theta)$, $\mathbf{M}_X(\theta)$, and $\mathbf{P}_\theta$ depend on the design X.

6.1.2 A Classification of Nonlinear Regression Models

Intrinsically Linear Models

The model (6.1) is intrinsically linear if

$$\mathbf{P}_\theta \mathbf{H}^{\boldsymbol{\cdot}}_{ij}(\theta) = \mathbf{H}^{\boldsymbol{\cdot}}_{ij}(\theta)\,,\ \forall i,j=1,\ldots,p \ \text{ and } \ \forall\theta\in \mathrm{int}(\Theta)\,, \tag{6.5}$$

where

$$\mathbf{H}^{\boldsymbol{\cdot}}_{ij}(\theta) = \frac{\partial^2\eta(\theta)}{\partial\theta_i\partial\theta_j}\,. \tag{6.6}$$

Such a model has a planar expectation surface $\mathbb{S}_\eta$. In terms of the Bates and Watts (1980) intrinsic measure of nonlinearity,

$$C_{int}(X,\theta;\mathbf{u}) = \frac{\|[\mathbf{I}_N - \mathbf{P}_\theta]\sum_{i,j=1}^p u_i \mathbf{H}^{\boldsymbol{\cdot}}_{ij}(\theta) u_j\|}{\mathbf{u}^\top \mathbf{M}_X(\theta)\mathbf{u}}\,, \tag{6.7}$$

see Pázman (1993b) for a detailed derivation of this expression, we have $C_{int}(X,\theta;\mathbf{u}) = 0$ at any point $\theta\in\mathrm{int}(\Theta)$ and in any direction $\mathbf{u}\in\mathbb{R}^p$. An intrinsically linear model can be reparameterized to become a linear model, with eventual boundaries on the parametric space. That means that there exists a one-to-one continuously differentiable mapping $\beta=\beta(\theta)$ having an inverse $\theta=\theta(\beta)$ such that for a matrix $\mathbf{F}$ and a vector $\mathbf{v}$ we can write

$$\eta[\theta(\beta)] = \mathbf{F}\beta + \mathbf{v}\,,\ \eta(\theta) = \mathbf{F}\beta(\theta) + \mathbf{v} \tag{6.8}$$

for every θ and β in the corresponding parameter spaces. Evidently, every linear model is intrinsically linear. More interestingly, when the design X consists of repetitions of trials at p distinct design points only, say $(x^{(1)},\ldots,x^{(p)})$, a nonlinear regression model becomes intrinsically linear, with the obvious reparameterization $\beta_i = \eta(x^{(i)},\theta)$, $i=1,\ldots,p$, making the model linear. One may refer to Ross (1990) for the use of such designs.

Remark 6.1. The curvature $C_{int}(X,\theta;\mathbf{u})$ varies like $1/\sqrt{N}$. More precisely

$$C_{int}(X,\theta) = \sup_{\mathbf{u}\in\mathbb{R}^p-\{\mathbf{0}\}} C_{int}(X,\theta;\mathbf{u}) = \frac{1}{\sqrt{N}}\,C_{int}(\xi_N,\theta)\,, \tag{6.9}$$

where the intrinsic curvature $C_{int}(\xi,\theta)$ is defined by (3.84) and where ξ_N is the empirical design measure associated with X. Denote by $X^{\otimes n}$ the design obtained by replicating n times each point of X, we thus have $C_{int}(X^{\otimes n},\theta) = C_{int}(X,\theta)/\sqrt{n}$. Now, the n replications of X have the same statistical effect as a reduction of the standard deviation σ of the errors by a factor $\sqrt{n}$, i.e., the same effect as a reduction of $\|\mathbf{y}-\eta(\bar\theta)\|$ by a factor $\sqrt{n}$. It is this phenomenon of getting closer to the surface $\mathbb{S}_\eta$ that reduces the effect of its curvature. □

Remark 6.2. It is often advocated that the inverse of the observed information matrix provides a better approximation of the covariance matrix of the maximum likelihood estimator than the inverse of the expected information matrix; see, e.g., Efron and Hinkley (1978) and Lindsay and Li (1997). The reason for nevertheless using the expected information matrix throughout this book is that the observed matrix generally depends on the observations. It happens, however, that for the nonlinear regression models (6.1), these two matrices coincide when evaluated at the LS estimator, provided that the model is intrinsically linear and the errors are normal; see Clyde and Chaloner (2002). Indeed, the observed matrix, equal to the opposite of the second derivative of the log-likelihood, can then be written as

$$\mathbf{M}_X^{obs}(\theta, \mathbf{y}) = \frac{1}{\sigma^2}\mathbf{M}_X(\theta) + \frac{1}{\sigma^2}\sum_{i=1}^{N} \frac{\partial^2 \eta(x_i, \theta)}{\partial\theta\partial\theta^\top}[y(x_i) - \eta(x_i, \theta)].$$

Using (6.5), we get

$$\{\mathbf{M}_X^{obs}(\theta, \mathbf{y})\}_{ij} = \frac{1}{\sigma^2}\{\mathbf{M}_X(\theta)\}_{ij} + \frac{1}{\sigma^2}[\mathbf{y} - \eta(\theta)]^\top \mathbf{P}_\theta \mathbf{H}_{ij}^{\cdot}(\theta)$$

where the second term is zero for $\theta = \hat{\theta}_{LS}^N$ since $[\mathbf{y} - \eta(\hat{\theta}_{LS}^N)]$ is orthogonal to the tangent plane to $\mathbb{S}_\eta$ at $\eta(\hat{\theta}_{LS}^N)$. □

Parametrically Linear Models

The model (6.1) is parametrically linear if

$$\mathbf{P}_\theta \mathbf{H}_{ij}^{\cdot}(\theta) = \mathbf{0}\,,\ \forall i, j = 1, \ldots, p \ \text{ and } \ \forall \theta \in \mathrm{int}(\Theta)\,. \tag{6.10}$$

According to the definition of $\mathbf{P}_\theta$, this is equivalent to $\mathbf{J}^\top(\theta)\mathbf{H}_{ij}^{\cdot}(\theta) = \mathbf{0}$ for all $i, j = 1, \ldots, p$ and all $\theta \in \mathrm{int}(\Theta)$ and holds evidently if $\mathbf{M}_X(\theta) = \mathbf{M}$ constant in $\mathrm{int}(\Theta)$. These conditions are in fact equivalent, so that parametrically linear models are those with a constant information matrix. Further, (6.10) holds if and only if the parametric-effect measure of nonlinearity (parametric curvature) of Bates and Watts (1980)

$$C_{par}(X, \theta; \mathbf{u}) = \frac{\|\mathbf{P}_\theta \sum_{i,j=1}^{p} u_i \mathbf{H}_{ij}^{\cdot}(\theta) u_j\|}{\mathbf{u}^\top \mathbf{M}_X(\theta)\mathbf{u}} \tag{6.11}$$

is zero for any $\theta \in \mathrm{int}(\Theta)$ and in any direction $\mathbf{u} \in \mathbb{R}^p$.

Linear Models

The model (6.1) is linear, i.e., $\eta(x_i, \theta) = \mathbf{f}^\top(x_i)\theta + c(x_i)$ for some $\mathbf{f}(x_i) \in \mathbb{R}^p$ and $c(x_i) \in \mathbb{R}$, if and only if it is intrinsically and parametrically linear. Geometrically, the model is intrinsically but not parametrically linear if $\mathbb{S}_\eta$ is planar, but the parametric curves $\{\eta(\theta_1,\ldots,\theta_{i-1},t,\theta_{i+1},\ldots,\theta_p) : (\theta_1,\ldots,\theta_{i-1},t,\theta_{i+1},\ldots,\theta_p) \in \Theta\}$, parameterized in t, do not correspond to parallel lines when θ varies. Conversely, the model is parametrically linear but not intrinsically linear when the surface $\mathbb{S}_\eta$ is curved, but the parametric curves above are regularly spaced over $\mathbb{S}_\eta$ when θ varies.

Flat Models

The notion of flat models is related to the existence of a regular reparameterization $\theta = \theta(\beta)$ such that the model $\mathbf{y} = \nu(\beta) + \varepsilon$ with $\nu(\beta) = \eta[\theta(\beta)]$ has a constant information matrix.[1] Contrary to a rather common opinion, such a reparameterization does not always exist. The problem is related to an old issue in differential geometry which was solved by Riemann; see Eisenhart (1960, p. 25): such a reparameterization exists if and only if the Riemannian curvature tensor $\mathbf{R}(\theta)$ with components

$$R_{hijk}(\theta) = T_{hjik}(\theta) - T_{hkij}(\theta)\,,$$

where

$$T_{hjik}(\theta) = [\mathbf{H}^{\cdot}_{hj}(\theta)]^\top [\mathbf{I}_N - \mathbf{P}_\theta]\mathbf{H}^{\cdot}_{ik}(\theta)\,,$$

is identically zero:

$$R_{hijk}(\theta) = 0 \;\text{ for all } i,j,h,k = 1,\ldots,p \;\text{ and for all } \theta \in \text{int}(\Theta)\,,$$

which we shall denote $\mathbf{R}(\theta) \equiv 0$. Any parametrically linear model is such that $\mathbf{R}(\theta) = 0$ for any $\theta \in \text{int}(\Theta)$. Also, from the definition of $\mathbf{R}(\theta)$, we have $\mathbf{R}(\theta) \equiv 0$ in any intrinsically linear model.

Regression models of the form (6.1) can thus be classified as shown in Fig. 6.1; see Pázman (1992a).

Only some particular models can be at the same time parametrically linear and not intrinsically linear; a typical example is $\eta(\theta) = (\cos(\theta),\ \sin(\theta))^\top$. On the other hand, any regression model (6.1) with $\dim(\theta) = 1$ is flat. Also, all models that are linear in all components of θ but one are flat. An example of intrinsically and parametrically nonlinear but flat model is the Michaelis–Menten model, defined by

$$\eta(x,\theta) = \frac{\theta_1\, x}{\theta_2 + x}\,,$$

which is nonlinear in θ_2 only. Notice that in the case $\dim(\theta) = 1$, the reparameterization $\beta = \beta(\theta)$ that makes the information matrix constant has the form

$$\beta(\theta) = \beta_0 + \int_{\theta_0}^{\theta} \left\| \frac{\mathrm{d}\eta(t)}{\mathrm{d}t} \right\| \mathrm{d}t$$

for some $\theta_0 \in \Theta$ and $\beta_0 \in \mathbb{R}$.

[1]Our definition of flat models thus differs from that of intrinsically linear models with which it is sometimes confounded in the literature.

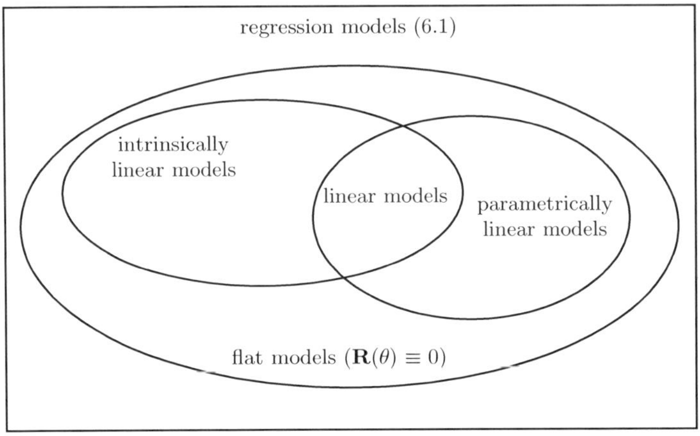

Fig. 6.1. A classification of regression models

6.1.3 Avoiding Failures of LS Estimation

A nonzero parametric curvature influences the statistical properties of the LS estimator $\hat{\theta}$ in that its p.d.f. may be nonsymmetrical, biased, and even bimodal; see, e.g., Example 3.10. A small intrinsic curvature does not have such an influence, but a large value of $\sigma C_{int}(X, \theta, \mathbf{u})$ can lead to a very unstable estimator; see Example 7.5. Also, when the expectation surface is almost overlapping, i.e., when $\min_{\|\theta-\theta'\|^2=\delta} \|\eta(\theta) - \eta(\theta')\|^2$ is small for a large δ, the estimator can be obtained at a large distance from $\bar{\theta}$, which is statistically not justified; see Example 7.6 in Sect. 7.3. Here we formulate some assumptions on the model (6.1) and on $\bar{\theta}$ that remove such difficulties.

For any $r > 0$, denote $\mathcal{G}(r) = \{\mathbf{y} \in \mathbb{R}^N : \|\mathbf{y} - \eta(\bar{\theta})\| < r\}$. We call θ a r-projection of $\mathbf{y}$ if $\|\mathbf{y} - \eta(\theta)\| < r$ and

$$\frac{\partial \|\mathbf{y} - \eta(\theta)\|^2}{\partial \theta} = \mathbf{0} \quad \text{or} \quad \theta = \hat{\theta}(\mathbf{y}) \text{ is on the boundary of } \Theta\,.$$

Denote by $\mathscr{B}(r)$ the set of all r-projections of points of $\mathcal{G}(r)$ and consider the tube $\mathcal{T}(r)$ around the expectation surface $\mathbb{S}_\eta$,

$$\mathcal{T}(r) = \{\mathbf{y} \in \mathbb{R}^N : \exists \theta \in \mathscr{B}(r) \text{ such that } \theta \text{ is an } r\text{-projection of } \mathbf{y}\}\,,$$

see Fig. 6.2. The assumption is as follows:

$\mathbf{H}_\mathcal{S}$: There exists $r > 0$ such that:

a) $\text{Prob}_{\bar{\theta}}[\mathcal{G}(r)] = \text{Prob}(\|\mathbf{y} - \eta(\bar{\theta})\| < r) \geq 1 - \epsilon\,.$
b) Every $\mathbf{y} \in \mathcal{T}(r)$ has one r-projection only.

Note that this assumption implies that $C_{int}(X, \theta, \mathbf{u}) \leq 1/r$ for every $\theta \in \mathscr{B}(r)$ and any $\mathbf{u} \in \mathbb{R}^p$.

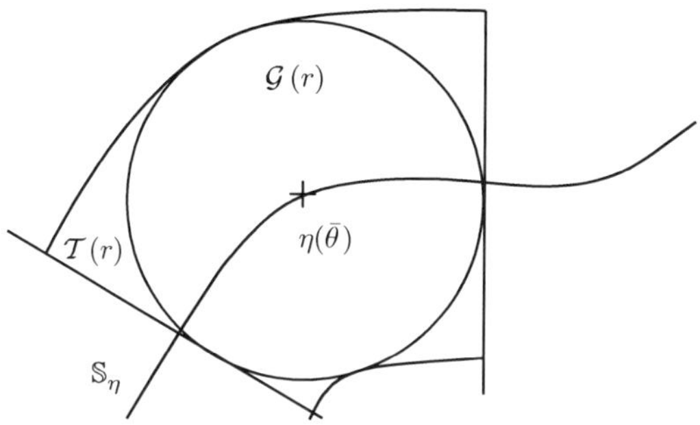

Fig. 6.2. r-projections and tube $\mathcal{T}(r)$ around $\mathbb{S}_\eta$

6.2 The Probability Density of the LS Estimator in Nonlinear Models with Normal Errors

We suppose throughout this section that the errors ε are normally distributed $\mathscr{N}(\mathbf{0}, \sigma^2 \mathbf{I}_N)$. When the regression model (6.1) is linear, i.e., when $\eta(x,\theta) = \mathbf{f}^\top(x)\theta + c(x)$, then the LS estimator $\hat{\theta}$ has the probability density

$$q(\theta|\bar{\theta}) = \frac{\det^{1/2}(\mathbf{F}^\top\mathbf{F})}{(2\pi)^{p/2}\,\sigma^p} \exp\left\{-\frac{1}{2\sigma^2}\|\mathbf{F}(\theta-\bar{\theta})\|^2\right\} \tag{6.12}$$

in $\text{int}(\Theta)$, where $\{\mathbf{F}\}_{i,\cdot} = \mathbf{f}^\top(x_i)$. As noticed in the next remark, the assumption of normal errors is justified when the design consists of repetitions of observations.

Remark 6.3. Consider the design $X^{\otimes n}$ consisting of n repetitions of each of the N points of the design X. Denote by $y_j(x_i)$ the j-th observation at x_i for $j = 1, \ldots, n$. The corresponding LS criterion (3.1) can be written as

$$\begin{aligned} J_N(\theta) &= \frac{1}{N}\sum_{i=1}^{N}\frac{1}{n}\sum_{j=1}^{n}[y_j(x_i) - \eta(x_i,\theta)]^2 \\ &= \frac{1}{N}\sum_{i=1}^{N}[\bar{y}(x_i) - \eta(x_i,\theta)]^2 + \frac{1}{N}\sum_{i=1}^{N}\left[\frac{1}{n}\sum_{j=1}^{n}y_j^2(x_i) - \bar{y}^2(x_i)\right] \end{aligned} \tag{6.13}$$

where $\bar{y}(x_i) = (1/n)\sum_{j=1}^{n} y_j(x_i)$ is the empirical mean of the observations at x_i and the last term within square brackets in (6.13) is their empirical variance. Since this part does not depend on θ, minimizing $J_N(\theta)$ is equivalent to minimizing $(1/N)\sum_{i=1}^{N}[\bar{y}(x_i) - \eta(x_i,\theta)]^2$, i.e., an LS criterion for the design X, the n observations at x_i being replaced by their empirical mean. This generalizes to the case of unequal numbers of repetitions of observations. Indeed, let n_i be the number of observations at x_i; the associated LS criterion is

$$J_N(\theta) = \sum_{i=1}^{N} w_i \, [\bar{y}(x_i) - \eta(x_i, \theta)]^2 + \sum_{i=1}^{N} w_i \left[\frac{1}{n_i} \sum_{j=1}^{n_i} y_j^2(x_i) - \bar{y}^2(x_i) \right]$$

where $w_i = n_i / \left(\sum_{i=1}^{N} n_i \right)$ and $\bar{y}(x_i) = (1/n_i) \sum_{j=1}^{n_i} y_j(x_i)$. Again, the second term does not depend on θ, and minimizing $J_N(\theta)$ is thus equivalent to minimizing the weighted LS criterion $\sum_{i=1}^{N} w_i [\bar{y}(x_i) - \eta(x_i, \theta)]^2$.

From the central limit theorem, $\bar{y}(x_i)$ is asymptotically normal when n_i tends to infinity provided that the errors have a finite variance, whatever their distribution is. The assumption of normality of the errors is thus reasonable for designs consisting of repetitions when the number of repetitions of each design points is large enough. □

6.2.1 Intrinsically Linear Models

For such models there exists a reparameterization $\beta = \beta(\theta)$ with the property (6.8). Hence, from (6.12) we have for the p.d.f. of the LS estimator of θ

$$q(\theta|\bar{\theta}) = \frac{\det^{1/2}(\mathbf{F}^\top \mathbf{F})}{(2\pi)^{p/2}\, \sigma^p} \exp\left\{ -\frac{1}{2\sigma^2} \|\mathbf{F}[\beta(\theta) - \beta(\bar{\theta})]\|^2 \right\} \times \left| \det \left(\frac{\partial \beta(\theta)}{\partial \theta} \right) \right| .$$

Using (6.8), we obtain $\mathbf{F}[\beta(\theta) - \beta(\bar{\theta})] = \eta(\theta) - \eta(\bar{\theta})$ and

$$\frac{\partial \beta^\top(\theta)}{\partial \theta} \mathbf{F}^\top \mathbf{F} \frac{\partial \beta(\theta)}{\partial \theta^\top} = \mathbf{J}^\top(\theta) \mathbf{J}(\theta) = \mathbf{M}_X(\theta) .$$

Hence,

$$q(\theta|\bar{\theta}) = \frac{\det^{1/2} \mathbf{M}_X(\theta)}{(2\pi)^{p/2}\, \sigma^p} \exp\left\{ -\frac{1}{2\sigma^2} \|\eta(\theta) - \eta(\bar{\theta})\|^2 \right\} . \tag{6.14}$$

When the design X consists of repetitions of trials at p distinct points only, the model is intrinsically linear (see Sect. 6.1.2) and (6.14) gives the exact density of the LS estimator. Note that D-optimal designs (based on the information matrix) are often supported on p points only; see Sect. 5.2.3.

6.2.2 Models with $\dim(\theta) = 1$

The case where θ is unidimensional is presented separately since accurate approximations of the p.d.f. of $\hat{\theta}$ are obtained rather easily. The expectation surface $\mathbb{S}_\eta$ is then a curve in $\mathbb{R}^N$, and we can approximate the distribution function of $\hat{\theta}$ at a point $t \in \mathrm{int}(\Theta)$ by

$$\mathbb{F}(t|\bar{\theta}) = \mathrm{Prob}_{\bar{\theta}}\{\mathbf{y} \in \mathbb{R}^N : \hat{\theta}(\mathbf{y}) < t\} = \mathrm{Prob}_{\bar{\theta}}\{\mathcal{Q}_t\} ,$$

where $\mathcal{Q}_t$ is the set of all samples of $\mathbb{R}^N$ having orthogonal projections onto $\mathbb{S}_\eta$ on that side of $\eta(t)$ which is opposite to the direction of the vector $\mathrm{d}\eta(t)/\mathrm{d}t$; that is,

$$\mathcal{Q}_t = \{\mathbf{y} \in \mathbb{R}^N : [\mathbf{y} - \eta(t)]^\top \frac{\mathrm{d}\eta(t)}{\mathrm{d}t} < 0\}.$$

This involves an approximation since not all projections from points from $\mathcal{Q}_t$ correspond in reality to solutions of (6.3)—they may correspond to other stationary points of the function $t \longrightarrow \|\mathbf{y} - \eta(t)\|^2$. Also, some points $\mathbf{y}$ not in $\mathcal{Q}_t$ may yield LS estimates $\hat{\theta}(\mathbf{y}) < t$. However, if the model (6.1) is such that $\mathrm{H}_{\mathcal{S}}$ (p. 172) is satisfied, this approximation is reasonable. Then we have

$$\begin{aligned}
\mathbb{F}(t|\bar{\theta}) &= \mathrm{Prob}_{\bar{\theta}}\{\mathcal{Q}_t\} \\
&= \mathrm{Prob}_{\bar{\theta}}\left\{\mathbf{y} \in \mathbb{R}^N : [\mathbf{y} - \eta(\bar{\theta})]^\top \frac{\mathrm{d}\eta(t)/\mathrm{d}t}{\|\mathrm{d}\eta(t)/\mathrm{d}t\|} < [\eta(t) - \eta(\bar{\theta})]^\top \frac{\mathrm{d}\eta(t)/\mathrm{d}t}{\|\mathrm{d}\eta(t)/\mathrm{d}t\|}\right\} \\
&= \int_{-\infty}^{a(t)} \frac{1}{\sqrt{2\pi}\,\sigma} \exp\left\{-\frac{x^2}{2\sigma^2}\right\} \mathrm{d}x\,,
\end{aligned}$$

with

$$a(t) = [\eta(t) - \eta(\bar{\theta})]^\top \frac{\mathrm{d}\eta(t)/\mathrm{d}t}{\|\mathrm{d}\eta(t)/\mathrm{d}t\|}\,,$$

since

$$[\mathbf{y} - \eta(\bar{\theta})]^\top \frac{\mathrm{d}\eta(t)/\mathrm{d}t}{\|\mathrm{d}\eta(t)/\mathrm{d}t\|} \sim \mathscr{N}(0, \sigma^2)\,.$$

The p.d.f. of $\hat{\theta}$ is then obtained by differentiating $\mathbb{F}(t|\bar{\theta})$ with respect to t, which gives

$$q(t|\bar{\theta}) = \frac{1}{\sqrt{2\pi}\,\sigma} \exp\left\{-\frac{a^2(t)}{2\sigma^2}\right\} \frac{\mathrm{d}a(t)}{\mathrm{d}t}\,.$$

After rearrangements, we obtain

$$\begin{aligned}
q(\theta|\bar{\theta}) = \; & \frac{\mathbf{M}_X(\theta) + [\eta(\theta) - \eta(\bar{\theta})]^\top [\mathbf{I}_N - \mathbf{P}_\theta](\mathrm{d}^2\eta(\theta)/\mathrm{d}\theta^2)}{\sqrt{2\pi}\,\sigma\,\mathbf{M}_X^{1/2}(\theta)} \\
& \times \exp\left\{-\frac{1}{2\sigma^2}\|\mathbf{P}_\theta[\eta(\theta) - \eta(\bar{\theta})]\|^2\right\}. \qquad (6.15)
\end{aligned}$$

Notice that

$$\mathbf{M}_X^{1/2}(\theta) = \left\|\frac{\mathrm{d}\eta(\theta)}{\mathrm{d}\theta}\right\| \quad \text{and} \quad \mathbf{P}_\theta = \frac{\mathrm{d}\eta(\theta)}{\mathrm{d}\theta}\frac{\mathrm{d}\eta^\top(\theta)}{\mathrm{d}\theta} \Big/ \left\|\frac{\mathrm{d}\eta(\theta)}{\mathrm{d}\theta}\right\|^2$$

since $\dim(\theta) = 1$.

6.2.3 Flat Models

This corresponds to a generalization of (6.15) to the case $\dim(\theta) > 1$ but still with the Riemannian curvature tensor $\mathbf{R}(\theta) = 0$ for all $\theta \in \mathrm{int}(\Theta)$. The p.d.f. of $\hat{\theta}$ can be obtained either as a direct extension of (6.15) to the multivariate case, as done in (Pázman, 1984b), or via a more algebraic approach. The later introduces new coordinates in the sample space $\mathbb{R}^N$, some of which being equal to the components of $\hat{\theta}$, and the normal density of $\mathbf{y}$ is then transferred to those new coordinates; see Pázman (1993b) for details. The resulting (approximate) p.d.f. of $\hat{\theta}$ takes the form

$$q(\theta|\bar{\theta}) = \frac{\det[\mathbf{Q}(\theta, \bar{\theta})]}{(2\pi)^{p/2}\,\sigma^p\,\det^{1/2}\mathbf{M}_X(\theta)}\,\exp\left\{-\frac{1}{2\sigma^2}\|\mathbf{P}_\theta[\eta(\theta) - \eta(\bar{\theta})]\|^2\right\}, \quad (6.16)$$

where

$$\{\mathbf{Q}(\theta, \bar{\theta})\}_{ij} = \{\mathbf{M}_X(\theta)\}_{ij} + [\eta(\theta) - \eta(\bar{\theta})]^\top[\mathbf{I}_N - \mathbf{P}_\theta]\mathbf{H}_{ij}^{\cdot\cdot}(\theta)\,. \quad (6.17)$$

The approximation is as precise as (6.15), i.e., is very accurate, if $\mathrm{H}_{\mathcal{S}}$ (p. 172) can be satisfied with a small ϵ. Then, we can say that $q(\theta|\bar{\theta})$ is "almost exact" in the region $\mathscr{B}(r)$ defined in Sect. 6.1.3. The p.d.f. (6.16) can also be obtained asymptotically for any model (6.1) through the saddle-point approximation technique, sometimes called the "small-sample asymptotics" (Hougaard 1985); see also Pázman (1990) for a discussion.

6.2.4 Models with Riemannian Curvature Tensor $\mathbf{R}(\boldsymbol{\theta}) \not\equiv 0$

When $\mathbf{R}(\theta) \not\equiv 0$, under $\mathrm{H}_{\mathcal{S}}$ (p. 172), an accurate approximation of the p.d.f. of $\hat{\theta}$ can be obtained, similar to (6.16) but with $\det[\mathbf{Q}(\theta, \bar{\theta})]$ replaced by an expression depending on the components of $\mathbf{Q}(\theta, \bar{\theta})$ and $\mathbf{R}(\theta)$. In the special case $\dim(\theta) = 2$, $\det[\mathbf{Q}(\theta, \bar{\theta})]$ is simply replaced by $\det[\mathbf{Q}(\theta, \bar{\theta})] + R_{1212}(\theta)$; see Pázman (1993b, p. 186). The expression becomes much more complicated for larger values of $\dim(\theta)$; see Pázman (1993a) and Gauchi and Pázman (2006).

6.2.5 Density of the Penalized LS Estimator

We consider here the p.d.f. of the penalized LS estimator defined by

$$\tilde{\theta} = \arg\min_{\theta \in \Theta}\left\{\|\mathbf{y} - \eta(\theta)\|^2 + 2w(\theta)\right\} \quad (6.18)$$

where $w(\theta)$ is the penalty term. The introduction of the penalty function $w(\cdot)$ allows us to take the influence of the boundary $\partial\Theta$ of Θ on the p.d.f. of the LS estimator into account. Consider a domain $\mathcal{D} \subset \Theta$, close to the boundary of Θ, say a tube defined by

$$\mathcal{D} = \{\theta \in \Theta : \min_{\theta^* \in \partial\Theta} \|\theta - \theta^*\| < \epsilon\}\,.$$

Take a twice continuously differentiable penalty $w(\theta)$ which is zero on $\Theta \setminus \mathcal{D}$, $+\infty$ on $\partial\Theta$, positive on $\mathcal{D} \setminus \partial\Theta$, and increasing when θ approaches $\partial\Theta$. The p.d.f. of $\tilde{\theta}$ in the case $\mathbf{R}(\theta) \equiv 0$ is given by

$$\tilde{q}(\theta|\bar{\theta}) = \frac{\det\left[\mathbf{Q}(\theta,\bar{\theta}) - \mathbf{A}(\theta) + \frac{\partial^2 w(\theta)}{\partial\theta\partial\theta^\top}\right]}{(2\pi)^{p/2}\,\sigma^p\,\det^{1/2}\mathbf{M}_X(\theta)} \times \exp\left\{-\frac{1}{2\sigma^2}\|\mathbf{P}_\theta[\eta(\theta) - \eta(\bar{\theta}) + \mathbf{u}(\theta)]\|^2\right\}, \qquad (6.19)$$

where

$$\{\mathbf{A}(\theta)\}_{ij} = \mathbf{u}^\top(\theta)\mathbf{H}^{\boldsymbol{\cdot}}_{ij}(\theta) \;\text{ and }\; \mathbf{u}(\theta) = \mathbf{J}(\theta)\,\mathbf{M}_X^{-1}(\theta)\,\frac{\partial w(\theta)}{\partial\theta},$$

see Pázman and Pronzato (1992) and Pázman (1992b). Note that (6.19) coincides with (6.16) when $\theta \in \Theta\backslash\mathcal{D}$; the values of the LS estimator $\hat{\theta}$ that should lie on $\partial\Theta$ are simply shifted inside $\mathcal{D}$.

Remark 6.4. In a Bayesian setting where θ has the prior density $\pi(\cdot)$, the maximum a posteriori estimator $\tilde{\theta}$ is

$$\begin{aligned}\tilde{\theta} &= \arg\max_{\theta\in\Theta}\log\pi(\mathbf{y}|\theta)\\ &= \arg\min_{\theta\in\Theta}\left\{\|\mathbf{y} - \eta(\theta)\|^2 - 2\sigma^2\log\pi(\theta)\right\}\end{aligned}$$

and its density in the case $\mathbf{R}(\theta) \equiv 0$ is also given by (6.19), with $w(\theta) = -\sigma^2\log\pi(\theta)$. □

6.2.6 Marginal Densities of the LS Estimator

Marginal densities, i.e., densities of the components of $\hat{\theta}$ or the density of a scalar function $h(\hat{\theta})$, are much more difficult to approximate than the density of the whole vector $\hat{\theta}$. The difficulty comes from the fact that the set $\mathcal{R}_\gamma$ of samples that give the same estimator of $h(\theta)$, $\mathcal{R}_\gamma = \{\mathbf{y} \in \mathbb{R}^N : h[\hat{\theta}(\mathbf{y})] = \gamma\}$, is composed of hyperplanes in $\mathbb{R}^N$ which intersect the expectation surface $\mathbb{S}_\eta$ orthogonally along the subsurface $\mathcal{C}_\gamma = \{\eta(\theta) : \theta \in \Theta,\ h(\theta) = \gamma\}$ of $\mathbb{S}_\eta$. An approximation of the density of $h(\hat{\theta})$ at the point γ is proposed in (Pázman and Pronzato, 1996), based on a local approximation of $\mathcal{R}_\gamma$ by a $(N-p)$-dimensional hyperplane in $\mathbb{R}^N$ with $p = \dim(\theta)$,

$$q(\gamma|\bar{\theta}) = \frac{1}{\sqrt{2\pi}\sigma\|\mathbf{b}_\gamma\|}\exp\left\{-\frac{1}{2\sigma^2}\|\mathbf{P}_\gamma[\eta(\theta_\gamma) - \eta(\bar{\theta})]\|^2\right\},$$

where

$$\begin{aligned}\theta_\gamma &= \arg\min_{\theta:h(\theta)=\gamma}\|\eta(\theta) - \eta(\bar{\theta})\|^2,\\ \mathbf{b}_\gamma &= \mathbf{J}(\theta_\gamma)\,\mathbf{M}_X^{-1}(\theta_\gamma)\left.\frac{\partial h(\theta)}{\partial\theta}\right|_{\theta_\gamma},\\ \mathbf{P}_\gamma &= \frac{\mathbf{b}_\gamma\mathbf{b}_\gamma^\top}{\|\mathbf{b}_\gamma\|^2}.\end{aligned}$$

A more precise approximation, taking the curvature of $\mathcal{R}_\gamma$ into account, is also proposed in (Pázman and Pronzato, 1996). The extension to bias-corrected LS estimators, see Remark 6.5, can be found in (Pázman and Pronzato, 1998). Marginal densities can easily be plotted and can thus be used to compare visually different experiments in terms of precision of the estimation of each individual component of θ; see Pronzato and Pázman (2001).

6.3 Optimality Criteria Based on the p.d.f. of the LS Estimator

The mean-squared error matrix of $\hat{\theta}$ is equal to

$$\mathbf{S}(X,\bar{\theta}) = \mathbb{E}_{X,\bar{\theta}}\left\{(\hat{\theta}-\bar{\theta})(\hat{\theta}-\bar{\theta})^\top\right\}.$$

The numerical evaluation of this quantity, or the optimization of a function of $\mathbf{S}(X,\bar{\theta})$ with respect to the design X, will be based on integration over a bounded domain Θ, possibly through a Monte Carlo technique. This requires that the boundary of Θ is taken into account. Indeed, ignoring the vectors $\hat{\theta}$ falling outside Θ would falsify the evaluation of $\mathbf{S}(X,\bar{\theta})$ and, when optimizing a function of $\mathbf{S}(X,\bar{\theta})$ with respect to X, would enforce the choice of a singular design ensuring that the distribution of $\hat{\theta}$ is widely spread outside Θ. The definition of Θ should also account for mathematical or physical constraints on the model parameters when such constraints exist: e.g.the response function $\eta(\theta)$ may be not defined beyond some limit on θ, or physical constraints may be imposed on θ for the response function to make sense. We must therefore consider the density of the penalized LS estimator (6.18) when expressing the components of $\mathbf{S}(X,\bar{\theta})$ as integrals. When the errors are normal $\mathscr{N}(\mathbf{0},\sigma^2\mathbf{I}_N)$, the approximation (6.19) of this density can be used, which yields

$$S_{ij}(X,\bar{\theta}) = \int_{\mathrm{int}(\Theta)} (\theta-\bar{\theta})_i(\theta-\bar{\theta})_j\,\tilde{q}_X(\theta|\bar{\theta})\,\mathrm{d}\theta\,,$$

where the dependence of the p.d.f. on the design X is mentioned explicitly.

Classical optimality criteria (to be maximized) can be applied to $\mathbf{S}(X,\bar{\theta})$, for instance

$$\Phi_A(X;\bar{\theta}) = -\mathrm{trace}[\mathbf{S}(X,\bar{\theta})] = -\int_{\mathrm{int}(\Theta)} \|\theta-\bar{\theta}\|^2\,\tilde{q}_X(\theta|\bar{\theta})\,\mathrm{d}\theta$$

for A-optimality,

$$\Phi_D(X;\bar{\theta}) = [\det\mathbf{S}(X,\bar{\theta})]^{-1/p}$$

for D-optimality, and

$$\Phi_{q,\mathbf{I}}(X;\bar{\theta}) = -\left\{\frac{1}{p}\,\mathrm{trace}[\mathbf{S}^q(X,\bar{\theta})]\right\}^{1/q}$$

for criteria from the Φ_q-class; see Sect. 5.1.2. When we are interested in the estimation of a function $h(\theta)$, a natural extension of the c-optimality criterion is

$$\Phi_c(X;\bar{\theta}) = -\int_{\mathrm{int}(\Theta)} [h(\theta) - h(\bar{\theta})]^2 \, \tilde{q}_X(\theta|\bar{\theta}) \, \mathrm{d}\theta \, .$$

As noticed in Sect. 5.1.2, D-optimality is related to the minimization of the Shannon entropy of the asymptotic normal distribution of the estimator. When the p.d.f. (6.19) is used, we get the following entropy criterion (to be minimized):

$$H_1(X;\bar{\theta}) = -\int_{\mathrm{int}} \tilde{q}_X(\theta|\bar{\theta}) \, \log[\tilde{q}_X(\theta|\bar{\theta})] \, \mathrm{d}\theta \, ,$$

see (5.6).

Writing a criterion $\Phi(X;\bar{\theta})$ in the form of *one* multivariate integral of a real-valued function is adequate for its minimization through a stochastic approximation technique (see, for instance, Kushner and Clark (1978), Ermoliev and Wets (1988), and Kushner and Yin (1997)), stochastic approximation being used for optimum design in (Pronzato and Walter, 1985), (Pázman and Pronzato, 1992) and presented in details in (Gauchi and Pázman, 2006); see also Sect. 9.4. This concerns directly $\Phi_A(X;\bar{\theta})$, $\Phi_c(X;\bar{\theta})$ and $H_1(X;\bar{\theta})$; we show below that it also applies to $\Phi_D(X;\bar{\theta})$ and $\Phi_{q,\mathbf{I}}(X;\bar{\theta})$. Notice that maximizing $\Phi_D(X;\bar{\theta})$ is equivalent to minimizing $\det \mathbf{S}(X,\bar{\theta})$ and maximizing $\Phi_{q,\mathbf{I}}(X;\bar{\theta})$ is equivalent to minimizing $\mathrm{trace}[\mathbf{S}^q(X,\bar{\theta})]$. We can write, see Gauchi and Pázman (2006),

$$\begin{aligned}
\det \mathbf{S}(X,\bar{\theta}) &= \sum_{\pi} \left[\mathrm{sign}(\pi) \prod_{k=1}^{p} \{\mathbf{S}(X,\bar{\theta})\}_{k,\pi(k)} \right] \\
&= \sum_{\pi} \left[\mathrm{sign}(\pi) \prod_{k=1}^{p} \int_{\mathrm{int}(\Theta)} \{\theta^{(k)} - \bar{\theta}\}_k \{\theta^{(k)} - \bar{\theta}\}_{\pi(k)} \, \tilde{q}_X(\theta^{(k)}|\bar{\theta}) \mathrm{d}\theta^{(k)} \right] \\
&= \int_{\{\mathrm{int}(\Theta)\}^{\otimes p}} L(\theta^{(1)},\ldots,\theta^{(p)},\bar{\theta},X) \, \mathrm{d}\theta^{(1)} \cdots \mathrm{d}\theta^{(p)} \, ,
\end{aligned}$$

where the sum is taken over all permutations of the set $\{1,\ldots,p\}$ (with signs ± 1),

$$L(\theta^{(1)},\ldots,\theta^{(p)},\bar{\theta},X) = \sum_{\pi} \left[\mathrm{sign}(\pi) \prod_{k=1}^{p} \{\theta^{(k)} - \bar{\theta}\}_k \{\theta^{(k)} - \bar{\theta}\}_{\pi(k)} \, \tilde{q}_X(\theta^{(k)}|\bar{\theta}) \right],$$

and where we use the notation $\theta^{(k)}$ to distinguish between the p variables in the multivariate integral. Similarly, we get for the criterion $\Phi_{q,\mathbf{I}}(X;\bar{\theta})$

$$\mathrm{trace}[\mathbf{S}^q(X,\bar{\theta})] = \int_{\{\mathrm{int}(\Theta)\}^{\otimes q}} L'(\theta^{(1)},\ldots,\theta^{(q)},\bar{\theta},X) \, \mathrm{d}\theta^{(1)} \cdots \mathrm{d}\theta^{(q)}$$

with

$$L'(\theta^{(1)}, \ldots, \theta^{(q)}, \bar{\theta}, X) = \text{trace}\left[\prod_{k=1}^{q} (\theta^{(k)} - \bar{\theta})(\theta^{(k)} - \bar{\theta})^\top \tilde{q}_X(\theta^{(k)}|\bar{\theta})\right].$$

6.4 Higher-Order Approximations of Optimality Criteria

The criteria presented in Sect. 6.3 are adequate even for small-sample experiments. However, their optimization may be difficult, even if it is carried through the use of stochastic approximation, hence the motivation for deriving analytical expressions. This happens to be feasible, at least approximately, when the errors ε in (6.1) have finite moments up to order three, with σ small, and when $\Theta = \mathbb{R}^p$, that is, when we can neglect the influence of the boundary of Θ on the estimation—which is justified if σ is small enough since we always assume that $\bar{\theta} \in \text{int}(\Theta)$.

We first rewrite the criteria as integrals with respect to the density $\varphi_{X,\bar{\theta}}(\mathbf{y})$ of the observations. We obtain

$$\int_{\mathbb{R}^N} \{h[\hat{\theta}(\mathbf{y})] - h(\bar{\theta})\}^2 \, \varphi_{X,\bar{\theta}}(\mathbf{y}) \, \mathrm{d}\mathbf{y}$$

for c-optimality and

$$\int_{\mathbb{R}^N} \{\hat{\theta}(\mathbf{y}) - \bar{\theta}\}_i \{\hat{\theta}(\mathbf{y}) - \bar{\theta}\}_j \, \varphi_{X,\bar{\theta}}(\mathbf{y}) \, \mathrm{d}\mathbf{y}$$

for the components of $\mathbf{S}(X, \bar{\theta})$. The entropy of the p.d.f. $q_X(\cdot|\bar{\theta})$ of $\hat{\theta}$ can be approximated by

$$\text{Ent}[q_X(\cdot|\bar{\theta})] = -\int_{\mathbb{R}^N} \log\{q_X[\hat{\theta}(\mathbf{y})|\bar{\theta}]\} \, \varphi_{X,\bar{\theta}}(\mathbf{y}) \, \mathrm{d}\mathbf{y}\,. \tag{6.20}$$

It gives the exact entropy of $q_X(\cdot|\bar{\theta})$, and thus the exact entropy of the p.d.f. of $\hat{\theta}$, in intrinsically linear models where $q_X(\cdot|\bar{\theta})$ is exact; see Sect. 6.2.1. More generally, we obtain an integral of the form

$$\int_{\mathbb{R}^N} F[\hat{\theta}(\mathbf{y}), \bar{\theta}, X] \, \varphi_{X,\bar{\theta}}(\mathbf{y}) \, \mathrm{d}\mathbf{y} \tag{6.21}$$

where $F(\cdot, \bar{\theta}, X)$ is a given differentiable function. Compared to Sect. 6.3, we have now integrals over $\mathbb{R}^N$ instead of $\mathbb{R}^p$. However, we can approximate the function $F(\cdot, \bar{\theta}, X)$ to be integrated by using a Taylor development in the neighborhood of the point $\eta(\bar{\theta})$.

Since $\hat{\theta}(\mathbf{y}) = \arg\min_\theta \|\mathbf{y} - \eta(\theta)\|^2$, we have $\hat{\theta}[\eta(\bar{\theta})] = \bar{\theta}$ and

$$
\begin{aligned}
F[\hat{\theta}(\mathbf{y}),\bar{\theta},X] &= F(\bar{\theta},\bar{\theta},X) + \left.\frac{\partial F(\theta,\bar{\theta},X)}{\partial\theta^\top}\right|_{\bar{\theta}} \left.\frac{\partial\hat{\theta}(\mathbf{y})}{\partial\mathbf{y}^\top}\right|_{\eta(\bar{\theta})} \varepsilon \\
&\quad + \frac{1}{2}\varepsilon^\top \left\{ \sum_{i=1}^{p} \left.\frac{\partial F(\theta,\bar{\theta},X)}{\partial\theta_i}\right|_{\bar{\theta}} \left.\frac{\partial^2[\hat{\theta}(\mathbf{y})]_i}{\partial\mathbf{y}\partial\mathbf{y}^\top}\right|_{\eta(\bar{\theta})} \right. \\
&\quad \left. + \left.\frac{\partial\hat{\theta}^\top(\mathbf{y})}{\partial\mathbf{y}}\right|_{\eta(\bar{\theta})} \left.\frac{\partial^2 F(\theta,\bar{\theta},X)}{\partial\theta\partial\theta^\top}\right|_{\bar{\theta}} \left.\frac{\partial\hat{\theta}(\mathbf{y})}{\partial\mathbf{y}^\top}\right|_{\eta(\bar{\theta})} \right\} \varepsilon + \ldots
\end{aligned}
$$

where $\varepsilon = \mathbf{y} - \eta(\bar{\theta})$ has zero mean and variance–covariance matrix $\sigma^2\mathbf{I}_N$. Therefore, the criterion (6.21) equals

$$
\begin{aligned}
\int_{\mathbb{R}^N} F[\hat{\theta}(\mathbf{y}),\bar{\theta},X]\,\varphi_{X,\bar{\theta}}(\mathbf{y})\,\mathrm{d}\mathbf{y} &= F(\bar{\theta},\bar{\theta},X) \\
&\quad + \frac{\sigma^2}{2}\sum_{i=1}^{N} \left\{ \left.\frac{\partial F(\theta,\bar{\theta},X)}{\partial\theta^\top}\right|_{\bar{\theta}} \left.\frac{\partial^2\hat{\theta}(\mathbf{y})}{\partial y_i^2}\right|_{\eta(\bar{\theta})} \right. \\
&\quad \left. + \left.\frac{\partial\hat{\theta}^\top(\mathbf{y})}{\partial y_i}\right|_{\eta(\bar{\theta})} \left.\frac{\partial^2 F(\theta,\bar{\theta},X)}{\partial\theta\partial\theta^\top}\right|_{\bar{\theta}} \left.\frac{\partial\hat{\theta}(\mathbf{y})}{\partial y_i}\right|_{\eta(\bar{\theta})} \right\} + \mathcal{O}(\sigma^3)\,.
\end{aligned}
$$

The only remaining difficulty is to express the derivatives of $\hat{\theta}(\mathbf{y})$ with respect to the components y_i. These derivatives can be obtained by using the implicit definition of $\hat{\theta}(\mathbf{y})$ through the equations

$$
\left.\frac{\partial}{\partial\theta}\|\mathbf{y}-\eta(\theta)\|^2\right|_{\theta=\hat{\theta}(\mathbf{y})} = \mathbf{0}\,.
$$

Denote $\mathbf{g}(\theta,\mathbf{y}) = [\partial\eta^\top(\theta)/\partial\theta]\,[\eta(\theta)-\mathbf{y}]$. We then obtain $\mathbf{g}[\bar{\theta},\eta(\bar{\theta})] = \mathbf{0}$, $\mathbf{g}[\hat{\theta}(\mathbf{y}),\mathbf{y}] = \mathbf{0}$,

$$
\begin{aligned}
\frac{\partial\mathbf{g}(\theta,\mathbf{y})}{\partial\theta^\top} &= \left\{ \mathbf{M}_X(\theta) + \sum_{i=1}^{N}\frac{\partial^2\eta(x_i,\theta)}{\partial\theta\partial\theta^\top}[\eta(x_i,\theta)-y(x_i)] \right\}, \\
\frac{\partial\mathbf{g}(\theta,\mathbf{y})}{\partial\mathbf{y}^\top} &= -\frac{\partial\eta^\top(\theta)}{\partial\theta}\,.
\end{aligned}
$$

Hence, from the implicit function theorem,

$$
\frac{\partial\hat{\theta}(\mathbf{y})}{\partial\mathbf{y}^\top} = -\left[\left.\frac{\partial\mathbf{g}(\theta,\mathbf{y})}{\partial\theta^\top}\right|_{\theta=\hat{\theta}(\mathbf{y})}\right]^{-1} \left.\frac{\partial\mathbf{g}(\theta,\mathbf{y})}{\partial\mathbf{y}^\top}\right|_{\theta=\hat{\theta}(\mathbf{y})}. \tag{6.22}
$$

Since $\hat{\theta}[\eta(\bar{\theta})] = \bar{\theta}$, we get

$$
\left.\frac{\partial\hat{\theta}(\mathbf{y})}{\partial\mathbf{y}^\top}\right|_{\eta(\bar{\theta})} = -\mathbf{M}_X^{-1}(\bar{\theta})\,\mathbf{J}^\top(\bar{\theta})\,. \tag{6.23}
$$

The higher-order derivatives of $\hat{\theta}(\mathbf{y})$ are obtained by differentiating (6.22) with respect to the components of $\mathbf{y}$. For the second-order derivatives we obtain

$$\frac{\partial^2\hat{\theta}(\mathbf{y})}{\partial y_i^2}\bigg|_{\eta(\bar{\theta})} = -\mathbf{M}_X^{-1}(\bar{\theta})\,\frac{\partial}{\partial y_i}\left[\frac{\partial \mathbf{g}(\theta,\mathbf{y})}{\partial\theta^\top}\bigg|_{\theta=\hat{\theta}(\mathbf{y})}\right]\bigg|_{\mathbf{y}=\eta(\bar{\theta})} \mathbf{M}_X^{-1}(\bar{\theta})\,\frac{\partial\eta(x_i,\theta)}{\partial\theta}\bigg|_{\bar{\theta}}$$
$$+\mathbf{M}_X^{-1}(\bar{\theta})\,\frac{\partial^2\eta(x_i,\theta)}{\partial\theta\partial\theta^\top}\bigg|_{\bar{\theta}}\,\frac{\partial\hat{\theta}(\mathbf{y})}{\partial y_i}\bigg|_{\eta(\bar{\theta})}. \tag{6.24}$$

We have

$$\frac{\partial}{\partial y_i}\left[\frac{\partial \mathbf{g}(\theta,\mathbf{y})}{\partial\theta^\top}\bigg|_{\theta=\hat{\theta}(\mathbf{y})}\right]\bigg|_{\mathbf{y}=\eta(\bar{\theta})} = \sum_{j=1}^{p}\frac{\partial^2\mathbf{g}[\theta,\eta(\bar{\theta})]}{\partial\theta_j\partial\theta^\top}\bigg|_{\bar{\theta}}\,\frac{\partial[\hat{\theta}(\mathbf{y})]_j}{\partial y_i}\bigg|_{\eta(\bar{\theta})}$$
$$+\frac{\partial}{\partial y_i}\left[\frac{\partial \mathbf{g}(\theta,\mathbf{y})}{\partial\theta^\top}\bigg|_{\bar{\theta}}\right]\bigg|_{\mathbf{y}=\eta(\bar{\theta})}$$
$$=\sum_{j=1}^{p}\left\{\frac{\partial^2\eta^\top(\theta)}{\partial\theta_j\partial\theta}\bigg|_{\bar{\theta}}\,\frac{\partial\eta(\theta)}{\partial\theta^\top}\bigg|_{\bar{\theta}}+\frac{\partial\eta^\top(\theta)}{\partial\theta}\bigg|_{\bar{\theta}}\,\frac{\partial^2\eta(\theta)}{\partial\theta_j\partial\theta^\top}\bigg|_{\bar{\theta}}\right.$$
$$\left.+\sum_{k=1}^{N}\frac{\partial^2\eta(x_k,\theta)}{\partial\theta\partial\theta^\top}\bigg|_{\bar{\theta}}\,\frac{\partial\eta(x_k,\theta)}{\partial\theta_j}\bigg|_{\bar{\theta}}\right\}\left\{\mathbf{M}_X^{-1}(\bar{\theta})\frac{\partial\eta(x_i,\theta)}{\partial\theta}\bigg|_{\bar{\theta}}\right\}_j$$
$$-\frac{\partial^2\eta(x_i,\theta)}{\partial\theta\partial\theta^\top}\bigg|_{\bar{\theta}},$$

which, together with (6.24), gives after simplification,

$$\sum_{i=1}^{N}\frac{\partial^2\hat{\theta}(\mathbf{y})}{\partial y_i^2}\bigg|_{\eta(\bar{\theta})} = -\mathbf{M}_X^{-1}(\bar{\theta})\,\mathbf{J}^\top(\bar{\theta})\sum_{i,j=1}^{p}\mathbf{H}_{\cdot ij}(\bar{\theta})\{\mathbf{M}_X^{-1}(\bar{\theta})\}_{ij}\,, \tag{6.25}$$

with $\mathbf{H}_{\cdot ij}(\theta)$ defined in (6.6).

Notice that, in the case of normal errors $\varepsilon\sim\mathscr{N}(\mathbf{0},\sigma^2\mathbf{I}_N)$, using the first-order approximation for $\hat{\theta}(\mathbf{y})$ amounts to using the asymptotic normal approximation for $\hat{\theta}$. Indeed, we have

$$\hat{\theta}[\eta(\bar{\theta})]+\frac{\partial\hat{\theta}(\mathbf{y})}{\partial\mathbf{y}^\top}\bigg|_{\eta(\bar{\theta})}\varepsilon = \bar{\theta}+\mathbf{M}_X^{-1}(\bar{\theta})\,\mathbf{J}^\top(\bar{\theta})\,\varepsilon\sim\mathscr{N}(\bar{\theta},\sigma^2\mathbf{M}_X^{-1}(\bar{\theta}))\,.$$

6.4.1 Approximate Bias and Mean-squared Error

Using the developments above we obtain that the bias of $\hat{\theta}$ is given by

$$\begin{aligned}\mathbf{b}(X,\bar{\theta}) &= \mathrm{I\!E}_{X,\bar{\theta}}\{\hat{\theta}\}-\bar{\theta}\\ &= \mathrm{I\!E}_{X,\bar{\theta}}\left\{\mathbf{M}_X^{-1}(\bar{\theta})\,\mathbf{J}^\top(\bar{\theta})\,\varepsilon+\frac{1}{2}\sum_{i,j=1}^{N}\left[\varepsilon_i\varepsilon_j\,\frac{\partial^2\hat{\theta}(\mathbf{y})}{\partial y_i\partial y_j}\bigg|_{\eta(\bar{\theta})}\right]\right\}+\mathcal{O}(\sigma^3)\\ &= \mathbf{b}_2(\theta)+\mathcal{O}(\sigma^3)\,,\end{aligned} \tag{6.26}$$

with

$$\mathbf{b}_2(\theta) = \mathbf{b}_2(X, \bar{\theta}) = -\frac{\sigma^2}{2}\, \mathbf{M}_X^{-1}(\bar{\theta})\, \mathbf{J}^\top(\bar{\theta}) \sum_{i,j=1}^{p} \mathbf{H}_{;ij}^{\cdot}(\bar{\theta})\, \{\mathbf{M}_X^{-1}(\bar{\theta})\}_{ij}\,, \qquad (6.27)$$

see (6.25), which corresponds to the expression obtained by Box (1971) using a different technique. When the errors ε_i are normal, the third-order moments $\mathbb{E}\{\varepsilon_i\varepsilon_j\varepsilon_k\}$ equal zero for all i, j and k, and the first neglected term in (6.26) is of order $\mathcal{O}(\sigma^4)$.

Remark 6.5. The approximate expression (6.27) for the bias of the LS estimator is at the origin of two methods for constructing bias-corrected estimators. The first one (Firth 1993) corresponds to the score-corrected estimator $\hat{\theta}_{\rm sc}$ obtained by solving the modified normal equations

$$\mathbf{v}(\theta) - \mathbf{M}_X(\theta)\mathbf{b}_2(X, \theta) = \mathbf{0}$$

for θ, where $\mathbf{v}(\theta) = \mathbf{J}^\top(\theta)\,[\mathbf{y} - \eta(\theta)]$ is the score function of the LS estimator $\hat{\theta}$. The second one (Pronzato and Pázman 1994a) is based on the identity $\mathbb{E}_{X,\bar{\theta}}\{\hat{\theta}(\mathbf{y}) - \bar{\theta} - \mathbf{b}(X, \bar{\theta})\} = \mathbf{0}$, and the estimator $\hat{\theta}_{\rm ts}$ is obtained by solving the equations $\theta + \mathbf{b}_2(\theta) = \hat{\theta}$ for θ. It thus corresponds to a two-stage procedure, with the LS estimator $\hat{\theta}$ determined first and $\hat{\theta}_{\rm ts}$ constructed from $\hat{\theta}$. The approximate joint and marginal densities of $\hat{\theta}_{\rm sc}$ and $\hat{\theta}_{\rm ts}$ are derived in (Pázman and Pronzato, 1998). □

From the developments above for the bias of the LS estimator, we see that the second-order approximation of mean-squared error matrix $\mathbf{S}(X, \bar{\theta})$ is

$$\mathbf{S}(X, \bar{\theta}) = \sigma^2 \mathbf{M}_X^{-1}(\bar{\theta}) + \mathcal{O}(\sigma^3)\,.$$

The first neglected term is of order $\mathcal{O}(\sigma^4)$ when the errors ε_i are normal. The calculation of this next term in the development requires the expression of fourth-order derivatives of $\hat{\theta}(\mathbf{y})$ with respect to $\mathbf{y}$. An approximation of $\mathbf{S}(X, \bar{\theta})$ is derived in (Clarke, 1980) using a different approach. However, the expression obtained is very complicated and thus difficult to use for the construction of design criteria.

6.4.2 Approximate Entropy of the p.d.f. of the LS Estimator

When the errors are normal $\mathcal{N}(\mathbf{0}, \sigma^2\mathbf{I}_N)$, the (Shannon) entropy of the p.d.f. $q_X(\cdot|\bar{\theta})$ of $\hat{\theta}$ can be approximated by (6.20), which gives

$$\begin{aligned}
\mathrm{Ent}[q_X(\cdot|\bar{\theta})] &= -\log q_X(\bar{\theta}|\bar{\theta}) - \frac{\sigma^2}{2} \sum_{i=1}^{N} \left.\frac{\partial^2 \log q_X[\hat{\theta}(\mathbf{y})|\bar{\theta}]}{\partial y_i^2}\right|_{\eta(\bar{\theta})} + \mathcal{O}(\sigma^4) \\
&= -\log q_X(\bar{\theta}|\bar{\theta}) - \frac{\sigma^2}{2} \sum_{i=1}^{N} \left\{ \left.\frac{\partial \hat{\theta}^\top(\mathbf{y})}{\partial y_i}\right|_{\eta(\bar{\theta})} \left.\frac{\partial^2 \log q_X(\theta|\bar{\theta})}{\partial\theta\partial\theta^\top}\right|_{\bar{\theta}} \left.\frac{\partial \hat{\theta}(\mathbf{y})}{\partial y_i}\right|_{\eta(\bar{\theta})} \right.
\end{aligned}$$

$$+\frac{\partial \log q_X(\theta|\bar{\theta})}{\partial \theta^\top}\bigg|_{\bar{\theta}} \frac{\partial^2 \hat{\theta}(\mathbf{y})}{\partial y_i^2}\bigg|_{\eta(\bar{\theta})}\Bigg\} + \mathcal{O}(\sigma^4)\,.$$

Direct calculation gives

$$-\log q_X(\bar{\theta}|\bar{\theta}) = -\frac{1}{2}\log\det \mathbf{M}_X(\bar{\theta}) + \frac{p}{2}\log(2\pi\sigma^2)\,.$$

Calculating the first- and second-order derivatives of $\log q_X(\theta|\bar{\theta})$ with respect to θ and using (6.23), (6.25), we obtain after simplification

$$\begin{aligned}\mathrm{Ent}[q_X(\cdot|\bar{\theta})] &= \mathrm{Ent}_1[q_X(\cdot|\bar{\theta})] \\ &\quad -\frac{\sigma^2}{2}\sum_{h,i,j,k=1}^{p} \left(\{\mathbf{M}_X^{-1}(\bar{\theta})\}_{ij}\left[\{\mathbf{M}_X^{-1}(\bar{\theta})\}_{kh}[R_{kjhi}(\bar{\theta}) + U^h_{kij}(\bar{\theta})]\right.\right. \\ &\quad \left.\left. -G^h_{ki}(\bar{\theta})\,G^k_{hj}(\bar{\theta}) - G^k_{kh}(\bar{\theta})\,G^h_{ij}(\bar{\theta})\right]\right) + \mathcal{O}(\sigma^4)\,,\end{aligned}$$

where

$$\mathrm{Ent}_1[q_X(\cdot|\bar{\theta})] = \frac{p}{2}[1+\log(2\pi\sigma^2)] - \frac{1}{2}\log\det \mathbf{M}_X(\bar{\theta})\,,$$

$$U^h_{kij}(\theta) = \frac{\partial \eta^\top(\theta)}{\partial\theta_k\partial\theta_i\partial\theta_j}\,\frac{\partial\eta(\theta)}{\partial\theta_h}$$

$$G^k_{ij}(\theta) = \sum_{h=1}^{p} \frac{\partial \eta^\top(\theta)}{\partial\theta_h}\,\mathbf{H}_{ij}^{\boldsymbol{\cdot}}\,\{\mathbf{M}_X^{-1}(\bar{\theta})\}_{hk}$$

and $R_{hijk}(\theta)$ is defined in Sect. 6.1.2, see Pronzato and Pázman (1994b). Notice that the approximation $\mathrm{Ent}_1[q_X(\cdot|\bar{\theta})]$ coincides with the entropy of the asymptotic normal distribution of $\hat{\theta}$, as can be checked by substituting $\mathbf{M}_X(\bar{\theta})/[N\sigma^2]$ for $\mathbf{M}(\xi,\bar{\theta})$ in (5.7).

6.5 Bibliographic Notes and Further Remarks

Design criteria based on confidence regions

Define $\mathbf{e}(\theta) = \mathbf{y} - \eta(\theta)$; one may observe that $\mathbf{e}^\top(\bar{\theta})\,\mathbf{P}_{\bar{\theta}}\,\mathbf{e}(\bar{\theta})/\sigma^2$ and $\mathbf{e}^\top(\bar{\theta})\,[\mathbf{I}_N - \mathbf{P}_{\bar{\theta}}]\,\mathbf{e}(\bar{\theta})/\sigma^2$, with $\mathbf{P}_\theta$ the projector given by (6.4), are, respectively, distributed like χ^2_p and χ^2_{N-p} random variables under the assumption of normal errors. Moreover, these variables are independent. Therefore, the region

$$\left\{\theta \in \mathbb{R}^p : \mathbf{e}^\top(\theta)\,\mathbf{P}_\theta\,\mathbf{e}(\theta)/\sigma^2 < \chi^2_p(1-\alpha)\right\}\,,$$

where $\chi^2_p(1-\alpha)$ is the $(1-\alpha)$ quantile of the χ^2_p distribution, is an exact confidence region of level $1-\alpha$ when σ^2 is known, whereas the region

$$\left\{\theta \in \mathbb{R}^p : \frac{N-p}{p} \frac{\mathbf{e}^\top(\theta)\, \mathbf{P}_\theta\, \mathbf{e}(\theta)}{\mathbf{e}^\top(\theta)\, [\mathbf{I}_N - \mathbf{P}_\theta]\, \mathbf{e}(\theta)} < F_{p,N-p}(1-\alpha)\right\}, \qquad (6.28)$$

where $F_{p,N-p}(1-\alpha)$ is the $(1-\alpha)$ quantile of the F distribution with p and $N-p$ degrees of freedom, is an exact confidence region of level $1-\alpha$ for the case where σ^2 is unknown; see, e.g., Halperin (1963), Hartley (1964), and Sundaraij (1978). Although the regions above are exact, in general they are not of minimum volume. Also, they can be composed of disconnected subsets when $\mathbb{S}_\eta$ is curved.

Approximate confidence regions can be obtained from the likelihood ratio. When σ^2 is known,

$$\left\{\theta \in \mathbb{R}^p : \|\mathbf{e}(\theta)\|^2 - \|\mathbf{e}(\hat{\theta})\|^2 < \sigma^2 \chi_p^2(1-\alpha)\right\}$$

has confidence level approximately $1-\alpha$ and, when σ^2 is unknown, the confidence level of

$$\left\{\theta \in \mathbb{R}^p : \|\mathbf{e}(\theta)\|^2 / \|\mathbf{e}(\hat{\theta})\|^2 < 1 + \frac{p}{N-p} F_{p,N-p}(1-\alpha)\right\}$$

is also approximately $1-\alpha$. These regions are usually connected.

Hamilton et al. (1982) use a quadratic approximation of $\mathbb{S}_\eta$ and approximate projections of confidence regions like those above on the tangent plane to $\mathbb{S}_\eta$ at $\eta(\hat{\theta})$. They obtain in this way ellipsoids on this tangent plane (instead of the spheres obtained when a linear approximation of $\mathbb{S}_\eta$ is used), which can then be mapped into the parameter space. A nice quadratic approximation of the volume of the resulting regions is used in (Hamilton and Watts, 1985) as a substitute to the D-optimality criterion, which is related to the volume of the asymptotic ellipsoid of concentration; see Sect. 5.1. Vila (1990) considers the design criterion defined by the exact expected volume of confidence regions (6.28) for a given α, which he then optimizes by stochastic approximation techniques; see also Vila and Gauchi (2007).

When σ^2 is unknown, instead of using an estimate of σ^2 based on residuals, like in (6.28), we can estimate σ^2 from replications of observations at some design points. Let n_i denote the number of observations at $x^{(i)}$, $i = 1, \ldots, m$, with $\sum_{i=1}^m n_i = N$, and define $\bar{y}_i = (1/n_i) \sum_{j=1}^{n_i} y_j(x^{(i)})$ with $y_j(x^{(i)})$ the j-th observation at $x^{(i)}$. Suppose that $n_i > 1$ for $i = 1, \ldots, m' \leq m$. Then, the estimator $\hat{\sigma}^2 = [1/(N-m')] \sum_{i=1}^{m'} \sum_{j=1}^{n_i} [y_j(x^{(i)}) - \bar{y}_i]^2$ is such that $(N-m')\hat{\sigma}^2/\sigma^2$ has the $\chi^2_{N-m'}$ distribution. One may thus consider confidence regions given by

$$\left\{\theta \in \mathbb{R}^p : \mathbf{e}^\top(\theta)\, \mathbf{P}_\theta\, \mathbf{e}(\theta) / (p\, \hat{\sigma}^2) < F_{p,N-m'}(1-\alpha)\right\}. \qquad (6.29)$$

Let $\bar{\mathbf{y}}$ denote the vector obtained by replacing each component $y_j(x^{(i)})$ by $\bar{y}_i$ in $\mathbf{y}$. One can easily check that $(\mathbf{y} - \bar{\mathbf{y}})^\top \mathbf{P}_{\bar{\theta}}\, \mathbf{e}(\bar{\theta}) = 0$, so that, under the assumption of normal errors, $\mathbf{e}^\top(\bar{\theta})\, \mathbf{P}_{\bar{\theta}}\, \mathbf{e}(\bar{\theta})$ and $\hat{\sigma}^2$ are independent and the regions (6.29) are exact. When the intrinsic curvature of the model is not too large, these regions are larger than the regions (6.28); on the other hand, they are not so much affected by a large intrinsic curvature. Since different designs allow different degrees of freedom $N - m'$ for the estimation of σ^2, the value of m' should enter the definition of design criteria based on the confidence regions (6.29), a problem considered in (Gilmour and Trinca, 2012) for the case of linear models.

7

Identifiability, Estimability, and Extended Optimality Criteria

Among the major difficulties that one may encounter when estimating parameters in a nonlinear model are the nonuniqueness of the estimator, its instability with respect to small perturbations of the observations, and the presence of local optimizers of the estimation criterion. Classically, those issues are ignored at the design stage: the designs of Chap. 5 are based on asymptotic local properties of the estimator; the approaches of Chap. 6 make use of an assumption ($H_{\mathcal{S}}$, p. 172) which allows us to avoid these difficulties. The main message of this chapter is that estimability issues can be taken into account at the design stage, through the definition of suitable design criteria. This forms a difficult area, still under development. Several new notions will be introduced, and a series of examples will illustrate the importance of the geometry of the model.

The qualitative notions of identifiability and estimability are introduced in Sects. 7.1 and 7.2, respectively. Numerical difficulties for the estimation of the model parameters, such as the presence of local optima and the instability of the estimator with respect to small perturbations of the observations, are considered in Sect. 7.3. They are related to a notion of estimability which is more quantitative and which we suggest to measure through the construction of an estimability function (Sect. 7.4) or an extended measure of nonlinearity (Sect. 7.5). The advantages and drawbacks of using p-point design in this context are exposed in Sect. 7.6. Section 7.7 suggests optimality criteria for designing experiments that take potential numerical difficulties into account. The notion of estimability is related to the estimator that is used; we focus our attention on LS estimation throughout the chapter, and the presentation of estimability for estimators other than LS is postponed to Sect. 7.8.

Throughout the chapter we consider the same framework as in Chap. 6, that is, we assume that the observations satisfy

$$y_i = y(x_i) = \eta(x_i, \bar{\theta}) + \varepsilon_i \,,\ i = 1, \ldots, N \,, \tag{7.1}$$

L. Pronzato and A. Pázman, *Design of Experiments in Nonlinear Models*, Lecture Notes in Statistics 212, DOI 10.1007/978-1-4614-6363-4_7,

with $\mathbb{E}(\varepsilon_i) = 0$, $\text{var}(\varepsilon_i) = \sigma^2$, $\text{cov}(\varepsilon_i, \varepsilon_j) = 0$ if $i \neq j$, $i, j = 1, \ldots, N$. We write in a vector notation

$$\mathbf{y} = \eta(\bar{\theta}) + \varepsilon\,, \quad \text{with } \mathbb{E}(\varepsilon) = \mathbf{0}\,, \ \text{Var}(\varepsilon) = \sigma^2 \mathbf{I}_N\,, \tag{7.2}$$

where $\eta(\theta) = \eta_X(\theta) = (\eta(x_1, \theta), \ldots, \eta(x_N, \theta))^\top$, $\mathbf{y} = (y_1, \ldots, y_N)^\top$, $\varepsilon = (\varepsilon_1, \ldots, \varepsilon_N)^\top$, and $X = (x_1, \ldots, x_N)$. Also, we suppose that the assumptions H_Θ, H1_η, and H2_η of Sect. 3.1 are satisfied: Θ is a compact subset of $\mathbb{R}^p$ such that $\Theta \subset \overline{\text{int}(\Theta)}$, and $\eta(x, \theta)$ is twice continuously differentiable with respect to $\theta \in \text{int}(\Theta)$ for any $x \in \mathscr{X}$. We denote

$$J_N(\theta) = \frac{1}{N} \|\eta(\theta) - \mathbf{y}\|^2 = \frac{1}{N} \sum_{i=1}^{N} [y(x_i) - \eta(x_i, \theta)]^2 \tag{7.3}$$

the LS criterion, and the LS estimator corresponds to the global minimizer of $\|\eta(\theta) - \mathbf{y}\|^2$ for $\theta \in \Theta$,

$$\hat{\theta}^N_{LS}(\mathbf{y}) = \arg\min_{\theta \in \Theta} \|\eta(\theta) - \mathbf{y}\|^2\,.$$

7.1 Identifiability

Given a parametric model describing how observations are generated, identifiability addresses the following question: is it possible to estimate the model parameters uniquely from a (possibly infinite) set of observations? The answer depends on the model structure itself, given by mathematical equations, on the choice of the design and on the set Θ in which the model parameters are looked for. One may refer to Koopmans and Reiersøl (1950) for an illuminating introduction to these ideas for general statistical models (design issues being not considered in that paper, however).

In the case of regression models such as (7.2), the issue of identifiability reduces to that of the uniqueness of a vector θ of model parameters associated with a given response vector $\eta(\theta) = (\eta(x_1, \theta), \ldots, \eta(x_N, \theta))^\top$ when N is arbitrarily large and $X = (x_1, \ldots, x_N)$ is arbitrary. One may refer, e.g., to Walter (1982, 1987) for definitions in a more general context. An important notion there is that of global identifiability: the model is *globally identifiable* when for almost any θ (in the sense of zero Lebesgue measure on Θ) it satisfies

$$\theta' \in \Theta \text{ and } \eta(\theta) = \eta(\theta') \text{ for all designs } X \Longrightarrow \theta' = \theta\,.$$

The restriction to almost all θ is to avoid pathological situations; the consideration of all designs X makes the property depend on the model equations only. On the other hand, when the property is true in some neighborhood of θ (i.e., for almost any θ, there exists $\mathcal{V}(\theta)$ such that $\eta(\theta) = \eta(\theta')$ for all designs X and $\theta' \in \mathcal{V}(\theta)$ implies $\theta' = \theta$), the model is said to be *locally identifiable*. When the result $\theta' = \theta$ is weakened into the equality of some component of θ, $\{\theta'\}_i = \{\theta\}_i$ say, this component is said to be globally (or locally) identifiable.

7.2 LS Estimability of Regression Models

We call estimability a notion, related to identifiability, that concerns the uniqueness of the estimate of the model parameters θ *for a given design*. The notion is thus related to the estimator which is considered—notice that the estimability conditions given in the consistency theorems of Chaps. 3 and 4 depend on the model and estimator. We focus here on LS estimation; other estimators will be considered in Sect. 7.8.

LS estimability can be understood as identifiability for a given design X: a regression model is *globally LS estimable at* $\theta \in \Theta$ *for* X if

$$\theta' \in \Theta \text{ and } \eta(\theta') = \eta(\theta) \Longrightarrow \theta' = \theta \,.$$

When the property is valid for almost all θ, in the sense of the Lebesgue measure on Θ, again in order to avoid pathological situations, the model is said to be *globally LS estimable* for X. Geometrically, this corresponds to a nonoverlapping expectation surface; see Sect. 6.1.1. A justification of this notion is as follows. Suppose that the experiment characterized by X is repeated n times. The global LS estimability at $\bar{\theta}$ for X guarantees the strong consistency of the LS estimator of θ when n tends to infinity. This is in accordance with the LS estimability condition (3.6) used in Theorem 3.1.

The definitions of local LS estimability at θ and local LS estimability for X follow similarly from the notion of local identifiability: the model is *locally LS estimable at* $\theta \in \Theta$ *for* X if there exists some neighborhood $\mathcal{V}(\theta)$ of θ such that

$$\theta' \in \mathcal{V}(\theta) \text{ and } \eta(\theta') = \eta(\theta) \Longrightarrow \theta' = \theta \,;$$

if this is true for almost any θ, then the model is *locally LS estimable for* X. Notice that the existence of at least one design X such that the model is globally (respectively, locally) LS estimable for X implies the global (respectively, local) identifiability of the model.

Example 7.1. Consider a one-parameter model with one observation at x for which the response $\eta(x,\theta)$ is the continuous function of θ plotted in Fig. 7.1. Then the model is globally LS estimable for x at any $\theta < \theta_a$ or $\theta > \theta_d$, locally LS estimable for x at any $\theta \in [\theta_a, \theta_b) \cup (\theta_c, \theta_d]$, and not estimable for x at a $\theta \in [\theta_b, \theta_c]$. □

Let $X = (x_1, \ldots, x_N)$ be any given design and consider the associated responses $\eta(\theta) = (\eta(x_1,\theta), \ldots, \eta(x_N,\theta))^\top$. The next theorem (see Pázman 1984a) states that $\eta(\hat{\theta}_{LS}^N) = \min_{\mathbf{z}\in\mathbb{S}_\eta} \|\mathbf{z}-\mathbf{y}\|^2$, with $\mathbb{S}_\eta = \{\eta(\theta) : \theta \in \Theta\}$ the expectation surface, is unique with probability 1 under fairly general conditions. We do not reproduce the proof, which requires long differential-geometric arguments presented in (Pázman, 1993b).

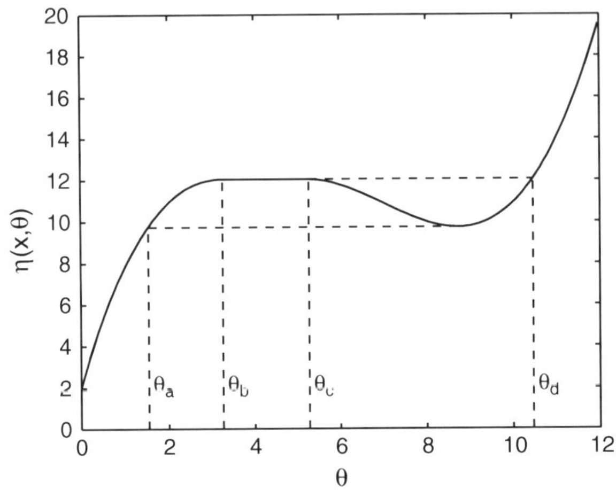

Fig. 7.1. Local and global estimability at θ

Theorem 7.2. *When the probability measure of the observation errors* ε_i *in the regression model (7.2) has a density with respect to the Lebesgue measure, the value of* $\mathbf{z} = \eta[\hat{\theta}^N_{LS}(\mathbf{y})]$ *on the expectation surface* $\mathbb{S}_\eta$ *closest (for the Euclidian distance) to* $\mathbf{y}$ *is unique w.p.1.*

This does not mean in general that $\hat{\theta}^N_{LS}(\mathbf{y})$ is unique, but due to Theorem 7.2, this uniqueness is ensured w.p.1 if the mapping $\theta \in \Theta \longrightarrow \eta(\theta) \in \mathbb{R}^N$ is unique, i.e., if the model is globally LS estimable for X. This evidences the practical importance of estimability.

Local LS estimability at a given θ for the experimental design X is easily tested by computing the information matrix

$$\mathbf{M}_X(\theta) = \frac{\partial \eta^\top(\theta)}{\partial \theta} \frac{\partial \eta(\theta)}{\partial \theta^\top} \,. \tag{7.4}$$

Indeed, $\text{rank}[\mathbf{M}_X(\theta)] = \dim(\theta)$ implies that the model is locally LS estimable at θ for X. The situation is particularly simple for linear models where $\eta(\theta)$ is linear in θ, $\eta(\theta) = \mathbf{F}(X)\theta + \mathbf{v}(X)$: local and global LS estimabilities are then equivalent. Testing global LS estimability for models nonlinear in θ is more complicated; a numerical approach will be proposed in Sect. 7.4 through the construction of an estimability function.

7.3 Numerical Issues Related to Estimability in Regression Models

Examples of difficulties with LS estimation concern the uniqueness of the global minimizer $\hat{\theta}^N_{LS}(\mathbf{y})$ of $J_N(\cdot)$ given by (7.3), the possible existence of values of θ far from $\hat{\theta}^N_{LS}$ but with almost similar values of $J_N(\cdot)$, the stability of $\hat{\theta}^N_{LS}(\mathbf{y})$ with respect to small perturbations of $\mathbf{y}$, the possible presence of local minimizers for $J_N(\cdot)$, etc.

Local Minimizers and Instability of $\hat{\theta}^N_{LS}$

A major issue in parameter estimation concerns the presence of local minimizers of the mapping $\theta \in \Theta \longrightarrow \|\eta(\theta) - \mathbf{y}\|^2$ or of the mapping $\mathbf{z} \in \mathbb{S}_\eta \longrightarrow \|\mathbf{z} - \mathbf{y}\|^2$. Indeed, most algorithms for LS estimation only perform a local search, and it is then difficult to certify that the minimizer obtained is the global one. In fact, when the probability measure of the observations $y(x_i)$ has a density with respect to the Lebesgue measure and $\mathbb{S}_\eta$ is curved, i.e., $C_{int}(X,\theta) > 0$ for some θ (see (6.9)), there is a strictly positive probability that the LS criterion $J_N(\cdot)$ has local minimizers that differ from the global one; see Demidenko (1989, 2000).

In the extreme situation where the expectation surface $\mathbb{S}_\eta$ is curved and $\mathbf{y}$ is close to its center of curvature, small perturbations of $\mathbf{y}$ may change drastically the values of $\eta[\hat{\theta}^N_{LS}(\mathbf{y})]$ and $\hat{\theta}^N_{LS}(\mathbf{y})$. If such a situation may occur with non-negligible probability for a design X, then either we should be sure to repeat observations at X a large enough number of times, see Remarks 6.1 and 6.3, or we should use another design.

Although it is intuitively clear that local minima will not exist if (i) the feasible domain Θ is not too large and (ii) the observations $\mathbf{y}$ are sufficiently close to the expectation surface $\mathbb{S}_\eta$, few precise results exist in this domain. The following property, stated in (Chavent, 1983), defines a tube $\mathcal{T}$ around $\mathbb{S}_\eta$ such that, for any $\mathbf{y} \in \mathcal{T}$, the LS criterion (7.3) has a unique global minimizer and no other local minimizer. See also Chavent (1987); more precise developments are presented in (Chavent, 1990, 1991).

Theorem 7.3. *Assume that $\eta(\cdot)$ is twice continuously differentiable in the interior of Θ, a compact subset of $\mathbb{R}^p$, and that $\partial\eta^\top(\theta)/\partial\theta$ has full rank p for any $\theta \in \mathrm{int}(\Theta)$. Define*

$$\alpha_\eta = \alpha_\eta(X,\Theta) = \left(\min_{\theta\in\Theta} \lambda_{\min}[\mathbf{M}_X(\theta)]\right)^{1/2} \tag{7.5}$$

and

$$\beta_\eta = \beta_\eta(X,\Theta) = \max_{\theta\in\Theta} \max_{\mathbf{u}\in\mathbb{R}^p,\ \|\mathbf{u}\|=1} \left(\sum_{i=1}^{N} \left[\mathbf{u}^\top \frac{\partial^2\eta(x_i,\theta)}{\partial\theta\partial\theta^\top}\,\mathbf{u}\right]^2\right)^{1/2} . \tag{7.6}$$

Assume that

$$\mathrm{diam}(\Theta) < 2\sqrt{2}\,\frac{\alpha_\eta}{\beta_\eta} \tag{7.7}$$

with $\text{diam}(\Theta) = \max_{\theta,\theta' \in \Theta} \|\theta' - \theta\|$. *Then, for any* $\mathbf{y}$ *such that*

$$d(\mathbf{y}, \mathbb{S}_\eta) = \min_{\theta\in\Theta} \|\mathbf{y} - \eta(\theta)\| = \sqrt{N} \min_{\theta\in\Theta} \sqrt{J_N(\theta)} < \frac{\alpha_\eta^2}{\beta_\eta} - \frac{\beta_\eta}{8} [\text{diam}(\Theta)]^2 \quad (7.8)$$

the LS criterion $J_N(\cdot)$ *given by (7.3) has a unique global minimizer* $\hat{\theta}_{LS}^N(\mathbf{y})$ *and no other local minimizer in* Θ*, with* $\hat{\theta}_{LS}^N(\mathbf{y})$ *depending continuously on* $\mathbf{y}$.

Convexity of the LS Criterion

From (7.7), local minimizers of the LS criterion may exist in Θ when its diameter is large. In general, having determined a minimizer $\hat{\theta}^N$ in a given set Θ, one may thus still wonder if it corresponds to the global optimum. The following property (Demidenko, 2000) gives a partial answer. The proof, based on simple convexity arguments, is reproduced below. Define the *total radius of curvature* of the model at θ as

$$R_\eta(\theta) = R_\eta(X, \theta) = \inf_{\mathbf{u}\in\mathbb{R}^p - \{\mathbf{0}\}} \frac{\mathbf{u}^\top \mathbf{M}_X(\theta)\mathbf{u}}{\left(\sum_{i=1}^N \left[\mathbf{u}^\top \frac{\partial^2 \eta(x_i,\theta)}{\partial\theta\partial\theta^\top} \mathbf{u}\right]^2\right)^{1/2}} . \quad (7.9)$$

Notice that $\alpha_\eta^2/\beta_\eta \leq \underline{R}_\eta = \min_{\theta\in\Theta} R_\eta(\theta)$.

Theorem 7.4. *Assume that* $\eta(\cdot)$ *is twice continuously differentiable in the interior of* Θ*, a compact set of* $\mathbb{R}^p$. *Let* $\hat{\theta}^N$ *be a local minimizer for the criterion (7.3). If*

$$N\, J_N(\hat{\theta}^N) < \min_{\theta\in\Theta} R_\eta^2(\theta) = \underline{R}_\eta^2 \quad (7.10)$$

and

$$\Theta_\eta = \{\theta \in \Theta : N\, J_N(\theta) < \underline{R}_\eta^2\} \text{ is a convex set}, \quad (7.11)$$

then $\hat{\theta}^N$ *is the global minimizer of* $J_N(\cdot)$. *Moreover,* $J_N(\cdot)$ *is convex on* Θ_η.

Proof. We have

$$\nabla_\theta^2 J_N(\theta) = \frac{\partial^2 J_N(\theta)}{\partial\theta\partial\theta^\top} = \frac{2}{N}\mathbf{M}_X(\theta) - \frac{2}{N}\sum_{i=1}^N [y(x_i) - \eta(x_i,\theta)] \frac{\partial^2\eta(x_i,\theta)}{\partial\theta\partial\theta^\top} .$$

Therefore, for any $\mathbf{u} \in \mathbb{R}^p$,

$$\begin{aligned}
\frac{1}{2}\mathbf{u}^\top \nabla_\theta^2 J_N(\theta)\mathbf{u} &\geq \frac{1}{N}\mathbf{u}^\top \mathbf{M}_X(\theta)\mathbf{u} - \frac{1}{N}\left(\sum_{i=1}^N [y(x_i) - \eta(x_i,\theta)]^2\right)^{1/2} \\
&\quad \times \left(\sum_{i=1}^N \left[\mathbf{u}^\top \frac{\partial^2\eta(x_i,\theta)}{\partial\theta\partial\theta^\top}\mathbf{u}\right]^2\right)^{1/2} \\
&= \frac{1}{N}\mathbf{u}^\top \mathbf{M}_X(\theta)\mathbf{u} - J_N^{1/2}(\theta)\left(\frac{1}{N}\sum_{i=1}^N \left[\mathbf{u}^\top \frac{\partial^2\eta(x_i,\theta)}{\partial\theta\partial\theta^\top}\mathbf{u}\right]^2\right)^{1/2}
\end{aligned}$$

and $N\,J_N(\theta) < \underline{R}_\eta^2$ implies $\mathbf{u}^\top \nabla_\theta^2 J_N(\theta)\mathbf{u} > 0$ so that $J_N(\cdot)$ is convex on Θ_η. If $\hat{\theta}^N$ is a local minimizer of $J_N(\cdot)$ that belongs to Θ_η, when Θ_η is convex, $\hat{\theta}^N$ is necessarily the global minimizer of $J_N(\cdot)$. ■

One may notice that the conditions (7.10), (7.11) of Theorem 7.4 can be replaced by $\Theta_\eta(t) = \{\theta \in \Theta : N\,J_N(\theta) < t\}$ is a convex set for some t satisfying $N\,J_N(\hat{\theta}^N) < t \leq \underline{R}_\eta^2$.

The two theorems above somewhat complete each other: under the assumption that the model structure used for estimation is correct, i.e., $\mathbb{E}\{\mathbf{y}\} = \eta(\bar{\theta})$ for some $\bar{\theta}$, Theorem 7.3 can be used to bound the probability that the LS criterion will have a unique global minimizer and no other local minimizer on a given set Θ; Theorem 7.4 gives a sufficient condition for a local minimum to be the global one, once the data have been collected.

However, some issues remain open. A first reason is that the computation of α_η, β_η, or $\underline{R}_\eta$ is not easy. An algorithm for computing $R_\eta(\theta)$ can be found in (Bates and Watts, 1980) and (Demidenko, 2000), and a lower bound on $\underline{R}_\eta$ can be obtained analytically in particular examples; see Demidenko (2000). A second reason, even more serious, is that the results of both theorems often cover limited regions of the parameter space only. Theorem 7.3 puts a condition on diam(Θ) (see (7.7)) that can be very conservative; the condition (7.11) of Theorem 7.4 is generally extremely difficult to check, and in general situations, the set Θ_η can be non-convex or even disconnected. In fact the examples in (Demidenko, 2000) require a case-by-case specific construction of convex sets, as large as possible, on which the LS criterion is convex. The following examples, originated from (Demidenko, 2000), illustrate the difficulties. We call *convexity region* the set of $\mathbf{y} \in \mathbb{R}^N$ such that the function $\theta \longrightarrow J_N(\theta)$ is convex on Θ and *multimodality region* the set of $\mathbf{y}$ such that this function admits several local minimizers.

Example 7.5. Consider the following one-parameter model:

$$\eta(\mathbf{x}, \theta) = \theta\{\mathbf{x}\}_1 + \theta^2\{\mathbf{x}\}_2$$

with $\Theta \subset \mathbb{R}$ and two observations at the design points $\mathbf{x}_1 = (1, 0)$, $\mathbf{x}_2 = (0, 1)$. Direct calculations give

$$\mathbf{M}_X(\theta) = 2\,R_\eta(\theta) = 1 + 4\theta^2\,,\ \underline{R}_\eta = \frac{1}{2}\,,\ \alpha_\eta = 1\,,\ \beta_\eta = 2\,.$$

Also, the intrinsic curvature defined by (6.9) equals

$$C_{int}(X, \theta) = \frac{2}{(1 + 4\theta^2)^{3/2}}$$

and is maximum at $\theta = 0$. The expectation surface $\mathbb{S}_\eta = \{(\theta, \theta^2)^\top : \theta \in \mathbb{R}\}$ is shown in Fig. 7.2 (solid-line parabola). The LS criterion (7.3) is a convex function of θ for $y_2 < 1/2$; the derivative $\nabla_\theta J_2(\theta)$ has three real roots, and $J_2(\theta)$

has thus two minima for $y_2 > 1/2+(3/4)(2|y_1|)^{2/3}$, $y_1 \in \mathbb{R}$. The corresponding regions of convexity and multimodality in the (y_1, y_2) plane are presented on the figure. Consider the tubes $\mathcal{T}(\epsilon) = \{\mathbf{y} : d(\mathbf{y}, \mathbb{S}_\eta) \leq \epsilon\}$ around $\mathbb{S}_\eta$. The condition $\epsilon \leq 1/2$ is necessary to guarantee that the LS criterion has a unique global minimizer for any $\mathbf{y} \in \mathcal{T}(\epsilon)$. Indeed, for any $\epsilon' > 1/2$, $\mathbf{y} = (0, \epsilon')^\top$ belongs to the multimodality region, and the LS criterion $J_2(\theta)$ has then two global minimizers at $\pm\sqrt{\epsilon' - 1/2}$ (note that the probability measure of the set of such points is zero according to Theorem 7.2).

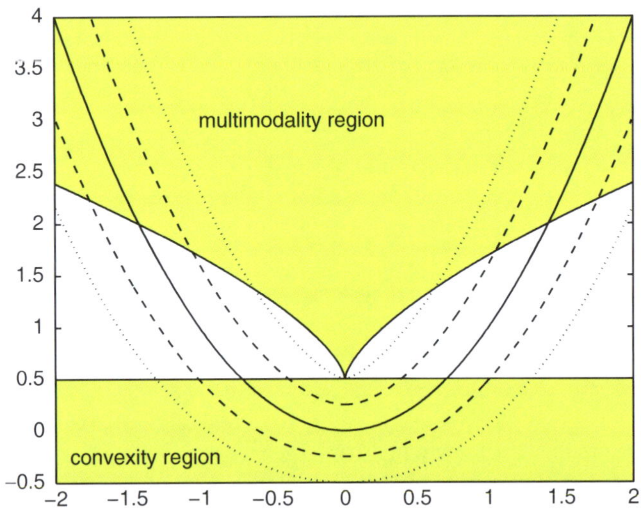

Fig. 7.2. Expectation surface (*solid-line parabola*), regions of convexity and multimodality of the LS criterion in the (y_1, y_2) plane (*colored*), and limits on $\mathbf{y}$ given by (7.8) for $\text{diam}(\Theta) = 1$ (*dashed line*) and $\text{diam}(\Theta) = 0$ (*dotted line*)

The bound on $d(\mathbf{y}, \mathbb{S}_\eta)$ given by (7.8) equals $b = b(\Theta) = 1/2 - [\text{diam}(\Theta)]^2/4$, which, for any value of $\text{diam}(\Theta)$, defines a tube $\mathcal{T}(b)$ around $\mathbb{S}_\eta$ (see the dashed-line parabolas on Fig. 7.2 obtained for $\text{diam}(\Theta) = 1$).

As we shall show now, this example illustrates that the bound (7.8) is rather pessimistic. Indeed, let $\hat{\theta}_1$ and $\hat{\theta}_2$ denote the two minimizers of $J_2(\theta)$ when $\mathbf{y}$ belongs to the multimodality region. One can check that $|\hat{\theta}_2 - \hat{\theta}_1|$ is minimum when $\mathbf{y}$ is on the boundary of the region, i.e., when it satisfies $y_2 = 1/2 + (3/4)(2|y_1|)^{2/3}$. The two minimizers are then $\hat{\theta}_1 = (2y_1)^{1/3}$ and $\hat{\theta}_2 = -(y_1/4)^{1/3}$, and the minimum of $|\hat{\theta}_2 - \hat{\theta}_1|$ when $d(\mathbf{y}, \mathbb{S}_\eta) \leq 1/2$ is obtained at $\mathbf{y} \simeq (0.6222\,,\ 1.3677)^\top$ and approximately equals $\Delta = 1.6134$. This means that for any Θ such that $\text{diam}(\Theta) < \Delta$ and any $\mathbf{y}$ such that $d(\mathbf{y}, \mathbb{S}_\eta) < 1/2$, there is one unique minimizer of $J_2(\theta)$ in Θ, a situation much more favorable than indicated by the bound (7.8).

Consider now the condition given by (7.10). One can check that when $\mathbf{y}$ is in the multimodality region, $J_2(\hat{\theta}_i) \geq \underline{R}_\eta^2/2 = 1/8$ for at least one of the two minimizers $\hat{\theta}_1, \hat{\theta}_2$. Any minimizer $\hat{\theta}_i$ satisfying (7.10) is thus the global minimizer of $J_2(\theta)$. The results given in Theorem 7.4 are more partial than that. The set Θ_η defined in (7.11) is not empty when $(y_1-\theta)^2+(y_2-\theta^2)^2 < 1/4$ for some θ, which defines the tube $\mathcal{T}(1/2)$ around $\mathbb{S}_\eta$; see the dotted parabolas on Fig. 7.2. Theorem 7.4 only covers the situation where $\mathbf{y}$ lies between these lines: it indicates that the local minimizer $\hat{\theta}$ such that $\|\mathbf{y} - \eta(\hat{\theta})\| < 1/2$ is the global one and that the LS criterion is convex for θ in the set $\Theta_\eta = \{\theta : \|\mathbf{y} - \eta(\theta)\| < 1/2\}$. One may notice that convexity is in fact satisfied in a larger set. Indeed, for any $\mathbf{y}$, the LS criterion is convex for θ in $\mathcal{C}(\mathbf{y}) = \{\theta : \theta^2 > (2y_2 - 1)/6\}$, whereas $\theta \in \Theta_\eta$ implies $\theta^2 > y_2 - 1/2$, so that $\Theta_\eta \subset \mathcal{C}(\mathbf{y})$. □

Example 7.6. We modify previous example by changing $\eta(\mathbf{x}, \theta)$ for negative θ, $\eta(\mathbf{x}, \theta) = (\theta\{\mathbf{x}\}_1 + \theta^2\{\mathbf{x}\}_2)\, \mathbb{I}_{\mathbb{R}^+}(\theta) + (\sin(\theta)\{\mathbf{x}\}_1 + 2[1 - \cos(\theta)]\{\mathbf{x}\}_2)\, \mathbb{I}_{\mathbb{R}^-}(\theta)$, with $\Theta = [\gamma, \infty)$, $\gamma > -2\pi$. Again, we make two observations at the design points $\mathbf{x}_1 = (1, 0)$ and $\mathbf{x}_2 = (0, 1)$, and $\eta(\theta)$ is twice continuously differentiable in the interior of Θ. We get $\mathbf{M}_X(\theta) = \cos^2\theta + 4\sin^2\theta$ for $\theta \leq 0$, with the same values as in Example 7.5 for α_η, β_η, and $\underline{R}_\eta$. For $\theta \leq 0$, the intrinsic curvature (6.9) is now

$$C_{int}(X, \theta) = \frac{2}{(4 - 3\cos^2\theta)^{3/2}}$$

and is maximum at $\theta = 0$. The situation is thus not worse than in Example 7.5 in terms of intrinsic curvature of the model.

Figure 7.3 presents the expectation surface $\mathbb{S}_\eta$. Note that $\gamma > -2\pi$ implies that the model is globally LS estimable at any $\theta \in \Theta$ for this design: to any $\mathbf{z} \in \mathbb{S}_\eta$ corresponds a unique $\theta \in \Theta$ such that $\mathbf{z} = \eta(\theta)$.

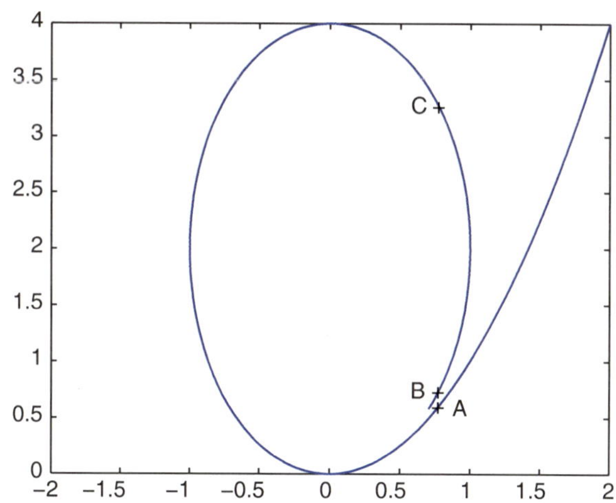

Fig. 7.3. Expectation surface $\mathbb{S}_\eta$ in Example 7.6 for $\gamma = -5.5$

Although the expectation surface $\mathbb{S}_\eta$ almost overlaps, the situation is similar to previous example concerning Theorem 7.3: when $\mathbf{y}$ belongs to a tube $\mathcal{T}(\epsilon) = \{\mathbf{y} : d(\mathbf{y}, \mathbb{S}_\eta) \leq \epsilon\}$ with $\epsilon < 1/2$, if there exist two minimizers $\hat{\theta}_1$ and $\hat{\theta}_2$, they are well separated: $|\hat{\theta}_1 - \hat{\theta}_2|$ is larger than some computable bound. At the same time, if γ is close enough to -2π, we can find observations $\mathbf{y}$ that yield two local minimizers $\hat{\theta}_1$ and $\hat{\theta}_2$ such that $J_2(\hat{\theta}_1)$ and $J_2(\hat{\theta}_2)$ are smaller than $\underline{R}_\eta^2/2$ (and $J_2(\hat{\theta}_1)$, $J_2(\hat{\theta}_2)$ can be arbitrarily close if γ approaches -2π), that is, condition (7.11) of Theorem 7.4 may not be satisfied. The reason is that Θ_η can be disconnected. □

Examples 7.5 and 7.6 show that the information provided by the curvature of the model is clearly not enough to measure the difficulty of the estimation of its parameters caused by an expectation surface that folds over itself. It is therefore extremely important to keep in view the possibility that $\mathbb{S}_\eta$ may overlap when choosing the experimental design X, i.e., before collecting the observations. In the next section we construct a function that gives a quantitative information on the LS estimability of the model for a given design X.

7.4 Estimability Function

7.4.1 Definition

Define the (local) *estimability function at* θ

$$E_{\eta,\theta}(\cdot) : \delta \in \mathbb{R}^+ \longrightarrow E_{\eta,\theta}(\delta) = \min_{\theta' \in \Theta,\, \|\theta' - \theta\|^2 = \delta} \|\eta(\theta') - \eta(\theta)\|^2 \tag{7.12}$$

and the (global) estimability function

$$E_\eta(\cdot) : \delta \in \mathbb{R}^+ \longrightarrow E_\eta(\delta) = \min_{\theta \in \Theta} E_{\eta,\theta}(\delta) = \min_{(\theta,\theta') \in \Theta^2,\, \|\theta' - \theta\|^2 = \delta} \|\eta(\theta') - \eta(\theta)\|^2$$

with the minimum over an empty set taken as $+\infty$. Notice that $E_{\eta,\theta}(\delta)$ and $E_\eta(\delta)$ depend on the design X. Although $E_{\eta,\theta}(\cdot)$ is defined at a particular θ, its construction involves the consideration of $\|\eta(\theta') - \eta(\theta)\|$ for θ' far from θ; it therefore carries information on the global LS estimability of the model; see in particular Theorem 7.10. Note that by repeating n times the experiment characterized by X, we multiply $E_{\eta,\theta}(\delta)$ and $E_\eta(\delta)$ by n. Also notice that $E_{\eta,\theta}(\cdot)$ and $E_\eta(\cdot)$ are one-dimensional curves that can be plotted whatever the values of N and $p = \dim(\theta)$, whereas the expectation surface can only be visualized for $p = 1$ or 2 and $N \leq 3$. An example with $p = 3$ and $N = 16$ will be given in Sect. 7.7.4.

7.4.2 Properties

Relation with *E*-optimality

When the model is linear in θ with $\Theta = \mathbb{R}^p$ and $\eta(\theta) = \mathbf{F}(X)\theta + \mathbf{v}(X)$, the function $E_\eta(\delta)$ is linear,

$$E_\eta(\delta) = \lambda_{\min}[\mathbf{F}^\top(X)\mathbf{F}(X)]\,\delta = \lambda_{\min}(\mathbf{M}_X)\,\delta\,,$$

which is related to the criterion of E-optimality; see Sect. 5.1.2.

For nonlinear models, we can use a second-order expansion of $\eta(\theta')$ for θ' close to $\theta \in \text{int}(\Theta)$,

$$\begin{aligned}\eta(\theta') &= \eta(\theta) + (\theta' - \theta)^\top \frac{\partial \eta(\theta)}{\partial \theta^\top} \\ &\quad + \frac{1}{2} \sum_{i,j=1}^{p} (\theta' - \theta)_i\, (\theta' - \theta)_j\, \frac{\partial^2 \eta(\theta)}{\partial \theta_i \partial \theta_j} + \mathcal{O}(\|\theta' - \theta\|^3)\,,\end{aligned}$$

to obtain

$$\|\eta(\theta') - \eta(\theta)\|^2 = (\theta' - \theta)^\top \mathbf{M}_X(\theta)(\theta' - \theta) + R_X(\theta, \theta') + \mathcal{O}(\|\theta' - \theta\|^4)\,,$$

where

$$R_X(\theta, \theta') = \sum_{i,j,k=1}^{p} (\theta' - \theta)_i\, (\theta' - \theta)_j\, (\theta' - \theta)_k\, \frac{\partial \eta^\top(\theta)}{\partial \theta_i}\, \frac{\partial^2 \eta(\theta)}{\partial \theta_j \partial \theta_k}\,.$$

For $\|\theta' - \theta\|^2 = \delta$, we get

$$\|\eta(\theta') - \eta(\theta)\|^2 = \delta\, \mathbf{u}^\top \mathbf{M}_X(\theta)\mathbf{u} + \delta^{3/2}\, A_X(\theta; \mathbf{u}) + \mathcal{O}(\delta^2)\,, \tag{7.13}$$

with $\mathbf{u} = (\theta' - \theta)/\|\theta' - \theta\|$ and

$$A_X(\theta; \mathbf{u}) = \sum_{i,j,k=1}^{p} u_i\, u_j\, u_k\, \frac{\partial \eta^\top(\theta)}{\partial \theta_i}\, \frac{\partial^2 \eta(\theta)}{\partial \theta_j \partial \theta_k}\,.$$

Therefore, for small δ,

$$\begin{aligned}E_{\eta,\theta}(\delta) &= \lambda_{\min}[\mathbf{M}_X(\theta)]\,\delta + \mathcal{O}(\delta^{3/2})\,, \qquad (7.14)\\ E_\eta(\delta) &= \min_{\theta \in \Theta} \lambda_{\min}[\mathbf{M}_X(\theta)]\,\delta + \mathcal{O}(\delta^{3/2})\,,\end{aligned}$$

and more precisely

$$E_{\eta,\theta}(\delta) = \lambda_{\min}[\mathbf{M}_X(\theta)]\,\delta + A_X(\theta)\,\delta^{3/2} + \mathcal{O}(\delta^2) \tag{7.15}$$

for some scalar $A_X(\theta)$ depending on X and θ.

Remark 7.7. The term $A_X(\theta; \mathbf{u})$ can be rewritten as

$$A_X(\theta; \mathbf{u}) = \sum_{i,j,k=1}^{p} u_i\, u_j\, u_k\, \frac{\partial \eta^\top(\theta)}{\partial \theta_i}\, \mathbf{P}_\theta\, \frac{\partial^2 \eta(\theta)}{\partial \theta_j \partial \theta_k}\,,$$

with $\mathbf{P}_\theta$ the projector defined by (6.4). Using Cauchy–Schwarz inequality, we obtain $|A_X(\theta;\mathbf{u})| \leq [\mathbf{u}^\top \mathbf{M}_X(\theta)\mathbf{u}]^{3/2}\, C_{par}(X,\theta;\mathbf{u})$, with $C_{par}(X,\theta;\mathbf{u})$ defined by (6.11), and therefore

$$\max_{\|\mathbf{u}\|=1} |A_X(\theta;\mathbf{u})| \leq \lambda_{\max}^{3/2}[\mathbf{M}_X(\theta)]\, C_{par}(X,\theta)\,,$$

with $C_{par}(X,\theta) = \sup_{\mathbf{u}\in\mathbb{R}^p-\{\mathbf{0}\}} C_{par}(X,\theta;\mathbf{u})$ the parametric curvature of the model. The major effect of nonlinearity on $E_{\eta,\theta}(\delta)$ for small δ is thus parametric and related to $C_{par}(X,\theta)$. □

Remark 7.8. Suppose that $\eta(\cdot)$ is continuous in θ. $E_{\eta,\theta}(\cdot)$ is then a continuous function of δ if a minimizing point $\theta'(\delta)$ in (7.12) belongs to $\mathrm{int}(\Theta)$—the case when all points $\theta'(\delta)$ are on the boundary of Θ is more delicate and may lead to left or right discontinuity of $E_{\eta,\theta}(\cdot)$. Similarly, denote by $(\theta(\delta),\theta'(\delta))$ a minimizing pair (θ,θ') in $E_\eta(\cdot)$; $E_\eta(\cdot)$ is then continuous in δ when $\theta(\delta)$ or $\theta'(\delta)$ belong to $\mathrm{int}(\Theta)$. When Θ is convex, $E_\eta(\delta)$ is thus continuous at any $\delta < \mathrm{diam}(\Theta)$.

$E_{\eta,\theta}(\cdot)$ is lower semicontinuous since its lower sets $\mathcal{L}(\alpha) = \{\delta \in \mathbb{R}^+ : E_{\eta,\theta}(\delta) \leq \alpha\}$ are closed for all α. Indeed, $\mathcal{L}(\alpha) = \{\delta \in \mathbb{R}^+ : \mathcal{D}_\theta(\alpha) \neq \varnothing\}$ where $\mathcal{D}_\theta(\alpha) = \{(\theta',\delta) \in \Theta \times \mathbb{R}^+ : \|\theta'-\theta\|^2 = \delta \text{ and } \|\eta(\theta')-\eta(\theta)\|^2 \leq \alpha\}$ is a closed set. $E_\eta(\cdot)$ is lower semicontinuous from similar arguments. As a consequence, $E_{\eta,\theta}(\cdot)$ has a minimum on any compact subset of $\mathbb{R}^+$. Also, $\omega_\theta^* = \min_{\delta\in\mathbb{R}^+} E_{\eta,\theta}(\delta)/\delta$ is well defined (see (7.16)) and so is $\Phi_{eE}(X;\theta) = \min_{\delta\in\mathbb{R}^+} E_{\eta,\theta}(\delta)(K+1/\delta)/N$ for some $K>0$, see Sect. 7.7.1. □

Relation with the Localization of the LS Estimator

The model is locally LS estimable at θ for X if $E_{\eta,\theta}(\delta) > 0$ for all $\delta \in (0,\Delta)$ for some $\Delta > 0$; it is globally LS estimable at θ for X if $E_{\eta,\theta}(\delta) > 0$ for all $\delta > 0$. Similarly, the function $E_\eta(\cdot)$ provides information about the global LS estimability for the design X (see Sect. 7.2), the model being globally LS estimable for X when $E_\eta(\delta) > 0$ for any $\delta > 0$. We shall see in Example 7.14 that $E_\eta(\delta)$ may be null for all δ in some interval $\mathcal{I}_\eta$ when the model is only locally LS estimable for X.

More importantly, we show hereafter that the functions $E_{\eta,\theta}(\cdot)$ and $E_\eta(\cdot)$ also provide information on the numerical properties of estimators, in connection with Theorems. 7.3 and 7.4. The first part of the results given below (Lemma 7.9) relates the estimability function $E_\eta(\cdot)$ to the results in Theorem 7.4. Those results are rather of the negative sort: in particular, it is shown that the set Θ_η defined by (7.11) is disconnected with positive probability when there exists $\delta' < \delta$ with $E_\eta(\delta) < \underline{R}_\eta^2 < E_\eta(\delta')$. The second part (Theorem 7.10) is more positive: we show that, for any θ, the search for $\hat{\theta}_{LS}^N$ can be restricted to a ball centered at θ with radius related to $\|\mathbf{y}-\eta(\theta)\|$.

For any given $\mathbf{y} \in \mathbb{R}^N$, consider the sets

$$\Theta_\eta(t) = \{\theta \in \Theta : \|\mathbf{y} - \eta(\theta)\|^2 < t\}\,, \; t > 0\,.$$

Note that $\Theta_\eta(t)$ depends on X and $\mathbf{y}$ and that the set Θ_η of Theorem 7.4 corresponds to $t = \underline{R}_\eta^2$. The following lemma relates the size and connectivity of $\Theta_\eta(t)$ to properties of the function $E_\eta(\cdot)$ and gives indications on the possibility for condition (7.11) not to be satisfied. The proof is given in Appendix C.

Lemma 7.9. *Assume that $\eta(\theta)$ is continuous for $\theta \in \Theta$, a compact subset of $\mathbb{R}^p$. We have:*

(i) For any $\theta, \theta' \in \Theta_\eta(t)$, $E_\eta(\|\theta - \theta'\|^2) < 4t$, and the maximum diameter $\overline{D}(t)$ of any connected part of $\Theta_\eta(t)$ satisfies $\overline{D}^2(t) \le \inf\{\delta : E_\eta(\delta) \ge 4t\}$.
(ii) Suppose that the probability measure of the observations $\mathbf{y}$ has a density with respect to the Lebesgue measure in $\mathbb{R}^N$. If there exists $\delta' < \delta$ such that $E_\eta(\delta) < t < E_\eta(\delta')$, then the probability that the set $\Theta_\eta(t)$ is not connected is strictly positive.

Next theorem shows that the calculation of

$$\omega_\theta^* = \min_{\delta \in \mathbb{R}^+} \frac{E_{\eta,\theta}(\delta)}{\delta} = \min_{\theta' \in \Theta} \frac{\|\eta(\theta') - \eta(\theta)\|^2}{\|\theta' - \theta\|^2} \tag{7.16}$$

allows us to draw conclusions about the location of $\hat{\theta}_{LS}^N$, thereby extending the results of Theorem 7.3.

Theorem 7.10. *For any $\theta \in \Theta$, $\hat{\theta}_{LS}^N$ belongs to the set Θ_θ defined by*

$$\Theta_\theta = \Theta \cap \mathscr{B}(\theta, 2d_\theta/\sqrt{\omega_\theta^*})\,,$$

where $d_\theta = \|\mathbf{y} - \eta(\theta)\|$ denotes the distance between $\mathbf{y}$ and $\eta(\theta)$. Moreover, Θ_θ contains no other local minimizer of (7.3) when

$$d_\theta < d_\theta^* = \frac{1}{4\beta_{\eta,\theta}} \left[\sqrt{\omega_\theta^*(\omega_\theta^* + 8\alpha_{\eta,\theta}^2)} - \omega_\theta^*\right] \tag{7.17}$$

where $\alpha_{\eta,\theta} = \alpha_\eta(X, \Theta_\theta)$, $\beta_{\eta,\theta} = \beta_\eta(X, \Theta_\theta)$, and ω_θ^* are, respectively, given by (7.5), (7.6), and (7.16).

Proof. The global minimizer $\hat{\theta}_{LS}^N$ of (7.3) satisfies $\|\mathbf{y} - \eta(\hat{\theta}_{LS}^N)\| \le d_\theta$, so that

$$\hat{\theta}_{LS}^N \in \{\theta' \in \Theta : \|\mathbf{y} - \eta(\theta')\| \le d_\theta\} \subset \{\theta' \in \Theta : \|\eta(\theta') - \eta(\theta)\| \le 2d_\theta\}\,. \tag{7.18}$$

Now, from the definition of ω_θ^*, $\|\theta' - \theta\|^2 = \delta > 4d_\theta^2/\omega_\theta^*$ implies that $E_{\eta,\theta}(\delta) > 4d_\theta^2$, and thus $\|\eta(\theta') - \eta(\theta)\|^2 > 4d_\theta^2$. Therefore, $\hat{\theta}_{LS}^N \in \Theta_\theta$.

For the second part of the theorem we simply apply Theorem 7.3 to the set Θ_θ, which indicates that Θ_θ contains no other local optimizer of (7.3) when $d_\theta + 2\beta_{\eta,\theta} d_\theta^2/\omega_\theta^* < \alpha_{\eta,\theta}^2/\beta_{\eta,\theta}$, i.e., when d_θ satisfies (7.17). ∎

Remark 7.11. The results in Theorem 7.10 can be improved as follows. Define $\overleftarrow{E}_{\eta,\theta}(s)$ as the solution of the following optimization problem:

$$\overleftarrow{E}_{\eta,\theta}(s) = \max_{\theta' \in \Theta,\, \|\eta(\theta') - \eta(\theta)\|^2 \leq s} \|\theta' - \theta\|^2 \,.$$

Then, for all $\theta' \in \Theta$, $\|\theta' - \theta\|^2 > \delta_\theta^* = \overleftarrow{E}_{\eta,\theta}(4d_\theta^2)$ implies that $\|\eta(\theta') - \eta(\theta)\|^2 > 4d_\theta^2$. From (7.18), we can thus substitute $\Theta'_\theta = \Theta \cap \mathscr{B}(\theta, \sqrt{\delta_\theta^*})$ for Θ_θ in Theorem 7.10. Notice that $\delta_\theta^* \leq 4d_\theta^2/\omega_\theta^*$, so that $\Theta'_\theta \subseteq \Theta_\theta$, which gives a better localization of $\hat{\theta}^N_{LS}$. Moreover, $\alpha_\eta(X, \Theta'_\theta) \geq \alpha_\eta(X, \Theta_\theta)$ and $\beta_\eta(X, \Theta'_\theta) \leq \beta_\eta(X, \Theta_\theta)$ (see the definitions (7.5), (7.6)), which gives a larger upper bound of admissible d_θ in (7.17). However, this requires the computation $\overleftarrow{E}_{\eta,\theta}(4d_\theta^2)$, which is more difficult than that of ω_θ^* used in Theorem 7.10. □

Theorem 7.10 shows the potential interest of having the estimability function increasing as fast as possible at points θ where $\|\mathbf{y} - \eta(\theta)\|$ is small. In practice, we can select a nominal value θ^0, such that we think that $\mathbf{y}$ will not be too far from $\eta(\theta^0)$ and then use a local design approach and construct a design X that makes $E_{\eta,\theta_0}(\delta)$ increase fast when δ increases. This type of experimental design will be considered in Sect. 7.7.1.

One important feature of $E_\eta(\delta)$ is that it accounts for the global effect of the intrinsic curvature of the model for θ varying in Θ. Indeed, the shape of the function $E_\eta(\cdot)$ is very different depending on whether the curvature is high only locally or over a large portion of Θ, yielding quite different situations concerning the presence of local minimizers. This is illustrated below by a continuation of Examples 7.5 and 7.6.

Example 7.5 (continued). Straightforward calculation gives

$$E_{\eta,\theta}(\delta) = \delta \min\{1 + (2\theta + \sqrt{\delta})^2,\, 1 + (2\theta - \sqrt{\delta})^2\}\,,$$

and thus $E_\eta(\delta) = \delta$. The model is globally LS estimable for X. The maximum diameter $\overline{D}_\eta$ of Θ_η (with $\Theta_\eta = \Theta_\eta(t)$ for $t = \underline{R}_\eta^2$) equals 1 and is obtained for $\mathbf{y} = (0,\ 1/4)^\top$, which coincides with the bound $\sqrt{4\underline{R}_\eta^2}$ given by Lemma 7.9-(i). The value of ω_θ^* defined by (7.16) equals 1 for all θ. Using the bounds $\beta_{\eta,\theta} \leq \beta_\eta = 2$, $\alpha_{\eta,\theta} \geq \alpha_\eta = 1$, we get the underestimated value 1/4 for d_θ^* in (7.17). The balls $\mathscr{B}(\theta, 2d_\theta^*/\sqrt{\omega_\theta^*})$ thus have diameter at least 1. This indicates that if we find a θ such that $\|\mathbf{y} - \eta(\theta)\| < 1/4$, then we can be sure that $|\hat{\theta}^N_{LS} - \theta| \leq 1/2$ and that there is no other local minimum of (7.3) in $[\theta - 1/2, \theta + 1/2]$. Notice that Theorem 7.3 alone indicates that for any θ, there is a unique local minimizer of the LS criterion in $[\theta - 1/2, \theta + 1/2]$ when $\|\mathbf{y} - \eta(\theta)\| < 1/4$, but it does not guarantee that the global minimizer over $\Theta = \mathbb{R}$ is in $[\theta - 1/2, \theta + 1/2]$. □

Example 7.6 (continued). The behavior of the estimability function $E_{\eta,\theta}(\cdot)$ can be inferred directly from Fig. 7.3 and is presented on Fig. 7.4, in log scale, for $\theta = 1.5$ (solid line) and $\theta = 0.75$ (dashed line). The presence of local minimas for $\theta = 1.5$ indicates that the expectation surface is curved; the situation is more extreme for $\theta = 0.75$ (which corresponds to point A on Fig. 7.3), where the presence of a local minimum with small value indicates that the expectation surface is almost folded over itself. Note that $E_{\eta,\theta}(\cdot)$ is discontinuous at $\delta = (\theta - \gamma)^2$.

The global estimability function $E_\eta(\delta)$ is plotted in Fig. 7.5 for δ between 0 and 50, showing a local minimum at $\delta^* \simeq 39.007$ where $E_\eta(\delta^*) \simeq 2.32 \cdot 10^{-3}$. Parameter estimation is clearly more difficult in this case than in Example 7.5. Easy calculations indicate that $E_\eta(\delta) = 4\sin^2(\sqrt{\delta}/2)$ for $\delta \in [0, -2(\gamma + \pi)]$ when $\gamma < -\pi$ with $\Theta = [\gamma, \infty)$ and that, from Lemma 7.9, the maximum diameter of any connected part of Θ_η satisfies $\overline{D}_\eta \leq \pi/3$. When the probability measure of $\mathbf{y}$ has a density with respect to the Lebesgue measure, there is a strictly positive probability that Θ_η is not connected. □

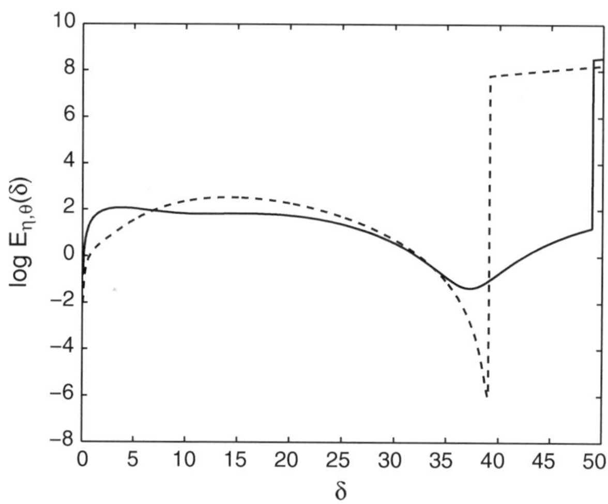

Fig. 7.4. Estimability function $E_{\eta,\theta}(\delta)$ (log scale) in Example 7.6 for $\theta = 1.5$ (*solid line*) and $\theta = 0.75$ (*dashed line*); $\Theta = [\gamma, \infty)$ with $\gamma = -5.5$

7.4.3 Replications and Design Measures

Denote by $X^{\otimes n}$ the design obtained by replicating n times each point of X. From Remark 6.3, the estimation problem with the design $X^{\otimes n}$ is equivalent to an estimation problem with the design X with, for $i = 1, \ldots, N$, the observation $\bar{y}(x_i)$ at x_i given by the empirical mean of the n observations at the same point. In this equivalent problem, the quantities α_η, β_η, and R_η are not modified, but the variance of $\bar{y}(x_i)$ is reduced by a factor n compared to var$[y(x_i)]$. Replications thus increase the probability that the LS criterion has a unique global minimizer and no other local minimizer; see (7.8) in Theorem 7.3. The benefit of replications is also revealed in Theorem 7.4 by considering $J_N(\theta) = (1/N)\sum_{i=1}^{N}[\bar{y}(x_i) - \eta(x_i,\theta)]^2$ for the replicated design.

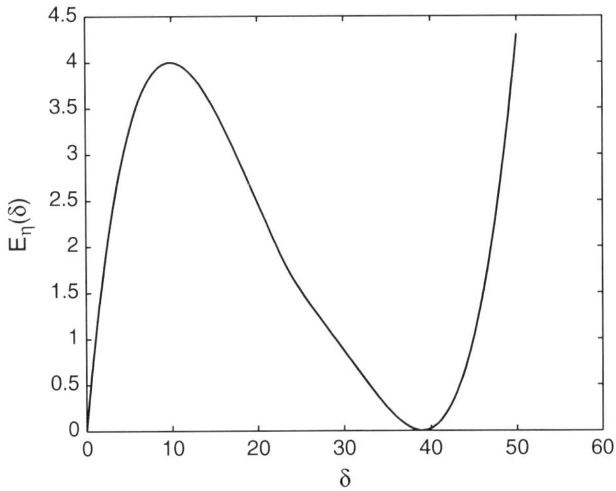

Fig. 7.5. Estimability function $E_\eta(\delta)$ in Example 7.6 for $\Theta = [\gamma, \infty)$ with $\gamma = -5.5$

The direct benefit of using $X^{\otimes n}$ in place of X is further evidenced by considering the estimability function $E_{\eta,\theta}(\delta)$. For small δ, the estimability function for the design X approximately equals $\delta\,\lambda_{\min}[\mathbf{M}_X(\theta)]$ (provided that the model is locally LS estimable at θ for X), see (7.14), and we have for the replicated design $X^{\otimes n}$

$$\lambda_{\min}[\mathbf{M}_{X^{\otimes n}}(\theta)] = n \lim_{\delta\to 0} E_{\eta,\theta}(\delta)/\delta\,.$$

Note that $\lambda_{\min}[\mathbf{M}_{X^{\otimes n}}(\theta)]$ can be related to the size of (asymptotic) confidence ellipsoids for θ when $n \to \infty$; see Sect. 5.1.1.

Since the estimability function for the replicated design $X^{\otimes n}$ depends linearly on n, the number of replications, it is natural to define estimability functions for design measures. Denoting $\|\eta(\cdot,\theta') - \eta(\cdot,\theta)\|_\xi^2 = \int_{\mathscr{X}}[\eta(x,\theta') - \eta(x,\theta)]^2\,\xi(\mathrm{d}x)$, with the estimability functions $E_{\eta,\theta}(\cdot)$, we associate a normalized version, denoted $E_{\eta,\theta}^{\xi}(\cdot)$, where the norm $\|\eta(\cdot,\theta') - \eta(\cdot,\theta)\|_\xi^2$ is used instead of $\|\eta(\theta') - \eta(\theta)\|$, that is,

$$E_{\eta,\theta}^{\xi}(\delta) = \min_{\theta'\in\Theta,\, \|\theta'-\theta\|^2=\delta} \|\eta(\cdot,\theta') - \eta(\cdot,\theta)\|_\xi^2\,. \qquad (7.19)$$

A model is globally LS estimable at θ for ξ, i.e., globally LS estimable at θ for the design X corresponding to the support of ξ, if $\theta' \in \Theta$ and $\|\eta(\cdot,\theta') -$

$\eta(\cdot, \theta)\|_\xi^2 = 0$ imply $\theta' = \theta$. An equivalent condition is thus $E_{\eta,\theta}^\xi(\delta) > 0$ for all $\delta > 0$. In the same way, the model is locally LS estimable at θ for ξ if $E_{\eta,\theta}^\xi(\delta) > 0$ for all $\delta \in (0, \Delta)$ for some $\Delta > 0$. Similarly to what was obtained in Sect. 7.4 (see (7.13)), we have

$$\|\eta(\cdot, \theta') - \eta(\cdot, \theta)\|_\xi^2 = \delta\, \mathbf{u}^\top \mathbf{M}(\xi, \theta)\mathbf{u} + \delta^{3/2}\, A(\xi, \theta; \mathbf{u}) + \mathcal{O}(\delta^2)\,, \tag{7.20}$$

with $\|\theta' - \theta\|^2 = \delta$, $\mathbf{u} = (\theta' - \theta)/\|\theta' - \theta\|$ and

$$A(\xi, \theta; \mathbf{u}) = \sum_{i,j,k=1}^{p} u_i\, u_j\, u_k \int_{\mathscr{X}} \frac{\partial^\top \eta(x, \theta)}{\partial \theta_i} \frac{\partial^2 \eta(x, \theta)}{\partial \theta_j \partial \theta_k}\, \xi(\mathrm{d}x)\,.$$

Therefore, for small δ,

$$E_{\eta,\theta}^\xi(\delta) = \lambda_{\min}[\mathbf{M}(\xi, \theta)]\, \delta + A(\xi, \theta)\, \delta^{3/2} + \mathcal{O}(\delta^2)\,,$$

where the scalar $A(\xi, \theta)$ depends on ξ and θ. Similarly to the developments in Remark 7.7, we also get that

$$\max_{\|\mathbf{u}\|=1} |A(\xi, \theta; \mathbf{u})| \le \lambda_{\max}^{3/2}[\mathbf{M}(\xi, \theta)]\, C_{par}(\xi, \theta)\,,$$

with

$$C_{par}(\xi, \theta) = \sup_{\mathbf{u} \in \mathbb{R}^p - \{\mathbf{0}\}} \frac{\|P_\theta \sum_{i,j=1}^p u_i [\partial^2 \eta(\cdot, \theta)/\partial \theta_i \partial \theta_j] u_j\|_\xi}{\mathbf{u}^\top \mathbf{M}(\xi, \theta)\mathbf{u}}$$

the parametric curvature of the model for the design measure ξ, where P_θ is the projector defined by (3.83).

7.4.4 Estimability for Parametric Functions

Consider the situation where we are interested in estimating a scalar function $h(\theta)$ of the model parameters. We shall suppose that $h(\cdot)$ satisfies assumption $\mathrm{H1}_h$ of Sect. 3.1.4, that is, $h(\cdot)$ is twice continuously differentiable in $\mathrm{int}(\Theta)$.

We say that $h(\cdot)$ is *globally LS estimable on Θ at θ for ξ* if

$$\theta' \in \Theta \text{ and } \|\eta(\cdot, \theta') - \eta(\cdot, \theta)\|_\xi^2 = 0 \Longrightarrow h(\theta') = h(\theta)\,,$$

in agreement with the estimability condition (3.12), and that $h(\cdot)$ is *locally LS estimable on Θ at θ for ξ* if there exists some $\epsilon > 0$ such that

$$\theta' \in \Theta\,, |h(\theta') - h(\theta)| < \epsilon \text{ and } \|\eta(\cdot, \theta') - \eta(\cdot, \theta)\|_\xi^2 = 0 \Longrightarrow h(\theta') = h(\theta)\,.$$

The definitions are similar for an exact design X, with the condition $\eta(\theta') = \eta(\theta)$ substituted for $\|\eta(\cdot, \theta') - \eta(\cdot, \theta)\|_\xi^2 = 0$. For a linear model with $\eta(\theta) = \mathbf{F}(X)\theta + \mathbf{v}(X)$ and a linear function of interest $h(\theta) = \mathbf{c}^\top\theta$, the condition of global LS estimability of $h(\cdot)$ is equivalent to $\mathbf{c} \in \mathcal{M}(\mathbf{M}_X)$ with $\mathbf{M}_X = \mathbf{F}^\top(X)\mathbf{F}(X)$, which is in agreement with the definition of the c-optimality criterion; see Sect. 5.1.2.

Consider the following generalizations of (7.12) and (7.19), respectively, for an exact design X and a design measure ξ:

$$E_{\eta,\theta,h}(\delta) = \min_{\theta'\in\Theta,\,|h(\theta')-h(\theta)|=\sqrt{\delta}} \|\eta(\theta') - \eta(\theta)\|^2 \,,$$

$$E^{\xi}_{\eta,\theta,h}(\delta) = \min_{\theta'\in\Theta,\,|h(\theta')-h(\theta)|=\sqrt{\delta}} \|\eta(\cdot,\theta') - \eta(\cdot,\theta)\|^2_{\xi} \,.$$

Only the case of design measures is considered below; similar developments can be made for exact designs.

The function $h(\cdot)$ is globally LS estimable on Θ at θ for ξ in the model $\eta(x,\cdot)$ if $E^{\xi}_{\eta,\theta,h}(\delta) > 0$ for all $\delta > 0$; when the model itself is globally LS estimable at θ for ξ, then $h(\cdot)$ is globally LS estimable at θ for ξ. In that case, using

$$|h(\theta') - h(\theta)|^2 = \epsilon \left(\mathbf{u}^\top \frac{\partial h(\theta)}{\partial\theta}\right)^2 + \epsilon^{3/2} \left(\mathbf{u}^\top \frac{\partial h(\theta)}{\partial\theta}\right) \left(\mathbf{u}^\top \frac{\partial^2 h(\theta)}{\partial\theta\partial\theta^\top}\mathbf{u}\right) + \mathcal{O}(\epsilon^2)$$

and (7.20) with $\epsilon = \|\theta' - \theta\|^2$ and $\mathbf{u} = (\theta' - \theta)/\sqrt{\epsilon}$, we obtain

$$E^{\xi}_{\eta,\theta,h}(\delta) = \frac{\delta}{\mathbf{c}^\top \mathbf{M}^{-1}(\xi,\theta)\mathbf{c}} + \mathcal{O}(\delta^{3/2}) \qquad (7.21)$$

where $\mathbf{c} = \partial h(\theta)/\partial\theta$.

When ξ is singular, $h(\cdot)$ is globally LS estimable on Θ at any θ for ξ if $h(\cdot)$ satisfies $\mathrm{H2}''_h$ (p. 44). Below is a simple example of a function $h(\cdot)$ that does not satisfy this condition and is not LS estimable.

Example 7.12. Consider the model $\eta(x,\theta) = x\,\theta_1^2 + \theta_2$, with $\theta = (\theta_1\,,\ \theta_2)^\top \in \mathbb{R}^2$, $x \in [-1,1]$, and the function of interest $h(\theta) = 2\theta_1 - \theta_2$. The Elfving's set $\mathscr{F}_{\theta^0}$ for $\theta^0 = (1\,,\ 1)^\top$ (see Sect. 5.2.4) corresponds to the rectangle $\{\mathbf{z} \in \mathbb{R}^2 : -2 \le z_1 \le 2 \text{ and } -1 \le z_2 \le 1\}$. Elfving's theorem (see Sect. 5.3.1) then indicates that the c-optimal design ξ_c^* for $h(\cdot)$ at θ^0 is the delta measure δ_{x_*} at $x_* = -1$. Figure 7.6 presents the sets $\{\theta \in \mathbb{R}^2 : \|\eta(\cdot,\theta) - \eta(\cdot,\theta^0)\|^2_{\xi_c^*} = 0\}$ and $\{\theta \in \mathbb{R}^2 : h(\theta) = h(\theta^0)\}$. For any $\delta > 0$, there exist two values of θ such that $h(\theta) = h(\theta^0) - \sqrt{\delta}$ and $\eta(x_*,\theta) = \eta(x_*,\theta^0)$. These two values tend to θ^0 as $\delta \to 0$. This means that $E^{\xi_c^*}_{\eta,\theta^0,h}(\delta) = 0$ and $h(\cdot)$ is not LS estimable, on any arbitrarily small ball centered at θ^0, at θ^0 for the c-optimal design $\xi_c^* = \delta_{x_*}$.

This estimability problem is due to the nonlinearity of $\eta(x,\theta)$ in θ_1. Indeed, consider the linearized version of $\eta(x,\theta)$ at θ^0, $\eta_L(x,\theta) = 2x(\theta_1 - 1) + \theta_2 - 1$. It satisfies $\eta_L(x_*,\theta) = \theta_2 - 2\theta_1 + 1 = 1 - h(\theta)$, so that $h(\cdot)$ is clearly LS estimable on $\mathbb{R}^2$ for ξ_c^* in this linearized model. □

The example above illustrates the fact that since c-optimum design is based on a linearization (here of the model response, but more generally of the model response and function of interest), in general it does not guarantee the LS estimability of $h(\cdot)$ due to the curvatures of the sets $\Theta_{\eta,\theta^0} = \{\theta \in$

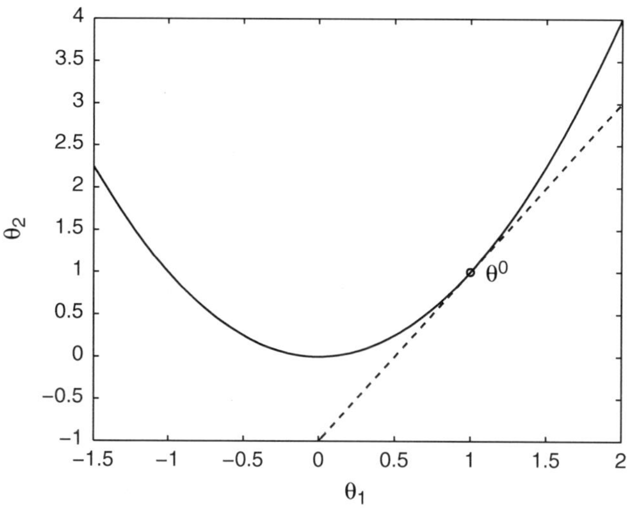

Fig. 7.6. $\{\theta \in \mathbb{R}^2 : \|\eta(\cdot,\theta) - \eta(\cdot,\theta^0)\|^2_{\xi^*_c} = 0\}$ (*solid parabola*) and $\{\theta \in \mathbb{R}^2 : h(\theta) = h(\theta^0)\}$ (*dashed line*)

$\mathbb{R}^p : \|\eta(\cdot,\theta) - \eta(\cdot,\theta^0)\|^2_{\xi} = 0\}$ and $\Theta_{h,\theta^0} = \{\theta \in \mathbb{R}^p : h(\theta) = h(\theta^0)\}$. In some rather pathological situations, however, it may happen that Θ_{η,θ^0} is included in Θ_{h,θ^0} although ξ is singular (with less than $p = \dim(\theta)$ support points) and $h(\theta)$ does not depend explicitly on $\eta(\cdot,\theta)$. Example 3.17 provides such a situation.

Example 3.17 (continued). We have $\eta(x,\theta) = \theta_1 x + \theta_2 x^2$ and $h(\theta) = -\theta_1/(2\theta_2)$.

Take any θ such that $\theta_2 \neq 0$ and any $x \neq x_* = -\theta_1/\theta_2$ and set $\xi = \delta_x$ the delta measure at x. When $\theta_1 + x\theta_2 \neq 2\theta_2\sqrt{\delta}$, the line $\{\theta' \in \mathbb{R}^2 : h(\theta') = h(\theta) + \sqrt{\delta}\}$ is not parallel to the line $\Theta_{\eta,\theta} = \{\theta' \in \mathbb{R}^2 : \eta(x,\theta') = \eta(x,\theta)\}$ and they both intersect at

$$\theta'_1 = \frac{(\theta_1 + x\theta_2)(\theta_1 - 2\theta_2\sqrt{\delta})}{\theta_1 + x\theta_2 - 2\theta_2\sqrt{\delta}}\,, \quad \theta'_2 = \frac{\theta_2(\theta_1 + x\theta_2)}{\theta_1 + x\theta_2 - 2\theta_2\sqrt{\delta}}\,.$$

When the two lines are parallel, i.e., when $\theta_1 + x\theta_2 = 2\theta_2\sqrt{\delta}$, then $\Theta_{\eta,\theta}$ intersects the line $\{\theta' \in \mathbb{R}^2 : h(\theta') = h(\theta) - \sqrt{\delta}\}$. Therefore, for any $\delta > 0$, there exists θ' such that $\eta(x,\theta') = \eta(x,\theta)$ and $|h(\theta') - h(\theta)| = \sqrt{\delta}$ so that $E^{\xi}_{\eta,\theta,h}(\delta) = 0$. In other words, $h(\cdot)$ is not LS estimable at θ for ξ, due to the fact that $\Theta_{\eta,\theta}$ is not included in $\Theta_{h,\theta} = \{\theta' \in \mathbb{R}^2 : h(\theta') = h(\theta)\}$.

Take now $x = x_* = -\theta_1/\theta_2$. The intersection between $\Theta_{\eta,\theta}$ and the two lines $\{\theta' \in \mathbb{R}^2 : |h(\theta') - h(\theta)| = \sqrt{\delta}\}$ is at $\theta' = \mathbf{0}$ for any δ; $h(\cdot)$ is thus globally

LS estimable on any compact set excluding $\mathbf{0}$ at θ for $\xi = \delta_{x_*}$. The reason is that $\Theta_{\eta,\theta} \subset \Theta_{h,\theta}$. However, $h(\cdot)$ is not LS estimable at any θ'' such that $\theta''_1 + x_*\theta''_2 \neq 0$; see Examples 3.13 and 3.17. □

If we discard pathological situations like the one in previous example, we are typically in one of the following two configurations:

1. $h(\cdot)$ depends directly on the model response on the support of some ξ (see H2$''_h$ (p. 44)); a typical case is when the function of interest is the response itself at some particular x_0 and $h(\theta) = \eta(x_0, \theta)$); $h(\cdot)$ is then globally LS estimable for ξ although ξ may be singular. In particular, this is the case in a linear model $\eta(x, \theta) = \mathbf{f}^\top(x)\theta$ with a linear function of interest $h(\theta) = \mathbf{c}^\top\theta$: $h(\cdot)$ remains globally LS estimable when ξ is singular but $\mathbf{c} \in \mathcal{M}[\mathbf{M}(\xi)]$ since $\mathbf{c} = \mathbf{M}(\xi)\mathbf{u}$ for some $\mathbf{u} \in \mathbb{R}^p$ and thus $h(\theta) = \int_{\mathscr{X}} \mathbf{u}^\top\mathbf{f}(x)\mathbf{f}^\top(x)\theta\,\xi(\mathrm{d}x) = \int_{\mathscr{X}} [\mathbf{u}^\top\mathbf{f}(x)]\,\eta(x, \theta)\,\xi(\mathrm{d}x)$.
2. $h(\cdot)$ does not depend directly on the model response. When ξ is singular (typically, it has less than p support points), there is no reason why the manifold $\Theta_{\eta,\theta} = \{\theta' \in \mathbb{R}^p : \|\eta(\cdot, \theta') - \eta(\cdot, \theta)\|^2_\xi = 0\}$ passing through θ, typically $(p-k)$-dimensional when ξ has k support points, should be included in the $(p-1)$-dimensional manifold $\Theta_{h,\theta} = \{\theta' \in \mathbb{R}^p : h(\theta') = h(\theta)\}$, and $h(\cdot)$ is not LS estimable at θ. When ξ is such that the two manifolds are tangent at θ, which is the case when ξ is a c-optimal design for $h(\cdot)$ at θ, a linearization of the model and function of interest may hide the difficulty; see Example 7.12.

Remark 7.13. When the interest is in some specific components $\{\theta\}_i$ of the parameter vector θ, i.e., $h(\theta) = \theta_i$ for some $i = 1, \ldots, \dim(\theta)$, for computational reasons one may prefer to use the following alternative definition:

$$E'_{\eta,\theta,i}(\delta) = \min_{\theta' \in \Theta_{\theta,i}(\delta)} \|\eta(\theta') - \eta(\theta)\|^2 ,$$

where $\Theta_{\theta,i}(\delta) = \{\theta' \in \Theta : \{\theta'\}_i = \{\theta\}_i \pm \sqrt{\delta} \text{ and } \{\theta'\}_j = \{\theta\}_j \text{ for } j \neq i\}$. Using the approximation (7.13) for small δ, we obtain $E'_{\eta,\theta,i}(\delta) = \delta\{\mathbf{M}_X(\theta)\}_{ii} + \mathcal{O}(\delta^{3/2})$—which is not related to any of the criteria of Chap. 5. In the case of a design measure we can define similarly

$$E'^{\xi}_{\eta,\theta,i}(\delta) = \min_{\theta' \in \Theta_{\theta,i}(\delta)} \|\eta(\cdot, \theta') - \eta(\cdot, \theta)\|^2_\xi \,. \tag{7.22}$$

Also, when the function of interest is linear in θ we can use

$$E'_{\eta,\theta,\mathbf{c}}(\delta) = \min_{\theta' \in \Theta_{\theta,\mathbf{c}}(\delta)} \|\eta(\theta') - \eta(\theta)\|^2 , \tag{7.23}$$

with $\Theta_{\theta,\mathbf{c}}(\delta) = \{\theta' \in \Theta : \theta' = \theta \pm \sqrt{\delta}\mathbf{c}/\|\mathbf{c}\|\}$, $\mathbf{c} \in \mathbb{R}^p$, which gives $E'_{\eta,\theta,\mathbf{c}}(\delta) = \delta\,\mathbf{c}^\top\mathbf{M}_X(\theta)\mathbf{c}/(\mathbf{c}^\top\mathbf{c}) + \mathcal{O}(\delta^{3/2})$ for small δ. For a design measure we can define

$$E'^{\xi}_{\eta,\theta,\mathbf{c}}(\delta) = \min_{\theta' \in \Theta_{\theta,\mathbf{c}}(\delta)} \|\eta(\cdot, \theta') - \eta(\cdot, \theta)\|^2_\xi \,, \ \mathbf{c} \in \mathbb{R}^p \,. \tag{7.24}$$

Notice that $E'_{\eta,\theta,i}(\cdot)$ and $E'_{\eta,\theta,\mathbf{c}}(\cdot)$ can easily be computed for any p since each one of the sets $\Theta_{\theta,i}(\delta)$ and $\Theta_{\theta,\mathbf{c}}(\delta)$ contains at most two points; see Sect. 7.7.4 for examples. □

7.5 An Extended Measure of Intrinsic Nonlinearity

As illustrated in Examples 7.5 and 7.6, a major source of difficulties for the determination of the LS estimator (existence of local minimizers of the LS criterion, instability with respect to small perturbations of the observations) comes from the intrinsic curvature $C_{int}(X,\theta)$ of the model, i.e., the curvature of $\mathbb{S}_\eta$. At the same time, controlling the value of $C_{int}(X,\theta)$ is not enough, as long as it can be positive: for instance, the curvatures are identical in Examples 7.5 and 7.6 although the situation is clearly more difficult in the latter. In this section we present an extended measure of intrinsic nonlinearity that takes the global behavior of the model more deeply into account.

The intrinsic curvature defined by (6.9) can be extended as follows. We define

$$K_{int,\alpha}(X,\theta) = \max_{\theta' \in \mathrm{int}(\Theta)} K_{int,\alpha}(X,\theta;\theta') \tag{7.25}$$

with

$$K_{int,\alpha}(X,\theta;\theta') = 2\,\frac{\|[\mathbf{I}_N - \mathbf{P}_\theta][\eta(\theta)-\eta(\theta')]\|}{\|\eta(\theta)-\eta(\theta')\|^2}\left(\frac{(\theta-\theta')^\top \mathbf{M}_X(\theta)(\theta-\theta')}{\|\eta(\theta)-\eta(\theta')\|^2}\right)^\alpha,$$

for some $\alpha \in \mathbb{R}^+$, where $\mathbf{P}_\theta$ is the projector (6.4) and $\mathbf{M}_X(\theta)$ is the information matrix (7.4).

When the model is intrinsically linear (see Sect. 6.1.2), then $\eta(\theta) = \mathbf{F}\beta(\theta) + \mathbf{v}$ for some invertible continuously differentiable reparameterization (see (6.8)) and $[\mathbf{I}_N - \mathbf{P}_\theta][\eta(\theta)-\eta(\theta')] = \mathbf{0}$ for all θ,θ'. Conversely, if X is such that $\mathbf{M}_X(\theta)$ has full rank for all θ (regular model), then $K_{int,\alpha}(X,\theta;\theta') = 0$ for all θ,θ' implies that the model is intrinsically linear.

Moreover, when Θ in (7.25) is restricted to a small neighborhood of θ, we show below that $K_{int,\alpha}(X,\theta)$ corresponds to the intrinsic curvature $C_{int}(X,\theta)$ defined by (6.9) so that $K_{int,\alpha}(X,\theta)$ can be considered as an extended, globalized version of $C_{int}(X,\theta)$. Indeed, for θ' close to $\theta \in \mathrm{int}(\Theta)$, define $\delta = \|\theta-\theta'\|$ and $\mathbf{u} = (\theta-\theta')/\delta$. We then have, using a Taylor development of $\eta(\theta')$ at the point θ,

$$\|[\mathbf{I}_N - \mathbf{P}_\theta][\eta(\theta)-\eta(\theta')]\| = \frac{\delta^2}{2}\left\|[\mathbf{I}_N - \mathbf{P}_\theta]\left(\sum_{i,j=1}^p u_i\,\frac{\partial^2\eta(\theta)}{\partial\theta_i\partial\theta_j}\,u_j\right)\right\| + \mathcal{O}(\delta^3)$$

and $\|\eta(\theta)-\eta(\theta')\|^2 = \delta^2[\mathbf{u}^\top \mathbf{M}_X(\theta)\mathbf{u}] + \mathcal{O}(\delta^3)$, which gives $K_{int,\alpha}(X,\theta;\theta') = C_{int}(X,\theta;\mathbf{u}) + \mathcal{O}(\delta)$, with $C_{int}(X,\theta;\mathbf{u})$ the intrinsic measure of nonlinearity given by (6.7). On the other hand, $C_{int}(X, \theta; \mathbf{u})$ only gives a local measure of nonlinearity, while $K_{int,\alpha}(X, \theta; \theta')$ is also affected by the behavior of $\eta(\theta')$ for θ' far from θ.

Consider, for instance, the situation in Fig. 7.7 where the expectation surface $\mathbb{S}_\eta$ is almost overlapping. For θ' such that $\eta(\theta) - \eta(\theta')$ is orthogonal to the tangent plane to $\mathbb{S}_\eta$ at $\eta(\theta)$, we have $[\mathbf{I}_N - \mathbf{P}_\theta][\eta(\theta) - \eta(\theta')] = \eta(\theta) - \eta(\theta')$, so that

$$K_{int,\alpha}(X, \theta; \theta') = 2\frac{[(\theta - \theta')^\top \mathbf{M}_X(\theta)(\theta - \theta')]^\alpha}{\|\eta(\theta) - \eta(\theta')\|^{1+2\alpha}} = 2\frac{[(\theta - \theta')^\top \mathbf{M}_X(\theta)(\theta - \theta')]^\alpha}{\epsilon^{1+2\alpha}}$$

which tends to infinity when $\epsilon \to 0$. Suppose now that $\dim(\theta) = 1$ and that $\mathbb{S}_\eta$ overlaps because $\eta(\theta)$ is a periodic function of θ with period T. Then direct calculations give

$$K_{int,\alpha}(X, \theta; \theta') = C_{int}(X, \theta; \mathbf{u}) \, \frac{T^{2\alpha}}{t^{2\alpha}} \, [1 + \mathcal{O}(t)]$$

for small $t = \theta' - (\theta + T)$.

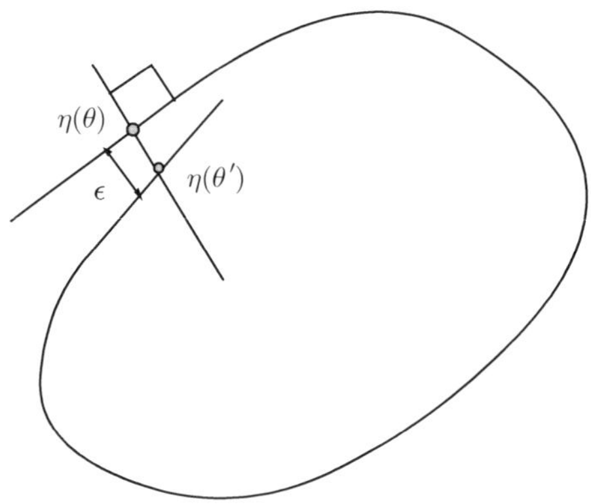

Fig. 7.7. An almost overlapping expectation surface

A continuation of Examples 7.5 and 7.6 will illustrate the different behaviors of $K_{int,\alpha}(X, \theta)$ and $C_{int}(X, \theta)$ and will serve to give some hints on how to choose α.

Example 7.5 (continued). Direct calculations yield

$$K_{int,\alpha}(X, \theta; \theta') = \frac{2(1 + 4\theta^2)^{\alpha - 1/2}}{[1 + (\theta + \theta')^2]^{1+\alpha}}$$

which is maximum for $\theta' = -\theta$. This gives

$$K_{int,\alpha}(X, \theta) = 2(1 + 4\theta^2)^{\alpha - 1/2} \,,$$

to be compared with $C_{int}(X,\theta) = 2/(1+4\theta^2)^{3/2}$. Figure 7.8 presents a plot of $C_{int}(X,\theta)$ and $K_{int,1}(X,\theta)$ as functions of θ. Although $\mathbb{S}_\eta$ is curved (see Fig. 7.2), there is no reason to consider that the effect of nonlinearity becomes more severe as $|\theta|$ increases. It thus seems reasonable to choose α such that

$$\max_{\theta' \in \Theta} \frac{[(\theta-\theta')^\top \mathbf{M}_X(\theta)(\theta-\theta')]^\alpha}{\|\eta(\theta)-\eta(\theta')\|^{1+2\alpha}}$$

remains bounded when $\|\theta\| \to \infty$. When taking $\alpha = 1/2$ in Example 7.5, we obtain $K_{int,1/2}(X,\theta) = 2$ for all θ. □

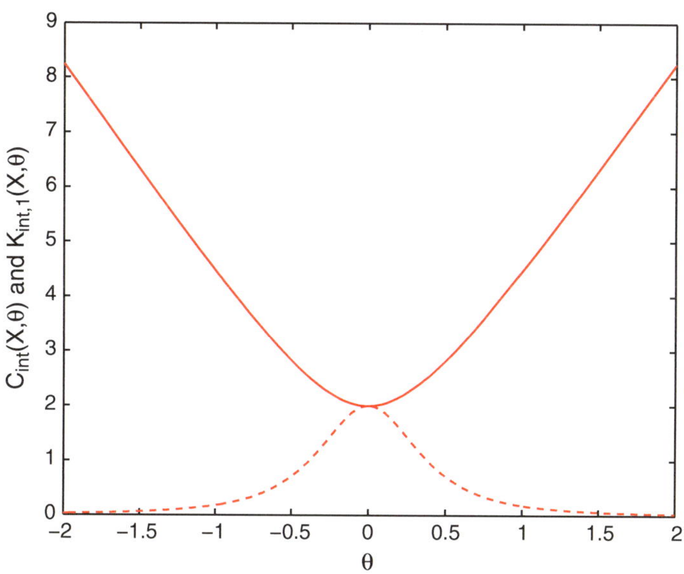

Fig. 7.8. $C_{int}(X,\theta)$ (*dashed line*) and $K_{int,1}(X,\theta)$ (*solid line*) as functions of θ in Example 7.5

Example 7.6 (continued). The expression of $K_{int,\alpha}(X,\theta;\theta')$ is more complicated than in Example 7.5. Consider the situation at $\theta = 0$. $K_{int,\alpha}(X,0;\theta')$ has the same expression as in Example 7.5 for $\theta' \geq 0$, i.e., $K_{int,\alpha}(X,0;\theta') = 2/(1+\theta'^2)^{1+\alpha}$, and equals

$$K_{int,\alpha}(X,0;\theta') = \frac{4\theta'^{2\alpha}}{(1-\cos\theta')^\alpha(5-3\cos\theta')^{1+\alpha}}$$

for $\theta' \leq 0$, which gives, for instance, $K_{int,1}(X,0;\theta') = 8\pi^2/(\theta'+2\pi)^2 - 8\pi/(\theta'+2\pi) + \mathcal{O}(1)$ and $K_{int,1/2}(X,0;\theta') = 4\pi/(\theta'+2\pi) - 2 + \mathcal{O}(\theta'+2\pi)$ when θ' approaches -2π. Figure 7.9 presents $K_{int,1}(X,0;\theta')$ as a function of θ'; the curve increases to infinity as θ' approaches -2π due to the overlapping of $\mathbb{S}_\eta$ caused by the periodicity of $\eta(\theta)$ for $\theta < 0$. For $\Theta = [\gamma,\infty)$ with $\gamma \lesssim -4.4768$ $K_{int,1}(X,0;\gamma) > K_{int,1}(X,0;0) = 2$.

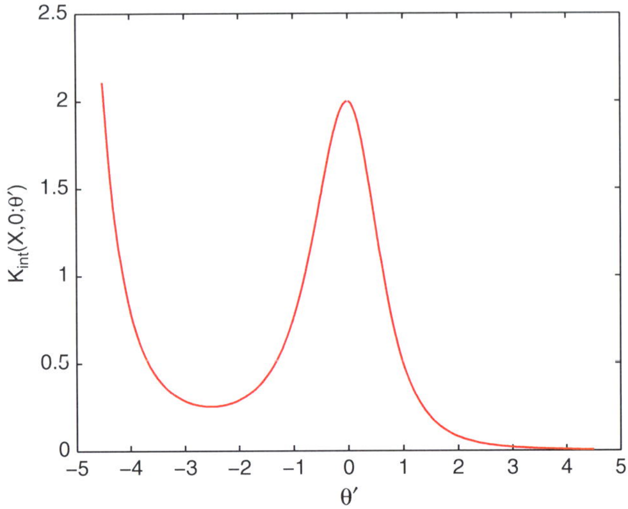

Fig. 7.9. $K_{int,1}(X, 0; \theta')$ as a function of θ' in Example 7.6

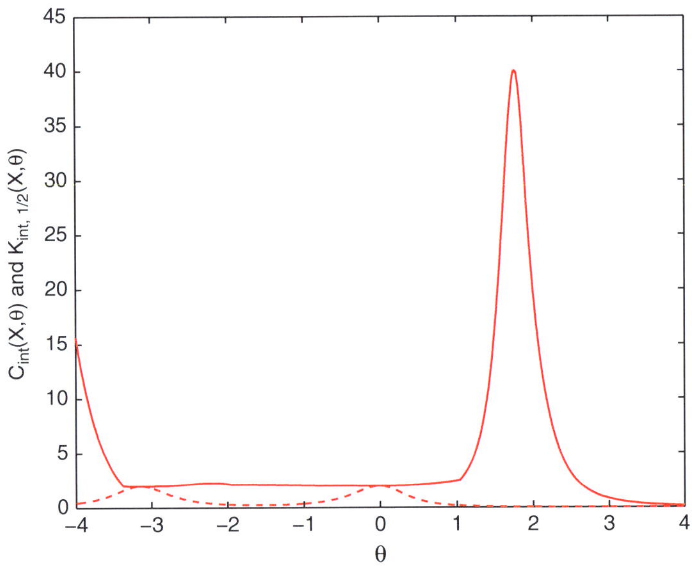

Fig. 7.10. $C_{int}(X, \theta)$ (*dashed line*) and $K_{int,1/2}(X, \theta)$ (*solid line*) as functions of θ in Example 7.6 for $\Theta = [-4, 4]$

Figure 7.10 shows $K_{int,1/2}(X, \theta)$ as a function of θ when $\Theta = [-4, 4]$. The intrinsic curvature

$$C_{int}(X,\theta) = \begin{cases} \frac{2}{(1+4\theta^2)^{3/2}} & \text{if } \theta \geq 0 \\ \frac{2}{(4-3\cos^2\theta)^{3/2}} & \text{if } \theta \leq 0 \end{cases}$$

is plotted in dashed line on the same figure. The fact that $\mathbb{S}_\eta$ tends to overlap (see Fig. 7.3) has no effect on $C_{int}(X,\theta)$ but has a strong influence on $K_{int,1/2}(X,\theta)$; in particular, $\eta(-4)$ is close to $\eta(\theta)$ for θ around 1.75 which produces large values of $K_{int,1/2}(X,\theta)$ for θ close to -4 or 1.75.

One may also calculate $K_{int,\alpha}(X,\theta)$ in a neighborhood of θ only, i.e., $K_{int,\alpha}(X,\theta) = \max_{\theta' \in \text{int}(\Theta_\theta)} K_{int,\alpha}(X,\theta;\theta')$ with $\Theta_\theta = \mathscr{B}(\theta,\delta) \cap \Theta$, $\delta > 0$. As Fig. 7.11 illustrates (for $\delta = 2$), this may be enough to detect a nonlinearity effect which is not revealed by $C_{int}(X,\theta)$. □

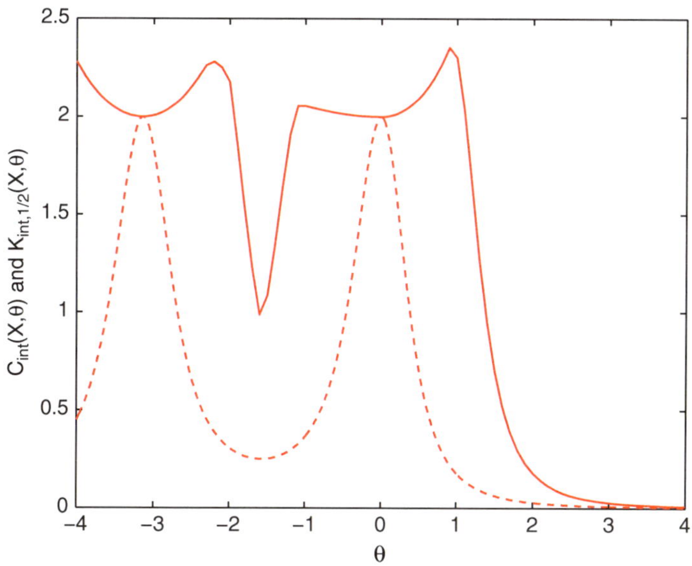

Fig. 7.11. $K_{int,1/2}(X,\theta) = \max_{\theta' \in \text{int}(\Theta_\theta)} K_{int,1/2}(X,\theta;\theta')$ with $\Theta_\theta = \mathscr{B}(\theta,2) \cap \Theta$ (*solid line*) as a function of θ in Example 7.6; $C_{int}(X,\theta)$ is in *dashed line*, $\Theta = [-4,4]$

7.6 Advantages and Drawbacks of Using p-point Designs

Since the intrinsic curvature of the model may cause numerical difficulties for the optimization of the LS criterion, the recommendation was made in (Pronzato and Walter, 2001) to choose an experimental design that ensures $C_{int}(X,\theta) = 0$ for (almost) all θ. The advantages and drawbacks of this approach are investigated below.

Consider a design X that consists of repetitions of trials at p different points only, denoted by $x^{(1)},\ldots,x^{(p)}$; see, e.g., Ross (1990) for the use of such designs. The model is then intrinsically linear (see Sect. 6.1.2), and $\|\mathbf{y}-\mathbf{z}\|^2$ has a unique minimum with respect to $\mathbf{z} \in \mathbb{S}_\eta$. Furthermore, using the reparameterization $\beta_i = \eta(x^{(i)}, \theta)$, $i = 1, \ldots, p$, we obtain a linear model $\eta(\beta) = \mathbf{F}\beta$ with $\mathbf{F} \in \mathbb{R}^{N \times p}$ and $\{\mathbf{F}\}_{ij} \in \{0, 1\}$ for all i, j. The minimizer $\hat{\beta} = \arg\min_{\beta \in \mathbb{R}} \|\mathbf{y} - \mathbf{F}\beta\|^2$ is given by

$$\hat{\beta} = (\mathbf{F}^\top \mathbf{F})^{-1} \mathbf{F}^\top \mathbf{y}$$

and the LS estimator is obtained by solving the equations

$$\hat{\beta}_i = \eta(x^{(i)}, \theta)\,,\ i = 1, \ldots, p\,, \tag{7.26}$$

for $\theta \in \Theta$, or by minimizing $\sum_{i=1}^p n_i [\hat{\beta}_i - \eta(x^{(i)}, \theta)]^2$, with n_i the number of repetitions at $x^{(i)}$, when (7.26) has no solution. Using a p-point design thus seems helpful.

However, it may happen that (7.26) has several solutions. That is, choosing a design that makes the model intrinsically linear may entail difficulties concerning global LS estimability. This drawback is illustrated by the next example.

Example 7.14. We take the same model as in Example 7.6, but change the second design point $\mathbf{x}_2$. When $\mathbf{x}_2 = \mathbf{x}_1 = (1, 0)$, the expectation surface $\mathbb{S}_\eta$ is included in the diagonal $y_1 = y_2$, and its intrinsic curvature $C_{int}(X, \theta)$ is zero for almost all θ: for $\Theta = [\gamma, \infty)$ and $\gamma < -\pi/2$, $\mathbb{S}_\eta = \{\mathbf{z} = (-1, -1)^\top + \alpha(1, 1)^\top\,,\ \alpha > 0\}$. Therefore, for any $\mathbf{y}$ in $\mathbb{R}^2$, $\|\mathbf{y} - \mathbf{z}\|$ has a unique minimum with respect to $\mathbf{z} \in \mathbb{S}_\eta$. However, several values of θ may correspond to this minimizing $\hat{\mathbf{z}}$, that is, global LS estimability may be lost.

Figure 7.12 presents the expectation surface $\mathbb{S}_\eta$ for $\mathbf{x}_2 = (1, 0.1)$, $\gamma = -5.5$ and shows the difficulty: when $\mathbf{x}_2$ gets close to $\mathbf{x}_1 = (1, 0)$, the intrinsic curvature of $\mathbb{S}_\eta$ tends to infinity for $\theta = -\pi/2$ and $\theta = -3\pi/2$.

When $\mathbf{x}_2 = \mathbf{x}_1 = (1, 0)$, having determined $\hat{\mathbf{z}}$ (unique) that minimizes $\|\mathbf{y} - \mathbf{z}\|$, $\mathbf{z} \in \mathbb{S}_\eta$, for $\gamma > -2\pi$ there may exist one, two, or three values of $\hat{\theta}$ such that $\eta(\hat{\theta}) = \hat{\mathbf{z}}$. For instance, when $\gamma = -7\pi/4$, if $\{\hat{\mathbf{z}}\}_1 > 1$, $\hat{\theta} = \{\hat{\mathbf{z}}\}_1$ is the unique solution; if $1/2 \le \{\hat{\mathbf{z}}\}_1 \le 1$, there are three solutions $\hat{\theta} = \{\hat{\mathbf{z}}\}_1$ and $\hat{\theta} = \arcsin(\{\hat{\mathbf{z}}\}_1)$ (which gives two values in $[-7\pi/4, 0]$); if $0 \le \{\hat{\mathbf{z}}\}_1 < 1/2$, there are two solutions $\hat{\theta} = \{\hat{\mathbf{z}}\}_1$ and $\hat{\theta} = \arcsin(\{\hat{\mathbf{z}}\}_1)$ (which gives one value in $[-7\pi/4, 0]$); for $-1 \le \{\hat{\mathbf{z}}\}_1 < 0$, there are two solutions again, satisfying $\hat{\theta} = \arcsin(\{\hat{\mathbf{z}}\}_1)$.

Figure 7.13 shows a plot of the estimability function $E_{\eta,\theta}(\cdot)$ when $\mathbf{x}_2 = \mathbf{x}_1 = (1, 0)$ for $\theta = 1.5$ (solid line) and $\theta = 0.75$ (dashed line); the possible existence of three solutions for $\hat{\theta}$ is revealed by $E_{\eta,\theta}(\delta)$ being equal to zero for three distinct values of δ: 0 and approximately 22.464 and 38.256. Figure 7.14 presents $E_\eta(\delta)$ for δ between 0 and 50. The fact that the model is only locally LS estimable results in $E_\eta(\delta)$ being equal to 0 for δ in the interval $[0, (\sin(\gamma) - \gamma)^2]$, i.e., approximately $[0, 38.51]$ for $\gamma = -5.5$. □

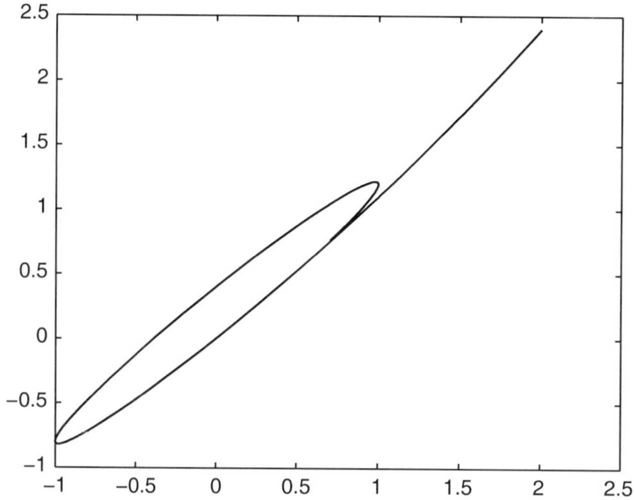

Fig. 7.12. Expectation surface $\mathbb{S}_\eta$ in Example 7.14 for $\gamma = -5.5$ and $\mathbf{x}_2 = (1, 0.1)$

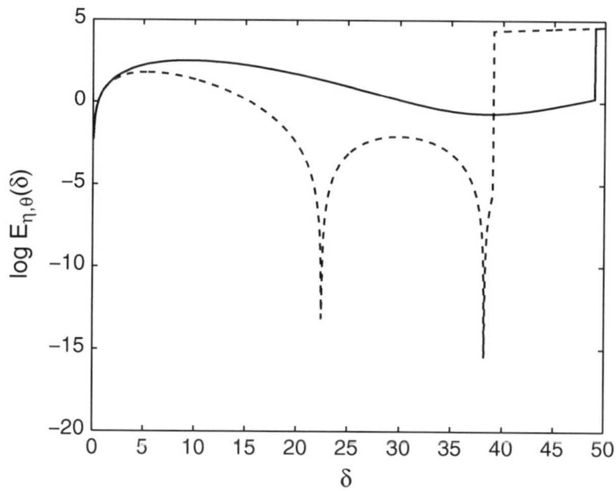

Fig. 7.13. Estimability function $E_{\eta,\theta}(\delta)$ (log scale) in Example 7.14 for $\theta = 1.5$ (*solid line*) and $\theta = 0.75$ (*dashed line*); $\gamma = -5.5$ and $\mathbf{x}_2 = \mathbf{x}_1 = (1, 0)$

Despite the estimability problems it may cause, choosing a design X that makes $\mathbb{S}_\eta$ plane can help to determine all local minima in a situation where $\mathbb{S}_\eta$ is curved. Indeed, having determined the unique $\hat{\mathbf{z}} \in \mathbb{S}_\eta$ that minimizes $\|\mathbf{y} - \mathbf{z}\|$ in the intrinsically linear model, the determination of all values of $\hat{\theta}$ such that $\eta(\hat{\theta}) = \hat{\mathbf{z}}$ is purely algebraic. Let $\hat{\theta}^{(i)}$, $i = 1, \ldots, s$, denote these

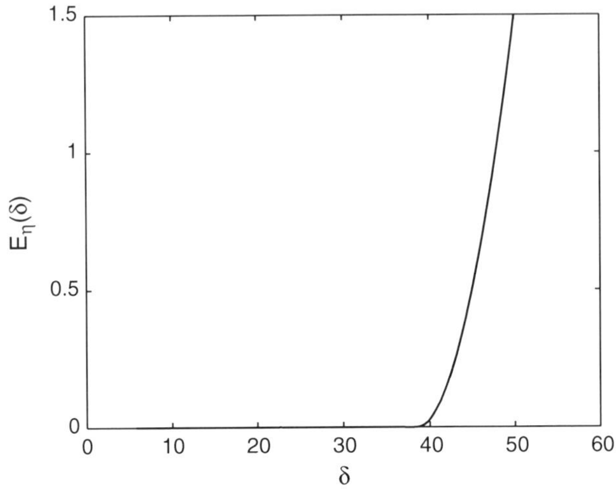

Fig. 7.14. Function $E_\eta(\delta)$ in Example 7.14 for $\gamma = -5.5$ and $\mathbf{x}_2 = \mathbf{x}_1 = (1, 0)$

solutions. When performing n repetitions of the design X, $\hat{\theta}^{(i)}$ tends to $\bar{\theta}$ for some $i \in \{1, \ldots, s\}$ as $n \to \infty$ provided that the model structure is locally LS estimable for X (and under suitable regularity conditions, see Chap. 3).

Augment now the design $X^{\otimes n}$ by adding new design points, in order to form a design X' such that the model is globally LS estimable for X'. The estimates $\hat{\theta}^{(i)}$ obtained with $X^{\otimes n}$ can then be used as initial values for the minimization of the LS criterion in the new problem with the design X' and its associated observations $\mathbf{y}'$.

Example 7.14 (continued). Take a design consisting of n repetitions of observations at $\mathbf{x}_1 = (1, 0)$ and consider the situation where there are three solutions $\hat{\theta}^{(i)}$ for $\hat{\theta}$, $i = 1, 2, 3$. For n large enough, $\hat{\theta}^{(i)}$ is close to $\bar{\theta}$ for some i. Consider the design X' obtained by augmenting $X^{\otimes n} = \mathbf{x}_1^{\otimes n}$ by $\mathbf{x}_2 = (0, 1)$ and suppose that the responses $[\eta(\mathbf{x}_1, \hat{\theta}^{(i)}), \eta(\mathbf{x}_2, \hat{\theta}^{(i)})]$ are indicated by the points A, B, and C on Fig. 7.3. Initializing the search for the LS estimator at each $\hat{\theta}^{(i)}$ successively then facilitates the determination of the global minimum of the LS criterion for the design X'.

Notice that when some ambiguity remains about which local minimizer to consider as the LS estimator of θ, because the local and global minimum values of the LS criterion are close, this ambiguity can be removed by repeating m observations at $\mathbf{x}_2$. Indeed, $[\bar{y}(\mathbf{x}_1), \bar{y}(\mathbf{x}_2)]$, with $\bar{y}(\mathbf{x}_i)$, the empirical mean of the observations at $\mathbf{x}_i$, converges to a point in the neighborhood of A, B, or C when n and m tend to infinity. More precisely, suppose that $n \to \infty$, $m \to \infty$, and $m/n \geq \alpha > 0$; then, the global minimum of the LS criterion

$$[n/(n+m)][\bar{y}(\mathbf{x}_1) - \eta(\mathbf{x}_1, \theta)]^2 + [m/(n+m)][\bar{y}(\mathbf{x}_2) - \eta(\mathbf{x}_1, \theta)]^2 ,$$

tends to zero (see Remark 6.3), whereas the other local minima remain bounded away from zero. A related test was proposed in (Pázman et al., 1969) in the context of elementary particle physics. □

7.7 Design of Experiments for Improving Estimability

Natural objectives for choosing a design X that ensures good estimability properties include the maximization of α_η and the minimization of β_η in Theorem 7.3 and the maximization of $\underline{R}_\eta$ in Theorem 7.4; note that $\alpha_\eta^2/\beta_\eta \leq \underline{R}_\eta$. The maximization of α_η corresponds to maximin E-optimal design; see Sects. 5.1.2 and 8.2. Additional requirements on the radius of curvature $R_\eta(\theta)$ can be taken into account by setting constraints on the curvature of the model while optimizing the design criterion; see, for instance, Clyde and Chaloner (2002). However, as the examples above illustrated, taking care of the curvature of the model is not enough to guarantee satisfying estimability properties. Also, as shown in Sect. 7.6, forcing the model to be intrinsically linear by using repetitions of a p-point design may entail difficulties concerning global LS estimability. For that reason, we consider below new design criteria that can be related to the estimability function defined in Sect. 7.4.[1]

7.7.1 Extended (Globalized) E-Optimality

Definition

Denote

$$D_{\eta,\theta,\delta}(X) = \frac{E_{\eta,\theta}(\delta)}{N}\,(K + 1/\delta)\,,$$

with K some positive number, and for a design measure ξ,

$$\begin{aligned} D_{\eta,\theta,\delta}(\xi) &= E^{\xi}_{\eta,\theta}(\delta)\,(K + 1/\delta)\,, \\ &= \min_{\theta'\in\Theta,\, \|\theta'-\theta\|^2=\delta} \left\{ \|\eta(\cdot,\theta') - \eta(\cdot,\theta)\|^2_\xi \,(K + 1/\delta) \right\}\,, \end{aligned} \tag{7.27}$$

see (7.19). Consider the design criterion defined by

$$\begin{aligned} \phi_{eE}(X) = \phi_{eE}(X;\theta) &= \min_{\delta\geq 0} D_{\eta,\theta,\delta}(X)\,, \\ &= \min_{\theta'\in\Theta} \left\{ \|\eta(\theta') - \eta(\theta)\|^2 \,(K + \|\theta'-\theta\|^{-2}) \right\}\,, \end{aligned}$$

[1] Another possibility, not explored here, would be to use a more classical optimality criterion (see Chap. 5) while imposing a constraint on the extended measure of nonlinearity $K_{int,\alpha}(X,\theta)$ defined in (7.25).

to be maximized with respect to X, and similarly

$$\phi_{eE}(\xi) = \phi_{eE}(\xi;\theta) = \min_{\delta\geq 0} D_{\eta,\theta,\delta}(\xi)\,,$$
$$= \min_{\theta'\in\Theta} \left\{ \|\eta(\cdot,\theta') - \eta(\cdot,\theta)\|_\xi^2 \,(K + \|\theta'-\theta\|^{-2}) \right\}\,,$$

to be maximized with respect to the design measure ξ. Notice that in a nonlinear regression model $\phi_{eE}(\cdot)$ depends on the value chosen for θ and can thus be considered as a local optimality criterion, in the same sense as in Chap. 5 when θ is set to a nominal value θ^0. On the other hand, the criterion is global in the sense that it depends on the behavior of $\eta(\cdot,\theta')$ for θ' far from θ. We can remove this (limited) locality by using $\phi_{MeE}(\xi) = \min_{\theta\in\Theta} \phi_{eE}(\xi)$, which corresponds to maximin-optimum design as considered in Sect. 8.2.

For a linear regression model with $\eta(\theta) = \mathbf{F}(X)\theta + \mathbf{v}(X)$ and $\Theta = \mathbb{R}^p$, we have $E_{\eta,\theta}(\delta) = E_\eta(\delta) = \delta\,\lambda_{\min}(\mathbf{M}_X) = N\,\delta\,\lambda_{\min}[\mathbf{M}(X)]$, with $\mathbf{M}(X) = \mathbf{M}_X/N = \mathbf{F}^\top(X)\mathbf{F}(X)/N$, and $E_{\eta,\theta}^\xi(\delta) = \delta\,\lambda_{\min}[\mathbf{M}(\xi)]$, so that

$$\phi_{eE}(X) = \lambda_{\min}[\mathbf{M}(X)] \quad\text{and}\quad \phi_{eE}(\xi) = \lambda_{\min}[\mathbf{M}(\xi)] \text{ for any } K \geq 0\,,$$

which corresponds to the E-optimality criterion of Sect. 5.1.2. $\phi_{eE}(\cdot)$ can thus be considered as an *extended E-optimality criterion.*

Consider now the case of nonlinear regression. When the model is locally LS estimable at θ for X, $E_{\eta,\theta}(\delta)/(N\delta)$ tends to $\lambda_{\min}[\mathbf{M}_X(\theta)]/N = \lambda_{\min}[\mathbf{M}(X,\theta)]$ as δ tends to zero; see (7.14). By continuity, we can thus define $D_{\eta,\theta,0}(X) = \lambda_{\min}[\mathbf{M}(X,\theta)]$. Denote $\delta^* = \arg\min_{\delta\geq 0} D_{\eta,\theta,\delta}(X)$. We have

$$\phi_{eE}(X;\theta) = D_{\eta,\theta,\delta^*}(X) = \frac{E_{\eta,\theta}(\delta^*)}{N}\,(K + 1/\delta^*) \leq \lambda_{\min}[\mathbf{M}(X,\theta)]$$

and thus

$$\frac{E_{\eta,\theta}(\delta^*)}{N} < \frac{\lambda_{\min}[\mathbf{M}(X,\theta)]}{K}\,.$$

Therefore, when the model is globally LS estimable for X at θ, so that $E_{\eta,\theta}(\delta) > 0$ for all $\delta > 0$, δ^* can be made arbitrarily small by taking K large enough. We show that this implies that $\phi_{eE}(X;\theta)$ can be made arbitrarily close to $\lambda_{\min}[\mathbf{M}(X,\theta)]$ by increasing K. Using the approximation (7.15) when $A_X(\theta) \geq 0$, we obtain that $D_{\eta,\theta,\delta}(X)$ is an increasing function of δ for δ small enough, which yields $\delta^* = 0$ and $\phi_{eE}(X;\theta) = \lambda_{\min}[\mathbf{M}(X,\theta)]$ for K large enough. When $A_X(\theta) < 0$, direct calculations give

$$\delta^* = \frac{A^2(X,\theta)}{4K^2\,\lambda_{\min}^2[\mathbf{M}(X,\theta)]} + \mathcal{O}(1/K^3)$$
$$\phi_{eE}(X;\theta) = \lambda_{\min}[\mathbf{M}(X,\theta)] - \frac{A^2(X,\theta)}{4K\,\lambda_{\min}[\mathbf{M}(X,\theta)]} + \mathcal{O}(1/K^2)$$

for large K, with $A(X,\theta) = A_X(\theta)/N$. Therefore, $\phi_{eE}(X;\theta)$ can always be made arbitrarily close to $\lambda_{\min}[\mathbf{M}(X,\theta)]$ by choosing K large enough. The same is true for $\phi_{eE}(\xi;\theta)$ which can be made arbitrarily close to $\lambda_{\min}[\mathbf{M}(\xi,\theta)]$ for K sufficiently large when the model is globally LS estimable for ξ at θ. Hence, the name of extended E-optimality criterion is justified for nonlinear models as well as for linear ones.

In the following we shall only consider the case of design measures and omit the dependence in θ for nonlinear models; the criterion is considered as evaluated at a nominal value θ^0. We have just seen that choosing K large makes $\phi_{eE}(\xi;\theta^0)$ approach the E-optimality criterion $\phi_E(\xi;\theta^0) = \Phi_E[\mathbf{M}(\xi,\theta^0)]$ of Sect. 5.1.2. This can also be enforced by taking Θ as (a subset of) a small ball around θ^0. At the same time, choosing K not too large ensures some protection against $\|\eta(\theta^0) - \eta(\theta)\|_\xi$ being small for some θ far from θ^0. A reasonable choice consists in taking K proportional to N/σ^2, so that for σ^2 sufficiently small, or N sufficiently large, the criterion $\phi_{eE}(\cdot)$ is close to $\phi_E(\cdot)$.

Example 7.15. We consider again Example 7.6 with different designs supported at $\mathbf{x}_1 = (1,0)$ and $\mathbf{x}_2 = (0,1)$ and defined by $\xi_\alpha = \alpha\delta_{\mathbf{x}_1} + (1-\alpha)\delta_{\mathbf{x}_2}$, $\alpha \in [0,1]$. For $\theta = 0.75$, $K = 1$, we have $\phi_{eE}(\xi_0) \simeq 1.4 \cdot 10^{-5}$, $\phi_{eE}(\xi_1) \simeq 9.7 \cdot 10^{-8}$, and the optimal design for $\phi_{eE}(\cdot)$ in this family is obtained for $\alpha^* \simeq 0.67$, with $\phi_{eE}(\xi_{0.67}) = 1.4 \cdot 10^{-3}$. Figure 7.15 gives a plot of $D_{\eta,\theta,\delta}(\xi_\alpha)$ defined by (7.27) for $\alpha = 0$, 1, and 0.67. Notice that the E-optimal design, maximizing $\mathbf{M}(X,\theta) = \alpha + (1-\alpha)4\theta^2$, is obtained for $\alpha = 0$. Other choices for K yield other optimal designs for $\phi_{eE}(\cdot)$, the optimal α approaching 0 as K increases ($\alpha^* \simeq 0.32$ for $K = 2,000$, $\alpha^* \simeq 0.01$ for $K = 5,000$, and $\alpha^* = 0$ for $K \gtrsim 5 \cdot 10^5$). □

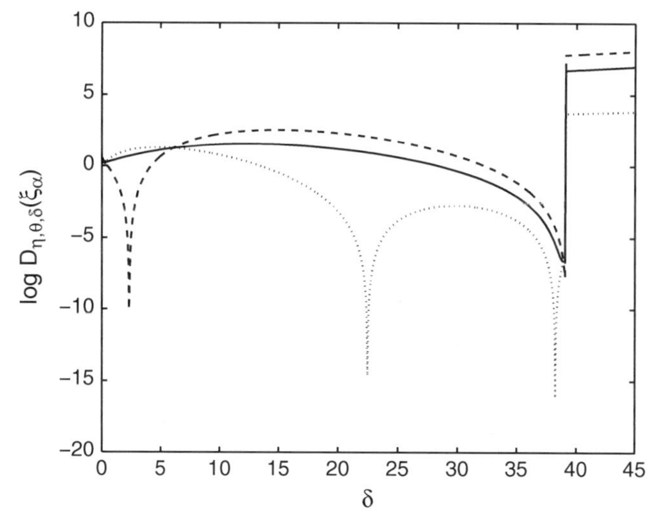

Fig. 7.15. $D_{\eta,\theta,\delta}(\xi)$ (log scale) as a function of δ in Example 7.15 for $\theta = 0.75$, $K = 1$, $\gamma = -5.5$ and different designs $\xi_\alpha = \alpha\delta_{\mathbf{x}_1} + (1-\alpha)\delta_{\mathbf{x}_2}$: $\alpha = 0$ (*dashed line*), $\alpha = 1$ (*dotted line*), and $\alpha = \alpha^* = 0.67$ (*solid line*)

Properties of $\phi_{eE}(\cdot)$

We show below that $\phi_{eE}(\cdot)$ shares some properties with criteria considered in Chap. 5 (concavity, positive homogeneity, existence of a directional derivative...).

As the minimum of linear functions of ξ, $\phi_{eE}(\cdot)$ is concave: for all $\xi, \nu \in \Xi$, the set of design measures on $\mathscr{X}$, for all $\alpha \in [0,1]$, and for all $\theta \in \Theta$, $\phi_{eE}[(1-\alpha)\xi+\alpha\nu] \geq (1-\alpha)\phi_{eE}(\xi)+\alpha\phi_{eE}(\nu)$. It is also positively homogeneous: $\phi_{eE}(a\xi) = a\,\phi_{eE}(\xi)$ for all $\xi \in \Xi$, $\theta \in \Theta$, and $a > 0$. Concavity implies the existence of directional derivatives; see Lemma 5.16. We can write

$$\phi_{eE}(\xi) = \min_{\theta' \in \Theta} \left\{ \|\eta(\cdot,\theta') - \eta(\cdot,\theta)\|_\xi^2 \left(K + \|\theta' - \theta\|^{-2}\right) \right\} , \tag{7.28}$$

and Lemma 5.18 gives the following (we suppose that $\eta(x,\theta)$ is continuous in θ over Θ, a compact subset of $\mathbb{R}^p$, for all $x \in \mathscr{X}$):

Theorem 7.16. *For any $\xi, \nu \in \Xi$, the directional derivative of the criterion $\phi_{eE}(\cdot)$ at ξ in the direction ν (see (5.33)) is given by*

$$F_{\phi_{eE}}(\xi;\nu) = \min_{\theta' \in \Theta_\theta(\xi)} \left\{ \|\eta(\cdot,\theta') - \eta(\cdot,\theta)\|_\nu^2 \left(K + \|\theta' - \theta\|^{-2}\right) \right\} - \phi_{eE}(\xi) ,$$

where

$$\Theta_\theta(\xi) = \left\{ \theta' \in \Theta : \|\eta(\cdot,\theta') - \eta(\cdot,\theta)\|_\xi^2 \left(K + \|\theta' - \theta\|^{-2}\right) = \phi_{eE}(\xi) \right\} .$$

A necessary-and-sufficient condition for optimality, based on the directional derivatives $F_{\phi_{eE}}(\xi;\nu)$, can be expressed in the same form as Theorem 5.21. We can write

$$F_{\phi_{eE}}(\xi;\nu) = \min_{\theta' \in \Theta_\theta(\xi)} \int_{\mathscr{X}} \Psi_{eE}(x,\theta',\xi)\,\nu(\mathrm{d}x)$$

where

$$\begin{aligned} \Psi_{eE}(x,\theta',\xi) = {} & (K + \|\theta' - \theta\|^{-2}) \\ & \times \left\{ [\eta(x,\theta') - \eta(x,\theta)]^2 - \|\eta(\cdot,\theta') - \eta(\cdot,\theta)\|_\xi^2 \right\} \end{aligned} \tag{7.29}$$

and ξ^* is optimal for $\phi_{eE}(\cdot)$ if and only if $\sup_{\nu \in \Xi} F_{\phi_{eE}}(\xi^*;\nu) \leq 0$.

One should notice that $\sup_{\nu \in \Xi} F_{\phi_{eE}}(\xi^*;\nu)$ is generally not obtained for ν equal to a one-point (delta) measure (see Remark 5.26-(*ii*)), which prohibits the usage of the classical vertex-direction algorithms of Sect. 9.1.1 for optimizing $\phi_{eE}(\cdot)$. A regularized version $\phi_{eE,\lambda}(\cdot)$ of $\phi_{eE}(\cdot)$ will be considered in Sect. 7.7.3, with the property that $\sup_{\nu \in \Xi} F_{\phi_{eE,\lambda}}(\xi;\nu)$ is obtained when ν is the delta measure δ_{x^*} at some $x^* \in \mathscr{X}$ (depending on ξ). Alternatively, one may notice that the maximization of $\phi_{eE}(\cdot)$ corresponds to a linear program.

Indeed, when Θ is finite, i.e., $\Theta = \{\theta^{(1)}, \theta^{(2)}, \ldots, \theta^{(m)}\}$, $\phi_{eE}(\xi)$ can be written as $\phi_{eE}(\xi) = \min_{j=1,\ldots,m} H_E(\xi, \theta^{(j)})$, where

$$H_E(\xi,\theta') = \|\eta(\cdot,\theta') - \eta(\cdot,\theta)\|_\xi^2 (K + \|\theta' - \theta\|^{-2}), \tag{7.30}$$

which is linear in ξ. If the design space $\mathscr{X}$ is finite too, that is, $\mathscr{X} = \{x^{(1)}, x^{(2)}, \ldots, x^{(\ell)}\}$, then the determination of an optimal design measure for $\phi_{eE}(\cdot)$ amounts to the determination of a scalar γ and of a vector of weights $\mathbf{w} = (w_1, w_2, \ldots, w_\ell)^\top$ such that γ is maximized with $\mathbf{w}$ and γ satisfying the linear constraints

$$\sum_{i=1}^{\ell} w_i = 1,$$

$$w_i \geq 0, \; i = 1, \ldots, \ell,$$

$$\sum_{i=1}^{\ell} w_i [\eta(x^{(i)}, \theta^{(j)}) - \eta(x^{(i)}, \theta)]^2 (K + \|\theta^{(j)} - \theta\|^{-2}) \geq \gamma, \; j = 1, \ldots, m.$$

The finite set Θ can be enlarged iteratively (see the relaxation algorithm of Sect. 9.3.2); solving the corresponding sequence of LP problems then corresponds to the method of cutting planes of Sect. 9.5.3.

7.7.2 Extended (Globalized) c-Optimality

We have seen in Sects. 5.1.2 and 5.4 that a c-optimal design for the estimation of a function $h(\theta)$ may be singular and that this raises specific difficulties in a nonlinear situation. Here we shall use a construction similar to that in Sect. 7.7.1 and define an extended version of c-optimality, with the objective to circumvent those difficulties and at the same time ensure some protection against $\|\eta(\cdot,\theta') - \eta(\cdot,\theta)\|_\xi$ being small for some θ' such that $|h(\theta') - h(\theta)|$ is large. The criterion that we construct is based on $\eta(x,\theta)$ and $h(\theta)$ but does not rely on a linearization with respect to θ, although it corresponds to c-optimum design for a linear model and a linear function of interest $h(\theta) = \mathbf{c}^\top\theta$.

Definition

We define

$$\phi_{ec}(\xi) = \min_{\delta \geq 0} \left\{ E^{\xi}_{\eta,\theta,h}(\delta)\,(K + 1/\delta) \right\} \tag{7.31}$$

$$= \min_{\theta' \in \Theta} \left\{ \|\eta(\cdot,\theta') - \eta(\cdot,\theta)\|_\xi^2 \,\left(K + |h(\theta') - h(\theta)|^{-2}\right) \right\}, \tag{7.32}$$

with K some positive constant.

When $\eta(x,\theta)$ and $h(\theta)$ are both linear in θ, we get

$$\phi_{ec}(\xi) = \min_{\theta' \in \Theta,\; \mathbf{c}^\top(\theta' - \theta) \neq 0} \frac{(\theta' - \theta)^\top \mathbf{M}(\xi)(\theta' - \theta)}{[\mathbf{c}^\top(\theta' - \theta)]^2}$$

and therefore, using Lemma 5.6, $\phi_{ec}(\xi) = [\mathbf{c}^\top\mathbf{M}^-(\xi)\mathbf{c}]^{-1} = \Phi_c^+[\mathbf{M}(\xi,\theta)]/(\mathbf{c}^\top\mathbf{c})$, which justifies that we consider $\phi_{ec}(\xi)$ as an *extended c-optimality criterion.*

In a nonlinear situation, when the model is globally LS estimable at θ for ξ, we obtain

$$\lim_{\delta \to 0} E^{\xi}_{\eta,\theta,h}(\delta)\,(K + 1/\delta) = \frac{1}{\mathbf{c}^\top \mathbf{M}^{-1}(\xi,\theta)\mathbf{c}}\,,$$

with $\mathbf{c} = \partial h(\theta)/\partial\theta$; see (7.21). Therefore, $\phi_{ec}(\xi) \leq [\mathbf{c}^\top \mathbf{M}^{-}(\xi,\theta)\mathbf{c}]^{-1}$. Let δ^* denote the minimizing δ in (7.31), we obtain

$$E^{\xi}_{\eta,\theta,h}(\delta^*) \leq \frac{1}{K\,\mathbf{c}^\top \mathbf{M}^{-1}(\xi,\theta)\mathbf{c}}\,,$$

which implies that δ^* can be made arbitrarily small by increasing K and, similarly to Sect. 7.7.1, $\phi_{ec}(\xi)$ can be approximated by $[\mathbf{c}^\top \mathbf{M}^{-1}(\xi,\theta)\mathbf{c}]^{-1}$ for large K. On the other hand, choosing K not too large ensures some protection against $\|\eta(\cdot,\theta') - \eta(\cdot,\theta)\|^2_\xi$ being small for some θ' such that $h(\theta')$ is significantly different from $h(\theta)$.

Properties of $\phi_{ec}(\cdot)$

Similarly to the case of the extended E-optimality criterion, $\phi_{ec}(\cdot)$ is concave and positively homogeneous. Also, concavity implies the existence of directional derivatives, and Lemma 5.18 gives the following:

Theorem 7.17. *For any $\xi, \nu \in \Xi$, the directional derivative of the criterion $\phi_{ec}(\cdot)$ at ξ in the direction ν is given by*

$$F_{\phi_{ec}}(\xi;\nu) = \min_{\theta' \in \Theta_{\theta,c}(\xi)} \left\{ \|\eta(\cdot,\theta') - \eta(\cdot,\theta)\|^2_\nu \left(K + |h(\theta') - h(\theta)|^{-2}\right)\right\} - \phi_{ec}(\xi)\,,$$

where

$$\Theta_{\theta,c}(\xi) = \left\{\theta' \in \Theta : \|\eta(\cdot,\theta') - \eta(\cdot,\theta)\|^2_\xi \left(K + |h(\theta') - h(\theta)|^{-2}\right) = \phi_{ec}(\xi)\right\}\,.$$

A necessary-and-sufficient condition for optimality based on the directional derivatives $F_{\phi_{ec}}(\xi;\nu)$ can be expressed in the same form as Theorem 5.21, and ξ^* is optimal for ϕ_{ec} if and only if $\sup_{\nu \in \Xi} F_{\phi_{ec}}(\xi^*;\nu) \leq 0$. A regularized version $\phi_{ec,\lambda}(\cdot)$ of $\phi_{ec}(\cdot)$ will be considered in Sect. 7.7.3, with the property that $\sup_{\nu\in\Xi} F_{\phi_{ec,\lambda}}(\xi;\nu)$ is obtained for $\nu = \delta_{x^*}$, the delta measure at some $x^* \in \mathscr{X}$ (depending on ξ). Also, when both Θ and $\mathscr{X}$ are finite, an optimal design maximizing $\phi_{ec}(\cdot)$ can be obtained by LP, following the same approach as for $\phi_{eE}(\cdot)$.

Remark 7.18 (Extended G-Optimality). Following the same lines as above, we can also define an extended G-optimality criterion by

$$\phi_{GG}(\xi) = \min_{\theta' \in \Theta} \left[\|\eta(\cdot,\theta') - \eta(\cdot,\theta)\|^2_\xi \left\{ K + \frac{1}{\max_{x\in\mathscr{X}} [\eta(x,\theta') - \eta(x,\theta)]^2} \right\}\right]$$

with K some positive constant. The fact that it corresponds to the G-optimality criterion of Sect. 5.1.2 for a linear model can easily be seen, noticing that in the model (7.2) with $\eta(x,\theta) = \mathbf{f}^\top(x)\theta + v(x)$, we have

$$\left\{\max_{x\in\mathscr{X}} \frac{N}{\sigma^2} \operatorname{var}\left[\mathbf{f}^\top(x)\hat{\theta}^N_{LS}\right]\right\}^{-1} = \inf_{x\in\mathscr{X}} \inf_{\mathbf{u}\in\mathbb{R}^p,\, \mathbf{u}^\top\mathbf{f}(x)\neq 0} \frac{\mathbf{u}^\top\mathbf{M}(X)\mathbf{u}}{[\mathbf{f}^\top(x)\mathbf{u}]^2}$$

$$= \inf_{\mathbf{u}\in\mathbb{R}^p,\, \mathbf{u}^\top\mathbf{f}(x)\neq 0} \left[\mathbf{u}^\top\mathbf{M}(X)\mathbf{u}\left\{K + \frac{1}{\max_{x\in\mathscr{X}}[\mathbf{f}^\top(x)\mathbf{u}]^2}\right\}\right],$$

see Lemma 5.6. Directional derivatives can be computed similarly to the cases of extended E and c-optimality; an optimal design can be obtained by LP when both Θ and $\mathscr{X}$ are finite. □

7.7.3 Maximum-Entropy Regularization of Estimability Criteria

Extended E-optimality

Applying the results of Sect. 8.3.2 on maximum-entropy regularization to the criterion $\phi_{eE}(\cdot)$ given by (7.28), we obtain the regularized version

$$\phi_{eE,\lambda}(\xi) = -\frac{1}{\lambda}\log\int_\Theta \exp\{-\lambda H_E(\xi,\theta')\}\,\mathrm{d}\theta' \tag{7.33}$$

where $H_E(\xi,\theta')$ is given by (7.30). The regularized criterion $\phi_{eE,\lambda}(\cdot)$ is concave; its directional derivative at ξ in the direction ν is

$$F_{\phi_{eE,\lambda}}(\xi;\nu) = \frac{\int_{\mathscr{X}}\int_\Theta \exp\{-\lambda H_E(\xi,\theta')\}\,\Psi_{eE}(x,\theta',\xi)\,\mathrm{d}\theta'\,\nu(\mathrm{d}x)}{\int_\Theta \exp\{-\lambda H_E(\xi,\theta')\}\,\mathrm{d}\theta'} \tag{7.34}$$

with $\Psi_{eE}(x,\theta',\xi)$ given by (7.29). Moreover, $\phi_{eE,\lambda}(\cdot)$ is differentiable, contrary to $\phi_{eE}(\cdot)$, and in the same vein as Lemma 5.20 and Theorem 5.21, we obtain a necessary-and-sufficient condition of optimality for $\phi_{eE,\lambda}(\cdot)$: ξ^* is optimal for $\phi_{eE,\lambda}(\cdot)$ if an only if

$$\max_{x\in\mathscr{X}}\int_\Theta \exp\{-\lambda H_E(\xi^*,\theta')\}\,\Psi_{eE}(x,\theta',\xi^*)\,\mathrm{d}\theta' \leq 0\,.$$

Note that (7.33) depends on the choice made for the set Θ. Also note that in order to make the computations easier, the integrals on θ' in (7.33), (7.34) can be replaced by finite sums. This will be illustrated by the examples of Sect. 7.7.4.

Extended c-optimality

The situation is similar to the case of extended E-optimality. Applying the results of Sect. 8.3.2 to the criterion $\phi_{ec}(\cdot)$ given by (7.32), we obtain the regularized version

$$\phi_{ec,\lambda}(\xi) = -\frac{1}{\lambda} \log \int_{\Theta} \exp\{-\lambda H_c(\xi,\theta')\} \, \mathrm{d}\theta' \tag{7.35}$$

where

$$H_c(\xi,\theta') = \|\eta(\cdot,\theta') - \eta(\cdot,\theta)\|_\xi^2 (K + |h(\theta') - h(\theta)|^{-2}) \, .$$

The regularized criterion $\phi_{ec,\lambda}(\cdot)$ is concave and differentiable with respect to ξ. Its directional derivative at ξ in the direction ν is

$$F_{\phi_{ec,\lambda}}(\xi;\nu) = \frac{\int_{\mathscr{X}} \int_{\Theta} \exp\{-\lambda H_c(\xi,\theta')\} \, \Psi_{ec}(x,\theta',\xi) \, \mathrm{d}\theta' \, \nu(\mathrm{d}x)}{\int_{\Theta} \exp\{-\lambda H_c(\xi,\theta')\} \, \mathrm{d}\theta'} \, , \tag{7.36}$$

where

$$\begin{aligned} \Psi_{ec}(x,\theta',\xi) = {} & (K + |h(\theta') - h(\theta)|^{-2}) \\ & \times \left\{ [\eta(x,\theta') - \eta(x,\theta)]^2 - \|\eta(\cdot,\theta') - \eta(\cdot,\theta)\|_\xi^2 \right\} \, . \end{aligned}$$

A necessary-and-sufficient condition for optimality is as follows: ξ^* is optimal for $\phi_{ec,\lambda}(\cdot)$ if an only if

$$\max_{x \in \mathscr{X}} \int_{\Theta} \exp\{-\lambda H_c(\xi^*,\theta')\} \, \Psi_{ec}(x,\theta',\xi^*) \, \mathrm{d}\theta' \leq 0 \, .$$

Again, in order to make the computations easier, the integrals on θ' in (7.35), (7.36) can be replaced by finite sums.

7.7.4 Numerical Examples

Three examples are presented below that illustrate the influence of the design on LS estimability and serve to indicate some implementation details for the construction of extended E and c-optimal designs.

Example 7.19. We consider the same example as in (Kieffer and Walter, 1998), see also Pronzato and Walter (2001), with

$$\eta(x,\theta) = \theta_1 [\exp(-\theta_2 x) - \exp(-\theta_3 x)] \, , \; \theta = (\theta_1, \theta_2, \theta_3)^\top \, , \; x \in \mathbb{R}^+ \, . \tag{7.37}$$

The analysis is made at $\theta^0 = (0.773\,, \; 0.214\,, \; 2.09)^\top$, the value of the LS estimator obtained from the data in (Kieffer and Walter, 1998) with the 16-point design $X_0 = (1, 2, \ldots, 16)$. We shall denote by ξ_0 the associated design measure

$$\xi_0 = \left\{ \begin{array}{cccc} 1 & 2 & \cdots & 16 \\ 1/16 & 1/16 & \cdots & 1/16 \end{array} \right\} ,$$

where the first row gives the support points and the second one their respective weights. The design space $\mathscr{X}$ is given by a regular grid of 1601 points in $[0, 16]$ spaced by $1/100$.

The D and E-optimal designs for θ^0 in $\mathscr{X}$ are supported on three points and are, respectively, given by

$$\xi_D^*(\theta^0) \simeq \begin{Bmatrix} 0.42 & 1.82 & 6.80 \\ 1/3 & 1/3 & 1/3 \end{Bmatrix},$$
$$\xi_E^*(\theta^0) \simeq \begin{Bmatrix} 0.29 & 1.83 & 9.0 \\ 0.4424 & 0.3318 & 0.2258 \end{Bmatrix}.$$

The values of $\det[\mathbf{M}(\xi,\theta^0)]$ and $\lambda_{\min}[\mathbf{M}(\xi,\theta^0)]$ for the three designs ξ_0, ξ_D^* and ξ_E^* are indicated in Table 7.1. We also indicate in the same table the values of the parametric, intrinsic, and total curvatures, computed at θ^0 using the algorithm from Bates and Watts (1980), for the different designs. Since they have different numbers of support points, normalized curvatures—for design measures—are considered (see Remark 6.1), that is,

$$\begin{aligned}
C_{int}(\xi,\theta) &= \sup_{\mathbf{u}\in\mathbb{R}^p-\{\mathbf{0}\}} \frac{\|[I-P_\theta]\sum_{i,j=1}^p u_i[\partial^2\eta(\cdot,\theta)/\partial\theta_i\partial\theta_j]u_j\|_\xi}{\mathbf{u}^\top\mathbf{M}(\xi,\theta)\mathbf{u}},\\
C_{par}(\xi,\theta) &= \sup_{\mathbf{u}\in\mathbb{R}^p-\{\mathbf{0}\}} \frac{\|P_\theta\sum_{i,j=1}^p u_i[\partial^2\eta(\cdot,\theta)/\partial\theta_i\partial\theta_j]u_j\|_\xi}{\mathbf{u}^\top\mathbf{M}(\xi,\theta)\mathbf{u}},\\
C_{tot}(\xi,\theta) &= \sup_{\mathbf{u}\in\mathbb{R}^p-\{\mathbf{0}\}} \frac{\|\sum_{i,j=1}^p u_i[\partial^2\eta(\cdot,\theta)/\partial\theta_i\partial\theta_j]u_j\|_\xi}{\mathbf{u}^\top\mathbf{M}(\xi,\theta)\mathbf{u}},\\
&\le C_{int}(\xi,\theta)+C_{par}(\xi,\theta),
\end{aligned}$$

with P_θ the projector defined by (3.83).

Table 7.1. Performances of different designs in Example 7.19

ξ	$\det[\mathbf{M}(\xi,\theta^0)]$	$\lambda_{\min}[\mathbf{M}(\xi,\theta^0)]$	$\phi_{eE,\lambda}(\xi)$	$C_{par}(\xi,\theta^0)$	$C_{int}(\xi,\theta^0)$	$C_{tot}(\xi,\theta^0)$
ξ_0	$6.38\cdot10^{-6}$	$1.92\cdot10^{-4}$	$3.12\cdot10^{-4}$	180.7	15.73	181.3
ξ_D^*	$1.40\cdot10^{-4}$	$1.69\cdot10^{-3}$	$1.88\cdot10^{-3}$	58.03	0	58.03
ξ_E^*	$9.19\cdot10^{-5}$	$2.04\cdot10^{-3}$	$2.41\cdot10^{-3}$	50.72	0	50.72
$\xi_{eE,\lambda}^*$	$9.73\cdot10^{-5}$	$2.00\cdot10^{-3}$	$2.52\cdot10^{-3}$	55.35	0	55.35

The design $\xi_{eE,\lambda}^*$ is optimal for the criterion $\phi_{eE,\lambda}(\cdot)$ given by (7.33); with $K=1$, $\lambda=10^5$ and with the integral replaced by a discrete sum over 800 points uniformly distributed over the sphere centered at θ^0 and of radius 10^{-2}

One may notice that the model (7.37) is only locally identifiable[2] if $\theta\in\mathbb{R}^3$. Indeed, exchanging the value of θ_2 and θ_3 and changing θ_1 into

[2]This is true for almost any θ (in the sense of zero Lebesgue measure on compact subsets on $\mathbb{R}^3$); notice that the model is not identifiable for $\theta_1=0$.

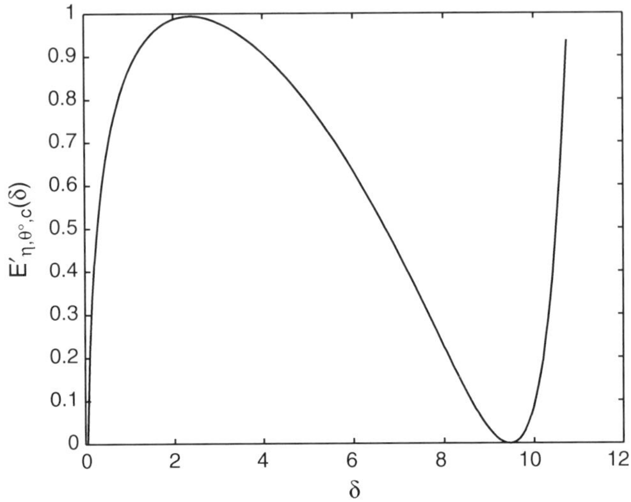

Fig. 7.16. Estimability function $E'_{\eta,\theta^0,\mathbf{c}}(\cdot)$ given by (7.23) for the design X_0 and $\mathbf{c} = (1\,,\ -1.2135\,,\ 1.2135)^\top$ in Example 7.19

$-\theta_1$ leaves $\eta(x,\theta)$ unmodified for any x. This is revealed when plotting the function $E'_{\eta,\theta^0,\mathbf{c}}(\cdot)$ given by (7.23) for the design X_0 and $\mathbf{c}$ close to $(1\,,\ -1.2135\,,\ 1.2135)^\top$ (with $1.2135 \simeq (\theta_3^0 - \theta_2^0)/(2\theta_1^0)$); see Fig. 7.16.

This problem of identifiability can be avoided by restricting Θ to positive θ_1 only. Taking $\Theta = [0,5] \times [0,5] \times [0,5]$, Kieffer and Walter (1998) find that for the observations $\mathbf{y}$ given in their Table 13.1, the LS criterion (7.3) has a global minimizer (the value we have taken here for θ^0) and two other local minimizers in Θ. This is due to the intrinsic curvature of the expectation surface for ξ_0; see Table 7.1. Figure 7.17 (resp. 7.18) presents the function $E'^{\xi}_{\eta,\theta^0,2}(\cdot)$ (resp. $E'^{\xi}_{\eta,\theta^0,3}(\cdot)$) defined by (7.22) for the three designs ξ_0, ξ_D^*, and ξ_E^*. The influence of the curvatures given in Table 7.1 for the three designs can be seen on those two figures. We construct below an optimal design for $\phi_{eE,\lambda}(\cdot)$ (see (7.33)) and show that it also yields $C_{int}(\xi,\theta) = 0$ for all θ.

To replace integrals by finite sums in (7.33), (7.34), we consider regular grids of points uniformly distributed on spheres centered at θ^0. Denote by $\mathcal{G}(\rho, M)$ such a grid, formed of M points at distance ρ from θ^0. We take $K = 1$. Note that the choice of K does not affect the optimal design when $\|\theta' - \theta\|$ is kept constant. The regularizing parameter λ in $\phi_{eE,\lambda}(\cdot)$ is set to 10^5. The optimal design for $\phi_{eE,\lambda}(\cdot)$ may be singular, i.e., supported on less than 3 points, when M is too small. When ρ tends to zero and M is large enough, the optimal design for the grid $\mathcal{G}(\rho, M)$ tends to ξ_E^*; we obtain

$$\xi^*_{eE,\lambda}(\theta^0) \simeq \left\{ \begin{array}{ccc} 0.3 & 1.7 & 7.9 \\ 0.491 & 0.307 & 0.202 \end{array} \right\}$$

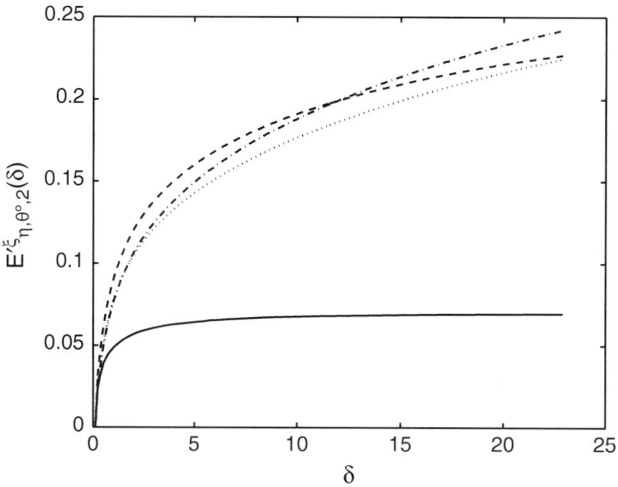

Fig. 7.17. Estimability function $E'^{\xi}_{\eta,\theta^0,2}(\cdot)$ given by (7.22) for the designs ξ_0 (*solid line*), ξ_D^* (*dashed line*), ξ_E^* (*dotted line*), and $\xi^*_{eE,\lambda}$ (*dash-dotted line*) in Example 7.19

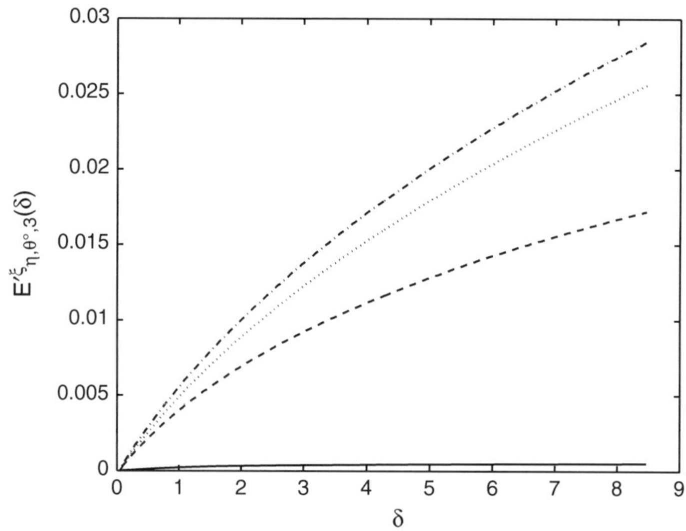

Fig. 7.18. Same as Fig. 7.17 but for $E'^{\xi}_{\eta,\theta^0,3}(\cdot)$

for $\rho = 0.01$ and $M = 800$. Its performances are indicated in Table 7.1. The normalized functions $E'^{\xi}_{\eta,\theta^0,2}(\cdot)$ and $E'^{\xi}_{\eta,\theta^0,3}(\cdot)$ obtained for $\xi^*_{eE,\lambda}$ are plotted in Figs. 7.17 and 7.18 (dash-dotted lines) and demonstrate the superiority of $\xi^*_{eE,\lambda}$ over the three other designs considered in terms of estimability properties. A direct optimization of $\phi_{eE}(\cdot)$ by LP for θ' on the same grid $\mathcal{G}(0.01, 800)$ yields a design ξ_{eE}^* close to $\xi_{eE,\lambda}^*$:

$$\xi_{eE}^*(\theta^0) \simeq \left\{ \begin{array}{ccc} 0.32 & 1.76 & 8.22 \\ 0.481 & 0.309 & 0.210 \end{array} \right\},$$

with $\phi_{eE}(\xi_{eE}^*) \simeq 2.55 \cdot 10^{-3}$ (evaluated on $\mathcal{G}(0.01, 800)$). □

Example 7.20. The model response is given by

$$\eta(\mathbf{x}, \theta) = \theta_1\{\mathbf{x}\}_1 + \theta_1^3(1 - \{\mathbf{x}\}_1) + \theta_2\{\mathbf{x}\}_2 + \theta_2^2(1 - \{\mathbf{x}\}_2)\,,\ \theta = (\theta_1, \theta_2)^\top\,,$$

for $\mathbf{x} \in \mathscr{X} = [0, 1]^2$. We consider local designs for $\theta^0 = (1/8, 1/8)^\top$. One may notice that the set $\{\partial\eta(\mathbf{x}, \theta)/\partial\theta\big|_{\theta^0} : \mathbf{x} \in \mathscr{X}\}$ is the rectangle $[3/64, 1] \times [1/4, 1]$, so that optimal designs for any isotonic criterion $\Phi(\cdot)$ (see Definition 5.3) are supported on the vertices (0, 1), (1, 0), and (1, 1) of $\mathscr{X}$. The D- and E-optimal designs are supported on three and two points, respectively,

$$\xi_D^*(\theta^0) \simeq \left\{ \begin{array}{ccc} \begin{pmatrix} 0 \\ 1 \end{pmatrix} & \begin{pmatrix} 1 \\ 0 \end{pmatrix} & \begin{pmatrix} 1 \\ 1 \end{pmatrix} \\ 0.4134 & 0.3184 & 0.2682 \end{array} \right\},\quad \xi_E^*(\theta^0) \simeq \left\{ \begin{array}{cc} \begin{pmatrix} 0 \\ 1 \end{pmatrix} & \begin{pmatrix} 1 \\ 0 \end{pmatrix} \\ 0.5113 & 0.4887 \end{array} \right\}.$$

When only the design points $\mathbf{x}_1 = (0\ 1)$ and $\mathbf{x}_2 = (1\ 0)$ are used, the model is only locally LS estimable. Indeed, the equations in θ'

$$\begin{aligned} \eta(\mathbf{x}_1, \theta') &= \eta(\mathbf{x}_1, \theta) \\ \eta(\mathbf{x}_2, \theta') &= \eta(\mathbf{x}_2, \theta) \end{aligned}$$

give not only the trivial solutions $\theta_1' = \theta_1$ and $\theta_2' = \theta_2$ but also θ_1' and θ_2' as roots of two univariate polynomials of the fifth degree, with coefficients depending on θ. Since these polynomials always admit at least one real root, at least one solution exists for θ' that is different from θ. For $\theta = \theta^0$, the vector $\theta^{0\prime} = (-0.9760\,,\ 0.3094)^\top$ gives approximately the same values as θ^0 for the responses at $\mathbf{x}_1$ and $\mathbf{x}_2$. This is confirmed by plotting the function $E'^{\xi}_{\eta,\theta^0,\mathbf{c}}(\cdot)$ given by (7.24) for $\xi = \xi_E^*$ and $\mathbf{c} = (1\,,\ -0.8462)^\top$ (with $-0.8462 \simeq (\theta^{0\prime}_2 - \theta_2^0)/(\theta^{0\prime}_1 - \theta_1^0)$); see Fig. 7.19. See also Fig. 7.21 for a plot of $E^{\xi_E^*}_{\eta,\theta^0}(\cdot)$ given by (7.19).

The model is globally LS estimable when the design points $\mathbf{x}_1 = (0\ 1)$, $\mathbf{x}_2 = (1\ 0)$, and $\mathbf{x}_3 = (1\ 1)$ are used, that is, the polynomial equations in θ' given by

$$\begin{aligned} \eta(\mathbf{x}_1, \theta') &= \eta(\mathbf{x}_1, \theta) \\ \eta(\mathbf{x}_2, \theta') &= \eta(\mathbf{x}_2, \theta) \\ \eta(\mathbf{x}_3, \theta') &= \eta(\mathbf{x}_3, \theta) \end{aligned}$$

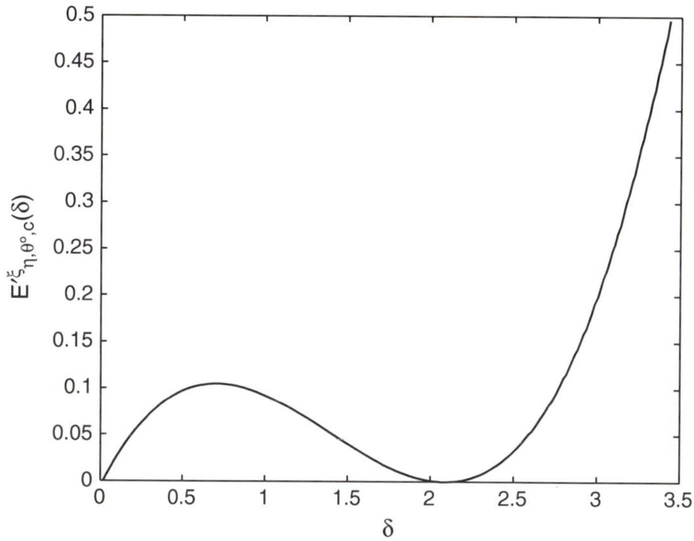

Fig. 7.19. Estimability function $E'^{\xi}_{\eta,\theta^0,\mathbf{c}}(\cdot)$ given by (7.24) for $\xi = \xi_E^*$ and $\mathbf{c} = (1\,,\; -0.8462)^\top$ in Example 7.20

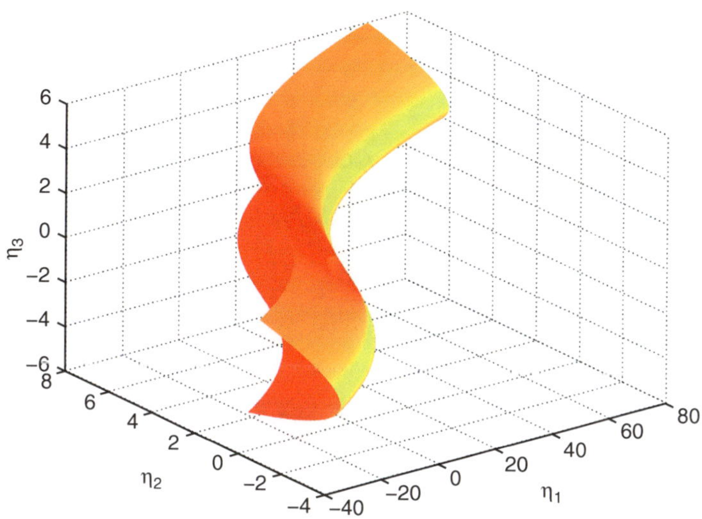

Fig. 7.20. Expectation surface $\mathbb{S}_\eta$ in Example 7.20 for the design $(\mathbf{x}_1, \mathbf{x}_2, \mathbf{x}_3)$ and $\theta \in [-3, 4] \times [-2, 2]$ ($\eta_i = \eta(\mathbf{x}_i, \theta)$, $i = 1, 2, 3$)

only have the solution $\theta' = \theta$. However, the expectation surface $\mathbb{S}_\eta$ for the design $(\mathbf{x}_1, \mathbf{x}_2, \mathbf{x}_3)$ has a rather complicated and strongly curved shape; see Fig. 7.20. The situation is thus favorable to the presence of local minima for the LS criterion; see Pronzato and Walter (2001) and Walter and Pronzato (1997,

Table 7.2. Performances of different designs in Example 7.20

ξ	$\det[\mathbf{M}(\xi,\theta^0)]$	$\lambda_{\min}[\mathbf{M}(\xi,\theta^0)]$	$\phi_{eE}(\xi)$	$C_{par}(\xi,\theta^0)$	$C_{int}(\xi,\theta^0)$	$C_{tot}(\xi,\theta^0)$
ξ_D^*	0.277	0.273	$3.30 \cdot 10^{-3}$	1.096	0.541	1.221
ξ_E^*	0.244	0.367	$3.23 \cdot 10^{-4}$	1.191	0	1.191
$\xi_{eE,\lambda}^{*,1}$	$7.68 \cdot 10^{-3}$	$5.06 \cdot 10^{-3}$	$2.36 \cdot 10^{-5}$	43.33	0	43.33
$\xi_{eE,\lambda}^{*,2}$	$8.63 \cdot 10^{-2}$	$7.98 \cdot 10^{-2}$	$7.28 \cdot 10^{-3}$	0.581	2.277	2.350
ξ_{eE}^*	$7.25 \cdot 10^{-2}$	$4.57 \cdot 10^{-2}$	$9.80 \cdot 10^{-3}$	0.776	3.201	3.290

$K = 0.01$ and $\lambda = 10^3$ for $\xi_{eE,\lambda}^{*,i}$, $i = 1, 2$; $\phi_{eE}(\cdot)$ is evaluated by a discrete sum (40,000 points) over $\cup_{i=1}^{200}\mathcal{G}(0.01\,i, 200)$

Chap. 4). The values of $\det[\mathbf{M}(\xi,\theta^0)]$ and $\lambda_{\min}[\mathbf{M}(\xi,\theta^0)]$ and the curvatures $C_{int}(\xi,\theta)$, $C_{int}(\xi,\theta)$, and $C_{int}(\xi,\theta)$ for the designs ξ_D^* and ξ_E^* are indicated in Table 7.2.

We have seen that estimability criterion (7.33) can produce designs close to an E-optimal design. Since here the design ξ_E^* makes the model only locally LS estimable, a design optimal for (7.33) can be expected to be significantly different. Similarly to what was done in Example 7.19, we replace integrals by finite sums in (7.33), (7.34) and consider regular grids $\mathcal{G}(\rho, M)$ formed of M points uniformly distributed on a circle centered at θ^0 with radius ρ. We take $K = 0.01$ (to enforce the protection against $\eta(\theta')$ being close to $\eta(\theta^0)$ for θ' far from θ^0) and $\lambda = 10^3$.

We first take a single grid $\mathcal{G}(1.25, 100)$; the choice $\rho = 1.25$ is to allow us to detect the presence of $\theta^{0'} = (-0.9760\,,\ 0.3094)^\top$ at approximate distance 1.116 from θ^0. The optimal design is then

$$\xi_{eE,\lambda}^{*,1}(\theta^0) \simeq \left\{ \begin{matrix} \begin{pmatrix} 0 \\ 0 \end{pmatrix} & \begin{pmatrix} 1 \\ 1 \end{pmatrix} \\ 0.2472 & 0.7528 \end{matrix} \right\}.$$

This design ensures that $\eta(\theta')$ is significantly different from $\eta(\theta^0)$ for all θ' at a distance about 1.25 from θ^0, but it does not protect against $\eta(\theta')$ being close to $\eta(\theta^0)$ for some θ closer or further from θ^0. In fact, the model is still only locally LS estimable for $\xi_{eE,\lambda}^{*,1}$: the two vectors of parameters $\theta^{0''} = (1/16)\,(-9-\sqrt{173}\,,\ 13+\sqrt{173})^\top$ and $\theta^{0'''} = (1/16)\,(-9+\sqrt{173}\,,\ 13-\sqrt{173})^\top$, respectively at approximate distances 2.135 and 0.190 from θ^0, yield the same model responses as θ^0. This shows that the choice of the set over which is computed the integral (or the discrete sum) in $\phi_{eE,\lambda}(\cdot)$ must be large enough to cover the region of interest around θ^0.

Take now a collection of grids $\mathcal{G}(\rho, 100)$ to optimize $\phi_{eE,\lambda}(\cdot)$, with ρ varying from 0.1 to 2, i.e., $\mathcal{G} = \cup_{i=1}^{20}\mathcal{G}(0.1\,i, 100)$. The corresponding optimal design is

$$\xi^{*,2}_{eE,\lambda}(\theta^0) \simeq \left\{ \begin{array}{ccc} \begin{pmatrix} 0 \\ 0 \end{pmatrix} & \begin{pmatrix} 1 \\ 0 \end{pmatrix} & \begin{pmatrix} 1 \\ 1 \end{pmatrix} \\ 0.2600 & 0.3575 & 0.3825 \end{array} \right\} .$$

When we use the finer grid $\mathcal{G} = \cup_{i=1}^{200} \mathcal{G}(0.01\,i, 200)$ and optimize $\phi_{eE}(\cdot)$ by LP, we get the optimal design

$$\xi^{*}_{eE}(\theta^0) \simeq \left\{ \begin{array}{cccc} \begin{pmatrix} 0 \\ 0 \end{pmatrix} & \begin{pmatrix} 0 \\ 1 \end{pmatrix} & \begin{pmatrix} 1 \\ 0 \end{pmatrix} & \begin{pmatrix} 1 \\ 1 \end{pmatrix} \\ 0.1131 & 0.0143 & 0.1377 & 0.7349 \end{array} \right\} .$$

The model is globally LS estimable for $\xi^{*,2}_{eE,\lambda}$ and ξ^*_{eE}. Figure 7.21 shows the estimability function $E^{\xi}_{\eta,\theta^0}(\cdot)$ given by (7.19) for the four designs ξ^*_D, ξ^*_E, $\xi^{*,2}_{eE,\lambda}$, and ξ^*_{eE}. The values of $\det[\mathbf{M}(\xi,\theta^0)]$, $\lambda_{\min}[\mathbf{M}(\xi,\theta^0)]$, and $\phi_{eE}(\xi)$ and of the curvatures $C_{int}(\xi,\theta)$, $C_{int}(\xi,\theta)$, and $C_{int}(\xi,\theta)$ for $\xi^{*,1}_{eE,\lambda}$, $\xi^{*,2}_{eE,\lambda}$, and ξ^*_{eE} are indicated in Table 7.2. The intrinsic and total curvatures for $\xi^{*,2}_{eE,\lambda}$ and ξ^*_{eE} are significantly larger than those for ξ^*_D and ξ^*_E. However, Fig. 7.21 indicates that the minimum of $\|\eta(\cdot,\theta') - \eta(\cdot,\theta^0)\|_{\xi}$, say for $\|\theta' - \theta^0\|^2 > 1.5$, is 0.149 for ξ^*_{eE} and 0.131 for $\xi^{*,2}_{eE,\lambda}$, but only 0.082 for ξ^*_D; it is zero for ξ^*_E since the model is only locally LS estimable for this design. Since K is small ($K = 0.01$), the computed values of $\phi_{eE}(\cdot)$ and $\omega^*_{\theta^0}$ given by (7.16) for the different designs are rather close; we get $\omega^*_{\theta^0} \simeq 3.23 \cdot 10^{-3}$, $3.16 \cdot 10^{-4}$, $2.36 \cdot 10^{-5}$, $7.11 \cdot 10^{-3}$, and $9.55 \cdot 10^{-3}$ for ξ^*_D, ξ^*_E, $\xi^{*,1}_{eE,\lambda}$, $\xi^{*,2}_{eE,\lambda}$, and ξ^*_{eE}, respectively. □

Example 7.21. Consider again the model (7.37) of Example 7.19, with now $\theta^0 = (21.80\,,\ 0.05884\,,\ 4.298)^\top$, the nominal value used in (Atkinson et al., 1993). The D- and E-optimal designs for θ^0 are, respectively, given by

$$\xi^*_D(\theta^0) \simeq \left\{ \begin{array}{ccc} 0.229 & 1.389 & 18.42 \\ 1/3 & 1/3 & 1/3 \end{array} \right\} ,$$

$$\xi^*_E(\theta^0) \simeq \left\{ \begin{array}{ccc} 0.170 & 1.398 & 23.36 \\ 0.199 & 0.662 & 0.139 \end{array} \right\} .$$

The c-optimal design for the estimation of the area under the curve

$$h(\theta) = \int_0^\infty \eta(x,\theta)\,\mathrm{d}x = \theta_1\ (1/\theta_2 - 1/\theta_3) \tag{7.38}$$

is the two-point design

$$\xi^*_c(\theta^0) \simeq \left\{ \begin{array}{cc} 0.233 & 17.63 \\ 0.0135 & 0.9865 \end{array} \right\} .$$

Consider now the regularized criterion $\phi_{ec,\lambda}(\cdot)$ given by (7.35). We take $K = 1$ and $\lambda = 10^3$ and replace the integral in (7.35) by a discrete sum on a grid

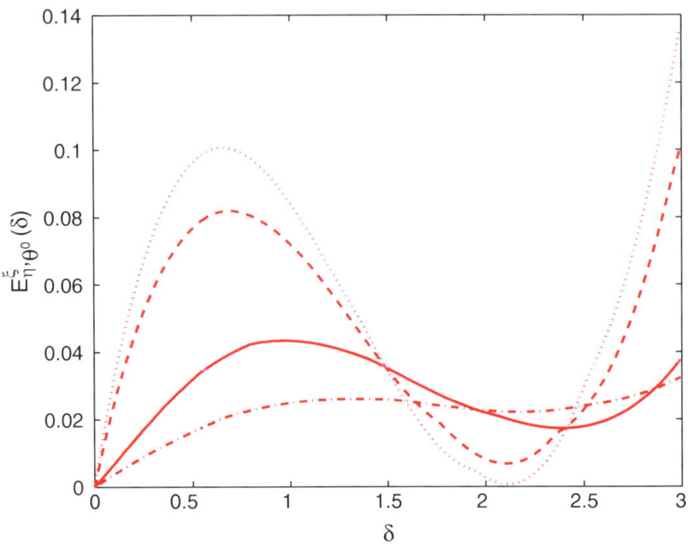

Fig. 7.21. Estimability function $E^{\xi}_{\eta,\theta^0}(\cdot)$ given by (7.19) for the four designs ξ_D^* (*dashed line*), ξ_E^* (*dotted line*), $\xi_{eE,\lambda}^{*,2}$ (*solid line*), and ξ_{eE}^* (*dash-dotted line*) in Example 7.20

Table 7.3. Performances of different designs in Example 7.21

ξ	$\det[\mathbf{M}(\xi,\theta^0)]$	$\lambda_{\min}[\mathbf{M}(\xi,\theta^0)]$	$\mathbf{c}^\top\mathbf{M}^-(\xi,\theta^0)\mathbf{c}$	$\phi_{ec,\lambda}(\xi)$
ξ_D^*	$1.62\cdot 10^3$	0.191	$6.39\cdot 10^3$	0.315
ξ_E^*	$6.87\cdot 10^2$	0.316	$1.65\cdot 10^4$	0.156
ξ_c^*	0	0	$2.19\cdot 10^3$	$\simeq 0$
ξ_{ec}^*	$1.28\cdot 10^3$	0.220	$5.79\cdot 10^3$	0.363

$\mathbf{c} = \partial h(\theta)/\partial\theta|_{\theta^0}$, with $h(\theta)$ given by (7.38); $K=1$ and $\lambda=10^3$ in $\phi_{ec,\lambda}(\xi)$, see (7.35), with the integral replaced by a discrete sum (5,000 points on the ellipsoid defined by $(\theta-\theta^0)^\top\mathbf{\Omega}(\theta-\theta^0)=1$ with $\mathbf{\Omega}=\mathrm{diag}(1/4,10^4,1)$)

of 5,000 points on the ellipsoid defined by $(\theta-\theta^0)^\top\mathbf{\Omega}(\theta-\theta^0)=1$ with $\mathbf{\Omega}=\mathrm{diag}(1/4,10^4,1)$, thus allowing a variation of ± 2, ± 0.01 and ± 1 on θ_1, θ_2, and θ_3, respectively. We restrict the design space to the eight points corresponding to the union of the supports of $\xi_D^*(\theta^0)$, $\xi_E^*(\theta^0)$, and $\xi_c^*(\theta^0)$. We then obtain the three-point optimal design

$$\xi_{ec}^*(\theta^0) \simeq \left\{ \begin{array}{ccc} 0.170 & 1.389 & 23.36 \\ 0.233 & 0.368 & 0.399 \end{array} \right\}.$$

The performances of these different designs are indicated in Table 7.3. One may notice that the design $\xi_{ec}^*(\theta^0)$, optimal for $\phi_{ec,\lambda}(\cdot)$, is second best for each other criterion. □

7.8 Remarks on Estimability for Estimators Other than LS

As mentioned in Sect. 7.2, the notion of estimability depends on the estimator that is used. Only the case of LS estimation has been considered throughout this chapter. This is not necessarily a strong limitation since the global LS estimability at $\bar{\theta}$ for X also guarantees that for the replicated design $X^{\otimes n}$ with n tending to infinity, the WLS estimator with strictly positive weights (see (3.10)), the TSLS estimator (see (3.55) in Theorem 3.26), and the M and ML estimators (see Theorems. 4.1 and 4.5) are all strongly consistent. It seems useful, however, to give precise definitions of estimability in regression models for the estimators considered in Chaps. 3 and 4 and to construct their specific estimability functions and extended optimality criteria. This is the objective of this section.

Take any exact design $X = (x_1, \ldots, x_k)$ and consider the replicated design $X^{\otimes n}$. Consider one of the estimators $\hat{\theta}^N$ of Chaps. 3 and 4, with $\hat{\theta}^N$ minimizing a criterion $J_N(\theta)$, with $N = n \times k$ and k the number of design points in X. The notion of estimability for the estimator $\hat{\theta}^N$ is then based on the property $J_N(\theta) \overset{\theta}{\rightsquigarrow} J_{\bar{\theta}}(\theta)$ a.s. as $n \to \infty$ with $J_{\bar{\theta}}(\cdot)$ having a unique minimum at $\theta = \bar{\theta}$: the model is globally estimable at $\bar{\theta}$ for X if

$$\theta' \in \Theta \text{ and } J_{\bar{\theta}}(\theta') = J_{\bar{\theta}}(\bar{\theta}) \Longrightarrow \theta' = \bar{\theta}\,,$$

in agreement with the estimability conditions used in the consistency theorems of Chaps. 3 and 4. Notice that the estimator is defined by the criterion $J_N(\cdot)$ and that estimability is defined by a property of the limiting function $J_{\bar{\theta}}(\cdot)$.

Now, to any limiting function $J_\theta(\cdot)$, we associate the estimability function

$$E_{J_\theta,\theta}(\cdot) : \delta \in \mathbb{R}^+ \longrightarrow E_{J_\theta,\theta}(\delta) = \min_{\theta' \in \Theta, \|\theta'-\theta\|^2 = \delta} J_\theta(\theta') - J_\theta(\theta)\,.$$

For instance, for LS estimation we have $J_\theta(\theta') - J_\theta(\theta) = \|\eta(\cdot, \theta') - \eta(\cdot, \theta)\|_\xi^2$ with ξ the design measure that put weight $1/k$ at each point of X (see the proof of Theorem 3.1), and we recover the definition (7.19) of $E_{\eta,\theta}^{\xi}(\cdot)$. When $J_\theta(\theta')$ is a smooth function of θ', we have $J_\theta(\theta') - J_\theta(\theta) = \frac{1}{2}(\theta' - \theta)^\top \mathbf{H}(\xi, \theta)(\theta' - \theta) + \mathcal{O}(\|\theta' - \theta\|^3)$ and obtain

$$E_{J_\theta,\theta}(\delta) = \frac{\delta}{2}\,\lambda_{\min}[\mathbf{H}(\xi, \theta)] + \mathcal{O}(\delta^{3/2})$$

with $\mathbf{H}(\xi, \theta)$ the Hessian matrix $\mathbf{H}(\xi, \theta) = \nabla_\theta^2 J_\theta(\theta) = \partial^2 J_\theta(\theta')/\partial\theta' \partial\theta'^\top \big|_{\theta'=\theta}$.

An extended E-optimality criterion can be defined similarly to what was done in Sect. 7.7.1,

$$\phi_{eE}(\xi; \theta) = \min_{\delta \geq 0} \{E_{J_\theta,\theta}(\delta)(K + 1/\delta)\}$$

with K some positive number. By letting K tend to infinity, we obtain optimal designs that tend to be E-optimal for $\mathbf{H}(\xi,\theta)$, whereas by taking K not too large, we enable some protection against $J_\theta(\theta') - J_\theta(\theta)$ being small for some θ' far from θ. Other extended optimality criteria can be defined in a similar way; see Sect. 7.7.2 for c-optimality and Remark 7.18 for G-optimality.

We have already used the Hessian matrix $\mathbf{H}(\xi,\theta)$ when investigating the asymptotic distribution of estimators, and several situations encountered in Chaps. 3 and 4 are considered below. It appears that constructing an optimal design based on the matrix $\mathbf{H}(\xi,\theta)$ corresponds to designing under the same idealistic assumptions as those considered in Sect. 5.5 for situations where the asymptotic variance–covariance matrix of the estimator is in product form.

The WLS estimator

The estimator minimizes $J_N(\theta)$ given by (3.5) and

$$J_\theta(\theta') - J_\theta(\theta) = \int_{\mathscr{X}} w(x)[\eta(x,\theta') - \eta(x,\theta)]^2\, \xi(\mathrm{d}x)\,,$$

see Remark 3.2-(ii). Direct calculation gives $\mathbf{H}(\xi,\theta) = 2\,\mathbf{M}_1(\xi,\theta)$ given by (3.23); see also Sect. 5.5.1.

The penalized WLS estimator

The estimator minimizes $J_N(\theta)$ given by (3.47) and

$$\begin{aligned} J_\theta(\theta') - J_\theta(\theta) &= \int_{\mathscr{X}} \lambda^{-1}(x,\theta')[\eta(x,\theta') - \eta(x,\theta)]^2\, \xi(\mathrm{d}x) \\ &\quad + \bar\beta \int_{\mathscr{X}} \left[\frac{\lambda(x,\theta)}{\lambda(x,\theta')} - \log\frac{\lambda(x,\theta)}{\lambda(x,\theta')} - 1\right] \xi(\mathrm{d}x)\,, \end{aligned}$$

see the proof of Theorem 3.22. Direct calculation gives $\mathbf{H}(\xi,\theta) = 2\,\mathbf{M}_1(\xi,\theta)$ given by (3.52); see also Sect. 5.5.2.

When $\bar\beta$ is considered as a nuisance parameter in the variance function (3.45), the estimator minimizes (3.49), see Remark 3.23, and

$$\begin{aligned} J_\theta(\theta') - J_\theta(\theta) &= \int_{\mathscr{X}} \log\frac{\lambda(x,\theta')}{\lambda(x,\theta)}\, \xi(\mathrm{d}x) \\ &\quad + \log\left\{\int_{\mathscr{X}} \frac{\lambda(x,\theta)}{\lambda(x,\theta')}\, \xi(\mathrm{d}x) + \int_{\mathscr{X}} \frac{[\eta(x,\theta') - \eta(x,\theta)]^2}{\beta\lambda(x,\theta')}\, \xi(\mathrm{d}x)\right\}, \end{aligned}$$

which gives

$$\mathbf{H}(\xi,\theta) = 2\left[\mathbf{M}_{1,\theta}(\xi,\bar\gamma) - \frac{\mathbf{v}_1(\xi,\bar\gamma)\mathbf{v}_1^\top(\xi,\bar\gamma)}{m_1(\bar\beta)}\right],$$

see (3.69).

The M estimator

The estimator minimizes $J_N(\theta)$ given by (4.11). We obtain

$$J_\theta(\theta')-J_\theta(\theta) = \int_{\mathscr{X}} \left(\int_{-\infty}^{\infty} \{\rho_x[\eta(x,\theta) - \eta(x,\theta') + \varepsilon] - \rho_x(\varepsilon)\} \, \bar{\varphi}_x(\varepsilon)\mathrm{d}\varepsilon \right) \xi(\mathrm{d}x) \, ,$$

see the proof of Theorem 4.1, and $\mathbf{H}(\xi,\theta) = \mathbf{M}_1(\xi,\theta)$ given by (4.18); see also Sect. 5.5.4.

The ML estimator

The estimator minimizes $J_N(\theta)$ given by (4.12). Direct calculations give $J_\theta(\theta') - J_\theta(\theta) = D(P_0 \| P_{\eta(\cdot,\theta')-\eta(\cdot,\theta)})$, the Kullback–Leibler divergence between $P_0(\mathrm{d}\varepsilon,\mathrm{d}x) = \bar{\varphi}_x(\varepsilon)\mathrm{d}\varepsilon\,\xi(\mathrm{d}x)$ and $P_{\eta(\cdot,\theta')-\eta(\cdot,\theta)}(\mathrm{d}\varepsilon,\mathrm{d}x) = \bar{\varphi}_x[\varepsilon + \eta(x,\theta) - \eta(x,\theta')]\mathrm{d}\varepsilon\,\xi(\mathrm{d}x)$, see Theorem 4.5, and $\mathbf{H}(\xi,\theta) = \mathbf{M}(\xi,\theta)$ given by (4.16).

7.9 Bibliographic Notes and Further Remarks

Distinguishability and discrimination

Distinguishability refers to the possibility of selecting one model structure among several candidates, the best one in terms of reproduction (and prediction) of the observations, see, e.g., Walter (1982, 1987) and Walter and Pronzato (1995). Although the notions of identifiability and distinguishability are not related (identifiability of two model structures is neither necessary nor sufficient for their distinguishability), distinguishability can be tested with methods similar to those used for testing identifiability. The quantitative counterpart to identifiability is experimental design for parameter estimation, which is the subject of this book. The quantitative counterpart to distinguishability is experimental design for model discrimination, where we try to maximize the power of statistical tests for choosing one model structure, see, e.g., Atkinson and Fedorov (1975a,b) for T-optimum design and the survey papers (Atkinson and Cox, 1974, Hill, 1978).

Experimental design for discrimination will not be considered here. However, as indicated in Sect. 1.3, parameter estimation can be used to test the validity of a model. D_s-optimum design, which focuses on the estimation of a subset of the parameter vector θ, see Chap. 5, can therefore also be used for model discrimination.

8

Nonlocal Optimum Design

The design criteria considered in Chap. 5 for nonlinear models are local, in the sense that they depend on the choice of a prior nominal value θ^0 for the model parameters θ. Similarly, the criteria of Chap. 6 for LS estimation in nonlinear regression models depend on the true value $\bar{\theta}$ of the parameters, and at the design stage, the unknown $\bar{\theta}$ must be replaced by some nominal value θ^0. The design criteria of Chap. 7 are based on a local estimability function and, in a certain sense, are thus local too.

This dependence of criteria on the location of θ in Θ is fundamental for nonlinear models since the amount of information collected from an experiment may very much depend on the true value $\bar{\theta}$ of the parameters. Therefore, choosing a design criterion is not enough in a nonlinear situation; we must also guess a value θ^0 for θ. Nothing forces this choice of θ^0 to be unique, and we may (should) consider a family of criteria $\xi \longrightarrow \phi(\xi;\theta^{(i)})$ for $\theta^{(i)} \in \Theta$. A first intuitive approach is to consider a finite set $\{\theta^{(1)},\ldots,\theta^{(M)}\}$, solve M local design problems, one for each $\theta^{(i)}$, and compare the values $\phi(\xi^j;\theta^{(i)})$ for each pair (i,j), with ξ^j the optimal design for $\theta^{(j)}$. Alternatively, comparing the performances of a procedure on a set of examples is usually done through mean, extremal, median, or quantile values. It is the purpose of this chapter to formalize such approaches and present their main properties. We shall always suppose that the criterion $\phi(\xi;\theta)$ for a given θ is to be maximized with respect to ξ. The presentation is for design measures, but the transposition to exact designs X is straightforward.

Section 8.1 is devoted to average-optimum design, where a prior probability measure $\mu(\cdot)$ is set on Θ. The average-optimal version of $\phi(\xi;\theta)$ is then $\phi_{AO}(\xi) = \int_\Theta \phi(\xi;\theta)\,\mu(\mathrm{d}\theta)$. Maximin-optimum design is considered in Sect. 8.2, with $\phi(\xi;\theta)$ replaced by $\phi_{MmO}(\xi) = \min_{\theta\in\Theta}\phi(\xi;\theta)$. Section 8.3 shows that a smooth transition between average and maximin optimality is possible, which allows us to avoid some algorithmic difficulties inherent to the maximin approach. Section 8.4 considers quantiles and probability level criteria, with the substitution of the median for the mean value for $\mu(\cdot)$ as a particular case. Sequential design is briefly considered in Sect. 8.5 in two

L. Pronzato and A. Pázman, *Design of Experiments in Nonlinear Models*, Lecture Notes in Statistics 212, DOI 10.1007/978-1-4614-6363-4_8,

particular situations: (*i*) the experiment is constructed in two stages, (*ii*) the experiment is fully sequential, i.e., we choose one design point at a time, and the design space $\mathscr{X}$ is finite.

8.1 Average-Optimum Design

When $\phi(\xi;\theta)$ is the design criterion for parameters θ, the associated average-optimality criterion is

$$\phi_{AO}(\xi) = \mathbb{E}_\mu\{\phi(\xi;\theta)\} = \int_\Theta \phi(\xi;\theta)\,\mu(\mathrm{d}\theta) \tag{8.1}$$

with $\mu(\cdot)$ a prior probability measure on Θ. Of course, when $\mu(\cdot)$ is the delta measure δ_{θ^0}, we recover the locally optimal design approach considered in previous chapters. We shall consider in particular the case where $\phi(\xi;\theta) = \Phi[\mathbf{M}(\xi,\theta)]$ with $\Phi(\cdot)$ one of the criteria considered in Chap. 5 and $\mathbf{M}(\xi,\theta)$ an information matrix.

One may wonder if it is legitimate to use a prior $\mu(\cdot)$ on θ for designing the experiment but not for estimating θ. This can be justified by asymptotic considerations (see, e.g., Theorem 4.16) or by the substitution of the prior on θ for the predictive distribution of the estimator, see Sect. 8.1.2. Another heuristic justification is that the harm caused by a bad prior at the design stage is usually less severe than the one that may result from using the same prior for estimation. Moreover, in many situations objective conclusions are required that should be based on standardized procedures not depending on subjective priors, whereas the use of all possible sources of information about the possible location of $\bar{\theta}$, even of subjective origin, is recommended to increase the chance of selecting an informative experiment.

8.1.1 Properties

When $\phi(\cdot;\theta)$ is positively homogeneous for all $\theta \in \Theta$, i.e., $\phi(a\xi;\theta) = a\phi(\xi;\theta)$ for any $a > 0$, any $\theta \in \Theta$, and any $\xi \in \Xi$, the set of probability measures on $\mathscr{X}$ (see Definition 5.3) then $\phi_{AO}(\cdot)$ is positively homogeneous too. An optimal design for $\phi(\cdot;\theta)$ is invariant with respect to the composition by a strictly increasing differentiable function $\psi(\cdot)$, i.e., $\arg\max_\xi \phi(\xi;\theta) = \arg\max_\xi \psi[\phi(\xi;\theta)]$; see Sect. 5.2.1 and Remark 5.22-(*i*). However, optimal designs for $\int_\Theta \psi[\phi(\cdot;\theta)]\,\mu(\mathrm{d}\theta)$ and $\psi\left[\int_\Theta \phi(\cdot;\theta)\,\mu(\mathrm{d}\theta)\right]$ generally differ when $\psi(\cdot)$ is a nonlinear function; see, e.g., Fedorov (1980). Also, maximizing $\mathbb{E}_\mu\{\phi(\xi;\theta)\}$ is different from maximizing the average efficiency

$$\mathscr{E}_{\phi_{AO}}(\xi) = \mathbb{E}_\mu\{\mathscr{E}_\phi(\xi;\theta)\} = \mathbb{E}_\mu\left\{\frac{\phi(\xi;\theta)}{\phi(\xi_\theta^*;\theta)}\right\},$$

where $\xi_\theta^* = \arg\max_{\xi\in\Xi} \phi(\xi;\theta)$; see Sect. 5.1.8. Note that using an efficiency criterion may prove quite useful when the magnitude of $\phi(\xi;\theta)$ depends strongly on the value of $\theta \in \Theta$.

The criterion $\phi_{AO}(\cdot)$ is concave when $\phi(\cdot;\theta)$ is concave for any $\theta \in \Theta$. Its directional derivative $F_{\phi_{AO}}(\xi;\nu)$ then exists for any $\xi, \nu \in \Xi$ such that $\phi_{AO}(\xi) > -\infty$ (see Lemma 5.16) and is given by

$$F_{\phi_{AO}}(\xi;\nu) = \int_\Theta F_{\phi_\theta}(\xi;\nu)\, \mu(\mathrm{d}\theta),$$

with $F_{\phi_\theta}(\xi;\nu)$ the directional derivative of $\phi(\cdot;\theta)$ at ξ in the direction ν. We thus obtain the following equivalence theorem for average-optimum design.

Theorem 8.1. *Let $\phi(\cdot;\theta)$ be a concave functional on the set Ξ of probability measures on $\mathscr{X}$ for any $\theta \in \Theta$. A design ξ^* is optimal for the criterion $\phi_{AO}(\cdot)$ defined by (8.1) (or ϕ_{AO}-optimal) if and only if*

$$\sup_{\nu\in\Xi} \int_\Theta F_{\phi_\theta}(\xi^*;\nu)\, \mu(\mathrm{d}\theta) = 0.$$

The theorem above takes a simple form when $\phi(\xi;\theta) = \Phi[\mathbf{M}(\xi,\theta)]$ with $\Phi(\cdot)$ differentiable; see Lemma 5.20: ξ^* is ϕ_{AO}-optimal if and only if

$$\max_{x\in\mathscr{X}} \int_\Theta \mathrm{trace}\left\{[\mathbf{M}_\theta(x) - \mathbf{M}(\xi,\theta)]\nabla_{\mathbf{M}}\Phi[\mathbf{M}(\xi,\theta)]\right\}\, \mu(\mathrm{d}\theta) = 0.$$

In that case, the determination of an average-optimal design ξ_{AO}^* does not raise any special difficulty apart from heavier computations than for locally optimum design due to evaluations of expected values for $\mu(\cdot)$. Numerical calculations are facilitated by taking $\mu(\cdot)$ as a discrete measure over a finite subset of Θ. When $\mu(\cdot)$ has a density with respect to the Lebesgue measure, the computations of integrals can be avoided by using stochastic approximation techniques; see Sect. 9.4.

Note that the bound on the number of support points indicated in Sect. 5.2.3 does not apply to $\phi_{AO}(\xi) = \int_\Theta \Phi[\mathbf{M}(\xi,\theta)]\, \mu(\mathrm{d}\theta)$. It has been reported by many authors that the number of support points of an average-optimal design ξ^* increases when the prior $\mu(\cdot)$ becomes less informative (see, e.g., Chaloner and Larntz (1989)); a sufficient condition is given in (Braess and Dette, 2007) under which the number of support points of ξ^* can be made arbitrarily large. On the other hand, average-optimal design measures generally have a moderate number of support points when Θ is not too big.

Example 8.2. Consider the simple case of the one-parameter regression model $\eta(x,\theta) = \exp(-\theta x)$ with homoscedastic errors having variance 1, $\theta > 0$, $x \geq 0$. The information matrix (here a scalar) for LS estimation with a design ξ on $\mathscr{X}$ is given by (3.32) which yields here $M(\xi,\theta) = \int_{\mathscr{X}} x^2 \exp(-2\theta x)\, \xi(\mathrm{d}x)$.

For any isotonic criterion $\Phi(\cdot)$, the optimal design ξ_θ^* on $\mathscr{X} = \mathbb{R}^+$ is the delta measure at $1/\theta$, and the associated value of the information matrix is $M(\xi_\theta^*, \theta) = 1/(\mathrm{e}\,\theta)^2$. We consider two different situations.

(A) Suppose that θ has the prior $\mu(\cdot)$ on a subset Θ of $\mathbb{R}^+$ and consider

$$\phi_{AO}(\xi) = \int_\Theta M(\xi,\theta)\,\mu(\mathrm{d}\theta) = \int_{\mathscr{X}}\int_\Theta x^2 \exp(-2\theta x)\,\mu(\mathrm{d}\theta)\,\xi(\mathrm{d}x).$$

Due to its linearity in ξ, we directly obtain that the optimal design is the delta measure at $x^* = \arg\max_{x\in\mathscr{X}} \int_\Theta x^2 \exp(-2\theta x)\,\mu(\mathrm{d}\theta)$; see Fig. 8.5 for an illustration.

Consider now the criterion $\phi_{ED}(\xi) = \int_\Theta \log[M(\xi,\theta)]\,\mu(\mathrm{d}\theta)$. For ξ the delta measure at x we obtain $\phi_{ED}(\delta_x) = \int_\Theta \log[x^2 \exp(-2\theta x)]\,\mu(\mathrm{d}\theta)$, and one can easily check that the optimal design among one-point design measures on $\mathscr{X} = \mathbb{R}^+$ is supported at $x^* = 1/\mathbb{E}_\mu(\theta)$. One can then check numerically that δ_{x^*} is optimal among all design measures on $\mathscr{X}$ when Θ remains reasonably small by computing

$$F_{\phi_{ED}}(\delta_{x^*}, x) = \int_\Theta \left(x^2\mathbb{E}_\mu^2(\theta)\exp\{-2\theta[x - 1/\mathbb{E}_\mu(\theta)]\} - 1\right)\,\mu(\mathrm{d}\theta)$$

and checking that $\max_{x\in\mathscr{X}} F_{\phi_{ED}}(\delta_{x^*}, x) \le 0$. When $\mu(\cdot)$ is uniform on $[\theta^* - \epsilon, \theta^* + \epsilon]$, $\epsilon \le \theta^*$, this is true if ϵ is not too large and numerical calculations indicate that δ_{x^*} is optimal on $\mathscr{X} = \mathbb{R}^+$ when $\epsilon/\theta^* \lesssim 0.9344$. It remains optimal for larger ϵ on a smaller design space $\mathscr{X}$. For instance, with $\mathscr{X} = [0,1]$, δ_{x^*} is optimal when $\epsilon \lesssim 4.680$ for $\theta^* = 5$, when $\epsilon \lesssim 3.768$ for $\theta^* = 4$, and when $\epsilon \lesssim 2.881$ for $\theta^* = 3$ and is optimal for $\mu(\cdot)$ uniform on $\Theta = [0,4]$.

(B) Consider now the efficiency criterion

$$\mathscr{E}_{\phi_{AO}}(\xi) = \mathbb{E}_\mu\left\{\frac{M(\xi,\theta)}{M(\xi_\theta^*,\theta)}\right\} = \int_\Theta\int_{\mathscr{X}} \mathrm{e}^2\theta^2 x^2 \exp(-2\theta x)\,\xi(\mathrm{d}x)\,\mu(\mathrm{d}\theta).$$

Due to its linearity in ξ, we directly obtain that the optimal design is the delta measure at $x^* = \arg\max_{x\in\mathscr{X}} \int_\Theta \theta^2 x^2 \exp(-2\theta x)\,\mu(\mathrm{d}\theta)$. □

The next section shows that some average-optimality criteria, which may be considered as ad hoc modifications of local criteria aimed at taking the dependence in θ^0 into account, correspond in fact to approximations of a Bayesian formulation of the design problem.

8.1.2 A Bayesian Interpretation

Maximizing the Expected Information Provided by an Experiment

Suppose that $\mu(\cdot)$ has a density $\pi(\cdot)$ with respect to the Lebesgue measure on Θ and denote by $\pi_{\mathbf{X},\mathbf{y}}(\cdot)$ the posterior p.d.f. of θ for the observations

$\mathbf{y} = [y(x_1), \ldots, y(x_N)]^\top$. The amount of information provided by the experiment characterized by the design X is defined by the decrease in the (Shannon) entropy of the distribution of θ,

$$\mathcal{J}(X; \mathbf{y}) = -\int_\Theta \pi(\theta) \log[\pi(\theta)] \, \mathrm{d}\theta + \int_\Theta \pi_{X,\mathbf{y}}(\theta) \log[\pi_{X,\mathbf{y}}(\theta)] \, \mathrm{d}\theta,$$

see Lindley (1956). Since $\mathcal{J}(X; \mathbf{y})$ depends on the observations $\mathbf{y}$, it must be averaged with respect to $\mathbf{y}$ to form a design criterion. The expected gain in information provided by X is then

$$\mathcal{I}(X) = \mathbb{E}\left\{\int_\Theta (\pi_{X,\mathbf{y}}(\theta) \log[\pi_{X,\mathbf{y}}(\theta)] - \pi(\theta) \log[\pi(\theta)]) \, \mathrm{d}\theta\right\}, \tag{8.2}$$

where the expectation $\mathbb{E}\{\cdot\}$ is with respect to the marginal distribution of $\mathbf{y}$, the p.d.f. of which is denoted by $\varphi_X^*(\cdot)$.

Denote by $D(\varphi(\cdot)\|\psi(\cdot))$ the Kullback–Leibler divergence (or information divergence, or relative entropy) between the p.d.f. $\varphi(\cdot)$ and $\psi(\cdot)$,

$$D(\varphi(\cdot)\|\psi(\cdot)) = \int \varphi(t) \log \frac{\varphi(t)}{\psi(t)} \, \mathrm{d}t.$$

Since $\mathbb{E}\{\pi_{X,\mathbf{y}}(\theta)\} = \pi(\theta)$, we can write

$$\begin{aligned}
\mathcal{I}(X) &= \mathbb{E}\left\{\int_\Theta \pi_{X,\mathbf{y}}(\theta) \log\left[\frac{\pi_{X,\mathbf{y}}(\theta)}{\pi(\theta)}\right] \mathrm{d}\theta\right\} = \mathbb{E}\{D(\pi_{X,\mathbf{y}}(\cdot)\|\pi(\cdot))\} \\
&= \int_{\mathbb{R}^N} \int_\Theta \varpi_X(\theta, \mathbf{y}) \log\left[\frac{\varpi_X(\theta, \mathbf{y})}{\pi(\theta)\varphi_X^*(\mathbf{y})}\right] \mathrm{d}\theta \, \mathrm{d}\mathbf{y} \\
&= D(\varpi_X(\cdot, \cdot)\|\pi(\cdot)\varphi_X^*(\cdot)) && (8.3) \\
&= \int_\Theta D(\varphi_{X,\theta}(\cdot)\|\varphi_X^*(\cdot)) \, \pi(\theta) \, \mathrm{d}\theta, && (8.4)
\end{aligned}$$

with $\varpi_X(\theta, \mathbf{y}) = \pi_{X,\mathbf{y}}(\theta) \, \varphi_X^*(\mathbf{y})$ the joint density of θ and $\mathbf{y}$ and $\varphi_{X,\theta}(\mathbf{y}) = \pi_{X,\mathbf{y}}(\theta) \, \varphi_X^*(\mathbf{y})/\pi(\theta)$. Note that (8.3) shows a full symmetry between θ and $\mathbf{y}$.

One can show that $\mathcal{I}(X) \geq 0$ with equality if and only if $\varphi_{X,\theta}(\mathbf{y})$ does not depend on θ (except possibly on a set of zero Lebesgue measure for θ); see Lindley (1956). This does not mean, however, that the gain in information provided by an experiment $\mathcal{J}(X; \mathbf{y})$ is necessarily positive: it is so only on the average; surprising results may lead to the posterior for θ being a severe revision of the prior and to a decrease of the amount of information; one may refer to Wynn (2004, 2007) for generalizations.

Average-Optimality Criteria

Only the first term of $\mathcal{I}(X)$ depends on X in (8.2), which we write

$$\phi_{\mathcal{I}}(X) = \int_{\mathbb{R}^N} \int_\Theta \pi_{X,\mathbf{y}}(\theta) \log[\pi_{X,\mathbf{y}}(\theta)] \, \varphi_X^*(\mathbf{y}) \, \mathrm{d}\theta \, \mathrm{d}\mathbf{y}.$$

Using the approximation

$$\pi_{X,\mathbf{y}}(\theta) \approx \mathscr{N}(\hat{\theta}^N_{ML}(\mathbf{y}), \mathbf{M}^{-1}(X, \hat{\theta}^N_{ML}(\mathbf{y}))/N), \tag{8.5}$$

see Sect. 4.5 and, e.g., Cox and Hinkley (1974, p. 399), we obtain from (5.7) that maximizing $\phi_{\mathcal{I}}(X)$ is approximately equivalent to maximizing

$$\int_{\mathbb{R}^N} \log\det[\mathbf{M}(X, \hat{\theta}^N_{ML}(\mathbf{y}))]\, \varphi^*_X(\mathbf{y})\, \mathrm{d}\mathbf{y}.$$

Under the assumption that the experiment is informative enough, the predictive distribution of $\hat{\theta}^N_{ML}(\mathbf{y})$ can be approximated by the prior, which then yields the average (or expected) D-optimality criterion

$$\phi_{ED}(X) = \int_{\Theta} \log\det[\mathbf{M}(X,\theta)]\, \mu(\mathrm{d}\theta), \tag{8.6}$$

see D'Argenio (1990) and Chaloner and Verdinelli (1995). Another approximation can alternatively be used for $\pi_{X,\mathbf{y}}(\theta)$:

$$\pi_{X,\mathbf{y}}(\theta) \approx \mathscr{N}(\hat{\theta}^N(\mathbf{y}), [N\, \mathbf{M}(X, \hat{\theta}^N(\mathbf{y})) + \mathbf{\Omega}^{-1}]^{-1}), \tag{8.7}$$

with $\hat{\theta}^N(\mathbf{y})$ the mode of $\pi_{X,\mathbf{y}}(\cdot)$, i.e., the maximum a posteriori estimator, and $\mathbf{\Omega}$ the prior covariance matrix of θ, or $\mathbb{E}_\pi\{[\partial \log \pi(\theta)/\partial\theta]\, [\partial \log \pi(\theta)/\partial\theta^\top]\}$; see Remark 4.18-($i$). Supposing again that the predictive distribution of $\hat{\theta}^N$ is close to the prior, we obtain the criterion

$$\phi_{BED}(X) = \int_{\Theta} \log\det[\mathbf{M}(X,\theta) + \mathbf{\Omega}^{-1}/N]\, \mu(\mathrm{d}\theta).$$

In the case where $\mu(\cdot)$ is the delta measure at some θ^0, or in a linear situation where $\mathbf{M}(X,\theta)$ does not depend on θ, we obtain a Bayesian version of the D-optimality criterion; see Sect. 5.6.

Consider now a utility function based on the (weighted) variance of the posterior rather than on its entropy, i.e., $-\mathbb{E}\{\int_\Theta (\theta - \tilde{\theta}^N)^\top \mathbf{Q}\mathbf{Q}^\top (\theta - \tilde{\theta}^N)\, \pi_{X,\mathbf{y}}(\theta)\, \mathrm{d}\theta\}$, with $\tilde{\theta}^N$ the posterior mean of θ, $\tilde{\theta}^N = \int_\Theta \theta\, \pi_{X,\mathbf{y}}(\theta)\, \mathrm{d}\theta$, and $\mathbf{Q}$ some $p \times m$ matrix. Using the approximation (8.5) and assuming that the predictive distribution of $\hat{\theta}^N_{ML}$ is close to the prior, we obtain the average (or expected) L-optimality criterion

$$\phi_{EL}(X) = -\int_{\Theta} \mathrm{trace}[\mathbf{Q}\mathbf{Q}^\top \mathbf{M}^{-1}(X,\theta)]\, \mu(\mathrm{d}\theta).$$

Similarly, using the approximation (8.7), we obtain

$$\phi_{BEL}(X) = -\int_{\Theta} \mathrm{trace}\{\mathbf{Q}\mathbf{Q}^\top [\mathbf{M}(X,\theta) + \mathbf{\Omega}^{-1}/N]^{-1}\}\, \mu(\mathrm{d}\theta),$$

which can be related to Bayesian L-optimum design; see Sect. 5.6 and Pilz (1983). Other utility functions are considered in the survey (Chaloner and Verdinelli 1995); one may also refer to Eaton et al. (1996) for a decision theoretic formulation of the Bayesian design problem.

The connection between the information provided by an experiment and average-optimality criteria can be made more rigorous through asymptotic considerations when the experiment characterized by X is replicated n times with $n \to \infty$. Clarke and Barron (1994) interpret $D(\varphi_{X,\theta}(\cdot)\|\varphi_X^*(\cdot))$ in (8.4) as the risk of a Bayesian strategy in a game against nature and $\mathcal{I}(X)$ is then the corresponding Bayes risk. It is shown in the same paper that, under suitable regularity conditions, we have for the design $X^{\otimes n}$ consisting of n independent replications of the experiment at X

$$D(\varphi_{X,\theta}(\cdot)\|\varphi_X^*(\cdot)) = \frac{p}{2}\log\frac{n}{2\pi\mathrm{e}} + \frac{1}{2}\log\det[\mathbf{M}(X,\theta)] + \log\frac{1}{\pi(\theta)} + o(1)$$

as $n \to \infty$, uniformly on compact sets in the interior of the support of $\pi(\cdot)$. Integration with respect to $\pi(\cdot)$ then gives

$$\mathcal{I}(X^{\otimes n}) = \frac{p}{2}\log\frac{n}{2\pi\mathrm{e}} + \frac{1}{2}\int_\Theta \log\det[\mathbf{M}(X,\theta)]\,\pi(\theta)\,\mathrm{d}\theta + H_1(\pi) + o(1) \quad (8.8)$$

with $H_1(\pi) = -\int_\Theta \pi(\theta)\,\log\pi(\theta)\,\mathrm{d}\theta$ the Shannon entropy of the prior for θ. The criterion to be maximized thus takes the form of the average D-optimality criterion given by (8.6).

Non-informative Priors

The expected gain in information (8.8) can be rewritten as

$$\mathcal{I}(X^{\otimes n}) = \frac{p}{2}\log\frac{n}{2\pi\mathrm{e}} - D(\pi(\cdot)\|\pi^*(\cdot)) + \log\left\{\int_\Theta \det{}^{1/2}[\mathbf{M}(X,\theta)]\,\mathrm{d}\theta\right\} + o(1)$$

with $\pi^*(\cdot)$ given by Jeffrey's prior,

$$\pi^*(\theta) = \frac{\det^{1/2}[\mathbf{M}(X,\theta)]}{\int_\Theta \det^{1/2}[\mathbf{M}(X,\theta)]\,\mathrm{d}\theta}. \quad (8.9)$$

Neglecting the terms that tend to zero as $n \to \infty$, the maximum of $\mathcal{I}(X^{\otimes n})$ with respect to $\pi(\cdot)$ is obtained for $\pi(\cdot) = \pi^*(\cdot)$ and equals

$$\mathcal{I}^*(X^{\otimes n}) = \frac{p}{2}\log\frac{n}{2\pi\mathrm{e}} + \log\left\{\int_\Theta \det{}^{1/2}[\mathbf{M}(X,\theta)]\,\mathrm{d}\theta\right\} + o(1).$$

To summarize, we should thus maximize $\int_\Theta \det^{1/2}\mathbf{M}(X,\theta)\,\mathrm{d}\theta$ when using Jeffrey's prior and maximize the average D-optimality criterion (8.6) when the prior density $\pi(\cdot)$ is given (and does not depend on X).

Other choices than (8.9) for a non-informative prior $\pi(\cdot)$ can be motivated by the objective of obtaining a uniform distribution of responses $\eta(x,\theta)$ for regression models, or more generally of (expected) log-likelihoods, over (a part of) $\mathscr{X}$. This is the approach proposed in (Bornkamp, 2011), which we follow in the rest of this section.

For a regression model (3.2), (3.3) we can define a new metric in Θ by considering the sets

$$\mathcal{A}_\epsilon(\nu,\theta^0) = \{\theta \in \Theta : \|\eta(\cdot,\theta) - \eta(\cdot,\theta^0)\|_\nu \leq \epsilon\},$$

with $\nu(\cdot)$ a probability measure on $\mathscr{X}$ related to the importance given to different regions of interest, θ^0 a point in $\mathrm{int}(\Theta)$ and $\|\cdot\|_\nu$ the norm in $\mathscr{L}_2(\nu)$; see (3.41). Let $m_\epsilon(\nu,\theta^0)$ denote the mass allocated to $\mathcal{A}_\epsilon(\nu,\theta^0)$ when θ has the prior $\pi(\cdot)$. Uniformity in the response space is then obtained when all $\mathcal{A}_\epsilon(\nu,\theta^0)$ receive the same mass as ϵ tends to zero, i.e., when $m_\epsilon(\nu,\theta^0)/\epsilon^p$ tends to a constant not depending on θ^0 as $\epsilon \to 0$; see, e.g., Dembski (1990) for precise statements. For small ϵ, the set $\mathcal{A}_\epsilon(\nu,\theta^0)$ can be approximated by

$$\mathcal{E}_\epsilon(\nu,\theta^0) = \{\theta' \in \Theta : [(\theta-\theta^0)^\top \mathbf{M}(\nu,\theta^0)(\theta-\theta^0)]^{1/2} \leq (\epsilon/\sigma)\},$$

with $\mathbf{M}(\nu,\theta^0)$ the information matrix (3.32). The volume of the ellipsoid $\mathcal{E}_\epsilon(\nu,\theta^0)$ equals $\epsilon^p V_p \det^{-1/2}\mathbf{M}(\nu,\theta^0)/\sigma^p$, with V_p the volume of the unit ball $\mathscr{B}(\mathbf{0},1)$ of $\mathbb{R}^p$; see Lemma 5.1-(i). The prior $\pi(\cdot)$ that yields a uniform distribution for the new metric is thus proportional to $\det^{1/2}\mathbf{M}(\nu,\theta)$, that is,

$$\pi_\nu^*(\theta) = \frac{\det^{1/2}[\mathbf{M}(\nu,\theta)]}{\int_\Theta \det^{1/2}[\mathbf{M}(\nu,\theta)]\,\mathrm{d}\theta}. \tag{8.10}$$

Bornkamp (2011) considers the case where $\nu(\cdot)$ corresponds to the uniform distribution on $\mathscr{X}$. One may notice that taking $\nu = \xi$, the design measure corresponding to the design X itself, gives Jeffrey's prior (8.9).

More generally, within the framework of Sect. 4.4, one may define a new metric on Θ by considering the sets

$$\mathcal{A}_\epsilon(Z,\theta^0) = \left\{\theta \in \Theta : \frac{2}{m}\,\mathbb{E}_{Z,\theta^0}\left[\log\left(\frac{\mathrm{L}_{Z,\mathbf{y}}(\theta^0)}{\mathrm{L}_{Z,\mathbf{y}}(\theta)}\right)\right] \leq \epsilon^2\right\},$$

with $Z = (z_1,\ldots,z_m)$ an m-point design on $\mathscr{X}$ and $\mathrm{L}_{Z,\mathbf{y}}(\theta^0)$ the likelihood of parameters θ^0 for the design Z at observations $\mathbf{y}$. The expected log-likelihood ratio in the definition of $\mathcal{A}_\epsilon(Z,\theta^0)$ corresponds to Kullback–Leibler divergence between the distributions of $\mathbf{y}$ for θ^0 and θ, a similar result is obtained when using Hellinger distance. For small ϵ, the set $\mathcal{A}_\epsilon(Z,\theta^0)$ can be approximated by

$$\mathcal{E}_\epsilon(Z,\theta^0) = \left\{\theta' \in \Theta : -\frac{1}{m}(\theta-\theta^0)^\top \mathbb{E}_{Z,\theta^0}\left[\left.\frac{\partial^2 \log \mathrm{L}_{Z,\mathbf{y}}(\theta^0)}{\partial\theta\partial\theta^\top}\right|_{\theta^0}\right](\theta-\theta^0) \leq \epsilon^2\right\};$$

that is, from Remark 4.14-(iv),

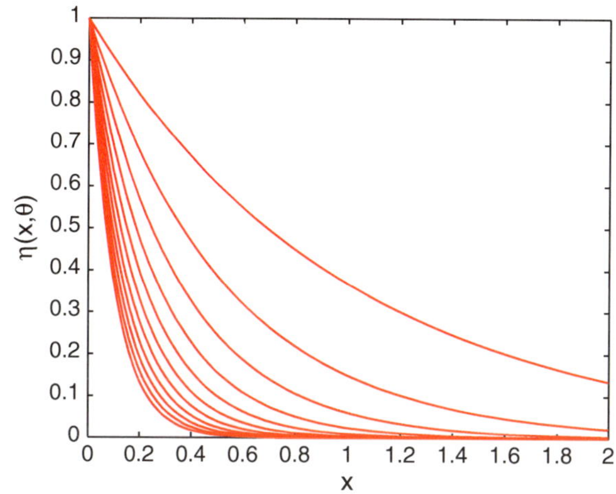

Fig. 8.1. α-quantiles for the distribution of $\eta(x,\theta)$ as functions of x for α=0, 0.1, 0.2 ..., 1; π is uniform on Θ

$$\mathcal{E}_\epsilon(Z,\theta^0) = \{\theta' \in \Theta : [(\theta - \theta^0)^\top \mathbf{M}(Z,\theta^0)(\theta - \theta^0)]^{1/2} \leq \epsilon\},$$

with $\mathbf{M}(Z,\theta^0)$ the information matrix (4.46). The volume of $\mathcal{E}_\epsilon(Z,\theta^0)$ equals $\epsilon^p V_p \det^{-1/2}\mathbf{M}(Z,\theta^0)$, which yields the prior

$$\pi_Z^*(\theta) = \frac{\det^{1/2}[\mathbf{M}(Z,\theta)]}{\int_\Theta \det^{1/2}[\mathbf{M}(Z,\theta)]\,\mathrm{d}\theta}.$$

Again, the choice of Z can be guided by regions of interest in $\mathscr{X}$ where we wish the expected log-likelihood to be uniformly distributed.

Example 8.3. Consider again the one-parameter regression model $\eta(x,\theta) = \exp(-\theta x)$ with homoscedastic errors, $x \geq 0$, $\theta \in \Theta = [\theta_{\min}, \theta_{\max}]$, $\theta_{\min} > 0$.

We have $M(\xi,\theta) = \int_{\mathscr{X}} x^2 \exp(-2\theta x)\,\xi(\mathrm{d}x)$. Suppose that ξ is the delta measure at x. When the prior p.d.f. $\pi(\cdot)$ corresponds to Jeffrey's prior (8.9), we obtain $\int_\Theta M^{1/2}(\delta_x,\theta)\,\mathrm{d}\theta = \exp(-\theta_{\min}x) - \exp(-\theta_{\max}x)$ so that the optimal value of x is $x^* = [\log\theta_{\max} - \log\theta_{\min}]/[\theta_{\max} - \theta_{\min}]$, with $x^* \simeq 0.2558$ for $\Theta = [1, 10]$.

Figures 8.1–8.3 present the α-quantiles for the distribution of $\eta(x,\theta)$ as functions of x for $\Theta = [1, 10]$, $\alpha = 0, 0.1, 0.2, \ldots, 1$ and different priors $\pi(\cdot)$: $\pi(\cdot)$ are uniform in Fig. 8.1; $\pi(\cdot) = \pi_\nu^*(\cdot)$ given by (8.10) in Fig. 8.2 (with ν uniform on $[0,2]$) and Fig. 8.3 (ν is uniform on $[1.5,2]$). The value of x that maximizes $\phi_{ED}(\delta_x) = \int_\Theta \log[x^2\exp(-2\theta x)]\,\pi(\theta)\,\mathrm{d}\theta$ is given by $x^* = 1/\mathbb{E}_\pi(\theta)$; see Example 8.2-A. We obtain $x^* \simeq 0.1818$, $x^* \simeq 0.3118$ and $x^* \simeq 0.6279$, respectively, for $\pi(\cdot)$ uniform on Θ, $\pi(\cdot) = \pi_\nu^*(\cdot)$ with ν uniform on $[0,2]$ and $\pi(\cdot) = \pi_\nu^*(\cdot)$ with ν is uniform on $[1.5,2]$. $\square$

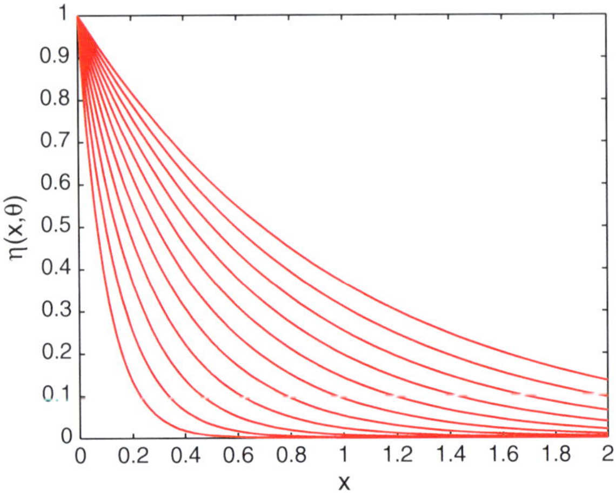

Fig. 8.2. α-quantiles for the distribution of $\eta(x,\theta)$ as functions of x for α=0, 0.1, 0.2 ..., 1; $\pi(\cdot) = \pi^*_\nu(\cdot)$ given by (8.10) with ν uniform on $[0, 2]$

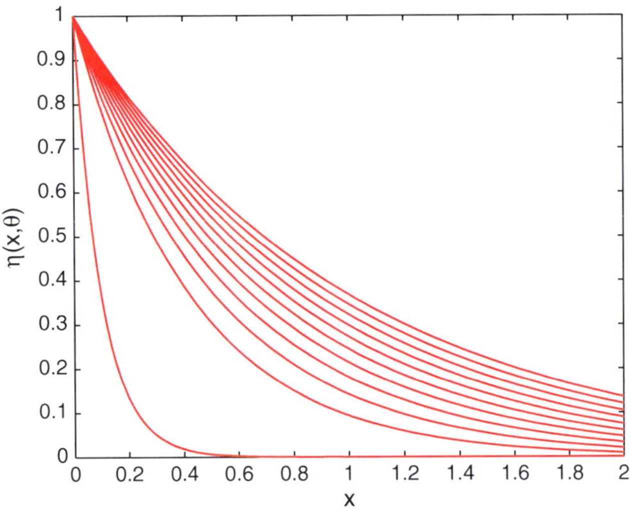

Fig. 8.3. α-quantiles for the distribution of $\eta(x,\theta)$ as functions of x for α=0, 0.1, 0.2 ..., 1; $\pi(\cdot) = \pi^*_\nu(\cdot)$ given by (8.10) with ν uniform on $[1.5, 2]$

8.2 Maximin-Optimum Design

Let $\phi(\xi;\theta)$ denote the criterion of interest for a given θ, to be maximized with respect to ξ. Its associated maximin-optimality version is

$$\phi_{MmO}(\xi) = \min_{\theta \in \Theta} \phi(\xi;\theta), \tag{8.11}$$

where Θ is a finite set or a compact subset of $\mathbb{R}^p$ with nonempty interior.

The bound on the number of support points of an optimal design given in Sect. 5.2.3 does not apply to $\phi_{MmO}(\xi) = \min_{\theta\in\Theta} \Phi[\mathbf{M}(\xi,\theta)]$. Braess and Dette (2007) give a sufficient condition under which the number of support points of a maximin D-optimal design can be made arbitrarily large by increasing the size of Θ. An example where this happens is given in (Rojas et al., 2007).

The maximin criterion $\phi_{MmO}(\cdot)$ is positively homogeneous when $\phi(\xi;\theta)$ is positively homogeneous for all $\theta \in \Theta$; see Definition 5.3. When the magnitude of $\phi(\xi;\theta)$ depends strongly on the value of $\theta \in \Theta$, it is recommended to consider the maximin-efficiency criterion

$$\mathscr{E}_{\phi_{MmO}}(\xi) = \min_{\theta\in\Theta} \mathscr{E}_{\phi}(\xi;\theta) = \min_{\theta\in\Theta} \left\{ \frac{\phi(\xi;\theta)}{\phi(\xi_\theta^*;\theta)} \right\},$$

where $\xi_\theta^* = \arg\max_{\xi\in\Xi} \phi(\xi;\theta)$.

If $\phi(\xi;\theta)$ is concave for all $\theta \in \Theta$, then $\phi_{MmO}(\cdot)$ is concave (as the minimum over a family of concave functionals). It thus admits a directional derivative $F_{\phi_{MmO}}(\xi;\nu)$ at any ξ such that $\phi_{MmO}(\xi) > -\infty$; see Lemma 5.16. When $\phi(\cdot;\theta)$ is differentiable for all θ,

$$F_{\phi_{MmO}}(\xi;\nu) = \min_{\theta\in\Theta(\xi)} F_{\phi_\theta}(\xi;\nu),$$

with $F_{\phi_\theta}(\xi;\nu)$ the directional derivative of $\phi(\cdot;\theta)$ at ξ in the direction ν and

$$\Theta(\xi) = \{\theta \in \Theta : \phi(\xi;\theta) = \phi_{MmO}(\xi)\}, \tag{8.12}$$

see (5.36). We thus obtain an equivalence theorem for $\phi_{MmO}(\cdot)$, which is a simple reformulation of Theorem 5.25 in Sect. 5.2.2.

Theorem 8.4. *Let $\phi(\cdot;\theta)$ be a concave and differentiable functional on the set Ξ of probability measures on $\mathscr{X}$ for any $\theta \in \Theta$. A design ξ^* is optimal for the criterion $\phi_{MmO}(\cdot)$ defined by (8.11) (or ϕ_{MmO}-optimal) if and only if*

$$\sup_{\nu\in\Xi} \min_{\theta\in\Theta(\xi^*)} F_{\phi_\theta}(\xi^*;\nu) = 0, \tag{8.13}$$

with $\Theta(\xi)$ defined by (8.12). An equivalent condition is

$$\max_{x\in\mathscr{X}} \int_{\Theta(\xi^*)} F_{\phi_\theta}(\xi^*,x)\, \mu^*(\mathrm{d}\theta) = 0 \quad \textit{for some } \mu^* \in \mathscr{M}_{\xi^*}, \tag{8.14}$$

with $\mathscr{M}_\xi$ the set of probability measures on $\Theta(\xi)$, $F_{\phi_\theta}(\xi,x) = F_{\phi_\theta}(\xi;\delta_x)$ and δ_x the delta measure at x.

Consider, for instance, the following global version of the estimability criterion $\phi_{eE}(X;\theta)$ of Sect. 7.7.1,

$$\phi_{MeE}(X) = \min_{\delta\geq 0} \frac{E_\eta(\delta)}{N}(K + 1/\delta) = \min_{\theta\in\Theta} \phi_{eE}(X,\theta),$$

to be maximized with respect to X. We can define similarly

$$\phi_{MeE}(\xi) = \min_{(\theta',\theta)\in\Theta^2} \|\eta(\cdot,\theta') - \eta(\cdot,\theta)\|_\xi^2 \left(K + \|\theta' - \theta\|^{-2}\right),$$

see (7.28), the directional derivative of which is given by

$$F_{\phi_{MeE}}(\xi;\nu) = \min_{\theta,\theta'\in\Theta^2(\xi)} \|\eta(\cdot,\theta') - \eta(\cdot,\theta)\|_\nu^2 \left(K + \|\theta' - \theta\|^{-2}\right) - \phi_{MeE}(\xi)$$

where

$$\Theta^2(\xi) = \left\{(\theta,\theta') \in \Theta^2 : \|\eta(\cdot,\theta') - \eta(\cdot,\theta)\|_\xi^2 \left(K + \|\theta' - \theta\|^{-2}\right) = \phi_{MeE}(\xi)\right\}.$$

The conditions of Theorem 8.4 are satisfied, and the necessary-and-sufficient condition (8.14) applies.

Example 8.5. Consider again the model of Example 8.2, for which $M(\xi,\theta) = \int_{\mathscr{X}} x^2 \exp(-2\theta x)\,\xi(\mathrm{d}x)$ with $\theta > 0$ and $\mathscr{X} \subseteq \mathbb{R}^+$, in the same two situations as in p. 238.

(A) The optimal design ξ_θ^* on $\mathscr{X} = \mathbb{R}^+$ for any isotonic criterion $\Phi(\cdot)$ maximizes $M(\xi,\theta)$ and is given by the delta measure at $1/\theta$. For Θ any compact subset of $\mathbb{R}^+$, the ϕ_{MmO}-optimal design that maximizes

$$\phi_{MmO}(\xi) = \min_{\theta\in\Theta} \Phi[M(\xi,\theta)]$$

is then the delta measure at $1/\theta_{\max}$ with $\theta_{\max} = \max(\theta)$; see Fig. 8.5 for an illustration.

(B) Consider now the efficiency criterion $\mathscr{E}(\xi;\theta) = M(\xi,\theta)/M(\xi_\theta^*,\theta)$ and its maximin version

$$\mathscr{E}_{\phi_{MmO}}(\xi) = \min_{\theta\in\Theta} \left\{\frac{M(\xi,\theta)}{M(\xi_\theta^*,\theta)}\right\} = \min_{\theta\in\Theta} \int_{\mathscr{X}} \mathrm{e}^2\theta^2 x^2 \exp(-2\theta x)\,\xi(\mathrm{d}x).$$

Suppose that $\Theta = [\theta_{\min}, \theta_{\max}]$. For $\theta_{\max} - \theta_{\min}$ small enough, the optimal design maximizing $\mathscr{E}_{\phi_{MmO}}(\xi)$ is the delta measure at the point

$$x^* = \frac{\log(\theta_{\max}) - \log(\theta_{\min})}{\theta_{\max} - \theta_{\min}}$$

satisfying $\mathscr{E}(\delta_{x^*};\theta_{\max}) = \mathscr{E}(\delta_{x^*};\theta_{\min})$. The optimality of this one-point design can be verified by using Theorem 8.4. Denote $\phi_\theta(\xi) = \phi(\xi;\theta) = \mathscr{E}(\xi;\theta)$, then $F_{\phi_\theta}(\xi,x) = \mathrm{e}^2\theta^2x^2 \exp(-2\theta x) - \phi(\xi;\theta)$. Take $\mu^* = (1-\beta)\delta_{\theta_{\min}} + \beta\delta_{\theta_{\max}}$. Numerical calculations show that the condition (8.14) is satisfied, for instance, when $\beta \simeq 0.4427$ for $\theta_{\min} = 1/2$, $\theta_{\max} = 1$ and when $\beta \simeq 0.4102$ for $\theta_{\min} = 1$, $\theta_{\max} = 3$. The one-point design δ_{x^*} ceases to be optimal if $\theta_{\max} - \theta_{\min}$ exceeds some threshold; for instance, it is already not optimal for $\theta_{\min} = 1$ and $\theta_{\max} = 4$. One can refer to Rojas et al. (2007) for a detailed analysis in another example with $\dim(\theta) = 1$. □

Example 8.6. We consider the maximin D-efficiency criterion $\mathscr{E}_{MmD}(\xi) = \min_{\theta\in\Theta} \phi(\xi;\theta)$, with

$$\phi(\xi;\theta) = \mathscr{E}_D(\xi;\theta) = \frac{\det^{1/p}[\mathbf{M}(\xi,\theta)]}{\det^{1/p}\{\mathbf{M}[\xi_D^*(\theta),\theta]\}}$$

and $\xi_D^*(\theta)$ a D-optimal design measure for θ (see Sect. 5.1.8), for the two-parameter regression model $\eta(x,\theta) = \theta_1 \exp(-\theta_2 x)$, with $\theta_2 > 0$ and $x \in \mathscr{X} = [0,1]$. Since the response $\eta(x,\theta)$ is linear in θ_1, $\mathscr{E}_D(\xi;\theta)$ does not depend on θ_1 and $\xi_D^*(\theta) = (1/2)\delta_0 + (1/2)\delta_{1/\theta_2}$. Suppose that $\theta_2 \in [0, \theta_{\max}]$. The optimal design measure ξ^* for $\mathscr{E}_{MmD}(\cdot)$ depends on $\theta_{\max}$, and the number of its support points increases with $\theta_{\max}$; see Braess and Dette (2007). The cutting-plane algorithm of Sect. 9.5.3, with Θ replaced by a regular grid with steps 10^{-3}, gives

$$\xi^* = \left\{\begin{matrix} 0 & \log 2 \simeq 0.6931 \\ 1/2 & 1/2 \end{matrix}\right\},\ \Theta(\xi^*) = \{1,\ 2\},\ \text{for } \theta_{\max} = 2,$$

$$\xi^* \simeq \left\{\begin{matrix} 0 & 0.1405 & 0.7345 \\ 0.4470 & 0.3447 & 0.2083 \end{matrix}\right\},\ \Theta(\xi^*) \simeq \{1,\ 3.491,\ 10\},\ \text{for } \theta_{\max} = 10,$$

and

$$\xi^* \simeq \left\{\begin{matrix} 0 & 0.0635 & 0.274 & 0.9 \\ 0.4201 & 0.3020 & 0.1484 & 0.1295 \end{matrix}\right\},\ \Theta(\xi^*) \simeq \{1,\ 3.604,\ 7.115,\ 20\},$$

for $\theta_{\max} = 20$, where $\Theta(\xi^*) = \{\theta \in \Theta : \phi(\xi;\theta) = \mathscr{E}_{MmD}(\xi)\}$; see (8.12). The optimality of ξ^* can be checked via condition (8.14): the measure μ^* that minimizes $\max_{x\in\mathscr{X}} \int_{\Theta(\xi^*)} F_{\phi_\theta}(\xi^*,x)\,\mu^*(\mathrm{d}\theta)$ is obtained as solution of an LP problem and is given by

$$\mu^* \simeq \left\{\begin{matrix} 1 & 2 \\ 0.5553 & 0.4447 \end{matrix}\right\} \text{ for } \theta_{\max} = 2,$$

$$\mu^* \simeq \left\{\begin{matrix} 1 & 3.491 & 10 \\ 0.3916 & 0.3140 & 0.2944 \end{matrix}\right\} \text{ for } \theta_{\max} = 10,$$

$$\mu^* \simeq \left\{\begin{matrix} 1 & 3.604 & 7.115 & 20 \\ 0.3272 & 0.2107 & 0.1905 & 0.2716 \end{matrix}\right\} \text{ for } \theta_{\max} = 20,$$

with the first line indicating the support points (the points in $\Theta(\xi^*)$) and the second their respective weights. Figure 8.4 shows $\int_{\Theta(\xi^*)} F_{\phi_\theta}(\xi^*,x)\,\mu^*(\mathrm{d}\theta)$ as a function of x when $\theta_{\max} = 2$ (solid line) and $\theta_{\max} = 20$ (dashed line). □

The supremum of $F_{\phi_{MmO}}(\xi;\nu)$ with respect to ν in (8.13) is generally not obtained for ν^* equal to a one-point (delta) measure; see Remark 5.26-(*ii*). Therefore, a maximin-optimal design cannot be obtained by using one of the vertex-direction algorithms of Sect. 9.1.1, and we should either use a specific algorithm (see Sects. 9.3.1 and 9.5) or smooth the design criterion. The latter is considered in the next section.

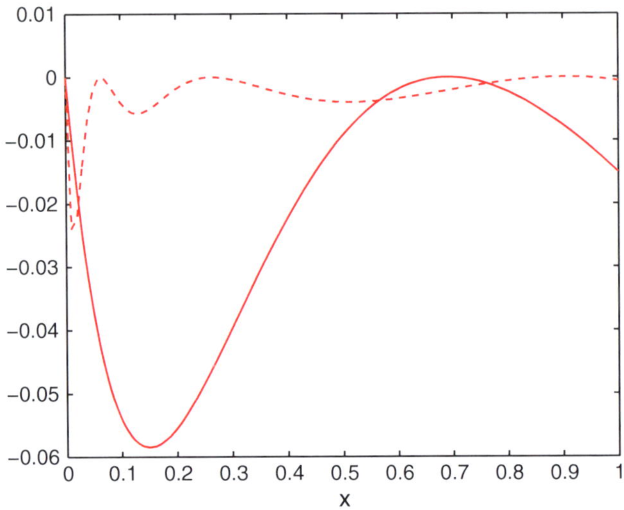

Fig. 8.4. Plot of $\int_{\Theta(\xi^*)} F_{\phi_\theta}(\xi^*, x)\, \mu^*(\mathrm{d}\theta)$ as a function of x in Example 8.6 for $\theta_{\max} = 2$ (*solid line*) and $\theta_{\max} = 20$ (*dashed line*)

8.3 Regularization of Maximin Criteria via Average Criteria

Two approaches are considered that ensure a smooth transition between the average-optimality criterion (8.1) and the maximin criterion (8.11), which is usually non-differentiable. The first one simply relies on properties of $\mathscr{L}_q$ norms; the second one relies on a maximum-entropy principle. In both cases the criterion that is constructed is concave and differentiable, so that its optimization can be carried out by using the algorithms of Sect. 9.1.

8.3.1 Regularization via $\mathscr{L}_q$ Norms

Consider the design criterion

$$\overline{\phi}_{MmO,q}(\xi) = \left[\int_{\Theta} \phi^{-q}(\xi;\theta)\, \mu(\mathrm{d}\theta) \right]^{-1/q} . \qquad (8.15)$$

We suppose that $\phi_{MmO}(\xi) = \min_{\theta\in\Theta} \phi(\xi;\theta) \geq 0$ and $\phi_{MmO}(\xi) > 0$ for some $\xi \in \Xi$, the set of probability measures on $\mathscr{X}$. We denote by Ξ^+ the set $\{\xi \in \Xi : \phi_{MmO}(\xi) > 0\}$. Either Θ is a finite set $\{\theta^{(1)}, \ldots, \theta^{(M)}\}$ and the probability measure $\mu(\cdot)$ for θ is such that $\min_{i=1,\ldots,M} \mu(\theta^{(i)}) > 0$ or Θ satisfies H_Θ (p. 22), $\mu(\cdot)$ has a density with respect to the Lebesgue measure which is bounded away from zero on Θ, and we suppose that $\phi(\xi;\theta)$ is continuous in $\theta \in \Theta$ for any $\xi \in \Xi^+$.

For any $\xi \in \Xi^+$ and any $q > 0$ we can write

$$\phi_{MmO}(\xi) = [\max_{\theta \in \Theta} \phi^{-q}(\xi;\theta)]^{-1/q} \tag{8.16}$$

and $\max_{\theta \in \Theta} \phi^{-q}(\xi;\theta) \geq \int_\Theta \phi^{-q}(\xi;\theta)\, \mu(\mathrm{d}\theta)$ implies that

$$\phi_{MmO}(\xi) \leq \overline{\phi}_{MmO,q}(\xi). \tag{8.17}$$

Similarly, $\phi_{MmO}(\xi) = [\min_{\theta \in \Theta} \phi^{-q}(\xi;\theta)]^{-1/q}$ for $q < 0$ and $\min_{\theta \in \Theta} \phi^{-q}(\xi;\theta) \leq \int_\Theta \phi^{-q}(\xi;\theta)\, \mu(\mathrm{d}\theta)$ implies (8.17).

Pointwise Convergence of $\overline{\phi}_{MmO,q}(\xi)$ to $\phi_{MmO}(\xi)$ as $q \to \infty$

For any given $\xi \in \Xi^+$, $\overline{\phi}_{MmO,q}(\xi) \to \phi_{MmO}(\xi)$ as q tends to ∞. Indeed, consider, for instance, the case where Θ is a compact set satisfying H_Θ. From the continuity in θ of $\phi(\xi;\theta)$ and the assumptions on Θ and $\mu(\cdot)$, for any $\xi \in \Xi^+$ and any $\epsilon > 0$, the set $\mathcal{A}_\epsilon(\xi) = \{\theta \in \Theta : \phi(\xi;\theta) < \phi_{MmO}(\xi) + \epsilon\}$ has positive measure $\mu[\mathcal{A}_\epsilon(\xi)] = c_\epsilon(\xi) > 0$. For $q > 0$, $\phi^{-q}(\xi;\theta) > [\phi_{MmO}(\xi) + \epsilon]^{-q}$ on $\mathcal{A}_\epsilon(\xi)$, so that $\overline{\phi}_{MmO,q}(\xi) \leq [\phi_{MmO}(\xi) + \epsilon]\, c_\epsilon^{-1/q}(\xi)$ and therefore $\limsup_{q \to \infty} \overline{\phi}_{MmO,q}(\xi) \leq \phi_{MmO}(\xi) + \epsilon$. Together with (8.17), it implies that $\lim_{q \to \infty} \overline{\phi}_{MmO,q}(\xi) = \phi_{MmO}(\xi)$.

The computation of the derivative $\partial \overline{\phi}_{MmO,q}(\xi)/\partial q$ gives

$$\begin{aligned}
\frac{\partial \overline{\phi}_{MmO,q}(\xi)}{\partial q} &= \frac{\overline{\phi}_{MmO,q}(\xi)}{q^2 \int_\Theta \phi^{-q}(\xi;\theta)\, \mu(\mathrm{d}\theta)} \\
&\times \left\{ \left[\int_\Theta \phi^{-q}(\xi;\theta)\, \mu(\mathrm{d}\theta) \right] \log \left[\int_\Theta \phi^{-q}(\xi;\theta)\, \mu(\mathrm{d}\theta) \right] \right. \\
&\left. - \int_\Theta \phi^{-q}(\xi;\theta)\, \log[\phi^{-q}(\xi;\theta)]\, \mu(\mathrm{d}\theta) \right\} \leq 0 \quad \text{for any } q,
\end{aligned} \tag{8.18}$$

where the inequality follows from Jensen's inequality (the function $x \longrightarrow x \log x$ being convex). Equality is obtained when $\phi(\xi;\cdot)$ is μ-a.s. constant; the inequality is strict in other situations, and $\overline{\phi}_{MmO,q}(\xi)$ decreases monotonically from $\max_{\theta \in \Theta} \phi(\xi;\theta)$ for $q \to -\infty$ to $\phi_{MmO}(\xi)$ as q increases to $+\infty$.

Notice that $\overline{\phi}_{MmO,-1}(\xi) = \phi_{AO}(\xi)$ (see (8.1)) and that for any $\xi \in \Xi^+$ we have $\lim_{q \to 0} \overline{\phi}_{MmO,q}(\xi) = \exp\left\{\int_\Theta \log[\phi(\xi;\theta)]\, \mu(\mathrm{d}\theta)\right\}$; we can thus define by continuity

$$\overline{\phi}_{MmO,0}(\xi) = \exp\left\{ \int_\Theta \log[\phi(\xi;\theta)]\, \mu(\mathrm{d}\theta) \right\}. \tag{8.19}$$

We thus obtain from (8.17) that $\overline{\phi}_{MmO,q}(\xi) \geq \phi_{MmO}(\xi)$ for all q.

Uniform Convergence of $\overline{\phi}_{MmO,q}(\cdot)$ to $\phi_{MmO}(\cdot)$ as $q \to \infty$ when Θ is Finite

Suppose now that Θ is finite, $\Theta = \{\theta^{(1)}, \ldots, \theta^{(M)}\}$, and consider

$$\underline{\phi}_{MmO,q}(\xi) = \left[\sum_{i=1}^{M} \phi^{-q}(\xi;\theta^{(i)})\right]^{-1/q}, \; q > 0. \tag{8.20}$$

For $q > 0$ and any $\xi \in \Xi^+$, (8.16) and $\max_{\theta\in\Theta} \phi^{-q}(\xi;\theta) \le \sum_{i=1}^{M} \phi^{-q}(\xi;\theta^{(i)})$ (since $\phi(\xi;\theta^{(i)}) > 0$ for all i) imply that $\phi_{MmO}(\xi) \ge \underline{\phi}_{MmO,q}(\xi)$.

From a property of $\mathscr{L}_q$ norms, $\underline{\phi}_{MmO,q_2}(\xi) \le \underline{\phi}_{MmO,q_1}(\xi)$ for any $q_1 > q_2 > 0$ and any $\xi \subset \Xi^+$, so that $\underline{\phi}_{MmO,q}(\xi)$ with $q > 0$ forms a lower bound on $\phi_{MmO}(\xi)$ which tends to $\phi_{MmO}(\xi)$ as $q \to \infty$. Notice that $\underline{\phi}_{MmO,0}(\xi)$ is not defined, with $\lim_{q\to 0^-} \underline{\phi}_{MmO,q}(\xi) = \infty$ and $\lim_{q\to 0^+} \underline{\phi}_{MmO,q}(\xi) = 0$, and that $\underline{\phi}_{MmO,q}(\xi)$ tends to $\max_{\theta\in\Theta} \phi(\xi;\theta)$ as $q \to -\infty$.

Denote by $\mu_i = \mu(\theta^{(i)})$ the weight given by the measure $\mu(\cdot)$ of (8.15) to $\theta^{(i)}$ for $i = 1, \ldots, M$ and $\underline{\mu} = \min_{i=1,\ldots,M} \mu_i$. We have, for any $\xi \in \Xi^+$ and any $q > 0$,

$$\underline{\phi}_{MmO,q}(\xi) \le \phi_{MmO}(\xi) \le \overline{\phi}_{MmO,q}(\xi) \le \underline{\mu}^{-1/q} \underline{\phi}_{MmO,q}(\xi), \tag{8.21}$$

so that

$$0 \le \phi_{MmO}(\xi) - \underline{\phi}_{MmO,q}(\xi) \le (\underline{\mu}^{-1/q} - 1)\phi_{MmO}(\xi^*)$$

and

$$0 \le \overline{\phi}_{MmO,q}(\xi) - \phi_{MmO}(\xi) \le (\underline{\mu}^{-1/q} - 1)\phi_{MmO}(\xi^*),$$

where ξ^* is optimal for $\phi_{MmO}(\cdot)$ and $\underline{\mu}^{-1/q}$ tends to 1 as $q \to \infty$.

The inequalities (8.21) can be used to obtain lower bounds on the maximin efficiency of designs optimal for $\underline{\phi}_{MmO,q}(\xi)$ or $\overline{\phi}_{MmO,q}(\xi)$. Indeed, let ξ_1^*, ξ_2^* be, respectively, optimal for $\underline{\phi}_{MmO,q}(\cdot)$ and $\overline{\phi}_{MmO,q}(\cdot)$ and let ϕ_{MmO}^* denote the optimal value of $\phi_{MmO}(\cdot)$, obtained for some design ξ_{MmO}^*. Then, using (8.21) and the optimality of ξ_1^* and ξ_2^*, we obtain

$$\frac{\phi_{MmO}(\xi_1^*)}{\phi_{MmO}^*} \ge \frac{\underline{\phi}_{MmO,q}(\xi_1^*)}{\phi_{MmO}^*} \ge \frac{\underline{\phi}_{MmO,q}(\xi_{MmO}^*)}{\phi_{MmO}^*} \ge \underline{\mu}^{1/q}$$

and

$$\frac{\phi_{MmO}(\xi_2^*)}{\phi_{MmO}^*} \ge \frac{\underline{\mu}^{1/q}\overline{\phi}_{MmO,q}(\xi_2^*)}{\phi_{MmO}^*} \ge \frac{\underline{\mu}^{1/q}\overline{\phi}_{MmO,q}(\xi_{MmO}^*)}{\phi_{MmO}^*} \ge \underline{\mu}^{1/q}.$$

Note that the best efficiencies are obtained for μ the uniform measure with $\underline{\mu} = 1/M$; in that case $\overline{\phi}_{MmO,q}(\cdot) = M^{1/q}\underline{\phi}_{MmO,q}(\cdot)$, and the maximin efficiency of optimal designs for $\underline{\phi}_{MmO,q}(\xi)$ or $\overline{\phi}_{MmO,q}(\xi)$ is at least $M^{-1/q}$.

Remark 8.7. In the case where Θ has a nonempty interior and satisfies H_Θ (p. 22), we can also define for $\xi \in \Xi^+$, similarly to (8.15),

$$\underline{\phi}_{MmO,q}(\xi) = \left[\int_\Theta \phi^{-q}(\xi;\theta)\,\mathrm{d}\theta\right]^{-1/q}, \; q > 0.$$

However, $\underline{\phi}_{MmO,q}(\xi)$ does not necessarily form a lower bound on $\phi_{MmO}(\xi)$, and the convergence of $\underline{\phi}_{MmO,q}(\xi)$ to $\phi_{MmO}(\xi)$ as $q \to \infty$ is not necessarily monotone in q, Example 8.10 will give an illustration. Taking $\mu(\theta)$ as the Lebesgue measure on Θ, i.e., $\mu(\mathrm{d}\theta) = \mathrm{d}\theta/V$ with $V = \mathrm{vol}(\Theta)$, we obtain that the derivative $\partial\underline{\phi}_{MmO,q}(\xi)/\partial q$ is given by an expression like (8.18) plus a term $\underline{\phi}_{MmO,q}(\xi)\log(V)/q^2$ which may be positive enough to counter Jensen's inequality. This criterion nevertheless satisfies the following; see, e.g., Rudin (1987): for any $\xi \in \Xi^+$ and $q_1 > q > q_2 > 0$, $\underline{\phi}_{MmO,q}(\xi) \geq \min\{\underline{\phi}_{MmO,q_1}(\xi), \underline{\phi}_{MmO,q_2}(\xi)\}$. □

Concavity of $\overline{\phi}_{MmO,q}(\cdot)$ for $q \geq -1$

Suppose that $\phi(\cdot;\theta)$ is twice differentiable and concave for any θ. Defining $\xi = (1-\alpha)\xi_0 + \alpha\nu$, we obtain

$$\frac{\partial\overline{\phi}_{MmO,q}(\xi)}{\partial\alpha} = \left[\int_\Theta \phi^{-q}(\xi;\theta)\,\mu(\mathrm{d}\theta)\right]^{-1/q-1} \int_\Theta \phi^{-q-1}(\xi;\theta)\,\frac{\partial\phi(\xi;\theta)}{\partial\alpha}\,\mu(\mathrm{d}\theta).$$

Direct calculations then give

$$\begin{aligned}\frac{\partial^2\overline{\phi}_{MmO,q}(\xi)}{\partial\alpha^2} &= \left[\int_\Theta \phi^{-q}(\xi;\theta)\,\mu(\mathrm{d}\theta)\right]^{-1/q-1} \int_\Theta \phi^{-q-1}(\xi;\theta)\,\frac{\partial^2\phi(\xi;\theta)}{\partial\alpha^2}\,\mu(\mathrm{d}\theta)\\ &\quad +(q+1)\left[\int_\Theta \phi^{-q}(\xi;\theta)\,\mu(\mathrm{d}\theta)\right]^{-1/q-2}\\ &\times\Bigg\{\left[\int_\Theta \phi^{-q-1}(\xi;\theta)\,\frac{\partial\phi(\xi;\theta)}{\partial\alpha}\,\mu(\mathrm{d}\theta)\right]^2\\ &-\left(\int_\Theta \phi^{-q}(\xi;\theta)\,\mu(\mathrm{d}\theta)\right)\left(\int_\Theta \phi^{-q-2}(\xi;\theta)\left[\frac{\partial\phi(\xi;\theta)}{\partial\alpha}\right]^2\mu(\mathrm{d}\theta)\right)\Bigg\}.\end{aligned}$$

Cauchy–Schwarz inequality implies that the term within curly brackets is smaller or equal than zero. For $q \geq -1$ we thus obtain

$$\begin{aligned}\frac{\partial^2\overline{\phi}_{MmO,q}(\xi)}{\partial\alpha^2} &\leq \left[\int_\Theta \phi^{-q}(\xi;\theta)\,\mu(\mathrm{d}\theta)\right]^{-1/q-1} \int_\Theta \phi^{-q-1}(\xi;\theta)\,\frac{\partial^2\phi(\xi;\theta)}{\partial\alpha^2}\,\mu(\mathrm{d}\theta)\\ &\leq 0.\end{aligned}$$

Directional Derivative

Denote by $F_{\phi_\theta}(\xi;\nu)$ the directional derivative of $\phi(\cdot;\theta)$ at ξ in the direction ν. Using the expression (8.15) of the criterion $\overline{\phi}_{MmO,q}(\cdot)$, we directly obtain that its directional derivative is given by

$$F_{\overline{\phi}_{MmO,q}}(\xi;\nu) = \frac{\int_\Theta \phi^{-q-1}(\xi;\theta)\, F_{\phi_\theta}(\xi;\nu)\, \mu(\mathrm{d}\theta)}{\left[\overline{\phi}_{MmO,q}(\xi)\right]^{-q-1}}.$$

Example 8.8. This is a continuation of Examples 8.2 and 8.5.

(A) Take first $\phi(\xi;\theta) = M(\xi,\theta) = \int_{\mathscr{X}} x^2 \exp(-2\theta x)\,\xi(\mathrm{d}x)$. The associated average-optimal design maximizing (8.1) is the delta measure at $x^*_{AO} = \arg\max_{x\in\mathscr{X}} \int_\Theta x^2 \exp(-2\theta x)\,\mu(\mathrm{d}\theta)$, and the maximin-optimal design maximizing (8.11) is the delta measure at $x^*_{MmO} = 1/\theta_{\max}$. For this particular situation where $M(\xi,\theta)$ is scalar, the criterion $\overline{\phi}_{MmO,0}(\cdot)$ given by (8.19) is given by $\overline{\phi}_{MmO,0}(\cdot) = \exp[\int_\Theta \log[M(\xi,\theta)]\,\mu(\mathrm{d}\theta)] = \exp[\phi_{ED}(\xi)]$. When Θ is not too large, the associated optimal design is the delta measure at $x^*_{MmO,0} = 1/\mathbb{E}_\mu(\theta)$; see Example 8.2. Figure 8.5 presents those criteria as functions of x when $\xi = \delta_x$ and $\mu(\cdot)$ is the uniform measure on $\Theta = [1,3]$; in that case, $x^*_{AO} \simeq 0.6149$, $x^*_{MmO} = 1/3$ and $x^*_{MmO,0} = 1/2$. The graph of $\overline{\phi}_{MmO,q}(\delta_x)$ as a function of x is presented on the same figure for several values of q, illustrating the decrease of $\overline{\phi}_{MmO,q}(\xi)$ as q increases, the closeness to $\phi_{AO}(\xi)$ for q close to -1, the closeness to $\overline{\phi}_{MmO,0}(\xi) = \exp[\phi_{ED}(\xi)]$ for q close to 0, and the closeness to $\phi_{MmO}(\xi)$ for q large enough. Notice that we are more interested into the position of the maxima of the criteria than in their magnitude; also note that a concave function of ξ is not necessarily concave when expressed as a function of the support points of ξ.

(B) Take now $\phi(\xi;\theta) = M(\xi,\theta)/M(\xi^*_\theta,\theta) = \int_{\mathscr{X}} \mathrm{e}^2\theta^2 x^2 \exp(-2\theta x)\,\xi(\mathrm{d}x)$. The associated average-optimal design maximizing (8.1) is $\delta_{x^*_{AO}}$ with $x^*_{AO} = \arg\max_{x\in\mathscr{X}} \int_\Theta \theta^2 x^2 \exp(-2\theta x)\,\mu(\mathrm{d}\theta)$. When $\Theta = [\theta_{\min},\theta_{\max}]$ and $\theta_{\max} - \theta_{\min}$ is small enough, the optimal design maximizing (8.11) is the delta measure at $x^*_{MmO} = [\log(\theta_{\max}) - \log(\theta_{\min})]/[\theta_{\max} - \theta_{\min}]$; see Example 8.5. We have $\phi_{MmO}(\delta_x) = \mathrm{e}^2 x^2 \min\{\theta^2_{\min} \exp(-2\theta_{\min}),\, \theta^2_{\max} \exp(-2\theta_{\max})\}$. The maximization of the criterion $\overline{\phi}_{MmO,0}(\cdot)$ given by (8.19) is equivalent to the maximization of $\overline{\phi}_{MmO,0}(\cdot)$ in the situation considered above where $\phi(\xi;\theta)$ was equal to $M(\xi,\theta)$; the associated optimal design is the delta measure at $x^*_{MmO,0} = 1/\mathbb{E}_\mu(\theta)$ when Θ is not too large. Figure 8.6 presents those criteria as functions of x when $\xi = \delta_x$ and $\mu(\cdot)$ is the uniform measure on $\Theta = [1,3]$; in that case, $x^*_{AO} \simeq 0.4961$, $x^*_{MmO} = \log(9)/4 \simeq 0.5493$, and $x^*_{MmO,0} = 1/2$. The graph of $\overline{\phi}_{MmO,q}(\delta_x)$ as a function of x is also presented for $q = -0.8, 0.4$ and 40. Note the closeness of the maximizers of $\overline{\phi}_{MmO,40}(\delta_x)$ and $\phi_{MmO}(\delta_x)$ although the former is differentiable with respect to x and the latter is not differentiable at x^*_{MmO}. □

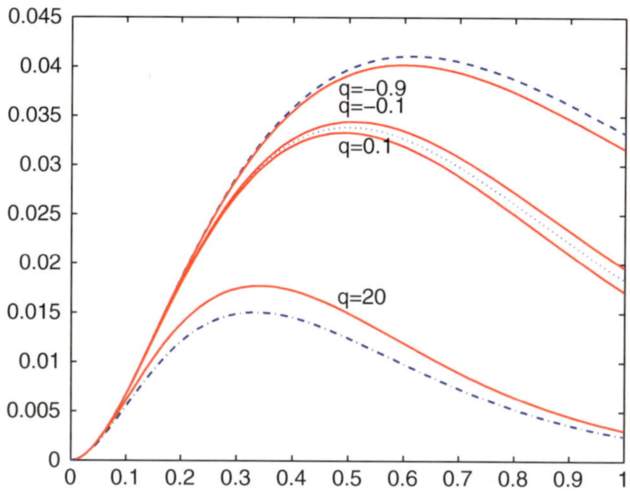

Fig. 8.5. $\phi_{AO}(\delta_x)$ (*dashed line*), $\overline{\phi}_{MmO,0}(\delta_x)$ (*dotted line*), $\phi_{MmO}(\delta_x)$ (*dash-dotted line*), and $\overline{\phi}_{MmO,q}(\delta_x)$ (*solid lines*, $q = -0.9, -0.1, 0.1, 20$) as functions of x in Example 8.8-A; $\mu(\cdot)$ is the uniform measure on $\Theta = [1, 3]$

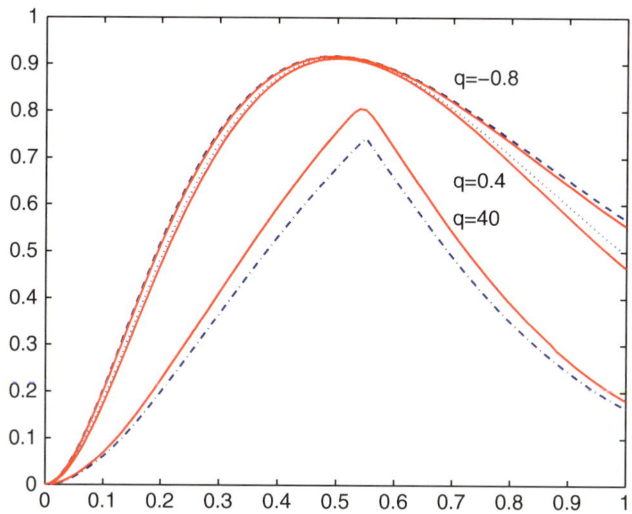

Fig. 8.6. $\phi_{AO}(\delta_x)$ (*dashed line*), $\overline{\phi}_{MmO,0}(\delta_x)$ (*dotted line*), $\phi_{MmO}(\delta_x)$ (*dash-dotted line*), and $\overline{\phi}_{MmO,q}(\delta_x)$ (*solid lines*, $q = -0.8, 0.4, 40$) as functions of x in Example 8.8-B; $\mu(\cdot)$ is the uniform measure on $\Theta = [1, 3]$

A direct generalization of regularization by $\mathscr{L}_q$ norm is as follows. Let $\psi(\cdot)$ be a strictly increasing function and $\overleftarrow{\psi}(\cdot)$ denote its inverse. Then,

$$\phi_{MmO}(\xi) = \overleftarrow{\psi}\{\min_{\theta \in \Theta} \psi[\phi(\xi;\theta)]\}.$$

Applying the $\mathscr{L}_q$ regularization above to the min function in this formulation of the maximin-optimality criterion, we can define

$$\overline{\phi}_{MmO,q,\psi}(\xi) = \overleftarrow{\psi}\left\{\left[\int_{\Theta} \{\psi[\phi(\xi;\theta)]\}^{-q}\, \mu(\mathrm{d}\theta)\right]^{-1/q}\right\} \tag{8.22}$$

and

$$\underline{\phi}_{MmO,q,\psi}(\xi) = \overleftarrow{\psi}\left\{\left[\int_{\Theta} \{\psi[\phi(\xi;\theta)]\}^{-q}\, \mathrm{d}\theta\right]^{-1/q}\right\}. \tag{8.23}$$

Concavity is not necessarily preserved, however. The regularization method presented in the next section corresponds to choosing $\psi(\cdot) = \exp(\cdot)$, which can be justified by maximum-entropy arguments in the case of (8.23) and is appealing in situations where $\phi(\xi;\theta)$ can take negative values. In that particular case, concavity is preserved.

8.3.2 Maximum-Entropy Regularization

The criterion $\phi_{MmO}(\cdot)$ can be equivalently defined by

$$\phi_{MmO}(\xi) = \min_{\mu \in \mathscr{M}(\Theta)} \int_{\Theta} \phi(\xi;\theta)\, \mu(\mathrm{d}\theta),$$

where $\mathscr{M}(\Theta)$ denotes the set of probability measures on Θ. The minimum is obtained for $\mu(\cdot)$ the delta measure at some $\theta \in \Theta$. The idea used by Li and Fang (1997) in the finite case is to regularize $\phi_{MmO}(\cdot)$ through a penalization of measures having small (Shannon) entropy, with a penalty coefficient that sets the amount of regularization introduced.[1] Consider the situation where Θ is a compact subset of $\mathbb{R}^p$ with nonempty interior that satisfies H_{Θ} (p. 22) and $\mu(\cdot)$ is a probability measure having the density $\pi(\cdot)$ with respect to the Lebesgue measure. We suppose that $\phi(\xi;\theta)$ is continuous in $\theta \in \Theta$ for any $\xi \in \Xi$, the set of probability measures on $\mathscr{X}$. Define

$$\phi_{ME,\lambda}(\xi) = \min_{\pi \in \mathscr{D}(\Theta)} \left\{\int_{\Theta} \phi(\xi;\theta)\, \pi(\theta)\, \mathrm{d}\theta + \frac{1}{\lambda}\int_{\Theta} \pi(\theta)\, \log[\pi(\theta)]\, \mathrm{d}\theta\right\}, \ \lambda > 0,$$

where $\mathscr{D}(\Theta)$ is the set of p.d.f. on Θ. This minimization problem has the solution

$$\pi^*(\theta) = \frac{\exp[-\lambda\phi(\xi;\theta)]}{\int_{\Theta} \exp[-\lambda\phi(\xi;\theta)]\, \mathrm{d}\theta},$$

[1] In a finite setting, $\sum_{i=1}^{M} \mu_i \phi(\xi;\theta^{(i)})$ can be interpreted as the Lagrange function for the maximization of $\phi_{MmO}(\cdot)$ with $\mu = (\mu_1,\ldots,\mu_M)$ the vector of Lagrange multipliers, restricted to sum to one; see Li and Fang (1997).

which gives after straightforward calculation

$$\phi_{ME,\lambda}(\xi) = -\frac{1}{\lambda} \log \left\{ \int_{\Theta} \exp[-\lambda \phi(\xi;\theta)] \, d\theta \right\} . \tag{8.24}$$

This exactly corresponds to the specialization of (8.23) to $\psi(\cdot) = \exp(\cdot)$.

When Θ is finite, $\Theta = \{\theta^{(1)}, \ldots, \theta^{(M)}\}$, which corresponds to the situation considered in (Li and Fang, 1997), we simply replace integrals by finite sums and obtain

$$\phi_{ME,\lambda}(\xi) = -\frac{1}{\lambda} \log \left\{ \sum_{i=1}^{M} \exp[-\lambda \phi(\xi;\theta^{(i)})] \right\} .$$

Remark 8.9. This method could be used to smooth some of the criteria considered in Sect. 5.1.2 that are not differentiable everywhere, for instance, those of E and MV-optimality, see Sect. 5.2.

A smoothing method for minimizing the sum of the r largest functions among m, or equivalently for maximizing the sum of the r smallest functions among m, is presented in (Pan et al., 2007), which could be used to smooth the E_k-optimality criteria, see Sect. 5.1.2. □

Pointwise Convergence of $\phi_{ME,\lambda}(\xi)$ to $\phi_{MmO}(\xi)$ as $\lambda \to \infty$

For any $\xi \in \Xi$ we have

$$\begin{aligned} \phi_{ME,\lambda}(\xi) &= \log \left\{ \int_{\Theta} \exp[-\lambda \phi(\xi;\theta)] \, d\theta \right\}^{-1/\lambda} \\ &= -\log \left\{ \int_{\Theta} (\exp[-\phi(\xi;\theta)])^{\lambda} \, d\theta \right\}^{1/\lambda} \end{aligned} \tag{8.25}$$

and thus $\lim_{\lambda\to\infty} \phi_{ME,\lambda}(\xi) = -\log \max_{\theta\in\Theta} \exp[-\phi(\xi;\theta)] = \phi_{MmO}(\xi)$.

Uniform Convergence of $\phi_{ME,\lambda}(\cdot)$ to $\phi_{MmO}(\cdot)$ as $\lambda \to \infty$ when Θ is Finite

Suppose that $\Theta = \{\theta^{(1)}, \ldots, \theta^{(M)}\}$ and take any $\lambda_2 \geq \lambda_1 > 0$. By a property of $\mathscr{L}_q$ norms,

$$\left\{ \sum_{i=1}^{M} (\exp[-\phi(\xi;\theta^{(i)})])^{\lambda_2} \right\}^{1/\lambda_2} \leq \left\{ \sum_{i=1}^{M} (\exp[-\phi(\xi;\theta^{(i)})])^{\lambda_1} \right\}^{1/\lambda_1}$$

for any $\xi \in \Xi$ and therefore $\phi_{ME,\lambda_2}(\xi) \geq \phi_{ME,\lambda_1}(\xi)$; see (8.25). This implies that $\phi_{ME,\lambda}(\xi)$ is a lower bound on $\phi_{MmO}(\xi)$ for any $\xi \in \Xi$ and any $\lambda > 0$ and

$$\begin{aligned}
0 \leq \phi_{MmO}(\xi) - \phi_{ME,\lambda}(\xi) &= \log\{\exp[\phi_{MmO}(\xi)]\} \\
&\quad + \frac{1}{\lambda} \log\left\{\sum_{i=1}^{M} \exp[-\lambda\phi(\xi;\theta^{(i)})]\right\} \\
&= \frac{1}{\lambda} \log\left\{\sum_{i=1}^{M} \frac{\exp[-\lambda\phi(\xi;\theta^{(i)})]}{\exp[-\lambda\phi_{MmO}(\xi)]}\right\} \\
&= \frac{1}{\lambda} \log\left\{\sum_{i=1}^{M} \exp(-\lambda[\phi(\xi;\theta) - \phi_{MmO}(\xi)])\right\} \\
&\leq \frac{1}{\lambda} \log M
\end{aligned}$$

which tends to zero as λ tends to infinity. Notice that this tells us how far an optimal design ξ^* for $\phi_{ME,\lambda}(\cdot)$ is from being maximin optimal. Indeed, let ξ^*_{MmO} denote an optimal design for $\phi_{MmO}(\cdot)$; we have

$$\phi_{MmO}(\xi^*) \geq \phi_{ME,\lambda}(\xi^*) \geq \phi_{ME,\lambda}(\xi^*_{MmO}) \geq \phi_{MmO}(\xi^*_{MmO}) - \frac{1}{\lambda} \log M,$$

so that

$$\frac{\phi_{MmO}(\xi^*)}{\phi_{MmO}(\xi^*_{MmO})} \geq 1 - \frac{\log M}{\lambda\, \phi_{MmO}(\xi^*_{MmO})}.$$

Concavity of $\phi_{ME,\lambda}(\cdot)$

If $\phi(\cdot;\theta)$ is concave for all $\theta \in \Theta$, then $\phi_{ME,\lambda}(\cdot)$ is concave for any $\lambda > 0$. Indeed, for any $\xi_0, \xi_1 \in \Xi$ and any $\alpha \in (0,1)$,

$$\begin{aligned}
\phi_{ME,\lambda}[(1-\alpha)\xi_0 + \alpha\xi_1] &= -\frac{1}{\lambda} \log\left\{\int_\Theta \exp(-\lambda\phi[(1-\alpha)\xi_0 + \alpha\xi_1;\theta])\, \mathrm{d}\theta\right\} \\
&\geq -\frac{1}{\lambda} \log\left\{\int_\Theta (\exp[-\lambda\phi(\xi_0;\theta)])^{1-\alpha} (\exp[-\lambda\phi(\xi_1;\theta)])^{\alpha}\, \mathrm{d}\theta\right\},
\end{aligned}$$

since $\phi[(1-\alpha)\xi_0 + \alpha\xi_1;\theta] \geq (1-\alpha)\phi(\xi_0;\theta) + \alpha\phi(\xi_1;\theta)$ for all $\theta \in \Theta$. Hölder's inequality then gives

$$\begin{aligned}
\phi_{ME,\lambda}[(1-\alpha)\xi_0 + \alpha\xi_1] &\geq -\frac{1}{\lambda} \log\left\{\left(\int_\Theta \exp[-\lambda\phi(\xi_0;\theta)]\, \mathrm{d}\theta\right)^{1-\alpha}\right. \\
&\qquad \left.\left(\int_\Theta \exp[-\lambda\phi(\xi_1;\theta)]\, \mathrm{d}\theta\right)^{\alpha}\right\} \\
&= (1-\alpha)\phi_{ME,\lambda}(\xi_0) + \alpha\phi_{ME,\lambda}(\xi_1).
\end{aligned}$$

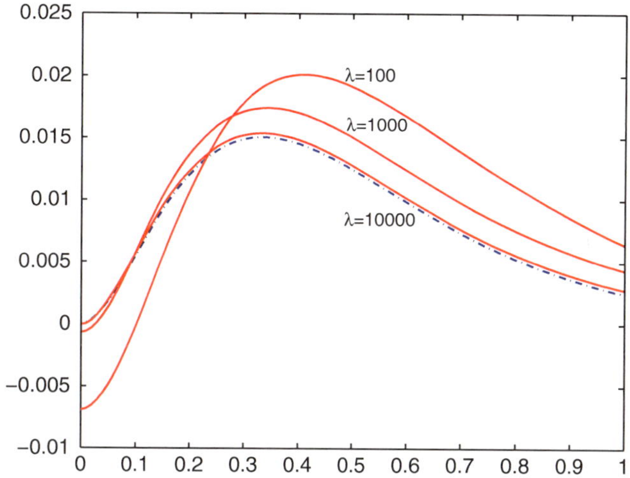

Fig. 8.7. $\phi_{MmO}(\delta_x)$ (*dash-dotted line*) and $\phi_{ME,\lambda}(\delta_x)$ (*solid lines*, $\lambda = 10^2, 10^3, 10^4$) as functions of x in Example 8.10; $\mu(\cdot)$ is the uniform measure on $\Theta = [1,3]$

Since the criterion $\phi_{ME,\lambda}(\cdot)$ given by (8.24) is concave, it enjoys the same properties as the local optimality criterion $\phi(\xi;\theta)$ on which it is based, that is, an equivalence theorem can be formulated (see Sect. 5.2.2) and optimization algorithms that converge globally to an optimum design measure are thus available; see Sect. 9.1.

Directional Derivative

Denote by $F_{\phi_\theta}(\xi;\nu)$ the directional derivative of $\phi(\cdot;\theta)$ at ξ in the direction ν. From the expression (8.24) of the criterion $\phi_{ME,\lambda}(\cdot)$, its directional derivative is

$$F_{\phi_{ME,\lambda}}(\xi;\nu) = \frac{\int_\Theta \exp[-\lambda\phi(\xi;\theta)]\, F_{\phi_\theta}(\xi;\nu)\, \mathrm{d}\theta}{\int_\Theta \exp[-\lambda\phi(\xi;\theta)]\, \mathrm{d}\theta}. \tag{8.26}$$

Note that it corresponds to a weighted average of $F_{\phi_\theta}(\xi;\nu)$ that gives more weight to θ such that $\phi(\xi;\theta)$ is small.

Example 8.10. This is a continuation of Examples 8.2, 8.5, and 8.8. We only consider situation A where $\phi(\xi;\theta) = M(\xi,\theta) = \int_{\mathscr{X}} x^2 \exp(-2\theta x)\,\xi(\mathrm{d}x)$. Figure 8.7 presents $\phi_{MmO}(\delta_x)$ and $\phi_{ME,\lambda}(\delta_x)$ as functions of x, for $\lambda = 10^2$, 10^3, and 10^4, when $\mu(\cdot)$ is the uniform measure on $\Theta = [1,3]$. Notice that the convergence of $\phi_{ME,\lambda}(\delta_x)$ to $\phi_{MmO}(\delta_x)$ as λ increases is not monotonic. □

An Upper Bound on $\phi_{MmO}(\xi)$

A criterion of the form (8.24) has been used in Sect. 7.7.3 to regularize the estimability criteria of Sect. 7.7. However, as the example above illustrates, the convergence of $\phi_{ME,\lambda}(\xi)$ to $\phi_{MmO}(\xi)$ as λ tends to infinity is not monotonic when $\mu(\cdot)$ is not finitely supported. Also, $\phi_{ME,\lambda}(\cdot)$ being constructed with the Lebesgue measure, we cannot make connections with average-optimum design of Sect. 8.1 for particular values of λ. This motivates the introduction of an alternative regularization, related to $\overline{\phi}_{MmO,q}(\cdot)$ of Sect. 8.3.1.

Define

$$\overline{\phi}_{ME,\lambda}(\xi) = -\frac{1}{\lambda}\log\left\{\int_{\Theta} \exp[-\lambda\phi(\xi;\theta)]\,\mu(\mathrm{d}\theta)\right\},\ \lambda > 0,$$

with $\mu(\cdot)$ a probability measure on Θ satisfying the same conditions as in Sect. 8.3.1. It corresponds to the specialization of (8.22) to $\psi(\cdot) = \exp(\cdot)$ and enjoys the same properties of convergence to $\phi_{MmO}(\xi)$ when $\lambda \to \infty$ as $\overline{\phi}_{MmO,q}(\cdot)$ when $q \to \infty$. Also, for all $\xi \in \Xi$ and all $\lambda > 0$, $\overline{\phi}_{ME,\lambda}(\xi) \geq \phi_{MmO}(\xi)$ and the convergence to $\phi_{MmO}(\xi)$ is monotonic as λ increases. Note that when Θ has a nonempty interior, by taking $\mu(\mathrm{d}\theta) = \mathrm{d}\theta/\mathrm{vol}(\Theta)$ we obtain $\phi_{ME,\lambda}(\xi) + \{\log[\mathrm{vol}(\Theta)]\}/\lambda \geq \phi_{MmO}(\xi)$. Since $\lim_{\lambda\to 0}\overline{\phi}_{ME,\lambda}(\xi) = \phi_{AO}(\xi)$ defined by (8.1), we can define by continuity

$$\overline{\phi}_{ME,0}(\xi) = \phi_{AO}(\xi).$$

The concavity of $\overline{\phi}_{ME,\lambda}(\cdot)$ follows from the same arguments as those used for $\phi_{ME,\lambda}(\cdot)$: we use Hölder's inequality with now integration with respect to the measure $\mu(\cdot)$. Its directional derivative takes the same form as (8.26), with $\mu(\mathrm{d}\theta)$ substituted for $\mathrm{d}\theta$. When Θ is finite, with $\Theta = \{\theta^{(1)},\ldots,\theta^{(M)}\}$, and μ gives weight μ_i to $\theta^{(i)}$ with $\underline{\mu} = \min_{i=1,\ldots,M}\mu_i > 0$, we have $\overline{\phi}_{ME,\lambda}(\xi) \leq \phi_{ME,\lambda}(\xi) - (1/\lambda)\log\underline{\mu}$ and thus

$$\overline{\phi}_{ME,\lambda}(\xi) + \frac{\log\underline{\mu}}{\lambda} \leq \phi_{ME,\lambda}(\xi) \leq \phi_{MmO}(\xi) \leq \overline{\phi}_{ME,\lambda}(\xi).$$

This implies that an optimal design ξ^* for $\overline{\phi}_{ME,\lambda}(\cdot)$ satisfies

$$\frac{\phi_{MmO}(\xi^*)}{\phi^*_{MmO}} \geq 1 + \frac{\log\underline{\mu}}{\lambda\,\phi^*_{MmO}},$$

where ϕ^*_{MmO} is the optimal value of $\phi^*_{MmO}(\cdot)$; the best bound is obtained for μ being the uniform measure and $\underline{\mu} = 1/M$.

Example 8.11. Consider again the situation of Example 8.10. Figure 8.8 presents $\phi_{AO}(\delta_x)$, $\phi_{MmO}(\delta_x)$, and $\overline{\phi}_{ME,\lambda}(\delta_x)$ as functions of x, for $\lambda = 1$, 10^2, 10^3, and 10^4, when $\mu(\cdot)$ is the uniform measure on $\Theta = [1,3]$. Note the closeness of $\overline{\phi}_{ME,\lambda}(\delta_x)$ to $\phi_{AO}(\delta_x)$ when λ is small and the monotonicity of the convergence to $\phi_{MmO}(\delta_x)$ when λ increases (compare with Fig. 8.7). □

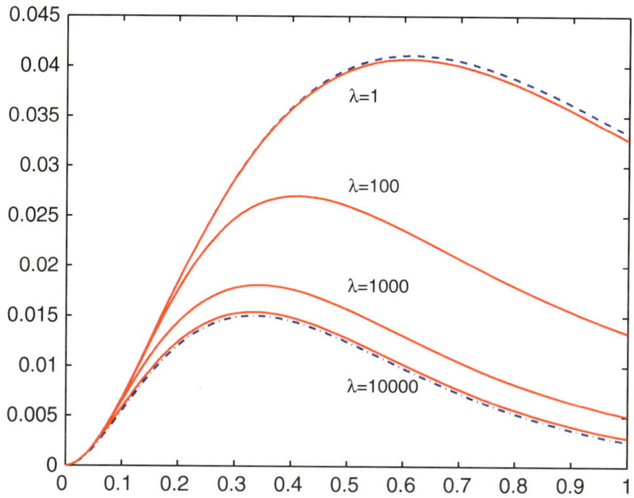

Fig. 8.8. $\phi_{AO}(\delta_x)$ (*dashed line*), $\phi_{MmO}(\delta_x)$ (*dash-dotted line*), and $\overline{\phi}_{ME,\lambda}(\delta_x)$ (*solid lines*, $\lambda = 1, 10^2, 10^3, 10^4$) as functions of x in Example 8.11; $\mu(\cdot)$ is the uniform measure on $\Theta = [1, 3]$

8.4 Probability Level and Quantile Criteria

Average and maximin-optimum designs can be considered as attempts to protect against a bad choice of the prior value θ^0 in locally optimum design. However, as shown below, the protection they provide is not totally satisfactory, and some difficulties remain.

Difficulties with Average and Maximin-Optimum Design

(*i*) Consider a design ξ^*_{A0} optimal for the criterion $\phi_{AO}(\cdot)$ given by (8.1). Although optimal in the average sense, ξ^*_{A0} may perform poorly for "many" values of θ, in the sense that $\mu\{\phi(\xi^*_{AO};\theta) < u\}$ may be large for some unacceptably low value of u.

(*ii*) For $\psi(\cdot)$ an increasing real function, the maximization of $\psi[\phi(\xi;\theta)]$ is equivalent to that of $\phi(\xi;\theta)$, but maximizing $\mathbb{E}_\mu\{\psi[\phi(\xi;\theta)]\}$ is not equivalent to maximizing $\mathbb{E}_\mu\{\phi(\xi;\theta)\}$ in general. Therefore, a single design criterion for local optimality yields infinitely many criteria for average optimality; see, e.g., Fedorov (1980). For instance, with the D-optimality criterion we can associate $\mathbb{E}_\mu\{\log\det[\mathbf{M}(\xi,\theta)]\}$ and $\mathbb{E}_\mu\{\det^{1/p}[\mathbf{M}(\xi,\theta)]\}$, both being concave on $\mathbb{M}^>$.

(*iii*) Consider a design ξ^*_{MmO} optimal for $\phi_{MmO}(\cdot)$ given by (8.11) with Θ a compact subset of $\mathbb{R}^p$ with nonempty interior. It frequently happens that, in fact,

$$\xi^*_{MmO} = \arg\max_{\xi\in\Xi}\min_{\theta\in\partial\Theta}\phi(\xi;\theta),$$

with $\partial\Theta$ the boundary of Θ. The dependency of a locally optimal design in the value chosen for θ^0 is then simply replaced by the dependency of ξ^*_{MmO} in the choice of some extreme points of Θ. Another (sometimes related) difficulty is that Θ may contain values of θ such that $\mathbf{M}(\xi,\theta)$ is close to being singular for all $\xi \in \Xi$ and the optimal design ξ^*_{MmO} then focuses on such θ; one may even encounter pathological situations where there exists some $\theta \in \Theta$ such that $\mathbf{M}(\xi,\theta)$ is singular for all $\xi \in \Xi$ and no maximin-optimal design exists.

Probability Level and Quantile Criteria

We consider here new stochastic design criteria based on the distribution of $\phi(\xi;\theta)$ when θ is distributed with some prior probability measure $\mu(\cdot)$ on $\Theta \subset \mathbb{R}^p$. In particular, we shall consider the probability levels

$$P_u(\xi) = \mu\{\phi(\xi;\theta) \geq u\} \tag{8.27}$$

and quantiles

$$Q_\alpha(\xi) = \max\{u : P_u(\xi) \geq 1-\alpha\}, \quad \alpha \in [0,1], \tag{8.28}$$

with u and α considered as free parameters, chosen by the user. When the range of possible values for $\phi(\xi;\theta)$ is known, which is the case, for instance, when $\phi(\cdot;\theta)$ is an efficiency criterion $\mathscr{E}_\phi(\cdot;\theta)$ with values in $[0,1]$ (see Sect. 5.1.8), we can specify a target level u and then maximize the probability $P_u(\xi)$ that the target is reached, or equivalently minimize the risk $1-P_u(\xi)$ that it is not. In other situations, we can specify a probability level α that defines an acceptable risk and maximize the value of u such that the probability that $\phi(\xi;\theta)$ is smaller than u is less than α, which corresponds to maximizing $Q_\alpha(\xi)$. We shall assume that $\phi[(1-\gamma)\xi+\gamma\nu;\theta]$ is continuously differentiable in $\gamma \in [0,1)$ for any θ and any probability measures ξ, ν on $\mathscr{X}$ such that $\mathbf{M}(\xi,\theta)$ is nonsingular. We also assume that $\phi(\xi;\theta)$ is continuous in θ and that the measure $\mu(\cdot)$ has a positive density on every open subset of Θ. This implies that $Q_\alpha(\xi)$ is defined as the solution in u of the equation $1-P_u(\xi)=\alpha$; see Fig. 8.9.

One may notice that the difficulties (*i*–*iii*) mentioned above for average and maximin-optimum design are explicitly taken into account by the proposed approach: the probability indicated in (*i*) is precisely $1-P_u(\xi)$ which is minimized, (*ii*) substituting $\psi[\phi(\xi;\theta)]$ for $\phi(\xi;\theta)$ with $\psi(\cdot)$ increasing leaves (8.27) and (8.28) unchanged, (*iii*) the role of the boundary of Θ is negligible when a small probability is attached to it, and moreover, probability measures with infinite support are allowed. It will be shown below that kernel smoothing can be used to make $P_u(\cdot)$ and $Q_\alpha(\cdot)$ differentiable. When $\phi(\xi;\theta)$ is concave in ξ for all θ, $\phi_{AO}(\cdot)$ and $\phi_{MmO}(\cdot)$ are also concave. Unfortunately,

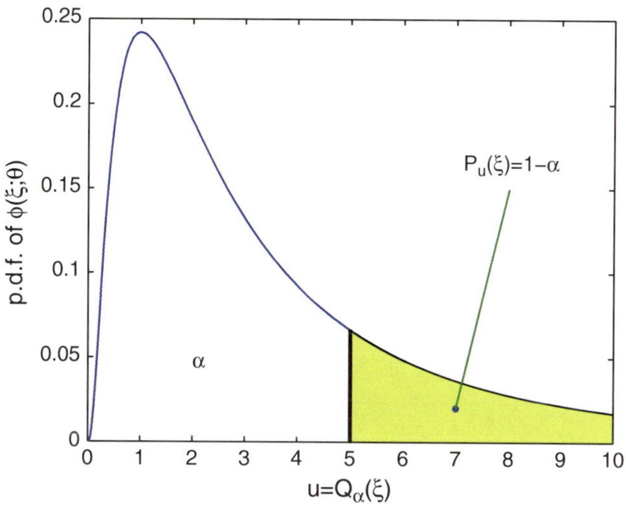

Fig. 8.9. Probability levels and quantiles for a design criterion $\phi(\xi;\theta)$

$P_u(\cdot)$ and $Q_\alpha(\cdot)$ are generally not, which is probably the main drawback of the approach. However, Q_α satisfies the following: suppose that the support Θ of $\mu(\cdot)$ is compact, then, $Q_\alpha(\xi) \to \phi_{MmO}(\xi)$ when $\alpha \to 0$, and a design optimal for Q_α will tend to be optimal for $\phi_{MmO}(\cdot)$ and vice versa.

Example 8.12. Consider again the situation of Examples 8.10 and 8.11 for which we have $\phi(\xi;\theta) = \int_{\mathscr{X}} x^2 \exp(-2\theta x)\,\xi(\mathrm{d}x)$, $x \geq 0$. Take $\mu(\cdot)$ as the uniform measure on $\Theta = [\theta_{\min}, \theta_{\max}]$. Direct calculations give

$$P_u(\delta_x) = \frac{\log[x^2 \exp(-2\theta_{\min} x)] - \log(u)}{2x(\theta_{\max} - \theta_{\min})}$$

and

$$Q_\alpha(\delta_x) = \phi_{MmO}(\delta_x) \exp[2x\alpha(\theta_{\max} - \theta_{\min})],$$

where $\phi_{MmO}(\delta_x) = x^2 \exp(-2\theta_{\max} x)$ is defined by (8.11). $Q_\alpha(\delta_x) \geq \phi_{MmO}(\delta_x)$ and tends to $\phi_{MmO}(\delta_x)$ when $\alpha \to 0$. Figure 8.10 presents $\phi_{MmO}(\delta_x)$ and $Q_\alpha(\delta_x)$ as functions of x for different values of α when $\mu(\cdot)$ is the uniform measure on $\Theta = [1, 3]$. □

In the next section we show how to compute the directional derivatives of $P_u(\xi)$ and $Q_\alpha(\xi)$. A steepest-ascent optimization algorithm can then be used to optimize $P_u(\cdot)$ or $Q_\alpha(\cdot)$, with guaranteed convergence to a local maximum only since the criteria are not necessarily concave.

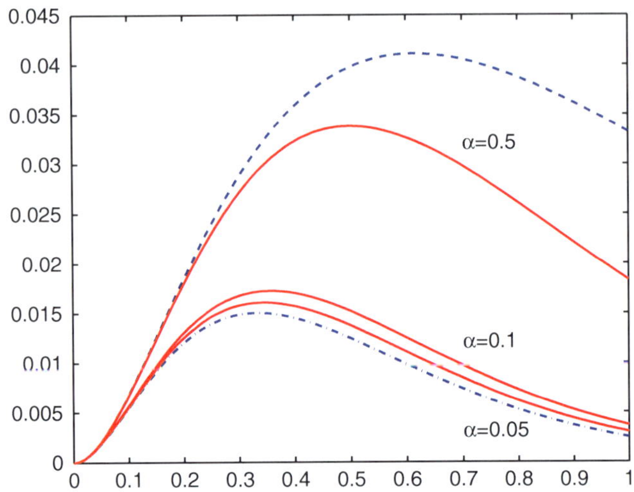

Fig. 8.10. $\phi_{AO}(\delta_x)$ (*dashed line*), $\phi_{MmO}(\delta_x)$ (*dash-dotted line*), and $Q_\alpha(\delta_x)$ (*solid lines*, $\alpha = 0.5, 0.1, 0.05$) as functions of x in Example 8.12; $\mu(\cdot)$ is the uniform measure on $\Theta = [1,3]$

Computation of Derivatives

Let $\xi = (1-\gamma)\xi_0 + \gamma\nu$ and consider the directional derivatives

$$F_{P_u}(\xi;\nu) = \left.\frac{\partial P_u(\xi)}{\partial\gamma}\right|_{\gamma=0}, \tag{8.29}$$

$$F_{Q_\alpha}(\xi;\nu) = \left.\frac{\partial Q_\alpha(\xi)}{\partial\gamma}\right|_{\gamma=0}. \tag{8.30}$$

Since $Q_\alpha(\xi)$ satisfies the implicit equation $P_{Q_\alpha(\xi)}(\xi) = 1-\alpha$, we can write

$$\left.\frac{\partial P_u(\xi)}{\partial\gamma}\right|_{u=Q_\alpha(\xi)} + \left.\frac{\partial P_u(\xi)}{\partial u}\right|_{u=Q_\alpha(\xi)} \frac{\partial Q_\alpha(\xi)}{\partial\gamma} = 0,$$

which gives

$$\frac{\partial Q_\alpha(\xi)}{\partial\gamma} = -\left.\left(\frac{\partial P_u(\xi)}{\partial\gamma}\Big/\frac{\partial P_u(\xi)}{\partial u}\right)\right|_{u=Q_\alpha(\xi)}. \tag{8.31}$$

To compute the derivatives $\partial P_u(\xi)/\partial\gamma$ and $\partial P_u(\xi)/\partial u$, we write $P_u(\xi)$ as

$$P_u(\xi) = \int_\Theta \mathbb{I}_{[u,\infty)}[\phi(\xi;\theta)]\,\mu(\mathrm{d}\theta) = \int_\Theta \mathbb{I}_{(-\infty,\phi(\xi;\theta)]}(u)\,\mu(\mathrm{d}\theta),$$

with $\mathbb{I}_{\mathcal{A}}(\cdot)$ the indicator function of the set $\mathcal{A}$. When approximating the indicator step function by a normal distribution function with small variance σ^2, the two expressions above respectively become

$$P_u(\xi) \approx \int_\Theta \mathbb{F}_{u,\sigma^2}[\phi(\xi;\theta)]\,\mu(\mathrm{d}\theta) = \int_\Theta [1 - \mathbb{F}_{\phi(\xi;\theta),\sigma^2}(u)]\,\mu(\mathrm{d}\theta)$$

with $\mathbb{F}_{a,\sigma^2}(\cdot)$ the distribution function of the normal $\mathscr{N}(a,\sigma^2)$. Differentiating these approximations, respectively, with respect to γ and u, we get

$$\left.\frac{\partial P_u(\xi)}{\partial \gamma}\right|_{\gamma=0} \approx \int_\Theta n_{u,\sigma^2}[\phi(\xi_0;\theta)]\left.\frac{\partial \phi(\xi;\theta)}{\partial \gamma}\right|_{\gamma=0} \mu(\mathrm{d}\theta), \tag{8.32}$$

$$\frac{\partial P_u(\xi)}{\partial u}_{|\gamma=0} \approx -\int_\Theta n_{\phi(\xi_0;\theta),\sigma^2}(u)\,\mu(\mathrm{d}\theta), \tag{8.33}$$

with $n_{a,\sigma^2}(\cdot)$ the density of $\mathbb{F}_{a,\sigma^2}(\cdot)$. These expressions can be substituted in (8.31) to form an approximation of $\partial Q_\alpha(\xi)/\partial\gamma\big|_{\gamma=0}$. As shown below, this type of approximation can be related to kernel smoothing.

$P_u(\xi)$ and $Q_\alpha(\xi)$ for a Normal Prior with Small Variance

Suppose that θ has the prior normal distribution $\mathscr{N}(\theta^0,\mathbf{\Omega})$ with $\mathbf{\Omega}$ small, which may be considered as a slight relaxation of locally optimum design. Define $\theta_u = \theta_u(\xi) = \arg\min_{\{\theta:\phi(\theta,\xi)=u\}} \|\theta-\theta^0\|^2_{\mathbf{\Omega}^{-1}}$, with $\|\mathbf{a}\|^2_{\mathbf{\Omega}^{-1}} = \mathbf{a}^\top\mathbf{\Omega}^{-1}\mathbf{a}$. Replacing $\phi(\xi;\theta)$ by its linear approximation around θ_u in the definition of $P_u(\xi)$, we get

$$\begin{aligned}
P_u(\xi) &\approx \mu\left\{\phi(\xi;\theta_u) + (\theta-\theta_u)^\top \left.\frac{\partial\phi(\xi;\theta)}{\partial\theta}\right|_{\theta_u} \geq u\right\} \\
&= \mu\left\{(\theta-\theta_u)^\top \left.\frac{\partial\phi(\xi;\theta)}{\partial\theta}\right|_{\theta_u} \geq 0\right\} \\
&= \mu\left\{\frac{(\theta-\theta^0)^\top\mathbf{\Omega}^{-1}\mathbf{h}_\xi(\theta_u)}{[\mathbf{h}_\xi^\top(\theta_u)\mathbf{\Omega}^{-1}\mathbf{h}_\xi(\theta_u)]^{1/2}} \geq \frac{(\theta_u-\theta^0)^\top\mathbf{\Omega}^{-1}\mathbf{h}_\xi(\theta_u)}{[\mathbf{h}_\xi^\top(\theta_u)\mathbf{\Omega}^{-1}\mathbf{h}_\xi(\theta_u)]^{1/2}}\right\}
\end{aligned}$$

where $\mathbf{h}_\xi(t) = \mathbf{\Omega}\partial\phi(\xi;\theta)/\partial\theta\big|_{\theta=t}$. The term on the left-hand side of the inequality is normally distributed $\mathscr{N}(0,1)$, and direct calculation shows that the term on the right-hand side equals $z\|\theta_u-\theta^0\|_{\mathbf{\Omega}^{-1}}$, with $z = \mathrm{sign}[u-\phi(\xi;\theta^0)]$. This gives the approximation

$$P_u(\xi) \approx \int_{z\|\theta_u-\theta^0\|_{\mathbf{\Omega}^{-1}}}^{\infty} n_{0,1}(t)\,\mathrm{d}t.$$

The value α in (8.28) satisfies $\alpha = \int_{-\infty}^{z\|\theta_u-\theta^0\|_{\mathbf{\Omega}^{-1}}} n_{0,1}(t)\,\mathrm{d}t$. Let q_α denote the α-quantile of the standard normal $\mathscr{N}(0,1)$, i.e., $\int_{-\infty}^{q_\alpha} n_{0,1}(t)\,\mathrm{d}t = \alpha$, $Q_\alpha(\xi)$ is thus defined by the value of u such that $z\|\theta_u-\theta^0\|_{\mathbf{\Omega}^{-1}} = q_\alpha$.

These expressions become simpler when the linear approximation of $\phi(\xi;\theta)$ is around θ^0, and we get

$$P_u(\xi) \approx \int_{s_u(\xi)}^{\infty} n_{0,1}(t)\,\mathrm{d}t,$$
$$Q_\alpha(\xi) \approx \phi(\xi;\theta^0) + q_\alpha[\mathbf{h}_\xi^\top(\theta^0)\boldsymbol{\Omega}^{-1}\mathbf{h}_\xi(\theta^0)]^{1/2},$$

with $s_u(\xi) = [u - \phi(\xi;\theta^0)]/[\mathbf{h}_\xi^\top(\theta^0)\boldsymbol{\Omega}^{-1}\mathbf{h}_\xi(\theta^0)]^{1/2}$. However, using a linear approximation of the boundary of the set $\mathcal{S}_u(\xi) = \{\theta \in \mathbb{R}^p : \phi(\xi;\theta) \geq u\}$ may give rather approximate results, and other approximations based on kernel smoothing are considered below.

Kernel Smoothing

The idea is to estimate $P_u(\xi)$, $Q_\alpha(\xi)$, and their derivatives through an approximation of the p.d.f. of $\phi(\xi;\theta)$ obtained by a standard kernel estimator

$$\varphi_{M,\xi}(z) = 1/(Mh_M)\sum_{i=1}^{M} \mathscr{K}\left\{[z - \phi(\xi;\theta^{(i)})]/h_M\right\}.$$

Here $\mathscr{K}(\cdot)$ is a symmetric kernel function, typically the p.d.f. of a probability measure on $\mathbb{R}$ with $\mathscr{K}(z) = \mathscr{K}(-z)$, e.g., $\mathscr{K}(\cdot) = n_{0,1}(\cdot)$, and $\theta^{(i)}$, $i = 1,\ldots,M$, is a sample of values of θ independently randomly generated with the prior measure $\mu(\cdot)$. The bandwidth h_M tends to zero as $M \to \infty$. From this we obtain directly

$$P_u(\xi) \approx \hat{P}_u^M(\xi) = \int_{-\infty}^{\infty} \mathbb{I}_{[u,\infty)}(z)\,\varphi_{M,\xi}(z)\,dz, \tag{8.34}$$

which is easily computed when $\int_u^\infty \mathscr{K}(z)dz$ has a simple form. The value of $Q_\alpha(\xi)$ can then be estimated by

$$\hat{Q}_\alpha^M(\xi) = \{u : \hat{P}_u^M(\xi) = 1 - \alpha\}, \tag{8.35}$$

which is also easily computed numerically. Alternatively, we may use kernel smoothing again and compute

$$\tilde{Q}_\alpha^M(\xi) = \frac{1}{h_M}\sum_{i=1}^{M} \phi_i^M(\xi)\int_{(i-1)/M}^{i/M} \mathscr{K}\left[(z-\alpha)/h_M\right]dz,$$

with $\phi_1^M(\xi) \leq \phi_2^M(\xi) \leq \cdots \leq \phi_M^M(\xi)$ the order statistics obtained from the $\phi(\xi;\theta^{(i)})$; see, e.g., Parzen (1979) and Yang (1985). However, this latter form seems to be less precise for values of α close to zero or one.

Consider now the computation of directional derivatives, with again $\xi = (1-\gamma)\xi_0 + \gamma\nu$. Direct calculations give

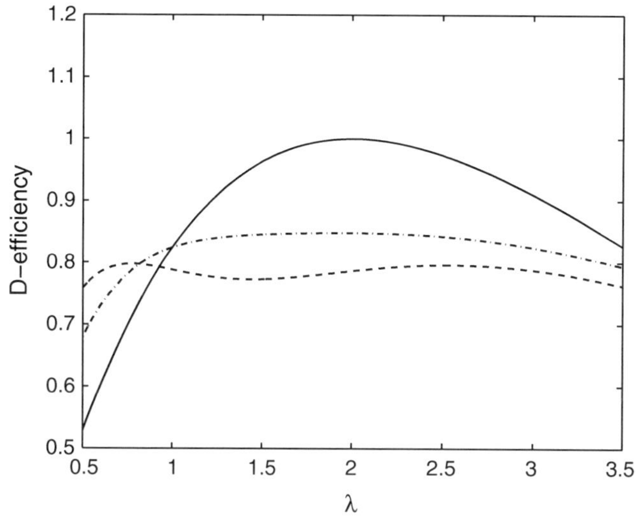

Fig. 8.11. D-efficiencies as function of λ for different designs; *solid line*, local D-optimal ξ_2^*; *dashed line*, optimal for $P_{0.75}$; *dash-dotted line*, optimal for $Q_{0.1}$

$$\left.\frac{\partial \hat{P}_u^M(\xi)}{\partial \gamma}\right|_{\gamma=0} = \frac{1}{Mh_M}\sum_{i=1}^{M} \left.\frac{\partial \phi(\xi;\theta^{(i)})}{\partial \gamma}\right|_{\gamma=0} \mathscr{K}\left(\frac{u-\phi(\xi_0;\theta^{(i)})}{h_M}\right), \quad (8.36)$$

$$\left.\frac{\partial \hat{P}_u^M(\xi)}{\partial u}\right|_{\gamma=0} = -\frac{1}{Mh_M}\sum_{i=1}^{M} \mathscr{K}\left(\frac{u-\phi(\xi_0;\theta^{(i)})}{h_M}\right), \quad (8.37)$$

to be used in (8.29)–(8.31). Notice that when taking $\sigma^2 = h_M$ and $\mu(\cdot)$ as the discrete measure with mass $1/M$ at each $\theta^{(i)}$, (8.32) and (8.33) respectively give (8.36) and (8.37) with $\mathscr{K}(\cdot) = n_{0,1}(\cdot)$.

The accuracy of the kernel approximations (8.34)–(8.37) improves as M increases with, on the other hand, a computational cost that increases with M.

Example 8.13. Consider the nonlinear regression model $\eta(x,\theta) = \beta e^{-\lambda x}$, with $\theta = (\beta,\ \lambda)^\top$ the vector of parameters to be estimated. We suppose that the errors are i.i.d. with variance 1. The information matrix $\mathbf{M}(\xi,\theta)$ for LS estimation with a design measure ξ then takes the form (3.32). We suppose that $\beta > 0$ and take $\mathscr{X} = [0,\infty)$. The local D-optimal experiment ξ_θ^* that maximizes $\det \mathbf{M}(\xi,\theta)$ puts mass 1/2 at $x = 0$ and $x = 1/\lambda$; the associated value of $\det \mathbf{M}(\xi,\theta)$ is $\det \mathbf{M}(\xi_\theta^*,\theta) = \beta^2/(4e^2\lambda^2)$. We consider the D-efficiency criterion defined by $\mathscr{E}_D(\xi;\theta) = \{\det \mathbf{M}(\xi,\theta)/\det \mathbf{M}(\xi_\theta^*,\theta)\}^{1/2} \in [0,1]$. Due to the linear dependency of $\eta(x,\theta)$ in β, ξ_θ^* and $\mathscr{E}_D(\xi;\theta)$ only depend on λ, and we shall write ξ_λ^*, $\mathscr{E}_D(\xi;\lambda)$ instead of ξ_θ^*, $\mathscr{E}_D(\xi;\theta)$. Supposing that $\lambda = 2$ when designing the experiment, the efficiency $\mathscr{E}_D(\xi_2^*;\lambda)$ is plotted as a function of λ (solid line) in Fig. 8.11.

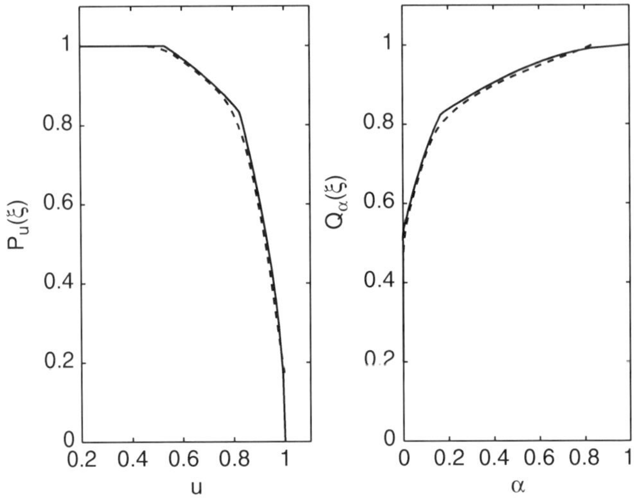

Fig. 8.12. *Left*: $\hat{P}_u^n(\xi)$ (*dashed line*) and $P_u(\xi)$ (*solid line*) as functions of u. *Right*: $\hat{Q}_\alpha^n(\xi)$ (*dashed line*) and $Q_\alpha(\xi)$ (*solid line*) as functions of α; $\xi = \xi_2^*$, $M = 100$

Suppose now that we only know that $\lambda \in [1/2, 7/2]$ and put a uniform prior for λ on that interval; ξ_2^* is then optimal for the midpoint, but its efficiency is less than 53 % for the endpoint $\lambda = 1/2$. We approximate $P_u(\xi)$ and $Q_\alpha(\xi)$ by kernel smoothing with $\mathscr{K}(\cdot) = n_{0,1}(\cdot)$ for $M = 100$ values $\hat{\lambda}^{(i)}$ equally spaced in $[0.5, 3.5]$. No special care is taken for the choice of h_M, and we simply use the rule $h_M = \hat{\sigma}_M(\mathscr{E}_D)M^{-1/5}$ (see, e.g., Scott 1992 p. 152), with $\hat{\sigma}_M(\mathscr{E}_D)$ the empirical standard deviation of the values $\mathscr{E}_D(\xi; \lambda^{(i)})$, $i = 1, \ldots, n$. Figure 8.12 shows the estimated values $\hat{P}_u^M$ (left) and $\hat{Q}_\alpha^M$ (right), in dashed lines, as functions of u and α, respectively, for $\xi = \xi_2^*$. One can check the reasonably good agreement with the exact values of P_u and Q_α, plotted in solid lines; increasing M to 1,000 makes the curves almost indistinguishable.

The optimization of $\hat{P}_{0.75}^n$ and $\hat{Q}_{0.10}^n$ with a vertex-direction (steepest-ascent) algorithm on the finite design space $\{0, 0.1, 0.2, \ldots, 5\}$ (see Sect. 9.1.1) gives the four-point designs

$$\xi^*(P_{0.75}) \simeq \left\{ \begin{matrix} 0 & 0.3 & 0.4 & 1.7 \\ 0.4523 & 0.0977 & 0.2532 & 0.1968 \end{matrix} \right\},$$

and

$$\xi^*(Q_{0.10}) \simeq \left\{ \begin{matrix} 0 & 0.3 & 0.4 & 1.3 \\ 0.4688 & 0.1008 & 0.2634 & 0.1670 \end{matrix} \right\},$$

where the first row indicates the support points and the second one their respective weights. They satisfy $\hat{P}_{0.75}^M[\xi^*(P_{0.75})] \simeq 0.9999$ and $\hat{Q}_\alpha^M[\xi^*(Q_{0.10})] \simeq 0.783$. The efficiencies of these designs are plotted in Fig. 8.11. The exact value $P_u[\xi^*(P_{0.75})]$ equals one, indicating that the efficiency is larger than 75 % for all possible values of λ.

We decrease now α to the value 0.01 in $Q_\alpha(\cdot)$. The optimization of $\hat{Q}^n_{0.01}$ gives a design very close to $\xi^*(P_{0.75})$ which, together with the shape of the curve in dashed line on Fig. 8.11, suggests that $\xi^*(P_{0.75})$ is almost maximin optimal. The comparison with the curve in dash-dotted line on the same figure, obtained for $\alpha = 0.1$, indicates that accepting a small loss of efficiency for about 10 % of the values of λ produces a significant increase of efficiency on most of the interval $[1/2, 7/2]$. □

8.5 Sequential Design

A radically different approach can sometimes be used to tackle the problem of dependency of the optimal design into the unknown value of θ. It consists in constructing the design sequentially, step by step, using at each step the information available for choosing an experiment adapted for next step. In *full-sequential design*, a single design point x_{k+1} is chosen after each observation $y(x_k)$. In *batch-sequential design*, design points are chosen m by m, with m the size of the batches. The number of design stages can also be fixed a priori, with the number of observations to be collected at each stage considered as a tuning parameter. In a standard two-stage strategy, for instance, a first experiment is run, e.g., optimal for some nominal value θ^0, with n observations collected; the model parameters are then estimated, say by $\hat{\theta}^n$; the remaining $N - n$ observations are collected with an experiment optimal for $\hat{\theta}^n$.

The literature on sequential experimental design is vast and rich, the applications are abundant. Sequential design is directly related to sequential analysis and testing, but such issues will not be considered here, and we shall only focus on the parameter estimation problem.[2] Also, we shall not touch to the field of adaptive design in clinical trials, to which many papers and books are devoted.

A Bayesian point of view is well adapted to formalize the increase of information on the location of θ or, equivalently, the decrease of uncertainty on θ, when the number of observations collected increases. When the horizon N is fixed, the optimal sequential experiment is then obtained as the solution of a stochastic dynamic programming problem. This is usually untractable, and approximations must be used, such as approximations of the posterior and pre-posterior distributions of θ (see Gautier and Pronzato 1998), together with Laplace approximations for integrals or stochastic approximation to avoid the calculation of such integrals; see Gautier and Pronzato (2000). However, numerical simulations indicate that there is not much to gain compared to traditional sequential design, where future planning is ignored and each design step

[2] The term *adaptive design* would thus be more appropriate to describe the kind of problem we shall deal with, which has strong connections with adaptive control; see, e.g., Pronzato (2008). Sequential design remains the usual denomination in the literature, however.

is considered as being the last one. There, the m design points at stage k, with $m = 1$ in full-sequential design, are chosen at best using the current estimated value $\hat{\theta}^k$ of θ, which corresponds to a forced-certainty-equivalence control in terms of stochastic control. We shall restrict our attention to such adaptive designs.

The Bayesian paradigm can also be used in asymptotic considerations. This is adopted, for instance, by Hu (1998) for full-sequential design and by Spokoinyi (1992) for a two-stage strategy, optimal in the asymptotically minimax sense. In particular, it facilitates the investigation of consistency issues, since the sequence of posterior means for θ forms a martingale which converges a.s. from the martingale convergence theorem. Hu (1998) also shows that the sequence of posterior variances of $f(\theta)$, with $f(\cdot)$ any bounded measurable function, forms a positive supermartingale which thus also converges a.s. However, estimating the parameters by their posterior mean is not practical in nonlinear situations, and we shall focus our attention on sequential design for LS estimation in regression models with stationary errors; see (3.2), (3.3). Similar developments can be made for other estimators and models considered in Chaps. 3 and 4. We shall denote by $\hat{\theta}^N$ the estimator based on N observations.

Our objective here will be modest: we simply want to construct experiments that are asymptotically optimal for the true value of the model parameters. More precisely, we want the asymptotic covariance matrix of the estimator to be the same as for an optimal experiment designed for the true (unknown) value of the model parameters. Two situations will be considered. In the first one (Sect. 8.5.1), the horizon is fixed, and the experiment is designed in two stages; we show that one should allocate $n = \mathcal{O}(\sqrt{N})$ observations at the first stage—and thus $N-n$ at the second. The second situation considered (Sect. 8.5.2) corresponds to full-sequential design. There are two major difficulties there. First, the sequential construction of the experiment destroys the independence among the observations $y(x_1), \ldots, y(x_N)$, which raises specific difficulties for proving the consistency of the estimator of θ. Second, the design of the experiment is usually based on a matrix $\mathbf{M}$ constructed similarly to the information matrices encountered in Chaps. 3 and 4; however, $\mathbf{M}$ is no longer the information matrix when the experiment is built sequentially. Therefore, this construction only makes sense if we can prove that the asymptotic covariance matrix of the estimator corresponds to $\mathbf{M}^{-1}$.

8.5.1 Two-Stage Allocation

Let $\Phi(\cdot)$ denote a strictly concave and differentiable criterion and ξ_θ^* denote the optimal design for θ, that is, ξ_θ^* maximizes $\Phi[\mathbf{M}(\xi, \theta)]$, with $\mathbf{M}(\xi, \theta)$ the information matrix for the design ξ and parameters θ in the model considered. We consider a two-stage strategy, where n observations are collected at stage 1 with some design $\xi_{(1)}$ and $N-n$ at stage 2 with the optimal design $\xi_{\hat{\theta}^n}^*$, where $\hat{\theta}^n$ denotes the estimated value of θ obtained from the n observations of stage 1. Such a problem is considered in (Gautier and Pronzato, 1999) within a Bayesian framework: the optimal proportion n/N can be obtained by solving a stochastic dynamic programming problem; calculations are facilitated by using various approximations for integrals or stochastic approximation to avoid the calculation of integrals. We shall adopt here a non-Bayesian point of view.

Asymptotic Considerations

There are no particular difficulties with the asymptotic behavior of $\hat{\theta}^N$; see, for instance, Chaudhuri and Mykland (1993) for a similar situation with ML estimation. We suppose that the design $\{x_k\}$ at first stage, for $k = 1, \ldots, n$, is asymptotically discrete with limiting measure $\xi_{(1)}$ (discrete) and that the design at second stage is also asymptotically discrete with limiting measure $\xi^*_{\hat{\theta}^n}$ (conditionally on stage 1). This is not restrictive: in practice we can choose the design points at each stage among the support points of the corresponding design measure and adjust the frequencies in order to guarantee convergence of the sampling measure to the target one, $\xi_{(1)}$ or $\xi^*_{\hat{\theta}^n}$.

Suppose that the number n of observations at stage 1 tends to infinity. For a $\xi_{(1)}$ ensuring usual estimability conditions (see (3.6)), $\hat{\theta}^n$ then satisfies $\hat{\theta}^n \overset{\text{a.s.}}{\to} \bar{\theta}$ as $n \to \infty$; see Theorem 3.1. The strong consistency of $\hat{\theta}^N$ follows from arguments similar to those used in the same theorem.

Since $\hat{\theta}^n \overset{\text{a.s.}}{\to} \bar{\theta}$ as $n \to \infty$, $\Phi[\mathbf{M}(\xi^*_{\hat{\theta}^n}, \bar{\theta})] \overset{\text{a.s.}}{\to} \Phi[\mathbf{M}(\xi^*_{\bar{\theta}}, \bar{\theta})]$; that is, $\xi^*_{\hat{\theta}^n}$ tends to be optimal for $\bar{\theta}$ and $\mathbf{M}(\xi^*_{\hat{\theta}^n}, \bar{\theta}) \to \mathbf{M}(\xi^*_{\bar{\theta}}, \bar{\theta})$, which is unique due to the strict concavity of $\Phi(\cdot)$—notice, however, that the optimal design measure for $\bar{\theta}$ is not necessarily unique. Suppose now that $n \to \infty$ and $n/N \to 0$ as $N \to \infty$. Consider the proof of Theorem 3.8. A Taylor development gives an equation similar to (3.27), where again $\nabla^2_\theta J_N(\beta^N_i) \overset{\text{a.s.}}{\to} 2\,\mathbf{M}_1(\xi, \bar{\theta})$, with $\mathbf{M}_1(\xi, \theta)$ given by (3.23) with $w(x) \equiv 1$ (ordinary LS). We decompose $-\sqrt{N}\nabla_\theta J_N(\bar{\theta})$ into

$$-\sqrt{N}\nabla_\theta J_N(\bar{\theta}) = \frac{2\sqrt{n}}{\sqrt{N}} \frac{1}{\sqrt{n}} \sum_{k=1}^{n} \varepsilon_k \left.\frac{\partial \eta(x_k, \theta)}{\partial \theta}\right|_{\bar{\theta}} + \frac{2\sqrt{N-n}}{\sqrt{N}} \frac{1}{\sqrt{N-n}} \sum_{k=n+1}^{N} \varepsilon_k \left.\frac{\partial \eta(x_k, \theta)}{\partial \theta}\right|_{\bar{\theta}} .$$

We then have $(1/\sqrt{n}) \sum_{k=1}^n \varepsilon_k\, \partial\eta(x_k,\theta)/\partial\theta\big|_{\bar{\theta}} \overset{\text{d}}{\to} \mathbf{v} \sim \mathscr{N}(\mathbf{0}, \mathbf{M}_2(\xi_{(1)}, \bar{\theta}))$ and $(1/\sqrt{N-n}) \sum_{k=n+1}^N \varepsilon_k\, \partial\eta(x_k,\theta)/\partial\theta\big|_{\bar{\theta}} \overset{\text{d}}{\to} \mathbf{w} \sim \mathscr{N}(\mathbf{0}, \mathbf{M}_2(\xi^*_{\hat{\theta}^n}, \bar{\theta}))$ conditionally on stage 1, with $\mathbf{M}_2(\xi, \theta)$ given by (3.25) with $w(x) \equiv 1$ and $\sigma^2(x) \equiv \sigma^2$. Therefore, $\sqrt{N}(\hat{\theta}^N - \bar{\theta}) \overset{\text{d}}{\to} \mathbf{z} \sim \mathscr{N}(\mathbf{0}, \mathbf{M}^{-1}(\xi^*_{\bar{\theta}}, \bar{\theta}))$, $N \to \infty$, with $\mathbf{M}(\xi, \theta)$ given by (3.26) with $\sigma^2(x) \equiv \sigma^2$. The two-stage experiment is thus asymptotically optimal for $\bar{\theta}$.

One can easily go further and give an indication of the optimal choice for n as $N \to \infty$; this is considered below.

Optimal Sample-Size for Stage 1

Suppose that $n \to \infty$ and $n/N \to 0$ as $N \to \infty$ and let ξ_N denote the two-stage design. We have $\xi_N = (n/N)\,\xi_{(1)} + (1 - n/N)\,\xi^*_{\hat{\theta}^n}$ so that

$$\Phi[\mathbf{M}(\xi_N,\bar{\theta})] = \Phi[\mathbf{M}(\xi^*_{\hat{\theta}^n},\bar{\theta})] + \frac{n}{N}\,T(\xi^*_{\hat{\theta}^n},\xi_{(1)},\bar{\theta}) + \mathcal{O}(n^2/N^2)$$

where

$$T(\xi^*_{\theta},\xi_{(1)},\bar{\theta}) = \mathrm{trace}\left\{\nabla_{\mathbf{M}}\Phi[\mathbf{M}(\xi^*_{\theta},\bar{\theta})] \times [\mathbf{M}(\xi_{(1)},\bar{\theta}) - \mathbf{M}(\xi^*_{\theta},\bar{\theta})]\right\}.$$

Now, the first two terms in the development above for $\Phi[\mathbf{M}(\xi_N,\bar{\theta})]$ are random since $\hat{\theta}^n$ is a random variable that satisfies $\sqrt{n}(\hat{\theta}^n - \bar{\theta}) \overset{\mathrm{d}}{\to} \mathbf{z} \sim \mathscr{N}(\mathbf{0},\mathbf{M}^{-1}(\xi_{(1)},\bar{\theta}))$, $n \to \infty$. Denote $\mathbf{A}(\bar{\theta})$ the Hessian matrix

$$\mathbf{A}(\bar{\theta}) = \left.\frac{\partial^2\Phi[\mathbf{M}(\xi^*_{\theta},\bar{\theta})]}{\partial\theta\partial\theta^\top}\right|_{\theta=\bar{\theta}},$$

which is nonpositive definite since $\Phi[\mathbf{M}(\xi^*_{\theta},\bar{\theta})]$ as a function of θ is maximum at $\theta = \bar{\theta}$. We obtain

$$\begin{aligned}\mathbb{E}\{\Phi[\mathbf{M}(\xi_N,\bar{\theta})]\} &= \Phi[\mathbf{M}(\xi^*_{\bar{\theta}},\bar{\theta})] + \frac{1}{2n}\,\mathrm{trace}[\mathbf{A}(\bar{\theta})\mathbf{M}^{-1}(\xi_{(1)},\bar{\theta})] \\ &\quad + \frac{n}{N}\,T(\xi^*_{\bar{\theta}},\xi_{(1)},\bar{\theta}) + \mathcal{O}(1/N) + \mathcal{O}(n^2/N^2) + \mathcal{O}(1/n^{3/2}).\end{aligned}$$

Note that $T(\xi^*_{\bar{\theta}},\xi_{(1)},\bar{\theta}) \leq 0$ with equality when $\xi_{(1)} = \xi^*_{\bar{\theta}}$. Neglecting the terms of order $1/N$, n^2/N^2, and $1/n^{3/2}$, we choose n that minimizes

$$t(n) = \left|\frac{1}{2n}\,\mathrm{trace}[\mathbf{A}(\bar{\theta})\mathbf{M}^{-1}(\xi_{(1)},\bar{\theta})] + \frac{n}{N}\,T(\xi^*_{\bar{\theta}},\xi_{(1)},\bar{\theta})\right|,$$

i.e., collect $n^* = \lceil \rho\sqrt{N} \rceil$ observations at stage 1 with

$$\rho = \left(\frac{\mathrm{trace}[\mathbf{A}(\bar{\theta})\mathbf{M}^{-1}(\xi_{(1)},\bar{\theta})]}{2\,T(\xi^*_{\bar{\theta}},\xi_{(1)},\bar{\theta})}\right)^{1/2}. \tag{8.38}$$

Since the value of ρ depends on $\bar{\theta}$, it cannot be used as a precise indication for the number of observations that should be collected at stage 1. It may serve as a guideline, however.

For instance, if a prior measure $\mu(\cdot)$ is set on $\bar{\theta}$, we may choose $\xi_{(1)}$ as the optimal design for the prior mean $\theta^0 = \mathbb{E}_\mu\{\theta\}$ and then choose n that maximizes $\mathbb{E}_\mu\{\mathrm{trace}[\mathbf{A}(\theta)\mathbf{M}^{-1}(\xi_{(1)},\theta)]/(2n) + (n/N)\,T(\xi^*_{\theta},\xi_{(1)},\theta)\}$, with $\mathbb{E}_\mu\{\cdot\}$ the expectation with respect to θ. This gives $n^* = \lceil \rho_\mu\sqrt{N} \rceil$ with

$$\rho_\mu = \left(\frac{\mathbb{E}_\mu\{\mathrm{trace}[\mathbf{A}(\theta)\mathbf{M}^{-1}(\xi_{(1)},\theta)]\}}{2\,\mathbb{E}_\mu\{T(\xi^*_{\theta},\xi_{(1)},\theta)\}}\right)^{1/2}. \tag{8.39}$$

Example 8.14. Consider LS estimation in the two-parameter model $\eta(x,\theta) = \theta_1 \exp(-\theta_2 x)$, $x \in \mathbb{R}^+$, with homoscedastic errors having variance $\sigma^2 = 1$. We take $\Phi(\mathbf{M}) = \log\det(\mathbf{M})$. Easy calculations indicate that the optimal design ξ_θ^* is given by $\xi_\theta^* = (1/2)\,\delta_0 + (1/2)\,\delta_{1/\theta_2}$ with δ_z the delta measure at z, which gives

$$\mathbf{A}(\bar{\theta}) = \begin{pmatrix} 0 & 0 \\ 0 & -2/\bar{\theta}_2^2 \end{pmatrix}.$$

We take $\xi_{(1)}$, the design measure at stage one, as the two-point measure $\xi_{(1)} = (1/2)\,\delta_0 + (1/2)\,\delta_x$. This gives

$$\mathrm{trace}[\mathbf{A}(\bar{\theta})\mathbf{M}^{-1}(\xi_{(1)},\bar{\theta})] = \frac{-4[1+\exp(2\bar{\theta}_2\,x)]}{\bar{\theta}_1^2\bar{\theta}_2^2\,x^2},$$
$$T(\xi_{\bar{\theta}}^*,\xi_{(1)},\bar{\theta}) = [\mathrm{e}^2\bar{\theta}_2^2\,x^2 + (1-\bar{\theta}_2\,x)^2]\,\exp(-2\bar{\theta}_2\,x) - 1.$$

Notice that $\mathrm{trace}[\mathbf{A}(\bar{\theta})\mathbf{M}^{-1}(\xi_{(1)},\bar{\theta})] < 0$ and $T(\xi_{\bar{\theta}}^*,\xi_{(1)},\bar{\theta}) \leq 0$ for any $x > 0$, with $T(\xi_{\bar{\theta}}^*,\xi_{(1)},\bar{\theta}) = 0$ when $x = 1/\bar{\theta}_2$. When considered as a function of x, ρ given by (8.38) tends to infinity when x tends to zero or $1/\bar{\theta}_2$, with a unique minimizer in $(0, 1/\bar{\theta}_2)$; it also tends to infinity when $x \to \infty$ with a unique minimizer in $(1/\bar{\theta}_2, \infty)$. The situation $x = 1/\bar{\theta}_2$ corresponds to $\xi_{(1)} = \xi_{\bar{\theta}}^*$, and all observations should then be taken at stage 1 (if we know the value of $\bar{\theta}$, we can design optimally for $\bar{\theta}$ at stage 1, and there is no reason for using a second stage); on the other hand, the sample size at stage 1 should increase to compensate the lack of information carried by $\xi_{(1)}$ when x is close to zero or very large.

Suppose now that θ_1 and θ_2 are independently distributed, with θ_2 uniformly distributed in $[2-\delta, 2+\delta]$. We take $x = 1/2$, so that $\xi_{(1)}$ is D-optimal for $\mathbb{E}_\mu\{\theta_2\} = 2$. The number of observations to collect at stage 1 to maximize $\mathbb{E}_\mu\{\Phi[\mathbf{M}(\xi_N,\theta)]\}$ then satisfies approximately (for N large enough) $n^* \simeq \lceil \rho_\mu \sqrt{N} \rceil$, where ρ_μ given by (8.39) is proportional to $1/\sqrt{\mathbb{E}_\mu\{\theta_1^2\}}$. Figure 8.13 presents ρ_μ as a function of δ (for $1/\sqrt{\mathbb{E}_\mu\{\theta_1^2\}} = 1$). □

8.5.2 Full-Sequential D-Optimum Design for LS Estimation in Nonlinear Regression Models

Full-sequential D-optimum design obeys the following rule. Take $x_1,\ldots,x_{k_0}$, $k_0 \geq p$, such that the associated information matrix $\mathbf{M}(\xi_{k_0},\theta)$ is nonsingular for any $\theta \in \Theta$, with ξ_k the empirical measure for $x_1,\ldots,x_k$. Then, for $k \geq k_0$, select x_{k+1} so as to maximize $\det\mathbf{M}(\xi_{k+1},\hat{\theta}^k)$, with $\hat{\theta}^k$ the estimated value

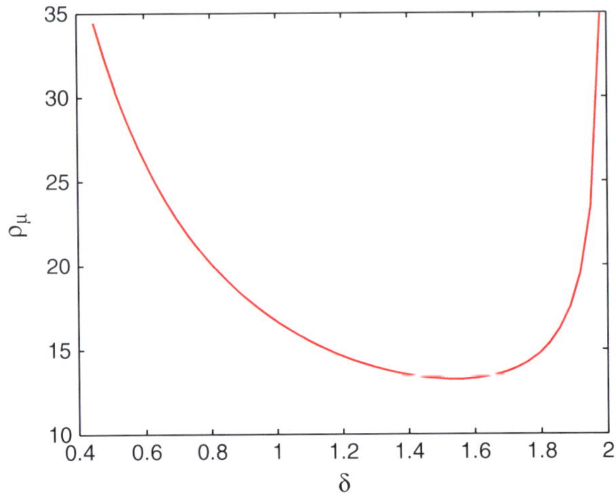

Fig. 8.13. Normalized ρ_μ $(1/\sqrt{\mathbb{E}_\mu\{\theta_1^2\}} = 1)$ as a function of δ in Example 8.14 when θ_1 and θ_2 are independent and θ_2 is uniformly distributed in $[2-\delta, 2+\delta]$

of θ obtained for the observations $y(x_1), \ldots, y(x_k)$. For the regression model (3.2), (3.3), direct calculations show that x_{k+1} is equivalently given by

$$x_{k+1} = \arg\max_{x \in \mathscr{X}} \mathbf{f}_{\hat{\theta}^k}^\top(x)\mathbf{M}^{-1}(\xi_k, \hat{\theta}^k)\mathbf{f}_{\hat{\theta}^k}(x) \tag{8.40}$$

with $\mathbf{f}_\theta(x) = \partial\eta(x,\theta)/\partial\theta$.

One may notice that when a fixed nominal value θ^0 is substituted for $\hat{\theta}^k$, (8.40) corresponds to one iteration of a vertex-direction algorithm for the maximization of $\log\det[\mathbf{M}(\xi, \theta^0)]$, with stepsize $1/(k+1)$; see (9.2), (9.3). One may thus expect that the almost sure convergence of $\hat{\theta}^k$ to some $\hat{\theta}^\infty$ would ensure the convergence of ξ_k to a design measure maximizing $\det[\mathbf{M}(\xi, \hat{\theta}^\infty)]$. Conversely, if the empirical measure ξ_k converges to a nonsingular design, we may expect $\hat{\theta}^k$ to be strongly consistent, and ξ_k should thus converge to a measure maximizing $\det[\mathbf{M}(\xi, \bar{\theta})]$. To obtain consistency results for $\hat{\theta}^k$ and convergence of ξ_k to an optimal design for $\bar{\theta}$, we need to get out of this circular argument.

Using a Bayesian estimator is a possible option, since the sequence of posterior means for θ forms a martingale which converges almost surely; this type of argumentation is used by Hu (1998). Few results are available for LS estimation: Ford and Silvey (1980) and Müller and Pötscher (1992) consider sequential c-optimum design in a particular model; Lai (1994) and Chaudhuri and Mykland (1995) introduce a subsequence of nonadaptive design points to ensure the consistency of the estimator; Chaudhuri and Mykland (1993) require that the size k_0 of the (nonadaptive) initial experiment grows with the increase in size of the total experiment, similarly to the situation where $n \to \infty$ in the two-stage approach of Sect. 8.5.1.

As already mentioned, the difficulty is due to the fact that the sequential construction of the experiment destroys the independence among the observations $y(x_1), \ldots, y(x_k)$, so that usual arguments for proving consistency do not apply. We shall thus resort to Theorem 3.5 to obtain the strong consistency of the estimator $\hat{\theta}^k$, using the assumption that the design space $\mathscr{X}$ is finite.

Suppose that

$$\mathscr{X} = \{x^{(1)}, x^{(2)}, \ldots, x^{(\ell)}\},\ \ell < \infty .$$

The idea used in (Pronzato, 2010a) is to consider separately the properties of the iterations (8.40) and the asymptotic behavior of $\hat{\theta}^k$ by considering the situation where $\{\hat{\theta}^k\}$ used in (8.40) is any sequence in Θ. We shall use the following assumption on the model

H$_{\mathscr{X}}$-*(i)*: $\inf_{\theta\in\Theta} \lambda_{\min}\left[\sum_{i=1}^{\ell} \mathbf{f}_\theta(x^{(i)})\mathbf{f}_\theta^\top(x^{(i)})\right] > \gamma > 0$.

The following property is proved in (Pronzato, 2010a).

Theorem 8.15. *Let $\{\hat{\theta}^k\}$ be an arbitrary sequence in Θ used to generate design points according to (8.40) in a finite design space $\mathscr{X}$ with an initialization such that $\mathbf{M}(\xi_{k_0}, \theta)$ is nonsingular for all θ in Θ. Let $r_{k,i} = r_k(x^{(i)})$ denote the number of times $x^{(i)}$ appears in the sequence $x_1, \ldots, x_k$, $i = 1, \ldots, \ell$, and consider the associated order statistics $r_{k,1:\ell} \geq r_{k,2:\ell} \geq \cdots \geq r_{k,\ell:\ell}$. Define*

$$q^* = \max\{j : \text{ there exists } \alpha > 0 \text{ such that } \liminf_{k\to\infty} r_{k,j:\ell}/k > \alpha\},$$

Then $H_{\mathscr{X}}$-(i) implies $q^* \geq p$.

For any sequence $\{\hat{\theta}^k\}$ used in (8.40), the conditions of Theorem 8.15 ensure the existence of k_1 and $\alpha > 0$ such that $r_{k,j:\ell} > \alpha k$ for all $k > k_1$ and all $j = 1, \ldots, p$. Under the additional assumption:

H$_{\mathscr{X}}$-*(ii)*: For all $\delta > 0$, there exists $\epsilon(\delta) > 0$ such that for any subset $\{i_1, \ldots, i_p\}$ of distinct elements of $\{1, \ldots, \ell\}$, $\inf_{\|\theta-\bar{\theta}\|\geq\delta} \sum_{j=1}^{p} [\eta(x^{(i_j)}, \theta) - \eta(x^{(i_j)}, \bar{\theta})]^2 > \epsilon(\delta)$

we thus obtain that $D_k(\theta, \bar{\theta})$ given by (3.16) satisfies $\inf_{\|\theta-\bar{\theta}\|\geq\delta} D_k(\theta, \bar{\theta}) > \alpha\, k\, \epsilon(\delta)$, $k > k_1$. Therefore, when $\{\hat{\theta}^k\}$ in (8.40) is the sequence of LS estimates, $\hat{\theta}^k \overset{\text{a.s.}}{\to} \bar{\theta}$ $(k \to \infty)$ from Theorem 3.5.

Having proved the strong consistency of $\hat{\theta}^k$, the next step is to show that $\mathbf{M}(\xi_k, \bar{\theta}) \overset{\text{a.s.}}{\to} \mathbf{M}(\xi^*_{\bar{\theta}}, \bar{\theta})$, with $\xi^*_{\bar{\theta}}$ maximizing $\det[\mathbf{M}(\xi, \bar{\theta})]$. This is proved in (Pronzato, 2010a) under the following additional assumption on $\mathscr{X}$:

H$_{\mathscr{X}}$-*(iii)*: $\lambda_{\min}\left[\sum_{j=1}^{p} \mathbf{f}_{\bar{\theta}}(x^{(i_j)})\mathbf{f}_{\bar{\theta}}^\top(x^{(i_j)})\right] \geq \bar{\gamma} > 0$ for any subset $\{i_1, \ldots, i_p\}$ of distinct elements of $\{1, \ldots, \ell\}$.

One difficulty remains: $\mathbf{M}(\xi_k, \theta)$ is not the information matrix for θ, due to the sequential construction of the design. One may refer to Ford and Silvey

(1980) for an empirical justifications for using $\mathbf{M}^{-1}(\xi_k, \hat{\theta}^k)$ to characterize the precision of the estimation in a (particular) sequential context; see also Ford et al. (1985) and Wu (1985) who make use of results in (Lai and Wei, 1982) on linear stochastic regression models. Other asymptotic considerations show that $\mathbf{M}(\xi_k, \hat{\theta}^k)$ can be used although it is not the information matrix, since

$$\sqrt{k}\,\mathbf{M}^{1/2}(\xi_k, \hat{\theta}^k)(\hat{\theta}^k - \bar{\theta}) \overset{\mathrm{d}}{\to} \mathbf{v} \sim \mathscr{N}(\mathbf{0}, \mathbf{I}_p), \tag{8.41}$$

a consequence of Pronzato (2009a, Theorem 2).

Example 8.16. Suppose that $y(x_i) = \bar{\theta}_1 x_i/(\bar{\theta}_2 + x_i) + \varepsilon_i$, with $\{\varepsilon_i\}$ satisfying (3.2), (3.3), $\Theta = [L_1, U_1] \times [L_2, U_2]$, $0 < L_j < \bar{\theta}_j < U_j$, $j = 1, 2$. When $\mathscr{X} = (0, \bar{x}]$, the D-optimal measure for θ on $\mathscr{X}$ is

$$\xi_D^*(\theta) = (1/2)\,\delta_{x_1^*(\theta)} + (1/2)\,\delta_{x_2^*} \tag{8.42}$$

with $x_1^*(\theta) = \theta_2 \bar{x}/(2\theta_2 + \bar{x}) < x_2^* = \bar{x}$. For $\hat{\theta}^k$ the LS estimator, Lai (1994) suggests the following design sequence:

$$\begin{cases} x_k = x_1^*(\hat{\theta}^{k-1}) & \text{if } k \text{ is even and } k \notin \{k_1, k_2 \ldots\} \\ x_k = \bar{x} & \text{if } k \text{ is odd and } k \notin \{k_1, k_2 \ldots\} \\ c/(1 + \log k) & \text{if } k \in \{k_1, k_2 \ldots\} \end{cases}$$

where $k_i \sim i^\alpha$ as $i \to \infty$, for some $c > 0$ and $1 < \alpha < 2$, in order to obtain the strong convergence of $\hat{\theta}^k$; see also Sect. 3.1.2. Hu (1998) shows that the introduction of the perturbations $x_k = c/(1 + \log k)$ if $k \in \{k_1, k_2 \ldots\}$ is not necessary when $\hat{\theta}^k$ is the posterior mean of θ given $y(x_1), \ldots, y(x_k)$ and that the sequence

$$\begin{cases} x_k = x_1^*(\hat{\theta}^{k-1}) & \text{if } k \text{ is even} \\ x_k = \bar{x} & \text{if } k \text{ is odd} \end{cases} \tag{8.43}$$

ensures $\hat{\theta}^k \overset{\text{a.s.}}{\to} \bar{\theta}$, $k \to \infty$.

Suppose now that $\mathscr{X}$ is finite, with $0 < \min(\mathscr{X}) < \max(\mathscr{X}) = \bar{x}$. One can check that $\mathrm{H}_{\mathscr{X}}$-(*ii*) is satisfied for $\theta \in \Theta$. Indeed, $\eta(x, \theta) = \eta(x, \bar{\theta})$ and $\eta(z, \theta) = \eta(z, \bar{\theta})$ for $\theta, \bar{\theta} \in \Theta$, $x > 0$, $z > 0$, and $x \neq z$ imply $\theta = \bar{\theta}$. Also, $\det[\mathbf{f}_\theta(x)\mathbf{f}_\theta^\top(x) + \mathbf{f}_\theta(z)\mathbf{f}_\theta^\top(z)] = x^2 z^2 \theta_1^2 (x - z)^2/[(\theta_2 + x)^4(\theta_2 + z)^4]$ so that $\mathrm{H}_{\mathscr{X}}$-(*i*), $\mathrm{H}_{\mathscr{X}}$-(*iii*) are satisfied, and the results above apply: when $\hat{\theta}^k$ in (8.40) is the LS estimator, $\hat{\theta}^k$ is strongly consistent and satisfies (8.41). Note that the LS estimator is much easier to obtain than the posterior mean of θ and that we do not need to know the form (8.42) of the D-optimal design to construct a design ξ_k asymptotically optimal for $\bar{\theta}$ through (8.40). □

Example 8.17. Consider again the model of Example 8.14 with homoscedastic errors satisfying (3.2), (3.3) and with Θ as in Example 8.16. Take $\mathscr{X}$ finite with $\min(\mathscr{X}) = \underline{x} \geq 0$. The D-optimal design measure is then $\xi_D^*(\theta) = (1/2)\,\delta_{\underline{x}} + (1/2)\,\delta_{\underline{x}+1/\theta_2}$. One can check that $\mathrm{H}_{\mathscr{X}}$-(*ii*) is satisfied for $\theta \in \Theta$, since $\eta(x,\theta) = \eta(x,\bar{\theta})$ and $\eta(z,\theta) = \eta(z,\bar{\theta})$ for $x \neq z$ imply $\theta = \bar{\theta}$; $\mathrm{H}_{\mathscr{X}}$-(*i*), $\mathrm{H}_{\mathscr{X}}$-(*iii*) are satisfied too since $\det[\mathbf{f}_\theta(x)\mathbf{f}_\theta^\top(x) + \mathbf{f}_\theta(z)\mathbf{f}_\theta^\top(z)] = \theta_1^2(x - z)^2\exp[-2\theta_2(x+z)]$. Therefore, the results above apply again: $\hat{\theta}^k$ is strongly consistent and satisfies (8.41) when we use (8.40). □

9

Algorithms: A Survey

9.1 Maximizing a Concave Differentiable Functional of a Probability Measure

We consider the maximization of a design criterion $\phi(\cdot)$ with respect to $\xi \in \Xi$, the set of probability measures on $\mathscr{X}$ compact. The algorithms we consider often rely on a discretization of $\mathscr{X}$ into a finite set $\mathscr{X}_\ell = \{x^{(1)}, \ldots, x^{(\ell)}\}$, which may be necessary for their practical implementation. This discretization can be progressively refined; see, e.g., Sect. 9.1.2 and Wu (1978a). For ξ a discrete design measure on the finite set $\mathscr{X}_\ell$, we shall denote by $\mathbf{w}$ the vector formed by its weights, $\mathbf{w} = (w_1, \ldots, w_\ell)^\top$ with $w_i = \xi(x^{(i)})$ for all i. We shall keep the same notation for the design criterion in terms of probability measure ξ or vector of weights $\mathbf{w}$ and write $\phi(\mathbf{w}) = \phi(\xi)$; ϕ^* will denote the optimal value of $\phi(\cdot)$, and we shall denote by $\nabla\phi(\mathbf{w}) = \partial\phi(\mathbf{w}')/\partial\mathbf{w}'\big|_{\mathbf{w}'=\mathbf{w}}$ the gradient of $\phi(\cdot)$ at $\mathbf{w}$.

The construction of a ϕ-optimal design on $\mathscr{X}_\ell$ amounts to minimizing $-\phi(\cdot)$, convex and differentiable, with respect to $\mathbf{w}$ belonging to the convex set

$$\mathscr{P}_{\ell-1} = \left\{ \mathbf{w} \in \mathbb{R}^\ell : w_i \geq 0\,,\ \sum_{i=1}^{\ell} w_i = 1 \right\} . \qquad (9.1)$$

In the optimization literature, such a problem is now considered as easy; one may refer, for instance, to the books (Hiriart-Urruty and Lemaréchal 1993; den Hertog 1994; Nesterov and Nemirovskii 1994; Ben-Tal and Nemirovskii 2001; Boyd and Vandenberghe 2004; Nesterov 2004) for recent developments on convex optimization. Two classical methods for convex programming (the ellipsoid and the cutting-plane methods), which are rather straightforward to implement and adapt to the optimization of design criteria (differentiable or not), are presented in Sect. 9.5. A peculiarity of design problems, however, is that the cardinality ℓ of the discretized set $\mathscr{X}_\ell$ may be quite large, making the convex optimization problem high dimensional. For that reason, some algorithms specifically dedicated to design problems are competitive compared

L. Pronzato and A. Pázman, *Design of Experiments in Nonlinear Models*, Lecture Notes in Statistics 212, DOI 10.1007/978-1-4614-6363-4_9,

to general-purpose convex-programming algorithms. It is the purpose of this section to present such design-specific algorithms and their connection with more classical optimization methods.

We shall not detail the convergence properties of all algorithms presented. Also, we shall neither indicate the required tuning of the various constants involved, in particular in the definitions of stopping rules, that makes an algorithm efficient—or simply makes it work—nor the ad hoc rules that allow us to remove support points with negligible mass or to merge support points that are close enough. The assumptions we shall make on $\phi(\cdot)$ are not necessarily the weakest possible, and only basic arguments justifying the principles of the algorithmic constructions will be given, supposing that calculations are performed with infinite precision. These arguments rely in particular on Theorem 5.21, which forms a most useful tool to check the optimality of a given design, and on Lemma 5.20 which indicates how far a given design is from the optimum.

A key idea is to keep the iterations simple, due to the large dimension of $\mathbf{w}$. Although we try to keep the presentation as general as possible, in particular, in order to cover the case of criteria considered in Chap. 8, we shall often refer to the situation where $\phi(\xi) = \Phi[\mathbf{M}(\xi,\theta)]$, with $\Phi(\cdot)$ one of the criteria of Sect. 5.1.2 and $\mathbf{M}(\xi,\theta)$ a $p \times p$ information matrix. We shall always suppose that $\Phi(\cdot)$ is bounded from above and concave and differentiable on the set $\mathbb{M}^{>}$ of positive-definite $p \times p$ matrices. When $\mathbf{M}(\xi,\theta) = \int_{\mathscr{X}} \mathbf{g}_\theta(x)\mathbf{g}_\theta^\top(x)\,\xi(\mathrm{d}x)$, $\mathbf{g}_\theta \in \mathbb{R}^p$, and $\Phi(\cdot)$ is isotonic (see Definition 5.3), an optimal design is then supported on points x such that $\mathbf{g}_\theta(x)$ is on the boundary of the Elfving's set $\mathscr{F}_\theta$, see Lemma 5.28. When $\mathscr{X}$ is discretized into $\mathscr{X}_\ell$, only the design points corresponding to vertices of the set $\mathscr{F}_\theta$ have to be considered. This may drastically reduce the cardinality of $\mathscr{X}_\ell$. One may refer, e.g., to Boissonnat and Yvinec (1998) and Cormen et al. (2001) for algorithms for the determination of the convex hull of a finite set. Notice that for the average optimality criterion $\int_\Theta \Phi[\mathbf{M}(\xi,\theta)]\,\mu(\mathrm{d}\theta)$, see Sect. 8.1, or the maximin-optimality criterion $\min_{\theta\in\Theta} \Phi[\mathbf{M}(\xi,\theta)]$ (see Sect. 8.2), when Θ is finite only the $x^{(i)}$ corresponding to a finite union of sets formed by vertices of $\mathscr{F}_\theta$, $\theta \in \Theta$, have to be considered.

Although the maximization of $\Phi(\mathbf{M})$ with respect to $\mathbf{M} \in \mathcal{M}_\theta(\Xi)$ given by (5.3) is a $p(p+1)/2$-dimensional problem, the optimal design ξ^* rather than the associated information matrix $\mathbf{M}(\xi^*,\theta)$ is usually the main concern. This excludes the use of duality theory (see Sect. 5.2.4) since we wish to use iterations that update design measures.

The design measure at the iteration k of an algorithm will be denoted by ξ_k and $\mathbf{w}^k$ will be the associated vector of weights. All the algorithms considered are such that $\xi_k \in \Xi$ and $\mathbf{w}^k \in \mathscr{P}_{\ell-1}$ for each k. The case of D-optimum design plays a special role in terms of optimization, due in particular to its connection with an optimal-ellipsoid problem (see Sect. 5.6) and to the simplifications it allows in some algorithms; Sect. 9.1.4 is devoted to the case $\phi(\xi) = \log\det[\mathbf{M}(\xi,\theta)]$.

9.1.1 Vertex-Direction Algorithms

Suppose that $\phi(\xi) = \Phi[\mathbf{M}(\xi,\theta)]$ with $\Phi(\cdot)$ concave and differentiable on $\mathbb{M}^{>}$ and that ξ_0 is a probability measure on $\mathscr{X}$ such that $\mathbf{M}(\xi_0,\theta)$ has full rank. Consider an algorithm that updates ξ_k at iteration k into

$$\xi_{k+1} = (1-\alpha_k)\xi_k + \alpha_k \delta_{x_{k+1}^+} \tag{9.2}$$

where

$$x_{k+1}^+ = \arg\max_{x\in\mathscr{X}} F_\phi(\xi_k, x)\,, \tag{9.3}$$

$F_\phi(\xi,x) = F_\phi(\xi;\delta_x)$, with δ_x the delta measure at x and $F_\phi(\xi;\nu)$ the directional derivative of $\phi(\cdot)$ at ξ in the direction ν; see Sect. 5.2.1. Several choices are possible for the stepsize α_k. It can be set to the value α_k^+ in $[0,1]$ that maximizes $\phi(\xi_{k+1})$; this corresponds to a one-dimensional concave optimization problem for which many line-search methods can be used; see in particular den Boeff and den Hertog (2007). It can also be taken as the k-th point in a sequence $\{\alpha_n\}$ that satisfies $0 \le \alpha_n \le 1$, $\alpha_n \to 0$, and $\sum_{n=0}^{\infty} \alpha_n = \infty$. A simple algorithm is thus as follows:

0. Start from ξ_0, a probability measure of $\mathscr{X}$ such that $\phi(\xi_0) > -\infty$; choose $\epsilon_0 > 0$; set $k=0$.
1. Compute x_{k+1}^+ given by (9.3).
2. If $F_\phi(\xi_k, x_{k+1}^+) < \epsilon_0$ stop; otherwise, perform one iteration (9.2), $k \leftarrow k+1$, return to step 1.

Although the algorithm can in principle be applied when $\mathscr{X}$ is a compact subset of $\mathbb{R}^d$ with nonempty interior, $F_\phi(\xi_k,x)$ in (9.3) is generally a multimodal function, and the determination of x_{k+1}^+ is often obtained by discretizing $\mathscr{X}$ into $\mathscr{X}_\ell$; local search in $\mathscr{X}$ with multistart can be used too. The extreme points of the set $\mathcal{M}_\theta(\Xi_\ell)$, with Ξ_ℓ the set of probability measures on $\mathscr{X}_\ell$, are of the form $\mathbf{M}_\theta(x^{(i)})$, see (5.1), and (9.2) adjusts ξ_k along the direction of one vertex of $\mathcal{M}_\theta(\Xi_\ell)$, hence the name of vertex-direction algorithm. When there are several solutions for x_{k+1}^+ in (9.3), it is enough to select one of them.[1] This method corresponds in fact to the method proposed by Frank and Wolfe (1956) in a more general context.

The example below is extremely simple but rich enough to illustrate the behavior of the various algorithms to be considered for differentiable criteria.

Example 9.1. We take $\ell = 3$ and $\phi(\cdot)$ a quadratic function with a maximum at $\mathbf{w}^* = (1/2\,,\ 1/2\,,\ 0)^\top$, on the boundary of the simplex $\mathscr{P}_2$ given by (9.1).

[1] It may happen that the maximum in (9.3) is reached at several x, in particular when the model possesses some symmetry. In that case, a faster convergence is obtained if the multiple maximizers $x_{k+1,i}^+$ $(i=1,2,\ldots,q)$ are introduced in one single step, using $\xi_{k+1} = (1-\alpha_k)\xi_k + (\alpha_k/q)\sum_i \delta_{x_{k+1,i}^+}$; see Atwood (1973).

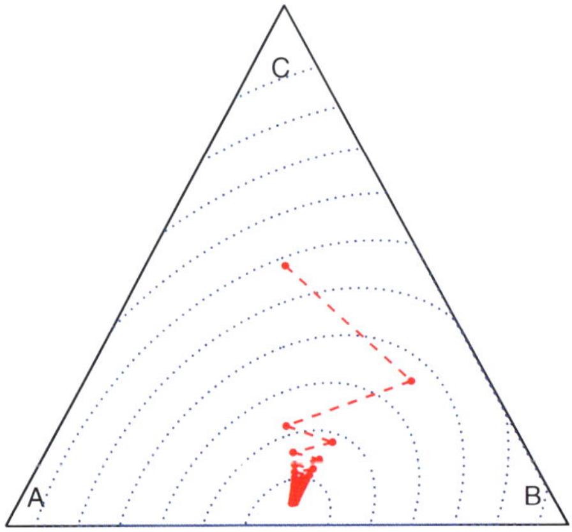

Fig. 9.1. Behavior of the vertex-direction algorithm (9.2), (9.3) with optimal stepsize in Example 9.1

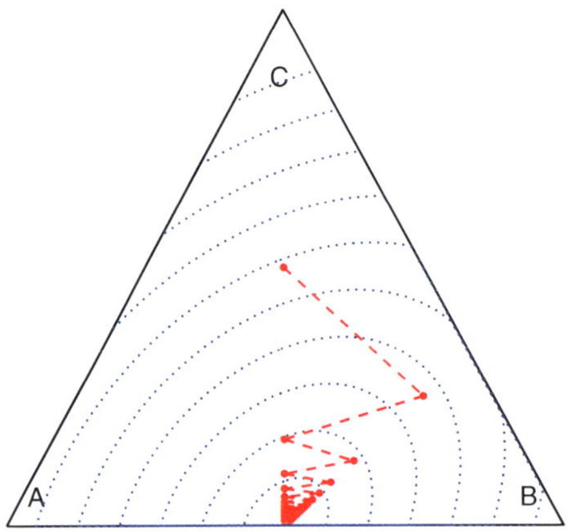

Fig. 9.2. Behavior of the vertex-direction algorithm (9.2), (9.3) with predefined stepsize sequence $\alpha_k = 1/(2+k)$ in Example 9.1

Figure 9.1 presents the evolution in $\mathscr{P}_2$ of 50 iterates $\mathbf{w}^k$ corresponding to ξ_k generated by the algorithm above, initialized at $\mathbf{w}^0 = (1/4\,,\ 1/4\,,\ 1/2)^\top$, when the stepsize α_k is chosen optimally. The corners A, B, and C respectively correspond to the vertices $(1\,,\ 0\,,\ 0)$, $(0\,,\ 1\,,\ 0)$, and $(0\,,\ 0\,,\ 1)$; the level sets of the function $\phi(\cdot)$ are indicated by dotted lines. As it can be seen on Fig. 9.1, the convergence to the optimum (the middle point between A and B) is rather slow. Figure 9.2 shows the path followed by the iterates for the predefined stepsize sequence $\alpha_k = 1/(2+k)$. The behavior is very similar to Fig. 9.1, and convergence to the optimum is also quite slow. □

Remark 9.2. Instead of choosing x_{k+1}^+ given by (9.3), it may be easier to simply use any x_{k+1}^+ satisfying

$$F_\phi(\xi_k, x_{k+1}^+) \geq \min\{\gamma, r \max_{x\in\mathscr{X}} F_\phi(\xi_k, x)\}$$

for some small $\gamma > 0$ and some $r \in (0, 1]$; see Wu (1978a). □

In the case of D-optimality, the proof of convergence to an optimal design can be found, respectively, in (Wynn, 1970) and (Fedorov, 1971, 1972) for α_k the k-th element of a non-summable sequence and α_k chosen optimally; the extension to D_s-optimum design is considered in (Wynn, 1972) and (Pázman, 1986, Proposition 5.1) when $\{\alpha_k\}$ is a predefined non-summable sequence and in (Atwood, 1980) when α_k is chosen optimally. Basic arguments justifying convergence to an optimal design in more general situations are given below.

Denote

$$\nabla^2\phi(\xi;\nu) = \left.\frac{\partial^2\phi[(1-\alpha)\xi + \alpha\nu]}{\partial\alpha^2}\right|_{\alpha=0}$$

(possibly equal to $-\infty$) and define

$$B(t) = \sup\left\{\left|\nabla^2\phi(\xi;\delta_x)\right| : \xi \in \Xi\,,\ x \in \mathscr{X}\,,\ \phi(\xi) \geq t\right\}$$

with δ_x the delta measure at x. When $\phi(\xi) = \Phi[\mathbf{M}(\xi,\theta)]$ with $\Phi(\cdot)$ a positively homogeneous, isotonic, and global criterion (see Definitions 5.3 and 5.8), we assume that

$$B(\epsilon) < \infty \text{ for all } \epsilon > 0\,. \tag{9.4}$$

Would the criterion not be written in a positively homogeneous form, so that $\phi(\cdot)$ may reach the value $-\infty$ (think in particular of the case $\phi(\xi) = \Phi[\mathbf{M}(\xi,\theta)]$ with $\Phi(\mathbf{M}) = -\infty$ for a singular $\mathbf{M}$), we would replace (9.4) by

$$B(t) < \infty \text{ for all } t > -\infty\,,$$

which is satisfied in particular by $\phi(\xi) = \log\det[\mathbf{M}(\xi,\theta)]$ and $\phi(\xi) = -\text{trace}[\mathbf{M}^{-1}(\xi,\theta)]$.

Convergence with Optimal Stepsize

Define $x^+(\xi) = \arg\max_{x\in\mathscr{X}} F_\phi(\xi, x)$, $\alpha^+(\xi) = \arg\max_{\alpha\in[0,1]} \phi[(1-\alpha)\xi + \alpha\delta_{x^+(\xi)}]$ and suppose that $\alpha_k = \alpha_k^+ = \alpha^+(\xi_k)$. The idea is to construct a lower bound on $h_k(\alpha) = \phi[(1-\alpha)\xi_k + \alpha\delta_{x_{k+1}^+}]$, quadratic in α, in order to obtain a lower bound on $\phi(\xi_{k+1}) - \phi(\xi_k)$ when $\phi(\xi_k)$ is bounded away from the optimal value ϕ^*. By that we shall prove the convergence of $\phi(\xi_k)$ to ϕ^*.

Suppose that $\phi(\xi) = \Phi[\mathbf{M}(\xi,\theta)]$ with $\Phi(\cdot)$ a positively homogeneous, isotonic, and global criterion. Then, $\phi[(1-\alpha)\xi + \alpha\delta_x] \to 0$ when $\alpha \to 1$ for any x and any ξ, so that for any $\epsilon > 0$, there exists $C(\epsilon) < 1$ such that $\alpha^+(\xi) < C(\epsilon)$ for all ξ satisfying $\phi(\xi) \geq \epsilon$. The developments are similar when $\Phi(\cdot)$ is not written in a positively homogeneous form, so that, for instance, $\phi[(1-\alpha)\xi + \alpha\delta_x] \to -\infty$ when $\alpha \to 1$ for any x and any ξ; then $\alpha^+(\xi) < C(t) < 1$ for all ξ such that $\phi(\xi) \geq t > -\infty$. Since the sequence

$\{\phi(\xi_k)\}$ is nondecreasing, we have $\alpha_k^+ < C_0 = C[\phi(\xi_0)]$, $k \geq 0$. For $\alpha \in [0, C_0]$ we have

$$h_k(\alpha) = \phi(\xi_k) + \alpha\, F_\phi(\xi_k, x_{k+1}^+) \\ + \frac{\alpha^2}{2}\, \nabla^2\phi[(1-\beta)\xi_k + \beta\delta_{x_{k+1}^+}; \delta_{x_{k+1}^+}]\,,\ k \geq 0\,,$$

for some $\beta \in [0, \alpha]$. Since $\beta < C_0$ and $\phi(\xi_k) \geq \phi(\xi_0)$, $\phi[(1-\beta)\xi_k + \beta\delta_{x_{k+1}^+}] \geq \epsilon_0$ for some $\epsilon_0 > 0$; therefore, $\nabla^2\phi[(1-\beta)\xi_k + \beta\delta_{x_{k+1}^+}; \delta_{x_{k+1}^+}] \geq -B(\epsilon_0)$. We thus obtain, for all $\alpha \in [0, C_0]$,

$$h_k(\alpha) \geq \phi(\xi_k) + \alpha\, F_\phi(\xi_k, x_{k+1}^+) - \frac{\alpha^2\, B(\epsilon_0)}{2}\,,\ k \geq 0\,.$$

Since $\{\phi(\xi_k)\}$ is nondecreasing and bounded, $\phi(\xi_k)$ converges to some constant ϕ_∞. Suppose that $\phi^* \geq \phi_\infty + \Delta$ with ϕ^* the optimal value of $\phi(\cdot)$ and $\Delta > 0$. From Lemma 5.20, it implies that $F_\phi(\xi_k, x_{k+1}^+) \geq \Delta$ for all k. Let $\Delta' = \min\{\Delta, C_0\, B(\epsilon_0)\}$. We obtain

$$\forall \alpha \in [0, C_0]\,,\ h_k(\alpha) \geq \phi(\xi_k) + \alpha\, \Delta' - \frac{\alpha^2\, B(\epsilon_0)}{2}\,,\ k \geq 0\,.$$

The right-hand side reaches its maximum value $\phi(\xi_k) + \Delta'^2/[2B(\epsilon_0)]$ at $\alpha^* = \Delta'/B(\epsilon_0) \leq C_0$, so that

$$\phi(\xi_{k+1}) = \max_{\alpha\in[0,1]} h_k(\alpha) = \max_{\alpha\in[0,C_0]} h_k(\alpha) \geq \phi(\xi_k) + \Delta'^2/[2B(\epsilon_0)]\,.$$

This implies that $\phi(\xi_k) \to \infty$, contradicting the fact that $\phi(\cdot)$ is bounded. Therefore, $\phi_\infty = \phi^*$. The result remains valid when x_{k+1}^+ is chosen as in Remark 9.2.

Remark 9.3.

(*i*) Using Armijo type arguments, we may relax the assumption of second-order differentiability and only suppose that $\nabla\Phi(\cdot)$ is continuous on the set $\{\mathbf{M} \in \mathcal{M}_\theta(\Xi) : \Phi(\mathbf{M}) \geq \Phi[\mathbf{M}(\xi_0, \theta)]\}$; see Wu (1978a, Theorem 1).

(*ii*) When $\Phi(\cdot)$ is a singular (or partial; see Sect. 5.1.6) criterion, an optimal design ξ^* may be singular, i.e., such that $\mathbf{M}(\xi^*, \theta)$ is singular. Also, $\Phi(\cdot)$ is generally not differentiable and not continuous at a singular $\mathbf{M}$; see Example 5.19 and Sect. 5.1.7. We may then use a regularized version of the criterion, $\phi(\xi) = \Phi\{\mathbf{M}[(1-\gamma)\xi + \gamma\tilde{\xi}, \theta]\}$ with γ a small positive number and $\tilde{\xi}$ such that $\mathbf{M}(\tilde{\xi}, \theta)$ has full rank; see (5.55) for regularized c-optimality. In that case, $\left|\nabla^2\phi(\xi; \delta_x)\right| < B < \infty$ for all $\xi \in \Xi$ and $x \in \mathscr{X}$, and we have for any $\alpha \in [0, 1]$

$$h_k(\alpha) \geq \phi(\xi_k) + \alpha\, F_\phi(\xi_k, x_{k+1}^+) - \frac{\alpha^2\, B}{2}\,,\ k \geq 0\,.$$

The convergence of $\phi(\xi_k)$ to ϕ^* then follows from arguments identical to those used above.

(*iii*) Suppose that the algorithm (9.2), (9.3) is used for the optimization of $\phi(\cdot) = \Phi[\mathbf{M}(\cdot,\theta)]$ with $\Phi(\cdot)$ singular (and not regularized). Notice that we may have $\alpha_k^+ = 1$ at some iterations since an optimal design ξ^* may be singular. By imposing $\alpha_k < 1$ in (9.2), we can force ξ_k to remain nonsingular (provided that ξ_0 is nonsingular), which plays the same role as enforcing a form of regularization. The idea is to prevent ξ_k from becoming close to singularity before it is close enough from being optimal and to force the matrix $\mathbf{M}(\xi_k,\theta)$ to remain well inside the cone $\mathbb{M}^>$; see Sect. 5.1.7. One may refer to Atwood (1980) for a detailed exposition on optimization algorithms for singular criteria. □

Convergence with a Non-Summable Stepsize Sequence

Suppose now that $\{\alpha_k\}$ satisfies $\alpha_k \geq 0$, $\alpha_k \to 0$, and $\sum_{i=0}^k \alpha_i \to \infty$ as $k \to \infty$ and that $\phi(\xi) = \Phi[\mathbf{M}(\xi,\theta)]$ with $\Phi(\cdot)$ a positively homogeneous, isotonic, and global criterion satisfying (9.4). Also suppose that $\liminf_{k\to\infty} \phi(\xi_k) > \epsilon$ for some $\epsilon > 0$—it is satisfied, for instance, when the criterion is regularized; see (5.55) for c-optimality. Using (9.4) and following the same approach as above, we obtain that, for k large enough,

$$\phi(\xi_{k+1}) = \phi[(1-\alpha_k)\xi_k + \alpha_k \delta_{x_{k+1}^+}] > \phi(\xi_k) + \alpha_k\, F_\phi(\xi_k, x_{k+1}^+) - \frac{\alpha_k^2 B(\epsilon')}{2} \tag{9.5}$$

for some $\epsilon' > 0$. Suppose that $\phi^* > \limsup_{k\to\infty} \phi(\xi_k) + \Delta$ for some $\Delta > 0$. From Lemma 5.20, it implies that $F_\phi(\xi_k, x_{k+1}^+) > \Delta$ for k large enough and (9.5) shows that $\phi(\xi_k) \to \infty$, contradicting the fact that $\phi(\cdot)$ is bounded from above. Therefore, $\limsup_{k\to\infty} \phi(\xi_k) = \phi^*$. Moreover, (9.5) implies that, for any $\Delta > 0$, $\phi(\xi_{k+1}) > \phi(\xi_k) - \Delta$ for k large enough. Also, $\phi(\xi_k) < \phi^* - \Delta$ implies $F_\phi(\xi_k, x_{k+1}^+) > \Delta$ and thus $\phi(\xi_{k+1}) > \phi(\xi_k)$ for large k. Altogether, it implies that $\phi(\xi_{k+1}) > \phi(\xi_k) - 2\Delta$ for k large enough. Since Δ is arbitrary, we obtain that $\phi(\xi_k) \to \phi^*$ as $k \to \infty$. This remains true when x_{k+1}^+ is chosen as in Remark 9.2.

We have thus obtained a dichotomous property, see Wu and Wynn (1978): either there exists a subsequence $\{\xi_{k_n}\}$ such that $\phi(\xi_{k_n}) \to 0$ or $\phi(\xi_k) \to \phi^*$, $k \to \infty$. Precise condition for eliminating the first possibility can be found, e.g., in (Wu and Wynn, 1978); see also Pázman (1986, Chap. 5) and Wu (1978a, Theorems 4 and 5). They are satisfied in particular for D-optimality, with $\phi(\xi) = \log\det[\mathbf{M}(\xi,\theta)]$; A-optimality, with $\phi(\xi) = -\mathrm{trace}[\mathbf{M}^{-1}(\xi,\theta)]$; and for Φ_q-optimality, where $\phi(\xi) = -\left\{(1/p)\mathrm{trace}[(\mathbf{Q}^\top\mathbf{M}^{-1}(\xi,\theta)\mathbf{Q})^q]\right\}^{1/q}$ with $q > 0$ and $\mathbf{Q}$ a full-rank $p \times p$ matrix—which can be taken equal to identity by applying a linear transformation to $\mathbf{M} \in \mathcal{M}_\theta(\Xi)$ given by (5.3).

Although the weights of some support points of ξ_k may decrease continuously along iterations (9.2), when $\alpha_k \in (0,1)$, they always stay strictly positive; i.e., support points of ξ_0 cannot be totally removed. The convergence of $\phi(\xi_k)$ to the optimal value ϕ^* is thus inevitably slow when we get close to the optimum. The method presented next allows us to set some weights to zero if necessary and is thus able to reduce the support of the initial design measure; this will also be the case for the algorithms of Sect. 9.1.2.

Vertex-Exchange Algorithm

When ξ_0, and therefore ξ_i for all $i \geq 0$, has finite support, Atwood (1973) has suggested that α_k in (9.2) could sometimes be taken negative in order to remove support points from ξ_k if necessary. We then compute

$$\alpha_k^+ = \arg\max_{\alpha\in[0,1]} \phi[(1-\alpha)\xi_k + \alpha\delta_{x_{k+1}^+}] \tag{9.6}$$

with x_{k+1}^+ given by (9.3), and

$$x_{k+1}^- = \arg\min_{x\in\mathcal{S}_{\xi_k}} F_\phi(\xi_k, x) \tag{9.7}$$

$$\alpha_k^- = \arg\max_{\alpha\in[-\xi_k(x_{k+1}^-)/[1-\xi_k(x_{k+1}^-)],0]} \phi[(1-\alpha)\xi_k + \alpha\delta_{x_{k+1}^-}], \tag{9.8}$$

with $\mathcal{S}_{\xi_k}$ the support of ξ_k. The admissible interval for α in α_k^- is chosen so that the weight at x_{k+1}^- remains nonnegative. When there are several solutions for x_{k+1}^-, any of them can be selected. The choice between $(1-\alpha_k^+)\xi_k + \alpha_k^+\delta_{x_{k+1}^+}$ and $(1-\alpha_k^-)\xi_k + \alpha_k^-\delta_{x_{k+1}^-}$ is made by comparing the associated values of $\phi(\cdot)$. See also St. John and Draper (1975) for other suggestions, in particular concerning the distribution of the weight removed from x_{k+1}^- on the other support points of ξ_k. We may alternatively decide earlier between the two types of iteration by taking

$$\begin{cases} x_{k+1} = x_{k+1}^+ \text{ and } \alpha_k = \alpha_k^+ \text{ if } F_\phi(\xi_k, x_{k+1}^+) \geq -F_\phi(\xi_k, x_{k+1}^-) \\ x_{k+1} = x_{k+1}^- \text{ and } \alpha_k = \alpha_k^- \text{ otherwise,} \end{cases} \tag{9.9}$$

and then

$$\xi_{k+1} = (1-\alpha_k)\xi_k + \alpha_k\delta_{x_{k+1}}. \tag{9.10}$$

In that case, only one optimization with respect to α is required at each iteration. This corresponds to the method suggested by Wolfe (1970) in a more general context.

A related but slightly different approach is proposed by Böhning (1985, 1986) and Molchanov and Zuyev (2001, 2002). By considering the set $\tilde{\Xi}$ of signed measures on $\mathscr{X}$ instead of the set Ξ, the directional derivative $\lim_{\alpha\to 0^+}[\phi(\xi+\alpha\nu) - \phi(\alpha)]/\alpha$, $\nu \in \tilde{\Xi}$, instead of $F_\phi(\xi;\nu)$, $\nu \in \Xi$, and using the total variation norm for ν, they obtain that the steepest-ascent direction at ξ_k corresponds to adding some weight α to x_{k+1}^+ given by (9.3) and removing the same weight from x_{k+1}^- given by (9.7), with $\alpha \to 0$. The iteration k of the corresponding algorithm is thus

$$\xi_{k+1} = \xi_k + \alpha_k(\delta_{x_{k+1}^+} - \delta_{x_{k+1}^-}) \tag{9.11}$$

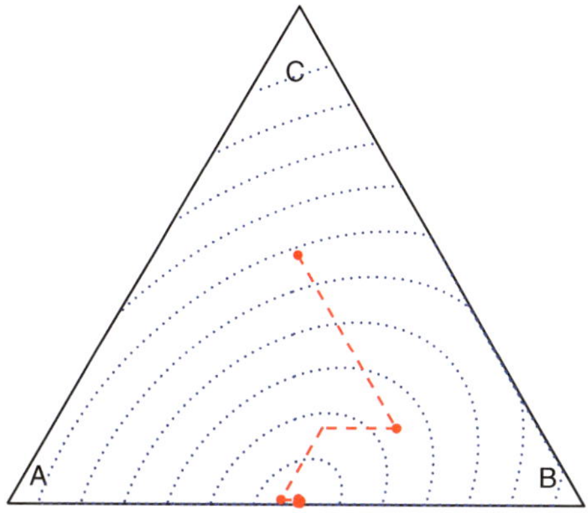

Fig. 9.3. Behavior of the vertex-direction algorithm (9.11), (9.12) in Example 9.1. Notice that compared with Fig. 9.1 the trajectory of the iterates is now parallel to the edges of $\mathscr{P}_{\ell-1}$

with

$$\alpha_k = \arg \max_{\alpha \in [0, \xi_k(x_{k+1}^-)]} \phi[\xi_k + \alpha(\delta_{x_{k+1}^+} - \delta_{x_{k+1}^-})] \,. \tag{9.12}$$

Note that when $\alpha_k = \xi_k(x_{k+1}^-)$, the support point x_{k+1}^- is removed from ξ_k, whereas a new support point x_{k+1}^+ is introduced, hence the name of vertex-exchange algorithm.

Example 9.1 (continued). We consider the same problem as in Example 9.1. Figure 9.3 presents the evolution in $\mathscr{P}_2$ of iterates $\mathbf{w}^k$ corresponding to ξ_k generated by (9.11), (9.12), initialized at $\mathbf{w}^0 = (1/4\,,\ 1/4\,,\ 1/2)^\top$; compare with Fig. 9.1. □

A method is suggested in (Böhning, 1985) for the construction of α_k, with quadratic convergence to the optimum. A simple modification allows us to consider stepsizes $\alpha_k \in [0,1]$ larger than $\xi_k(x_{k+1}^-)$:

$$\left\{ \begin{array}{ll} 1) & \text{Sort the support points of } \xi_k \text{ by increasing values of } F_\phi(\xi_k, x), \\ & \text{with } \{x_{k+1}^-, x_{k+1}^{2-}, x_{k+1}^{3-} \ldots\} \text{ the corresponding ordered set.} \\ 2) & \text{Transfer weight successively from } x_{k+1}^-,\, x_{k+1}^{2-} \cdots \text{ to } x_{k+1}^+. \end{array} \right. \tag{9.13}$$

Both the method of Atwood and (9.11) allow the suppression of support points from ξ_k, which yields a significant improvement in the speed of convergence to the optimum compared to (9.2); see Example 9.14 for an illustration. Clearly, when a support point is suppressed from ξ_k, it would be important to know whether it can be considered as suppressed for ever or not, i.e., whether it can be removed from $\mathscr{X}_\ell$ or not. We shall see in Sect. 9.1.4 how to answer this question in the case of D-optimum design.

9.1.2 Constrained Gradient and Gradient Projection

We suppose that $\phi(\cdot)$ is concave and differentiable for all $\xi \in \Xi$, that $\mathscr{X}$ is discretized into $\mathscr{X}_\ell$, and consider the problem of maximizing $\phi(\mathbf{w})$ with respect to $\mathbf{w} \in \mathscr{P}_{\ell-1}$; see (9.1). We denote by $\nabla\phi(\mathbf{w})$ the gradient vector $\partial\phi(\mathbf{w})/\partial\mathbf{w} \in \mathbb{R}^\ell$.

Constrained Gradient

The constrained-gradient (or conditional gradient) method transforms $\mathbf{w}^k$ at iteration k into

$$\mathbf{w}^{k+1} = \mathbf{w}^k + \alpha_k(\mathbf{u}_+^k - \mathbf{w}^k)$$

with $\mathbf{u}_+^k \in \mathscr{P}_{\ell-1}$ and $(\mathbf{u}_+^k - \mathbf{w}^k)^\top \nabla\phi(\mathbf{w}^k)$ as large as possible, i.e.,

$$\mathbf{u}_+^k = \arg\max_{\mathbf{u}\in\mathscr{P}_{\ell-1}} \mathbf{u}^\top \nabla\phi(\mathbf{w}^k) = \mathbf{e}_i \,,$$

the i-th unit vector, with

$$i = \arg\max_{j\in\{1,\ldots,\ell\}} \{\nabla\phi(\mathbf{w}^k)\}_j = \arg\max_{j\in\{1,\ldots,\ell\}} F_\phi(\xi_k, \delta_{x^{(j)}}) \,.$$

This is also called the iterative barycentric coordinate method; see Khachiyan (1996). This algorithm thus coincides with (9.2), (9.3). The constrained-gradient algorithm is known to sometimes converge quite slowly; see, e.g., Polyak (1987); see also Example 9.1. The gradient-projection algorithm considered below generally yields significantly faster convergence to the optimum.

Gradient Projection

Suppose that $\mathbf{w}^k$ has strictly positive components. We then project $\nabla\phi(\mathbf{w}^k)$ orthogonally on the linear space $\mathcal{L} = \{\mathbf{z} \in \mathbb{R}^\ell : \mathbf{1}^\top\mathbf{z} = 0\}$ with $\mathbf{1} = (1, \ldots, 1)^\top$, to form

$$\mathbf{d}^k = \nabla\phi(\mathbf{w}^k) - [\mathbf{1}^\top\nabla\phi(\mathbf{w}^k)]\,\mathbf{1}/\ell \,. \tag{9.14}$$

The next point $\mathbf{w}^{k+1}$ is then constructed according to

$$\mathbf{w}^{k+1} = \mathbf{w}^k + \alpha_k\mathbf{d}^k \tag{9.15}$$

where

$$\alpha_k = \arg\max_{\alpha\in[0,\bar{\alpha}_k]} \phi(\mathbf{w}^k + \alpha\mathbf{d}^k) \;\text{ with }\; \bar{\alpha}_k = \min\{w_i^k/|d_i^k| : d_i^k < 0\} \tag{9.16}$$

and d_i^k the i-th component of $\mathbf{d}^k$. Notice that $\mathbf{1}^\top\mathbf{d}^k = 0$, so that $\sum_{i=1}^\ell w_i^{k+1} = 1$, and that $\alpha \in [0, \bar{\alpha}_k]$ ensures that $w_i^{k+1} \geq 0$ for all i. Also,

$$\left.\frac{\mathrm{d}\phi(\mathbf{w}^{k+1})}{\mathrm{d}\alpha_k}\right|_{\alpha_k=0^+} = \nabla^\top\phi(\mathbf{w}^k)\mathbf{d}^k$$
$$= \|\nabla\phi(\mathbf{w}^k)\|^2 - [\mathbf{1}^\top\nabla\phi(\mathbf{w}^k)]^2/\ell \geq 0$$

(from Cauchy–Schwarz inequality), with equality if and only if $\nabla\phi(\mathbf{w}^k)$ is proportional to $\mathbf{1}$, i.e., $\mathbf{e}_i^\top\nabla\phi(\mathbf{w}^k) = c$ for all i and some scalar c, and thus $\mathbf{e}_i^\top\nabla\phi(\mathbf{w}^k) = {\mathbf{w}^k}^\top\nabla\phi(\mathbf{w}^k)$, or equivalently $F_\phi(\xi_k, x^{(i)}) = 0$ for all i; see Sect. 5.2.1. From the equivalence theorem, see the discussion following Remark 5.22, this is equivalent to $\mathbf{w}^k$ being optimal for $\phi(\cdot)$. For a non-optimal $\mathbf{w}^k$ with positive components, $\mathbf{d}^k$ is thus a direction of increase for $\phi(\cdot)$.

More generally, when some components of $\mathbf{w}^k$ equal zero, the gradient-projection method constructs $\mathbf{w}^{k+1}$ according to

$$\mathbf{w}^{k+1} = P_{\mathscr{P}_{\ell-1}}\left[\mathbf{w}^k + \alpha_k\nabla\phi(\mathbf{w}^k)\right] \tag{9.17}$$

with $P_{\mathscr{P}_{\ell-1}}(\mathbf{z})$ the orthogonal projection of $\mathbf{z} \in \mathbb{R}^\ell$ onto $\mathscr{P}_{\ell-1}$ given by (9.1) and α_k a suitably chosen stepsize. It can be chosen to maximize $\phi[\mathbf{w}_+^k(\alpha)]$ with respect to $\alpha \geq 0$, with $\mathbf{w}_+^k(\alpha) = P_{\mathscr{P}_{\ell-1}}\left[\mathbf{w}^k + \alpha\nabla\phi(\mathbf{w}^k)\right]$; see McCormick and Tapia (1972) for a proof of convergence. One may refer to Calamai and Moré (1987) for an analysis of the algorithm when using inexact line search. Note that a non-monotone method may yield faster convergence to the optimum; see Birgin et al. (2000) and Dai and Fletcher (2006). In Sect. 9.3.1 (subgradient algorithm), we shall see that, under suitable assumptions on $\phi(\cdot)$, choosing α_k as the k-th element of a positive sequence satisfying $\alpha_k \to 0$ and $\sum_{i=0}^k \alpha_i \to \infty$ also yields (non-monotonic) convergence to an optimal design.

Remark 9.4. The method requires the computation of a projection onto $\mathscr{P}_{\ell-1}$ for each value of α. This projection can be obtained as the solution of the following quadratic-programming (QP) problem: $\mathbf{w}_+^k(\alpha)$ minimizes $\|\mathbf{w}-[\mathbf{w}^k + \alpha\nabla\phi(\mathbf{w}^k)]\|^2$ with respect to $\mathbf{w}$ satisfying the linear constraints $\mathbf{w} \in \mathscr{P}_{\ell-1}$. One may also use the following property: for any $\mathbf{z} \in \mathbb{R}^\ell$, the projection of $\mathbf{z}$ onto $\mathscr{P}_{\ell-1}$ is given by $P_{\mathscr{P}_{\ell-1}}(\mathbf{z}) = \mathbf{w}(\mathbf{z}, t^*)$, where $\mathbf{w}(\mathbf{z}, t) = \max\{\mathbf{z} - t\mathbf{1}, \mathbf{0}\}$ (componentwise) and t^* maximizes $L[\mathbf{w}(\mathbf{z}, t), t]$ with respect to t, with $L(\mathbf{w}, t)$ the partial Lagrangian $L(\mathbf{w}, t) = (1/2)\|\mathbf{w} - \mathbf{z}\|^2 + t\,[\mathbf{1}^\top\mathbf{w} - 1]$. Indeed, $L(\mathbf{w}, t)$ can be written as

$$L(\mathbf{w}, t) = (1/2)\|\mathbf{w} - (\mathbf{z} - t\mathbf{1})\|^2 + t\,[\mathbf{1}^\top\mathbf{z} - 1] - \ell\, t^2/2\,,$$

which reaches its minimum with respect to $\mathbf{w} \geq \mathbf{0}$ for $\mathbf{w} = \mathbf{w}(\mathbf{z}, t)$. One may notice that $\max\{\max_i(z_i) - t^*, 0\} \leq \mathbf{1}^\top\mathbf{w}(\mathbf{z}, t^*) = 1 \leq \ell\max\{\max_i(z_i) - t^*, 0\}$, so that the search for t^* can be restricted to the interval $[\max_i(z_i) - 1, \max_i(z_i) - 1/\ell]$.

When additional constraints to $\mathbf{w} \in \mathscr{P}_{\ell-1}$ are present that define a convex set $\mathscr{P}' \subset \mathscr{P}_{\ell-1}$ (see Sects. 5.1.9 and 5.1.10), one should consider the orthogonal projection onto $\mathscr{P}'$. One may refer, e.g., to Dai and Fletcher (2006) for an efficient algorithm for computing such projections. □

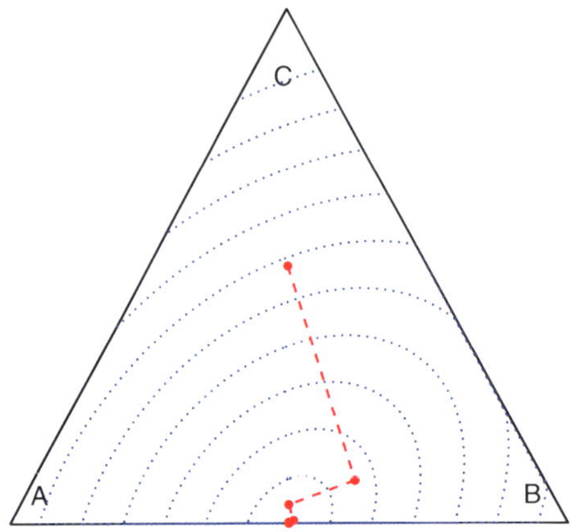

Fig. 9.4. Behavior of the gradient-projection algorithm (9.15), (9.16) in Example 9.1

Example 9.1 (continued). We consider the same problem as in Example 9.1. Figure 9.4 presents the evolution in $\mathscr{P}_2$ of iterates $\mathbf{w}^k$ generated by (9.15), (9.16), initialized at $\mathbf{w}^0 = (1/4\,,\ 1/4\,,\ 1/2)^\top$; compare with Figs. 9.1 and 9.3. □

The high dimension of $\mathbf{w}$ may set a limitation on the use of this method. It is thus recommended to combine vertex-direction iterations with iterations of the type (9.17). A simple prototype algorithm is as follows:

0. Start from ξ_0, a discrete probability measure on $\mathscr{X}$ with finite support $\mathcal{S}_{\xi_0}$ ($\xi_0(x) > 0$ for all $x \in \mathcal{S}_{\xi_0}$) and such that $\phi(\xi_0) > -\infty$; choose $\epsilon_1 > \epsilon_0 > 0$, set $k = 0$.
1. Set $k_0 = k$, optimize the weights $\mathbf{w}^k$ for designs supported on $\mathcal{S}_{\xi_{k_0}}$ using iterations of the type (9.17) initialized at ξ_{k_0}, increment k, and stop iterating at the first k for which $\max_{x\in\mathcal{S}_{\xi_{k_0}}} F_\phi(\xi_k, x) < \epsilon_0$.
2. If $\max_{x\in\mathscr{X}} F_\phi(\xi_k, x) < \epsilon_1$ stop; otherwise perform one iteration of a vertex-direction algorithm initialized at ξ_k, $k \leftarrow k+1$, return to step 1.

Notice that the dimension of $\mathbf{w}^{k_0}$ at step 1 may vary. We do not detail the choice of constants ϵ_0, ϵ_1 (and of those hidden in the definitions of stopping rules for determining the stepsizes of the gradient-projection and vertex-direction iterations of steps 1 and 2) ensuring a fast enough convergence of the algorithm; one may refer, e.g., to Polak (1971) for a general exposition.

In the algorithm proposed by Wu (1978a,b), which we reproduce below, gradient-projection iterations are only used with vectors $\mathbf{w}^k$ having strictly positive components; they can thus be based on (9.14), (9.15) and do not require computations of projections (9.17). The switching criterion between gradient projection and vertex direction is based on the observation that, due to the concavity of $\phi(\cdot)$, $\mathbf{w}^{k+1}$ given by (9.15) satisfies $\phi(\mathbf{w}^{k+1}) \leq$

$\phi(\mathbf{w}^k) + \bar{\alpha}_k \nabla^\top \phi(\mathbf{w}^k)\mathbf{d}^k$ with $\bar{\alpha}_k$ defined in (9.16). When $\phi(\mathbf{w}^k)$ cannot be very much improved, the search for a suitable $\alpha_k \in [0, \bar{\alpha}_k]$ is thus futile, and it is advantageous to switch to the vertex-direction method (9.2), (9.3). Notice that this is the case in particular when $\bar{\alpha}_k$ equals zero, i.e., when some components of $\mathbf{w}^k$ equal zero. One vertex-direction iteration with (9.2), (9.3) may then reset all components of the next $\mathbf{w}$ to positive values if necessary or leave some weights equal to zero, depending on which vertex direction is used. The algorithm is as follows. Every time a new vector of weights $\mathbf{w}^k$ is formed from a measure ξ_k (at steps 0 and 3), only support points with positive weights are considered, so that the dimension of $\mathbf{w}^k$ may vary along the iterations.

0. Start from ξ_0, a discrete probability measure on $\mathscr{X}$ with finite support $\mathcal{S}_{\xi_0}$ and such that $\phi(\xi^{(0)}) > -\infty$, choose ϵ_0 and $\Delta > 0$, set $k = 0$, and form $\mathbf{w}^0$.
1. Compute $\mathbf{d}^k$ and $\bar{\alpha}_k$ of (9.14) and (9.16); if $\bar{\alpha}_k \nabla^\top \phi(\mathbf{w}^k)\mathbf{d}^k \leq \epsilon_0$, go to step 3; otherwise, go to step 2.
2. Update $\mathbf{w}^k$ according to (9.15) with α_k given by (9.16), $k \leftarrow k+1$; go to step 1.
3. If $\max_{x \in \mathscr{X}} F_\phi(\xi_k, x) < \Delta$ stop; otherwise perform one iteration of the vertex-direction algorithm (9.2), (9.3) with α_k maximizing $\phi[(1-\alpha)\xi_k + \alpha\delta_{x_{k+1}^+}]$ with respect to $\alpha \in [0, 1]$; $k \leftarrow k+1$, form $\mathbf{w}^k$, and go to step 1.

When $\phi(\cdot) = \Phi[\mathbf{M}(\cdot, \theta)]$ with $\Phi(\cdot)$ a criterion from Chap. 5, the algorithm performs a sequence of optimizations over a sequence of polyhedra inscribed in $\mathcal{M}_\theta(\Xi)$ given by (5.3). Detailed explanations and proof of convergence are given in (Wu, 1978a); results of numerical experiments are presented in (Wu, 1978b).

Several improvements are suggested in the same papers, which we reproduce below:

(i) At step 3 we may choose x_{k+1}^+ as suggested in Remark 9.2.
(ii) Using arguments similar to those in Sect. 9.1.1, one can show that step 2 is eventually skipped for δ small enough. In order to benefit for the efficiency of the gradient-projection iterations of step 2, we may thus use a sequence of constants ϵ_i decreasing to zero at step 1, instead of using a fixed value ϵ_0. In that case ϵ_i is decreased to ϵ_{i+1} whenever a value $\max_{x \in \mathscr{X}} F_\phi(\xi_k, x)$ smaller than all previous ones is observed at step 3.
(iii) In order to favor reduction of support size, we may start step 3 by an attempt to move from $\mathbf{w}^k$ to $\mathbf{w}^k + \bar{\alpha}_k \mathbf{d}^k$ and accept the move (which reduces the support size) if $\phi(\mathbf{w}^k + \bar{\alpha}_k \mathbf{d}^k) \geq \phi(\mathbf{w}^k)$.

Step 3 of the algorithm above requires the determination of $\max_{x \in \mathscr{X}} F_\phi(\xi_k, x)$, which is difficult to implement when $\mathscr{X}$ is not finite. The usual practice is then to refine progressively a discretization of $\mathscr{X}$, i.e., to consider a sequence of imbedded finite sets $\mathscr{X}_\ell \subset \mathscr{X}_{\ell+1} \subset \cdots \subset \mathscr{X}$, and to use a positive sequence $\{\Delta_j\}$ decreasing to zero instead of a fixed Δ at step 3. A new version of step 3 is then as follows (with j initialized at ℓ at step 0); see Wu (1978a).

3') If $\max_{x\in\mathscr{X}_j} F_\phi(\xi_k, x) \geq \Delta_j$, choose $x^+_{k+1} = \arg\max_{x\in\mathscr{X}_j} F_\phi(\xi_k, x)$ and update ξ_k into $\xi_{k+1} = (1-\alpha_k)\xi_k + \alpha_k \delta_{x^+_{k+1}}$ with α_k maximizing $\phi[(1-\alpha)\xi_k + \alpha\delta_{x^+_{k+1}}]$ with respect to $\alpha \in [0,1]$, $k \leftarrow k+1$, form $\mathbf{w}^k$, and go to step 1; otherwise $j \leftarrow j+1$, repeat step 3'.

When step 3' is repeated, the search for x^+_{k+1} is made in $\mathscr{X}_j \setminus \mathscr{X}_{j-1}$. Note that $\{\phi(\xi_k)\}$ forms a monotone increasing sequence. Suppose that the sets $\mathscr{X}_j$ are such that

$$\max_{x\in\mathscr{X}} F_\phi(\xi, x) \leq \max_{x\in\mathscr{X}_j} F_\phi(\xi, x) + \Delta'_j \tag{9.18}$$

for all $j \geq \ell$ and all ξ such that $\phi(\xi) \geq \phi(\xi_0)$, with $\{\Delta'_j\}$ a positive sequence decreasing to zero. Then, an infinite loop with ξ_k at step 3' means that $\max_{x\in\mathscr{X}} F_\phi(\xi_k, x) < \Delta_j + \Delta'_j$ for all j; therefore, $\max_{x\in\mathscr{X}} F_\phi(\xi_k, x) = 0$, and ξ_k is optimal on $\mathscr{X}$. If an infinite loop does not occur, step 2 is eventually skipped (see point (*ii*) above) and all iterations are eventually of the vertex-direction type (9.2), (9.3). Suppose that the sets $\mathscr{X}_j$ are such that $\Delta_j \geq \gamma\Delta'_j$ for some $\gamma > 0$, then (9.18) implies that $F_\phi(\xi_k, x^+_{k+1}) \geq [\gamma/(1+\gamma)] \max_{x\in\mathscr{X}} F_\phi(\xi, x)$ at step 3'; the choice of x^+_{k+1} is thus compatible with Remark 9.2, and the convergence to an optimal design follows from that of the vertex-direction algorithm; see Sect. 9.1.1. When $\phi(\cdot)$ is such that $F_\phi(\xi, \cdot)$ is Lipschitz on $\mathscr{X}$ with Lipschitz constant L for all ξ such that $\phi(\xi) \geq \phi(\xi_0)$, we have $\max_{x\in\mathscr{X}} F_\phi(\xi, x) \leq \max_{x\in\mathscr{X}_j} F_\phi(\xi, x) + L \max_{x\in\mathscr{X}} \min_{x^{(t)}\in\mathscr{X}_j} \|x - x^{(t)}\|$, so that we may take $\Delta_j = C \max_{x\in\mathscr{X}} \min_{x^{(t)}\in\mathscr{X}_j} \|x - x^{(t)}\|$ for some constant C, and we only require that the sets $\mathscr{X}_j$ satisfy a space-filling property in $\mathscr{X}$ ($\Delta_j \to 0$ as $j \to \infty$).

Second-Order Methods

Suppose again that $\mathbf{w}^k$ has strictly positive components. For $\mathbf{\Lambda}$ a positive-definite $\ell \times \ell$ matrix, instead of using (9.14), we can also construct $\mathbf{d}^k$ from the projection of $\mathbf{\Lambda}\nabla\phi(\mathbf{w}^k)$ onto $\mathcal{L} = \{\mathbf{z} \in \mathbb{R}^\ell : \mathbf{1}^\top\mathbf{z} = 0\}$ for the norm induced by $\mathbf{\Lambda}^{-1}$, $\|\mathbf{z}\|_{\mathbf{\Lambda}^{-1}} = \mathbf{z}^\top\mathbf{\Lambda}^{-1}\mathbf{z}$. We thereby obtain the following generalization of (9.14):

$$\mathbf{d}^k = \mathbf{\Lambda}\nabla\phi(\mathbf{w}^k) - \frac{\mathbf{1}^\top\mathbf{\Lambda}\nabla\phi(\mathbf{w}^k)}{\mathbf{1}^\top\mathbf{\Lambda}\mathbf{1}}\mathbf{\Lambda}\mathbf{1}\,. \tag{9.19}$$

Note that (9.19) coincides with (9.14) for $\mathbf{\Lambda} = \mathbf{I}_\ell$. Again, $\mathbf{1}^\top\mathbf{d}^k = 0$ and

$$\left.\frac{\mathrm{d}\phi(\mathbf{w}^{k+1})}{\mathrm{d}\alpha_k}\right|_{\alpha_k=0^+} = \nabla^\top\phi(\mathbf{w}^k)\mathbf{\Lambda}\nabla\phi(\mathbf{w}^k) - \frac{[\mathbf{1}^\top\mathbf{\Lambda}\nabla\phi(\mathbf{w}^k)]^2}{\mathbf{1}^\top\mathbf{\Lambda}\mathbf{1}} \geq 0\,,$$

with equality if and only if $(\mathbf{e}_i - \mathbf{w}^k)^\top\nabla\phi(\mathbf{w}^k) = 0$ for all i, which means that $\mathbf{w}^k$ is optimal for $\phi(\cdot)$. The direction $\mathbf{d}^k$ is thus an ascent direction at $\mathbf{w}^k$, and we can use an iteration of the form (9.15). Notice that $\mathbf{\Lambda}$ may depend on the iteration number k. Taking $\mathbf{\Lambda} = \mathbf{\Lambda}_k = \mathbf{H}^{-1}(\mathbf{w}^k)$, the inverse of the Hessian matrix of second-order derivatives of $\phi(\cdot)$, we obtain the Newton method, quasi-Newton methods correspond to using an approximation of the inverse of the Hessian, etc. One can refer, e.g., to Polyak (1987) and Bonnans et al. (2006) for a general exposition on such second-order methods, including the popular sequential quadratic programming, and to Atwood (1976) for the implementation of the Newton method in a design setting. Particular choices of $\mathbf{\Lambda} = \mathbf{\Lambda}_k$ adapted to the design context are considered in the next section.

9.1.3 Multiplicative Algorithms

Suppose that $\mathscr{X}$ is discretized into $\mathscr{X}_\ell$. Taking $\mathbf{\Lambda} = \mathrm{diag}\{w_1^k, \ldots, w_\ell^k\}$ in (9.19) gives

$$\begin{aligned} d_i^k &= w_i^k \{\nabla\phi(\mathbf{w}^k)\}_i - [\nabla^\top \phi(\mathbf{w}^k)\mathbf{w}^k] w_i^k \\ &= w_i^k (\mathbf{e}_i - \mathbf{w}^k)^\top \nabla\phi(\mathbf{w}^k) = w_i^k F_\phi(\xi_k, x^{(i)}) \,, \end{aligned}$$

with d_i^k the i-th component of $\mathbf{d}^k$, and therefore, from (9.15),

$$w_i^{k+1} = w_i^k \left[1 + \alpha_k \, F_\phi(\xi_k, x^{(i)})\right] , \tag{9.20}$$

which increases the weights of points such that $F_\phi(\xi_k, x^{(i)}) > 0$ and decreases the weights of the others. As shown below, a simple choice of α_k may provide a simple globally convergent monotonic algorithm in some situations.

Consider the case of D-optimum design, where $\phi(\xi) = \log\det[\mathbf{M}(\xi, \theta)]$, with $\mathbf{M}(\xi, \theta) = \int_{\mathscr{X}} \mathbf{g}_\theta(x)\mathbf{g}_\theta^\top(x)\,\xi(\mathrm{d}x)$ and $\mathbf{g}_\theta(x) \in \mathbb{R}^p$, so that $F_\phi(\xi, x) = \mathbf{g}_\theta^\top(x)\mathbf{M}^{-1}(\xi, \theta)\mathbf{g}_\theta(x) - p$. The iteration (9.20) with $\alpha_k = 1/p$ then gives

$$w_i^{k+1} = w_i^k \, \frac{D(x^{(i)}; \xi_k)}{p} , \tag{9.21}$$

with

$$D(x; \xi) = \mathbf{g}_\theta^\top(x)\mathbf{M}^{-1}(\xi, \theta)\mathbf{g}_\theta(x) \tag{9.22}$$

This algorithm is shown in (Titterington, 1976) and (Pázman, 1986) to be globally monotonically convergent on $\mathscr{X}_\ell$ when $\xi_0(x^{(i)}) > 0$ for all $x^{(i)} \in \mathscr{X}_\ell$. More generally, by taking

$$\alpha_k = \frac{1}{\nabla^\top \phi(\mathbf{w}^k)\mathbf{w}^k}$$

in (9.20), we obtain

$$w_i^{k+1} = w_i^k \, \frac{\{\nabla\phi(\mathbf{w}^k)\}_i}{\nabla^\top \phi(\mathbf{w}^k)\mathbf{w}^k} , \tag{9.23}$$

or equivalently

$$w_i^{k+1} = w_i^k \frac{\mathbf{g}_\theta^\top(x^{(i)})\nabla_{\mathbf{M}}\Phi[\mathbf{M}(\xi_k,\theta)]\mathbf{g}_\theta(x^{(i)})}{\mathrm{trace}\{\mathbf{M}(\xi_k,\theta)\nabla_{\mathbf{M}}\Phi[\mathbf{M}(\xi_k,\theta)]\}}$$

when $\phi(\xi) = \Phi[\mathbf{M}(\xi,\theta)]$. This can be further generalized into

$$w_i^{k+1} = w_i^k \frac{\{\nabla\phi(\mathbf{w}^k)\}_i^\lambda}{\sum_{i=1}^\ell w_i^k \{\nabla\phi(\mathbf{w}^k)\}_i^\lambda}, \ \lambda > 0, \tag{9.24}$$

see Silvey et al. (1978) and Fellman (1989). Monotonicity is proved in (Torsney, 1983) for A-optimality and in (Fellman, 1974) for c-optimality, both with $\lambda = 1/2$. One may refer to Torsney (2009) for a historical review. Yu (2010a) shows that (9.24) is monotonic for any $\lambda \in (0,1]$ when $\phi(\xi) = \Phi[\mathbf{M}(\xi,\theta)]$ and the function $\Psi(\cdot)$ defined by $\Psi(\mathbf{M}) = -\Phi(\mathbf{M}^{-1})$ is isotonic and concave on $\mathbb{M}^>$; see Sect. 5.1.3. Global convergence to an optimal design is proved in the same paper for a large class of criteria, including D- and A-optimality.

Take now

$$\alpha_k = \frac{1}{\nabla^\top\phi(\mathbf{w}^k)\mathbf{w}^k - \beta_k}$$

which gives another generalization of (9.23), considered in (Dette et al., 2008):

$$w_i^{k+1} = w_i^k \frac{\{\nabla\phi(\mathbf{w}^k)\}_i - \beta_k}{\nabla^\top\phi(\mathbf{w}^k)\mathbf{w}^k - \beta_k}. \tag{9.25}$$

In the case of D-optimality, we get $\alpha_k = 1/(p-\beta_k)$ and

$$w_i^{k+1} = w_i^k \frac{D(x^{(i)};\xi_k) - \beta_k}{p - \beta_k},$$

with $D(x;\xi)$ given by (9.22). The condition $0 < \alpha_k \leq \bar{\alpha}_k = \min\{w_i^k/|d_i^k| : d_i^k < 0\}$ (see (9.16)) gives $-\infty < \beta_k \leq \bar{\beta}_k = \min_{i=1,\ldots,\ell} D(x^{(i)};\xi_k)$. Dette et al. (2008) prove global monotonic convergence to a D-optimal design when $-\infty < \beta_k \leq \bar{\beta}_k/2$ for all k. Numerical simulations indicate that the convergence to the optimum is significantly faster for $\beta_k = \bar{\beta}_k/2$ than for $\beta_k = 0$ which corresponds to (9.21).

Remark 9.5. When $\mathbf{g}_\theta(x) = [\mathbf{z}_\theta^\top(x)\ 1]^\top$, with $\mathbf{z}_\theta(x) \in \mathbb{R}^{p-1}$, $p \geq 2$, the dual of the D-optimal design problem on $\mathscr{X}_\ell$ corresponds to the construction of the minimum-volume ellipsoid, with free center, containing the points $\mathbf{z}_\theta(x^{(i)})$; see Sect. 5.6 and the section below. The algorithm (9.25) with $\beta_k = 1$ for all k then corresponds to a method proposed by Silvey et al. (1978) and Titterington (1976, 1978). The monotonicity and global convergence to the optimum when $p \geq 3$ was left as a conjecture, supported by numerical simulations, until the recent proof in (Yu, 2010a,b). It had been noticed in (Pronzato et al., 2000, Chap. 7) that when $p = 2$ and $\mathbf{g}(x) = [x\ 1]^\top$ with x in some interval, then the algorithm corresponds to a renormalized version of the steepest descent method for the minimization of a quadratic function and may oscillate between two non-optimal solutions. □

Remark 9.6. Multiplicative iterations similar to (9.21) suffer from the same defect as vertex-direction iterations (9.2) with $\alpha_k \in (0,1)$: all support points present in the initial design measure ξ_0 keep a strictly positive weight along iterations, and the convergence of $\phi(\xi_k)$ to the optimal ϕ^* is ineluctably slow when approaching the optimum. A mixture of multiplicative and vertex-exchange algorithms is proposed in (Yu, 2011) for D-optimum design; it includes a nearest-neighbor exchange strategy that helps apportioning weights between adjacent points and has the property that poor support points are quickly removed from the support of ξ_k. □

9.1.4 *D*-optimum Design

Besides motivations related to the properties of the D-optimality criterion in terms of characterization of the precision of the estimation (see Sects. 5.1.2 and 5.1.8), D-optimum design has attracted a lot of attention due to the simplifications it allows in vertex-direction algorithms and to the connection with a classical problem in geometry. Here $\phi(\xi) = \log\det[\mathbf{M}(\xi,\theta)]$, with $\mathbf{M}(\xi,\theta) = \int_{\mathscr{X}} \mathbf{g}_\theta(x)\mathbf{g}_\theta^\top(x)\,\xi(\mathrm{d}x)$.

Exploiting Submodularity

The submodularity property of the determinant function and of the function $X \longrightarrow D(x;\xi)$ given by (9.22), with ξ the design measure associated with the exact design X, can be used to simplify calculations in a D-optimal design algorithm based on (9.2), (9.3); see Robertazzi and Schwartz (1989). When the design space is the finite set $\mathscr{X}_\ell = \{x^{(1)},\ldots,x^{(\ell)}\}$, we have from (9.3) $x_{k+1}^+ = \arg\max_{x\in\mathscr{X}_\ell} D(x;\xi_k)$. One can show that it is not necessary to compute $D(x;\xi_k)$ for all $x \in \mathscr{X}_\ell$ at each iteration to obtain x_{k+1}^+. Indeed, we have for any $\mathbf{v} \in \mathbb{R}^p$ and any $j \geq 0$

$$\mathbf{v}^\top\mathbf{M}^{-1}(\xi_{j+1},\theta)\mathbf{v} = \frac{\mathbf{v}^\top\mathbf{M}^{-1}(\xi_j,\theta)\mathbf{v}}{1-\alpha_j} - \frac{\alpha_j\left[\mathbf{v}^\top\mathbf{M}^{-1}(\xi_j,\theta)\mathbf{g}_\theta(x_{j+1})\right]^2}{(1-\alpha_j)[(1-\alpha_j)+\alpha_j D(x_{j+1},\xi_j)]}\,.$$

Hence,

$$\mathbf{v}^\top\mathbf{M}^{-1}(\xi_{j+1},\theta)\mathbf{v} \leq \frac{\mathbf{v}^\top\mathbf{M}^{-1}(\xi_j,\theta)\mathbf{v}}{1-\alpha_j}\,,\ 0<\alpha_j<1\,,$$

and by induction

$$\forall n \geq 1\,,\ \mathbf{v}^\top\mathbf{M}^{-1}(\xi_{j+n},\theta)\mathbf{v} \leq \mathbf{v}^\top\mathbf{M}^{-1}(\xi_j,\theta)\mathbf{v}\prod_{i=j}^{j+n-1}(1-\alpha_i)^{-1}\,.$$

Suppose that at iteration $k = j + n > j$, the value $D(x^{(l)}; \xi_j)$ has been stored. Then, knowing that

$$D(x^{(i^*)}; \xi_k) > D(x^{(l)}; \xi_j) \prod_{i=j}^{k-1} (1 - \alpha_i)^{-1}$$

we can safely conclude that $D(x^{(i^*)}; \xi_k) > D(x^{(l)}; \xi_k)$ without having to compute $D(x^{(l)}; \xi_k)$. This is exploited in the algorithm below; see Robertazzi and Schwartz (1989). For any sequence $\{\gamma_i\}$, we define $\prod_{i=a}^{b} \gamma_i = 1$ if $b < a$:

0. Start from ξ_0, a discrete probability measure on the set $\mathscr{X}_\ell$ such that $\det[\mathbf{M}(\xi_0, \theta)] > 0$; set $k = 0$ and $s(i) = 0$ for all i; compute and store the values $\mathcal{D}(i; s(i)) = D(x^{(i)}; \xi_0)$ for all $x^{(i)} \in \mathscr{X}_\ell$; choose $\epsilon_0 > 0$.
1. At current iteration k:
   1a. Find $j^* = \arg\max_{j=1,\ldots,\ell} \mathcal{D}(j; s(j)) \prod_{i=s(j)}^{k-1} (1 - \alpha_i)^{-1}$; if $k = 0$, go to step 2.
   1b. Set $s(j^*) = k$; compute $\mathcal{D}(j^*; k) = D(x^{(j^*)}; \xi_k)$; if $\mathcal{D}(j^*; s(j^*)) \geq \mathcal{D}(j; s(j)) \prod_{i=s(j)}^{k-1} (1 - \alpha_i)^{-1}$ for all $j = 1, \ldots, \ell$, go to step 2; otherwise, return to step 1a.
2. If $\mathcal{D}(j^*; s(j^*)) < p + \epsilon_0$, stop ; otherwise take $x_{k+1}^+ = x^{(j^*)}$; perform one iteration (9.2) to update ξ_k into ξ_{k+1}, $k \leftarrow k + 1$ and return to step 1.

Explicit Optimal Stepsizes

One can easily check that α_k^+ and α_k^-, respectively defined by (9.6) and (9.8), are given explicitly as

$$\alpha_k^+ = h_k(x_{k+1}^+) \tag{9.26}$$

$$\alpha_k^- = \begin{cases} h_k(x_{k+1}^-) \text{ if } D(x_{k+1}^-; \xi_k) > \frac{p}{1+(p-1)\xi_k(x_{k+1}^-)} \\ \frac{-\xi_k(x_{k+1}^-)}{1-\xi_k(x_{k+1}^-)} \text{ otherwise}, \end{cases} \tag{9.27}$$

with

$$h_k(x) = \frac{D(x; \xi_k) - p}{p[D(x; \xi_k) - 1]}$$

and $D(x; \xi)$ given by (9.22). Similarly, direct calculations show that α_k given by (9.12) satisfies

$$\alpha_k = \min \left\{ \xi_k(x_{k+1}^-), \frac{D(x_{k+1}^+; \xi_k) - D(x_{k+1}^-; \xi_k)}{2[D(x_{k+1}^+; \xi_k) D(x_{k+1}^-; \xi_k) - D^2(x_{k+1}^+, x_{k+1}^-; \xi_k)]} \right\} \tag{9.28}$$

where

$$D(x_1, x_2; \xi) = \mathbf{g}_\theta^\top(x_1) \mathbf{M}^{-1}(\xi, \theta) \mathbf{g}_\theta(x_2) \,.$$

Reduction of the Design Space

When $\alpha_k = \alpha_k^- = -\xi_k(x_{k+1}^-)/[1-\xi_k(x_{k+1}^-)]$ in (9.27) or $\alpha_k = \xi_k(x_{k+1}^-)$ in (9.28), it means that the weight at x_{k+1}^- is set at 0, that is, x_{k+1}^- is removed from ξ_{k+1}. This does not mean that x_{k+1}^- may not reenter the support of $\xi_{k'}$ for some $k' > k$. However, it is shown in (Harman and Pronzato, 2007) that any design point x satisfying

$$D(x;\xi_k) \leq H(\epsilon_k) = p\left[1 + \frac{\epsilon_k}{2} - \frac{\sqrt{\epsilon_k(4+\epsilon_k-4/p)}}{2}\right], \tag{9.29}$$

with $\epsilon_k = \max_{x\in\mathscr{X}} D(x;\xi_k) - p$, cannot be support point for a D-optimal design and can thus be safely removed from $\mathscr{X}$. This can be used for any ξ_k and thus implemented as an additional ingredient to any optimization algorithm for D-optimum design. In the case of a finite design space $\mathscr{X}_\ell$, the cardinality of the space can thus be decreased during the progress of the algorithm; see, for instance, Examples 9.16 and 9.17 for an illustration. Notice that $H(\epsilon)$ is a decreasing function of ϵ, satisfying $H(0) = p$ and $\lim_{\epsilon\to\infty} H(\epsilon) = 1$, so that the inequality above becomes more and more powerful for removing points from $\mathscr{X}$ as ϵ_k gets smaller and smaller, i.e., during the progress of ξ_k towards a D-optimal design.

Minimum-Volume Covering Ellipsoids and the Complexity of D-optimum Design

An algorithm based on (9.2), (9.3) for the construction of the minimum-volume ellipsoid enclosing a given set of points $\mathscr{Z}_\ell = \{\mathbf{z}^{(1)},\ldots,\mathbf{z}^{(\ell)}\} \subset \mathbb{R}^d$ is analyzed in (Khachiyan, 1996). Denote

$$\mathbf{M}(\xi) = \int \begin{bmatrix}\mathbf{z}\\1\end{bmatrix} [\mathbf{z}^\top\ 1]\, \xi(\mathrm{d}\mathbf{z}),$$

with ξ a probability measure on $\mathbb{R}^d$, and denote by $\mathcal{E}(\xi)$ the ellipsoid

$$\mathcal{E}(\xi) = \{\mathbf{u}\in\mathbb{R}^{d+1} : \mathbf{u}^\top[(d+1)\mathbf{M}(\xi)]^{-1}\mathbf{u} \leq 1\}.$$

Let ξ_D^* be a D-optimal design measure for the linear regression model with regressors $[{\mathbf{z}^{(i)}}^\top\ 1]^\top$; that is, ξ_D^* maximizes $\log\det[\mathbf{M}(\xi)]$ with respect to ξ in the set of probability measures on $\mathscr{Z}_\ell$. The intersection of $\mathcal{E}(\xi_D^*)$ with the hyperplane $\mathcal{H} = \{\mathbf{u}\in\mathbb{R}^{d+1} : u_{d+1} = 1\}$ gives the minimum-volume ellipsoid containing $\mathscr{Z}_\ell$; see Shor and Berezovski (1992) and Khachiyan and Todd (1993). The method in (Khachiyan, 1996) uses iterations (9.2), (9.3), initialized at ξ_0 giving weight $1/\ell$ at each point of $\mathscr{Z}_\ell$, to construct a sequence of discrete design measures ξ_k on $\mathscr{Z}_\ell$ that converges to a D-optimal measure ξ_D^*. To the sequence of design measures ξ_k generated corresponds a sequence of ellipsoids $\mathcal{E}(\xi_k)$. It is shown in (Todd and Yildirim, 2007) that the polar ellipsoids

$$\mathcal{E}^o(\xi_k) = \{\mathbf{v} \in \mathbb{R}^{d+1} : \mathbf{v}^\top \mathbf{M}(\xi_k)\mathbf{v} \leq 1\}$$

all contain the polar polytope

$$\mathscr{Z}_\ell^o = \{\mathbf{v} \in \mathbb{R}^{d+1} : \left|[\mathbf{z}^{(i)\top} \; 1]\mathbf{v}\right| \leq 1 \,, \; i = 1, \ldots, \ell\}$$

and (surprisingly) exactly coincide with the ellipsoids generated by the parallel-cut version of the ellipsoid method[2] for enclosing $\mathscr{Z}_\ell^o$ into an ellipsoid; see Bland et al. (1981), and Todd (1982); see also Sect. 9.3.1. If the algorithm is stopped when $F_\phi(\xi_n, x_{n+1}^+) \leq (d+1)\epsilon$, then the convex hull of $\mathscr{Z}_\ell$ is included in $\sqrt{1+\epsilon}\,[\mathcal{E}(\xi_n) \cap \mathcal{H}]$ and $[1/\sqrt{(d+1)(1+\epsilon)}]\,\mathcal{E}^o(\xi_n)$ is contained in $\mathscr{Z}_\ell^o$.

Todd and Yildirim (2007) also show that an algorithm based on (9.9), (9.10), with α_k^+ and α_k^- given by (9.26) and (9.27), computes an approximate D-optimal design ξ^* on a finite space $\mathscr{X}_\ell$, satisfying

$$\begin{cases} \max_{i=1,\ldots,\ell} D(x^{(i)}; \xi^*) \leq (1+\epsilon)p \\ D(x^{(i)}; \xi^*) \geq (1-\epsilon)p \text{ for all } i \text{ such that } \xi^*(x^{(i)}) > 0 \,, \end{cases}$$

in $\mathcal{O}\,\{p[\log(p) + 1/\epsilon]\}$ iterations, with $D(x;\xi)$ given by (9.22); see Ahipasaoglu et al. (2008) for more precise results and extensions to other concave differentiable criteria.

9.2 Exact Design

The exact design problem corresponds to the situation where the number N of observations to be collected is fixed. We denote by X the collection of design points to be used, $X = (x_1, \ldots, x_N)$, with possible repetitions; that is, we may have $x_i = x_j$ for some $i \neq j$. We only consider here locally optimum design: when the criterion depends on the value of the model parameters θ, we assume that a nominal value θ^0 is used for θ. An optimal exact design is denoted by X^*. Maximin-optimum exact design will be considered in Sect. 9.3.2 and average-optimum exact design in Sect. 9.4.2.

In approximate design theory an optimal design measure ξ^* is characterized by its support points $x^{(i)}$ and associated weights w_i^*, $i = 1, \ldots, m$, which are positive reals that sum to one. When N is fixed, one may thus consider constructing an exact design by rounding the weights w_i^* of an optimal design measure ξ^*. The support points are then unchanged, each quota Nw_i^* being rounded to an integer r_i, the number of repetitions of observations at $x^{(i)}$, with the constraint that $\sum_{i=1}^m r_i = N$, which in general precludes using a simple numerical rounding of each Nw_i^* to its closest integer.

[2]This is the ellipsoid method used by Khachiyan (1979) to prove the polynomial complexity of LP.

Fedorov (1972, p. 157) suggests to allocate observations in several steps. First allocate $r_i^0 = \lceil (N-m)w_i^* \rceil$ observations at $x^{(i)}$, $i = 1, \ldots, m$, where $\lceil t \rceil$ is the smallest integer satisfying $\lceil t \rceil \geq t$. The remaining $N - \sum_{i=1}^m r_i^0$ observations can then be added one by one to points where $(N-m)w_i^* - r_i^0$ is larger than $-1/2$. This allocation rule gets close to optimality as N increases when $\phi(X) = \Phi[\mathbf{M}(X,\theta)]$ with $\Phi(\cdot)$ isotonic and positively homogeneous; see Definition 5.3. Indeed, denote by X^0 the exact design with r_i^0 observations at $X^{(i)}$, $i = 1, \ldots, m$. We have

$$\phi(\xi^*) \geq \phi(X^*) \geq \phi(X^0) \geq \frac{N-m}{N}\,\phi(\xi^*)\,,$$

so that

$$\phi(X^*) - \phi(X^0) \leq \frac{m}{N}\,\phi(\xi^*)\,,$$

see Fedorov (1972, Corollary 1, p. 157). Notice that this indicates that when the optimal design measure ξ^* is not unique, the one with fewer support points should be used in the rounding procedure.

Other rounding methods are discussed in (Pukelsheim and Reider, 1992), where it is shown that Adams apportionment, which consists in choosing the r_i that maximize $\min_{i=1,\ldots,m} r_i/w_i^*$, guarantees the best efficiency bounds for all isotonic and positively homogeneous criteria. Although this rounding method is the most efficient, one should note that an exact optimal design is not necessarily in the class of designs having the same support as an approximate optimal design—consider in particular the case where $p \leq N < m$. It is therefore of interest to consider techniques for constructing exact optimal designs directly, without passing through the intermediate step of an approximate design.[3]

When the design space $\mathscr{X}$ is a compact subset of $\mathbb{R}^d$ with nonempty interior and $X = (x_1, \ldots, x_N)$, the design problem is a continuous optimization problem in $\mathbb{R}^{N \times d}$ with N constraints ($x_i \in \mathscr{X}$ for all i). If $N \times d$ is not too large, general-purpose optimization algorithms can be used. However, since there exist local optima (notice in particular that any permutation of the x_i leaves the criterion value unchanged), global optimization algorithms must be used. A straightforward application of a global algorithm in dimension $N \times d$ is computationally prohibitive even for moderate values of N and d. It is thus recommended to use algorithms that take the specific structure

[3]There exists a situation, however, where an approximate design can be implemented directly, without requiring any rounding of its weights, whatever their value might be. This is when the design corresponds to the construction of the optimal input signal for a dynamical system, this input signal being characterized by its power spectral density (which plays the same role as the design measure ξ); see, e.g., Goodwin and Payne (1977, Chap. 6), Zarrop (1979), Ljung (1987, Chap. 14), Walter and Pronzato (1997, Chap. 6).

of the problem into account and attempt to decompose the optimization in $\mathbb{R}^{N\times d}$ into a series of optimizations in $\mathbb{R}^d$. This is the objective of exchange algorithms, the principles of which are presented in Sect. 9.2.1.

Consider now the case where $\mathscr{X}$ is a finite set, $\mathscr{X} = \{x^{(1)},\ldots,x^{(\ell)}\}$. Since there are $(\ell+N-1)!/[N!(\ell-1)!]$ possible designs with N points, exhaustive search is clearly computationally unfeasible even for moderate N and ℓ. Exchange algorithms can be used in this case too. However, when the design criterion is a function of the information matrix, $\phi(X) = \Phi[\mathbf{M}(X,\theta)]$, and $\Phi(\cdot)$ is concave, a continuous relaxation can be used in a branch-and-bound tree search, with guaranteed convergence to the optimum. This is considered in Sect. 9.2.2.

9.2.1 Exchange Methods

The presentation is for general criteria $\phi(\cdot)$, not necessarily of the form $\phi(X) = \Phi[\mathbf{M}(X,\theta)]$. No concavity property is exploited, and the algorithms can thus also be used to optimize the criteria considered in Chap. 6.

The central idea of exchange algorithms is to exchange *one* design point x_j^k at iteration k with a better one x^* according to $\phi(\cdot)$, which we represent as

$$X^k = (x_1^k,\ldots,\underset{\substack{\updownarrow\\ x^*}}{x_j^k},\ldots,x_N^k)\,.$$

There basically exist two variants.

In Fedorov's algorithm (1972), all N possible exchanges are considered successively, each time starting from X^k. At iteration k, for every $j=1,\ldots,N$ we solve the optimization problem (continuous if $\mathscr{X}$ has a nonempty interior, or grid-search problem if $\mathscr{X}$ is finite) $x_j^* = \max_{x\in\mathscr{X}}\phi([X_{-j}^k,x])$, where $[X_{-j}^k,x]$ denotes the N-point design X^k with x substituted for x_j^k. The exchange finally retained corresponds to j_0 such that $\phi([X_{-j_0}^k,x_{j_0}^*]) = \max_{j=1,\ldots,N}\phi([X_{-j}^k,x_j^*])$, and X^k is updated into $X^{k+1} = [X_{-j_0}^k,x_{j_0}^*]$.

In the simplest version of Mitchell's DETMAX algorithm (1974), at iteration k, we suppose that one additional observation is allowed, at some $x_{N+1}\in\mathscr{X}$. Denote $x_{N+1}^* = \arg\max_{x_{N+1}\in\mathscr{X}}\phi([X^k,X_{N+1}])$ and

$$X^{k+} = (x_1^k,\ldots,x_j^k,\ldots,x_N^k,x_{N+1}^*)$$

the resulting $(N+1)$-point design. We then return to a N-point design by removing one design point. All possible cancellations are considered; that is, we consider the N designs X_{-j}^{k+}, $j=1,\ldots,N$, with x_j^k removed from X^k. The less penalizing one in terms of $\phi(\cdot)$ is retained, which corresponds to j_0 such that $\phi(X_{-j_0}^{k+}) = \arg\max_{j=1,\ldots,N} X_{-j}^{k+}$. Globally, it means that the design point $x_{j_0}^k$ has been replaced by x_{N+1}^*. This corresponds to what is called an excursion of length 1, and longer excursions may also be considered; see also

Galil and Kiefer (1980) and Johnson and Nachtsheim (1983). One iteration of the basic algorithm only requires one optimization with respect to $x_{N+1} \in \mathscr{X}$, followed by N evaluations of the criterion $\phi(\cdot)$. The iterations are thus simpler than with Fedorov's algorithm; however, more iterations are usually required to reach convergence. One may refer to Cook and Nachtsheim (1980) for a comparison between these algorithms.

It must be noticed that dead ends are possible for both types of algorithms: the DETMAX algorithm is stopped when $\phi(X^{k+}_{-j_0}) < \phi(X^k)$ at some iteration; in Fedorov's algorithm, it may be impossible to improve $\phi(\cdot)$ when optimizing with respect to one point x_j^k at a time only. Since there is no guarantee of convergence to an optimal design, it is recommended to restart the algorithms several times using different initializations. Heuristics have been proposed to escape from local optima (see, for instance, Bohachevsky et al. (1986) for a method based on simulated annealing), however without guaranteed convergence to the neighborhood of an optimal design in a finite number of iterations.

9.2.2 Branch and Bound

Here we suppose that $\phi(X) = \Phi[\mathbf{M}(X,\theta)]$, with $\Phi(\cdot)$ concave and differentiable in $\mathbb{M}^>$ (see Chap. 5), and that $\mathbf{M}_\theta(x)$ in (5.1) has rank one, i.e., $\mathbf{M}_\theta(x) = \mathbf{g}_\theta(x)\mathbf{g}_\theta^\top(x)$. We also suppose that $\mathscr{X}$ is finite, $\mathscr{X} = \{x^{(1)},\ldots,x^{(\ell)}\}$. Denote by r_i the number of repetitions of observations at the point $x^{(i)}$. The exact design problem then consists in maximizing

$$\phi(\mathbf{r}/N) = \Phi\left[\frac{1}{N}\sum_{i=1}^{\ell} r_i\, \mathbf{g}_\theta(x^{(i)})\mathbf{g}_\theta^\top(x^{(i)})\right]$$

with respect to $\mathbf{r} = (r_1,\ldots,r_\ell) \in \mathbb{N}^\ell$ subject to

$$0 \le r_i \le N\,,\ i = 1,\ldots,N\,, \tag{9.30}$$

and $\sum_{i=1}^{\ell} r_i = N$. We shall consider generalizations of this problem, obtained by replacing the constraints (9.30) by

$$0 \le \underline{r}_i \le r_i \le \overline{r_i} \le N\,,\ i = 1,\ldots,\ell\,.$$

Notice that we must have $\sum_i \underline{r}_i \le N$ and $\sum_i \overline{r_i} \ge N$ to be consistent with the constraint $\sum_{i=1}^{\ell} r_i = N$. The vectors $\underline{\mathbf{r}} = (\underline{r}_1,\ldots,\underline{r}_\ell)$ and $\overline{\mathbf{r}} = (\overline{r}_1,\ldots,\overline{r}_\ell)$ then define a node $N(\underline{\mathbf{r}},\overline{\mathbf{r}})$ in a tree structure (see Welch 1982), with the following problem attached to it:

$$\begin{cases} \text{maximize } \phi(\mathbf{r}) \text{ with respect to } \mathbf{r} \in \mathbb{N}^\ell \\ \text{subject to } \underline{\mathbf{r}} \le \mathbf{r} \le \overline{\mathbf{r}} \text{ and } \sum_{i=1}^{\ell} r_i = N\,, \end{cases} \tag{9.31}$$

where inequalities for vectors should be interpreted componentwise. The root of the tree, at level 0, corresponds to $\underline{\mathbf{r}} = (0,\ldots,0)$ and $\overline{\mathbf{r}} = (N,\ldots,N)$;

the problem attached to the root is thus the original exact design problem. A node such that $\underline{\mathbf{r}} = \overline{\mathbf{r}}$ is called a leaf (a node with no descendants); solving the attached problem simply means evaluating $\phi(\underline{\mathbf{r}})$. A node $N(\underline{\mathbf{r}}, \overline{\mathbf{r}})$ which is not a leaf can be split into two descendant nodes (of higher level), obtained by partitioning the set $\{\mathbf{r} \in \mathbb{N}^\ell : \underline{\mathbf{r}} \leq \mathbf{r} \leq \overline{\mathbf{r}}\}$ into the two subsets

$$N(\underline{\mathbf{r}}^L(i_0, r_0), \overline{\mathbf{r}}^L(i_0, r_0)) \text{ and } N(\underline{\mathbf{r}}^R(i_0, r_0), \overline{\mathbf{r}}^R(i_0, r_0)), \tag{9.32}$$

respectively such that less than r_0 observations and at least r_0 observations are made at $x^{(i_0)}$. In order to obtain nonempty partitions, we must choose i_0 and r_0 such that

$$\begin{aligned} &\underline{r}_{i_0} + 1 \leq r_0 \leq \overline{r}_{i_0}, \\ &\sum_{i \neq i_0} \underline{r}_i + r_0 \leq N, \\ &N \leq \sum_{i \neq i_0} \overline{r}_i + r_0 - 1. \end{aligned}$$

The lower and upper bounds $\underline{\mathbf{r}}^L, \overline{\mathbf{r}}^L, \underline{\mathbf{r}}^R$, and $\overline{\mathbf{r}}^R$ are then given by

$$\begin{aligned} &\underline{r}_i^L(i_0, r_0) = \underline{r}_i^R(i_0, r_0) = \underline{r}_i \text{ and } \overline{r}_i^L(i_0, r_0) = \overline{r}_i^R(i_0, r_0) = \overline{r}_i \text{ for all } i \neq i_0 \\ &\underline{r}_{i_0}^L(i_0, r_0) = \underline{r}_{i_0}, \ \overline{r}_{i_0}^R(i_0, r_0) = \overline{r}_{i_0} \\ &\overline{r}_{i_0}^L(i_0, r_0) = r_0 - 1, \ \underline{r}_{i_0}^R(i_0, r_0) = r_0. \end{aligned}$$

Starting from the root and applying the branching process above to all nodes encountered, we reach all leafs of the tree. The central idea of branch-and-bound algorithms is that exhaustive inspection of all leafs can be avoided if an upper bound $\overline{B}(\underline{\mathbf{r}}, \overline{\mathbf{r}})$ on the optimal criterion value for the problem attached to the node $N(\underline{\mathbf{r}}, \overline{\mathbf{r}})$ is available. Indeed, let $\underline{B}$ denote a lower bound on the optimal value ϕ^* of $\phi(\mathbf{r})$ in the original problem (i.e., at the root of the tree); then, if $\overline{B}(\underline{\mathbf{r}}, \overline{\mathbf{r}}) < \underline{B}$, the node $N(\underline{\mathbf{r}}, \overline{\mathbf{r}})$ and all its descendant nodes need not be further considered and can thus be discarded. The lower bound $\underline{B}$ of ϕ^* may result from an evaluation at any guessed solution, e.g., obtained from rounding an optimal approximate design, or from the execution of an exchange algorithm. It can be improved, i.e., increased, during the branch-and-bound search by simply making use of the maximum value of $\phi(\cdot)$ encountered.

Four ingredients still have to be specified to define entirely the algorithm:

1. The construction of the upper bounds $\overline{B}(\underline{\mathbf{r}}, \overline{\mathbf{r}})$
2. The rule for choosing the nodes $N(\underline{\mathbf{r}}, \overline{\mathbf{r}})$ to be processed
3. The splitting rule, i.e., the choice of i_0 and r_0 in (9.32)

We consider them successively.

1. *Construction of an upper bound* $\overline{B}(\underline{\mathbf{r}}, \overline{\mathbf{r}})$. Classically, in discrete search problems, the construction relies on a continuous relaxation of the problem

attached to the node. A detailed construction is given in (Welch, 1982) for the case of D-optimum design. The relaxed version of (9.31) is

$$\begin{cases} \text{maximize } \phi(\mathbf{w}) = \Phi\left[\sum_{i=1}^{\ell} w_i\, \mathbf{g}_\theta(x^{(i)})\mathbf{g}_\theta^\top(x^{(i)})\right] \text{ with respect to } \mathbf{w} \in \mathbb{R}^\ell \\ \text{subject to } \mathbf{w} \in \mathscr{P}(\underline{\mathbf{r}}, \overline{\mathbf{r}})\,, \end{cases} \tag{9.33}$$

with

$$\mathscr{P}(\underline{\mathbf{r}}, \overline{\mathbf{r}}) = \{\mathbf{w} \in \mathbb{R}^\ell : \underline{\mathbf{r}}/N \leq \mathbf{w} \leq \overline{\mathbf{r}}/N \text{ and } \sum_{i=1}^{\ell} w_i = 1\}\,.$$

When $\Phi(\cdot)$ is concave and differentiable, we can invoke Lemma 5.20 to obtain an upper bound on $\phi(\mathbf{w})$ for $\mathbf{w} \in \mathscr{P}(\underline{\mathbf{r}}, \overline{\mathbf{r}})$. Take any $\mathbf{w}' \in \mathscr{P}(\underline{\mathbf{r}}, \overline{\mathbf{r}})$, denote by ξ and ξ' the design measures respectively associated with $\mathbf{w}$ and $\mathbf{w}'$ and denote by $\Xi(\underline{\mathbf{r}}, \overline{\mathbf{r}})$ the set of design measures associated with elements of $\mathscr{P}(\underline{\mathbf{r}}, \overline{\mathbf{r}})$. We then have

$$\max_{\mathbf{w} \in \mathscr{P}(\underline{\mathbf{r}}, \overline{\mathbf{r}})} \phi(\mathbf{w}) = \max_{\xi \in \Xi(\underline{\mathbf{r}}, \overline{\mathbf{r}})} \phi(\xi) \leq \phi(\xi') + \sup_{\nu \in \Xi(\underline{\mathbf{r}}, \overline{\mathbf{r}})} F_\phi(\xi'; \nu)\,,$$

which can be rewritten as

$$\max_{\mathbf{w} \in \mathscr{P}(\underline{\mathbf{r}}, \overline{\mathbf{r}})} \phi(\mathbf{w}) \leq \phi(\mathbf{w}') + \max_{\mathbf{w}'' \in \mathscr{P}(\underline{\mathbf{r}}, \overline{\mathbf{r}})} (\mathbf{w}'' - \mathbf{w}')^\top \nabla\phi(\mathbf{w}')\,.$$

The maximization of $(\mathbf{w}'' - \mathbf{w}')^\top \nabla\phi(\mathbf{w}')$ with respect to $\mathbf{w}'' \in \mathscr{P}(\underline{\mathbf{r}}, \overline{\mathbf{r}})$ is an LP problem which can be solved by standard procedures, such as the simplex algorithm, for instance. This yields the bound

$$\overline{B}(\underline{\mathbf{r}}, \overline{\mathbf{r}}) = \phi(\mathbf{w}') + \max_{\mathbf{w}'' \in \mathscr{P}(\underline{\mathbf{r}}, \overline{\mathbf{r}})} (\mathbf{w}'' - \mathbf{w}')^\top \nabla\phi(\mathbf{w}') \tag{9.34}$$

on the optimal criterion value for the problem attached to $N(\underline{\mathbf{r}}, \overline{\mathbf{r}})$, where $\mathbf{w}'$ is any point in $\mathscr{P}(\underline{\mathbf{r}}, \overline{\mathbf{r}})$. This bound can be improved, i.e., decreased, by constructing $\mathbf{w}'$ through a few optimization steps for the problem (9.33), using algorithms from Sect. 9.1. As usual in branch-and-bound algorithms, a compromise between the quality of the bound and the amount of calculations required needs to be made; see, e.g., Minoux (1983, Chap. 7, Vol. 2).

2. *Choice of the node to be processed.* The two classical rules are called "depth first" and "width first." In the depth-first method, the node selected among those awaiting treatment is the one with highest level. When several such nodes exist, the one with highest upper bound $\overline{B}(\underline{\mathbf{r}}, \overline{\mathbf{r}})$ is selected, i.e., the most promising one. This presents the advantage that leafs are reached quickly, so that a lower bound $\underline{B}$ on ϕ^* can be found or improved. This approach is used in (Uciński and Patan, 1982). In the width-first method, the node with highest upper bound $\overline{B}(\underline{\mathbf{r}}, \overline{\mathbf{r}})$ is always selected first. This usually requires more exploration before finding a solution, i.e., reaching a leaf, but the first solution obtained is generally of better quality than with the depth-first method.

Since there is no universal best rule, one usually resorts to heuristics. In (Welch, 1982), the next node to be evaluated after branching is the descendant $N(\underline{\mathbf{r}}^R(i_0, r_0), \overline{\mathbf{r}}^R(i_0, r_0))$, the other descendant $N(\underline{\mathbf{r}}^L(i_0, r_0), \overline{\mathbf{r}}^L(i_0, r_0))$ being placed on the top of a stack of nodes awaiting treatment. When there are no descendants, because the node is a leaf (which may be used to update $\underline{B}$) or because the inequality $\overline{B}(\underline{\mathbf{r}}, \overline{\mathbf{r}}) < \underline{B}$ indicates that it can be discarded, the next node to be processed in taken from the top of the stack.

3. *Splitting rule.* Many variants exist, mainly based on heuristics. Welch (1982) uses $r_0 = \underline{r}_{i_0} + 1$ and selects i_0 such that $\{\nabla\phi(\underline{\mathbf{r}} + \gamma\mathbf{1})\}_{i_0}$ is maximal among all components such that $\underline{r}_i < \overline{r}_i$, with γ a small positive constant and $\mathbf{1}$ the vector of ℓ ones. This is to avoid $\nabla\phi$ being infinite when $\underline{\mathbf{r}}$ has less than p positive values. One might also consider choosing i_0 such that $\overline{r}_i - \underline{r}_i$ is maximal. Another technique consists in choosing i_0 and r_0 such that the upper bounds associated with the descendants differ most. This choice is in some sense the most informative, one of the descendants being a much better candidate that the other for containing the optimum. It is rather computationally intensive, however, which means that a compromise between performance and cost has again to be made.

9.3 Maximin-Optimum Design

9.3.1 Non-Differentiable Optimization of a Design Measure

The methods presented in Sect. 9.1 cannot be used directly when $\phi(\cdot)$ is not differentiable. We shall only consider here the case where $\mathscr{X}$ is discretized into $\mathscr{X}_\ell$ having ℓ elements, so that the design problem corresponds to the maximization of $\phi(\mathbf{w})$ with respect to the vector of weights $\mathbf{w} \in \mathscr{P}_{\ell-1}$; see (9.1). We suppose that $\phi(\cdot)$ is concave and denote by $F_\phi(\xi;\nu)$ the directional derivative of $\phi(\cdot)$ at ξ in the direction ν; see Sect. 5.2.1. One may refer, e.g., to Minoux (1983, Vol. 1, Chap. 4), Shor (1985) and Bonnans et al. (2006, Chaps. 8–10) for a general exposition on methods for non-differentiable problems; see also Sect. 9.5.

We shall pay special attention to the case where $\phi(\xi) = \phi_{MmO}(\xi) = \min_{\theta\in\Theta}\phi(\xi;\theta)$, with $\phi(\cdot;\theta)$ concave and differentiable for all $\theta \in \Theta$; see Sect. 8.2. In practical situations, it is reasonable to suppose that Θ is a finite set. The grid method in (Dem'yanov and Malozemov, 1974, Chap. 6), based on the construction of a sequence of discretized sets $\Theta_j \subset \Theta_{j+1} \subset \cdots \subset \Theta$ satisfying $\lim_{j\to\infty}\max_{\theta\in\Theta}\min_{\theta'\in\Theta_j}\|\theta - \theta'\| = 0$, can be used when Θ is a compact subset of $\mathbb{R}^p$. The relaxation method of Sect. 9.3.2 allows us to iteratively add points to a finite Θ; see also Sect. 9.5.3 for the related method of cutting planes. The directional derivative $F_\phi(\xi;\nu)$ of $\phi(\cdot)$ at ξ in the direction ν is given by

$$F_\phi(\xi;\nu) = \min_{\theta\in\Theta(\xi)} F_{\phi_\theta}(\xi;\nu) \tag{9.35}$$

where

$$\Theta(\xi) = \{\theta \in \Theta : \phi(\xi;\theta) = \phi(\xi)\} \tag{9.36}$$

and $F_{\phi_\theta}(\xi;\nu)$ is the directional derivative of $\phi(\cdot;\theta)$, see Sect. 8.2.

A Multi-Vertex Direction Method

In general, the steepest-ascent direction ν^* maximizing $F_\phi(\xi;\nu)$ given by (9.35) is not equal to the delta measure at some x; see Remark 5.26-(*ii*). Moreover, it may happen that no such vertex direction is a direction of increase for $\phi(\cdot)$; see Example 9.8. This precludes the direct use of one of the vertex-direction algorithms of Sect. 9.1.1.

A natural (but rather unfortunate) idea is to nevertheless use an algorithm based on the multi-vertex steepest-ascent direction, that is, take

$$\xi_{k+1} = (1-\alpha_k)\xi_k + \alpha_k \nu_k^+ \tag{9.37}$$

with

$$\begin{aligned} \nu_k^+ = \arg\max_{\nu\in\Xi} F_\phi(\xi_k;\nu) &= \arg\max_{\nu\in\Xi} \min_{\theta\in\Theta(\xi_k)} F_{\phi_\theta}(\xi_k;\nu) \\ &= \arg\max_{\nu\in\Xi} \min_{\theta\in\Theta(\xi_k)} \sum_{i=1}^{\ell} w_i\, F_{\phi_\theta}(\xi_k, x^{(i)}) \,, \end{aligned}$$

where $w_i = \nu(x^{(i)})$ for all i, $F_{\phi_\theta}(\xi_k, x^{(i)}) = F_{\phi_\theta}(\xi_k;\delta_{x^{(i)}})$ with δ_x the delta measure at x and where we used the linearity in ν of $F_{\phi_\theta}(\xi_k;\nu)$. The stepsize α_k is obtained by maximizing $\phi[(1-\alpha)\xi_k + \alpha\nu_k^+]$ with respect to $\alpha \in [0,1]$. This line search corresponds to a non-differentiable but concave maximization problem for which a Golden-Section or Fibonacci type algorithm can be used; see also den Boeff and den Hertog (2007). Written in terms of an iteration on the vector of weights $\mathbf{w}^k$, with $w_i^k = \xi_k(x^{(i)})$ for all i, (9.37) becomes

$$\mathbf{w}^{k+1} = (1-\alpha_k)\mathbf{w}^k + \alpha_k \mathbf{w}_+^k \tag{9.38}$$

where

$$\mathbf{w}_+^k = \arg\max_{\mathbf{w}\in\mathscr{P}_{\ell-1}} \min_{\theta\in\Theta(\xi_k)} \nabla^\top \phi_\theta(\mathbf{w}^k)(\mathbf{w}-\mathbf{w}^k) \tag{9.39}$$

and $\nabla\phi_\theta(\mathbf{w})$ is the gradient of $\phi(\cdot;\theta)$ at $\mathbf{w}$, $\nabla\phi_\theta(\mathbf{w}) = \partial\phi(\mathbf{w}';\theta)/\partial\mathbf{w}'\big|_{\mathbf{w}'=\mathbf{w}}$.

Remark 9.7.

(*i*) $\mathbf{w}_+^k$ given by (9.39) is easily determined when Θ is a finite set since it corresponds to the solution of an (finite dimensional) LP problem. It may differ in general from

$$\mathbf{w}_+^k = \arg\max_{\mathbf{w}\in\mathscr{P}_{\ell-1}} \min_{\theta\in\Theta(\xi_k)} \nabla^\top \phi_\theta(\mathbf{w}^k)\mathbf{w}\,.$$

However, from Remark 5.26-(*iii*), they both coincide when $\phi_\theta(\xi) = \Phi[\mathbf{M}(\xi,\theta)]$ with $\Phi(\cdot)$ isotonic and differentiable.

(*ii*) The calculation of $F_\phi(\xi_k;\nu_k^+) = \min_{\theta\in\Theta(\xi_k)} \nabla^\top \phi_\theta(\mathbf{w}^k)(\mathbf{w}_+^k - \mathbf{w}^k)$ can be used to form a stopping criterion for the algorithm since Lemma 5.20 indicates that the maximum value ϕ^* of $\phi(\cdot)$ over Ξ satisfies $\phi^* \leq \phi(\xi_k) + F_\phi(\xi_k;\nu_k^+)$. □

Since $\{\phi(\xi_k)\}$ forms an increasing sequence bounded by ϕ^*, it converges to some ϕ_∞. However, a direct implementation of this method is not recommended since the sequence it generates may be such that $\phi_\infty < \phi^*$. This is illustrated by the following simple example, rich enough to reveal the difficulties caused by non-differentiable criteria. It will be used as a test-case for the algorithms to be presented for non-differentiable optimization.

Example 9.8. The example is derived from Example 9.1 in (Bonnans et al., 2006). We take $\ell = 3$ and consider a new orthonormal system of coordinates in $\mathbb{R}^3$ defined by the transformation $\mathbf{w} \in \mathbb{R}^3 \longrightarrow \mathbf{z} = \mathbf{T}\mathbf{w}$, with

$$\mathbf{T} = \begin{pmatrix} -\frac{\sqrt{2}}{2} & \frac{\sqrt{2}}{2} & 0 \\ -\frac{1}{\sqrt{6}} & -\frac{1}{\sqrt{6}} & \frac{\sqrt{2}}{\sqrt{3}} \\ \frac{1}{\sqrt{3}} & \frac{1}{\sqrt{3}} & \frac{1}{\sqrt{3}} \end{pmatrix}.$$

Then, $\mathbf{t} = (z_1\,,\ z_2)^\top = \mathbf{T}_2\mathbf{w}$, with

$$\mathbf{T}_2 = \begin{pmatrix} -\frac{\sqrt{2}}{2} & \frac{\sqrt{2}}{2} & 0 \\ -\frac{1}{\sqrt{6}} & -\frac{1}{\sqrt{6}} & \frac{\sqrt{2}}{\sqrt{3}} \end{pmatrix},$$

defines coordinates on the simplex $\mathscr{P}_2$ given by (9.1). The criterion $\phi(\cdot)$ is defined by $\phi(\mathbf{w}) = \min\{\phi_1(\mathbf{w}), \phi_2(\mathbf{w}), \phi_3(\mathbf{w}), \phi_4(\mathbf{w})\}$ with $\phi_i(\mathbf{w}) = \phi(\mathbf{w};\theta^{(i)}) = \mathbf{w}^\top\mathbf{T}_2^\top\theta^{(i)}$, $i = 1,\ldots,4$, and $\theta^{(1)} = (-5\,,\ -2)^\top$, $\theta^{(2)} = (-2\,,\ -3)^\top$, $\theta^{(3)} = (5\,,\ -2)^\top$, $\theta^{(4)} = (2\,,\ -3)^\top$.

Figure 9.5 presents the level sets of $\phi(\cdot)$ (dotted lines) on $\mathscr{P}_2$, together with the kinks (dash-dotted lines), i.e., the lines on $\mathscr{P}_2$ on which $\phi(\cdot)$ is not differentiable: when $\phi_1(\mathbf{w}) = \phi_2(\mathbf{w})$, or $\phi_1(\mathbf{w}) = \phi_3(\mathbf{w})$, or $\phi_2(\mathbf{w}) = \phi_4(\mathbf{w})$, or $\phi_3(\mathbf{w}) = \phi_4(\mathbf{w})$. The corners A, B, and C respectively correspond to the vertices $\mathbf{e}_1 = (1\,,\ 0\,,\ 0)^\top$, $\mathbf{e}_2 = (0\,,\ 1\,,\ 0)^\top$, and $\mathbf{e}_3 = (0\,,\ 0\,,\ 1)^\top$; the point O at the center of the triangle ABC corresponds to $\mathbf{w} = (1/3\,,\ 1/3\,,\ 1/3)^\top$ (and $\mathbf{t} = \mathbf{0}$ in new coordinates on $\mathscr{P}_2$). Note that all vertex directions originated from O are directions along which $\phi(\cdot)$ decreases, although the maximum value of $\phi(\cdot)$ in $\mathscr{P}_2$ is obtained at the midpoint between A and B, i.e., $\mathbf{w}^* = (1/2\,,\ 1/2\,,\ 0)^\top$, marked by a circle in Fig. 9.5.

Consider the kink defined by $\phi_3(\mathbf{w}) = \phi_4(\mathbf{w})$. Its equation on $\mathscr{P}_2$ is $t_2 = -3t_1$. One can easily check that for any $\mathbf{w}$ corresponding to a point on this line, the value $\mathbf{w}_+$ maximizing $\min_{i=3,4} \nabla^\top\phi_i(\mathbf{w})(\mathbf{w}' - \mathbf{w})$ with respect to $\mathbf{w}' \in \mathscr{P}_2$ is given by $\mathbf{e}_2$. Note that $\nabla^\top\phi_3(\mathbf{w})\mathbf{w} = \nabla^\top\phi_4(\mathbf{w})\mathbf{w}$ when $\mathbf{w}$ is such that $\phi_3(\mathbf{w}) = \phi_4(\mathbf{w}) = \phi(\mathbf{w})$; see Remark 9.7-(*i*). Similarly, the steepest-ascent direction at a $\mathbf{w}$ on the kink defined by $\phi_1(\mathbf{w}) = \phi_2(\mathbf{w})$ (the line

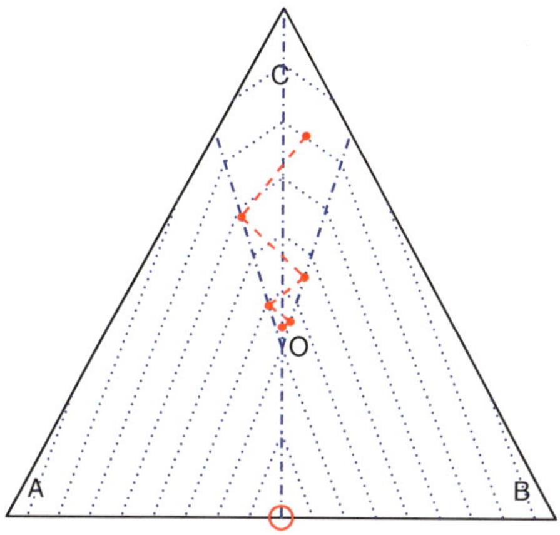

Fig. 9.5. Level sets (*dotted lines*) and kinks (*dash-dotted lines*) on $\mathscr{P}_2$ for the non-differentiable problem in Example 9.8, together with a sequence of iterates generated by the steepest-ascent method (9.38), (9.39) with optimal stepsize

$t_2 = 3t_1$ in new coordinates on $\mathscr{P}_2$) corresponds to $\mathbf{e}_1$. A sequence of iterates $\mathbf{w}^k$ for the multi-vertex direction algorithm with optimal stepsize, initialized at $\mathbf{w}^0 = (1/12\,,\ 2/12\,,\ 3/4)^\top$, is plotted in Fig. 9.5. The sequence converges (slowly) to $\mathbf{w}^\infty = (1/3\,,\ 1/3\,,\ 1/3)^\top$ which is not optimal. □

A reason for the failure of the multi-vertex direction algorithm (9.38), (9.39) is the fact that the set $\Theta(\xi)$ defined by (9.36) is too small. Consider instead

$$\Theta_\epsilon(\xi) = \{\theta \in \Theta : \phi(\xi;\theta) \leq \phi(\xi) + \epsilon\}\,,\ \epsilon > 0\,,$$

and use at iteration k

$$\nu_k^+ = \arg\max_{\nu\in\Xi}\ \min_{\theta\in\Theta_\epsilon(\xi_k)} F_{\phi_\theta}(\xi_k;\nu)\,.$$

The substitution of $\Theta_\epsilon(\xi)$ for $\Theta(\xi)$ in the necessary-and-sufficient condition for optimality

$$\max_{\nu\in\Xi}\ \min_{\theta\in\Theta(\xi)} F_{\phi_\theta}(\xi;\nu) = 0 \Longleftrightarrow \phi(\xi) = \phi^*\,,$$

see Theorem 5.21, is not without consequences. Indeed, we have for any $\theta \in \Theta_\epsilon(\xi)$ and any optimal design ξ^* such that $\phi(\xi^*) = \phi^*$,

$$\phi^* \leq \phi(\xi^*;\theta) \leq \phi(\xi;\theta) + F_{\phi_\theta}(\xi;\xi^*) \leq \phi(\xi) + \epsilon + F_{\phi_\theta}(\xi;\xi^*)\,.$$

Therefore,

$$\phi^* \leq \phi(\xi) + \epsilon + \min_{\theta\in\Theta_\epsilon(\xi)} F_{\phi_\theta}(\xi;\xi^*) \leq \phi(\xi) + \epsilon + \max_{\nu\in\Xi}\ \min_{\theta\in\Theta_\epsilon(\xi)} F_{\phi_\theta}(\xi;\nu)\,,$$

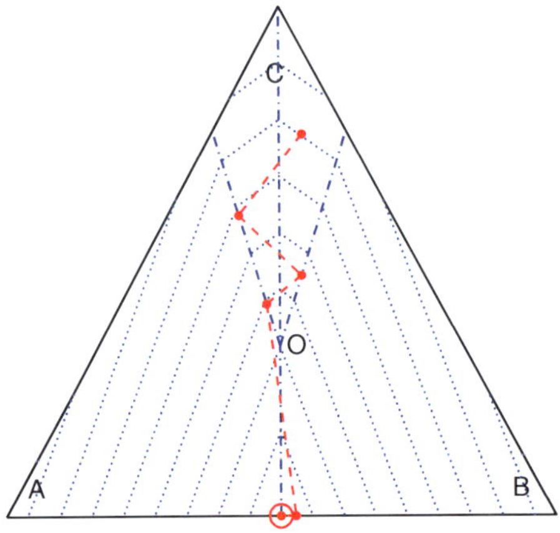

Fig. 9.6. Same as Fig. 9.5 but for iterations using $\Theta_\epsilon(\xi_k)$ in (9.39) with $\epsilon = 0.2$

so that $\max_{\nu\in\Xi}\min_{\theta\in\Theta_\epsilon(\xi)} F_{\phi_\theta}(\xi;\nu) = 0$ only implies that $\phi(\xi) \geq \phi^* - \epsilon$. It is thus recommended to take $\epsilon = \epsilon_k$ decreasing with k and use iterations of the form (9.38), (9.39) with $\Theta_{\epsilon_k}(\xi_k)$ substituted for $\Theta(\xi_k)$ in (9.39). This can be related to bundle methods used for general non-differentiable problems; see Lemaréchal et al. (1995); Bonnans et al. (2006, Chap. 10).

Example 9.8 (continued). We consider the same problem as in Example 9.8. Figure 9.6 presents the evolution in $\mathscr{P}_2$ of iterates $\mathbf{w}^k$ generated by the multi-vertex steepest-ascent method (9.38), (9.39) with optimal stepsize, initialized at $\mathbf{w}^0 = (1/12\,,\, 2/12\,,\, 3/4)^\top$, when $\Theta_{0.2}(\xi_k)$ is substituted for $\Theta(\xi_k)$ in (9.39); compare with Fig. 9.5. □

Subgradient Projection

Subgradient-projection methods form a direct extension of the method of Sect. 9.1.2. Denote by $\tilde{\nabla}\phi(\mathbf{w})$ a subgradient of $\phi(\cdot)$ at $\mathbf{w} \in \mathbb{R}^\ell$; see Appendix A for definitions and properties of subdifferentials and subgradients. We consider in particular the case of maximin-optimum design where $\phi(\mathbf{w}) = \phi_{MmO}(\mathbf{w}) = \min_{\theta\in\Theta}\phi(\mathbf{w};\theta)$, with $\phi(\cdot;\theta)$ concave and differentiable for all $\theta \in \Theta$ and Θ finite. A subgradient $\tilde{\nabla}\phi(\mathbf{w})$ is then directly available, and we may take $\tilde{\nabla}\phi(\mathbf{w}) = \nabla\phi_{\theta^*}(\mathbf{w})$, the gradient of $\phi_{\theta^*}(\cdot) = \phi(\cdot;\theta^*)$, with θ^* any point in

$$\Theta(\mathbf{w}) = \{\theta \in \Theta : \phi(\mathbf{w};\theta) = \phi(\mathbf{w})\}\,.$$

Notice that the whole subdifferential $\partial\phi(\mathbf{w})$ is available:

$$\partial\phi(\mathbf{w}) = \left\{ \int_\Theta \tilde{\nabla}\phi_\theta(\mathbf{w})\,\mu(\mathrm{d}\theta) : \mu \in \mathscr{M}_\mathbf{w} \right\}, \tag{9.40}$$

with $\mathscr{M}_\mathbf{w}$ the set of probability measures on $\Theta(\mathbf{w})$.

A direct substitution of a subgradient $\tilde{\nabla}\phi(\mathbf{w}^k)$ for $\nabla\phi(\mathbf{w}^k)$ in the gradient-projection method of Sect. 9.1.2 is bound to failure since $\tilde{\nabla}\phi(\mathbf{w}^k)$ does not necessarily correspond to a direction of increase for $\phi(\cdot)$. The steepest-ascent direction, given by

$$\bar{\nabla}\phi(\mathbf{w}^k) = \arg \min_{\mathbf{z}\in\partial\phi(\mathbf{w}^k)} \|\mathbf{z}\| \,, \tag{9.41}$$

see Dem'yanov and Malozemov (1974, Chap. 3), can be used instead. Note that the determination of $\bar{\nabla}\phi(\mathbf{w}^k)$ corresponds to the solution of a finite dimensional QP problem when Θ is finite. However, the method does not necessarily converge to the optimum, for the same reasons as for the multi-vertex steepest-ascent method presented above: the set $\Theta(\mathbf{w})$ is too small, and instead of $\partial\phi(\mathbf{w}^k)$, we should consider

$$\partial_\epsilon\phi(\mathbf{w}) = \left\{ \int_\Theta \tilde{\nabla}\phi_\theta(\mathbf{w})\,\mu(\mathrm{d}\theta) : \mu \in \mathscr{M}_{\mathbf{w},\epsilon} \right\} \tag{9.42}$$

in the construction of the steepest-ascent direction (9.41), with $\mathscr{M}_{\mathbf{w},\epsilon}$ the set of probability measures on

$$\Theta_\epsilon(\mathbf{w}) = \{\theta \in \Theta : \phi(\mathbf{w};\theta) \le \phi(\mathbf{w}) + \epsilon\} \,,\ \epsilon > 0 \,.$$

One may take ϵ decreasing with k; see the method of successive approximations in (Dem'yanov and Malozemov, 1974, Chap. 3). A definition of $\partial\phi(\mathbf{w})$ more general than (9.40) is

$$\partial\phi(\mathbf{w}) = \{\mathbf{z} \in \mathbb{R}^\ell : \phi(\mathbf{w}') \le \phi(\mathbf{w}) + \mathbf{z}^\top(\mathbf{w}' - \mathbf{w}) \text{ for all } \mathbf{w}' \in \mathbb{R}^\ell\} \,,$$

and we may define the ϵ-subdifferential $\partial\phi_\epsilon(\mathbf{w})$ accordingly by

$$\partial_\epsilon\phi(\mathbf{w}) = \{\mathbf{z} \in \mathbb{R}^\ell : \phi(\mathbf{w}') \le \phi(\mathbf{w}) + \mathbf{z}^\top(\mathbf{w}' - \mathbf{w}) + \epsilon \text{ for all } \mathbf{w}' \in \mathbb{R}^\ell\} \,.$$

Example 9.8 (continued). We consider the same problem as in Example 9.8. Figure 9.7 presents the evolution in $\mathscr{P}_2$ of iterates $\mathbf{w}^k$ generated by the projected-subgradient algorithm when using the steepest-ascent direction (9.41), with optimal stepsize, initialized at $\mathbf{w}^0 = (1/12\,,\ 2/12\,,\ 3/4)^\top$. The sequence converges (slowly) to $\mathbf{w}^\infty = (1/3\,,\ 1/3\,,\ 1/3)^\top$ which is not optimal; the behavior is similar to that in Fig. 9.5.

Figure 9.8 corresponds to the situation where (9.42) is substituted for $\partial\phi(\mathbf{w}^k)$ in (9.41). The algorithm now converges to the optimum in four iterations. □

One may refer, e.g., to Lemaréchal et al. (1995) and Bonnans et al. (2006, Chap. 10) for an exposition on more sophisticated methods, called bundle methods, which do not require the knowledge of the whole ϵ-subdifferential $\partial_\epsilon\phi(\mathbf{w})$. Methods that only require the knowledge of one arbitrary subgradient $\tilde{\nabla}\phi(\mathbf{w})$ instead of the whole subdifferential are called black-box methods. The (projected) subgradient algorithm presented below is probably the

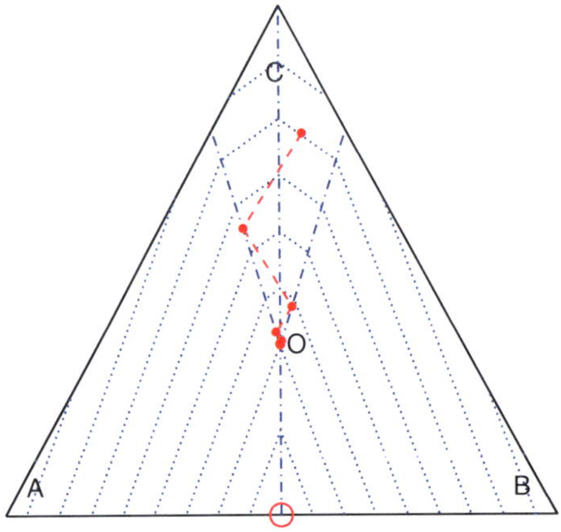

Fig. 9.7. Same as Fig. 9.5 but for the projected-subgradient algorithm using the steepest-ascent direction (9.41)

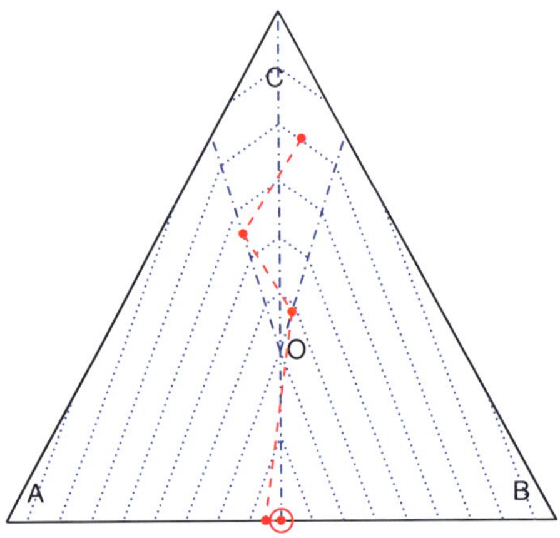

Fig. 9.8. Same as Fig. 9.5 but for the projected-subgradient algorithm using the steepest-ascent direction (9.41) with the ϵ-subdifferential (9.42) ($\epsilon = 0.2$)

simplest among them; two other black-box methods (ellipsoid and cutting planes) will be presented in Sect. 9.5; see especially Sect. 9.5.3 for a particular bundle method called the level method.

Instead of trying to construct an ascent direction from the subdifferentials $\partial\phi(\mathbf{w})$ or $\partial_\epsilon\phi(\mathbf{w})$ and then optimize the stepsize in that direction, we may abandon the objective of ensuring a monotonic increase of $\phi(\cdot)$ and directly use subgradient directions $\tilde{\nabla}\phi(\mathbf{w})$ with a predefined stepsize sequence $\{\alpha_k\}$. We may then simply substitute a subgradient $\tilde{\nabla}\phi(\mathbf{w}^k)$ for $\nabla\phi(\mathbf{w}^k)$ in (9.17) and choose a sequence $\{\alpha_k\}$ that satisfies

$$\alpha_k \geq 0\,,\ \alpha_k \to 0\,,\ \sum_{k=0}^{\infty} \alpha_k = \infty\,. \tag{9.43}$$

Other choices can be made for the stepsize α_k; see, for instance, Goffin and Kiwiel (1999) for a target-level approach. Since $\mathbf{w}^{k+1}$ is the orthogonal projection of $\mathbf{w}_+^k = \mathbf{w}^k + \alpha_k \tilde{\nabla}\phi(\mathbf{w}^k)$ onto $\mathscr{P}_{\ell-1}$, we have $\|\mathbf{w}^{k+1} - \mathbf{w}\| \leq \|\mathbf{w}_+^k - \mathbf{w}\|$ for any $\mathbf{w} \in \mathscr{P}_{\ell-1}$.

Suppose that $\phi(\cdot)$ is a positive criterion satisfying $\sup\{\|\tilde{\nabla}\phi(\mathbf{w})\| : \mathbf{w} \in \mathscr{P}_{\ell-1} \text{ and } \phi(\mathbf{w}) \geq \Delta\} = C(\Delta) < \infty$. Then, assuming that $\liminf_{k\to\infty} \phi(\mathbf{w}^k) > \epsilon$ for some $\epsilon > 0$, standard arguments, reproduced below, show that the algorithm satisfies $\limsup_{k\to\infty} \phi(\mathbf{w}^k) = \phi^* = \max_{\mathbf{w}\in\mathscr{P}_{\ell-1}} \phi(\mathbf{w})$ when α_k satisfies (9.43); see Correa and Lemaréchal (1993). Indeed, the subgradient inequality

$$\phi(\mathbf{w}') \leq \phi(\mathbf{w}) + \tilde{\nabla}^\top \phi(\mathbf{w})(\mathbf{w}' - \mathbf{w}) \text{ for all } \mathbf{w}'\,, \tag{9.44}$$

see (A.1), and the inequality

$$\begin{aligned} \|\mathbf{w}^{k+1} - \mathbf{w}\|^2 &\leq \|\mathbf{w}_+^k - \mathbf{w}\|^2 \\ &= \|\mathbf{w}^k - \mathbf{w}\|^2 + 2\alpha_k \tilde{\nabla}^\top \phi(\mathbf{w}^k)(\mathbf{w}^k - \mathbf{w}) + \alpha_k^2 \|\tilde{\nabla}\phi(\mathbf{w}^k)\|^2 \end{aligned}$$

imply that

$$\|\mathbf{w}^{k+1} - \mathbf{w}\|^2 \leq \|\mathbf{w}^k - \mathbf{w}\|^2 + 2\alpha_k[\phi(\mathbf{w}^k) - \phi(\mathbf{w})] + \alpha_k^2 \|\tilde{\nabla}\phi(\mathbf{w}^k)\|^2 \tag{9.45}$$

for any $\mathbf{w} \in \mathscr{P}_{\ell-1}$. Suppose that there exist $\Delta > 0$ and k_0 such that $\phi(\mathbf{w}^k) < \phi^* - \Delta$ for all $k \geq k_0$. We may assume that k_0 is large enough to ensure that $\|\tilde{\nabla}\phi(\mathbf{w}^k)\| < C(\epsilon)$ and $\alpha_k < \Delta/C^2(\epsilon)$ for any $k \geq k_0$, which implies that

$$\|\mathbf{w}^{k+1} - \mathbf{w}^*\|^2 \leq \|\mathbf{w}^k - \mathbf{w}^*\|^2 - 2\alpha_k \Delta + \alpha_k \Delta \text{ for all } k \geq k_0\,,$$

where $\mathbf{w}^* \in \mathscr{P}_{\ell-1}$ is such that $\phi(\mathbf{w}^*) = \phi^*$. We obtain by summation

$$0 \leq \|\mathbf{w}^k - \mathbf{w}^*\|^2 \leq \|\mathbf{w}^{k_0} - \mathbf{w}^*\|^2 - \Delta \sum_{j=k_0}^{k-1} \alpha_j\,,\ k > k_0\,,$$

which contradicts (9.43). Therefore, for any $\Delta > 0$ and any k_0, there exists some $k > k_0$ such that $\phi(\mathbf{w}^k) \geq \phi^* - \Delta$, that is, $\limsup_{k\to\infty} \phi(\mathbf{w}^k) = \phi^*$.

If we suppose, moreover, that $\{\alpha_k\}$ is square summable, i.e., $\sum_{k=0}^{\infty} \alpha_k^2 < \infty$, we obtain that $\mathbf{w}^k$ tends to a maximum point of $\phi(\cdot)$ as $k \to \infty$; see Correa and Lemaréchal (1993). The proof goes as follows. Since $\mathbf{w}^k \in \mathscr{P}_{\ell-1}$ for all k, $\{\mathbf{w}^k\}$ has a cluster point in $\mathscr{P}_{\ell-1}$. The continuity of $\phi(\cdot)$ and $\limsup_{k\to\infty} \phi(\mathbf{w}^k) = \phi^*$ imply that such a cluster point maximizes $\phi(\cdot)$; denote it by $\mathbf{w}^*$. Applying (9.45) to $\mathbf{w}^*$ we obtain, for k large enough,

$$\|\mathbf{w}^{k+1} - \mathbf{w}^*\|^2 \leq \|\mathbf{w}^k - \mathbf{w}^*\|^2 + \alpha_k^2\, C^2(\epsilon)\,.$$

Take any $\Delta > 0$ and choose k_0 large enough so that $\|\mathbf{w}^{k_0} - \mathbf{w}^*\|^2 < \Delta/2$ and $C^2(\epsilon)\sum_{k_0}^{\infty} \alpha_k^2 < \Delta/2$. Then, by summation,

$$\|\mathbf{w}^{k+1} - \mathbf{w}^*\|^2 \leq \|\mathbf{w}^{k_0} - \mathbf{w}^*\|^2 + C^2(\epsilon) \sum_{j=k_0}^{k} \alpha_j^2 < \Delta$$

for all $k \geq k_0$.

To summarize, we have thus obtained a dichotomous property similar to that in Sect. 9.1.1: either there exists a subsequence $\{\mathbf{w}^{k_n}\}$ such that $\phi(\mathbf{w}^{k_n}) \to 0$, or $\limsup_{k\to\infty} \phi(\mathbf{w}^k) = \phi^*$ when $\{\alpha_k\}$ satisfies (9.43) and, moreover, $\mathbf{w}^k$ tends to some maximizer $\mathbf{w}^*$ of $\phi(\cdot)$ when $\{\alpha_k\}$ is square summable in addition to (9.43).

Remark 9.9.

(*i*) The diverging case is eliminated when we can guarantee that $\|\tilde{\nabla}\phi(\mathbf{w})\| < C$ for all $\mathbf{w} \in \mathscr{P}_{\ell-1}$. Suppose that $\phi(\xi) = \min_{\theta\in\Theta} \Phi[\mathbf{M}(\xi,\theta)]$ with $\Phi(\cdot)$ one of the criteria of Sect. 5.1. This condition can be obtained by a regularization of the criterion (see Remark 9.3-(*ii*) and (*iii*)) or by a truncation of $\mathscr{P}_{\ell-1}$ that avoids singular designs. Indeed, we can force all designs to be supported on p points at least by projecting $\mathbf{w}_+^k$ onto the set obtained by truncating all $(p-1)$-dimensional faces from the simplex $\mathscr{P}_{\ell-1}$, i.e., by projecting $\mathbf{w}_+^k$ onto

$$\mathscr{P}'_{\ell-1} = \{\mathbf{w} \in \mathbb{R}^\ell : w_i \geq 0\,,\ \sum_{i=1}^{\ell} w_i = 1 \text{ and}$$
$$w_{i_1} + \cdots + w_{i_m} \leq 1 - \Delta \ \text{ for all } i_1 < i_2 < \cdots < i_m \text{ and } m < p\}$$

for some small $\Delta > 0$. For some particular criteria $\phi(\cdot)$, the diverging case can also be eliminated following arguments similar to those in (Wu and Wynn, 1978).

(*ii*) One can check that in order to obtain $\limsup_{k\to\infty} \phi(\mathbf{w}^k) = \phi^*$, it is enough to have $\sum_{k=0}^{\infty} \alpha_k = \infty$ and $\alpha_k \|\tilde{\nabla}\phi(\mathbf{w}^k)\|^2 \to 0$; see (9.45). When normalized stepsizes $\alpha_k = \gamma_k/\|\tilde{\nabla}\phi(\mathbf{w}^k)\|$ are used, it is thus enough to have $\gamma_k \|\tilde{\nabla}\phi(\mathbf{w}^k)\| \to 0$ and $\sum_{k=0}^{\infty} \gamma_k/\|\tilde{\nabla}\phi(\mathbf{w}^k)\| = \infty$. To obtain additionally that $\mathbf{w}^k$ tend to a maximizer of $\phi(\cdot)$ it is enough to have $\sum_{k=0}^{\infty} \alpha_k^2 \|\tilde{\nabla}\phi(\mathbf{w}^k)\|^2 = \sum_{k=0}^{\infty} \gamma_k^2 < \infty$.

(*iii*) Additional constraints that define a convex set $\mathscr{P}' \subset \mathscr{P}_{\ell-1}$ (see Sects. 5.1.9 and 5.1.10) can be taken into account by considering the projection of $\mathbf{w}_+^k = \mathbf{w}^k + \alpha_k \tilde{\nabla}\phi(\mathbf{w}^k)$ onto $\mathscr{P}'$ (see Remark 9.4). □

Example 9.8 (continued). We consider the same problem as in Example 9.8. Figure 9.9 presents the evolution in $\mathscr{P}_2$ of iterates $\mathbf{w}^k$ generated by the projected-subgradient algorithm with a predefined stepsize sequence satisfying (9.43). Notice the oscillations of the path followed by the iterates and the slow convergence to the optimum, a behavior similar to that in Fig. 9.2. □

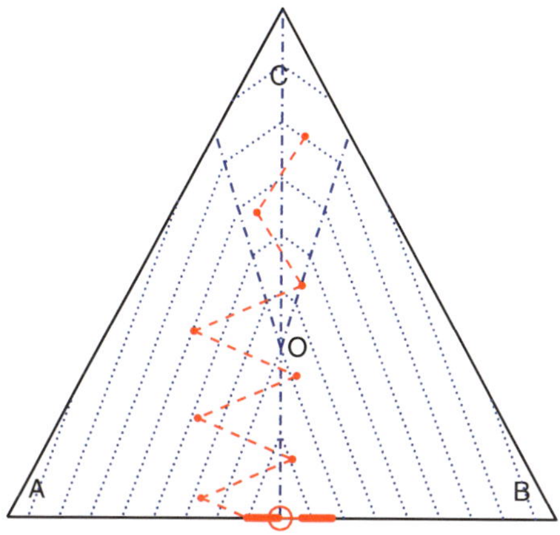

Fig. 9.9. Same as Fig. 9.5 but for the projected-subgradient algorithm using the stepsize sequence $\alpha_k = 1/(20+k)$

9.3.2 Maximin-Optimum Exact Design

Consider the maximization of the criterion

$$\phi_{MmO}(X) = \min_{\theta \in \Theta} \phi(X;\theta)\,,$$

with $X = (x_1, \ldots, x_N)$ an exact design of given size N. The algorithms of Sect. 9.2 can be used directly for this maximin-optimum design problem when the admissible parameter space Θ is finite, $\Theta = \{\theta^{(1)}, \ldots, \theta^{(M)}\}$, since the calculation of $\min_{\theta\in\Theta} \phi(X;\theta)$ simply amounts to M evaluations $\phi(X;\theta^{(i)})$, $i = 1, \ldots, M$. The situation is much more complicated when Θ is a compact subset of $\mathbb{R}^p$ with nonempty interior. A relaxation algorithm is proposed for this case in (Pronzato and Walter, 1988), based on a method by Shimizu and Aiyoshi (1980). It amounts to solving a series a maximin problems for an imbedded sequence of finite sets $\Theta^1 \subset \Theta^2 \subset \ldots$

0. Choose $\epsilon > 0$ and $\theta^{(1)} \in \Theta$; set $\Theta^1 = \{\theta^{(1)}\}$ and $i = 1$.
1. Compute $X_i^* = \arg\max_{X\in\mathscr{X}^N} \min_{\theta\in\Theta^i} \phi(X;\theta)$.
2. Compute $\theta^{(i+1)} = \arg\min_{\theta\in\Theta} \phi(X_i^*;\theta)$; if

$$\min_{\theta\in\Theta^i} \phi(X_i^*;\theta) - \phi(X_i^*;\theta^{(i+1)}) < \epsilon\,,$$

   stop, take X_i^* as an ϵ-optimal design for $\phi_{MmO}(\cdot)$; otherwise, set $\Theta^{i+1} = \Theta^i \cup \{\theta^{(i+1)}\}$, $i \leftarrow i+1$ and return to step 1.

Remark 9.10. A similar idea can be used to maximize a function of weights $\mathbf{w}$ allocated to given support points, $\phi(\mathbf{w}) = \phi_{MmO}(\mathbf{w}) = \min_{\theta\in\Theta} \phi(\mathbf{w};\theta)$, with $\phi(\cdot;\theta)$ concave and differentiable for all $\theta \in \Theta$, as in Sect. 9.3.1. Each passage through step 1 then requires the maximization of a non-differentiable function of the weights $\mathbf{w}$. We may also combine the augmentation of Θ^i with the search for the optimal $\mathbf{w}$, as in the method of cutting planes of Sect. 9.5.3. □

9.4 Average-Optimum Design

9.4.1 Average-Optimal Design Measures and Stochastic Approximation

The methods presented in Sects. 9.1.1 and 9.1.2 can be used directly when

$$\phi(\xi) = \phi_{AO}(\xi) = \int_{\Theta} \phi(\xi;\theta)\, \mu(\mathrm{d}\theta) \tag{9.46}$$

and all $\phi(\cdot;\theta)$, $\theta \in \Theta$, are concave and differentiable functions; see Sect. 8.1. Indeed, $\phi(\cdot)$ is then concave and differentiable; its directional derivative at ξ in the direction ν is

$$F_{\phi}(\xi;\nu) = \int_{\Theta} F_{\phi_\theta}(\xi;\nu)\, \mu(\mathrm{d}\theta)\,,$$

with $F_{\phi_\theta}(\xi;\nu)$ the directional derivative of $\phi(\cdot;\theta)$. Similarly, for ξ a discrete measure with weights $\mathbf{w}$, the gradient of $\phi(\cdot)$ at $\mathbf{w}$ is

$$\nabla\phi(\mathbf{w}) = \int_{\Theta} \nabla\phi(\mathbf{w};\theta)\, \mu(\mathrm{d}\theta)\,,$$

with $\nabla\phi(\mathbf{w};\theta)$ the gradient of $\phi(\cdot;\theta)$. However, each evaluation of $\phi(\cdot)$, of its directional derivative, or gradient, requires the computation of an expected value for the probability measure $\mu(\cdot)$ on Θ. This does not raise any special difficulty when the integral in (9.46) is reduced to a finite sum, i.e., when $\mu(\cdot)$ is a discrete measure on a finite set, but is computationally heavy in other situations. On the other hand, stochastic approximation techniques yield simple methods for the iterative maximization of (9.46) without the explicit computation of expected values for $\mu(\cdot)$.

A direct application of these techniques to the gradient-projection algorithm (9.15) yields the following stochastic algorithm. Define $\mathbf{w}_k^+ = \mathbf{w}_k + \alpha_k \nabla\phi(\mathbf{w}^k;\tilde{\theta}^k)$ with $\tilde{\theta}^k$ the k-th element of an i.i.d. sequence generated with the probability measure $\mu(\cdot)$. We then take $\mathbf{w}^{k+1} = P_{\mathscr{P}_{\ell-1}}(\mathbf{w}_+^k)$ with $P_{\mathscr{P}_{\ell-1}}(\cdot)$ the orthogonal projection onto $\mathscr{P}_{\ell-1}$; see Remark 9.4. The stepsize α_k is usually taken as the k-th element of a positive sequence satisfying $\sum_{k=0}^{\infty} \alpha_k = \infty$ and $\sum_{k=0}^{\infty} \alpha_k^2 < \infty$. General conditions for the convergence of such stochastic algorithms can be found, e.g., in (Kushner and Clark, 1978), (Ermoliev and Wets, 1988) and (Kushner and Yin, 1997). Polyak (1990) and Polyak and Juditsky (1992) have shown that by choosing a sequence $\{\alpha_k\}$ with slower convergence towards 0, such that $\alpha_k/\alpha_{k+1} = 1 + o(\alpha_k)$, for instance,

$\alpha_k = 1/(k+1)^\gamma$, $0 < \gamma < 1$, one can obtain the fastest possible convergence to the optimum when the successive iterates are averaged, i.e., when considering $\tilde{\mathbf{w}}^k = [1/(k+1)] \sum_{i=0}^{k} \mathbf{w}^i$; see also Kushner and Yang (1993) and Kushner and Yin (1997, Chap. 11).

9.4.2 Average-Optimum Exact Design

When X is an exact design, the algorithms of Sect. 9.2 can be used directly if $\mu(\cdot)$ is a discrete measure on a finite set since the evaluation of $\int_\Theta \phi(X;\theta)\,\mu(\mathrm{d}\theta)$ simply amounts to the calculation of a discrete sum. A stochastic approximation method can be used to maximize $\phi_{AO}(\xi) = \int_\Theta \phi(X;\theta)\,\mu(\mathrm{d}\theta)$ when $\mathscr{X}$ is a compact subset of $\mathbb{R}^d$ with nonempty interior and $\mu(\cdot)$ is any probability measure on Θ, possibly with a density with respect to the Lebesgue measure. Starting from some $X^0 = \{\mathbf{x}_1^0, \ldots, \mathbf{x}_N^0\} \in \mathscr{X}^N$, the k-th iteration of the algorithm has the form

$$\mathbf{x}_i^{k+1} = P_{\mathscr{X}} \left[\mathbf{x}_i^k + \alpha_k \left. \frac{\partial \phi(X;\tilde{\theta}^k)}{\partial \mathbf{x}_i} \right|_{X=X^k} \right], \; i = 1, \ldots, N,$$

where $\tilde{\theta}^k$ is the k-th element of a sequence of i.i.d. random variables distributed with the probability measure $\mu(\cdot)$ and $P_{\mathscr{X}}(\cdot)$ denotes the orthogonal projection onto $\mathscr{X}$. This projection corresponds to a simple truncation of each component of $\mathbf{x}$ when $\mathscr{X}$ is the hyper-rectangle $\{\mathbf{x} \in \mathbb{R}^d : \mathbf{a} \leq \mathbf{x} \leq \mathbf{b}\}$, where inequalities should be interpreted componentwise. A typical choice for the stepsize is $\alpha_k = \alpha/(k+1)$ for some $\alpha > 0$. In order to avoid the difficult choice of a suitable constant α, it is suggested in (Pronzato and Walter, 1985) to use

$$\mathbf{x}_i^{k+1} = P_{\mathscr{X}} \left[\mathbf{x}_i^k + \alpha_k \, \mathbf{D}_{i,k} \left. \frac{\partial \phi(X;\tilde{\theta}^k)}{\partial \mathbf{x}_i} \right|_{X=X^k} \right], \; i = 1, \ldots, N,$$

where $\mathbf{D}_{i,k}$ is diagonal, $\mathbf{D}_{i,k} = \mathrm{diag}\{\mathbf{u}_i^k\}$ with $\mathbf{u}_i^k$ the vector defined by

$$\{\mathbf{u}_i^k\}_j = \frac{\{\mathbf{x}_{\max}\}_j - \{\mathbf{x}_{\min}\}_j}{\left[\frac{1}{k+1} \sum_{n=0}^{k} \left(\left. \frac{\partial \phi(X;\tilde{\theta}^n)}{\partial \{\mathbf{x}_i\}_j} \right|_{X=X^n} \right)^2 \right]^{1/2}}$$

with $\{\mathbf{x}_{\max}\}_j = \max_{x \in \mathscr{X}}(\{\mathbf{x}\}_j)$, $\{\mathbf{x}_{\min}\}_j = \min_{x \in \mathscr{X}}(\{\mathbf{x}\}_j)$, $j = 1, \ldots, d$. Note that $\mathbf{x}_i^1 = P_{\mathscr{X}}(\mathbf{x}_{i,+}^0)$ with $\{\mathbf{x}_{i,+}^0\}_j = \{\mathbf{x}_i^0\}_j \pm \alpha_0 \{\mathbf{x}_{\max} - \mathbf{x}_{\min}\}_j$ for all i, which may serve as a guideline for choosing α_0. Again, one may consider averaging the iterates with $\alpha_k = \alpha/(k+1)^\gamma$ with $0 < \alpha < 1$; see Polyak (1990); Polyak and Juditsky (1992); and Kushner and Yang (1993). Since this corresponds to a stochastic version of a local search algorithm, one may only expect convergence to a local optimum. It is therefore recommended to repeat several optimizations initialized at different designs X^0. Average exact D-optimum design is considered in (Pronzato and Walter, 1985); an application of stochastic approximation to the criteria considered in Sect. 6.3 is presented in (Pázman and Pronzato, 1992) and (Gauchi and Pázman, 2006).

9.5 Two Methods for Convex Programming

The objective of this section is to present two classical methods for convex programming (ellipsoid and cutting planes) that can be used for the maximization of a concave design criterion $\phi(\cdot)$ with respect to $\mathbf{w}$ in the probability simplex $\mathscr{P}_{\ell-1}$ given by (9.1). Among the multitude of methods available, we single out those two because (i) they are easy to implement and (ii) they can optimize differentiable and non-differentiable criteria without requiring any specific adaptation. In particular, they can be applied rather straightforwardly to maximin-optimum design. The first one (the ellipsoid method) has the reputation of being fairly robust (see Ecker and Kupferschmid 1983) and does not rely on any external solver or optimizer; however, it is rather slow when the dimension ℓ gets large. The second one (cutting planes) is based on LP: similarly to the relaxation algorithm of Sect. 9.3.2, it amounts to solving a series of maximin problems, but with the particularity that each maximization (step 1) forms now an LP problem. The method thus requires the application of an LP solver; we believe, however, that this is not a strong handicap since such solvers are widely available. Its application to E-, c-, G-, and D-optimum design is detailed. The performance of the cutting-plane method can be improved by adding a QP step at each iteration. This yields a particular bundle method, called the level method, which is briefly presented.

Throughout the section we shall denote indifferently by $\nabla\phi(\mathbf{w})$ the gradient of $\phi(\cdot)$ at $\mathbf{w}$ when $\phi(\cdot)$ is differentiable or an arbitrary subgradient if $\phi(\cdot)$ is not differentiable.

9.5.1 Principles for Cutting Strategies and Interior-Point Methods

Suppose that $k+1$ gradients of $\phi(\cdot)$, or $k+1$ arbitrary subgradients if $\phi(\cdot)$ is not differentiable, have been evaluated at $\mathbf{w}^0, \mathbf{w}^1, \ldots, \mathbf{w}^k$, all in $\mathscr{P}_{\ell-1}$. The subgradient inequality (9.44) gives

$$\phi(\mathbf{w}') \leq \phi(\mathbf{w}) + \nabla^\top\phi(\mathbf{w})(\mathbf{w}' - \mathbf{w}) \quad \text{for any} \quad \mathbf{w},\ \mathbf{w}' \in \mathscr{P}_{\ell-1}\,, \tag{9.47}$$

which implies that the solution $\mathbf{w}^*$ that maximizes $\phi(\cdot)$ satisfies, for $j = 0, \ldots, k$,

$$0 \leq \overline{\phi}_k - \phi(\mathbf{w}^j) \leq \phi(\mathbf{w}^*) - \phi(\mathbf{w}^j) \leq \nabla^\top\phi(\mathbf{w}^j)(\mathbf{w}^* - \mathbf{w}^j)\,, \tag{9.48}$$

where

$$\overline{\phi}_k = \max_{i=0,\ldots,k} \phi(\mathbf{w}^i) \, .$$

Hence,

$$\overline{\phi}_k \le \phi(\mathbf{w}^*) \le \min_{j=0,\ldots,k} \left\{ \phi(\mathbf{w}^j) + \nabla^\top \phi(\mathbf{w}^j)(\mathbf{w}^* - \mathbf{w}^j) \right\} \, .$$

The k-th iteration of the cutting-plane method (Kelley, 1960), presented into more details in Sect. 9.5.3, chooses $\mathbf{w}^{k+1}$ that maximizes a piecewise linear (upper) approximation of $\phi(\cdot)$ based on previous evaluations; that is,

$$\mathbf{w}^{k+1} = \arg \max_{\mathbf{w} \in \mathscr{P}_{\ell-1}} \min_{j=0,\ldots,k} \left\{ \phi(\mathbf{w}^j) + \nabla^\top \phi(\mathbf{w}^j)(\mathbf{w} - \mathbf{w}^j) \right\} \, , \tag{9.49}$$

which corresponds to the solution of a linear program.

Instead of solving this linear program accurately at each iteration, we may alternatively choose $\mathbf{w}^{k+1}$ as an interior point of the polytope $\mathcal{P}_k \in \mathscr{P}_{\ell-1}$ defined by

$$\mathcal{P}_k = \{ \mathbf{w} \in \mathscr{P}_{\ell-1} : \nabla^\top \phi(\mathbf{w}^j)(\mathbf{w} - \mathbf{w}^j) \ge \overline{\phi}_k - \phi(\mathbf{w}^j) \, , \; j = 0, \ldots, k \} \, ,$$

see (9.48). Taking $\mathbf{w}^{k+1}$ at the center of gravity of $\mathcal{P}_k$ gives the fastest method (in the worst case) in terms of number of iterations, among all methods using only subgradient information; see Levin (1965). However, computing the center of gravity of a polytope in $\mathbb{R}^\ell$ is not an easy task, and alternative centers must be considered. Taking $\mathbf{w}^{k+1}$ as the center of the maximum-volume ellipsoid inscribed in $\mathcal{P}_k$ forms a very efficient method in terms of number of iterations required for a given accuracy on the location of $\mathbf{w}^*$. However, each iteration requires the computation of a maximum-volume inscribed ellipsoid for $\mathcal{P}_k$ (see Tarasov et al. 1988; Khachiyan and Todd 1993), which is computationally expensive, especially when the dimension ℓ is large. We may also take $\mathbf{w}^{k+1}$ as the analytic center $\mathbf{w}_c^k$ of $\mathcal{P}_k$; see Nesterov (1995). Rewriting the constraints that define $\mathcal{P}_k$ as $\mathbf{a}_i^\top \mathbf{w} \le b_i$, $i = 1, \ldots, m_k$, $\mathbf{w}_c^k$ is defined as the point in $\mathbb{R}^\ell$ that minimizes the logarithmic-barrier function $-\sum_{i=1}^{m_k} \log(b_i - \mathbf{a}_i^\top \mathbf{w})$. When an initial point in $\mathcal{P}_k$ is known, $\mathbf{w}_c^k$ can be obtained via a standard Newton method; techniques exist that allow us to use unfeasible starting points; see, e.g., Boyd and Vandenberghe (2004, Chap. 10). These ideas are key ingredients for sophisticated interior-point methods for convex programming, and one may refer to den Hertog (1994), Nesterov and Nemirovskii (1994), Ye (1997), Wright (1998), and Nesterov (2004) for general expositions on the subject.

The so-called ellipsoid method, presented below, constructs a sequence of outer ellipsoids for $\mathcal{P}_k$, with decreasing volumes, that shrink to $\mathbf{w}^*$.

9.5.2 The Ellipsoid Method

The method is known to be quite slow in high dimension (see Example 9.13 for an illustration), but it deserves particular attention due to the facility of its implementation, its very general applicability, and the simplicity of the geometric ideas on which it relies. Therefore, it may sometimes form a useful tool for solving design problems, differentiable or not, in small dimension (i.e., if the cardinality ℓ of $\mathscr{X}_\ell$ is small enough) when no other specialized algorithm is available.

The method exploits the subgradient inequality (9.47) which implies that the optimum $\mathbf{w}^*$ satisfies

$$\nabla^\top \phi(\mathbf{w}^k)(\mathbf{w}^* - \mathbf{w}^k) \geq \phi(\mathbf{w}^*) - \phi(\mathbf{w}^k) \geq 0 \tag{9.50}$$

for any choice of $\mathbf{w}^k$ in $\mathscr{P}_{\ell-1}$—and any subgradient $\nabla\phi(\mathbf{w}^k)$ if $\phi(\cdot)$ is not differentiable. If $\mathbf{w}^k$ is the center of some ellipsoid $\mathcal{E}_k$ that contains $\mathbf{w}^*$, then we know that $\mathbf{w}^*$ should belong to the half-ellipsoid $\{\mathbf{w} \in \mathcal{E}_k : \nabla^\top \phi(\mathbf{w}^k)(\mathbf{w} - \mathbf{w}^k) \geq 0\}$. It happens that the minimum-volume ellipsoid containing a half-ellipsoid is easy to determine, and we can then take $\mathbf{w}^{k+1}$ as the center of this new ellipsoid $\mathcal{E}_{k+1}$. $\mathcal{E}_{k+1}$ has smaller volume than $\mathcal{E}_k$ and necessarily contains $\mathbf{w}^*$, so that a sequence of such ellipsoids will shrink around $\mathbf{w}^*$ (provided that $\mathcal{E}_0$ contains $\mathbf{w}^*$). We shall see that the sequence of volumes $\mathrm{vol}(\mathcal{E}_k)$ decreases at the rate of a geometric progression. One may refer to Grötschel et al. (1980) and Bland et al. (1981) for a general exposition on the ellipsoid method for combinatorial and LP problems. The method, initially proposed by Shor (1977), has been used by Khachiyan (1979) to prove the polynomial complexity of LP. It can be interpreted as a modification of the subgradient algorithm, with a space dilation in the subgradient direction; see Bland et al. (1981). It is often considered as of theoretical interest only, since its convergence is rather slow when the dimension is large and, moreover, because handling ellipsoids in large dimensions is rather cumbersome. However, Ecker and Kupferschmid (1983, 1985) show that the method is competitive for nonlinear programming when the dimension is moderate—up to 36 in (Ecker and Kupferschmid, 1985). Its simplicity and quite general applicability make it attractive for design problems when the cardinality ℓ of the discretized set $\mathscr{X}_\ell$ is small enough.

In the lines above we have ignored the presence of the constraints $\mathbf{w} \in \mathscr{P}_{\ell-1}$ defined by (9.1). Suppose for the moment that only inequality constraints are present, in the form $\mathbf{A}\mathbf{w} \leq \mathbf{b}$, with $\mathbf{A} \in \mathbb{R}^{m\times\ell}$ and $\mathbf{b} \in \mathbb{R}^m$, where the inequality should be interpreted componentwise. Suppose that at iteration k, $\mathbf{w}^k$ constructed as above violates (at least one of) these constraints. Let j denote the index of a violated constraint, for instance, the most violated. Instead of cutting the ellipsoid $\mathcal{E}_k$ by the hyperplane $\mathcal{H}_k^o = \{\mathbf{w} : \nabla^\top \phi(\mathbf{w}^k)(\mathbf{w} - \mathbf{w}^k) = 0\}$ defined by the objective function to be maximized, we cut it by the hyperplane $\mathcal{H}_{k,j}^c = \{\mathbf{w} : \mathbf{a}_j^\top(\mathbf{w} - \mathbf{w}^k) = 0\}$ defined by a violated constraint, with $\mathbf{a}_j^\top$ the j-th row of $\mathbf{A}$. Note that this cut goes through the center of $\mathcal{E}_k$; for that reason it is called a central cut. Using a deeper cut is possible and produces a larger volume reduction for the next ellipsoid: the cut is then defined by the hyperplane $\{\mathbf{w} : \mathbf{a}_j^\top \mathbf{w} = b_j\}$.

The construction of the minimum-volume ellipsoid containing such a portion of an ellipsoid is also easy to construct; see below. Such cuts by constraints, followed by the construction of new ellipsoids with decreasing volumes, are repeated until a center $\mathbf{w}^{k'}$ is obtained that satisfies all constraints $\mathbf{A}\mathbf{w}^{k'} \leq \mathbf{b}$. A central cut by $\mathcal{H}_{k'}^o = \{\mathbf{w} : \nabla^\top \phi(\mathbf{w}^{k'})(\mathbf{w} - \mathbf{w}^{k'}) = 0\}$ is then operated, and the process is repeated.

The constraints $\mathbf{w} \in \mathscr{P}_{\ell-1}$ contain the equality constraint $\mathbf{w}^\top \mathbf{1} = 1$, with $\mathbf{1}$ the ℓ-dimensional vector of ones. Shah et al. (2000) use a projection method to take equality constraints into account within an ellipsoid algorithm. Due to the simplicity of the constraint $\mathbf{w}^\top \mathbf{1} = 1$, it is easier here to eliminate a variable, say the last one w_ℓ, and write admissible points in $\mathbb{R}^{\ell-1}$ as

$$\underline{\mathbf{w}} = (w_1, \ldots, w_{\ell-1})^\top \in \mathscr{W}_{\ell-1} = \left\{ \underline{\mathbf{w}} \in \mathbb{R}^{\ell-1} : \underline{w}_i \geq 0\,,\ \sum_{i=1}^{\ell-1} \underline{w}_i \leq 1 \right\}.$$

Note that $w_\ell = 1 - \sum_{i=1}^{\ell-1} \underline{w}_i$. Simulations indicate that the choice of the variable eliminated has marginal influence on the behavior of the algorithm. With a slight abuse of notation, we shall write $\phi(\underline{\mathbf{w}})$ for the value of $\phi(\cdot)$ at the associated $\mathbf{w}$. Denoting by $\nabla\phi_{\ell-1}(\cdot)$ a gradient (or subgradient) in these new coordinates $\underline{\mathbf{w}}$, we have

$$\nabla\phi_{\ell-1}(\underline{\mathbf{w}}) = [\mathbf{I}_{\ell-1} \quad -\mathbf{1}_{\ell-1}]\ \nabla\phi(\mathbf{w})\,,$$

with $\mathbf{I}_{\ell-1}$ and $\mathbf{1}_{\ell-1}$, respectively, the $(\ell-1)$-dimensional identity matrix and vector of ones; that is,

$$\{\nabla\phi_{\ell-1}(\underline{\mathbf{w}})\}_i = \{\nabla\phi(\mathbf{w})\}_i - \{\nabla\phi(\mathbf{w})\}_\ell\ ,\ i = 1, \ldots, \ell-1\,.$$

The inequality constraints $\underline{\mathbf{w}} \in \mathscr{W}_{\ell-1}$ are in the form $\mathbf{A}\underline{\mathbf{w}} \leq \mathbf{b}$, where

$$\mathbf{A} = \begin{bmatrix} -\mathbf{I}_{\ell-1} \\ \mathbf{1}_{\ell-1}^\top \end{bmatrix} \quad \text{and} \quad \mathbf{b} = \begin{bmatrix} \mathbf{0}_{\ell-1} \\ 1 \end{bmatrix}$$

with $\mathbf{0}_{\ell-1}$ the $(\ell-1)$-dimensional null vector.

The algorithm goes as follows. We denote by $\mathcal{E}(\mathbf{c}, \mathbf{E})$ the ellipsoid in $\mathbb{R}^{\ell-1}$ defined by

$$\mathcal{E}(\mathbf{c}, \mathbf{E}) = \{\underline{\mathbf{w}} \in \mathbb{R}^{\ell-1} : (\underline{\mathbf{w}} - \mathbf{c})^\top \mathbf{E}^{-1} (\underline{\mathbf{w}} - \mathbf{c}) \leq 1\}\,,$$

with $\mathbf{c} \in \mathbb{R}^{\ell-1}$ and $\mathbf{E}$ a $(\ell-1) \times (\ell-1)$ symmetric positive-definite matrix.

0. Start from $\underline{\mathbf{w}}^0$ and $\mathbf{E}_0$ such that $\mathscr{W}_{\ell-1} \subset \mathcal{E}_0 = \mathcal{E}(\underline{\mathbf{w}}^0, \mathbf{E}_0)$; choose $\epsilon > 0$; set $k = 0$.
1. If $\mathbf{A}\underline{\mathbf{w}}^k \leq \mathbf{b}$, go to step 3, otherwise go to step 2.
2. Construct $\mathcal{E}_{k+1} = \mathcal{E}(\underline{\mathbf{w}}^{k+1}, \mathbf{E}_{k+1})$, the minimum-volume ellipsoid containing the intersection between $\mathcal{E}_k$ and the half-space $\mathcal{H}_{k,j^*} = \{\underline{\mathbf{w}} \in \mathbb{R}^{\ell-1} : \mathbf{a}_{j^*}^\top (\underline{\mathbf{w}} - \underline{\mathbf{w}}^k) \leq 0\}$, where $\mathbf{a}_{j^*}^\top \underline{\mathbf{w}}^k - b_{j^*} = \max_j \mathbf{a}_j^\top \underline{\mathbf{w}}^k - b_j$ (j^* corresponds to the most violated constraint); $k \leftarrow k+1$, return to step 1.

3. If $\max_{\underline{\mathbf{w}}\in\mathcal{E}_k} \nabla^\top \phi_{\ell-1}(\underline{\mathbf{w}}^k)(\underline{\mathbf{w}}-\underline{\mathbf{w}}^k) < \epsilon$ stop; otherwise go to step 4.
4. Construct $\mathcal{E}_{k+1} = \mathcal{E}(\underline{\mathbf{w}}^{k+1}, \mathbf{E}_{k+1})$, the minimum-volume ellipsoid containing $\mathcal{E}_k \cap \{\underline{\mathbf{w}} \in \mathbb{R}^{\ell-1} : \nabla^\top \phi_{\ell-1}(\underline{\mathbf{w}}^k)(\underline{\mathbf{w}}-\underline{\mathbf{w}}^k) \geq 0\}$; $k \leftarrow k+1$, return to step 1.

A reasonable choice for $\mathcal{E}_0$ is the minimum-volume ellipsoid containing the simplex $\mathscr{W}_{\ell-1}$. Easy calculations show that it corresponds to

$$\begin{aligned} \underline{\mathbf{w}}^0 &= \mathbf{1}_{\ell-1}/\ell\,, \\ \mathbf{E}_0 &= \tfrac{\ell-1}{\ell}\left[\mathbf{I}_{\ell-1} - \mathbf{1}_{\ell-1}\mathbf{1}_{\ell-1}^\top/\ell\right]. \end{aligned} \tag{9.51}$$

Notice that $\max_{\underline{\mathbf{w}}\subset\mathcal{E}_k} \mathbf{a}^\top(\underline{\mathbf{w}}-\underline{\mathbf{w}}^k) = \sqrt{\mathbf{a}^\top\mathbf{E}_k\mathbf{a}}$ for any $\mathbf{a}\in\mathbb{R}^{\ell-1}$ and that $\max_{\underline{\mathbf{w}}\in\mathcal{E}_k} \nabla^\top \phi_{\ell-1}(\underline{\mathbf{w}}^k)(\underline{\mathbf{w}}-\underline{\mathbf{w}}^k) < \epsilon$ at step 3 implies that $\phi(\underline{\mathbf{w}}^k) > \phi(\mathbf{w}^*)-\epsilon$; see (9.50). The explicit construction of $\underline{\mathbf{w}}^{k+1}$ and $\mathbf{E}_{k+1}$ at steps 2 and 4 is as follows (see Grötschel et al. 1980 and Bland et al. 1981):

$$\begin{aligned} \underline{\mathbf{w}}^{k+1} &= \underline{\mathbf{w}}^k - \rho\,\frac{\mathbf{E}_k\mathbf{a}}{\sqrt{\mathbf{a}^\top\mathbf{E}_k\mathbf{a}}}\,, \\ \mathbf{E}_{k+1} &= s\left[\mathbf{E}_k - \tau\,\frac{\mathbf{E}_k\mathbf{a}\mathbf{a}^\top\mathbf{E}_k}{\mathbf{a}^\top\mathbf{E}_k\mathbf{a}}\right], \end{aligned} \tag{9.52}$$

where $\mathbf{a} = \mathbf{a}_{j^*}$ at step 2 and $\mathbf{a} = -\nabla\phi_{\ell-1}(\underline{\mathbf{w}}^k)$ at step 4 and where

$$\rho = \frac{1}{\delta+1}\,,\quad s = \frac{\delta^2}{\delta^2-1}\,,\quad \tau = \frac{2}{\delta+1}\,, \tag{9.53}$$

with $\delta = \ell - 1$ denoting the space dimension. Note that we may update the Cholesky factorization of $\mathbf{E}_k$ instead of updating $\mathbf{E}_k$ itself, which renders the computations simpler and numerically more robust. The ratio of the volume of $\mathcal{E}_{k+1}$ to the volume of $\mathcal{E}_k$ is

$$r(\delta) = \frac{\mathrm{vol}(\mathcal{E}_{k+1})}{\mathrm{vol}(\mathcal{E}_k)} = s^{\delta/2}\sqrt{1-\tau} = \frac{\delta}{\delta+1}\left(\frac{\delta^2}{\delta^2-1}\right)^{(\delta-1)/2} < 1\,. \tag{9.54}$$

The volumes of the ellipsoids constructed thus decrease as a geometric series with (constant) ratio $r(\delta)$. Figure 9.10 shows $r(\delta)$ as a function of δ; it is clear that the algorithm becomes very slow if $\ell = \delta + 1$ is very large.

As already noticed, a deeper cut can be used at step 2 by constructing $\mathcal{E}_{k+1} = \mathcal{E}(\underline{\mathbf{w}}^{k+1}, \mathbf{E}_{k+1})$ as the minimum-volume ellipsoid containing $\mathcal{E}_k \cap \mathcal{H}'_{k,j^*}$, with $\mathcal{H}'_{k,j^*} = \{\underline{\mathbf{w}} \in \mathbb{R}^{\ell-1} : \mathbf{a}_{j^*}^\top\underline{\mathbf{w}} \leq b_{j^*}\}$. In that case, denoting

$$\alpha = \frac{\mathbf{a}_{j^*}^\top\underline{\mathbf{w}}^k - b_{j^*}}{\sqrt{\mathbf{a}_{j^*}^\top\mathbf{E}_k\mathbf{a}_{j^*}}}\,,$$

which is strictly positive since the index j^* corresponds to a violated constraint, we compute ρ, s, and τ as

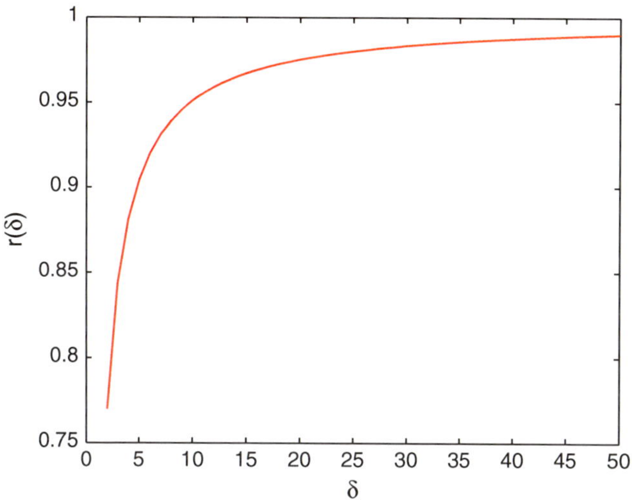

Fig. 9.10. Ratio of the volume of $\mathcal{E}_{k+1}$ to the volume of $\mathcal{E}_k$ in the ellipsoid algorithm with central cuts in dimension δ

$$\rho = \frac{1+\alpha\delta}{\delta+1}\,, \quad s = \frac{\delta^2(1-\alpha^2)}{\delta^2-1}\,, \quad \tau = \frac{2(1+\alpha\delta)}{(\alpha+1)(\delta+1)}\,. \tag{9.55}$$

Note that $\alpha \leq 1$; the value $\alpha = 1$ corresponds to the situation where the hyperplane $\{\underline{\mathbf{w}} \in \mathbb{R}^{\ell-1} : \mathbf{a}_{j^*}^\top \underline{\mathbf{w}} = b_{j^*}\}$ is tangent to the ellipsoid $\mathcal{E}_k$: the next ellipsoid $\mathcal{E}_{k+1}$ is then reduced to the single point $\underline{\mathbf{w}}^k - \mathbf{E}_k\mathbf{a}_{j^*}/\sqrt{\mathbf{a}_{j^*}^\top \mathbf{E}_k \mathbf{a}_{j^*}}$. The ratio of the volume of $\mathcal{E}_{k+1}$ to the volume of $\mathcal{E}_k$ is still given by $r(\delta,\alpha) = s^{\delta/2}\sqrt{1-\tau}$ and is now smaller than the value given in (9.54). Figure 9.11 shows $r(\delta,\alpha)$ as a function of δ for values of α ranging from 0 (top curve, identical to that in Fig. 9.10) to 0.9, by steps of 0.1.

One might think that the acceleration due to deep cuts compared to central cuts is important, especially in high dimension. However, it should be noted that constraint cuts with a large α can occur at early iterations only—and even not so if the initial ellipsoid $\mathcal{E}_0$ is well chosen. Indeed, consider the iteration following a central cut of $\mathcal{E}_k$ at step 4 with direction orthogonal to $\mathbf{a}$. Direct calculations show that when $\mathbf{a}_j^\top \underline{\mathbf{w}}^{k+1} > b_j$, the depth of a cut by the constraint $\mathbf{a}_j^\top \underline{\mathbf{w}} \leq b_j$ satisfies

$$\alpha = \frac{\sqrt{\mathbf{a}^\top \mathbf{E}_k \mathbf{a}}\,(\mathbf{a}_j^\top \underline{\mathbf{w}}^k - b_j) - \rho\, \mathbf{a}_j^\top \mathbf{E}_k \mathbf{a}}{\sqrt{s}\,\left[(\mathbf{a}_j^\top \mathbf{E}_k \mathbf{a}_j)(\mathbf{a}^\top \mathbf{E}_k \mathbf{a}) - \tau\,(\mathbf{a}_j^\top \mathbf{E}_k \mathbf{a})^2\right]^{1/2}}\,,$$

with ρ, s, and τ given by (9.53). Now, $\mathbf{a}_j^\top \underline{\mathbf{w}}^k - b_j$ is negative since step 4 was used the iteration before, so that, from Cauchy–Schwarz inequality,

$$\alpha \leq \alpha^* = \frac{\rho}{\sqrt{s(1-\tau)}} = \frac{1}{\delta}\,,$$

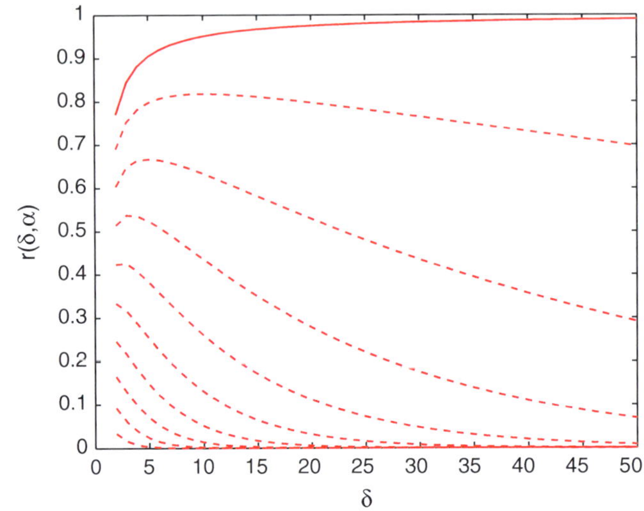

Fig. 9.11. Ratio of the volume of $\mathcal{E}_{k+1}$ to the volume of $\mathcal{E}_k$ in the ellipsoid algorithm with deep cuts in dimension δ: $r(\delta, \alpha)$ as a function of δ for $\alpha = 0$ (*top*), $0.1, \ldots, 0.9$ (*bottom*)

see Fig. 9.13 (left) for an illustration. Figure 9.12 shows $r(\delta, \alpha)$ as a function of δ for $\alpha = 0$ and $\alpha = \alpha^*$. As it can be seen, the improvement due to deep cuts cannot be important.

Another possible improvement consists in using cuts deeper than through the center of the ellipsoid also at step 4 of the algorithm. Indeed, from (9.48) the optimal weights $\mathbf{w}^*$ satisfy

$$\nabla^\top \phi_{\ell-1}(\underline{\mathbf{w}}^k)(\underline{\mathbf{w}}^* - \underline{\mathbf{w}}^k) \geq \gamma_k = \max_{\mathbf{w}^i \in \mathscr{P}_{\ell-1},\, i=0,\ldots,k} \phi(\mathbf{w}^i) - \phi(\mathbf{w}^k)$$

when $\mathbf{w}^k \in \mathscr{P}_{\ell-1}$, and we can thus use a deep cut with

$$\alpha = \frac{\gamma_k}{\sqrt{\nabla^\top \phi_{\ell-1}(\underline{\mathbf{w}}^k)\mathbf{E}_k \nabla \phi_{\ell-1}(\underline{\mathbf{w}}^k)}},$$

which is strictly positive at some iterations, rather than a central cut with $\alpha = 0$; ρ, s, and τ to be used in (9.52) are then given by (9.55). Although the algorithm usually goes less often through step 4 than step 2 (see, e.g., the figures in Table 9.1), cuts are often deeper in step 4 than in step 2, and the effects on the acceleration of the algorithm are then comparable. Figure 9.13 shows a typical picture of the evolution of the depth α_k of the cut along k, for step 2 on the left and step 4 on the right (notice the different vertical scales).

Remark 9.11. The algorithm can easily manage linear cost constraints like (5.26); they simply need to be added to those defining $\mathscr{W}_{\ell-1}$. Also, if linear

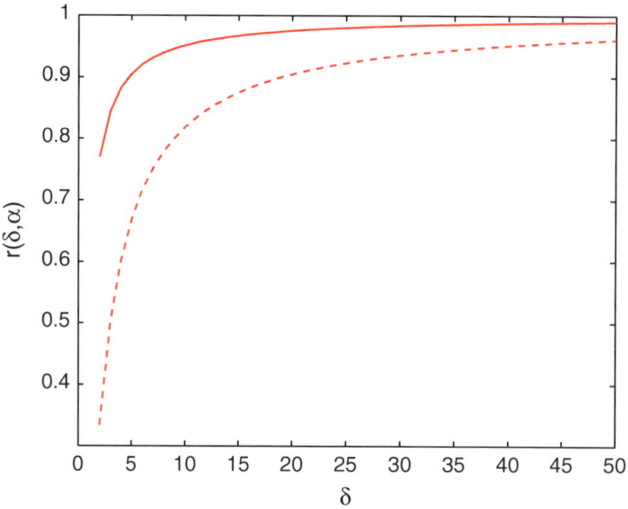

Fig. 9.12. Ratio of the volume of $\mathcal{E}_{k+1}$ to the volume of $\mathcal{E}_k$ in the ellipsoid algorithm with deep cuts in dimension δ: $r(\delta, \alpha)$ as a function of δ for $\alpha = 0$ (*solid line*) and $\alpha = 1/\delta$ (*dashed line*)

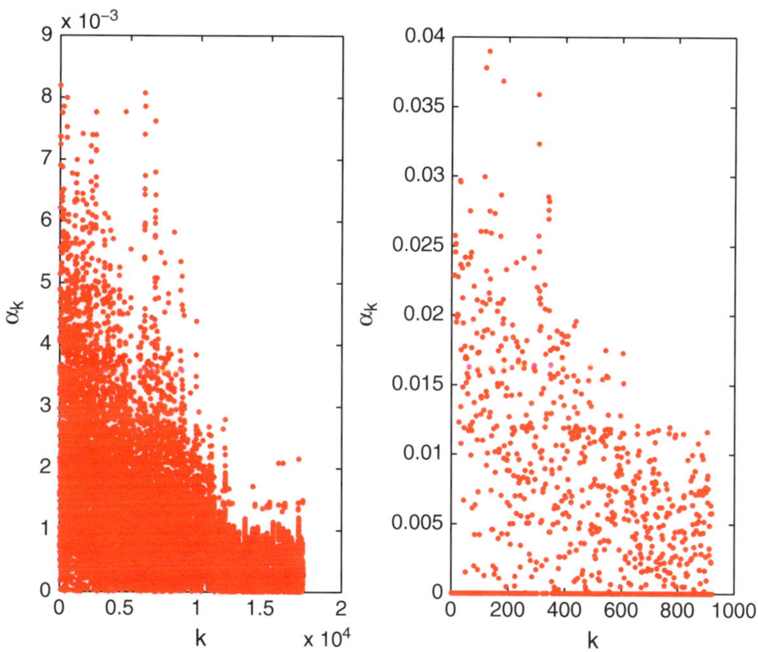

Fig. 9.13. Depth α_k of the k-th cut, at step 2 (*left*) and step 4 (*right*) (Example 9.13-B, $\eta(x, \theta) = \theta_1 + \theta_2 x + \theta_3 x^2$, $\ell = 41$)

equality constraints are present, we can eliminate more variables to define a vector $\underline{\mathbf{w}}$ with dimension less than $\ell - 1$ or use the projection approach of Shah et al. (2000).

The situation is only slightly more complicated when the constraints are nonlinear but define a convex set, as in Sect. 5.1.9. Suppose that to the linear constraints $\mathbf{A}\underline{\mathbf{w}} \leq \mathbf{b}$, we add a new constraint $\phi_1(\underline{\mathbf{w}}) \geq \Delta$, with $\phi_1(\cdot)$ a concave design criterion. Denote by $\nabla\phi_{1,\ell-1}(\underline{\mathbf{w}})$ a gradient (or subgradient) of $\phi_1(\cdot)$ at $\underline{\mathbf{w}} \in \mathscr{W}_{\ell-1}$. Step 1 of the ellipsoid algorithm is then changed into:

1') If $\mathbf{A}\underline{\mathbf{w}}^k \leq \mathbf{b}$ and $\phi_1(\underline{\mathbf{w}}^k) \geq \Delta$, go to step 3, otherwise go to step 2.

Step 2 is not modified if $\mathbf{a}_{j^*}^\top \underline{\mathbf{w}}^k - b_{j^*} \geq \Delta - \phi_1(\underline{\mathbf{w}}^k)$. If, on the other hand, $\Delta - \phi_1(\underline{\mathbf{w}}^k) > \mathbf{a}_{j^*}^\top \underline{\mathbf{w}}^k - b_{j^*}$, the constraint $\phi_1(\underline{\mathbf{w}}) \geq \Delta$ is the most violated, and we construct the minimum-volume ellipsoid containing the intersection between $\mathcal{E}_k$ and the half-space

$$\mathcal{H}_{\phi_1,k} = \{\underline{\mathbf{w}} \in \mathbb{R}^{\ell-1} : \nabla^\top \phi_{1,\ell-1}(\underline{\mathbf{w}}^k)(\underline{\mathbf{w}} - \underline{\mathbf{w}}^k) \geq 0\},$$

which corresponds to a central cut, or the half-space

$$\mathcal{H}'_{\phi_1,k} = \{\underline{\mathbf{w}} \in \mathbb{R}^{\ell-1} : \nabla^\top \phi_{1,\ell-1}(\underline{\mathbf{w}}^k)(\underline{\mathbf{w}} - \underline{\mathbf{w}}^k) \geq \Delta - \phi_1(\underline{\mathbf{w}}^k)\},$$

which corresponds to a deep cut with

$$\alpha = \frac{\Delta - \phi_1(\underline{\mathbf{w}}^k)}{\sqrt{\nabla^\top \phi_{1,\ell-1}(\underline{\mathbf{w}}^k)\mathbf{E}_k \nabla\phi_{1,\ell-1}(\underline{\mathbf{w}}^k)}}\,. \qquad \square$$

Example 9.12.

A. Consider the same non-differentiable problem as in Example 9.8. Figure 9.14 presents the evolution in the simplex $\mathscr{P}_2$ of the sequence of points $\mathbf{w}^k$ corresponding to the centers $\underline{\mathbf{w}}^k$ obtained at step 4 of the ellipsoid algorithm above (with central cuts), initialized according to (9.51). The first three ellipsoids constructed are plotted. The first constraint cut arrives at iteration 4. When ϵ is set to 10^{-3}, the algorithm stops after passing 46 times through step 4 (central cut by the objective) and 21 times through step 2 (central cut by a violated constraint). These figures become, respectively, 50 and 24 when ϵ is set to 10^{-6}. When deep cuts are used at step 2, the algorithm stops a little earlier and goes 45 times through step 4 and 14 times through step 2 for $\epsilon = 10^{-3}$ and 50 times through step 4 and 16 times through step 2 for $\epsilon = 10^{-6}$, indicating that the effect of using deep rather than central cuts at step 2 is limited.
B. We apply now the ellipsoid algorithm to the optimization of a differentiable criterion and consider the same problem as in Example 9.1. Figure 9.15 shows the sequence of centers and the first three ellipsoid generated by the algorithm with central cuts. Again, there is no big difference in performance between using deep and central cuts at step 2.

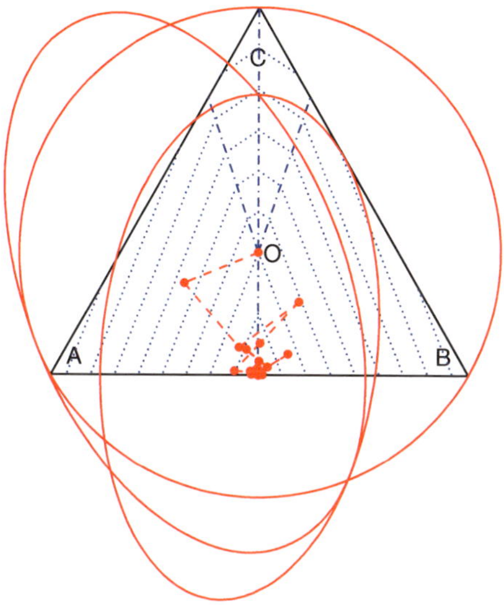

Fig. 9.14. Ellipsoids $\mathcal{E}_0, \mathcal{E}_1, \mathcal{E}_2$, and sequence of centers $\underline{\mathbf{w}}^k$ obtained at step 4 of the ellipsoid algorithm with central cuts in Example 9.12-A

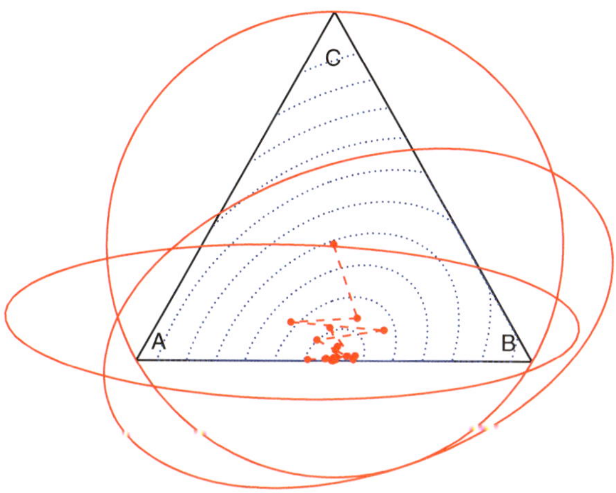

Fig. 9.15. Same as Fig. 9.14 but for Example 9.12-B

□

The simplicity of the ellipsoid algorithm can make it attractive when ℓ is small enough, but it must be stressed that it becomes desperately slow when ℓ is large, as shown in the next example.

Example 9.13 (Behavior in High Dimension).

A. We use the c-optimal design problem of Example 5.34 to illustrate the slow convergence of the ellipsoid algorithm in "high" dimension

(δ=dim($\underline{\mathbf{w}}$)=100). We take $\mathbf{c} = (0.5, \;\; 0.25)^\top$; the c-optimal design in $\mathscr{X} = [0,1]$ is then singular and puts mass one at $x^* = 0.5$, and the c-optimality criterion is non-differentiable at the optimum.

We discretize $\mathscr{X}$ into a regular grid $\mathscr{X}_\ell$ of 101 points $x^{(1)} = 0$, $x^{(2)} = 0.01, \ldots, x^{(100)} = 0.99$, $x^{(101)} = 1$. Let w_i denote the weight given at $x^{(i)}$. A subgradient at $\mathbf{w}$ is then given by $\{\nabla\phi_c(\mathbf{w})\}_i = [\mathbf{f}^\top(x^{(i)})\mathbf{M}^-(\mathbf{w})\mathbf{c}]^2$ for $i = 1, \ldots, \ell$ (see Sect. 5.2.1) with $\mathbf{f}(x) = \partial\eta(x,\theta)/\partial\theta$ and $\mathbf{M}^-(\mathbf{w})$ any g-inverse of the information matrix for the weights $\mathbf{w}$. When initialized according to (9.51), the ellipsoid algorithm with central cuts stops after going 2,288 times through step 4 and 112,270 times through step 2 for $\epsilon = 10^{-3}$. The mass at $x^* = 0.5$ is then approximately 0.980. Using deep cuts at step 2 does not help very much, since the algorithm still goes 2,213 times through step 4 and 102,598 times through step 2 before stopping, with a mass at x^* of about 0.981. It is clear that much more efficient techniques could be used to solve this problem: for instance, an LP solver (see Remark 5.32-(*ii*)) or the cutting-plane method, see Example 9.15 in Sect. 9.5.3.

B. We consider D-optimum design for two quadratic models; the optimal solutions are indicated in Remark 5.24.

The first model is $\eta(x,\theta) = \theta_1 + \theta_2 x + \theta_3 x^2$, with $x \in \mathscr{X} = [-1,1]$ which is discretized in a uniform grid of ℓ points. Table 9.1 indicates the number of times ellipsoid algorithms with central and deep cuts and $\epsilon = 10^{-3}$, initialized by (9.51), go through steps 2 and 4 for different choices of ℓ. The second model is given by the sum of two polynomial models as above, $\eta([x,y],\theta) = \theta_0 + \alpha_1 x + \alpha_2 x^2 + \beta_1 y + \beta_2 y^2$, with $\theta = (\theta_0, \alpha_1, \alpha_2, \beta_1, \beta_2)^\top$ and $x \in [-1,1]$, $y \in [-1,1]$. We discretize the square $[-1,1]^2$ into a uniform grid with ℓ points and use ellipsoid algorithms with central and deep cuts, initialized by (9.51). Table 9.1 indicates number of passages through steps 2 and 4 for $\ell = 5^2$ and $\ell = 9^2$ when $\epsilon = 10^{-3}$. □

Table 9.1. Behavior of ellipsoid algorithms with central and deep cuts for D-optimum design in Example 9.13-B: number of passages through steps 2 and 4 and total number of ellipsoids constructed for different discretizations of $\mathscr{X}$ ($\epsilon = 10^{-3}$ and $\mathscr{E}_0$ is given by (9.51))

		Central cuts			Deep cuts at step 2			Deep cuts at steps 2 and 4		
Model	ℓ	Step 2	Step 4	Total	Step 2	Step 4	Total	Step 2	Step 4	Total
$\eta(x,\theta)$	21	4 600	636	5 236	4 132	603	4 735	3 822	455	4 277
	41	19 998	1 308	21 306	18 299	1 247	19 546	17 287	922	18 209
	101	130 428	3 302	133 730	121 179	3 167	124 346	115 254	2 070	117 324
$\eta([x,y],\theta)$	25	5 624	2 148	7 772	5 304	2 088	7 392	4 821	1 588	6 509
	81	95 340	8 100	103 440	90 616	7 768	98 384	86 587	5 931	92 518

It is instructive to compare numerically the performance of the ellipsoid method with that of an algorithm of Sect. 9.1.1 on the same example.

We shall use the vertex-exchange algorithm defined by (9.11), (9.12) with the modification (9.13). This algorithm requires an initial measure ξ^0. The convergence to the optimum is ensured for all ξ^0 such that the initial information matrix is nonsingular. However, most design algorithms tend to converge slowly close to the optimum because they have difficulties in removing poor support points or in merging clusters around good ones; see Remark 9.6.[4] Choosing a ξ^0 with a small number of support points may thus help a vertex-exchange algorithm since it has the ability to exchange them for better points and may avoid introducing poor points along the iterations. In order to present results that do not depend on the particular ξ^0 selected, we choose a rather unfavorable configuration and take ξ^0 as the uniform measure on the ℓ points of the design space. If clusters of neighboring points are present after convergence of the algorithm, i.e., when the directional derivative $F_\phi(\xi_k, x)$ is less than some prescribed ϵ for all x, they are simply aggregated. Another possible option, more sophisticated, would be to make a few iterations with gradient projection (see Sect. 9.1.2) or with the cutting-plane method (see Sect. 9.5.3) while restricting the design space to the support of the design returned by the vertex-exchange algorithm.

Example 9.14. We continue Example 9.13-B, but with an algorithm based on (9.11)–(9.13); the design space $\mathscr{X}$ is discretized in a grid $\mathscr{X}_\ell$ with ℓ points; the initial measure ξ^0 is uniform on $\mathscr{X}_\ell$; the stopping rule is $\max_{x\in\mathscr{X}_\ell} F_\phi(\xi_k, x) < 10^{-4}$.

For the model $\eta(x,\theta) = \theta_1 + \theta_2 x + \theta_3 x^2$ with $\mathscr{X} = [-1,1]$ discretized in a uniform grid $\mathscr{X}_\ell$ of 2,001 points, the algorithm stops after 27 iterations only, requiring 488 evaluations of $\log\det \mathbf{M}(\cdot)$—most of them are used for the line search in (9.12), (9.13), performed with a golden-section type algorithm. The design obtained is supported on -1 and 1 (with weights approximately $1/3+1.3\cdot 10^{-7}$ and $1/3+1.3\cdot 10^{-6}$) and on 25 points around zero (at distance less than $1.2\cdot 10^{-2}$) that receive the rest of the mass. The aggregation of those points gives a measure supported on -1, 0, and 1 with a maximum error on the optimal weights of about $1.5\cdot 10^{-6}$.

For the second model $\eta([x,y],\theta) = \theta_0 + \alpha_2 x + \alpha_3 x^2 + \beta_2 y + \beta_3 y^2$ with $\mathscr{X} = [-1,1]^2$ discretized into a uniform grid with $\ell = 41\times 41 = 1,681$ points, the algorithm stops after 172 iterations, requiring 3,516 evaluations of $\log\det \mathbf{M}(\cdot)$; the 9 optimal support points are returned, no aggregation is required, and the optimal weights are determined with an error less than $3.0\cdot 10^{-6}$. □

---

[4]The cutting-plane method of Sect. 9.5.3 forms an exception due to the use of an LP solver providing precise solutions; however, it is restricted to moderate values of ℓ; see Examples 9.16 and 9.17.

9.5.3 The Cutting-Plane Method

We first consider the case of design criteria that can be written in the form

$$\phi_{\mathcal{U}}(\mathbf{w}) = \min_{u \in \mathcal{U}} \mathbf{u}^\top \mathbf{M}(\mathbf{w})\mathbf{u}\,, \tag{9.56}$$

with $\mathbf{M}(\mathbf{w})$ the $p \times p$ information matrix for a vector $\mathbf{w}$ of weights w_i allocated to design points $x^{(i)}$, $i = 1, \ldots, \ell$,

$$\mathbf{M}(\mathbf{w}) = \sum_{i=1}^{\ell} w_i \; \mathbf{M}_\theta(x^{(i)})\,, \tag{9.57}$$

see (5.1), and $\mathcal{U}$ some set of vectors in $\mathbb{R}^p$. This covers in particular the cases of E-, c-, and G-optimum design; see Sect. 5.1.2. In general, the set $\mathcal{U}$ is infinite, and the maximization of $\phi_{\mathcal{U}}(\cdot)$ involves an infinite number of constraints. We shall see that the application of the relaxation procedure of Sect. 9.3.2 then yields an algorithm that exactly coincides with the method of cutting planes introduced in Sect. 9.5.1. The algorithm will be detailed for E-, c-, and G-optimum design before we consider the maximization of a concave criterion $\phi(\cdot)$ in more general terms, with D-optimality as a particular case. A simple modification of the algorithm gives a particular bundle method (the level method) which is briefly presented.

E-, *c*-, and *G*-Optimum Design

Since the term $\mathbf{u}^\top \mathbf{M}(\mathbf{w})\mathbf{u}$ in (9.56) is linear in $\mathbf{w}$, maximizing $\phi_{\mathcal{U}}(\mathbf{w})$ with respect to $\mathbf{w} \in \mathscr{P}_{\ell-1}$ amounts to solving the following LP problem, with an infinite number of constraints if $\mathcal{U}$ is not finite:

$$\begin{aligned} \text{maximize} \quad & \mathbf{a}^\top \mathbf{z}\,, \text{ where } \mathbf{a} = (0, \ldots, 0, 1)^\top \in \mathbb{R}^{\ell+1}\,, \\ & \text{with respect to } \mathbf{z} = (\mathbf{w}^\top,\ t)^\top \in \mathbb{R}^{\ell+1} \\ \text{subject to} \quad & \mathbf{w} \in \mathscr{P}_{\ell-1} \\ & \text{and } \mathbf{u}^\top \mathbf{M}(\mathbf{w})\mathbf{u} \geq t \text{ for all } \mathbf{u} \in \mathcal{U}\,. \end{aligned}$$

Using a relaxation procedure similar to that of Shimizu and Aiyoshi (1980) (see Sect. 9.3.2), one may thus consider the solution of a series of relaxed LP problems, using at step k the finite set of constraints

$$\mathbf{w} \in \mathscr{P}_{\ell-1} \text{ and } \mathbf{u}^\top \mathbf{M}(\mathbf{w})\mathbf{u} \geq t \text{ for all } \mathbf{u} \in \mathcal{U}^k\,,$$

where $\mathcal{U}^k = \{\mathbf{u}^{(1)}, \ldots, \mathbf{u}^{(k)}\} \subset \mathcal{U}$. Once a solution $\mathbf{w}^k$ of this problem is obtained, using a standard LP solver, the set $\mathcal{U}^k$ is enlarged to $\mathcal{U}^{k+1} = \mathcal{U}^k \cup \{\mathbf{u}^{(k+1)}\}$ with $\mathbf{u}^{(k+1)}$ given by the constraint most violated by $\mathbf{w}^k$, i.e.,

$$\mathbf{u}^{(k+1)} = \arg\min_{\mathbf{u} \in \mathcal{U}} \mathbf{u}^\top \mathbf{M}(\mathbf{w}^k)\mathbf{u}\,. \tag{9.58}$$

This yields the following algorithm for the maximization of $\phi_{\mathcal{U}}(\mathbf{w})$.

0. Take any $\mathbf{w}^0 \in \mathscr{P}_{\ell-1}$, choose $\epsilon > 0$, and set $\mathbf{U}^0 = \varnothing$ and $k = 0$.
1. Compute $\mathbf{u}^{(k+1)}$ given by (9.58) and set $\mathcal{U}^{k+1} = \mathcal{U}^k \cup \{\mathbf{u}^{(k+1)}\}$.
2. Use an LP solver to determine $\hat{\mathbf{z}} = (\hat{\mathbf{w}}^\top,\ \hat{t})^\top$ solution of

$$\begin{aligned} \text{maximize} \quad & \mathbf{a}^\top \mathbf{z}\,, \text{ where } \mathbf{a} = (0,\ldots,0,1)^\top \in \mathbb{R}^{\ell+1}\,, \\ & \text{with respect to } \mathbf{z} = (\mathbf{w}^\top,\ t)^\top \in \mathbb{R}^{\ell+1} \\ \text{subject to} \quad & \mathbf{w} \in \mathscr{P}_{\ell-1} \text{ and } \mathbf{u}^\top \mathbf{M}(\mathbf{w})\mathbf{u} \geq t \text{ for all } \mathbf{u} \in \mathcal{U}^{k+1}\,. \end{aligned}$$

3. Set $\mathbf{w}^{k+1} = \hat{\mathbf{w}}$. If $\Delta_{k+1} = \hat{t} - \phi_{\mathcal{U}}(\mathbf{w}^{k+1}) < \epsilon$, take $\mathbf{w}^{k+1}$ as an ϵ-optimal solution and stop; otherwise $k \leftarrow k+1$, return to step 1.

One may observe that $\phi_{\mathcal{U}}(\mathbf{w}^j) = \mathbf{u}^{(j+1)^\top} \mathbf{M}(\mathbf{w}^j)\mathbf{u}^{(j+1)}$ for all $j = 0, 1, 2\ldots$ and that the vector with components

$$\{\nabla \phi_{\mathcal{U}}(\mathbf{w}^j)\}_i = \mathbf{u}^{(j+1)^\top} \mathbf{M}_\theta(x^{(i)})\mathbf{u}^{(j+1)}\,,\ i = 1,\ldots,\ell\,,$$

forms a subgradient of $\phi_{\mathcal{U}}(\cdot)$ at $\mathbf{w}^j$; see Appendix A and Lemma 5.18. It also satisfies $\mathbf{u}^{(j+1)^\top} \mathbf{M}(\mathbf{w})\mathbf{u}^{(j+1)} = \nabla^\top \phi_{\mathcal{U}}(\mathbf{w}^j)\mathbf{w}$ for all $\mathbf{w} \in \mathscr{P}_{\ell-1}$. Each of the constraints

$$\mathbf{u}^{(j+1)^\top} \mathbf{M}(\mathbf{w})\mathbf{u}^{(j+1)} \geq t$$

used at step 2, with $j = 0,\ldots,k$, can thus be written as

$$\nabla^\top \phi_{\mathcal{U}}(\mathbf{w}^j)\mathbf{w} = \nabla^\top \phi_{\mathcal{U}}(\mathbf{w}^j)(\mathbf{w} - \mathbf{w}^j) + \phi_{\mathcal{U}}(\mathbf{w}^j) \geq t\,.$$

Therefore, $\mathbf{w}^{k+1} = \hat{\mathbf{w}}$ determined at step 2 maximizes

$$\min_{j=0,\ldots,k} \{\phi_{\mathcal{U}}(\mathbf{w}^j) + \nabla^\top \phi_{\mathcal{U}}(\mathbf{w}^j)(\mathbf{w} - \mathbf{w}^j)\}$$

with respect to $\mathbf{w} \in \mathscr{P}_{\ell-1}$, and the algorithm corresponds to the cutting-plane method (9.49). Also, the optimal value $\phi_{\mathcal{U}}^*$ satisfies $\phi_{\mathcal{U}}(\mathbf{w}^{k+1}) \leq \phi_{\mathcal{U}}^* \leq \hat{t}$ at every iteration, and the value Δ_{k+1} of step 3 gives an upper bound on the distance to the optimum in terms of criterion value.

E-optimum Design

The E-optimality criterion

$$\phi_E(\mathbf{w}) = \lambda_{\min}[\mathbf{M}(\mathbf{w})] = \min_{\|\mathbf{u}\|=1} \mathbf{u}^\top \mathbf{M}(\mathbf{w})\mathbf{u}$$

corresponds to (9.56) with $\mathcal{U} = \{\mathbf{u} \in \mathbb{R}^p : \|\mathbf{u}\| = 1\}$. The vector $\mathbf{u}^{(k+1)}$ to be used at step 1 of the algorithm can be taken as any eigenvector of $\mathbf{M}(\mathbf{w}^k)$ associated with its minimum eigenvalue and normed to 1.

c-optimum Design

For $\mathbf{c}$ a given vector of $\mathbb{R}^p$, which we suppose of norm 1 without any loss of generality, the positively homogeneous form of the c-optimality criterion is

$$\phi_c^+(\mathbf{w}) = \{\mathbf{c}^\top \mathbf{M}^-(\mathbf{w})\mathbf{c}\}^{-1} \text{ if } \mathbf{c} \in \mathcal{M}(\mathbf{M}(\mathbf{w})) \text{ and } \phi_c^+(\mathbf{w}) = 0 \text{ otherwise},$$

with $\mathbf{M}^-$ any g-inverse of $\mathbf{M}$; see Sects. 5.1.2 and 5.1.4. Using Lemma 5.6, we can rewrite $\phi_c^+(\mathbf{w})$ as

$$\phi_c^+(\mathbf{w}) = \min_{\mathbf{u}^\top \mathbf{c}=1} \mathbf{u}^\top \mathbf{M}(\mathbf{w})\mathbf{u},$$

which has the form (9.56) with $\mathcal{U} = \{\mathbf{u} \in \mathbb{R}^p : \mathbf{u}^\top \mathbf{c} = 1\}$. Using Lemma 5.6 again, the vector $\mathbf{u}^{(k+1)}$ of (9.58) can be taken as

$$\mathbf{u}^{(k+1)} = \begin{cases} \frac{\mathbf{M}^-(\mathbf{w}^k)\mathbf{c}}{\mathbf{c}^\top \mathbf{M}^-(\mathbf{w}^k)\mathbf{c}} \text{ if } \mathbf{c} \in \mathcal{M}(\mathbf{M}(\mathbf{w}^k)) \\ \mathbf{v}/(\mathbf{v}^\top \mathbf{c}) \text{ for some } \mathbf{v} \in \mathcal{N}(\mathbf{M}(\mathbf{w}^k)) \text{ satisfying } \mathbf{v}^\top \mathbf{c} \neq 0 \text{ otherwise.} \end{cases}$$

Example 9.15. Consider the same situation as in Example 9.13-A. When $\mathscr{X} = [0, 1]$ is discretized into a regular grid with 101 points, the algorithm above with $\epsilon = 10^{-3}$ and initialized with $w_i^0 = 1/101$ for all i returns the optimal design measure (the delta measure at $x^* = 0.5$) in 6 iterations, thus requiring the solutions of 6 LP problems in $\mathbb{R}^{102}$. □

The behavior of the algorithm is similar when we replace the g-inverse $\mathbf{M}^-(\mathbf{w})$ by $[\mathbf{M}(\mathbf{w}) + \gamma \mathbf{I}]^{-1}$, with $\mathbf{I}_p$ the p-dimensional identity matrix and γ a small positive number (e.g., $\gamma = 10^{-6}$), thus simplifying the construction of $\mathbf{u}^{(k+1)}$ at step 1. Note that, using Remark 5.32-(*ii*), a c-optimal design can be obtained as the solution of *one* LP problem. However, the algorithm above can be generalized to any concave criterion $\phi(\cdot)$, whereas the LP formulation of Remark 5.32 is specific to c-optimality.

G-optimum Design

Suppose that the information matrix $\mathbf{M}(\mathbf{w})$ can be written as

$$\mathbf{M}(\mathbf{w}) = \sum_{i=1}^{\ell} w_i \, \mathbf{f}_\theta(x^{(i)}) \mathbf{f}_\theta^\top(x^{(i)})$$

and consider the following form of the G-optimality criterion (see Sect. 5.1.2):

$$\phi_G(\mathbf{w}) = \min_{x \in \mathscr{X}} \frac{1}{\mathbf{f}_\theta^\top(x)\mathbf{M}^-(\mathbf{w})\mathbf{f}_\theta(x)} = \min_{x \in \mathscr{X}} \min_{\mathbf{u}^\top \mathbf{f}_\theta(x)=1} \mathbf{u}^\top \mathbf{M}(\mathbf{w})\mathbf{u}, \tag{9.59}$$

where we used Lemma 5.6. Suppose that $\mathbf{M}(\mathbf{w})$ is nonsingular. The minimum on the right-hand side of (9.59) is then obtained for

$$x = x_*(\mathbf{w}) = \arg\max_{x \in \mathscr{X}} \mathbf{f}_\theta^\top(x)\mathbf{M}^{-1}(\mathbf{w})\mathbf{f}_\theta(x)$$

$$\text{and } \mathbf{u} = \mathbf{u}^*(\mathbf{w}) = \frac{\mathbf{M}^{-1}(\mathbf{w})\mathbf{f}_\theta[x_*(\mathbf{w})]}{\mathbf{f}_\theta^\top[x_*(\mathbf{w})]\mathbf{M}^{-1}(\mathbf{w})\mathbf{f}_\theta[x_*(\mathbf{w})]}$$

and equals $\{\mathbf{f}_\theta^\top[x_*(\mathbf{w})]\mathbf{M}^{-1}(\mathbf{w})\mathbf{f}_\theta[x_*(\mathbf{w})]\}^{-1}$. Therefore, the criterion $\phi_G(\mathbf{w})$ has the form (9.56), with $\mathscr{U} = \{\mathbf{u} \in \mathbb{R}^p : \mathbf{u}^\top \mathbf{f}_\theta(x) = 1 \text{ for some } x \in \mathscr{X}\}$. The vector $\mathbf{u}^{(k+1)}$ to be used at step 1 of the algorithm is given by $\mathbf{u}^*(\mathbf{w}^k)$.

At first iterations, when the set $\mathscr{U}^{k+1}$ is still not rich enough, the vector of weights $\hat{\mathbf{w}}$ constructed at step 2 has generally less than p nonzero components and the matrix $\mathbf{M}(\hat{\mathbf{w}})$ is thus singular. One may then use regularization and substitute $\mathbf{M}(\mathbf{w}^{k+1}) + \gamma\mathbf{I}$ for $\mathbf{M}(\mathbf{w}^{k+1})$ in the calculation of $\mathbf{u}^*(\mathbf{w}^{k+1})$, with γ a small positive number. Another possibility is to construct $\mathbf{w}^{k+1}$ at step 3 as $\mathbf{w}^{k+1} = (1-\alpha)\mathbf{w}^k + \alpha\hat{\mathbf{w}}$ when $\mathbf{M}(\hat{\mathbf{w}})$ is singular, or close to being singular, with α some number in $(0, 1)$. When $\mathbf{M}(\mathbf{w}^0)$ has full rank, this ensures that $\mathbf{M}(\mathbf{w}^k)$ has full rank for all k. This modified construction of $\mathbf{w}^{k+1}$ can be used at every iteration, but in that case, all weights that are initially strictly positive remain strictly positive in all subsequent iterations; see Remark 9.6.

Example 9.16. We consider G-optimum design (equivalent to D-optimum design) for the two models of Example 9.13-B: $\eta(x,\theta) = \theta_1 + \theta_2 x + \theta_3 x^2$, $x \in \mathscr{X} = [-1, 1]$, and $\eta([x,y],\theta) = \theta_0 + \alpha_1 x + \alpha_2 x^2 + \beta_1 y + \beta_2 y^2$, $(x,y) \in \mathscr{X} = [-1,1]^2$; $\mathscr{X}$ is discretized into a uniform grid $\mathscr{X}_\ell$ with ℓ points. The algorithm is initialized at the uniform measure on $\mathscr{X}_\ell$, and we take $\epsilon = 10^{-6}$. When $\lambda_{\min}[\mathbf{M}(\hat{\mathbf{w}})] < 10^{-6}$, we set $\mathbf{w}^{k+1} = (1-\alpha)\mathbf{w}^k + \alpha\hat{\mathbf{w}}$ at step 3, with $\alpha = 0.9$. Table 9.2 gives the number $k_{\max}$ of steps before the algorithm above stops. The maximum error between the weights returned by the algorithm and the optimal weights (1/3 for the first model, 1/9 for the second) is less than 10^{-16} for the first model and 10^{-6} for the second.

The inequality (9.29) can be used to remove from $\mathscr{X}_\ell$ points that cannot be support points of an optimal design measure, and Table 9.2 also indicates the number $\ell^*_{k_{\max}}$ of remaining points after $k_{\max}$ iterations. The reduction of the cardinality of $\mathscr{X}_\ell$ corresponds to a reduction of the dimension of the LP problem to be solved at each iteration and thus yields a significant acceleration of the algorithm—even if it generally does not reduce the number of iterations necessary to reach the required accuracy ϵ. However, when ℓ gets large, the first iterations require an important computational effort, and the method is then not competitive compared with the algorithm defined by (9.11)–(9.13); see Example 9.14. □

General Concave Criteria

For $\phi(\cdot)$ a general concave function of $\mathbf{w} \in \mathscr{P}_{\ell-1}$, differentiable or not, the cutting-plane algorithm is based on (9.49). For all $k \geq 0$, we define $\overline{\phi}_k = \max_{i=0,\ldots,k} \phi(\mathbf{w}^i)$.

Table 9.2. Behavior of the cutting-plane method for G-optimum design in Example 9.16 and D-optimum design in Example 9.17: number $k_{\max}$ of steps before $\Delta_{k+1} < 10^{-6}$ for different discretizations of $\mathscr{X}$ ($\mathbf{w}_i^0 = 1/\ell$ for all i); number $\ell^*_{k_{\max}}$ of remaining points in $\mathscr{X}_\ell$ after $k_{\max}$ iterations when (9.29) is used to reduce the design space; (9.60) is used with $\alpha = 0.9$ when $\lambda_{\min}[\mathbf{M}(\hat{\mathbf{w}})] < 10^{-6}$

		G-optimality		D-optimality $\log\det\mathbf{M}$		D-optimality $\det^{1/p}\mathbf{M}$	
Model	ℓ	$k_{\max}$	$\ell^*_{k_{\max}}$	$k_{\max}$	$\ell^*_{k_{\max}}$	$k_{\max}$	$\ell^*_{k_{\max}}$
$\eta(x,\theta)$	21	6	3	29	5	11	5
	41	7	3	32	7	25	11
	101	7	5	27	15	25	19
$\eta([x,y],\theta)$	25	28	9	198	9	24	9
	81	29	9	207	9	182	21

0. Take any $\mathbf{w}^0 \in \mathscr{P}_{\ell-1}$, choose $\epsilon > 0$, and set $k = 0$.
1. Use an LP solver to determine $\hat{\mathbf{z}} = (\hat{\mathbf{w}}^\top,\ \hat{t})^\top$ solution of

$$\begin{aligned} \text{maximize} \quad & \mathbf{a}^\top \mathbf{z}\,, \text{ where } \mathbf{a} = (0,\ldots,0,1)^\top \in \mathbb{R}^{\ell+1}\,, \\ & \text{with respect to } \mathbf{z} = (\mathbf{w}^\top,\ t)^\top \in \mathbb{R}^{\ell+1} \\ \text{subject to} \quad & \mathbf{w} \in \mathscr{P}_{\ell-1} \\ & \text{and } \nabla^\top \phi(\mathbf{w}^j)(\mathbf{w}-\mathbf{w}^j) + \phi(\mathbf{w}^j) \geq t \text{ for } j = 0,\ldots,k\,. \end{aligned}$$

2. Set $\mathbf{w}^{k+1} = \hat{\mathbf{w}}$. If $\Delta_{k+1} = \hat{t} - \overline{\phi}_{k+1} < \epsilon$, take $\mathbf{w}^{j^*}$ such that $j^* = \arg\max_{j=0,\ldots,k+1} \phi(\mathbf{w}^j)$ as an ϵ-optimal solution and stop; otherwise $k \leftarrow k+1$, return to step 1.

The vector $\hat{\mathbf{w}}$ constructed at step 1 has generally less than p nonzero components at first iterations. Regularization can then be used to avoid singular matrices $\mathbf{M}(\mathbf{w}^k)$ if necessary. One may also construct $\mathbf{w}^{k+1}$ at step 2 according to

$$\mathbf{w}^{k+1} = (1-\alpha)\mathbf{w}^k + \alpha\hat{\mathbf{w}}\,,\ \alpha \in (0,1)\,, \tag{9.60}$$

when $\mathbf{M}(\hat{\mathbf{w}})$ is close to singularity; $\mathbf{M}(\mathbf{w}^k)$ has then full rank for all k when $\mathbf{M}(\mathbf{w}^0)$ has full rank. This form of updating can be used at every iteration of the algorithm. However, if α is maintained strictly smaller than 1, then all nonzero initial weights remain strictly positive; see Remark 9.6. When $\phi(\cdot)$ is differentiable, $\hat{\mathbf{w}} - \mathbf{w}^k$ is an ascent direction (see Hearn and Lawphongpanich 1989), and one may thus determine an optimal stepsize in (9.60).

Example 9.17. We consider the maximization of $\phi_D(\mathbf{w}) = \log\det\mathbf{M}(\mathbf{w})$, with gradient $\{\nabla\phi_D(\mathbf{w})\}_i = \text{trace}[\mathbf{M}^{-1}(\mathbf{w})\mathbf{M}_\theta(x^{(i)})]$ when $\mathbf{M}(\mathbf{w})$ is given by (9.57), in the same situation as in Example 9.16. We use (9.60) at step 2, with $\alpha = 0.9$, when $\lambda_{\min}[\mathbf{M}(\hat{\mathbf{w}})] < 10^{-6}$. Table 9.2 gives the number $k_{\max}$ of steps before the algorithm above stops when initialized at the uniform measure on $\mathscr{X}_\ell$ and $\epsilon = 10^{-6}$ at step 2. The maximum error between the weights returned by the algorithm and the D-optimal weights (1/3 for the first model, 1/9 for the second) is less than $2.0 \cdot 10^{-4}$. The table also indicates the number $\ell^*_{k_{\max}}$ of remaining points in $\mathscr{X}_\ell$ after $k_{\max}$ iterations when (9.29) is used to remove points that cannot be support points of a D-optimal design measure.

The cutting-plane algorithm may behave differently when an equivalent form of the design criterion is used (see Sect. 5.1.4), and we also consider $\phi_D^+(\mathbf{w}) = \det^{1/p}[\mathbf{M}(\mathbf{w})]$, with gradient $\nabla\phi_D^+(\mathbf{w}) = (1/p)\,\phi_D^+(\mathbf{w})\,\nabla\phi_D(\mathbf{w})$. The performances obtained with this criterion are also indicated in Table 9.2. □

The cutting-plane algorithm can easily manage linear inequalities defined by cost constraints like (5.26), or linear equality constraints, since they can be directly taken into account by the LP solver. As shown in (Veinott, 1967) (see also Avriel 2003, Chap. 14), the algorithm can also be adapted to the presence of nonlinear constraints defining a convex set; see Sect. 5.1.9.

When $\phi(\cdot)$ is not differentiable, $\nabla\phi(\mathbf{w})$ can be taken as any subgradient of $\phi(\cdot)$ at $\mathbf{w}$. The method can thus be used in particular for maximin-optimum design where the criterion is $\phi_{MmO}(\mathbf{w}) = \min_{\theta\in\Theta}\phi(\mathbf{w};\theta)$, with Θ a compact subset of $\mathbb{R}^p$, finite of with nonempty interior, and $\phi(\cdot;\theta)$ is concave for all $\theta \in \Theta$; see Sect. 9.3.1. In that case, a subgradient of $\phi_{MmO}(\cdot)$ at $\mathbf{w}^j$, to be used at step 1, is given by $\nabla\phi_{MmO}(\mathbf{w}^j) = \nabla\phi_{\theta^*}(\mathbf{w}^j)$, with θ^* such that $\phi_{\theta^*}(\mathbf{w}^j) = \phi(\mathbf{w}^j;\theta^*) = \min_{\theta\in\Theta}\phi(\mathbf{w}^j;\theta)$. Example 8.6 gives an illustration.

The Level Method

Although the examples presented above indicate convincing performance when the value of ℓ remains moderate, the method of cutting planes is known to have sometimes rather poor convergence properties; see, e.g., Bonnans et al. (2006, Chap. 9), Nesterov (2004, Sect. 3.3.2). In particular, the search for $\hat{\mathbf{w}}$ in the whole simplex $\mathscr{P}_{\ell-1}$ at step 1 sometimes produces numerical instability. It is then helpful to restrict this search to some neighborhood of the best solution obtained so far. This is a central idea in bundle methods; see Lemaréchal et al. (1995), Bonnans et al. (2006, Chaps. 9–10). One may use in particular the level method (see Nesterov 2004, Sect. 3.3.3), which we briefly describe hereafter.

The level method adds to each iteration of the cutting-plane algorithm presented above a QP step. The solution of the LP problem of step 1, based on a piecewise linear approximation of $\phi(\cdot)$ (see (9.49)) is only used to compute an upper bound t^{k+1} on the optimal criterion value ϕ^*. The next weights $\mathbf{w}^{k+1}$ are then obtained as the orthogonal projection of $\mathbf{w}^k$ onto the polyhedron

$$\mathcal{P}_k = \{\mathbf{w} \in \mathscr{P}_{\ell-1} : \nabla^\top\phi(\mathbf{w}^j)(\mathbf{w} - \mathbf{w}^j) + \phi(\mathbf{w}^j) \geq L_k(\alpha) \text{ for } j = 0, \ldots, k\},$$

where

$$L_k(\alpha) = (1-\alpha)t^{k+1} + \alpha\overline{\phi}_k ,$$

with $\overline{\phi}_k = \max_{j=0,\ldots,k} \phi(\mathbf{w}^j)$, for some given $\alpha \in (0,1)$. The algorithm is as follows:

0. Take any $\mathbf{w}^0 \in \mathscr{P}_{\ell-1}$, choose $\epsilon > 0$, and set $k = 0$.
1. Use an LP solver to determine t^{k+1} solution of

$$\begin{aligned} &\text{maximize} \quad t \\ &\text{subject to} \quad \mathbf{w} \in \mathscr{P}_{\ell-1} \\ &\qquad \text{and } \nabla^\top \phi(\mathbf{w}^j)(\mathbf{w}-\mathbf{w}^j) + \phi(\mathbf{w}^j) \geq t \text{ for } j = 0,\ldots,k\,. \end{aligned}$$

2. If $\Delta_{k+1} = t^{k+1} - \overline{\phi}_k < \epsilon$, take $\mathbf{w}^{j^*}$ with $j^* = \arg\max_{j=0,\ldots,k} \phi(\mathbf{w}^j)$ as an ϵ-optimal solution and stop; otherwise go to step 3.
3. Find $\mathbf{w}^{k+1}$ solution of the QP problem

$$\begin{aligned} &\text{minimize} \quad \|\mathbf{w}-\mathbf{w}^k\|^2 \\ &\text{subject to} \quad \mathbf{w} \in \mathscr{P}_{\ell-1} \\ &\qquad \text{and } \nabla^\top \phi(\mathbf{w}^j)(\mathbf{w}-\mathbf{w}^j) + \phi(\mathbf{w}^j) \geq L_k(\alpha) \text{ for } j = 0,\ldots,k\,, \end{aligned}$$

   $k \leftarrow k+1$, return to step 1.

Note that the sequence $\{\overline{\phi}_k\}$ of record values is increasing while $\{t^k\}$ is decreasing, with the optimal value ϕ^* satisfying $\overline{\phi}_k \leq \phi^* \leq t^k$ for all k. The complexity analysis in (Nesterov, 2004) yields the optimal choice $\alpha^* = 1/(2+\sqrt{2}) \simeq 0.2929$ for α. The method can be used for differentiable or non-differentiable criteria; when $\phi(\cdot)$ is not differentiable, $\nabla\phi(\mathbf{w})$ can be any subgradient of $\phi(\cdot)$ at $\mathbf{w}$. When additional linear (equality or inequality) constraints are present (see Sect. 5.1.10), they can be directly taken into account at steps 1 (LP) and 3 (QP).

Compared with the method of cutting planes, each iteration of the level method requires the solutions of an LP and a QP problem. However, the theoretical performance of the method, measured through complexity bounds, is much better, and numerical experimentations indicate that in general significantly less iterations than with the method of cutting planes are required to reach a solution with similar accuracy on the criterion value. Next example gives an illustration.

Example 9.17 (continued). We consider D-optimum design for the two models of Example 9.16 with $\phi_D(\mathbf{w}) = \log\det \mathbf{M}(\mathbf{w})$ and $\phi_D^+(\mathbf{w}) = \det^{1/p}[\mathbf{M}(\mathbf{w})]$. We take $\epsilon = 10^{-6}$ and $\mathbf{w}^0$ corresponds to the uniform measure on $\mathscr{X}_\ell$. Table 9.3 reports the results obtained with the level method with $\alpha = \alpha^*$: number $k_{\max}$ of steps before the algorithm above stops and number $\ell^*_{k_{\max}}$ of remaining points in $\mathscr{X}_\ell$ after $k_{\max}$ iterations when (9.29) is used to remove points that cannot be support points of a D-optimal design measure. The designs

Table 9.3. Behavior of the level method with $\alpha = \alpha^*$ for D-optimum design in Example 9.17: number $k_{\max}$ of steps before $\Delta_{k+1} < 10^{-6}$ for different discretizations of $\mathscr{X}$ ($\mathbf{w}_i^0 = 1/\ell$ for all i,) and number $\ell^*_{k_{\max}}$ of remaining points in $\mathscr{X}_\ell$ after $k_{\max}$ iterations when (9.29) is used to reduce the design space

		$\log\det\mathbf{M}$		$\det^{1/p}\mathbf{M}$	
Model	ℓ	$k_{\max}$	$\ell^*_{k_{\max}}$	$k_{\max}$	$\ell^*_{k_{\max}}$
$\eta(x,\theta)$	21	15	5	15	5
	41	16	5	16	9
	101	18	15	17	15
$\eta([x,y],\theta)$	25	19	9	14	9
	81	22	9	16	9

generated along the iterations are nonsingular, so that we do not need to take care of singular information matrices. The method behaves similarly for the two criteria $\phi_D(\cdot)$ and $\phi_D^+(\cdot)$. Notice the smaller number of iterations required to reach convergence compared with Table 9.2. □

Appendix A

Subdifferentials and Subgradients

Let $\Phi(\cdot)$ be a concave criterion function defined on some set $\mathcal{M} \subset \mathbb{M}$, e.g., $\mathcal{M} = \mathbb{M}^{\geq}$. The definition of $\Phi(\cdot)$ can be extended to any $p \times p$ symmetric matrix in $\mathbb{M}$ by setting $\Phi(\mathbf{M}) = -\infty$ for $\mathbf{M} \notin \mathcal{M}$. This extension is then concave on $\mathbb{M}$; its effective domain is the set $\mathrm{dom}(\Phi) = \{\mathbf{M} \in \mathbb{M} : \Phi(\mathbf{M}) > -\infty\}$. Note that $\mathbb{M}^{\geq} \subset \mathrm{dom}(\Phi)$ when $\Phi(\cdot)$ positively homogeneous and isotonic; see Lemma 5.4-(*iii*). A concave function $\Phi(\cdot)$ is called proper when $\mathrm{dom}(\Phi) \neq \emptyset$ and $\Phi(\mathbf{M}) < \infty$ for all $\mathbf{M} \in \mathbb{M}$. As a rule all the criteria we consider are proper.

When $\Phi(\cdot) : \mathbb{M} \longrightarrow \mathbb{R}$ is non-differentiable, the notion of gradient can be generalized as follows. A matrix $\tilde{\mathbf{M}}$ is called a subgradient of $\Phi(\cdot)$ at $\mathbf{M}$ if

$$\Phi(\mathbf{A}) \leq \Phi(\mathbf{M}) + \mathrm{trace}[\tilde{\mathbf{M}}(\mathbf{A} - \mathbf{M})]\,,\ \forall \mathbf{A} \in \mathbb{M}\,. \tag{A.1}$$

Here $\mathrm{trace}(\mathbf{A}, \mathbf{B})$ is the usual scalar product between $\mathbf{A}$ and $\mathbf{B}$ in $\mathbb{M}$. The set of all subgradients of $\Phi(\cdot)$ at $\mathbf{M}$ is called the subdifferential[1] of $\Phi(\cdot)$ at $\mathbf{M}$ and is denoted by $\partial\Phi(\mathbf{M})$. The fact that these notions generalize that of gradient is due to the property $\partial\Phi(\mathbf{M}) = \{\nabla_{\mathbf{M}}\Phi(\mathbf{M})\}$ when $\Phi(\cdot)$ is differentiable at $\mathbf{M}$. In other situations $\partial\Phi(\mathbf{M})$ is not reduced to that singleton; it defines a convex set, closed if bounded, empty when $\mathbf{M} \notin \mathrm{dom}(\Phi)$, and satisfies the following properties. For any $\Phi(\cdot)$ concave on $\mathbb{M}$,

$$\partial(\alpha\Phi)(\mathbf{M}) = \alpha\partial\Phi(\mathbf{M})\,,\ \forall \mathbf{M} \in \mathbb{M}\,,\ \forall \alpha > 0\,. \tag{A.2}$$

For any $\Phi(\cdot)$ and $f(\cdot)$ concave on $\mathbb{M}$

$$\partial[\Phi + f](\mathbf{M}) = \partial\Phi(\mathbf{M}) + \partial f(\mathbf{M})\,,\ \forall \mathbf{M} \in \mathbb{M}\,, \tag{A.3}$$

[1]Subgradients and subdifferentials are usually defined for convex functions. We keep the same denomination here, although supergradients and superdifferentials might be more appropriate due to the upper-bound property (A.1); see Rockafellar (1970, p. 308).

L. Pronzato and A. Pázman, *Design of Experiments in Nonlinear Models*, Lecture Notes in Statistics 212, DOI 10.1007/978-1-4614-6363-4,

if there exists some $\mathbf{A} \in \mathbb{M}$ where $f(\mathbf{A})$ is finite and $\Phi(\cdot)$ is continuous; see Alexéev et al. (1987, Sect. 3). Another sufficient condition is that the effective domains of $\Phi(\cdot)$ and $f(\cdot)$ overlap sufficiently, i.e., that their relative interiors[2] have a point in common; see Rockafellar (1970, p. 223). Also, for any $\Phi_1(\cdot)$, $\Phi_2(\cdot)$ concave on $\mathbb{M}$, continuous at $\hat{\mathbf{M}}$ such that $\Phi_1(\hat{\mathbf{M}}) = \Phi_2(\hat{\mathbf{M}})$,

$$\partial[\min(\Phi_1, \Phi_2)](\hat{\mathbf{M}}) = \mathrm{conv}[\partial\Phi_1(\hat{\mathbf{M}}) \cup \partial\Phi_2(\hat{\mathbf{M}})] \qquad \text{(A.4)}$$

with $\mathrm{conv}(\mathcal{S})$ the convex hull of the set $\mathcal{S}$; see Alexéev et al. (1987, Sect. 3). For a continuous version of this property, consider a set of proper criteria functions $\Phi_\gamma(\cdot)$ from $\mathbb{M}$ to $\mathbb{R}$ (i.e., such that $\Phi_\gamma(\mathbf{M}) > -\infty$ for some $\mathbf{M}$ and $\Phi_\gamma(\mathbf{M}) < \infty$ for all $\mathbf{M} \in \mathbb{M}$) with $\gamma \in \Gamma$, a compact subset of $\mathbb{R}$, such that $\Phi_\gamma(\cdot)$ is concave and upper semicontinuous for all $\gamma \in \Gamma$ and the function $\gamma \longrightarrow \Phi_\gamma(\mathbf{M})$ is lower semicontinuous in γ for all $\mathbf{M}$. Suppose that $\Phi_\gamma(\cdot)$ is continuous at $\hat{\mathbf{M}}$ for all $\gamma \in \Gamma$ and define $\Phi^*(\mathbf{M}) = \min_{\gamma\in\Gamma} \Phi_\gamma(\mathbf{M})$ and $\Gamma^*(\mathbf{M}) = \{\gamma \in \Gamma : \Phi_\gamma(\mathbf{M}) = \Phi^*(\mathbf{M})\}$. Then, $\Phi^*(\cdot)$ is concave, and any element $\tilde{\mathbf{M}}$ of its subdifferential $\partial\Phi^*(\hat{\mathbf{M}})$ at $\hat{\mathbf{M}}$ can be written as

$$\tilde{\mathbf{M}} = \sum_{i=1}^{r} \alpha_i \tilde{\mathbf{M}}_i \qquad \text{(A.5)}$$

with $r \leq p(p+1)/2 + 1$, $\sum_{i=1}^{r} \alpha_i = 1$, $\alpha_i > 0$, and $\tilde{\mathbf{M}}_i \in \partial\Phi_{\gamma_i}(\hat{\mathbf{M}})$ for some $\gamma_i \in \Gamma^*(\hat{\mathbf{M}})$, $i = 1, \ldots, r$; see Alexéev et al. (1987, p. 67).

Subgradients can also be defined for indicator functions. Let $\mathcal{M}$ be a convex subset of $\mathbb{M}$ and define

$$\hat{\mathbb{I}}_{\mathcal{M}}(\mathbf{M}) = \begin{cases} 0 \text{ if } \mathbf{M} \in \mathcal{M} \\ -\infty \text{ otherwise.} \end{cases}$$

Then, $\tilde{\mathbf{M}} \in \partial\hat{\mathbb{I}}_{\mathcal{M}}(\mathbf{M})$ if and only if $\hat{\mathbb{I}}_{\mathcal{M}}(\mathbf{A}) \leq \hat{\mathbb{I}}_{\mathcal{M}}(\mathbf{M}) + \mathrm{trace}[\tilde{\mathbf{M}}(\mathbf{A} - \mathbf{M})]$ for all $\mathbf{A} \in \mathbb{M}$, see (A.1), and therefore $\mathbf{A} \in \mathcal{M}$ implies $\mathbf{M} \in \mathcal{M}$ and $\mathrm{trace}[\tilde{\mathbf{M}}(\mathbf{A} - \mathbf{M})] \geq 0$, which means that $-\tilde{\mathbf{M}}$ is normal to $\mathcal{M}$ at $\mathbf{M}$; see Rockafellar (1970, p. 215).

With the notions of subgradients and subdifferentials a large part of the results of differential calculus remain valid for non-differentiable functions. In particular, a necessary-and-sufficient condition for a concave criterion $\Phi(\cdot)$ to reach its maximum value on $\mathbb{M}$ at $\mathbf{M}^*$ is that $\mathbf{O} \in \partial\Phi(\mathbf{M}^*)$, with $\mathbf{O}$ the null matrix; see Rockafellar (1970, p. 264). From this we directly obtain the following; see Pukelsheim (1993, p. 162).

[2]The relative interior of a convex set $\mathcal{S}$ is the interior of $\mathcal{S}$ regarded as a subset of the smallest affine set containing $\mathcal{S}$ (i.e., the affine hull of $\mathcal{S}$).

Theorem A.1. *Let $\Phi(\cdot)$ be a concave criterion taking finite values on $\mathbb{M}^{>}$ and let $\mathcal{M}$ be a convex subset of $\mathbb{M}^{\geq}$ that intersects $\mathbb{M}^{>}$. Then $\mathbf{M}^*$ maximizes $\Phi(\cdot)$ over $\mathcal{M}$ if and only if there exists $\tilde{\mathbf{M}} \in \partial\Phi(\mathbf{M}^*)$ such that*

$$\text{trace}[\tilde{\mathbf{M}}(\mathbf{A} - \mathbf{M}^*)] \leq 0\,, \forall \mathbf{A} \in \mathcal{M}\,. \tag{A.6}$$

Indeed, using (A.3) the necessary-and-sufficient condition $\mathbf{O} \in \partial[\Phi + \hat{\mathbb{I}}_{\mathcal{M}}](\mathbf{M}^*)$ becomes: there exists $\tilde{\mathbf{M}} \in \partial\Phi(\mathbf{M}^*)$ such that $-\tilde{\mathbf{M}} \in \partial\hat{\mathbb{I}}_{\mathcal{M}}(\mathbf{M}^*)$, which gives (A.6).

In the particular case where $\Phi(\cdot)$ is differentiable with $\mathcal{M} = \mathcal{M}_\theta(\Xi)$, Theorem A.1 says that $\mathbf{M}^*$ is Φ-optimal on $\mathcal{M}_\theta(\Xi)$ if and only if $F_\Phi(\mathbf{M}^*, \mathbf{A}) = \text{trace}[\nabla_{\mathbf{M}}\Phi(\mathbf{M}^*)(\mathbf{A} - \mathbf{M}^*)] \leq 0$ for all $\mathbf{A} \in \mathcal{M}_\theta(\Xi)$. Writing $\mathbf{M}^* = \mathbf{M}(\xi^*)$ and $\mathbf{A} = \mathbf{M}(\nu)$ for some ξ^* and ν in Ξ, we obtain that ξ^* is ϕ-optimal on Ξ if and only if $F_\phi(\xi^*; \nu) \leq 0$ for all $\nu \in \Xi$, see (5.34), which corresponds to the equivalence theorem 5.21 (note that $F_\phi(\xi; \xi) = 0$ for all ξ).

More generally, consider the case where $\Phi(\cdot)$ is not differentiable everywhere. Then, the one-sided directional derivative $\Phi'(\mathbf{M}^*, \mathbf{A})$ defined by (5.30) is given by

$$\Phi'(\mathbf{M}^*, \mathbf{A}) = \inf\{\text{trace}(\tilde{\mathbf{M}}\mathbf{A}) : \tilde{\mathbf{M}} \in \partial\Phi(\mathbf{M}^*)\}\,, \tag{A.7}$$

see Rockafellar (1970, pp. 216–217), and the subgradient theorem says that $\mathbf{M}^*$ is Φ-optimal on $\mathcal{M}_\theta(\Xi)$ if and only if

$$\begin{aligned} F_\Phi(\mathbf{M}^*, \mathbf{A}) &= \Phi'(\mathbf{M}^*, \mathbf{A} - \mathbf{M}^*) \\ &= \inf_{\tilde{\mathbf{M}} \in \partial\Phi(\mathbf{M}^*)} \text{trace}[\tilde{\mathbf{M}}(\mathbf{A} - \mathbf{M}^*)] \leq 0\,,\ \forall \mathbf{A} \in \mathcal{M}_\theta(\Xi)\,, \end{aligned} \tag{A.8}$$

which again corresponds to the equivalence theorem. Since the subdifferential $\partial\Phi(\mathbf{M}^*)$ is convex, the minimax theorem applies (Dem'yanov and Malozemov 1974, Theorem 5.2, p. 218). The necessary-and-sufficient condition (A.8) for the Φ-optimality of $\mathbf{M}^*$ on $\mathcal{M}_\theta(\Xi)$ can be expressed as the existence of $\tilde{\mathbf{M}} \in \partial\Phi(\mathbf{M}^*)$ such that $\text{trace}[\tilde{\mathbf{M}}(\mathbf{A} - \mathbf{M}^*)] \leq 0$ for all $\mathbf{A} \in \mathcal{M}_\theta(\Xi)$. This type of condition has been used, for instance, in Theorem 5.38.

Finally, notice that the expressions for the directional derivatives of non-differentiable criteria obtained in Sect. 5.2.1 by using Lemmas 5.17 and 5.18 are direct consequences of (A.7) and (A.4), (A.5).

Appendix B

Computation of Derivatives Through Sensitivity Functions

The computation of the derivatives $\partial\eta(x,\theta)/\partial\theta_i$, $i = 1,\dots,p$, of the model response $\eta(x,\theta)$ with respect to the model parameters θ is a mandatory step for most of the developments presented throughout the book: they are required, for instance, to evaluate the information matrix, the curvatures of the model, etc. However, in many circumstances the analytic expression of $\eta(x,\theta)$ is unknown, and its derivatives can only be obtained numerically. This appendix shows that this does not raise any particular difficulty, apart perhaps the computational time required by numerical calculations performed on a computer.

Consider the case, often met in practical applications, when $\eta(x,\theta)$ is the solution of a differential equation (similar developments can be made for recurrence equations).[1] Then, the derivatives $\partial\eta(x,\theta)/\partial\theta_i$, also called *sensitivity functions*, are solutions of other differential equations, which can easily be derived from the original one; see, e.g., Rabitz et al. (1983) and Walter and Pronzato (1997, Chap. 4). Only first-order derivatives are considered hereafter, but the developments easily extend to higher-order derivation. One can refer to classical textbooks on numerical analysis for methods to solve initial-value problems; see, e.g., Stoer and Bulirsch (1993).

Consider, for instance, the following state-space representation for the equations that give the response $\eta(x,\theta)$:

$$\dot{\mathbf{v}}(x,\theta,t) = \frac{\mathrm{d}\mathbf{v}(x,\theta,t)}{\mathrm{d}t} = \mathbf{F}[\mathbf{v}(x,\theta,t),\theta], \quad \mathbf{v}(x,\theta,0) = \mathbf{v}_0(x,\theta)\,, \quad \text{(B.1)}$$

$$\eta(x,\theta,t) = \mathbf{H}[\mathbf{v}(x,\theta,t),\theta]\,. \quad \text{(B.2)}$$

Here t denotes the time and $\mathbf{v}(x,\theta,t)$ the vector of state variables at time t for experimental conditions x and model parameters θ. The dependence of the

[1] Other methods for exact differentiation, based on adjoint state or adjoint code approaches, are available for more general situations where $\eta(x,\theta)$ is not given by a differential or recurrence equation, see, e.g., Walter and Pronzato (1997, Chap. 4).

L. Pronzato and A. Pázman, *Design of Experiments in Nonlinear Models*, Lecture Notes in Statistics 212, DOI 10.1007/978-1-4614-6363-4,

right-hand side in some input signal $u(t)$ is omitted for the sake of simplicity of notations. Also, the (matrix) functions $\mathbf{F}$ and $\mathbf{H}$ might depend explicitly on t, which would correspond to a nonstationary system. The notation $\mathbf{v}_0(x, \theta)$ is to stress the fact that the initial conditions may be part of the unknown parameters to be estimated.

We wish to determine the values of $\partial\eta(x, \theta, t)/\partial\theta_i$, $i = 1, \ldots, p$, at some particular values of t given by the sampling times $t_1, t_2, \ldots, t_N$ at which the observations are performed. Note that, although we write $\eta(x, \theta, t)$, these sampling times may be part of the design variables x. Also, x may include some control variables that influence the input signal $u(t)$, in which case we would write $u(t) = u(x, t)$.

The derivation of (B.2) with respect to θ_i gives

$$\frac{\partial\eta(x, \theta, t)}{\partial\theta_i} = \left.\frac{\partial\mathbf{H}(\mathbf{v}, \theta)}{\partial\mathbf{v}^\top}\right|_{\mathbf{v}(x,\theta,t)} \frac{\partial\mathbf{v}(x, \theta, t)}{\partial\theta_i} + \left.\frac{\partial\mathbf{H}(\mathbf{v}, \theta)}{\partial\theta_i}\right|_{\mathbf{v}(x,\theta,t)} \tag{B.3}$$

which requires the evaluation of the derivative $\partial\mathbf{v}(x, \theta, t)/\partial\theta_i$. It is obtained by differentiating the evolution equations (B.1) of the system,

$$\frac{\mathrm{d}}{\mathrm{d}t}\frac{\partial\mathbf{v}(x, \theta, t)}{\partial\theta_i} = \left.\frac{\partial\mathbf{F}(\mathbf{v}, \theta)}{\partial\mathbf{v}^\top}\right|_{\mathbf{v}(x,\theta,t)} \frac{\partial\mathbf{v}(x, \theta, t)}{\partial\theta_i} + \left.\frac{\partial\mathbf{F}(\mathbf{v}, \theta)}{\partial\theta_i}\right|_{\mathbf{v}(x,\theta,t)}, \tag{B.4}$$

and the initial conditions

$$\frac{\partial\mathbf{v}(x, \theta, 0)}{\partial\theta_i} = \frac{\partial\mathbf{v}_0(x, \theta)}{\partial\theta_i}.$$

Therefore, the solution of the initial-value problem (B.1) gives $\eta(x, \theta, t)$, and the solution of p initial-value problems similar to (B.4) gives the sensitivity functions $\partial\eta(x, \theta, t)/\partial\theta_i$, $i = 1, \ldots, p$, through (B.3); see Valko and Vajda (1984) and Bilardello et al. (1993) for details. Notice that the differential equations (B.4) corresponding to θ_i and θ_j with $i \neq j$ are independent, i.e., the solutions can be obtained independently once the trajectory of $\mathbf{v}(x, \theta, t)$ has been obtained. Also note that (B.4) is linear in $\partial\mathbf{v}(x, \theta, t)/\partial\theta_i$ (but nonstationary since $\partial\mathbf{F}(\mathbf{v}, \theta)/\partial\mathbf{v}^\top|_{\mathbf{v}(x,\theta,t)}$ depends on t), and only the driving term $\partial\mathbf{F}(\mathbf{v}, \theta)/\partial\theta_i|_{\mathbf{v}(x,\theta,t)}$ and initial conditions $\partial\mathbf{v}_0(x, \theta)/\partial\theta_i$ depend on i.

On the other hand, although approximating the derivatives $\partial\eta(x, \theta, t)/\partial\theta_i$ by finite differences might seem simpler, it would require the solutions of $p + 1$ initial-value problems of the type (B.1) and would thus only produce approximate results for similar efforts. The situation is even more favorable to exact calculations when the state-space representation (B.1) is linear, i.e., when the differential equation that gives $\eta(x, \theta, t)$ is linear, with known initial conditions. Then, if $\eta(x, \theta, t)$ is solution of an m-th-order differential equation, $\eta(x, \theta, t)$ and its derivatives $\partial\eta(x, \theta, t)/\partial\theta_i$, $i = 1, \ldots, p$, can be obtained by solving an initial-value problem for a differential equation of order $2m$ only, whatever the number p of parameters. Indeed, consider the following m-th-order differential equation

$$\eta^{(m)}(x,\theta,t) + \sum_{i=0}^{m-1} \theta_i\, \eta^{(i)}(x,\theta,t) = \sum_{i=m}^{m+q} \theta_i\, u^{(i-m)}(t)\,, \tag{B.5}$$

where $\eta^{(k)}(x,\theta,t)$ and $u^{(k)}(t)$, $k \geq 0$, respectively denote the k-th-order derivatives of $\eta(x,\theta,t)$ and $u(t)$ with respect to t (with $\eta^{(0)}(x,\theta,t) = \eta(x,\theta,t)$ and $u^{(0)}(t) = u(t)$) and where the initial conditions $\eta^{(i)}(x,\theta,0) = \alpha_i$, $i = 0,\ldots,m-1$, are known. Denote by $s_j(x,\theta,t)$ the sensitivity functions

$$s_j(x,\theta,t) = \frac{\partial \eta(x,\theta,t)}{\partial \theta_j}\,,\; j = 0,\ldots,m+q\,.$$

They are solutions of differential equations of order m, obtained by differentiating (B.5) with respect to the $m+q+1$ parameters θ_i,

$$s_j^{(m)}(x,\theta,t) + \sum_{i=0}^{m-1} \theta_i\, s_j^{(i)}(x,\theta,t) = u^{(j-m)}(t)\,,\; j = m,\ldots,m+q\,, \tag{B.6}$$

$$s_j^{(m)}(x,\theta,t) + \sum_{i=0}^{m-1} \theta_i\, s_j^{(i)}(x,\theta,t) = -\eta^{(j)}(x,\theta,t)\,,\; j = 0,\ldots,m-1\,, \tag{B.7}$$

with zero initial conditions since the α_i are known. The computation of $\eta(x,\theta,t)$ and its derivatives $s_j(x,\theta,t)$ then seems to require the solution of an initial-value problem for $m+q+2$ differential equations of order m. However, one may notice that all these differential equations have the same homogeneous part (left-hand side) and only differ by their driving terms. The computations can thus be simplified as follows. First solve (B.6) for $j = m$ to obtain $s_m(x,\theta,t)$. Then, by linearity, we have $s_{m+k}(x,\theta,t) = \dot{s}_{m+k-1}(x,\theta,t)$ for $k = 1,\ldots,q$. Assume for the moment that the initial conditions α_i equal zero. Then, by linearity again, $\eta(x,\theta,t) = \sum_{j=m}^{m+q} \theta_j s_j(x,\theta,t)$. The solution of (B.7) for $j = 0$ gives $s_0(x,\theta,t)$, and by differentiation with respect to t we get $s_k(x,\theta,t) = \dot{s}_{k-1}(x,\theta,t)$ for $k = 1,\ldots,m-1$. The response $\eta(x,\theta,t)$ and the $m+q+1$ sensitivity functions are thus obtained by solving two initial-value problems for a differential equation of order m. When the α_i are not zero, the solution $\eta(x,\theta,t)$ must be corrected to take those initial conditions into account. This can be done through a state-space representation. Define the vector of state variables at time t by $\mathbf{w}(x,\theta,t) = [\eta^{(m-1)}(x,\theta,t), \eta^{(m-2)}(x,\theta,t), \ldots, \eta^{(0)}(x,\theta,t)]^\top$. It satisfies the differential equation

$$\dot{\mathbf{w}}(x,\theta,t) = \mathbf{A}(\theta)\mathbf{w}(x,\theta,t) + \sum_{i=m}^{m+q} \theta_i\, u^{(i-m)}(t)\; \mathbf{e}_1$$

with $\mathbf{e}_1 = (1,0,\ldots,0)^\top$ the first basis vector of $\mathbb{R}^m$ and $\mathbf{A} = \mathbf{A}(\theta)$ the $m \times m$ matrix

$$\mathbf{A} = \begin{pmatrix} -\theta_{m-1} & -\theta_{m-2} & \cdots & -\theta_1 & -\theta_0 \\ 1 & 0 & \cdots & 0 & 0 \\ 0 & 1 & \cdots & 0 & 0 \\ \vdots & \vdots & \ddots & \vdots & \vdots \\ 0 & 0 & \cdots & 1 & 0 \end{pmatrix}.$$

The response $\eta(x, \theta, t)$ is then given by

$$\eta(x, \theta, t) = \sum_{j=m}^{m+q} \theta_j s_j(x, \theta, t) + \sum_{i=1}^{n_\lambda} \sum_{j=1}^{n_{\lambda_i}} c_{i,j}\, t^{j-1} \exp(\lambda_i t)\,,$$

where n_λ denotes the number of distinct eigenvalues of $\mathbf{A}$, the eigenvalue λ_i having the multiplicity n_{λ_i}. The m constants $c_{i,j}$ are determined from the initial conditions $\eta^{(i)}(x, \theta, 0) = \alpha_i$, $i = 0, \ldots, m-1$.

Appendix C

Proofs

Lemma 2.5. *Let $\{x_i\}$ be an asymptotically discrete design with measure ξ. Assume that $a(x,\theta)$ is a bounded function on $\mathscr{X} \times \Theta$ and that to every $x \in \mathscr{X}$ we can associate a random variable $\varepsilon(x)$. Let $\{\varepsilon_i\}$ be a sequence of independent random variables, with ε_i distributed like $\varepsilon(x_i)$, and assume that for all $x \in \mathscr{X}$*

$$\mathbb{E}\{b[\varepsilon(x)]\} = m(x)\,,\ |m(x)| < \bar{m} < \infty\,,$$
$$\mathrm{var}\{b[\varepsilon(x)]\} = V(x) < \bar{V} < \infty\,,$$

with $b(\cdot)$ a Borel function on $\mathbb{R}$. Then we have

$$\frac{1}{N}\sum_{k=1}^{N} a(x_k,\theta)b(\varepsilon_k) \overset{\theta}{\rightsquigarrow} \sum_{x\in S_\xi} a(x,\theta)m(x)\xi(x)$$

as N tends to ∞, where $\overset{\theta}{\rightsquigarrow}$ means uniform convergence with respect to $\theta \in \Theta$, and the convergence is almost sure (a.s.), i.e., with probability one, with respect to the random sequence $\{\varepsilon_i\}$.

Proof. For any θ, we can write

$$\begin{aligned}
&\left|\frac{1}{N}\sum_{k=1}^{N} a(x_k,\theta)b(\varepsilon_k) - \sum_{x\in S_\xi} a(x,\theta)m(x)\xi(x)\right| \\
&\leq \left|\frac{1}{N}\sum_{k=1,\,x_k\notin S_\xi}^{N} a(x_k,\theta)b(\varepsilon_k)\right| \\
&\quad + \left|\frac{1}{N}\sum_{k=1,\,x_k\in S_\xi}^{N} a(x_k,\theta)b(\varepsilon_k) - \sum_{x\in S_\xi} a(x,\theta)m(x)\xi(x)\right|\,. \qquad (C.1)
\end{aligned}$$

Let $N(x)/N$ be the relative frequency of the point x in the sequence $x_1,\ldots,x_N$. The second term is bounded by

L. Pronzato and A. Pázman, *Design of Experiments in Nonlinear Models*, Lecture Notes in Statistics 212, DOI 10.1007/978-1-4614-6363-4,

$$\sum_{x \in \mathcal{S}_\xi} \sup_{\theta \in \Theta} |a(x,\theta)| \left| \frac{N(x)}{N} \frac{1}{N(x)} \sum_{k=1}^{N(x)} b(\varepsilon_k) - m(x)\xi(x) \right| ;$$

since $\sum_{k=1}^{N(x)} b(\varepsilon_k)/N(x)$ converges a.s. to $m(x)$ (SLLN) and $N(x)/N - \xi(x)$ tends to zero, this term tends a.s. to zero, uniformly in θ. Let A_N denote the first term on the right-hand side of (C.1) and $N(\mathscr{X} \setminus \mathcal{S}_\xi)$ denote the number of points among $x_1, \ldots, x_N$ that belong to the set $\mathscr{X} \setminus \mathcal{S}_\xi$. If $N(\mathscr{X} \setminus \mathcal{S}_\xi)$ is finite, the lemma is proved. Otherwise, A_N satisfies

$$|A_N| - \left| \frac{1}{N} \sum_{k=1,\, x_k \notin \mathcal{S}_\xi}^{N} a(x_k, \theta) b(\varepsilon_k) \right| \leq$$
$$\sup_{x \in \mathscr{X},\, \theta \in \Theta} |a(x,\theta)| \frac{N(\mathscr{X} \setminus \mathcal{S}_\xi)}{N} \frac{1}{N(\mathscr{X} \setminus \mathcal{S}_\xi)} \sum_{k=1,\, x_k \notin \mathcal{S}_\xi}^{N} |b(\varepsilon_k)| .$$

Now, the SLLN applied to the independent sequence of random variables $|b(\varepsilon_k)|$ gives

$$\frac{1}{N(\mathscr{X} \setminus \mathcal{S}_\xi)} \sum_{k=1,\, x_k \notin \mathcal{S}_\xi}^{N} |b(\varepsilon_k)| - \frac{1}{N(\mathscr{X} \setminus \mathcal{S}_\xi)} \sum_{k=1,\, x_k \notin \mathcal{S}_\xi}^{N} \mathbb{E}\{|b[\varepsilon(x_k)]|\} \stackrel{\text{a.s.}}{\to} 0$$

as $N \to \infty$. Moreover,

$$\frac{1}{N(\mathscr{X} \setminus \mathcal{S}_\xi)} \sum_{k=1,\, x_k \notin \mathcal{S}_\xi}^{N} \mathbb{E}\{|b[\varepsilon(x_k)]|\} \leq \sup_{x \in \mathscr{X}} \mathbb{E}\{|b[\varepsilon(x)]|\}$$
$$\leq \sup_{x \in \mathscr{X}} \sqrt{V(x) + m^2(x)} < \infty .$$

Since $N(\mathscr{X} \setminus \mathcal{S}_\xi)/N \to 0$, A_N tends to zero a.s. and uniformly in θ, which completes the proof. ■

Lemma 2.6. *Let $\{z_i\}$ be a sequence of i.i.d. random vectors of $\mathbb{R}^r$ and $a(z,\theta)$ be a Borel measurable real function on $\mathbb{R}^r \times \Theta$, continuous in $\theta \in \Theta$ for any z, with Θ a compact subset of $\mathbb{R}^p$. Assume that*

$$\mathbb{E}\{\max_{\theta \in \Theta} |a(z_1, \theta)|\} < \infty , \tag{C.2}$$

then $\mathbb{E}\{a(z_1,\theta)\}$ is continuous in $\theta \in \Theta$ and

$$\frac{1}{N} \sum_{i=1}^{N} a(z_i, \theta) \stackrel{\theta}{\rightsquigarrow} \mathbb{E}[a(z_1,\theta)] \text{ a.s. when } N \to \infty .$$

Proof. We use a construction similar to that in (Bierens, 1994, p. 43). Take some fixed $\theta^1 \in \Theta$ and consider the set

$$\mathscr{B}(\theta^1, \delta) = \left\{\theta \in \Theta : \ \left\|\theta - \theta^1\right\| \leq \delta\right\}.$$

Define $\bar{a}_\delta(z)$ and $\underline{a}_\delta(z)$ as the maximum and the minimum of $a(z, \theta)$ over the set $\mathscr{B}(\theta^1, \delta)$, which are properly defined random variables from Lemma 2.9. The expectations $\mathbb{E}\{|\underline{a}_\delta(z_1)|\}$ and $\mathbb{E}\{|\bar{a}_\delta(z_1)|\}$ are bounded by

$$\mathbb{E}\{\max_{\theta\in\Theta} |a(z_1, \theta)|\} < \infty .$$

Also, $\bar{a}_\delta(z) - \underline{a}_\delta(z)$ is an increasing function of δ. Hence, we can interchange the order of the limit and expectation in the following expression:

$$\lim_{\delta \searrow 0} \left[\mathbb{E}\{\bar{a}_\delta(z_1)\} - \mathbb{E}\{\underline{a}_\delta(z_1)\}\right] = \mathbb{E}\left\{\lim_{\delta \searrow 0} \left[\bar{a}_\delta(z_1) - \underline{a}_\delta(z_1)\right]\right\} = 0 ,$$

which proves the continuity of $\mathbb{E}\{a(z_1, \theta)\}$ at θ^1 and implies

$$\forall \beta > 0 , \ \exists \delta(\beta) > 0 \ \text{ such that } \ \left|\mathbb{E}\{\bar{a}_{\delta(\beta)}(z_1)\} - \mathbb{E}\{\underline{a}_{\delta(\beta)}(z_1)\}\right| < \frac{\beta}{2} .$$

Hence we can write for every $\theta \in \mathscr{B}(\theta^1, \delta(\beta))$

$$\begin{aligned}
\frac{1}{N}\sum_k \underline{a}_{\delta(\beta)}(z_k) - \mathbb{E}\{\underline{a}_{\delta(\beta)}(z_1)\} - \frac{\beta}{2} &\leq \frac{1}{N}\sum_k \underline{a}_{\delta(\beta)}(z_k) - \mathbb{E}\{\bar{a}_{\delta(\beta)}(z_1)\} \\
&\leq \frac{1}{N}\sum_k a(z_k, \theta) - \mathbb{E}\{a(z_1, \theta)\} \\
&\leq \frac{1}{N}\sum_k \bar{a}_{\delta(\beta)}(z_k) - \mathbb{E}\{\underline{a}_{\delta(\beta)}(z_1)\} \\
&\leq \frac{1}{N}\sum_k \bar{a}_{\delta(\beta)}(z_k) - \mathbb{E}\{\bar{a}_{\delta(\beta)}(z_1)\} + \frac{\beta}{2} .
\end{aligned}$$

From the SLLN, we have that $\forall \gamma > 0$, $\exists N_1(\beta, \gamma)$ such that

$$\text{Prob}\left\{\forall N > N_1(\beta, \gamma) , \ \left|\frac{1}{N}\sum_k \bar{a}_{\delta(\beta)}(z_k) - \mathbb{E}\{\bar{a}_{\delta(\beta)}(z_1)\}\right| < \frac{\beta}{2}\right\} > 1 - \frac{\gamma}{2} ,$$

$$\text{Prob}\left\{\forall N > N_1(\beta, \gamma) , \ \left|\frac{1}{N}\sum_k \underline{a}_{\delta(\beta)}(z_k) - \mathbb{E}\{\underline{a}_{\delta(\beta)}(z_1)\}\right| < \frac{\beta}{2}\right\} > 1 - \frac{\gamma}{2} .$$

Combining with previous inequalities, we obtain

$$\text{Prob}\left\{\forall N > N_1(\beta, \gamma), \max_{\theta\in\mathscr{B}(\theta^1,\delta(\beta))} \left|\frac{1}{N}\sum_k a(z_k, \theta) - \mathbb{E}\{a(z_1, \theta)\}\right| < \beta\right\} > 1 - \gamma .$$

It only remains to cover Θ with a finite numbers of sets $\mathscr{B}(\theta^i, \delta(\beta))$, $i = 1, \ldots, n(\beta)$, which is always possible from the compactness assumption. For any $\alpha > 0, \beta > 0$, take $\gamma = \alpha/n(\beta)$, $N(\beta) = \max_i N_i(\beta, \gamma)$. We obtain

$$\text{Prob}\left\{\forall N > N(\beta)\,,\ \max_{\theta \in \Theta}\left|\frac{1}{N}\sum_k a(z_k, \theta) - \mathbb{E}\{a(z_1, \theta)\}\right| < \beta\right\} > 1 - \alpha\,,$$

which completes the proof. ■

Lemma 2.7. *Let* $\{z_i\}$, θ, Θ *and* $a(z, \theta)$ *be defined as in Lemma 2.6. Assume that*

$$\sup_{\theta \in \Theta} \mathbb{E}\{|a(z_1, \theta)|\} < \infty$$

and that $a(z, \theta)$ *is continuous in* $\theta \in \Theta$ *uniformly in* z*. Then the conclusions of Lemma 2.6 apply.*

Proof. We only need to prove (C.2). The continuity of $a(z, \theta)$ with respect to θ being uniform in z, we have: $\forall \theta^1 \in \Theta\,,\ \forall \epsilon > 0\,,\ \exists \delta(\epsilon) > 0$ such that

$$\forall \theta \in \mathcal{C}(\theta^1, \delta(\epsilon)) = \mathscr{B}(\theta^1, \delta(\epsilon)) \cap \Theta\,,\ \ \sup_z |a(z, \theta) - a(z, \theta^1)| < \epsilon\,.$$

This implies that for all $\theta \in \mathcal{C}(\theta^1, \delta(\epsilon))$, $|a(z, \theta)| < |a(z, \theta^1)| + \epsilon\ \forall z$; that is,

$$\bar{a}^1_{\delta(\epsilon)}(z) = \max_{\theta \in \mathcal{C}(\theta^1, \delta(\epsilon))} |a(z, \theta)| < |a(z, \theta^1)| + \epsilon\ \forall z\,,$$

with $\mathbb{E}\{|a(z_1, \theta^1)|\} < \infty$ by assumption. Therefore, $\mathbb{E}\{|\bar{a}^1_{\delta(\epsilon)}(z_1)|\} < \infty$. Now, we can cover Θ by a finite number of balls $\mathscr{B}(\theta^k, \delta_k(\epsilon))$, $k = 1, \ldots, n(\epsilon)$ and

$$\max_{\theta \in \Theta} |a(z, \theta)| = \max_{k=1,\ldots,n(\epsilon)} \bar{a}^1_{\delta_k(\epsilon)}(z)$$

which implies (C.2). ■

Lemma 2.8. *Let* $\{x_i\}$ *be an asymptotically discrete design with measure* ξ*. Assume that to every* $x \in \mathscr{X}$ *we can associate a random variable* $\varepsilon(x)$*. Let* $\{\varepsilon_i\}$ *be a sequence of independent random variables, with* ε_i *distributed like* $\varepsilon(x_i)$*. Let* $a(x, \varepsilon, \theta)$ *be a Borel measurable function of* ε *for any* $(x, \theta) \in \mathscr{X} \times \Theta$*, continuous in* $\theta \in \Theta$ *for any* x *and* ε*, with* Θ *a compact subset of* $\mathbb{R}^p$*. Assume that*

$$\forall x \in \mathcal{S}_\xi\,,\ \ \mathbb{E}\{\max_{\theta \in \Theta} |a[x, \epsilon(x), \theta]|\} < \infty\,, \tag{C.3}$$

$$\forall x \in \mathscr{X} \setminus \mathcal{S}_\xi\,,\ \ \mathbb{E}\{\max_{\theta \in \Theta} |a[x, \epsilon(x), \theta]|^2\} < \infty\,. \tag{C.4}$$

Then we have

$$\frac{1}{N}\sum_{k=1}^{N} a(x_k,\epsilon_k,\theta) \overset{\theta}{\rightsquigarrow} \sum_{x\in S_\xi} \mathbb{E}\{a[x,\epsilon(x),\theta]\}\,\xi(x) \text{ a.s. when } N\to\infty\,,$$

where the function on the right-hand side is continuous in θ on Θ.

Proof. We have

$$\left|\frac{1}{N}\sum_{k=1}^{N} a(x_k,\epsilon_k,\theta) - \sum_{x\in S_\xi} \mathbb{E}\{a[x,\epsilon(x),\theta]\}\xi(x)\right| \le A_N + B_N$$

with

$$A_N = \left|\frac{1}{N}\sum_{k=1,\,x_k\notin\mathcal{S}_\xi}^{N} a(x_k,\epsilon_k,\theta)\right|$$

and

$$B_N = \left|\frac{1}{N}\sum_{k=1,\,x_k\in\mathcal{S}_\xi}^{N} a(x_k,\epsilon_k,\theta) - \sum_{x\in S_\xi} \mathbb{E}\{a[x,\epsilon(x),\theta]\}\xi(x)\right|.$$

Then

$$B_N \le \sum_{x\in S_\xi}\left|\left[\frac{N(x)}{N}\,\frac{1}{N(x)}\sum_{k=1,\,x_k=x}^{N} a(x,\epsilon_k,\theta)\right] - \mathbb{E}\{a[x,\epsilon(x),\theta]\}\xi(x)\right|$$

where, for each $x\in\mathcal{S}_\xi$, the $a(x,\epsilon_k,\theta)$ are i.i.d. random variables satisfying (C.3). Lemma 2.6 thus applies, and, since $N(x)/N$ tends to $\xi(x)$, $B_N \overset{\theta}{\rightsquigarrow} 0$ a.s. when $N\to\infty$. Also, from the same lemma, $\mathbb{E}\{a[x,\epsilon(x),\theta]\}$ is a continuous function of θ for any x.

$$A_N < \bar{A}_N = \frac{N(\mathscr{X}\setminus\mathcal{S}_\xi)}{N}\,\frac{1}{N(\mathscr{X}\setminus\mathcal{S}_\xi)}\sum_{k=1,\,x_k\notin\mathcal{S}_\xi}^{N} \max_{\theta\in\Theta}|a(x_k,\epsilon_k,\theta)|\,,$$

where $N(\mathscr{X}\setminus\mathcal{S}_\xi)$ denotes the number of points among $x_1,\ldots,x_N$ that belong to the set $\mathscr{X}\setminus\mathcal{S}_\xi$. If $N(\mathscr{X}\setminus\mathcal{S}_\xi)<\infty$, $A_N \overset{\theta}{\rightsquigarrow} 0$ a.s. when $N\to\infty$. Otherwise, the independent random variables $\max_{\theta\in\Theta}|a(x_k,\epsilon_k,\theta)|$ satisfy (C.4), and the SLLN then implies

$$\frac{1}{N(\mathscr{X}\setminus\mathcal{S}_\xi)}\sum_{k=1,\,x_k\notin\mathcal{S}_\xi}^{N}\left(\max_{\theta\in\Theta}|a(x_k,\epsilon_k,\theta)| - \mathbb{E}\{\max_{\theta\in\Theta}|a[x_k,\epsilon(x_k),\theta]|\}\right) \overset{\text{a.s.}}{\to} 0$$

when $N\to\infty$, which implies $\bar{A}_N \overset{\text{a.s.}}{\to} 0$ since $N(\mathscr{X}\setminus\mathcal{S}_\xi)/N\to 0$, and therefore $A_N \overset{\theta}{\rightsquigarrow} 0$ a.s. when $N\to\infty$. ■

Lemma 2.9 (Jennrich 1969). *Let Θ be a compact subset of $\mathbb{R}^p$, $\mathscr{Z}$ be a measurable subset of $\mathbb{R}^m$ and $J(z,\theta)$ be a Borel measurable real function on $\mathscr{Z}\times\Theta$, continuous in $\theta\in\Theta$ for any $z\in\mathscr{Z}$. Then there exists a mapping $\hat\theta$ from $\mathscr{Z}$ into Θ with Borel measurable components such that $J[z,\hat\theta(z)] = \min_{\theta\in\Theta} J(z,\theta)$, which therefore is also Borel measurable. If, moreover, $J(z,\theta)$ is continuous on $\mathscr{Z}\times\Theta$, then $\min_{\theta\in\Theta} J(z,\theta)$ is a continuous function on $\mathscr{Z}$.*

Proof. $J(z,\theta)$ is a measurable function of z for any $\theta\in\Theta$ and a continuous function of θ for any $z\in\mathscr{Z}$. Let $\{\Theta_k\}$ be an increasing sequence of finite subsets of Θ whose limit is dense in Θ. For any k, there exists a measurable function $\tilde\theta^k$ from $\mathscr{Z}$ into Θ_k such that

$$\forall z\in\mathscr{Z}\,,\ J(z,\tilde\theta^k) = \min_{\theta\in\Theta_k} J(z,\theta)\,.$$

Define $\hat\theta_1 = \hat\theta_1(z) = \liminf_{k\to\infty}\tilde\theta_1^k(z)$ (with $\tilde\theta_1^k$ the first component of $\tilde\theta^k$), and notice that $\hat\theta_1$ is measurable. For any $z\in\mathscr{Z}$, there exists a subsequence $\{\tilde\theta^{k_i}(z)\}$ of $\{\tilde\theta^k(z)\}$ that converges to a point $\tilde\theta\in\Theta$ such that

$$\tilde\theta = \tilde\theta(z) = (\hat\theta_1(z),\tilde\theta_2,\ldots,\tilde\theta_p)\,.$$

Now,

$$\begin{aligned}
\min_{(\theta_2,\ldots,\theta_p);(\hat\theta_1(z),\theta_2,\ldots,\theta_p)\in\Theta} J[z,(\hat\theta_1(z),\theta_2,\ldots,\theta_p)] &\le J[z,(\hat\theta_1(z),\tilde\theta_2,\ldots,\tilde\theta_p)]\\
&= J(z,\tilde\theta)\\
&= \lim_{i\to\infty} J[z,\tilde\theta^{k_i}(z)]\\
&= \lim_{i\to\infty}\min_{\theta\in\Theta_{k_i}} J(z,\theta)\\
&= \min_{\theta\in\Theta} J(z,\theta)
\end{aligned}$$

where the last equality follows from the fact that $\lim_{k\to\infty}\Theta_k$ is dense in Θ. Therefore, for any $z\in\mathscr{Z}$,

$$\min_{(\theta_2,\ldots,\theta_p);(\hat\theta_1(z),\theta_2,\ldots,\theta_p)\in\Theta} J[z,(\hat\theta_1(z),\theta_2,\ldots,\theta_p)] = \min_{\theta\in\Theta} J(z,\theta)\,.$$

Define $J_1[z,(\theta_2,\ldots,\theta_p)] = J[z,(\hat\theta_1(z),\theta_2,\ldots,\theta_p)]$. It is a continuous function of $(\theta_2,\ldots,\theta_p)$ for all $z\in\mathscr{Z}$ and a measurable function of z for all $(\theta_2,\ldots,\theta_p)$ such that $(\hat\theta_1(z),\theta_2,\ldots,\theta_p)\in\Theta$. Apply the same arguments to J_1 to obtain a measurable function $\hat\theta_2$ such that, for any $z\in\mathscr{Z}$,

$$\min_{(\theta_3,\ldots,\theta_p);(\hat\theta_1(z),\hat\theta_2(z),\theta_3,\ldots,\theta_p)\in\Theta} J[z,(\hat\theta_1(z),\hat\theta_2(z),\theta_3,\ldots,\theta_p)] = \min_{\theta\in\Theta} J(z,\theta)\,.$$

Continuing in this manner, we construct real-valued functions $\hat\theta_1,\ldots,\hat\theta_p$ such that, for any $z\in\mathscr{Z}$,

$$J[z, (\hat{\theta}_1(z), \ldots, \hat{\theta}_p(z))] = \min_{\theta \in \Theta} J(z, \theta) .$$

Hence, $\hat{\theta} = (\hat{\theta}_1(z), \ldots, \hat{\theta}_p(z))$ is a measurable function from $\mathscr{Z}$ into Θ with the desirable property.

We show now that the continuity of $J(z, \theta)$ on $\mathscr{Z} \times \Theta$, Θ compact, implies that $\min_{\theta \in \Theta} J(z, \theta)$ is continuous in z.

$J(\cdot, \cdot)$ is uniformly continuous on compact subsets of $\mathscr{Z} \times \Theta \subset \mathbb{R}^m \times \mathbb{R}^p$. Therefore, $\forall z_0 \in \mathscr{Z}$, $\forall \epsilon > 0$, $\exists \delta > 0$ such that

$$\forall \theta \in \Theta \,, \; \forall z \in \mathscr{B}(z_0, \delta) \,, \; J(z_0, \theta) - \epsilon < J(z, \theta) < J(z_0, \theta) + \epsilon \,,$$

where $\mathscr{B}(z_0, \delta) = \{z \in \mathbb{R}^m : \|z - z_0\| \leq \delta\}$. This implies

$$\forall z \in \mathscr{B}(z_0, \delta) \,, \; \min_{\theta \in \Theta} J(z_0, \theta) - \epsilon \leq \min_{\theta \in \Theta} J(z, \theta) \leq \min_{\theta \in \Theta} J(z_0, \theta) + \epsilon$$

and $\min_{\theta \in \Theta} J(z, \theta)$ is thus continuous at z_0. Since z_0 is arbitrary, it is continuous for all $z \in \mathscr{Z}$. ■

Lemma 2.10. *Assume that the sequence of functions $\{J_N(\theta)\}$ converges uniformly on Θ to the function $J_{\bar{\theta}}(\theta)$, with $J_N(\theta)$ continuous with respect to $\theta \in \Theta$ for any N, Θ a compact subset of $\mathbb{R}^p$, and $J_{\bar{\theta}}(\theta)$ such that*

$$\forall \theta \in \Theta \,, \; \theta \neq \bar{\theta} \Longrightarrow J_{\bar{\theta}}(\theta) > J_{\bar{\theta}}(\bar{\theta}) \,.$$

Then $\lim_{N \to \infty} \hat{\theta}^N = \bar{\theta}$, where $\hat{\theta}^N \in \arg\min_{\theta \in \Theta} J_N(\theta)$. When the functions $J_N(\cdot)$ are random, and the uniform convergence to $J_{\bar{\theta}}(\cdot)$ is almost sure, the convergence of $\hat{\theta}^N$ to $\bar{\theta}$ is also almost sure.

Proof. The function $J_{\bar{\theta}}(\cdot)$ is continuous, and therefore, $\forall \beta > 0$, $\exists \epsilon > 0$ such that $J_{\bar{\theta}}(\theta) < J_{\bar{\theta}}(\bar{\theta}) + \epsilon$ implies $\|\theta - \bar{\theta}\| < \beta$. Indeed, for any $\beta > 0$ define

$$\underline{J}(\beta) = \min_{\{\theta \in \Theta : \, \|\theta - \bar{\theta}\| \geq \beta\}} J_{\bar{\theta}}(\theta) \,, \; \epsilon = \epsilon(\beta) = \frac{\underline{J}(\beta) - J_{\bar{\theta}}(\bar{\theta})}{2} \,.$$

We have $\underline{J}(\beta) > J_{\bar{\theta}}(\bar{\theta})$ and thus $\epsilon(\beta) > 0$. Assume that $J_{\bar{\theta}}(\theta) < J_{\bar{\theta}}(\bar{\theta}) + \epsilon = [\underline{J}(\beta) + J_{\bar{\theta}}(\bar{\theta})]/2$. It implies $J_{\bar{\theta}}(\theta) < \underline{J}(\beta)$ and thus $\|\theta - \bar{\theta}\| < \beta$.

Now, the uniform convergence of $J_N(\cdot)$ implies that there exists N_0 such that $\forall N > N_0$ and $\forall \theta \in \Theta$, $|J_N(\theta) - J_{\bar{\theta}}(\theta)| < \epsilon/2$. Therefore, $|J_N(\bar{\theta}) - J_{\bar{\theta}}(\bar{\theta})| < \epsilon/2$, $|J_N(\hat{\theta}^N) - J_{\bar{\theta}}(\hat{\theta}^N)| < \epsilon/2$; hence, $J_{\bar{\theta}}(\hat{\theta}^N) < J_N(\hat{\theta}^N) + \epsilon/2 \leq J_N(\bar{\theta}) + \epsilon/2 < J_{\bar{\theta}}(\bar{\theta}) + \epsilon$, and thus $\|\hat{\theta}^N - \bar{\theta}\| < \beta$. Almost sure statements follow immediately. ■

Lemma 2.11. *Assume that the sequence of functions $\{J_N(\theta)\}$ converges uniformly on Θ to the function $J_{\bar{\theta}}(\theta)$, with $J_N(\theta)$ continuous with respect to*

$\theta \in \Theta$ for any N, Θ a compact subset of $\mathbb{R}^p$. Let $\Theta^\# = \arg\min_{\theta\in\Theta} J_{\bar\theta}(\theta)$ denote the set of minimizers of $J_{\bar\theta}(\theta)$. Then $\lim_{N\to\infty} d(\hat\theta^N, \Theta^\#) = 0$, where $\hat\theta^N \in \arg\min_{\theta\in\Theta} J_N(\theta)$. When the functions $J_N(\cdot)$ are random and the uniform convergence to $J_{\bar\theta}(\cdot)$ is almost sure, the convergence of $d(\hat\theta^N, \Theta^\#)$ to 0 is also almost sure.

Proof. The proof is similar to that of Lemma 2.10, we simply change $J_{\bar\theta}(\bar\theta)$ into $J_{\bar\theta}(\Theta^\#) = \min_{\theta\in\Theta} J_{\bar\theta}(\theta)$ and define

$$\underline{J}(\beta) = \min_{\{\theta\in\Theta:\, d(\theta,\Theta^\#)\geq\beta\}} J_{\bar\theta}(\theta)$$

with $d(\theta, \Theta^\#) = \min_{\theta'\in\Theta^\#} \|\theta - \theta'\|$. ■

Lemma 2.12 (Jennrich 1969). *Let Θ be a convex compact subset of $\mathbb{R}^p$, $\mathscr{Z}$ be a measurable subset of $\mathbb{R}^m$ and $J(z,\theta)$ be a Borel measurable real function on $\mathscr{Z}\times\Theta$, continuously differentiable in $\theta \in \mathrm{int}(\Theta)$ for any $z \in \mathscr{Z}$. Let $\theta^1(z)$ and $\theta^2(z)$ be measurable functions from $\mathscr{Z}$ into Θ. There exists a measurable function $\tilde\theta$ from $\mathscr{Z}$ into $\mathrm{int}(\Theta)$ such that for all $z \in \mathscr{Z}$ $\tilde\theta(z)$ lies on the segment joining $\theta^1(z)$ and $\theta^2(z)$ and*

$$J[z,\theta^1(z)] - J[z,\theta^2(z)] = \left.\frac{\partial J(z,\theta)}{\partial\theta^\top}\right|_{\tilde\theta(z)} [\theta^1(z) - \theta^2(z)] \,.$$

Proof. Let $d(z,\theta)$ denote the Euclidian distance from θ to the segment joining $\theta^1(z)$ and $\theta^2(z)$ and define

$$D(z,\theta) = \left| J[z,\theta^1(z)] - J[z,\theta^2(z)] - \frac{\partial J(z,\theta)}{\partial\theta^\top}[\theta^1(z) - \theta^2(z)] \right| + d(z,\theta) \,,$$

which is a measurable function of z for any $\theta \in \Theta$ and is continuous in θ for any $z \in \mathscr{Z}$. We can then apply Lemma 2.9: there exists a measurable function $\tilde\theta(z)$ from $\mathscr{Z}$ into Θ such that for any $z \in \mathscr{Z}$, $\tilde\theta(z)$ minimizes $D(z,\theta)$ with respect to $\theta \in \Theta$. From the (Taylor) mean value theorem, this $\tilde\theta(z)$ has the property that for any $z \in \mathscr{Z}$, $D[z,\tilde\theta(z)] = 0$, which completes the proof. ■

Lemma 3.4 (Wu 1981). *If for any $\delta > 0$*

$$\liminf_{N\to\infty} \inf_{\|\theta-\bar\theta\|\geq\delta} [S_N(\theta) - S_N(\bar\theta)] > 0 \ a.s. \tag{C.5}$$

then $\hat\theta^N_{LS} \overset{\text{a.s.}}{\to} \bar\theta$ as $N \to \infty$. If for any $\delta > 0$

$$\mathrm{Prob}\left\{ \inf_{\|\theta-\bar\theta\|\geq\delta} [S_N(\theta) - S_N(\bar\theta)] > 0 \right\} \to 1 \,,\ N \to \infty \,, \tag{C.6}$$

then $\hat\theta^N_{LS} \overset{\text{p}}{\to} \bar\theta$ as $N \to \infty$.

Proof. If $\hat{\theta}_{LS}^N \stackrel{\text{a.s.}}{\to} \bar{\theta}$ is not true, there exists some $\delta > 0$ such that

$$\text{Prob}(\limsup_{N\to\infty} \|\hat{\theta}_{LS}^N - \bar{\theta}\| \geq \delta) > 0 \,.$$

Now, $S_N(\hat{\theta}_{LS}^N) - S_N(\bar{\theta}) \leq 0$ from the definition of $\hat{\theta}_{LS}^N$, and $\|\hat{\theta}_{LS}^N - \bar{\theta}\| \geq \delta$ implies $\inf_{\|\theta-\bar{\theta}\|\geq\delta}[S_N(\theta) - S_N(\bar{\theta})] \leq 0$. Therefore, $\text{Prob}[\liminf_{N\to\infty} \inf_{\|\theta-\bar{\theta}\|\geq\delta} [S_N(\theta) - S_N(\bar{\theta})] \leq 0] > 0$, which contradicts (C.5).

When $\inf_{\|\theta-\bar{\theta}\|\geq\delta}[S_N(\theta) - S_N(\bar{\theta})] > 0$, then $\|\hat{\theta}_{LS}^N - \bar{\theta}\| < \delta$. Therefore, when (C.6) is satisfied, for any $\delta > 0$ and $\epsilon > 0$ there exists N_0 such that for all $N > N_0$ $\text{Prob}\{\|\hat{\theta}_{LS}^N - \bar{\theta}\| < \delta\} \geq 1 - \epsilon$, that is, $\hat{\theta}_{LS}^N \stackrel{\text{p}}{\to} \bar{\theta}$. ■

Lemma 3.7. *Let* $\mathbf{u}, \mathbf{v}$ *be two random vectors of* $\mathbb{R}^r$ *and* $\mathbb{R}^s$ *respectively defined on a probability space with measure* μ, *with* $\mathbb{E}(\|\mathbf{u}\|^2) < \infty$ *and* $\mathbb{E}(\|\mathbf{v}\|^2) < \infty$. *We have*

$$\mathbb{E}(\mathbf{u}\mathbf{u}^\top) \succeq \mathbb{E}(\mathbf{u}\mathbf{v}^\top)[\mathbb{E}(\mathbf{v}\mathbf{v}^\top)]^+\mathbb{E}(\mathbf{v}\mathbf{u}^\top)\,, \tag{C.7}$$

where $\mathbf{M}^+$ *denotes the Moore–Penrose g-inverse of* $\mathbf{M}$ *and* $\mathbf{A} \succeq \mathbf{B}$ *means that* $\mathbf{A} - \mathbf{B}$ *is nonnegative definite. Moreover, the equality is obtained in (C.7) if and only if* $\mathbf{u} = \mathbf{A}\mathbf{v}$ μ*-a.s. for some nonrandom matrix* $\mathbf{A}$.

Proof. Since $\mathbb{E}(\{\mathbf{u}\}_i^2) < \infty$, $i = 1, \ldots, r$ and $\mathbb{E}(\{\mathbf{v}\}_i^2) < \infty$, $i = 1, \ldots, s$, Cauchy–Schwarz inequality gives

$$\left\{\mathbb{E}\left[\begin{pmatrix}\mathbf{u}\\ \mathbf{v}\end{pmatrix} (\mathbf{u}^\top\ \mathbf{v}^\top)\right]\right\}_{ij} < \infty$$

for any $i, j = 1, \ldots, r + s$, so that $\mathbb{E}(\mathbf{u}\mathbf{u}^\top)$, $\mathbb{E}(\mathbf{u}\mathbf{v}^\top)$, and $\mathbb{E}(\mathbf{v}\mathbf{v}^\top)$ are well defined. Consider $\mathbb{E}[(\mathbf{x}^\top\mathbf{u} + \mathbf{y}^\top\mathbf{v})^2]$ for some nonrandom $(\mathbf{x}, \mathbf{y}) \in \mathbb{R}^r \times \mathbb{R}^s$. By direct expansion, we obtain

$$\mathbb{E}[(\mathbf{x}^\top\mathbf{u})^2] + 2\mathbf{x}^\top\mathbb{E}(\mathbf{u}\mathbf{v}^\top)\mathbf{y} + \mathbf{y}^\top\mathbb{E}(\mathbf{v}\mathbf{v}^\top)\mathbf{y} \geq 0 \tag{C.8}$$

which reaches its minimum value with respect to $\mathbf{y}$ when

$$\mathbb{E}(\mathbf{v}\mathbf{v}^\top)\mathbf{y} = -\mathbb{E}(\mathbf{v}\mathbf{u}^\top)\mathbf{x}\,.$$

This system is compatible, and thus consistent; see Harville (1997, p. 73). Indeed,

$$\mathbf{z}^\top\mathbb{E}(\mathbf{v}\mathbf{v}^\top) = \mathbf{0}^\top \Longrightarrow \mathbb{E}(\mathbf{z}^\top\mathbf{v}\mathbf{v}^\top\mathbf{z}) = 0 \Longrightarrow \mathbf{z}^\top\mathbf{v} = 0 \ \mu\text{-a.s.}$$
$$\Longrightarrow \mathbf{z}^\top\mathbb{E}(\mathbf{v}\mathbf{u}^\top)\mathbf{x} - \mathbb{E}(\mathbf{z}^\top\mathbf{v}\mathbf{u}^\top\mathbf{x}) = 0\,.$$

Therefore, the solution $\mathbf{y}^*$ is given by

$$\mathbf{y}^* = -[\mathbb{E}(\mathbf{v}\mathbf{v}^\top)]^-\mathbb{E}(\mathbf{v}\mathbf{u}^\top)\mathbf{x}$$

for any g-inverse of $\mathbb{E}(\mathbf{v}\mathbf{v}^\top)$; see Harville (1997, p. 108). Take

$$\mathbf{y}^* = -[\mathbb{E}(\mathbf{v}\mathbf{v}^\top)]^+ \mathbb{E}(\mathbf{v}\mathbf{u}^\top)\mathbf{x}$$

with $\mathbf{M}^+$ the Moore–Penrose g-inverse of $\mathbf{M}$, see Harville (1997, p. 493), and substitute $\mathbf{y}^*$ for $\mathbf{y}$ in (C.8). We obtain

$$\mathbf{x}^\top \mathbb{E}(\mathbf{u}\mathbf{u}^\top)\mathbf{x} \geq \mathbf{x}^\top \mathbb{E}(\mathbf{u}\mathbf{v}^\top)[\mathbb{E}(\mathbf{v}\mathbf{v}^\top)]^+ \mathbb{E}(\mathbf{v}\mathbf{u}^\top)\mathbf{x}$$

for any nonrandom vector $\mathbf{x} \in \mathbb{R}^r$, i.e., (C.7).

Assume that equality is attained. Taking $\mathbf{A} = \mathbb{E}(\mathbf{u}\mathbf{v}^\top)[\mathbb{E}(\mathbf{v}\mathbf{v}^\top)]^+$ we obtain $\mathbb{E}[(\mathbf{u} - \mathbf{A}\mathbf{v})(\mathbf{u} - \mathbf{A}\mathbf{v})^\top] = \mathbf{O}$ and thus $\mathbf{u} = \mathbf{A}\mathbf{v}$, μ-a.s. ■

Lemma 5.1. *Let $\mathbf{A}$ be a $p \times p$ positive-definite matrix and let $\mathcal{E}_A = \{\mathbf{t} \in \mathbb{R}^p : \mathbf{t}^\top \mathbf{A}\mathbf{t} \leq 1\}$. Then:*

(i) $\mathrm{vol}(\mathcal{E}_A) = V_p \det^{-1/2} \mathbf{A}$, *with* $V_p = \pi^{p/2}/\Gamma(p/2+1) = \mathrm{vol}[\mathscr{B}(\mathbf{0},1)]$, *the volume of the unit ball $\mathscr{B}(\mathbf{0},1)$ in $\mathbb{R}^p$.*

(ii) For any vector $\mathbf{c} \in \mathbb{R}^p$ we have

$$\max_{\mathbf{t} \in \mathcal{E}_A} (\mathbf{c}^\top \mathbf{t})^2 = \mathbf{c}^\top \mathbf{A}^{-1} \mathbf{c}\,;$$

in particular, when $\|\mathbf{c}\| = 1$, then $\mathbf{c}^\top \mathbf{A}^{-1}\mathbf{c}$ is the squared half-length of the orthogonal projection of $\mathcal{E}_A$ onto the straight line defined by $\mathbf{c}$.

(iii) $\max_{\|\mathbf{c}\|=1} \mathbf{c}^\top \mathbf{A}^{-1} \mathbf{c} = 1/\lambda_{\min}(\mathbf{A}) = R^2(\mathcal{E}_A)$, *with $\lambda_{\min}(\mathbf{A})$ the minimum eigenvalue of $\mathbf{A}$ and $R(\mathcal{E}_A)$ the radius of the smallest ball containing $\mathcal{E}_A$; the length of a principal axis of $\mathcal{E}_A$ equals $2/\sqrt{\lambda_i(\mathbf{A})}$ with $\lambda_i(\mathbf{A})$ an eigenvalue of $\mathbf{A}$.*

(iv) The squared length of the half-diagonal of the parallelepiped containing $\mathcal{E}_A$ and parallel to the coordinate axes of the Euclidean space $\mathbb{R}^p$ equals the sum of the squared half-lengths of the principal axes of $\mathcal{E}_A$ and is given by $\mathrm{trace}(\mathbf{A}^{-1})$.

(v) Let $\mathcal{E}_B$ be defined similarly to $\mathcal{E}_A$ but for the $p \times p$ positive-definite matrix $\mathbf{B}$, then the following statements are equivalent:

(a) $\mathcal{E}_A \subseteq \mathcal{E}_B$.

(b) $\mathbf{A} \succeq \mathbf{B}$, *i.e., the matrix $\mathbf{A} - \mathbf{B}$ is nonnegative definite.*

(c) For any $\mathbf{c} \in \mathbb{R}^p$, $\mathbf{c}^\top \mathbf{A}^{-1} \mathbf{c} \leq \mathbf{c}^\top \mathbf{B}^{-1}\mathbf{c}$, i.e., $\mathbf{B}^{-1} \succeq \mathbf{A}^{-1}$.

Proof.

(i) We can write

$$\begin{aligned}\mathrm{vol}(\mathcal{E}_A) &= \int_{\{\mathbf{t}\in\mathbb{R}^p : \mathbf{t}^\top \mathbf{A}\mathbf{t} \leq 1\}} d\mathbf{t} = \int_{\{\mathbf{u}\in\mathbb{R}^p : \mathbf{u}^\top \mathbf{u} \leq 1\}} \det{}^{-1/2}(\mathbf{A})\, d\mathbf{u} \\ &= [\det{}^{-1/2}\mathbf{A}]\, \mathrm{vol}[\mathscr{B}(\mathbf{0},1)]\,.\end{aligned}$$

(*ii*) From Cauchy–Schwarz inequality we have

$$\forall \mathbf{t} \in \mathbb{R}^p\,, \quad (\mathbf{c}^\top \mathbf{t})^2 = [(\mathbf{A}^{-1/2}\mathbf{c})^\top (\mathbf{A}^{1/2}\mathbf{t})]^2 \leq (\mathbf{c}^\top \mathbf{A}^{-1}\mathbf{c})(\mathbf{t}^\top \mathbf{A}\mathbf{t})\,.$$

Therefore,

$$\mathbf{c}^\top \mathbf{A}^{-1}\mathbf{c} \geq \sup_{\mathbf{t}\neq \mathbf{0}} \frac{(\mathbf{c}^\top \mathbf{t})^2}{\mathbf{t}\mathbf{A}\mathbf{t}}\,.$$

For $\mathbf{t} = \mathbf{A}^{-1}\mathbf{c}$, the ratio on right-hand side equals $\mathbf{c}^\top \mathbf{A}^{-1}\mathbf{c}$, so that

$$\mathbf{c}^\top \mathbf{A}^{-1}\mathbf{c} = \max_{\mathbf{t}\neq \mathbf{0}} \frac{(\mathbf{c}^\top \mathbf{t})^2}{\mathbf{t}\mathbf{A}\mathbf{t}} = \max_{\{\mathbf{t}\in\mathbb{R}^p:\mathbf{t}^\top\mathbf{A}\mathbf{t}\leq 1\,,\ \mathbf{t}\neq\mathbf{0}\}} \frac{(\mathbf{c}^\top \mathbf{t})^2}{\mathbf{t}\mathbf{A}\mathbf{t}} = \max_{\mathbf{t}\in\mathcal{E}_A} (\mathbf{c}^\top \mathbf{t})^2\,.$$

When $\|\mathbf{c}\| = 1$, then $\mathbf{c}\mathbf{c}^\top\mathbf{t}$ is the orthogonal projection of $\mathbf{t}$ onto the straight line defined by $\mathbf{c}$ and its squared length equals $\|\mathbf{c}\mathbf{c}^\top\mathbf{t}\|^2 = (\mathbf{c}^\top\mathbf{t})^2$.

(*iii*) The largest orthogonal projection of $\mathcal{E}_A$ onto the straight line defined by $\mathbf{c}$ is obtained when $\mathbf{c}$ goes in the direction of the main axis of $\mathcal{E}_A$. Let $\lambda_1 = \lambda_{\min}(\mathbf{A}) \leq \lambda_2 \leq \cdots \leq \lambda_p$ denote the eigenvalues of $\mathbf{A}$. In a basis of associated eigenvectors, $\mathcal{E}_A$ is defined by $\{\mathbf{y} \in \mathbb{R}^p : \sum_{i=1}^p y_i^2\,\lambda_i \leq 1\}$, with y_i the i-th component of $\mathbf{y}$. The half-length of the longest principal axis of $\mathcal{E}_A$ is thus $R(\mathcal{E}_A) = 1/\sqrt{\lambda_1}$. The length of the i-th principal axis is $2/\sqrt{\lambda_i}$.

(*iv*) From the same arguments as above, the sum of the squared half-lengths of the principal axes of $\mathcal{E}_A$ is $\sum_{i=1}^p \lambda_i^{-1} = \mathrm{trace}(\mathbf{A}^{-1})$. Let $\mathbf{e}_k$ denote the k-th basis vector of $\mathbb{R}^p$; then $\{\mathbf{A}^{-1}\}_{kk} = \mathbf{e}_k^\top\mathbf{A}^{-1}\mathbf{e}_k$ is the squared half-length of the orthogonal projection of $\mathcal{E}_A$ onto the k-th coordinate axis. By the Pythagorean relation in $\mathbb{R}^p$, we obtain that the squared length of the half-diagonal of the parallelepiped containing $\mathcal{E}_A$ and parallel to the coordinate axes equals $\sum_{k=1}^p \{\mathbf{A}^{-1}\}_{kk} = \mathrm{trace}(\mathbf{A}^{-1})$.

(*v*) The implication (b) $\Longrightarrow$ (a) is a direct consequence of the definitions of $\mathcal{E}_A$ and $\mathcal{E}_B$. The implication (a) $\Longrightarrow$ (c) follows from (*ii*). Suppose that (c) holds. Take any vector $\mathbf{v} \in \mathbb{R}^p$ and denote $\mathbf{s} = \mathbf{A}\mathbf{v}$, $\mathbf{z} = \mathbf{B}\mathbf{v}$. We have

$$\begin{aligned} 0 &\leq (\mathbf{s}-\mathbf{z})^\top\mathbf{A}^{-1}(\mathbf{s}-\mathbf{z}) = \mathbf{s}^\top\mathbf{A}^{-1}\mathbf{s} + \mathbf{z}^\top\mathbf{A}^{-1}\mathbf{z} - 2\mathbf{s}^\top\mathbf{A}^{-1}\mathbf{z} \\ &\leq \mathbf{s}^\top\mathbf{A}^{-1}\mathbf{s} + \mathbf{z}^\top\mathbf{B}^{-1}\mathbf{z} - 2\mathbf{v}^\top\mathbf{z} = \mathbf{s}^\top\mathbf{A}^{-1}\mathbf{s} + \mathbf{z}^\top\mathbf{B}^{-1}\mathbf{z} - 2\mathbf{z}^\top\mathbf{B}^{-1}\mathbf{z} \\ &= \mathbf{v}^\top\mathbf{A}\mathbf{v} - \mathbf{v}^\top\mathbf{B}\mathbf{v}\,; \end{aligned}$$

that is, $\mathbf{A} \succeq \mathbf{B}$. ■

Lemma 5.2. *Suppose that the estimator $\hat{\theta}^N$ in the regression model (3.2) satisfies $\sqrt{N}(\hat{\theta}^N - \bar{\theta}) \xrightarrow{\mathrm{d}} \mathbf{z} \sim \mathscr{N}(\mathbf{0}, \mathbf{M}^{-1}(\xi,\bar{\theta}))$ as $N \to \infty$. Then, for N large we have approximately*

$$\text{Prob}\left\{y(x_1),\ldots,y(x_N) : \forall x \in \mathscr{X}, |\eta(x,\hat{\theta}^N) - \eta(x,\bar{\theta})| \leq \frac{1}{\sqrt{N}}\left[\chi_p^2(1-\alpha)\frac{\partial\eta(x,\theta)}{\partial\theta^\top}\bigg|_{\bar{\theta}} \mathbf{M}^{-1}(\xi,\bar{\theta})\frac{\partial\eta(x,\theta)}{\partial\theta}\bigg|_{\bar{\theta}}\right]^{1/2}\right\} \geq 1-\alpha$$

where $\chi_p^2(1-\alpha)$ *is the* $(1-\alpha)$ *quantile of the* χ_p^2 *distribution.*

Proof. Since $\hat{\theta}^N - \bar{\theta}$ is approximately normal $\mathscr{N}(\mathbf{0}, \mathbf{M}^{-1}(\xi,\bar{\theta})/N)$, the quantity $N(\hat{\theta}^N - \bar{\theta})^\top \mathbf{M}(\xi,\bar{\theta})(\hat{\theta}^N - \bar{\theta})$ follows approximately the χ_p^2 distribution. Hence, for N large

$$\text{Prob}\left\{\mathbf{y} : (\hat{\theta}^N - \bar{\theta})^\top \mathbf{H}(\hat{\theta}^N - \bar{\theta}) \leq 1\right\} \simeq 1-\alpha$$

where $\mathbf{H} = N\mathbf{M}(\xi,\bar{\theta})/\chi_p^2(1-\alpha)$ and $\mathbf{y} = [y(x_1),\ldots,y(x_N)]^\top$. Since $\mathbf{u}^\top\mathbf{H}\mathbf{u} \leq 1$ is equivalent to $(\mathbf{v}^\top\mathbf{u})^2 \leq \mathbf{v}^\top\mathbf{H}^{-1}\mathbf{v}$ for all $\mathbf{v} \in \mathbb{R}^p$ (from Cauchy–Schwarz inequality), for large N we have

$$\begin{aligned}
&\text{Prob}\left\{\mathbf{y} : \forall x \in \mathscr{X}, |\eta(x,\hat{\theta}^N) - \eta(x,\bar{\theta})|^2 \leq \frac{1}{N}\chi_p^2(1-\alpha)\frac{\partial\eta(x,\theta)}{\partial\theta^\top}\bigg|_{\bar{\theta}} \mathbf{M}^{-1}(\xi,\bar{\theta})\frac{\partial\eta(x,\theta)}{\partial\theta}\bigg|_{\bar{\theta}}\right\} \\
&\simeq \text{Prob}\left\{\mathbf{y} : \forall x \in \mathscr{X}, \left|\frac{\partial\eta(x,\theta)}{\partial\theta^\top}\bigg|_{\bar{\theta}}(\hat{\theta}^N - \bar{\theta})\right|^2 \leq \frac{\partial\eta(x,\theta)}{\partial\theta^\top}\bigg|_{\bar{\theta}} \mathbf{H}^{-1}\frac{\partial\eta(x,\theta)}{\partial\theta}\bigg|_{\bar{\theta}}\right\} \\
&\geq \text{Prob}\left\{\mathbf{y} : \forall \mathbf{v} \in \mathbb{R}^p, \left|\mathbf{v}^\top(\hat{\theta}^N - \bar{\theta})\right|^2 \leq \mathbf{v}^\top\mathbf{H}^{-1}\mathbf{v}\right\} \\
&= \text{Prob}\left\{\mathbf{y} : (\hat{\theta}^N - \bar{\theta})^\top\mathbf{H}(\hat{\theta}^N - \bar{\theta}) \leq 1\right\} \simeq 1-\alpha\,. \qquad \blacksquare
\end{aligned}$$

Lemma 5.4 (Pukelsheim 1993, Sects. 5.2, 5.4). *Let* $\Phi(\cdot)$ *be a function from* $\mathbb{M}^{\geq}$ *to* $\mathbb{R}$. *Then,*

(i) When $\Phi(\cdot)$ *is positively homogeneous, it is concave if and only if it is superadditive, i.e.,* $\Phi(\mathbf{M}_1+\mathbf{M}_2) \geq \Phi(\mathbf{M}_1)+\Phi(\mathbf{M}_2)$ *for all* $\mathbf{M}_1,\mathbf{M}_2 \in \mathbb{M}^{\geq}$.
(ii) When $\Phi(\cdot)$ *is superadditive, nonnegativity implies isotonicity.*
(iii) When $\Phi(\cdot)$ *is positively homogeneous, isotonicity implies nonnegativity (i.e.,* $\Phi(\mathbf{M}) \geq 0$ *for all* $\mathbf{M}$ *in* $\mathbb{M}^{\geq}$*); moreover, either* Φ *is identically zero or* $\Phi(.)$ *is strictly positive on the open set* $\mathbb{M}^{>}$.

Proof.

(*i*) Take any $\mathbf{M}_1,\mathbf{M}_2 \in \mathbb{M}^{\geq}$, any $\alpha \in (0,1)$. Superadditivity gives $\Phi[(1-\alpha)\mathbf{M}_1 + \alpha\mathbf{M}_2] \geq \Phi[(1-\alpha)\mathbf{M}_1] + \Phi(\alpha\mathbf{M}_2) = (1-\alpha)\Phi(\mathbf{M}_1) + \alpha\Phi(\mathbf{M}_2)$ and thus implies concavity. Conversely, concavity implies $\Phi(\mathbf{M}_1+\mathbf{M}_2) = \Phi[(2\mathbf{M}_1+2\mathbf{M}_2)/2) \geq (1/2)\Phi(2\mathbf{M}_1) + (1/2)\Phi(2\mathbf{M}_2) = \Phi(\mathbf{M}_1)+\Phi(\mathbf{M}_2)$.

(*ii*) Take any $\mathbf{M}_1 \succeq \mathbf{M}_2 \in \mathbb{M}^{\geq}$. Superadditivity and nonnegativity imply $\Phi(\mathbf{M}_1) - \Phi(\mathbf{M}_2) = \Phi(\mathbf{M}_1 - \mathbf{M}_2 + \mathbf{M}_2) - \Phi(\mathbf{M}_2) \geq \Phi(\mathbf{M}_1 - \mathbf{M}_2) \geq 0$ so that $\Phi(\cdot)$ is isotonic.

(*iii*) Isotonicity implies $\Phi(\mathbf{M}) \geq \Phi(\mathbf{O})$ for any $\mathbf{M} \in \mathbb{M}^{\geq}$, and positive homogeneity gives $\Phi(\mathbf{O}) = \Phi(0\mathbf{M}) = 0$, so that $\Phi(\cdot)$ is nonnegative. If Φ is non identically zero, there exists some $\mathbf{M}^*$ in $\mathbb{M}^{\geq}$ such that $\Phi(\mathbf{M}^*) > 0$. Then, for any $\mathbf{M} \in \mathbb{M}^{>}$, there exists $\alpha > 0$ such that $\alpha\mathbf{M} - \mathbf{M}^* \succeq \mathbf{O}$ and isotonicity with positive homogeneity imply $\Phi(\mathbf{M}) = \Phi(\alpha\mathbf{M})/\alpha \geq \Phi(\mathbf{M}^*)/\alpha > 0$. ∎

Lemma 5.5. *For any $p \times p$ matrix $\mathbf{M}$ in $\mathbb{M}^{\geq}$ and any $\mathbf{c} \in \mathcal{M}(\mathbf{M})$ (i.e., such that $\mathbf{c} = \mathbf{M}\mathbf{u}$ for some $\mathbf{u} \in \mathbb{R}^p$) we have*

$$\Phi_c(\mathbf{M}) = -\mathbf{c}^\top \mathbf{M}^- \mathbf{c} = \min_{\mathbf{z} \in \mathbb{R}^p} [\mathbf{z}^\top \mathbf{M} \mathbf{z} - 2\mathbf{z}^\top \mathbf{c}] .$$

When $\mathbf{c} \notin \mathcal{M}(\mathbf{M})$, the right-hand side equals $-\infty$.

Proof. When $\mathbf{c} \in \mathcal{M}(\mathbf{M})$, we can write $[\mathbf{z}^\top \mathbf{M}\mathbf{z} - 2\mathbf{z}^\top \mathbf{c}] - \Phi_c(\mathbf{M}) = (\mathbf{M}^-\mathbf{c} - \mathbf{z})^\top \mathbf{M}(\mathbf{M}^-\mathbf{c} - \mathbf{z}) \geq 0$, with $\mathbf{M}^-$ any g-inverse of $\mathbf{M}$. When $\mathbf{c} \notin \mathcal{M}(\mathbf{M})$, take $\mathbf{z} = \gamma\mathbf{u}$ with $\gamma > 0$ and $\mathbf{u}$ any element of $\mathcal{N}(\mathbf{M}) = \{\mathbf{u} \in \mathbb{R}^p : \mathbf{M}\mathbf{u} = \mathbf{0}\}$ such that $\mathbf{c}^\top \mathbf{u} = s > 0$. Then, $\mathbf{z}^\top \mathbf{M}\mathbf{z} = 0$ and $\mathbf{z}^\top \mathbf{c} = \gamma s$ which can be made arbitrarily large. ∎

Lemma 5.6. *For any $p \times p$ matrix $\mathbf{M}$ in $\mathbb{M}^{\geq}$ and any $\mathbf{c} \in \mathcal{M}(\mathbf{M})$, we have*

$$\Phi_c^+(\mathbf{M}) = (\mathbf{c}^\top \mathbf{c})(\mathbf{c}^\top \mathbf{M}^- \mathbf{c})^{-1} = (\mathbf{c}^\top \mathbf{c}) \min_{\mathbf{z}^\top \mathbf{c} = 1} \mathbf{z}^\top \mathbf{M} \mathbf{z} .$$

When $\mathbf{c} \notin \mathcal{M}(\mathbf{M})$, the minimum on the right-hand side equals 0.

Proof. When $\mathbf{c} \in \mathcal{M}(\mathbf{M})$, Cauchy–Schwarz inequality gives

$$\begin{aligned}(\mathbf{c}^\top \mathbf{M}^- \mathbf{c})(\mathbf{z}^\top \mathbf{M}\mathbf{z}) &= (\mathbf{c}^\top \mathbf{M}^- \mathbf{M}\mathbf{M}^- \mathbf{c})(\mathbf{z}^\top \mathbf{M}\mathbf{z}) \\ &\geq (\mathbf{c}^\top \mathbf{M}^- \mathbf{M}\mathbf{z})^2 = (\mathbf{c}^\top \mathbf{z})^2\end{aligned}$$

for any $\mathbf{z} \in \mathbb{R}^p$. Therefore,

$$\mathbf{c}^\top \mathbf{M}^- \mathbf{c} \geq \sup_{\mathbf{z}^\top \mathbf{M}\mathbf{z} \neq 0} \frac{(\mathbf{c}^\top \mathbf{z})^2}{\mathbf{z}^\top \mathbf{M}\mathbf{z}} .$$

Taking $\mathbf{z} = \mathbf{M}^- \mathbf{c}$ gives equality since then $\mathbf{z}^\top \mathbf{M}\mathbf{z} = \mathbf{c}^\top \mathbf{M}^- \mathbf{c} > 0$. We can thus write

$$(\mathbf{c}^\top \mathbf{M}^- \mathbf{c}) = \sup_{\mathbf{z}^\top \mathbf{M}\mathbf{z} \neq 0} \frac{(\mathbf{c}^\top \mathbf{z})^2}{\mathbf{z}^\top \mathbf{M}\mathbf{z}} = \sup_{\mathbf{z}^\top \mathbf{c} = 1} \frac{1}{\mathbf{z}^\top \mathbf{M}\mathbf{z}} .$$

When $\mathbf{c} \notin \mathcal{M}(\mathbf{M})$, take $\mathbf{z}$ as any element of $\mathcal{N}(\mathbf{M})$ such that $\mathbf{c}^\top \mathbf{z} \neq 0$. Then, $\mathbf{z}^\top \mathbf{M}\mathbf{z} = 0$, and the supremum in the equation above is infinite. ∎

Lemma 5.7. *For any $p \times p$ matrix $\mathbf{M}$ in $\mathbb{M}^{\geq}$ partitioned as*

$$\mathbf{M} = \begin{pmatrix} \mathbf{M}_{11} & \mathbf{M}_{12} \\ \mathbf{M}_{21} & \mathbf{M}_{22} \end{pmatrix}$$

with $\mathbf{M}_{11}$ of dimension $s \times s$, we have

$$\log\det(\mathbf{M}_{11} - \mathbf{M}_{12}\mathbf{M}_{22}^{-}\mathbf{M}_{21}) \leq \log\det(\mathbf{M}_{11} + \mathbf{D}^\top\mathbf{M}_{22}\mathbf{D} - \mathbf{M}_{12}\mathbf{D} - \mathbf{D}^\top\mathbf{M}_{21})$$

for any $\mathbf{D} \in \mathbb{R}^{(p-s)\times s}$, with equality if and only if $\mathbf{M}_{22}\mathbf{D} = \mathbf{M}_{12}$.

Proof. Take any matrix $\mathbf{C}$ solution of $\mathbf{M}_{22}\mathbf{C} = \mathbf{M}_{21}$ (which is equivalent to $\mathbf{C} = \mathbf{M}_{22}^{-}\mathbf{M}_{21}$ for some g-inverse $\mathbf{M}_{22}^{-}$ of $\mathbf{M}_{22}$), and denote $\mathbf{M}^* = \mathbf{M}_{11} - \mathbf{M}_{12}\mathbf{M}_{22}^{-}\mathbf{M}_{21}$. Then, for any matrix $\mathbf{D}$ in $\mathbb{R}^{(p-s)\times s}$,

$$\mathbf{M}_{11} + \mathbf{D}^\top\mathbf{M}_{22}\mathbf{D} - \mathbf{M}_{12}\mathbf{D} - \mathbf{D}^\top\mathbf{M}_{21} - \mathbf{M}^* = (\mathbf{C} - \mathbf{D})^\top\mathbf{M}_{22}(\mathbf{C} - \mathbf{D})\,,$$

which is positive definite unless $\mathbf{M}_{22}\mathbf{D} = \mathbf{M}_{22}\mathbf{C} = \mathbf{M}_{21}$. When $\mathbf{M}^*$ is nonsingular, the strict isotonicity of the function $\log\det(\cdot)$ on $\mathbb{M}^{>}$ (see Sect. 5.1.5) concludes the proof. When $\mathbf{M}^*$ is singular, $\log\det(\mathbf{M}^*) = -\infty$, we also have $\log\det[\mathbf{M}_{11} + \mathbf{D}^\top\mathbf{M}_{22}\mathbf{D} - \mathbf{M}_{12}\mathbf{D} - \mathbf{D}^\top\mathbf{M}_{21}] = -\infty$ when $\mathbf{M}_{22}\mathbf{D} = \mathbf{M}_{21}$ since then $\mathbf{M}_{11} + \mathbf{D}^\top\mathbf{M}_{22}\mathbf{D} - \mathbf{M}_{12}\mathbf{D} - \mathbf{D}^\top\mathbf{M}_{21} = \mathbf{M}^*$. ■

Lemma 5.11. *The criterion $\phi_c(\cdot) = \Phi_c[\mathbf{M}(\cdot)]$, with $\Phi_c(\mathbf{M})$ given by (5.9), is upper semicontinuous at any $\xi_* \in \Xi_c = \{\xi \in \Xi : \mathbf{c} \in \mathcal{M}[\mathbf{M}(\xi)]\}$.*

Proof. Take $\xi_* \in \Xi_c$, and consider any sequence $\{\xi_n\}$ of measures in Ξ converging weakly to ξ_*. We have $\phi_c(\xi_n) = -\infty$ if $\xi_n \in \Xi \setminus \Xi_c$ and, from Lemma 5.5, $\phi_c(\xi_n) \leq \mathbf{z}^\top\mathbf{M}(\xi_n)\mathbf{z} - 2\mathbf{z}^\top\mathbf{c}$ for any $\mathbf{z} \in \mathbb{R}^p$ otherwise. Therefore, for any $\mathbf{z} \in \mathbb{R}^p$,

$$\limsup_{n\to\infty} \phi_c(\xi_n) \leq \limsup_{n\to\infty} [\mathbf{z}^\top\mathbf{M}(\xi_n)\mathbf{z} - 2\mathbf{z}^\top\mathbf{c}] = \mathbf{z}^\top\mathbf{M}(\xi_*)\mathbf{z} - 2\mathbf{z}^\top\mathbf{c}\,;$$

that is,

$$\limsup_{n\to\infty} \phi_c(\xi_n) \leq \min_{\mathbf{z}\in\mathbb{R}^p} [\mathbf{z}^\top\mathbf{M}(\xi_*)\mathbf{z} - 2\mathbf{z}^\top\mathbf{c}] = \phi_c(\xi_*)\,,$$

so that $\phi_c(\cdot)$ is upper semicontinuous at ξ_*. ■

Lemma 5.12. *Consider a sequence of matrices satisfying (5.19) and suppose that $\mathbf{c} \in \mathcal{M}(\mathbf{M}_0)$. Then, under the conditions*
C1: $\|\mathbf{R}_t\| = [\mathrm{trace}(\mathbf{R}_t^\top\mathbf{R}_t)]^{1/2} = o(t^\alpha)$ as $t \to 0^+$,
and
C2: $\mathbf{M}_0 + t^\alpha\mathbf{M}_\alpha \in \mathbb{M}^{>}$ for arbitrary small $t > 0$,
we have

$$\lim_{t\to 0^+} \Phi_c[\mathbf{M}(t)] = \Phi_c(\mathbf{M}_0)\,. \tag{C.9}$$

Proof. Define $r_t = \|\mathbf{R}_t\|/t^\alpha$. We first show that $\mathbf{M}(t) - (1-\sqrt{r_t})(\mathbf{M}_0 + t^\alpha \mathbf{M}_\alpha) \in \mathbb{M}^>$ for t small enough. Take any $\mathbf{z} \in \mathbb{R}^p$, $\mathbf{z} \neq \mathbf{0}$. We have

$$\begin{aligned}\mathbf{z}^\top \mathbf{M}(t)\mathbf{z} - (1-\sqrt{r_t})\,\mathbf{z}^\top(\mathbf{M}_0 + t^\alpha \mathbf{M}_\alpha)\mathbf{z} &= \sqrt{r_t}\,\mathbf{z}^\top(\mathbf{M}_0 + t^\alpha \mathbf{M}_\alpha)\mathbf{z} + \mathbf{z}^\top \mathbf{R}_t \mathbf{z} \\ &\geq \sqrt{r_t}\,\mathbf{z}^\top(\mathbf{M}_0 + t^\alpha \mathbf{M}_\alpha)\mathbf{z} - \|\mathbf{z}\|^2 \|\mathbf{R}_t\| \\ &= \sqrt{r_t}\, t^\alpha \,\|\mathbf{z}\|^2 \left(\frac{\mathbf{z}^\top(\mathbf{M}_0/t^\alpha + \mathbf{M}_\alpha)\mathbf{z}}{\|\mathbf{z}\|^2} - \sqrt{r_t} \right) \\ &\geq \sqrt{r_t}\, t^\alpha \,\|\mathbf{z}\|^2 \left(\frac{\mathbf{z}^\top(\mathbf{M}_0 + t_0^\alpha \mathbf{M}_\alpha)\mathbf{z}}{t_0^\alpha \|\mathbf{z}\|^2} - \sqrt{r_t} \right)\end{aligned}$$

for all $t < t_0$. $\mathbf{z}^\top(\mathbf{M}_0 + t_0^\alpha \mathbf{M}_\alpha)\mathbf{z}/[t_0^\alpha \|\mathbf{z}\|^2] > 0$ from C2, while $\sqrt{r_t}$ tends to zero as $t \to 0$ from C1. Therefore, there exists t_1 such that $\mathbf{M}(t) - (1-\sqrt{r_t})(\mathbf{M}_0 + t^\alpha \mathbf{M}_\alpha) \in \mathbb{M}^>$ for $0 < t < t_1$. We thus obtain $\Phi_c[\mathbf{M}(t)] \geq (1-\sqrt{r_t})^{-1}\Phi_c(\mathbf{M}_0 + t^\alpha \mathbf{M}_\alpha)$ for $0 < t < t_1$.

Next, we write $\mathbf{M}_0 + t^\alpha \mathbf{M}_\alpha$ as

$$\mathbf{M}_0 + t^\alpha \mathbf{M}_\alpha = (1-\gamma_t)\mathbf{M}_0 + \gamma_t \mathbf{M}_{0,\alpha}$$

with $\mathbf{M}_{0,\alpha} = \mathbf{M}_0 + t^\alpha \mathbf{M}_\alpha$ and $\gamma_t = (t/t_0)^\alpha$. Then, from the concavity of $\Phi_c(\cdot)$, $\Phi_c(\mathbf{M}_0 + t^\alpha \mathbf{M}_\alpha) \geq (1-\gamma_t)\Phi_c(\mathbf{M}_0) + \gamma_t \Phi_c(\mathbf{M}_{0,\alpha})$, which implies

$$\liminf_{t\to 0^+} \Phi_c[\mathbf{M}(t)] \geq \lim_{t\to 0^+} \frac{1}{1-\sqrt{r_t}}\,[(1-\gamma_t)\Phi_c(\mathbf{M}_0) + \gamma_t \Phi_c(\mathbf{M}_{0,\alpha})] = \Phi_c(\mathbf{M}_0)\,.$$

Finally, from the upper semicontinuity of $\Phi_c(\cdot)$ we have $\limsup_{t\to 0^+} \Phi_c[\mathbf{M}(t)] \leq \Phi_c(\mathbf{M}_0)$, which implies (C.9). ■

Corollary 5.13. *For a sequence of matrices $\mathbf{M}(t) \in \mathbb{M}^\geq$ satisfying (5.19) with $\mathbf{c} \in \mathcal{M}(\mathbf{M}_0)$ and the condition C1, either the continuity property (C.9) is satisfied or the convergence of $\mathbf{M}(t)$ to $\mathbf{M}_0$ is along a hyperplane tangent to the cone $\mathbb{M}^\geq$ at $\mathbf{M}_0$, i.e., $\mathbf{M}_\alpha$ belongs to a supporting hyperplane to $\mathbb{M}^\geq$ at $\mathbf{M}_0$.*

Proof. Let $\mathbf{A} \in \mathbb{M}$ define a supporting hyperplane $\mathcal{H}_A$ to the cone $\mathbb{M}^\geq$ at $\mathbf{M}_0$; it satisfies $\text{trace}(\mathbf{A}\mathbf{M}_0) = 0$ and $\text{trace}(\mathbf{A}\mathbf{M}) \geq 0$ for any $\mathbf{M} \in \mathbb{M}^\geq$ ($\mathbf{A}$ is thus normal to $\mathbb{M}^\geq$ at $\mathbf{M}_0$ and $\mathbf{A} \in \mathbb{M}^\geq$). We have $\text{trace}[\mathbf{A}\mathbf{M}(t)] = t^\alpha\,\text{trace}(\mathbf{A}\mathbf{M}_\alpha) + \text{trace}(\mathbf{A}\mathbf{R}_t) \geq 0$ (since $\mathbf{M}(t) \in \mathbb{M}^\geq$), and thus $\text{trace}(\mathbf{A}\mathbf{M}_\alpha) \geq -\|\mathbf{A}\|\,\|\mathbf{R}_t\|/t^\alpha$, which tends to zero from C1. This implies $\text{trace}(\mathbf{A}\mathbf{M}_\alpha) \geq 0$; that is, $\mathbf{M}_\alpha$ is on the same side of $\mathcal{H}_A$ as $\mathbb{M}^\geq$.

There are two alternatives. Either C2 is satisfied and Lemma 5.12 implies (C.9), or C2 in not satisfied. In the latter case, for any $t > 0$ there exists $\mathbf{z}_t$ with $\|\mathbf{z}_t\| = 1$ such that $\mathbf{z}_t^\top(\mathbf{M}_0 + t^\alpha \mathbf{M}_\alpha)\mathbf{z}_t \leq 0$. From any such sequence $\{\mathbf{z}_t\}$ we extract a subsequence converging to some $\mathbf{z}_*$, which thus satisfies $\mathbf{z}_*^\top \mathbf{M}_0 \mathbf{z}_* \leq 0$, and therefore $\mathbf{z}_*^\top \mathbf{M}_0 \mathbf{z}_* = 0$ since $\mathbf{M}_0 \in \mathbb{M}^\geq$. Also, $\mathbf{z}_t^\top \mathbf{M}_\alpha \mathbf{z}_t \leq 0$ (since $\mathbf{z}_t^\top \mathbf{M}_0 \mathbf{z}_t \geq 0$) and thus $\mathbf{z}_*^\top \mathbf{M}_\alpha \mathbf{z}_* \leq 0$. Take $\mathbf{A} = \mathbf{z}_*\mathbf{z}_*^\top$; it defines a supporting hyperplane $\mathcal{H}_A$ to $\mathbb{M}^{\geq}$ at $\mathbf{M}_0$. From the developments above we obtain $\text{trace}(\mathbf{A}\mathbf{M}_\alpha) = \mathbf{z}_*^\top \mathbf{M}_\alpha \mathbf{z}_* \geq 0$ and thus $\text{trace}(\mathbf{A}\mathbf{M}_\alpha) = 0$; that is, $\mathbf{M}_\alpha$ belongs to $\mathcal{H}_A$. ■

Remark.

(*i*) When the sequence of matrices $\mathbf{M}(t)$ satisfies (5.19) and C1 with $\mathbf{c} \in \mathcal{M}(\mathbf{M}_0)$, if $\lim_{t\to 0^+} \Phi_c[\mathbf{M}(t)] \neq \Phi_c(\mathbf{M}_0)$, it means that C2 is not satisfied and, from the proof of Corollary 5.13, that $\mathbf{M}_\alpha$ belongs to a supporting hyperplane to $\mathbb{M}^{\geq}$ at $\mathbf{M}_0$. Conversely, if $\mathbf{M}_\alpha$ does not belong to such a tangent hyperplane, C2 and thus (C.9) are satisfied.

(*ii*) The condition C1 in Corollary 5.13 can be replaced by $\mathbf{R}_t \in \mathbb{M}^{\geq}$. Indeed, in that case $\mathbf{M}(t) - (\mathbf{M}_0 + t^\alpha \mathbf{R}_t) \in \mathbb{M}^{\geq}$, $\Phi_c[\mathbf{M}(t)] \geq \Phi_c(\mathbf{M}_0 + t^\alpha \mathbf{R}_t)$, and the rest of the proof is similar to that of Corollary 5.13. □

Lemma 5.28. *When the design criterion $\Phi(\cdot)$ is isotonic, an optimal design is supported at values of x such that $\mathbf{g}_\theta(x)$ is on the boundary of the Elfving's set $\mathscr{F}_\theta$.*

Proof. Suppose that $\mathbf{M}(\xi,\theta) = \sum_{i=1}^m \xi_i\, \mathbf{g}_\theta(x^{(i)})\mathbf{g}_\theta^\top(x^{(i)})$, $m \leq p(p+1)/2+1$, see Sect. 5.2.3, with $x^{(1)}$ such that $\mathbf{g}_\theta(x^{(1)})$ lies in the interior of $\mathscr{F}_\theta$. We can then decompose $\mathbf{g}_\theta(x^{(1)})$ into

$$\mathbf{g}_\theta(x^{(1)}) = \sum_{j=1}^{p+1} \alpha_j \mathbf{g}_j \,,$$

with $\alpha_j \geq 0$, $\sum_{j=1}^{p+1} \alpha_j = 1$, and $\mathbf{g}_j = \pm \mathbf{g}_\theta(x^{(j)})$ for some $x^{(j)}$ with $\mathbf{g}_j$ belonging to the boundary of $\mathscr{F}_\theta$. For any $\mathbf{u} \in \mathbb{R}^p$, consider $\Delta(\mathbf{u}) = \mathbf{u}^\top \mathbf{M}'(\xi,\theta)\mathbf{u} - \mathbf{u}^\top \mathbf{M}(\xi,\theta)\mathbf{u}$, where $\mathbf{M}'(\xi,\theta)$ is obtained by substituting $\sum_{j=1}^{p+1} \alpha_j \mathbf{g}_j \mathbf{g}_j^\top$ for $\mathbf{g}_\theta(x^{(1)})\mathbf{g}_\theta^\top(x^{(1)})$ in $\mathbf{M}(\xi,\theta)$. We have

$$\begin{aligned}
\Delta(\mathbf{u}) &= \xi_1 \left\{ \sum_{j=1}^{p+1} \alpha_j \,[\mathbf{u}^\top \mathbf{g}_\theta(x^{(j)})]^2 - [\mathbf{u}^\top \mathbf{g}_\theta(x^{(1)})]^2 \right\} \\
&= \xi_1 \left\{ \sum_{j=1}^{p+1} \alpha_j \,[\mathbf{u}^\top \mathbf{g}_j]^2 - \left[\sum_{j=1}^{p+1} \alpha_j (\mathbf{u}^\top \mathbf{g}_j) \right]^2 \right\}
\end{aligned}$$

and $\Delta(\mathbf{u}) \geq 0$ from Cauchy–Schwarz inequality. Therefore $\mathbf{M}'(\xi,\theta) \succeq \mathbf{M}(\xi,\theta)$ and $\Phi[\mathbf{M}'(\xi,\theta)] \geq \Phi[\mathbf{M}(\xi,\theta)]$. ■

Lemma 7.9. *Assume that $\eta(\theta)$ is continuous for $\theta \in \Theta$, a compact subset of $\mathbb{R}^p$. We have:*

(*i*) *For any* $\theta, \theta' \in \Theta_\eta(t)$, $E_\eta(\|\theta - \theta'\|^2) < 4t$, *and the maximum diameter* $\overline{D}(t)$ *of any connected part of* $\Theta_\eta(t)$ *satisfies* $\overline{D}^2(t) \leq \inf\{\delta : E_\eta(\delta) \geq 4t\}$.

(*ii*) *Suppose that the probability measure of the observations* $\mathbf{y}$ *has a density with respect to the Lebesgue measure in* $\mathbb{R}^N$. *If there exists* $\delta' < \delta$ *such that* $E_\eta(\delta) < t < E_\eta(\delta')$, *then the probability that the set* $\Theta_\eta(t)$ *is not connected is strictly positive.*

Proof.

(*i*) For any $\theta, \theta' \in \Theta_\eta(t)$, $\|\eta(\theta) - \eta(\theta')\| \leq \|\eta(\theta) - \mathbf{y}\| + \|\mathbf{y} - \eta(\theta')\| < 2\sqrt{t}$, therefore $E_\eta(\|\theta - \theta'\|^2) < 4t$. Let $\mathcal{C}(t)$ denote any connected part of $\Theta_\eta(t)$; if θ and θ' are in $\mathcal{C}(t)$, for any $\delta \in [0, \|\theta - \theta'\|^2]$, there exists $\theta'' \in \mathscr{B}(\theta, \sqrt{\delta})$ such that $\theta'' \in \mathcal{C}(t)$, with $\mathscr{B}(\theta, \sqrt{\delta})$ the closed ball of center θ and radius $\sqrt{\delta}$. This implies $E_\eta(\delta) < 4t$ for any $\delta \in [0, \mathrm{diam}^2[\mathcal{C}(t)]]$, and therefore $\mathrm{diam}^2[\mathcal{C}(t)] \leq \inf\{\delta : E_\eta(\delta) \geq 4t\}$.

(*ii*) Define $\alpha = (1/2)\min\{E_\eta(\delta') - t, t - E_\eta(\delta)\}$, take θ_1 and θ_2 in Θ such that $\|\theta_1 - \theta_2\|^2 = \delta$ and $\|\eta(\theta_1) - \eta(\theta_2)\|^2 = E_\eta(\delta)$. Consider the set $\mathcal{A}_{\delta'} = \Theta \cap \{\theta : \|\theta - \theta_1\|^2 = \delta'\}$.
Suppose first that $\mathcal{A}_{\delta'}$ is empty. Suppose that $\|\mathbf{y} - \eta(\theta_1)\| \leq \sqrt{E_\eta(\delta)}$ and $\|\mathbf{y} - \eta(\theta_2)\| \leq \sqrt{E_\eta(\delta)}$, which happens with a strictly positive probability. Then $E_\eta(\delta) < t$ implies that $\theta_1 \in \Theta_\eta(t)$, $\theta_2 \in \Theta_\eta(t)$, and $\Theta_\eta(t)$ is not connected.
Suppose now that $\mathcal{A}_{\delta'}$ is not empty and that $\mathbf{y}$ satisfies $\|\mathbf{y} - \eta(\theta_2)\| \leq \sqrt{E_\eta(\delta)}$ and $\|\mathbf{y} - \eta(\theta_1)\| < \sqrt{t + \alpha} - \sqrt{t}$, which again happens with strictly positive probability. Since $\alpha < t$, $\sqrt{t + \alpha} - \sqrt{t} < \sqrt{t}$ and $\theta_1 \in \Theta_\eta(t)$. Also, $E_\eta(\delta) < t$ implies $\theta_2 \in \Theta_\eta(t)$. Any θ in $\mathcal{A}_{\delta'}$ satisfies $\|\eta(\theta) - \eta(\theta_1)\|^2 \geq E_\eta(\delta') > t + \alpha$; therefore,

$$\begin{aligned}\|\mathbf{y} - \eta(\theta)\| &\geq \|\eta(\theta) - \eta(\theta_1)\| - \|\mathbf{y} - \eta(\theta_1)\| \\ &> \sqrt{t + \alpha} - \|\eta(\theta_1) - \mathbf{y}\| > \sqrt{t}\end{aligned}$$

and $\theta \notin \Theta_\eta(t)$, which implies that $\Theta_\eta(t)$ is not connected. ■

Symbols and Notation

$\Rightarrow$	Convergence in general or weak convergence (of probability measures or distribution functions)
$\stackrel{\mathrm{d}}{\to}$	Convergence in distribution
$\stackrel{\mathrm{p}}{\to}$	Convergence in probability
$\stackrel{\mathrm{a.s.}}{\to}$	Almost sure convergence
$\stackrel{\theta}{\rightsquigarrow}$	Uniform convergence with respect to θ
$\sim$	Distributed
a, A	Scalars
$\mathcal{A}$, $\mathscr{A}$, $\mathbb{A}$	Sets
$\mathbf{a}$	Column vector
α	Scalar or column vector
$\mathbf{A}$	Matrix
$\mathbf{a}^\top$, $\mathbf{A}^\top$	Transposed of $\mathbf{a}$ and $\mathbf{A}$
$\mathbf{A}^-$	A generalized inverse of $\mathbf{A}$ (i.e., $\mathbf{A}\mathbf{A}^-\mathbf{A} = \mathbf{A}$)
$\mathbf{A}^+$	The Moore–Penrose g-inverse of $\mathbf{A}$ (i.e., $\mathbf{A}\mathbf{A}^+\mathbf{A} = \mathbf{A}$, $\mathbf{A}^+\mathbf{A}\mathbf{A}^+ = \mathbf{A}^+$, $(\mathbf{A}\mathbf{A}^+)^\top = \mathbf{A}\mathbf{A}^+$ and $(\mathbf{A}^+\mathbf{A})^\top = \mathbf{A}^+\mathbf{A}$)
$\\|\mathbf{a}\\| = \\|\mathbf{a}\\|_2$	Euclidian norm of $\mathbf{a} = (a_1, \ldots, a_d)^\top \in \mathbb{R}^d$, $\\|\mathbf{a}\\| = \left(\sum_{i=1}^d a_i^2\right)^{1/2}$
$\\|\mathbf{a}\\|_1$	$\mathscr{L}_1$ norm of $\mathbf{a} = (a_1, \ldots, a_d)^\top \in \mathbb{R}^d$, $\\|\mathbf{a}\\|_1 = \sum_{i=1}^d \lvert a_i \rvert$
$\\|\mathbf{a}\\|_{\boldsymbol{\Omega}}$	$(\mathbf{a}^\top \boldsymbol{\Omega} \mathbf{a})^{1/2}$, for some $\boldsymbol{\Omega} \in \mathbb{M}^{\geq}$
$\\|\cdot\\|_\xi$	Norm in $\mathscr{L}_2(\xi)$, $\\|\phi\\|_\xi = \left[\int_{\mathscr{X}} \phi^2(x)\, \xi(\mathrm{d}x)\right]^{1/2}$, $\phi \in \mathscr{L}_2(\xi)$
$\langle \cdot, \cdot \rangle_\xi$	Inner product in $\mathscr{L}_2(\xi)$, $\langle \phi, \psi \rangle_\xi = \int_{\mathscr{X}} \phi(x)\, \psi(x)\, \xi(\mathrm{d}x)$, $\phi, \psi \in \mathscr{L}_2(\xi)$
$\stackrel{\xi}{\equiv}$	Parameter equivalence for the design ξ in a regression model, $\theta \stackrel{\xi}{\equiv} \theta^*$ when $\\|\eta(\cdot, \theta) - \eta(\cdot, \theta^*)\\|_\xi = 0$
$\{\mathbf{a}_i\}_j$	j-th component of $\mathbf{a}_i$

L. Pronzato and A. Pázman, *Design of Experiments in Nonlinear Models*, Lecture Notes in Statistics 212, DOI 10.1007/978-1-4614-6363-4,

$\{\mathbf{A}\}_{ij}$	(i,j)-th entry of $\mathbf{A}$
$\{\mathbf{A}\}_{i,\cdot}$	i-th row of $\mathbf{A}$
$\mathbf{A} \succeq \mathbf{B}$	$\mathbf{A} - \mathbf{B} \in \mathbb{M}^{\geq}$ (is nonnegative definite), $\mathbf{A}, \mathbf{B} \in \mathbb{M}$
$\mathbf{A} \succ \mathbf{B}$	$\mathbf{A} - \mathbf{B} \in \mathbb{M}^{>}$ (is positive definite), $\mathbf{A}, \mathbf{B} \in \mathbb{M}$
$\nabla f(\cdot)$	Gradient vector of $f(\cdot)$, $\{\nabla f(\alpha)\}_i = \partial f(\theta)/\partial\theta_i\big\|_{\theta=\alpha}$
$\nabla^2 f(\cdot)$	Hessian matrix of $f(\cdot)$, $\{\nabla^2 f(\alpha)\}_{ij} = \partial^2 f(\theta)/\partial\theta_i\partial\theta_j\big\|_{\theta=\alpha}$
$\tilde{\nabla} f(\cdot)$	Subgradient of $f(\cdot)$
$\partial f(\cdot)$	Subdifferential of $f(\cdot)$
$x'(\cdot)$	Derivative of $x(\cdot)$
$x''(\cdot)$	Second-order derivative of $x(\cdot)$
$\mathbf{0}$	Null vector, $\{\mathbf{0}\}_i = 0$ for all i
$\mathbf{O}$	Null matrix, $\{\mathbf{O}\}_{i,j} = 0$ for all i, j
$\mathbf{1}$	Vector of ones, $\{\mathbf{1}\}_i = 1$ for all i
a.s.	Almost sure(ly)
$\mathscr{B}(\mathbf{c}, r)$	Closed ball $\{\mathbf{x} \in \mathbb{R}^d : \\|\mathbf{x} - \mathbf{c}\\| \leq r\}$
$C_{int}(\xi, \theta)$	Intrinsic curvature of a regression model at θ for the design measure ξ
$C_{int}(X, \theta)$	Intrinsic curvature of a regression model at θ for the exact design X
$C_{par}(\xi, \theta)$	Parametric curvature at θ for ξ
$C_{par}(X, \theta)$	Parametric curvature at θ for X
$C_{tot}(\xi, \theta)$	Total curvature at θ for ξ
$\mathrm{diag}(\mathbf{a})$	Diagonal matrix with vector $\mathbf{a}$ on its diagonal
d.f.	(Cumulative) distribution function
$\mathbf{e}_i$	i-th basis vector
$\mathscr{E}_\phi(\cdot)$	Efficiency criterion associated with $\phi(\cdot)$, $\mathscr{E}_\phi(\xi) = \frac{\phi^+(\xi)}{\phi^+(\xi^*)}$ with ξ^* optimal for $\phi(\cdot)$
$\mathbb{E}(\cdot)$	Expectation
$\mathbb{E}_\mu(\cdot)$	Expectation for the probability measure μ
$\mathbb{E}_\pi(\cdot)$	Expectation for the p.d.f. $\pi(\cdot)$
$\mathbb{E}_x(\cdot)$	Conditional expectation for a given x, $\mathbb{E}_x(\omega) = \mathbb{E}(\omega\|x)$
$\mathbb{F}(\cdot)$	Distribution function (d.f.)
$\mathbf{f}(\cdot)$	Regressor in a linear regression model, $\eta(x, \theta) = \mathbf{f}^\top(x)\theta$
$\mathbf{f}_\theta(\cdot)$	Derivative in a nonlinear regression model, $\mathbf{f}_\theta(x) = \partial\eta(x, \theta)/\partial\theta$
$F_\phi(\xi; \nu)$	Directional derivative of $\phi(\cdot)$ at ξ in the direction ν
$F_\phi(\xi, x)$	Directional derivative of $\phi(\cdot)$ at ξ in the direction δ_x
$\mathscr{F}_\theta$	Elfving's set, convex closure of the set $\{\mathbf{f}_\theta(x) : x \in \mathscr{X}\} \cup \{-\mathbf{f}_\theta(x) : x \in \mathscr{X}\}$
$\mathbf{I}_q$	q-dimensional identity matrix
i.i.d.	Independently and identically distributed
$\mathbb{I}_{\mathcal{A}}(\cdot)$	Indicator function of the set $\mathcal{A}$
$\mathrm{int}(\mathcal{A})$	Interior of the set $\mathcal{A}$

$J_N(\cdot)$	Estimation criterion
$J_{\bar{\theta}}(\cdot)$	Limiting value of $J_N(\cdot)$ as $N \to \infty$ (under uniform convergence conditions)
ℓ	Number of elements in the finite design space $\mathscr{X}_\ell$
LP	Linear programming
LS	Least squares
$\mathrm{L}_{X,\mathbf{y}}(\theta)$	Likelihood of parameters θ for the design X at observations $\mathbf{y}$
$\mathscr{L}_2(\xi)$	Hilbert space of square-integrable real-valued functions ϕ, $\mathscr{L}_2(\xi) = \{\phi(\cdot) : \mathscr{X} \longrightarrow \mathbb{R}, \int_{\mathscr{X}} \phi^2(x)\, \xi(\mathrm{d}x) < \infty\}$
ML	Maximum likelihood
MSE	Mean-squared error
$\mathbf{M}_X(\theta)$	Information matrix for the (exact) design X
$\mathbf{M}(X, \theta)$	Normalized information matrix for the (exact) design X, $\mathbf{M}(X, \theta) = \mathbf{M}_X(\theta)/N$
$\mathbf{M}(\xi, \theta)$	Normalized information matrix for the design measure ξ
$\mathbf{M}(\xi)$	Normalized information matrix $\mathbf{M}(\xi, \theta^0)$ (local design)
$\mathbf{M}_\theta(x)$	Normalized information matrix $\mathbf{M}(\delta_x, \theta)$
$\mathbb{M}$	Set of symmetric $p \times p$ matrices
$\mathbb{M}^{\geq}$	Subset of $\mathbb{M}$ formed by nonnegative-definite matrices
$\mathbb{M}^{>}$	Subset of $\mathbb{M}$ formed by positive-definite matrices
$\mathcal{M}(\mathbf{M})$	Column space of the matrix $\mathbf{M}$, $\mathcal{M}(\mathbf{M}) = \{\mathbf{Mu} : \mathbf{u} \in \mathbb{R}^p\}$
$\mathcal{M}_\theta(\mathscr{X})$	$\{\mathbf{M}_\theta(x) : x \in \mathscr{X}\}$
$\mathcal{M}_\theta(\Xi)$	$\{\mathbf{M}(\xi, \theta) : \xi \in \Xi\}$
$\mathscr{M}$	Set of probability measures
N	Number of observations
$\mathcal{N}(\mathbf{M})$	Null space of the matrix $\mathbf{M}$, $\mathcal{N}(\mathbf{M}) = \{\mathbf{u} \in \mathbb{R}^p : \mathbf{Mu} = \mathbf{0}\}$
$\mathscr{N}(\mathbf{a}, \mathbf{V})$	Normal distribution (mean $\mathbf{a}$, variance–covariance matrix $\mathbf{V}$)
$o_{\mathrm{p}}(\cdot)$	$\alpha_n = o_{\mathrm{p}}(\beta_n)$ if $\{\alpha_n/\beta_n\} \stackrel{\mathrm{p}}{\to} 0$, $n \to \infty$
$\mathcal{O}_{\mathrm{p}}(\cdot)$	$\alpha_n = \mathcal{O}_{\mathrm{p}}(\beta_n)$ if $\{\alpha_n/\beta_n\}$ is bounded in probability, $n \to \infty$
p	Dimension of the parameter vector θ
p.d.f.	Probability density function
P_θ, $\mathbf{P}_\theta$	Projectors
$\mathscr{P}_{\ell-1}$	Probability simplex $\{\mathbf{w} \in \mathbb{R}^\ell : w_i \geq 0\,,\ \sum_{i=1}^{\ell} w_i = 1\}$
QP	Quadratic programming
$\mathbb{R}^p$	p-dimensional Euclidian space of real column vectors
$\mathbf{R}(\theta)$	Riemannian curvature tensor
$s(x)$	Skewness of the p.d.f. $\bar{\varphi}_x(\cdot)$, $s(x) = \mathbb{E}_x\{\varepsilon^3(x)\}\sigma^{-3}(x)$
$\mathbb{S}_\eta$	Expectation surface, $\mathbb{S}_\eta = \{\eta(\theta) : \theta \in \Theta\}$
$\mathcal{S}_\xi$	Support of the design measure ξ
SLLN	Strong law of large numbers
TSLS	Two-stage least squares
$\mathrm{var}(\cdot)$	Variance
$\mathrm{var}_x(\cdot)$	Conditional variance for a given x

$\mathrm{Var}(\cdot)$	Variance–covariance matrix
w.p.1	With probability one
$\mathscr{W}_{\ell-1}$	$\{\underline{\mathbf{w}} \in \mathbb{R}^{\ell-1} : \underline{w}_i \geq 0,\ \sum_{i=1}^{\ell-1} \underline{w}_i \leq 1\}$
WLS	Weighted least squares
x_i	i-th design point, experimental variables for the i-th trial
$x^{(i)}$	i-th element in a finite design space $\mathscr{X}_\ell$
X	Exact design with fixed size N, $X = (x_1, \ldots, x_N)$
$\mathscr{X}$	Design space (in general a compact subset of $\mathbb{R}^d$)
$\mathscr{X}_\ell$	Finite design space with ℓ elements
$y(x)$	Observation (random variable) at x
$\mathbf{y}$	Vector of observations, $\mathbf{y} = [y(x_1), \ldots, y(x_N)]^\top$
δ_x	Delta measure with mass 1 at x
$\varepsilon_i = \varepsilon(x_i)$	Measurement error (with zero mean, $\mathbb{E}_x\{\varepsilon(x)\} = 0$)
$\eta(x, \theta)$	Mean (or expected) response at $x \in \mathscr{X}$ for parameters θ in a regression model
$\eta(\theta)$	Vector of responses, $\eta(\theta) = \eta_X(\theta) = [\eta(x_1, \theta), \ldots, \eta(x_N, \theta)]^\top$
θ	Vector of parameters $(\theta_1, \ldots, \theta_p)^\top \in \Theta \subset \mathbb{R}^p$
$\hat{\theta}^N$	Estimator of θ for N observations
$\bar{\theta}$	True value of θ
Θ	Parameter space, a subset of $\mathbb{R}^p$
$\overline{\Theta}$	Closure of Θ
$\partial\Theta$	Boundary of Θ
$\Theta^{\#}$	Set of global minimizers of $J_{\bar{\theta}}(\cdot)$
$\kappa(x)$	Kurtosis of the p.d.f. $\bar{\varphi}_x(\cdot)$, $\kappa(x) = \mathbb{E}_x\{\varepsilon^4(x)\}\sigma^{-4}(x) - 3$
$\lambda(x, \bar{\theta})$	Parameterized variance function $\mathbb{E}_x\{\varepsilon^2(x)\}$ in a (mixed) regression model
$\lambda_{\min}(\mathbf{A})$	Minimum eigenvalue of $\mathbf{A}$
$\lambda_{\max}(\mathbf{A})$	Maximum eigenvalue of $\mathbf{A}$
μ	Probability measure (e.g., prior measure for θ)
ξ	Design measure (a probability measure on $\mathscr{X}$)
ξ^*	Optimum design measure
Ξ	Set of design measures on $\mathscr{X}$
$\pi(\cdot)$	Prior p.d.f. for θ
$\pi_{X,\mathbf{y}}(\cdot)$	Posterior p.d.f. for θ given $\mathbf{y}$ for the design X
$\varpi_X(\cdot, \cdot)$	p.d.f. of the joint distribution of θ and $\mathbf{y}$ for the design X
$\sigma^2(x)$	Variance of the error $\varepsilon(x)$
$\bar{\varphi}(\cdot)$	p.d.f. of the errors ε (regression model with i.i.d. errors)
$\bar{\varphi}_x(\cdot)$	p.d.f. of the errors $\varepsilon(x)$ (regression model)
$\varphi_{x,\theta}(\cdot)$	p.d.f. of the observations $y(x)$ (e.g., exponential family with parameters θ)
$\varphi_{X,\theta}(\cdot)$	p.d.f. of $\mathbf{y}$ given θ for the design X
$\varphi^*_X(\cdot)$	p.d.f. of the marginal distribution of $\mathbf{y}$ for the design X
$\phi(\cdot)$	Design criterion, function of a design measure ξ
$\phi^+(\cdot)$	Positively homogenous form of $\phi(\cdot)$

ϕ^*	Optimum (i.e., maximum) value of $\phi(\xi)$, $\xi \in \Xi$
$\Phi(\cdot)$	Design criterion, function of an information matrix
$\Phi^+(\cdot)$	Positively homogenous form of $\Phi(\cdot)$

List of Labeled Assumptions

H$_\Theta$, page 22: Θ is a compact subset of $\mathbb{R}^p$ such that $\Theta \subset \overline{\mathrm{int}(\Theta)}$.

H1$_\eta$, page 22: $\eta(x,\theta)$ is bounded on $\mathscr{X} \times \Theta$ and $\eta(x,\theta)$ is continuous on Θ, $\forall x \in \mathscr{X}$.

H2$_\eta$, page 22: $\bar{\theta} \in \mathrm{int}(\Theta)$ and, $\forall x \in \mathscr{X}$, $\eta(x,\theta)$ is twice continuously differentiable with respect to $\theta \in \mathrm{int}(\Theta)$, and its first two derivatives are bounded on $\mathscr{X} \times \mathrm{int}(\Theta)$.

H1$_h$, page 36: The function $h(\cdot) : \Theta \longrightarrow \mathbb{R}$ is continuous and has continuous second-order derivatives in $\mathrm{int}(\Theta)$.

H3$_\eta$, page 43: Let $\mathscr{S}_\epsilon$ denote the set $\left\{\theta \in \mathrm{int}(\Theta) : \left\|\eta(\cdot,\theta) - \eta(\cdot,\bar{\theta})\right\|_\xi^2 < \epsilon\right\}$, then there exists $\epsilon > 0$ such that for every $\theta^\#$ and θ^* in $\mathscr{S}_\epsilon$ we have

$$\left[\frac{\partial}{\partial\theta}\left\|\eta(\cdot,\theta) - \eta(\cdot,\theta^\#)\right\|_\xi^2\right]_{\theta=\theta^*} = \mathbf{0} \implies \theta^\# \stackrel{\xi}{\equiv} \theta^* .$$

H4$_\eta$, page 43: For any point $\theta^* \stackrel{\xi}{\equiv} \bar{\theta}$ there exists a neighborhood $\mathcal{V}(\theta^*)$ such that

$$\forall \theta \in \mathcal{V}(\theta^*)\,,\ \mathrm{rank}[\mathbf{M}(\xi,\theta)] = \mathrm{rank}[\mathbf{M}(\xi,\theta^*)]\,.$$

H2$_h$, page 43: The function $h(\cdot)$ is defined and has a continuous nonzero vector of derivatives $\partial h(\theta)/\partial\theta$ on $\mathrm{int}(\Theta)$. Moreover, for any $\theta \stackrel{\xi}{\equiv} \bar{\theta}$, there exists a linear mapping A_θ from $\mathscr{L}_2(\xi)$ to $\mathbb{R}$ (a continuous linear functional on $\mathscr{L}_2(\xi)$), such that $A_\theta = A_{\bar{\theta}}$ and that

L. Pronzato and A. Pázman, *Design of Experiments in Nonlinear Models*, Lecture Notes in Statistics 212, DOI 10.1007/978-1-4614-6363-4,

$$\frac{\partial h(\theta)}{\partial \theta_i} = A_\theta \left[\{\mathbf{f}_\theta\}_i\right] \,, \; i = 1, \ldots, p \,,$$

where $\{\mathbf{f}_\theta\}_i$ is defined by (3.42).

H2$'_h$, page 44: There exists a function $\Psi(\cdot)$, with continuous gradient, such that $h(\theta) = \Psi[\eta(\theta)]$, with $\eta(\theta) = (\eta(x^{(1)}, \theta), \ldots, \eta(x^{(k)}, \theta))^\top$.

H2$''_h$, page 44: $h(\theta) = \Psi[h_1(\theta), \ldots, h_k(\theta)]$ with $\Psi(\cdot)$ a continuously differentiable function of k variables and with

$$h_i(\theta) = \int_{\mathscr{X}} g_i[\eta(x, \theta), x] \, \xi(\mathrm{d}x) \,, \; i = 1, \ldots, k \,,$$

for some functions $g_i(t, x)$ differentiable with respect to t for any x in the support of ξ.

H3$_h$, page 47: The vector function $\mathbf{h}(\theta)$ has a continuous Jacobian $\partial \mathbf{h}(\theta)/\partial \theta^\top$ on $\mathrm{int}(\Theta)$. Moreover, for each $\theta \overset{\xi}{\equiv} \bar{\theta}$ there exists a continuous linear mapping B_θ from $\mathscr{L}_2(\xi)$ to $\mathbb{R}^q$ such that $B_\theta = B_{\bar{\theta}}$ and that

$$\frac{\partial \mathbf{h}(\theta)}{\partial \theta_i} = B_\theta \left[\{\mathbf{f}_\theta\}_i\right] \,, \; i = 1, \ldots, p \,,$$

where $\{\mathbf{f}_\theta\}_i$ is given by (3.42).

H1$_\lambda$, page 48: $\lambda(x, \bar{\theta})$ is bounded and strictly positive on $\mathscr{X}$, $\lambda^{-1}(x, \theta)$ is bounded on $\mathscr{X} \times \Theta$, and $\lambda(x, \theta)$ is continuous on Θ for all $x \in \mathscr{X}$.

H2$_\lambda$, page 48: For all $x \in \mathscr{X}$, $\lambda(x, \theta)$ is twice continuously differentiable with respect to $\theta \in \mathrm{int}(\Theta)$, and its first two derivatives are bounded on $\mathscr{X} \times \mathrm{int}(\Theta)$.

H$_\mathcal{S}$, page 172: There exists $r > 0$ such that:

(a) $\mathrm{Prob}_{\bar{\theta}}[\mathcal{G}(r)] = \mathrm{Prob}(\|\mathbf{y} - \eta(\theta)\| < r) \geq 1 - \epsilon$.
(b) Every $\mathbf{y} \in \mathcal{T}(r)$ has one r-projection only.

H$_\mathscr{X}$-(*i*), page 273: $\inf_{\theta \in \Theta} \lambda_{\min} \left[\sum_{i=1}^{\ell} \mathbf{f}_\theta(x^{(i)}) \mathbf{f}_\theta^\top(x^{(i)})\right] > \gamma > 0$.

H$_\mathscr{X}$-(*ii*), page 273: For all $\delta > 0$ there exists $\epsilon(\delta) > 0$ such that for any subset $\{i_1, \ldots, i_p\}$ of distinct elements of $\{1, \ldots, \ell\}$, $\inf_{\|\theta - \bar{\theta}\| \geq \delta} \sum_{j=1}^{p} [\eta(x^{(i_j)}, \theta) - \eta(x^{(i_j)}, \bar{\theta})]^2 > \epsilon(\delta)$.

H$_\mathscr{X}$-(*iii*), page 273: $\lambda_{\min} \left[\sum_{j=1}^{p} \mathbf{f}_{\bar{\theta}}(x^{(i_j)}) \mathbf{f}_{\bar{\theta}}^\top(x^{(i_j)})\right] \geq \bar{\gamma} > 0$ for any subset $\{i_1, \ldots, i_p\}$ of distinct elements of $\{1, \ldots, \ell\}$.

References

Ahipasaoglu, S., P. Sun, and M. Todd (2008). Linear convergence of a modified Frank-Wolfe algorithm for computing minimum volume enclosing ellipsoids. *Optim. Methods Softw. 23*, 5–19.

Alexéev, V., E. Galéev, and V. Tikhomirov (1987). *Recueil de Problèmes d'Optimisation*. Moscou: MIR.

Andrews, D. (1987). Consistency in nonlinear econometric models: a generic uniform law of large numbers. *Econometrica 55*(6), 1465–1471.

Andrews, D. (1992). Generic uniform convergence. *Econometric Theory 8*, 241–257.

Atkinson, A. (2003). Transforming both sides and optimum experimental design for a nonlinear model arising from second-order chemical kinetics. In *Tatra Mountains Math. Pub.*, Volume 26, pp. 29–39.

Atkinson, A. (2004). Some Bayesian optimum designs for response transformation in nonlinear models with nonconstant variance. In A. Di Bucchianico, H. Läuter, and H. Wynn (Eds.), *mODa'7 – Advances in Model–Oriented Design and Analysis, Proc. 7th Int. Workshop, Heeze (Netherlands)*, Heidelberg, pp. 13–21. Physica Verlag.

Atkinson, A., K. Chaloner, A. Herzberg, and J. Juritz (1993). Optimal experimental designs for properties of a compartmental model. *Biometrics 49*, 325–337.

Atkinson, A. and R. Cook (1996). Designing for a response transformation parameter. *J. Roy. Statist. Soc. B59*, 111–124.

Atkinson, A. and D. Cox (1974). Planning experiments for discriminating between models (with discussion). *J. Roy. Statist. Soc. B36*, 321–348.

Atkinson, A. and A. Donev (1992). *Optimum Experimental Design*. Oxford Univ. Press.

Atkinson, A. and V. Fedorov (1975a). The design of experiments for discriminating between two rival models. *Biometrika 62*(1), 57–70.

Atkinson, A. and V. Fedorov (1975b). Optimal design: experiments for discriminating between several models. *Biometrika 62*(2), 289–303.

L. Pronzato and A. Pázman, *Design of Experiments in Nonlinear Models*, Lecture Notes in Statistics 212, DOI 10.1007/978-1-4614-6363-4,

Atwood, C. (1969). Optimal and efficient designs of experiments. *Ann. Math. Statist. 40*(5), 1570–1602.

Atwood, C. (1973). Sequences converging to D-optimal designs of experiments. *Ann. Statist. 1*(2), 342–352.

Atwood, C. (1976). Convergent design sequences for sufficiently regular optimality criteria. *Ann. Statist. 4*(6), 1124–1138.

Atwood, C. (1980). Convergent design sequences for sufficiently regular optimality criteria, II singular case. *Ann. Statist. 8*(4), 894–912.

Avriel, M. (2003). *Nonlinear Programming. Analysis and Methods.* New York: Dover. [Originally published by Prentice Hall, 1976].

Barndorff-Nielsen, O. (1978). *Information and Exponential Families in Statistical Theory.* Chichester: Wiley.

Bartlett, P. (2003). Prediction algorithms: complexity, concentration and convexity. In *Prep. 13th IFAC Symp. on System Identification, Rotterdam*, pp. 1507–1517.

Bates, D. and D. Watts (1980). Relative curvature measures of nonlinearity. *J. Roy. Statist. Soc. B42*, 1–25.

Bates, D. and D. Watts (Eds.) (1988). *Nonlinear regression Analysis and its Applications.* New York: Wiley.

Ben-Tal, A. and A. Nemirovskii (2001). *Lectures on Modern Convex Optimization: Analysis, Algorithms, and Engineering Applications.* Philadelphia: MPS/SIAM Series on Optim. 2.

Beran, R. (1974). Asymptotically efficient rank estimates in location models. *Ann. Statist. 2*, 63–74.

Berlinet, A. and C. Thomas-Agnan (2004). *Reproducing Kernel Hilbert Spaces in Probability and Statistics.* Boston: Kluwer.

Bickel, P. (1982). On adaptive estimation. *Ann. Statist. 10*, 647–671.

Bickel, P., C. Klassen, Y. Ritov, and J. Wellner (1993). *Efficient and Adaptive Estimation for Semiparametric Models.* Baltimore: Johns Hopkins Univ. Press.

Biedermann, S. and H. Dette (2001). Minimax optimal designs for nonparametric regression — a further optimality property of the uniform distribution. In A. Atkinson, P. Hackl, and W. Müller (Eds.), *mODa'6 – Advances in Model–Oriented Design and Analysis, Proc. 6th Int. Workshop, Puchberg/Schneberg (Austria)*, Heidelberg, pp. 13–20. Physica Verlag.

Bierens, H. (1994). *Topics in Advanced Econometrics.* Cambridge: Cambridge Univ. Press.

Bilardello, P., X. Joulia, J.-M. Le Lann, H. Delmas, and B. Koehret (1993). A general strategy for parameter estimation in differential-algebraic systems. *Comput. Chem. Eng. 17*, 517–525.

Billingsley, P. (1995). *Probability and Measure.* New York: Wiley. [3rd ed.].

Birgin, E., J. Martínez, and M. Raydan (2000). Nonmonotone spectral projected gradient methods on convex sets. *SIAM J. Optim. 10*(4), 1196–1211.

Bland, R., D. Goldfarb, and M. Todd (1981). The ellipsoid method: a survey. *Oper. Res. 29*(6), 1039–1091.

Bloomfield, P. and G. Watson (1975). The inefficiency of least squares. *Biometrika 62*(1), 121–128.

Bohachevsky, I., M. Johnson, and M. Stein (1986). Generalized simulated annealing for function optimization. *Technometrics 28*(3), 209–217.

Böhning, D. (1985). Numerical estimation of a probability measure. *J. Stat. Plann. Inference 11*, 57–69.

Böhning, D. (1986). A vertex-exchange-method in D-optimal design theory. *Metrika 33*, 337–347.

Boissonnat, J.-D. and M. Yvinec (1998). *Algorithmic Geometry.* Cambridge Univ. Press.

Bonnans, J., J. Gilbert, C. Lemaréchal, and C. Sagastizábal (2006). *Numerical Optimization. Theoretical and Practical Aspects.* Heidelberg: Springer. [2nd ed.].

Bornkamp, B. (2011). Functional uniform priors for nonlinear modelling. Technical Report arXiv:1110.4400v1.

Box, G. and D. Cox (1964). An analysis of transformations (with discussion). *J. Roy. Statist. Soc. B26*, 211–252.

Box, G. and H. Lucas (1959). Design of experiments in nonlinear situations. *Biometrika 46*, 77–90.

Box, M. (1971). Bias in nonlinear estimation. *J. Roy. Statist. Soc. B33*, 171–201.

Boyd, S. and L. Vandenberghe (2004). *Convex Optimization.* Cambridge: Cambridge Univ. Press.

Braess, D. and H. Dette (2007). On the number of support points of maximin and Bayesian optimal designs. *Ann. Statist. 35*(2), 772–792.

Caines, P. (1988). *Linear Stochastic Systems.* New York: Wiley.

Calamai, P. and J. Moré (1987). Projected gradient methods for linearly constrained problems. *Math. Programming 39*, 93–116.

Carroll, R. and D. Ruppert (1982). A comparison between maximum likelihood and generalized least squares in a heteroscedastic linear model. *J. Amer. Statist. Assoc. 77*(380), 878–882.

Casella, G. (2008). *Statistical Design.* New York: Springer.

Chaloner, K. and K. Larntz (1989). Optimal Bayesian design applied to logistic regression experiments. *J. Stat. Plann. Inference 21*, 191–208.

Chaloner, K. and I. Verdinelli (1995). Bayesian experimental design: a review. *Statist. Sci. 10*(3), 273–304.

Chaudhuri, P. and P. Mykland (1993). Nonlinear experiments: optimal design and inference based likelihood. *J. Amer. Statist. Assoc. 88*(422), 538–546.

Chaudhuri, P. and P. Mykland (1995). On efficiently designing of nonlinear experiments. *Statistica Sinica 5*, 421–440.

Chavent, G. (1983). Local stability of the output least square parameter estimation technique. *Matematicada Applicada e Computacional 2*(1), 3–22.

Chavent, G. (1987). Identifiability of parameters in the output least square formulation. In E. Walter (Ed.), *Identifiability of Parametric Models*, Chapter 6, pp. 67–74. Oxford: Pergamon.

Chavent, G. (1990). A new sufficient condition for the wellposedness of non-linear least-square problems arising in identification and control. In A. Bensoussan and J. Lions (Eds.), *Analysis and Optimization of Systems,* Lecture Notes in Control and Inform. Sci., vol. 144, pp. 452–463. Springer.

Chavent, G. (1991). New size × curvature conditions for strict quasiconvexity of sets. *SIAM J. Control Optim. 29*(6), 1348–1372.

Cheng, M.-Y., P. Hall, and D. Titterington (1998). Optimal design for curve estimation by local linear smoothing. *Bernoulli 4*(1), 3–14.

Clarke, B. and A. Barron (1994). Jeffrey's prior is asymptotically less favorable under entropy risk. *J. Stat. Plann. Inference 41*, 37–60.

Clarke, G. (1980). Moments of the least-squares estimators in a non-linear regression model. *J. Roy. Statist. Soc. B42*, 227–237.

Clyde, M. and K. Chaloner (2002). Constrained design strategies for improving normal approximations in nonlinear regression problems. *J. Stat. Plann. Inference 104*, 175–196.

Cochran, W. and G. Cox (1957). *Experimental Designs.* New York: Wiley. [2nd ed., 2005].

Cook, D. and V. Fedorov (1995). Constrained optimization of experimental design (invited discussion paper). *Statistics 26*, 129–178.

Cook, D. and W. Wong (1994). On the equivalence between constrained and compound optimal designs. *J. Amer. Statist. Assoc. 89*(426), 687–692.

Cook, R., D. Hawkins, and S. Weisberg (1993). Exact iterative computation of the robust multivariate minimum volume ellipsoid estimator. *Statist. Probab. Lett. 16*, 213–218.

Cook, R. and C. Nachtsheim (1980). A comparison of algorithms for constructing exact D-optimal designs. *Technometrics 22*(3), 315–324.

Cormen, T., C. Leiserson, R. Rivest, and C. Stein (2001). *Introduction to Algorithms.* MIT Press and McGraw-Hill. [2nd ed.].

Correa, R. and C. Lemaréchal (1993). Convergence of some algorithms for convex minimization. *Math. Programming 62*, 261–275.

Cox, D. and D. Hinkley (1974). *Theoretical Statistics.* London: Chapman & Hall.

Cramér, H. (1946). *Mathematical Methods of Statistics.* Princeton, N.J.: Princeton University Press.

Cucker, F. and S. Smale (2001). On the mathematical foundations of learning. *Bull. AMS 39*(1), 1–49.

Dai, Y.-H. and R. Fletcher (2006). New algorithms for singly linearly constrained quadratic programs subject to lower and upper bounds. *Math. Programming A 106*, 403–421.

D'Argenio, D. (1990). Incorporating prior parameter uncertainty in the design of sampling schedules for pharmacokinetic parameter estimation experiments. *Math. Biosci. 99*, 105–118.

Davies, L. (1992). The asymptotics of Rousseeuw's minimum volume ellipsoid estimator. *Ann. Statist. 20*(4), 1828–1843.

de la Garza, A. (1954). Spacing of information in polynomial regression. *Ann. Math. Statist. 25*, 123–130.

del Pino, G. (1989). The unifying role of iterative generalized least squares in statistical algorithms (with discussion). *Statist. Sci. 4*(4), 394–408.

Dembski, W. (1990). Uniform probability. *J. Theoret. Probab. 3*(4), 611–626.

Demidenko, E. (1989). *Optimization and Regression.* Moscow: Nauka. [In Russian].

Demidenko, E. (2000). Is this the least squares estimate? *Biometrika 87*(2), 437–452.

Dem'yanov, V. and V. Malozemov (1974). *Introduction to Minimax.* New York: Dover.

den Boeff, E. and D. den Hertog (2007). Efficient line search methods for convex functions. *SIAM J. Optim. 18*(1), 338–363.

den Hertog, D. (1994). *Interior Point Approach to Linear, Quadratic and Convex Programming.* Dordrecht: Kluwer.

Dette, H. (1993). Elfving's theorem for D-optimality. *Ann. Statist. 21*(2), 753–766.

Dette, H. and V. Melas (2011). A note on de la Garza phenomenon for locally optimal designs. *Ann. Statist. 39*(2), 1266–1281.

Dette, H., A. Pepelyshev, and A. Zhigljavsky (2008). Improving updating rules in multiplicative algorithms for computing D-optimal designs. *J. Stat. Plann. Inference 53*, 312–320.

Dette, H. and W. Studden (1993). Geometry of E-optimality. *Ann. Statist. 21*(1), 416–433.

Downing, D., V. Fedorov, and S. Leonov (2001). Extracting information from the variance function: optimal design. In A. Atkinson, P. Hackl, and W. Müller (Eds.), *mODa'6 – Advances in Model–Oriented Design and Analysis, Proc. 6th Int. Workshop, Puchberg/Schneberg (Austria)*, pp. 45–52. Heidelberg: Physica Verlag.

Dragalin, V. and V. Fedorov (2006). Adaptive designs for dose-finding based on efficacy-toxicity response. *J. Stat. Plann. Inference 136*, 1800–1823.

Eaton, M., A. Giovagnoli, and P. Sebastiani (1996). A predictive approach to the Bayesian design problem with application to normal regression models. *Biometrika 83*(1), 111–125.

Ecker, J. and M. Kupferschmid (1983). An ellipsoid algorithm for nonlinear programming. *Math. Programming 27*, 83–106.

Ecker, J. and M. Kupferschmid (1985). A computational comparison of the ellipsoidal algorithm with several nonlinear programming algorithms. *SIAM J. Control Optim. 23*(5), 657–674.

Efron, B. and D. Hinkley (1978). Assessing the accuracy of the maximum likelihood estimator: observed versus expected Fisher information. *Biometrika 65*(3), 457–487.

Eisenhart, L. (1960). *Riemannian Geometry.* Princeton, N.J.: Princeton Univ. Press.

Elfving, G. (1952). Optimum allocation in linear regression. *Ann. Math. Statist. 23*, 255–262.

Ermoliev, Y. and R.-B. Wets (Eds.) (1988). *Numerical Techniques for Stochastic Optimization Problems.* Berlin: Springer.

Fedorov, V. (1971). The design of experiments in the multiresponse case. *Theory Probab. Appl. 16*(2), 323–332.

Fedorov, V. (1972). *Theory of Optimal Experiments.* New York: Academic Press.

Fedorov, V. (1980). Convex design theory. *Math. Operationsforsch. Statist. Ser. Statist. 11*(3), 403–413.

Fedorov, V. (1989). Optimal design with bounded density: optimization algorithms of the exchange type. *J. Stat. Plann. Inference 22*, 1–13.

Fedorov, V. and P. Hackl (1997). *Model-Oriented Design of Experiments.* Berlin: Springer.

Fedorov, V. and A. Pázman (1968). Design of physical experiments. *Fortschritte der Physik 16*, 325–355.

Fellman, J. (1974). On the allocation of linear observations. *Comment. Phys. Math. 44*, 27–78.

Fellman, J. (1989). An empirical study of a class of iterative searches for optimal designs. *J. Stat. Plann. Inference 21*, 85–92.

Firth, D. (1993). Bias reduction of maximum likelihood estimates. *Biometrika 80*(1), 27–38.

Ford, I. and S. Silvey (1980). A sequentially constructed design for estimating a nonlinear parametric function. *Biometrika 67*(2), 381–388.

Ford, I., D. Titterington, and C. Wu (1985). Inference and sequential design. *Biometrika 72*(3), 545–551.

Fourgeaud, C. and A. Fuchs (1967). *Statistique.* Paris: Dunod.

Frank, M. and P. Wolfe (1956). An algorithm for quadratic programming. *Naval Res. Logist. Quart. 3*, 95–110.

Galil, Z. and J. Kiefer (1977). Comparison of simplex designs for quadratic mixture models. *Technometrics 19*(4), 445–453.

Galil, Z. and J. Kiefer (1980). Time- and space-saving computer methods, related to Mitchell's DETMAX, for finding D-optimum designs. *Technometrics 22*(3), 301–313.

Gallant, A. (1987). *Nonlinear Statistical Models.* New York: Wiley.

Gauchi, J.-P. and A. Pázman (2006). Designs in nonlinear regression by stochastic minimization of functionals of the mean square error matrix. *J. Stat. Plann. Inference 136*, 1135–1152.

Gautier, R. and L. Pronzato (1998). Sequential design and active control. In N. Flournoy, W. Rosenberger, and W. Wong (Eds.), *New Developments and Applications in Experimental Design*, Lecture Notes — Monograph Series, vol. 34, pp. 138–151. Hayward: IMS.

Gautier, R. and L. Pronzato (1999). Some results on optimal allocation in two–step sequential design. *Biometrical Lett. 36*(1), 15–30.

Gautier, R. and L. Pronzato (2000). Adaptive control for sequential design. *Discuss. Math. Probab. Stat. 20*(1), 97–114.

Gilmour, S. and L. Trinca (2012). Optimum design of experiments for statistical inference (with discussion). *J. Roy. Statist. Soc. C61*(3), 1–25.

Goffin, J.-L. and K. Kiwiel (1999). Convergence of a simple subgradient level method. *Math. Programming 85*, 207–211.

Goodwin, G. and R. Payne (1977). *Dynamic System Identification: Experiment Design and Data Analysis*. New York: Academic Press.

Gorman, J. and A. Hero (1990). Lower bounds for parametric estimation with constraints. *IEEE Trans. Information Theory 26*, 1285–1301.

Green, P. (1984). Iteratively reweighted least squares for maximum likelihood estimation, and some robust and resistant alternatives (with discussion). *J. Roy. Statist. Soc. B-46*(2), 149–192.

Grötschel, M., L. Lovász, and A. Schrijver (1980). *Geometric Algorithms and Combinatorial Optimization*. Berlin: Springer.

Hadjihassan, S., L. Pronzato, E. Walter, and I. Vuchkov (1997). Robust design for quality improvement by ellipsoidal bounding. In G. Yin and Q. Zhang (Eds.), *Proc. 1996 AMS–SIAM Summer Seminar, Math. of Stochastic Manufacturing Systems*, Lectures in Applied Math., vol. 33, Providence, Rhode Island, pp. 127–138. American Math. Soc.

Halperin, M. (1963). Confidence interval estimation in non-linear regression. *J. Roy. Statist. Soc. B25*, 330–333.

Hamilton, D. and D. Watts (1985). A quadratic design criterion for precise estimation in nonlinear regression models. *Technometrics 27*, 241–250.

Hamilton, D., D. Watts, and D. Bates (1982). Accounting for intrinsic nonlinearities in nonlinear regression parameter inference regions. *Ann. Statist. 10*, 386–393.

Harman, R. (2004a). Lower bounds on efficiency ratios based on ϕ_p-optimal designs. In A. Di Bucchianico, H. Läuter, and H. Wynn (Eds.), *mODa'7 – Advances in Model–Oriented Design and Analysis, Proc. 7th Int. Workshop, Heeze (Netherlands)*, Heidelberg, pp. 89–96. Physica Verlag.

Harman, R. (2004b). Minimal efficiency of designs under the class of orthogonally invariant information criteria. *Metrika 60*, 137–153.

Harman, R. and T. Jurík (2008). Computing c-optimal experimental designs using the simplex method of linear programming. *Comput. Statist. Data Anal. 53*, 247–254.

Harman, R. and L. Pronzato (2007). Improvements on removing non-optimal support points in D-optimum design algorithms. *Statist. Probab. Lett. 77*, 90–94.

Harman, R. and M. Trnovská (2009). Approximate D-optimal designs of experiments on the convex hull of a finite set of information matrices. *Math. Slovaca 59*(5), 693–704.

Hartley, H. (1964). Exact confidence regions for parameters in non-linear regression laws. *Biometrika 51*(3&4), 347–353.

Harville, D. (1997). *Matrix Algebra from a Statistician's Perspective.* Heidelberg: Springer.

Hastie, T., R. Tibshirani, and J. Friedman (2001). *The Elements of Statistical Learning. Data Mining, Inference and Prediction.* Heidelberg: Springer.

Hearn, D. and S. Lawphongpanich (1989). Lagrangian dual ascent by generalized linear programming. *Oper. Research Lett. 8*, 189–196.

Heyde, C. (1997). *Quasi-likelihood and its Application. A General Approach to Optimal Parameter Estimation.* New York: Springer.

Hill, P. (1978). A review of experimental design procedures for regression model discrimination. *Technometrics 20*, 15–21.

Hiriart-Urruty, J. and C. Lemaréchal (1993). *Convex Analysis and Minimization Algorithms, part 1 and 2.* Berlin: Springer.

Hougaard, P. (1985). Saddlepoint approximations for curved exponential families. *Statist. Probab. Lett. 3*, 161–166.

Hu, I. (1998). On sequential designs in nonlinear problems. *Biometrika 85*(2), 496–503.

Huber, P. (1981). *Robust Statistics.* New York: John Wiley.

Ibragimov, I. and R. Has'minskii (1981). *Statistical Estimation. Asymptotic Theory.* Heidelberg: Springer.

Ivanov, A. (1997). *Asymptotic Theory of Nonlinear Regression.* Dordrecht: Kluwer.

Jennrich, R. (1969). Asymptotic properties of nonlinear least squares estimation. *Ann. Math. Statist. 40*, 633–643.

Jobson, J. and W. Fuller (1980). Least squares estimation when the covariance matrix and parameter vector are functionally related. *J. Amer. Statist. Assoc. 75*(369), 176–181.

Johnson, M. and C. Nachtsheim (1983). Some guidelines for constructing exact D-optimal designs on convex design spaces. *Technometrics 25*, 271–277.

Jørgensen, B. (1997). Exponential dispersion models. *J. Roy. Statist. Soc. B49*, 127–167.

Karlin, S. and W. Studden (1966). Optimal experimental designs. *Ann. Math. Statist. 37*, 783–815.

Kelley, J. (1960). The cutting plane method for solving convex programs. *SIAM J. 8*, 703–712.

Khachiyan, L. (1979). A polynomial algorithm in linear programming. *Doklady Akademïa Nauk SSSR 244*, 1093–1096. [English transl. *Soviet Math. Doklady*, 20, 191–194].

Khachiyan, L. (1996). Rounding of polytopes in the real number model of computation. *Math. Oper. Res. 21*(2), 307–320.

Khachiyan, L. and M. Todd (1993). On the complexity of approximating the maximal inscribed ellipsoid for a polytope. *Math. Programming A61*(2), 137–159.

Kiefer, J. (1962). Two more criteria equivalent to D-optimality of designs. *Ann. Math. Statist. 33*(2), 792–796.

Kiefer, J. (1974). General equivalence theory for optimum designs (approximate theory). *Ann. Statist. 2*(5), 849–879.

Kiefer, J. and J. Wolfowitz (1960). The equivalence of two extremum problems. *Canad. J. Math. 12*, 363–366.

Kieffer, M. and E. Walter (1998). Interval analysis for guaranteed nonlinear parameter estimation. In A. Atkinson, L. Pronzato, and H. Wynn (Eds.), *MODA5 – Advances in Model-Oriented Data Analysis and Experimental Design, Marseilles (France), June 1998*, pp. 115–125. Heidelberg: Physica Verlag.

Kim, J. and D. Pollard (1990). Cube root asymptotics. *Ann. Statist. 18*(1), 191–219.

Koopmans, T. and O. Reiersøl (1950). The identification of structural characteristics. *Ann. Math. Statist. 21*(2), 163–181.

Kushner, H. and D. Clark (1978). *Stochastic Approximation Methods for Constrained and Unconstrained Systems*. Heidelberg: Springer.

Kushner, H. and J. Yang (1993). Stochastic approximation with averaging of the iterates: optimal asymptotic rate of convergence for general processes. *SIAM J. Control Optim. 31*(4), 1045–1062.

Kushner, H. and G. Yin (1997). *Stochastic Approximation Algorithms and Applications*. Heidelberg: Springer.

Lai, T. (1994). Asymptotic properties of nonlinear least squares estimates in stochastic regression models. *Ann. Statist. 22*(4), 1917–1930.

Lai, T., H. Robbins, and C. Wei (1978). Strong consistency of least squares estimates in multiple regression. *Proc. Nat. Acad. Sci. USA 75*(7), 3034–3036.

Lai, T., H. Robbins, and C. Wei (1979). Strong consistency of least squares estimates in multiple regression II. *J. Multivariate Anal. 9*, 343–361.

Lai, T. and C. Wei (1982). Least squares estimates in stochastic regression models with applications to identification and control of dynamic systems. *Ann. Statist. 10*(1), 154–166.

Le Cam, L. (1953). On some asymptotic properties of maximum likelihood estimates and related Bayes' estimates. *Univ. California Pub. in Stat. 1*, 277–230.

Le Cam, L. (1960). Local asymptotically normal families of distributions. *Univ. California Pub. in Stat. 3*, 37–98.

Lehmann, E. and G. Casella (1998). *Theory of Point Estimation*. Heidelberg: Springer.

Lemaréchal, C., A. Nemirovskii, and Y. Nesterov (1995). New variants of bundle methods. *Math. Programming 69*(1), 111–147.

Levin, A. (1965). On an algorithm for the minimization of convex functions. *Soviet Math. Dokl. 6*, 286–290.

Li, X.-S. and S.-C. Fang (1997). On the entropic regularization method for solving min-max problems with applications. *Math. Methods Oper. Res. 46*, 119–130.

Liang, K.-Y. and S. Zeger (1995). Inference based on estimating functions in the presence of nuisance parameters. *Statist. Sci. 10*(2), 158–173.

Lindley, D. (1956). On a measure of information provided by an experiment. *Ann. Math. Statist. 27*, 986–1005.

Lindsay, B. and B. Li (1997). On the second-order optimality of the observed Fisher information. *Ann. Statist. 25*(5), 2172–2199.

Liu, S. and H. Neudecker (1997). Kantorovich inequalities and efficiency comparisons for several classes of estimators in linear models. *Statistica Neerlandica 51*(3), 345–355.

Ljung, L. (1987). *System Identification, Theory for the User*. Englewood Cliffs: Prentice Hall.

López-Fidalgo, J. and J. Rodríguez-Díaz (1998). Characteristic polynomial criteria in optimal experimental design. In A. Atkinson, L. Pronzato, and H. Wynn (Eds.), *Advances in Model–Oriented Data Analysis and Experimental Design, Proc. MODA'5, Marseilles, June 22–26, 1998*, pp. 31–38. Heidelberg: Physica Verlag.

López-Fidalgo, J., B. Torsney, and R. Ardanuy (1998). MV-optimisation in weighted linear regression. In A. Atkinson, L. Pronzato, and H. Wynn (Eds.), *Advances in Model–Oriented Data Analysis and Experimental Design, Proc. MODA'5, Marseilles, June 22–26, 1998*, pp. 39–50. Heidelberg: Physica Verlag.

Magnus, J. and H. Neudecker (1999). *Matrix Differential Calculus, with Applications in Statistics and Econometrics*. New York: Wiley.

Manski, C. (1984). Adaptive estimation of nonlinear regression models. *Econometric Rev. 3*(2), 145–194.

McCormick, G. and R. Tapia (1972). The gradient projection method under mild differentiability conditions. *SIAM J. Control 10*(1), 93–98.

McCullagh, P. and J. Nelder (1989). *Generalized Linear Models*. London: Chapman & Hall. [2nd ed.].

Mikulecká, J. (1983). On a hybrid experimental design. *Kybernetika 19*(1), 1–14.

Minoux, M. (1983). *Programmation Mathématique, Théorie et Algorithmes, vol. 1 & 2*. Paris: Dunod.

Mitchell, T. (1974). An algorithm for the construction of "D-optimal" experimental designs. *Technometrics 16*, 203–210.

Molchanov, I. and S. Zuyev (2001). Variational calculus in the space of measures and optimal design. In A. Atkinson, B. Bogacka, and A. Zhigljavsky (Eds.), *Optimum Design 2000*, Chapter 8, pp. 79–90. Dordrecht: Kluwer.

Molchanov, I. and S. Zuyev (2002). Steepest descent algorithm in a space of measures. *Stat. Comput. 12*, 115–123.

Montgomery, D. (1976). *Design and Analysis of Experiments*. New York: Wiley. [6th ed., 2005].

Morris, M. and T. Mitchell (1995). Exploratory designs for computational experiments. *J. Stat. Plann. Inference 43*, 381–402.

Müller, C. and A. Pázman (1998). Applications of necessary and sufficient conditions for maximin efficient designs. *Metrika 48*, 1–19.

Müller, H.-G. (1984). Optimal designs for nonparametric kernel regression. *Statist. Probab. Lett. 2*, 285–290.

Müller, W. (2007). *Collecting Spatial Data. Optimum Design of Experiments for Random Fields*. Heidelberg: Physica-Verlag. [3rd ed.].

Müller, W. and A. Pázman (2003). Measures for designs in experiments with correlated errors. *Biometrika 90*(2), 423–434.

Müller, W. and B. Pötscher (1992). Batch sequential design for a nonlinear estimation problem. In V. Fedorov, W. Müller, and I. Vuchkov (Eds.), *Model-Oriented Data Analysis II, Proc. 2nd IIASA Workshop, St Kyrik (Bulgaria), May 1990*, pp. 77–87. Heidelberg: Physica Verlag.

Nadaraya, E. (1964). On estimating regression. *Theory Probab. Appl. 9*, 141–142.

Nesterov, Y. (1995). Complexity estimates of some cutting plane methods based on the analytic center. *Math. Programming 69*, 149–176.

Nesterov, Y. (2004). *Introductory Lectures to Convex Optimization: A Basic Course*. Dordrecht: Kluwer.

Nesterov, Y. and A. Nemirovskii (1994). *Interior-Point Polynomial Algorithms in Convex Programming*. Philadelphia: SIAM.

Newey, W. (1991). Uniform convergence in probability and stochastic equicontinuity. *Econometrica 9*(4), 1161–1167.

Pan, S., S. He, and X. Li (2007). Smoothing method for minimizing the sum of the r largest functions. *Optim. Methods Softw. 22*, 267–277.

Parzen, E. (1962). On estimation of a probability density function and mode. *Ann. Math. Statist. 35*, 1065–1076.

Parzen, E. (1979). Nonparametric statistical data modeling. *J. Amer. Statist. Assoc. 74*(365), 105–121.

Pázman, A. (1980). Singular experimental designs. *Math. Operationsforsch. Statist. Ser. Statist. 16*, 137–149.

Pázman, A. (1984a). Nonlinear least squares – uniqueness versus ambiguity. *Math. Operationsforsch. Statist. Ser. Statist. 15*, 323–336.

Pázman, A. (1984b). Probability distribution of the multivariate nonlinear least-squares estimates. *Kybernetika 20*, 209–230.

Pázman, A. (1986). *Foundations of Optimum Experimental Design*. Dordrecht (co-pub. VEDA, Bratislava): Reidel (Kluwer group).

Pázman, A. (1990). Small-sample distributional properties of nonlinear regression estimators. a geometric approach (with discussion). *Statistics 21*(3), 323–367.

Pázman, A. (1992a). A classification of nonlinear regression models and parameter confidence regions. *Kybernetika 28*(6), 444–453.

Pázman, A. (1992b). Geometry of the nonlinear regression with prior. *Acta Math. Univ. Comenianae LXI*, 263–276.

Pázman, A. (1993a). Higher dimensional nonlinear regression — a statistical use of the Riemannian curvature tensor. *Statistics 25*, 17–28.

Pázman, A. (1993b). *Nonlinear Statistical Models*. Dordrecht: Kluwer.

Pázman, A. (2001). Concentration sets, Elfving sets and norms in optimum design. In A. Atkinson, B. Bogacka, and A. Zhigljavsky (Eds.), *Optimum Design 2000*, Chapter 10, pp. 101–112. Dordrecht: Kluwer.

Pázman, A. (2002a). Optimal design of nonlinear experiments with parameter constraints. *Metrika 56*, 113–130.

Pázman, A. (2002b). Results on nonlinear least squares estimators under nonlinear equality constraints. *J. Stat. Plann. Inference 103*, 401–420.

Pázman, A., S. Bilenkaya, J. Bystrický, Z. Janout, and F. Lehar (1969). The problem of un-ambiguity of phase-shift analysis I: Description of a powerful discriminating criterion. *Czech J. Physics B19*, 882–890.

Pázman, A. and W. Müller (2001). Optimum design of experiments subject to correlated errors. *Statist. Probab. Lett. 52*, 29–34.

Pázman, A. and W. Müller (2010). A note on the relationship between two approaches to optimal design under correlation. In A. Giovagnoli, A. Atkinson, and B. Torsney (Eds.), *mODa9 — Advances in Model-Oriented Design and Analysis, Proc. 9th Int. Workshop in Model-Oriented Design and Analysis, Bertinoro, Italy, June 14–18, 2010*, pp. 145–148. Heidelberg: Physica Verlag.

Pázman, A. and L. Pronzato (1992). Nonlinear experimental design based on the distribution of estimators. *J. Stat. Plann. Inference 33*, 385–402.

Pázman, A. and L. Pronzato (1996). A Dirac function method for densities of nonlinear statistics and for marginal densities in nonlinear regression. *Statist. Probab. Lett. 26*, 159–167.

Pázman, A. and L. Pronzato (1998). Approximate densities of two bias–corrected nonlinear LS estimators. In A. Atkinson, L. Pronzato, and H. Wynn (Eds.), *MODA'5 – Advances in Model–Oriented Data Analysis and Experimental Design, Proc. 5th Int. Workshop, Marseille, 22–26 juin 1998*, pp. 145–152. Heidelberg: Physica Verlag.

Pázman, A. and L. Pronzato (2004). Simultaneous choice of design and estimator in nonlinear regression with parameterized variance. In A. Di Bucchianico, H. Läuter, and H. Wynn (Eds.), *mODa'7 – Advances in Model–Oriented Design and Analysis, Proc. 7th Int. Workshop, Heeze (Netherlands)*, Heidelberg, pp. 117–124. Physica Verlag.

Pázman, A. and L. Pronzato (2006a). Asymptotic criteria for designs in nonlinear regression with model errors. *Math. Slovaca 56*(5), 543–553.

Pázman, A. and L. Pronzato (2006b). On the irregular behavior of LS estimators for asymptotically singular designs. *Statist. Probab. Lett. 76*, 1089–1096.

Pázman, A. and L. Pronzato (2009). Asymptotic normality of nonlinear least squares under singular experimental designs. In L. Pronzato and A. Zhigljavsky (Eds.), *Optimal Design and Related Areas in Optimization and Statistics*, Chapter 8, pp. 167–191. Springer.

Pečarić, J., S. Puntanen, and G. Styan (1996). Some further matrix extensions of the Cauchy-Schwarz and Kantorovich inequalities, with some statistical applications. *Linear Algebra Appl. 237/238*, 455–476.

Phillips, R. (2002). Least absolute deviations estimation via the EM algorithm. *Stat. Comput. 12*, 281–285.

Pilz, J. (1983). *Bayesian Estimation and Experimental Design in Linear Regression Models*, Volume 55. Leipzig: Teubner-Texte zur Mathematik. [Also published by Wiley, New York, 1991].

Polak, E. (1971). *Computational Methods in Optimization, a Unified Approach.* New York: Academic Press.

Polak, E. (1987). *Optimization. Algorithms and Consistent Approximations.* New York: Springer.

Polyak, B. (1987). *Introduction to Optimization.* New York: Optimization Software.

Polyak, B. (1990). New stochastic approximation type procedures. *Automat. i Telemekh. 7*, 98–107.

Polyak, B. and A. Juditsky (1992). Acceleration of stochastic approximation by averaging. *SIAM J. Control Optim. 30*, 838–855.

Pronzato, L. (2004). A minimax equivalence theorem for optimum bounded design measures. *Statist. Probab. Lett. 68*, 325–331.

Pronzato, L. (2006). On the sequential construction of optimum bounded designs. *J. Stat. Plann. Inference 136*, 2783–2804.

Pronzato, L. (2008). Optimal experimental design and some related control problems. *Automatica 44*, 303–325.

Pronzato, L. (2009a). Asymptotic properties of nonlinear estimates in stochastic models with finite design space. *Statist. Probab. Lett. 79*, 2307–2313.

Pronzato, L. (2009b). On the regularization of singular c-optimal designs. *Math. Slovaca 59*(5), 611–626.

Pronzato, L. (2010a). One-step ahead adaptive D-optimal design on a finite design space is asymptotically optimal. *Metrika 71*(2), 219–238.

Pronzato, L. (2010b). Penalized optimal designs for dose-finding. *J. Stat. Plann. Inference 140*, 283–306.

Pronzato, L., C.-Y. Huang, and E. Walter (1991). Nonsequential T-optimal design for model discrimination: new algorithms. In A. Pázman and J. Volaufová (Eds.), *Proc. ProbaStat'91*, Bratislava, pp. 130–136.

Pronzato, L. and W. Müller (2012). Design of computer experiments: space filling and beyond. *Stat. & Comput. 22*(3), 681–701.

Pronzato, L. and A. Pázman (1994a). Bias correction in nonlinear regression via two-stages least-squares estimation. In M. Blanke and T. Söderström (Eds.), *Prep. 10th IFAC/IFORS Symp. on Identification and System Parameter Estimation*, Volume 1, Copenhagen, pp. 137–142. Danish Automation Society.

Pronzato, L. and A. Pázman (1994b). Second-order approximation of the entropy in nonlinear least-squares estimation. *Kybernetika 30*(2), 187–198. [*Erratum* 32(1):104, 1996].

Pronzato, L. and A. Pázman (2001). Using densities of estimators to compare pharmacokinetic experiments. *Comput. Biol. Med. 31*(3), 179–195.

Pronzato, L. and A. Pázman (2004). Recursively re-weighted least-squares estimation in regression models with parameterized variance. In *Proc. EUSIPCO'2004, Vienna, Austria*, pp. 621–624.

Pronzato, L. and E. Walter (1985). Robust experiment design via stochastic approximation. *Math. Biosci. 75*, 103–120.

Pronzato, L. and E. Walter (1988). Robust experiment design via maximin optimization. *Math. Biosci. 89*, 161–176.

Pronzato, L. and E. Walter (1994). Minimal-volume ellipsoids. *Internat. J. Adapt. Control Signal Process. 8*, 15–30.

Pronzato, L. and E. Walter (1996). Volume-optimal inner and outer ellipsoids. In M. Milanese, J.-P. Norton, H. Piet-Lahanier, and E. Walter (Eds.), *Bounding Approaches to System Identification*, pp. 119–138. London: Plenum.

Pronzato, L. and E. Walter (2001). Eliminating suboptimal local minimizers in nonlinear parameter estimation. *Technometrics 43*(4), 434–442.

Pronzato, L., H. Wynn, and A. Zhigljavsky (2000). *Dynamical Search.* Boca Raton: Chapman & Hall/CRC.

Pronzato, L., H. Wynn, and A. Zhigljavsky (2005). Kantorovich-type inequalities for operators via D-optimal design theory. *Linear Algebra Appl. 410*, 160–169.

Pukelsheim, F. (1993). *Optimal Experimental Design.* New York: Wiley.

Pukelsheim, F. and S. Reider (1992). Efficient rounding of approximate designs. *Biometrika 79*(4), 763–770.

Rabitz, H., M. Kramer, and D. Dacol (1983). Sensitivity analysis in chemical kinetics. *Ann. Rev. Phys. Chem. 34*, 414–461.

Raynaud, H.-F., L. Pronzato, and E. Walter (2000). Robust identification and control based on ellipsoidal parametric uncertainty descriptions. *Eur. J. Control 6*(3), 245–257.

Rényi, A. (1961). On measures of entropy and information. In *Proc. 4th Berkeley Symp. on Math. Statist. and Prob.*, pp. 547–561.

Robertazzi, T. and S. Schwartz (1989). An accelerated sequential algorithm for producing D-optimal designs. *SIAM J. Sci. Statist. Comput. 10*(2), 341–358.

Rockafellar, R. (1970). *Convex Analysis.* Princeton: Princeton Univ. Press.

Rojas, C., J. Welsh, G. Goodwin, and A. Feuer (2007). Robust optimal experiment design for system identification. *Automatica 43*, 993–1008.

Ross, G. (1990). *Nonlinear estimation.* New York: Springer.

Rousseeuw, P. (1984). Least median of squares regression. *J. Amer. Statist. Assoc. 79*, 871–880.

Rousseeuw, P. and A. Leroy (1987). *Robust Regression and Outlier Detection.* New York: Wiley.

Rudin, W. (1987). *Real and Complex Analysis.* New York: McGraw-Hill. [3rd ed.].

Sahm, M. and R. Schwabe (2001). A note on optimal bounded designs. In A. Atkinson, B. Bogacka, and A. Zhigljavsky (Eds.), *Optimum Design 2000*, Chapter 13, pp. 131–140. Dordrecht: Kluwer.

Schaback, R. (2003). Mathematical results concerning kernel techniques. In *Prep. 13th IFAC Symp. on System Identification, Rotterdam*, pp. 1814–1819.

Schlossmacher, E. (1973). An iterative technique for absolute deviations curve fitting. *J. Amer. Statist. Assoc. 68*(344), 857–859.

Schwabe, R. (1995). Designing experiments for additive nonlinear models. In C. Kitsos and W. Müller (Eds.), *MODA4 – Advances in Model-Oriented Data Analysis, Spetses (Greece), june 1995*, pp. 77–85. Heidelberg: Physica Verlag.

Schwabe, R. (1996). *Optimum Designs for Multi-Factor Models.* New York: Springer.

Schwabe, R. (1997). Maximin efficient designs. Another view at D-optimality. *Statist. Probab. Lett. 35*, 109–114.

Schwabe, R. and W. Wong (1999). Efficiency bounds for product designs in linear models. *Ann. Inst. Statist. Math. 51*, 723–730.

Scott, D. (1992). *Multivariate Density Estimation.* New York: Wiley.

Shah, S., J. Mitchell, and M. Kupferschmid (2000). An ellipsoid algorithm for equality-constrained nonlinear programs. *Comput. Oper. Res. 28*(1), 85–92.

Shimizu, K. and E. Aiyoshi (1980). Necessary conditions for min-max problems and algorithm by a relaxation procedure. *IEEE Trans. Automatic Control 25*, 62–66.

Shiryaev, A. (1996). *Probability.* Berlin: Springer.

Shor, N. (1977). Cut-off method with space extension in convex programming problems. *Kibernetika 13*(1), 94–95. [English Trans. *Cybernetics Syst. Anal.*, **13**(1), 94–96].

Shor, N. (1985). *Minimization Methods for Non-Differentiable Functions.* Berlin: Springer.

Shor, N. and O. Berezovski (1992). New algorithms for constructing optimal circumscribed and inscribed ellipsoids. *Optim. Methods Softw. 1*, 283–299.

Sibson, R. (1972). Discussion on a paper by H.P. Wynn. *J. Roy. Statist. Soc. B34*, 181–183.

Sibson, R. (1974). D_A-optimality and duality. *Progress in Stat. Colloq. Math. Soc. Janos Bolyai 9*, 677–692.

Silvey, S. (1980). *Optimal Design.* London: Chapman & Hall.

Silvey, S. and D. Titterington (1973). A geometric approach to optimal design theory. *Biometrika 60*(1), 21–32.

Silvey, S., D. Titterington, and B. Torsney (1978). An algorithm for optimal designs on a finite design space. *Comm. Statist. – Theory Methods A7*(14), 1379–1389.

Söderström, T. and P. Stoica (1981). Comparison of some instrumental variable methods—consistency and accuracy aspects. *Automatica 17*(1), 101–115.

Söderström, T. and P. Stoica (1983). *Instrumental Variable Methods for System Identification.* New York: Springer.

Söderström, T. and P. Stoica (1989). *System Identification.* New York: Prentice Hall.

Spokoinyi, V. (1992). On asymptotically optimal sequential experimental design. *Advances in Soviet Math. 12*, 135–150.

St. John, R. and N. Draper (1975). D-optimality for regression designs: a review. *Technometrics 17*(1), 15–23.

Stein, C. (1956). Efficient nonparametric testing and estimation. In *Proc. 3rd Berkeley Symp. Math. Stat. Prob.*, Volume 1, pp. 187–196. Berkeley: Univ. of California Press.

Stoer, J. and R. Bulirsch (1993). *Introduction to Numerical Analysis (2nd Edition).* Heidelberg: Springer.

Stoica, P. (1998). On the Cramér-Rao bound under parametric constraints. *IEEE Signal Proc. Lett. 5*(7), 177–179.

Stoica, P. (2001). Parameter estimation problems with singular information matrices. *IEEE Trans. Signal Proc. 49*, 87–90.

Stone, C. (1975). Adaptive maximum likelihood estimators of a location parameter. *Ann. Statist. 3*(2), 267–284.

Studden, W. (1971). Elfving's theorem and optimal designs for quadratic loss. *Ann. Math. Statist. 42*, 1613–1621.

Sun, P. and R. Freund (2004). Computation of minimum-volume covering ellipsoids. *Oper. Res. 52*(5), 690–706.

Sundaraij, N. (1978). A method for confidence regions for nonlinear models. *Austral. J. Statist. 20*(3), 270–274.

Tarasov, S., L. Khachiyan, and I. Erlich (1988). The method of inscribed ellipsoids. *Soviet Math. Dokl. 37*(1), 226–230.

Titterington, D. (1975). Optimal design: some geometrical aspects of D-optimality. *Biometrika 62*(2), 313–320.

Titterington, D. (1976). Algorithms for computing D-optimal designs on a finite design space. In *Proc. 1976 Conf. on Information Science and Systems*, Baltimore, pp. 213–216. Dept. of Electronic Engineering, John Hopkins Univ.

Titterington, D. (1978). Estimation of correlation coefficients by ellipsoidal trimming. *J. Roy. Statist. Soc. C27*(3), 227–234.

Todd, M. (1982). On minimum volume ellipsoids containing part of a given ellipsoid. *Math. Oper. Res. 7*(2), 253–261.

Todd, M. and E. Yildirim (2007). On Khachiyan's algorithm for the computation of minimum volume enclosing ellipsoids. *Discrete Appl. Math. 155*, 1731–1744.

Torsney, B. (1983). A moment inequality and monotonicity of an algorithm. In K. Kortanek and A. Fiacco (Eds.), *Proc. Int. Symp. on Semi-infinite Programming and Applications*, Heidelberg, pp. 249–260. Springer.

Torsney, B. (1986). Moment inequalities via optimal design theory. *Linear Algebra Appl. 82*, 237–253.

Torsney, B. (2009). W-iterations and ripples therefrom. In L. Pronzato and A. Zhigljavsky (Eds.), *Optimal Design and Related Areas in Optimization and Statistics*, Chapter 1, pp. 1–12. Springer.

Tsybakov, A. (2004). *Introduction à l'estimation non-paramétrique.* Heidelberg: Springer.

Uciński, D. and M. Patan (1982). *D*-optimal design of a monitoring network for parameter estimation of distributed systems. *J. Global Optim. 39*, 291–322.

Valko, P. and S. Vajda (1984). An extended ODE solver for sensitivity calculations. *Comput. Chem. 8*(4), 255–271.

van de Geer, S. (2000). *Empirical Processes in M-estimation.* Cambridge: Cambridge Univ. Press.

van der Vaart, A. (1998). *Asymptotic Statistics.* Cambridge: Cambridge Univ. Press.

van der Vaart, A. (2002). The statistical work of Lucien Le Cam. *Ann. Statist. 30*(3), 631–682.

van der Vaart, A. and J. Wellner (1996). *Weak Convergence and Empirical Processes, with Applications to Statistics.* New York: Springer.

Vandenberghe, L., S. Boyd, and S.-P. Wu (1998). Determinant maximisation with linear matrix inequality constraints. *SIAM J. Matrix Anal. Appl. 19*(2), 499–533.

Vapnik, V. (1998). *Statistical Learning Theory.* New York: Wiley.

Vapnik, V. (2000). *The Nature of Statistical Learning Theory.* New York: Springer. [2nd ed.].

Vazquez, E. (2005). Modélisation comportementale de systèmes nonlinéaires multivariables par méthodes à noyaux et applications. Ph.D. Thesis, Université Paris XI, Orsay, France.

Vazquez, E. and E. Walter (2003). Multi-output support vector regression. In *Prep. 13th IFAC Symp. on System Identification, Rotterdam*, pp. 1820–1825.

Veinott, A. (1967). The supporting hyperplane method for unimodal programming. *Oper. Research 15*, 147–152.

Vila, J.-P. (1990). Exact experimental designs via stochastic optimization for nonlinear regression models. In *Proc. Compstat, Int. Assoc. for Statistical Computing*, pp. 291–296. Heidelberg: Physica Verlag.

Vila, J.-P. and J.-P. Gauchi (2007). Optimal designs based on exact confidence regions for parameter estimation of a nonlinear regression model. *J. Stat. Plann. Inference 137*, 2935–2953.

Wald, A. (1949). Note on the consistency of the maximum likelihood estimate. *Ann. Math. Statist. 20*, 595–601.

Walter, E. (1982). *Identifiability of State-Space Models.* Berlin: Springer Verlag.

Walter, E. (Ed.) (1987). *Identifiability of Parametric Models.* Oxford: Pergamon Press.

Walter, E. and L. Pronzato (1995). Identifiabilities and nonlinearities. In A. Fossard and D. Normand-Cyrot (Eds.), *Nonlinear Systems. Modeling and Estimation*, Chapter 3, pp. 111–143. London: Chapman & Hall.

Walter, E. and L. Pronzato (1997). *Identification of Parametric Models from Experimental Data.* Heidelberg: Springer.

Watson, G. (1964). Smooth regression analysis. *Sankhya, Series A 26*, 359–372.

Welch, W. (1982). Branch-and-bound search for experimental designs based on D-optimality and other criteria. *Technometrics 24*(1), 41–28.

Welzl, E. (1991). Smallest enclosing disks (balls and ellipsoids). In *New Results and New Trends in Computer Science*, Lecture Notes in Comput. Sci., vol. 555, pp. 359–370. Heidelberg: Springer.

Wolfe, P. (1970). Convergence theory in nonlinear programming. In J. Abadie (Ed.), *Integer and Nonlinear Programming*, pp. 1–36. Amsterdam: North Holland.

Wright, M. (1998). The interior-point revolution in constrained optimization. Technical Report 98–4–09, Computing Sciences Research Center, Bell Laboratories, Murray Hill, New Jersey 07974.

Wu, C. (1978a). Some algorithmic aspects of the theory of optimal designs. *Ann. Statist. 6*(6), 1286–1301.

Wu, C. (1978b). Some iterative procedures for generating nonsingular optimal designs. *Comm. Statist. – Theory Methods A7*(14), 1399–1412.

Wu, C. (1980). Characterizing the consistent directions of least squares estimates. *Ann. Statist. 8*(4), 789–801.

Wu, C. (1981). Asymptotic theory of nonlinear least squares estimation. *Ann. Statist. 9*(3), 501–513.

Wu, C. (1983). Further results on the consistent directions of least squares estimators. *Ann. Statist. 11*(4), 1257–1262.

Wu, C. (1985). Asymptotic inference from sequential design in a nonlinear situation. *Biometrika 72*(3), 553–558.

Wu, C. and H. Wynn (1978). The convergence of general step–length algorithms for regular optimum design criteria. *Ann. Statist. 6*(6), 1273–1285.

Wynn, H. (1970). The sequential generation of D-optimum experimental designs. *Ann. Math. Statist. 41*, 1655–1664.

Wynn, H. (1972). Results in the theory and construction of D-optimum experimental designs. *J. Roy. Statist. Soc. B34*, 133–147.

Wynn, H. (1977). Optimum designs for finite populations sampling. In S. Gupta and D. Moore (Eds.), *Statistical Decision Theory and Related Topics II*, pp. 471–478. New York: Academic Press.

Wynn, H. (1982). Optimum submeasures with applications to finite population sampling. In S. Gupta and J. Berger (Eds.), *Statistical Decision Theory and Related Topics III. Proc. 3rd Purdue Symp., vol. 2*, pp. 485–495. New York: Academic Press.

Wynn, H. (2004). Maximum entropy sampling and general equivalence theory. In A. Di Bucchianico, H. Läuter, and H. Wynn (Eds.), *mODa'7 – Advances in Model–Oriented Design and Analysis, Proc. 7th Int. Workshop, Heeze (Netherlands)*, pp. 211–218. Heidelberg: Physica Verlag.

Wynn, H. (2007). Bayesian information-based learning and majorization. Technical report, http://mucm.group.shef.ac.uk/index.html.

Yang, M. (2010). On de la Garza phenomenon. *Ann. Statist. 38*(4), 2499–2524.

Yang, M. and J. Stufken (2009). Support points of locally optimal designs for nonlinear models with two parameters. *Ann. Statist. 37*(1), 518–541.

Yang, S.-S. (1985). A smooth nonparametric estimator of a quantile function. *J. Amer. Statist. Assoc. 80*(392), 1004–1011.

Ye, Y. (1997). *Interior-Point Algorithms: Theory and Analysis.* Chichester: Wiley.

Yu, Y. (2010a). Monotonic convergence of a general algorithm for computing optimal designs. *Ann. Statist. 38*(3), 1593–1606.

Yu, Y. (2010b). Strict monotonicity and convergence rate of Titterington's algorithm for computing D-optimal designs. *Comput. Statist. Data Anal. 54*, 1419–1425.

Yu, Y. (2011). D-optimal designs via a cocktail algorithm. *Stat. Comput. 21*, 475–481.

Zarrop, M. (1979). *Optimal Experiment Design for Dynamic System Identification.* Heidelberg: Springer.

Zhang, Y. and L. Gao (2003). On numerical solution of the maximum volume ellipsoid problem. *SIAM J. Optim. 14*(1), 53–76.

Zhu, Z. and H. Zhang (2006). Spatial sampling design under the infill asymptotic framework. *Environmetrics 17*(4), 323–337.

Author Index

L. Pronzato and A. Pázman, *Design of Experiments in Nonlinear Models*, Lecture Notes in Statistics 212, DOI 10.1007/978-1-4614-6363-4,

Subject Index

L. Pronzato and A. Pázman, *Design of Experiments in Nonlinear Models*, Lecture Notes in Statistics 212, DOI 10.1007/978-1-4614-6363-4,

Printed by Printforce, the Netherlands